Antiseptic Stewardship

Günter Kampf

Antiseptic Stewardship

Biocide Resistance and Clinical Implications

Springer

Günter Kampf
Institute of Hygiene and Environmental Medicine
University of Greifswald
Greifswald, Germany

ISBN 978-3-319-98784-2 ISBN 978-3-319-98785-9 (eBook)
https://doi.org/10.1007/978-3-319-98785-9

Library of Congress Control Number: 2018950922

This Springer imprint is published by the registered company Springer Nature Switzerland AG
The registered company address is: Gewerbestrasse 11, 6330 Cham, Switzerland

Foreword

Biocides (disinfectants, antiseptics, preservatives) usage has increased worldwide notably for applications that do not necessarily require the application of biocides, particularly in the home environment. The amount of biocides used in Europe is difficult to quantify as the number of products containing a biocide and biocide applications have increased dramatically in the last 10 years. It is thus logical to assume that microbial exposure to biocides has also increased. Parallel, but not separate, from the increase in biocidal products commercially available is the rise in antimicrobial resistance (AMR) in bacteria, which results primarily from the overuse and misuse of chemotherapeutic antibiotics for human and veterinary medicine, but also for industrial processes such as fermentation. Recent calculations from Lord O'Neil's AMR report to the British government predict human deaths caused by untreatable AMR to reach 10 million worldwide by 2050, well above other diseases including cancer. Biocidal products have a role to play in reducing AMR notably on hard and porous surfaces, with disinfection and antisepsis, and in products through preservation. The increase in biocidal products is most likely due to a better understanding by the public of hygiene concepts and AMR, and the absolute need to control infection, creating opportunities for the industry to meet the need for products that can inhibit or eliminate the risk of infection or spoilage. Although biocidal products play an essential part in controlling micro-organisms on surfaces and in products, the overuse of biocides and biocidal products has raised concerns among regulators, about environmental toxicity following product applications, and on risks associated with emerging bacterial resistance to specific biocidal agents, and cross-resistance to unrelated substances including chemotherapeutic antibiotics. In Europe, the Biocidal Products Regulation now mentioned the need for manufacturers to measure the impact of biocidal products on emerging resistance and cross-resistance, while the US Food and Drug Administration has recently published a rule to restrict the use of a number of cationic and phenolic biocides in certain products, based on potential toxicity and bacterial resistance issues.

Hence, if the use of biocides and biocidal products is necessary and beneficial on the one hand, overuse and misuse of biocides may be detrimental on the other hand. This book looks at the main biocides used in common formulations developed for healthcare applications. It provides useful information on biocide activity against bacteria and fungi, and evidence of emerging resistance and cross-resistance following biocide exposure. It also dedicates a number of chapters promoting appropriate biocidal product usage and good stewardship of biocidal products in healthcare settings. The subjects presented in this book are topical and of great interests. Overall, the information provided in this book provides a better understanding of the efficacy and limitations of commonly used biocides and their applications.

Cardiff, UK
June 2018

Prof. Jean-Yves Maillard
School of Pharmacy and Pharmaceutical Sciences
Cardiff University

Preface

A chemical that constantly stresses bacteria to adapt, and behaviour that promotes antibiotic resistance needs to be stopped immediately when the benefits are null.

Patrick J. McNamara and Stuart B. Levy (2016)

The indicated use of antiseptics and disinfectants is regarded as a major contribution to prevent the transmission of multidrug- or pan-resistant pathogens. Some antiseptic products, however, contain additional non-volatile active ingredients with a doubtful or sometimes even without a contribution to the overall antimicrobial efficacy. But these agents can at the same time cause adaptation and resistance, mainly among Gram-negative bacterial species. The resistance may even cover other biocidal agents or selected antibiotics. Chlorhexidine digluconate is such a biocidal agent used in different types of products such as alcohol-based hand rubs, antimicrobial soaps, alcohol-based skin antiseptics and antiseptic mouth rinses. In some of the applications, there is good evidence that it contributes to patient safety, e.g. when used in combination with alcohol as a skin antiseptic for the insertion of a central venous catheter or for puncture site care. Its effect in alcohol-based hand rubs, however, is at least doubtful.

After my publication in the Journal of Hospital Infection in 2016 on acquired resistance to chlorhexidine and the proposal to establish an antiseptic stewardship initiative, I have received some very encouraging emails from clinical colleagues who were grateful for the review and who supported the principal idea of an antiseptic stewardship based on their own clinical experience. This type of feedback was motivation enough to look at the entire topic in a broader perspective.

Although the evaluation of biocidal agents was done with a lot of care for completeness and experimental details, I may still have missed some studies. But the overall picture is probably quite complete and allows learning which of the biocidal agents has a higher risk for promoting resistance in which types of pathogens. Healthcare workers are invited to critically look at the product labels in the section "composition" and to find out which of the active agents is in the product even if not declared as an active agent. Regulative authorities are invited to ask the manufacturers about the evidence-based antimicrobial effects of specific substances which may even result in non-approval of specific products if the risks for selection pressure by a substance outweigh any possible benefits. And manufacturers are invited to take all the findings into account when formulating antiseptic products. At the end, I hope that the book contributes to reducing unnecessary selection pressure by the different types of antiseptic agents.

Greifswald, Germany Günter Kampf

About this Book

For this book, typical antiseptic substances have been selected which are used in various fields of applications (e.g. human medicine, veterinary medicine, food production and handling) and are used in at least two types of antiseptic products (e.g. hand disinfectants, surface disinfectants, skin antiseptics) by at least two manufacturers. Another prerequisite was to have published evidence available for each antiseptic agent allowing a best possible comprehensive review of its antimicrobial efficacy and resistance. One aspect is very important in this context. The summary on each biocidal agent aims to provide a neutral and complete picture but does not intend to favour or disadvantage specific biocidal agents. It does also not intend to favour or disadvantage specific manufacturers or companies.

In the first part of each chapter, the chemical is characterized followed by its typical applications including the regulatory frame in the European Union and the USA. A summary of the activity of each antiseptic agent against bacteria, fungi and mycobacteria is the next part. It includes an overview on MIC values to determine a microbiostatic activity and data from suspension tests to determine a microbiocidal activity obtained with culture collection strains and all other types of clinical and environmental isolates. It also includes a description of the efficacy against the micro-organisms in biofilms. Viruses were not included because adaptation and resistance were regarded as defence mechanisms of living cells. Bacterial spores were also not included because they are considered the most resistant form of a micro-organism anyway so that an adaptation or an acquired resistance is not expectable and is unlikely to change the use of antiseptics.

It is followed by all data on any type of adaptive response by micro-organisms to low-level exposure to the biocidal agent. This may be a change of susceptibility to the biocide itself, to other biocidal agents or antibiotics (e.g. measured by a higher MIC value), a change of biofilm formation, a change of efflux pump activity or a change of horizontal gene transfer. Taking all the information together will hopefully allow to see that some antiseptic agents have a higher risk for microbial adaptation and resistance, and other agents have a lower risk.

Finally, a description of the frequency of resistance can be found, e.g. isolates with high MIC values, contaminated biocidal products or even outbreaks or pseudo-outbreaks of infections caused by contaminated biocidal products. Possible mechanisms of resistance are reviewed such as specific resistance genes, plasmids and efflux pumps that may extrude deleterious compounds, such as antibiotics, drugs and solvents. Cross-resistance to other biocidal agents and antibiotics is also summarized in this part. Some studies have reported that a species became resistant to an antibiotic based on accepted break points and methods. In this case, an isolate will be described as resistant to the antibiotic. Other authors described an MIC change (e.g. by microdilution or Etest) or a change of the zone of inhibition (e.g. by disc diffusion test) without an assignment to "resistant" or "susceptible". This type of finding will be described as cross-tolerance. In addition, data on biofilm development, removal and fixation are summarized for each antiseptic agent. Based on the agents' summary, it should be possible to establish an antiseptic stewardship initiative.

Contents

About the Author

Günter Kampf is an associate professor for Hygiene and Environmental Medicine at the University of Greifswald, Germany. He has published more than 180 mostly international articles on various aspects of infection control, mainly hand hygiene and surface disinfection. In 2017, he published his second textbook on hand hygiene in German as an editor. He is a recognized medical specialist in Hygiene and Environmental Medicine and has worked for 18 years for a manufacturer of chemical disinfectants, in the last 5 years as a Director Science. Since 2016 he is self-employed and continues his scientific work together with infection control consultancy for hospitals, medical practices and companies with infection control issues (www.guenter-kampf-hygiene.de).

Abbreviations

3MRGN	Isolate with resistance to three of the following four antibiotic classes: acylureidopenicillins, third-generation and fourth-generation cephalosporins, carbapenems and fluoroquinolones
4MRGN	Isolate with resistance to all of the following four antibiotic classes: acylureidopenicillins, third-generation and fourth-generation cephalosporins, carbapenems and fluoroquinolones
A. acidoterrestris	Alicyclobacillus acidoterrestris
A. actinomycetemcomitans	Aggregatibacter actinomycetemcomitans
A. alternata	Alternaria alternata
A. anitratus	Acinetobacter anitratus
A. aphrophilus	Aggregatibacter aphrophilus
A. baumannii	Acinetobacter baumannii
A. calcoaceticus	Acinetobacter calcoaceticus
A. delafieldii	Acidovorax delafieldii
A. elegans	Actinomucor elegans
A. ferrooxidans	Acidithiobacillus ferrooxidans
A. flavipes	Aspergillus flavipes
A. flavus	Aspergillus flavus
A. fumigatus	Aspergillus fumigatus
A. gyllenbergii	Acinetobacter gyllenbergii
A. hydrophila	Aeromonas hydrophila
A. israelii	Actinomyces israelii
A. jandaei	Aeromonas jandaei
A. junii	Acinetobacter junii
A. laidlawii	Acheloplasma laidlawii
A. lwoffii	Acinetobacter lwoffii
A. naeslundii	Actinomyces naeslundii
A. nidulans	Aspergillus nidulans
A. niger	Aspergillus niger
A. nosocomialis	Acinetobacter nosocomialis
A. ochraceus	Aspergillus ochraceus

A. odontolyticus	Actinomyces odontolyticus
A. oleivorans	Acinetobacter oleivorans
A. parasiticus	Aspergillus parasiticus
A. proteolyticus	Aranicola proteolyticus
A. salmonicida	Aeromonas salmonicida
A. terreus	Aspergillus terreus
A. ustus	Aspergillus ustus
A. versicolor	Aspergillus versicolor
A. viscosus	Actinomyces viscosus
A. westerdijkiae	Aspergillus westerdijkiae
A. xylosoxidans	Achromobacter xylosoxidans
Ag-NP	Silver nanoparticles
ASTM	American Society for Testing and Materials
ATCC	American Type Culture Collection
B. abortus	Brucella abortus
B. adolescentis	Bifidobacterium adolescentis
B. afzelii	Borrelia afzelii
B. amyloliquefaciens	Bacillus amyloliquefaciens
B. animalis	Bifidobacterium animalis
B. bifidum	Bifidobacterium bifidum
B. breve	Bifidobacterium breve
B. burgdorferi	Borrelia burgdorferi
B. catenulatum	Bifidobacterium catenulatum
B. cenocepacia	Burkholderia cenocepacia
B. cepacia	Burkholderia cepacia
B. cereus	Bacillus cereus
B. diminuta	Brevundimonas diminuta
B. fragilis	Bacteroides fragilis
B. garinii	Borrelia garinii
B. gingivalis	Bacteroides gingivalis
B. infantis	Bifidobacterium infantis
B. intermedius	Bacteroides intermedius
B. licheniformis	Bacillus licheniformis
B. longum	Bifidobacterium longum
B. mallei	Burkholderia mallei
B. megaterium	Bacillus megaterium
B. melaninogenicus	Bacteroides melaninogenicus
B. melitensis	Brucella melitensis
B. petrii	Bordetella petrii
B. pseudocatenulatum	Bifidobacterium pseudocatenulatum
B. pseudolongum	Bifidobacterium pseudolongum
B. pseudomallei	Burkholderia pseudomallei
B. pumilus	Bacillus pumilus
B. sanguinis	Brevibacterium sanguinis
B. spicifera	Bipolaris spicifera

B. stearothermophilus	Bacillus stearothermophilus
B. subtilis	Bacillus subtilis
B. suis	Bifidobacterium suis
B. thailandensis	Burkholderia thailandensis
B. thermoacidophilum	Bifidobacterium thermoacidophilum
C. acidovorans	Comamonas acidovorans
C. albicans	Candida albicans
C. argentea	Candida argentea
C. ciferrii	Candida ciferrii
C. concisus	Campylobacter concisus
C. difficile	Clostridium difficile
C. diphtheriae	Corynebacterium diphtheriae
C. dubliniensis	Candida dubliniensis
C. famata	Candida famata
C. funicola	Chaetomium funicola
C. gingivalis	Capnocytophaga gingivalis
C. glabrata	Candida glabrata
C. globosum	Chaetomium globosum
C. guilliermondii	Candida guilliermondii
C. indologenes	Chryseobacterium indologenes
C. intermedia	Candida intermedia
C. intermedius	Citrobacter intermedius
C. jeikeium	Corynebacterium jeikeium
C. jejuni	Campylobacter jejuni
C. kefyr	Candida kefyr
C. koseri	Citrobacter koseri
C. krusei	Candida krusei
C. liriodendri	Cylindrocarpon liriodendri
C. lusitaniae	Candida lusitaniae
C. luteola	Chryseomonas luteola
C. macrodidymum	Cylindrocarpon macrodidymum
C. matruchotti	Corynebacterium matruchotti
C. melibiosica	Candida melibiosica
C. meningosepticum	Chryseobacterium meningosepticum
C. metallidurans	Cupriavidus metallidurans
C. neoformans	Cryptococcus neoformans
C. norvegensis	Candida norvegensis
C. novyi	Clostridium novyi
C. ochracea	Capnocytophaga ochracea
C. oleophila	Candida oleophila
C. orthopsilosis	Candida orthopsilosis
C. parapsilosis	Candida parapsilosis
C. pelliculosa	Candida pelliculosa
C. perfringens	Clostridium perfringens
C. piscicola	Carnobacterium piscicola

C. pseudogenitalium	Corynebacterium pseudogenitalium
C. pseudotropicalis	Candida pseudotropicalis
C. rectus	Campylobacter rectus
C. renale	Corynebacterium renale
C. rodentium	Citrobacter rodentium
C. sakazakii	Cronobacter sakazakii
C. sake	Candida sake
C. striatum	Corynebacterium striatum
C. trachomatis	Chlamydia trachomatis
C. tropicalis	Candida tropicalis
C. uniguttulatus	Cryptococcus uniguttulatus
C. utilis	Candida utilis
C. xerosis	Corynebacterium xerosis
CAS	Chemical abstracts service
CFU	Colony-forming units
CHA	Chlorhexidine diacetate
CHG	Chlorhexidine digluconate
CIP	Collection of Institut Pasteur
CMCC	National Center for Medical Culture Collections
CNS	Coagulase-negative staphylococci
D. acidovorans	Delftia acidovorans
D. hansenii	Debaryomyces hansenii
DSM	Deutsche Sammlung von Mikroorganismen
DT50	50% dissipation time
E. aerogenes	Enterobacter aerogenes
E. amylovora	Erwinia amylovora
E. asburiae	Enterobacter asburiae
E. avium	Enterococcus avium
E. casseliflavus	Enterococcus casseliflavus
E. cloacae	Enterobacter cloacae
E. coli	Escherichia coli
E. corrodens	Eikenella corrodens
E. durans	Enterococcus durans
E. faecalis	Enterococcus faecalis
E. gergoviae	Enterobacter gergoviae
E. hirae	Enterococcus hirae
E. ludwigii	Enterobacter ludwigii
E. nigrum	Epicoccum nigrum
E. nodatum	Eubacterium nodatum
E. raffinosus	Enterococcus raffinosus
E. repens	Eurotium repens
E. rhusiopathiae	Erysipelothrix rhusiopathiae
E. saccharolyticus	Enterococcus saccharolyticus
E. solitarius	Enterococcus solitarius
ECHA	European Chemicals Agency

ECOFF	Epidemiological cut-off value
EN	European norm
EPA	Environmental Protection Agency
ESBL	Extended spectrum β-lactamase
F. alocis	Filifactor alocis
F. indologenes	Flavobacterium indologenes
F. lichenicola	Fusarium lichenicola
F. noatunensis	Francisella noatunensis
F. nucleatum	Fusobacterium nucleatum
F. oryzihabitans	Flavimonas oryzihabitans
F. oxysporum	Fusarium oxysporum
F. proliferatum	Fusarium proliferatum
F. psychrophilum	Flavobacterium psychrophilum
F. solani	Fusarium solani
F. tularensis	Francisella tularensis
F. verticillioides	Fusarium verticillioides
FDA	Food and Drug Administration
G. haemolysans	Gemella haemolysans
G. vaginalis	Gardnerella vaginalis
h	Hour(s)
H. alvei	Hafnia alvei
H. anomala	Hansenula anomala
H. burtonii	Hyphopichia burtonii
H. flavidus	Humicoccus flavidus
H. gallinarum	Halonella gallinarum
H. influenzae	Haemophilus influenzae
H. parainfluenzae	Haemophilus parainfluenzae
H. parasuis	Haemophilus parasuis
H. pylori	Helicobacter pylori
H. valbyensis	Hanseniaspora valbyensis
ICU	Intensive care unit
IUPAC	International Union of Pure and Applied Chemistry
JCM	Japanese Collection of Microorganisms
K. aerogenes	Klebsiella aerogenes
K. apiculata	Kloeckera apiculata
K. oxytoca	Klebsiella oxytoca
K. planticola	Klebsiella planticola
K. pneumoniae	Klebsiella pneumoniae
K. quasipneumoniae	Klebsiella quasipneumoniae
K. terrigena	Klebsiella terrigena
L. acidophilus	Lactobacillus acidophilus
L. amylovorus	Lactobacillus amylovorus
L. brevis	Lactobacillus brevis
L. brunescens	Lysobacter brunescens
L. bulgaricus	Lactobacillus bulgaricus

L. coryniformis	Lactobacillus coryniformis
L. fermentum	Lactobacillus fermentum
L. garvieae	Lactococcus garvieae
L. grayi	Listeria grayi
L. helveticus	Lactobacillus helveticus
L. innocua	Listeria innocua
L. lactis	Lactococcus lactis
L. mesenteroides	Leuconostoc mesenteroides
L. monocytogenes	Listeria monocytogenes
L. odontolyticus	Lactobacillus odontolyticus
L. paracasei	Lactobacillus paracasei
L. pentosus	Lactobacillus pentosus
L. plantarum	Lactobacillus plantarum
L. pneumophila	Legionella pneumophila
L. pseudomesenteroides	Leuconostoc pseudomesenteroides
L. reuteri	Lactobacillus reuteri
L. rhamnosus	Lactobacillus rhamnosus
L. salivarius	Lactobacillus salivarius
L. seeligeri	Listeria seeligeri
L. welshimeri	Listeria welshimeri
M. abscessus	Mycobacterium abscessus
M. adhaesivum	Methylobacterium adhaesivum
M. aquaticum	Methylobacterium aquaticum
M. avium	Mycobacterium avium
M. bolletii	Mycobacterium bolletii
M. bovis	Mycobacterium bovis
M. canis	Microsporum canis
M. chelonae	Mycobacterium chelonae
M. circinelloides	Mucor circinelloides
M. fortuitum	Mycobacterium fortuitum
M. frederiksbergense	Mycobacterium frederiksbergense
M. fructicola	Metschnikowia fructicola
M. furfur	Malassezia furfur
M. gallisepticum	Mycoplasma gallisepticum
M. gypseum	Microsporum gypseum
M. kansasii	Mycobacterium kansasii
M. luteus	Micrococcus luteus
M. marinum	Mycobacterium marinum
M. massiliense	Mycobacterium massiliense
M. morganii	Morganella morganii
M. nonchromogenicum	Mycobacterium nonchromogenicum
M. osloensis	Moraxella osloensis
M. pachydermatis	Malassezia pachydermatis
M. phlei	Mycobacterium phlei
M. phyllosphaeriae	Microbacterium phyllosphaeriae

M. pneumoniae	Mycoplasma pneumoniae
M. racemosus	Mucor racemosus
M. rhodesianum	Methylobacterium rhodesianum
M. ruber	Monascus ruber
M. scrofulaceum	Mycobacterium scrofulaceum
M. slooffiae	Malassezia slooffiae
M. smegmatis	Mycobacterium smegmatis
M. suaveolens	Moniliella suaveolens
M. sympodialis	Malassezia sympodialis
M. terrae	Mycobacterium terrae
M. testaceum	Microbacterium testaceum
M. tuberculosis	Mycobacterium tuberculosis
M. xenopi	Mycobacterium xenopi
MBC	Minimum bactericidal concentration
MBEC	Minimum biofilm-eliminating concentration
MDR	Multidrug resistant
MIC	Minimum inhibitory concentration
MIC_{max}	Highest MIC value
min	Minute(s)
MRCNS	Methicillin-resistant coagulase-negative staphylococci
MRSA	Methicillin-resistant Staphylococcus aureus
MRSE	Methicillin-resistant Staphylococcus epidermidis
MRSP	Methicillin-resistant Staphylococcus pseudointermedius
MSCNS	Methicillin-susceptible coagulase-negative staphylococci
MSSA	Methicillin-susceptible Staphylococcus aureus
MSSP	Methicillin-susceptible Staphylococcus pseudointermedius
MTCC	Microbial Type Culture Collection and Gene Bank
N. asteroides	Nocardia asteroides
N. pseudofischeri	Neosartorya pseudofischeri
N. subflava	Neisseria subflava
NCIMB	National Collection of Industrial Food and Marine Bacteria
NCPF	National Collection of Pathogenic Fungi
NCTC	National Collection of Type Cultures
O. anthropi	Ochrobactrum anthropi
P	Commercial product
P. acnes	Propionibacterium acnes
P. aeruginosa	Pseudomonas aeruginosa
P. agglomerans	Pantoea agglomerans
P. alcalifaciens	Providencia alcalifaciens
P. aleophilum	Phaeoacremonium aleophilum

P. alkylphenolia	Pseudomonas alkylphenolia
P. anaerobius	Peptostreptococcus anaerobius
P. ananatis	Pantoea ananatis
P. anomala	Pichia anomala
P. aurantiogriseum	Penicillium aurantiogriseum
P. caseifulvum	Penicillium caseifulvum
P. chlamydospora	Phaeomoniella chlamydospora
P. chlororaphis	Pseudomonas chlororaphis
P. chrysogenum	Penicillium chrysogenum
P. citrinum	Penicillium citrinum
P. commune	Penicillium commune
P. corylophilum	Penicillium corylophilum
P. crustosum	Penicillium crustosum
P. denticola	Prevotella denticola
P. diminuta	Pseudomonas diminuta
P. discolor	Penicillium discolor
P. endodontalis	Porphyromonas endodontalis
P. expansum	Penicillium expansum
P. fluorescens	Pseudomonas fluorescens
P. fragi	Pseudomonas fragi
P. gingivalis	Porphyromonas gingivalis
P. intermedia	Prevotella intermedia
P. lundensis	Pseudomonas lundensis
P. marginalis	Pseudomonas marginalis
P. melaninogenica	Prevotella melaninogenica
P. mexicana	Pseudoxanthomonas mexicana
P. micra	Parvimonas micra
P. micros	Peptostreptococcus micros
P. mirabilis	Proteus mirabilis
P. morganii	Proteus morganii
P. multocida	Pasteurella multocida
P. nalgiovense	Penicillium nalgiovense
P. nigrescens	Prevotella nigrescens
P. nitroreducens	Pseudomonas nitroreducens
P. nitroreductans	Pseudomonas nitroreductans
P. norvegensis	Pichia norvegensis
P. ohmeri	Pichia ohmeri
P. paneum	Penicillium paneum
P. putida	Pseudomonas putida
P. pyocyanea	Pseudomonas pyocyanea
P. rettgeri	Proteus rettgeri
P. roqueforti	Penicillium roqueforti
P. solitum	Penicillium solitum
P. stutzeri	Pseudomonas stutzeri
P. verrucosum	Penicillium verrucosum

P. vesicularis	Pseudomonas vesicularis
P. vulgaris	Proteus vulgaris
PRSP	Penicillin-resistant Streptococcus pneumoniae
PTFE	Polytetrafluoroethylene
PVC	Polyvinyl chloride
QAC	Quaternary ammonium compound
R. dentocariosa	Rothia dentocariosa
R. erythropolis	Rhodococcus erythropolis
R. microsporus	Rhizopus microsporus
R. mucilaginosa	Rhodotorula mucilaginosa
R. nigricans	Rhizopus nigricans
R. pickettii	Ralstonia pickettii
R. planticola	Raoultella planticola
R. rubra	Rhodotorula rubra
R. rubrum	Rhodospirillum rubrum
S	Solution of antiseptic agent
s	Second(s)
S. Anatum	Salmonella Anatum
S. anginosus	Streptococcus anginosus
S. apiospermum	Scedosporium apiospermum
S. arboriculus	Saccharomyces arboriculus
S. aureus	Staphylococcus aureus
S. bayanus	Saccharomyces bayanus
S. brevicaulis	Scopulariopsis brevicaulis
S. capitis	Staphylococcus capitis
S. caprae	Staphylococcus caprae
S. cariocanus	Saccharomyces cariocanus
S. carlsbergensis	Saccharomyces carlsbergensis
S. cerevisiae	Saccharomyces cerevisiae
S. choleraesuis	Salmonella choleraesuis
S. chromogenes	Staphylococcus chromogenes
S. cohnii	Staphylococcus cohnii
S. constellatus	Streptococcus constellatus
S. delphini	Staphylococcus delphini
S. enterica	Salmonella enterica
S. Enteritidis	Salmonella Enteritidis
S. epidermidis	Staphylococcus epidermidis
S. equorum	Staphylococcus equorum
S. fleurettii	Staphylococcus fleurettii
S. flexneri	Shigella flexneri
S. gordonii	Streptococcus gordonii
S. Hadar	Salmonella Hadar
S. haemolyticus	Staphylococcus haemolyticus
S. hominis	Staphylococcus hominis
S. hyicus	Staphylococcus hyicus

S. Infantis	Salmonella Infantis
S. intermedius	Streptococcus intermedius
S. Kentucky	Salmonella Kentucky
S. kloosii	Staphylococcus kloosii
S. kudriavzevii	Saccharomyces kudriavzevii
S. lentus	Staphylococcus lentus
S. liquefaciens	Serratia liquefaciens
S. lugdunensis	Staphylococcus lugdunensis
S. maltophilia	Stenotrophomonas maltophilia
S. marcescens	Serratia marcescens
S. mikatae	Saccharomyces mikatae
S. mitis	Streptococcus mitis
S. mizutae	Sphingobacterium mizutae
S. multivorum	Sphingobacterium multivorum
S. mutans	Streptococcus mutans
S. oralis	Streptococcus oralis
S. paradoxus	Saccharomyces paradoxus
S. parasanguinis	Streptococcus parasanguinis
S. pasteuri	Staphylococcus pasteuri
S. paucimobilis	Sphingomonas paucimobilis
S. pneumoniae	Streptococcus pneumoniae
S. pombe	Schizosaccharomyces pombe
S. proteamaculans	Serratia proteamaculans
S. pseudintermedius	Staphylococcus pseudintermedius
S. putrefaciens	Shewanella putrefaciens
S. pyogenes	Streptococcus pyogenes
S. salivarius	Streptococcus salivarius
S. sanguinis	Streptococcus sanguinis
S. sanguis	Streptococcus sanguis
S. saprophyticus	Staphylococcus saprophyticus
S. schleiferi	Staphylococcus schleiferi
S. sciuri	Staphylococcus sciuri
S. Senftenberg	Salmonella Senftenberg
S. simulans	Staphylococcus simulans
S. sobrinus	Streptococcus sobrinus
S. soli	Sphingomonas soli
S. sonnei	Shigella sonnei
S. spiritivorum	Sphingobacterium spiritivorum
S. thermophilus	Streptococcus thermophilus
S. Thompson	Salmonella Thompson
S. Typhimurium	Salmonella Typhimurium
S. uvarum	Saccharomyces uvarum
S. viridians	Streptococcus viridians
S. warneri	Staphylococcus warneri
S. wittichii	Sphingomonas wittichii

S. xiamenensis	Shewanella xiamenensis
S. xylosus	Staphylococcus xylosus
S. yanoikuyae	Sphingobium yanoikuyae
SCCS	Scientific Committee on Consumer Safety
SEM	Scanning electron microscopy
t	Ton(s)
T. asahii	Trichosporon asahii
T. delbrueckii	Torulaspora delbrueckii
T. forsythia	Tannerella forsythia
T. harzianum	Trichoderma harzianum
T. longibrachiatum	Trichoderma longibrachiatum
T. mentagrophytes	Trichophyton mentagrophytes
T. rubrum	Trichophyton rubrum
T. viride	Trichoderma viride
T. whipplei	Tropheryma whipplei
V. alginolyticus	Vibrio alginolyticus
V. atypica	Veillonella atypica
V. cholerae	Vibrio cholerae
V. dispar	Veillonella dispar
V. indigofera	Vogesella indigofera
V. parahaemolyticus	Vibrio parahaemolyticus
V. parvula	Veillonella parvula
V. vulnificus	Vibrio vulnificus
v/v	Volume by volume
VBNC	Viable but non-culturable
VISA	Vancomycin intermediate-resistant Staphylococcus aureus
VISE	Vancomycin intermediate-resistant Staphylococcus epidermidis
VRE	Vancomycin-resistant Enterococcus spp.
w/w	Weight by weight
WD	Washer disinfector
WHO	World Health Organization
X. aerolatus	Xenophilus aerolatus
X. citri	Xanthomonas citri
X. maltophilia	Xanthomonas maltophilia
Y. enterocolitica	Yersinia enterocolitica
Y. pestis	Yersinia pestis
Y. pseudotuberculosis	Yersinia pseudotuberculosis
Y. ruckeri	Yersinia ruckeri

List of Figures

Introduction

1

1.1 Background

Antibiotic resistance is increasing worldwide. Since 2015, various measures are enforced by the WHO in a global action plan to at least slow down this development, e.g. to reduce the incidence of infection through effective sanitation, hygiene and infection prevention measures [28]. For this purpose, various antiseptic agents are used for different types of applications such as hand disinfection, skin antisepsis or surface disinfection. These antiseptic products usually comply with the necessary respective efficacy standards and aim to reduce the microbial load on the target organ or surface to a level that a transmission of the micro-organism to a patient or to food is no longer possible. The excessive use of antiseptics and biocides in health care, agriculture and the environment, however, can lead to resistance against these compounds and potentially also to cross-resistance to antibiotics [21, 25]. Concerns have been raised in recent years regarding co-selection for antibiotic resistance among bacteria exposed to biocides used as disinfectants, antiseptics and preservatives, and to heavy metals (particularly copper and zinc) used as growth promoters and therapeutic agents for some livestock species [26]. It is therefore worth to have a look at various antiseptic agents used for different types of application and to assess their ability to develop resistance to the compound itself and cross-resistance to other antiseptic compounds and antibiotics.

While biocides are generally not as well-studied as antibiotics, it is becoming clear that bacteria employ the same major resistance strategies against antiseptic agents such as inhibitor inactivation, target site alteration and target site exclusion [2]. The main mechanisms of biocidal resistance such as cellular impermeability, efflux pumps, plasmids and the presence of biofilms have been reviewed extensively in 2001 [22]. Although reports of resistance have often paralleled issues including inadequate cleaning, incorrect product use or ineffective infection control practices [17], it seems relevant to address specifically the potential of biocidal resistance for the most commonly used agents.

G. Kampf, *Antiseptic Stewardship*, https://doi.org/10.1007/978-3-319-98785-9_1

Bacterial populations exploit a range of mechanisms to survive challenges [3]. Bacteria can seek protected environments passively by avoiding deadly environments or actively by manipulating their phenotypic expression and gathering in structured biofilm communities [3]. In addition, an underappreciated healthcare-associated ecosystem exists and strongly suggests that effective control of the overall multidrug-resistant micro-organisms burden will require stewardship interventions that take into account both primary and secondary impacts of antibiotic treatments [27], which may also include antiseptic treatments. Water flow paths also trigger the formation of antibiotic resistance, since they transport antibiotics, multiresistant bacteria and free resistance genes through the soil, so that they function as hot spots for the accumulation of antibiotics and trigger the formation of resistance genes in soil [13].

Resistance or adaptation to biocidal agents has been described already 130 years ago, e.g. in 1887 by Kossiakoff when bacteria acquired the faculty of developing resistance to gradually increasing doses of some chemical agents such as boric acid or mercury chloride. Later in 1912, Regenstein studied the adaptation of bacteria to disinfectants including phenol [23]. The history of scientific evidence on bacterial adaptation to biocidal agents was reviewed by Russell in 2004 and is a valuable source of information for the interested reader [23].

Sudden awareness of the evaluation of risks and benefits was created in 2016 when the FDA banned 19 active ingredients including triclosan for antimicrobial soaps used by the general population at home [4]. The reason for the ban is rather simple but a milestone at the same time: "A risk must be balanced with the demonstration of a direct clinical benefit (i.e. *a reduction of infection*)—that the product is superior to washing with non-antibacterial soap and water in reducing infection". The decision has raised a lot of support in the scientific community: "We applaud this rule specifically because of the associated risks that triclosan poses to the spread of antibiotic resistance throughout the environment. This persistent chemical constantly stresses bacteria to adapt, and behaviour that promotes antibiotic resistance needs to be stopped immediately when the benefits are null" [18].

When we look at other antiseptic products containing two or more active ingredients, we can find similar scenarios, for example an alcohol-based hand rub containing chlorhexidine. Does a direct clinical benefit justify the supplementation of chlorhexidine taking into account that "this persistent chemical may constantly stress bacteria to adapt in the immediate patient surrounding"? The same questions can be raised for alcohol-based skin antiseptics supplemented with "persistent active ingredients". It can also be raised for antiseptic soaps based on one active ingredient typically used in food processing or veterinary medicine. And it can be raised for surface disinfectants, e.g. containing a composition of quaternary ammonium compounds. Is the risk for bacterial adaptation and resistance higher for the quaternary ammonium compounds compared to other biocidal agents such as hydrogen peroxide or sodium hypochlorite so that the promotion of resistance can be reduced with a product based on antiseptic compounds with a lower adaptation potential?

Selection and enrichment for antibiotic resistant bacteria is often a consequence of weak, non-lethal selective pressures—caused by low levels of antibiotics [1]. Adaptive tolerance is a specific class of non-mutational resistance that is characterized by its transient nature, although some adaptive changes may be stable. It occurs in response to certain environmental conditions or due to epigenetic phenomena like persistence [7]. That is why the effects of low-level exposure to antiseptics are addressed as well as any known resistance mechanisms, resistance genes and cross-resistance to other biocidal agents and antibiotics as they may all contribute to biocide tolerance [16]. For the final evaluation, the results on any adaptive MIC change are summarized in three general categories: no MIC change, weak MIC change ($\leq$4-fold) and strong MIC change (>4-fold) which may be unstable, stable or of unknown stability. The adaptive effect may be mediated by efflux pumps [6, 9]. Antibacterial biocides at low concentrations can also contribute to antibiotic resistance development by facilitating the spread of antibiotic resistance between bacteria [10].

Another aspect is the role of biofilms [14]. They are the predominant mode of microbial growth in drinking water systems. A dynamic exchange of individuals occurs between the attached and planktonic populations, while lateral gene transfer mediates genetic exchange in these bacterial communities [5]. In addition, the deciphering and control of anti-biofilm properties represent future challenges in human infection control [19] and in food processing [24]. Prevention of biofilm formation on implant surfaces is considered to be a key element for the successful prevention of implant infections caused by *S. aureus* and *S. epidermidis* [20]. Various types of infections associated with biofilms have been described [12]. Any effect of biocidal agents on biofilm formation, removal and fixation are therefore essential. Possible effects of low-level exposure also include inhibition or promotion of biofilm formation.

1.2 Dimensions of Antiseptic Stewardship

In 2005, it was suggested that the scientific community must weigh the risks and benefits of using biocides in clinical and community environments, to determine whether additional precautions are needed to guide biocide development and use [15]. In recent years, good stewardship programmes have been requested, not only for clinically used antibiotics, but also for antimicrobials used in agriculture, and for critically important antiseptics such as chlorhexidine [11, 25]. The Florence statement from 2017 suggests two types of actions that are highly relevant in an antiseptic stewardship programme.

1. Avoid chemicals except where they provide an evidence-based health benefit, and there is adequate evidence demonstrating they are safe [8].

Many antiseptic products contain two or more biocidal agents. For example, quite a number of alcohol-based hand rubs are supplemented with non-volatile biocidal agents such as chlorhexidine digluconate, mecetronium etilsulfate or octenidine dihydrochloride. The aim is to achieve a "persistent activity", e.g. during surgical hand disinfection. Skin disinfectants may also contain such agents in order to achieve a "persistent activity" with the aim to slow down bacterial regrowth and finally reduce surgical site infections. But what should be done if these agents do not contribute to the overall efficacy and do not provide an evidence-based health benefit but are prone to enhance the development of resistance and cross-resistance, e.g., by low-level exposure, or stimulate biofilm formation so that persistence of the target micro-organism and its adaptation are more likely? Alternative antiseptic products with the same spectrum of antimicrobial activity are often available, may be even with the same active agents except for the one claimed to have "persistent activity". In order to reduce unnecessary selection pressure in all fields of application, it seems mandatory to select these alternative antiseptic products provided that the overall efficacy, user acceptability, dermal tolerance or material compatibility are non-inferior. The WHO goal to reduce the incidence of infection through effective sanitation, hygiene and infection prevention measures can still be assured. At the same time, the selection pressure on biocidal agents and possible cross-resistance to other biocidal agents or antibiotics can be reduced.

2. Where antimicrobials are necessary, use safer alternatives that are not persistent and pose no risk to humans or ecosystems [8].

Especially for the disinfection of inanimate surfaces in health care, animal production, food processing or veterinary medicine, products often contain two or more active ingredients. Some of them have a low potential to adapt, others may lose some of their efficacy after a few applications so that the target micro-organisms have become tolerant to the commonly used concentration. If this happens, it has two relevant implications. First, the antiseptic is unlikely to exhibit its anticipated effect in real life, e.g. in health care. Second, adapted or resistant micro-organisms survive the antiseptic treatment and may persist on the surfaces. It will be easier for them to resist the next antiseptic treatment with the same product. If the adaptation or resistance is correlated with enhanced biofilm formation, then it may be even more difficult if not impossible to eliminate the pathogens from the surface. That is why it is relevant to know the potential of different antiseptic agents to adapt, to develop resistance or to enhance biofilm formation. Alternative antiseptic agents with the same spectrum of antimicrobial activity may be available, probably with other active agents. In order to reduce unnecessary selection pressure in all fields of application, it seems mandatory to select these alternative antiseptic products provided that the overall efficacy and material compatibility are comparable. The WHO goal to reduce the incidence of infection through effective sanitation, hygiene and infection prevention measures can still be assured. At the same time, the selection pressure on biocidal agents and possible cross-resistance to other biocidal agents or antibiotics can be reduced.

1.3 Antiseptic Stewardship Per Type of Application

This book will hopefully help to make informed decisions on the selection of biocidal products with the preference for those agents exhibiting a lower selection pressure. Alcohol-based products for hand disinfection or skin antisepsis may not need most of additional antiseptic agents unless there is an evidence-based health benefit for this specific antiseptic agent. Nevertheless, products supplemented with these antiseptic agents such as chlorhexidine digluconate, benzalkonium chloride or octenidine dihydrochloride are also often very effective, comply with the efficacy criteria and have good user acceptability. And yet they contain one of these substances that are likely to increase selection pressure without a direct clinical benefit.

In surface disinfection, some of the commonly used biocidal agents are compared regarding their potential for selection pressure. For this type of application, it may be an option to prefer those agents that are less prone for an adaptive response and inhibit rather than increase biofilm formation. Having this awareness is already the beginning of antiseptic stewardship in real life.

I am confident that opportunities to reduce selection pressure can be found in various settings by a more careful and responsible product selection without losing the required antimicrobial efficacy of the antiseptic. The magnitude of the contribution of antiseptic stewardship to slow down the global development of antimicrobial resistance is almost impossible to predict but it is nevertheless one component that will quite likely contribute.

References

1. Andersson DI, Hughes D (2012) Evolution of antibiotic resistance at non-lethal drug concentrations. Drug Resis Updat: Rev Comment Antimicro Anticancer Chemother 15 (3):162–172. https://doi.org/10.1016/j.drup.2012.03.005
2. Chapman JS (2003) Biocide resistance mechanisms. Int Biodeter Biodegr 51(2):133–138. https://doi.org/10.1016/S0964-8305(02)00097-5
3. Cogan NG (2013) Concepts in disinfection of bacterial populations. Math Biosci 245(2):111–125. https://doi.org/10.1016/j.mbs.2013.07.007
4. Department of Health and Human Services; Food and Drug Administration (2016) Safety and effectiveness of consumer antiseptics. Topical Antimicro Drug Products Over-the-Counter Human Use. Fed Reg 81(172):61106–61130
5. Farkas A, Butiuc-Keul A, Ciataras D, Neamtu C, Craciunas C, Podar D, Dragan-Bularda M (2013) Microbiological contamination and resistance genes in biofilms occurring during the drinking water treatment process. Sci Total Environ 443:932–938. https://doi.org/10.1016/j.scitotenv.2012.11.068
6. Fernandes P, Ferreira BS, Cabral JM (2003) Solvent tolerance in bacteria: role of efflux pumps and cross-resistance with antibiotics. Int J Antimicrob Agents 22(3):211–216
7. Fernandez L, Breidenstein EB, Hancock RE (2011) Creeping baselines and adaptive resistance to antibiotics. Drug Resis Updat: Rev Comment Antimicro Anticancer Chemother 14(1):1–21. https://doi.org/10.1016/j.drup.2011.01.001

8. Halden RU, Lindeman AE, Aiello AE, Andrews D, Arnold WA, Fair P, Fuoco RE, Geer LA, Johnson PI, Lohmann R, McNeill K, Sacks VP, Schettler T, Weber R, Zoeller RT, Blum A (2017) The florence statement on triclosan and triclocarban. Environ Health Perspect 125 (6):064501. https://doi.org/10.1289/ehp1788
9. Hernando-Amado S, Blanco P, Alcalde-Rico M, Corona F, Reales-Calderon JA, Sanchez MB, Martinez JL (2016) Multidrug efflux pumps as main players in intrinsic and acquired resistance to antimicrobials. Drug Resis Updat: Rev Comment Antimicro Anticancer Chemother 28:13–27. https://doi.org/10.1016/j.drup.2016.06.007
10. Jutkina J, Marathe NP, Flach CF, Larsson DGJ (2017) Antibiotics and common antibacterial biocides stimulate horizontal transfer of resistance at low concentrations. Sci Total Environ 616–617:172–178. https://doi.org/10.1016/j.scitotenv.2017.10.312
11. Kampf G (2016) Acquired resistance to chlorhexidine—is it time to establish an "antiseptic stewardship" initiative? J Hosp Infect 94(3):213–227. https://doi.org/10.1016/j.jhin.2016.08.018
12. Lebeaux D, Chauhan A, Rendueles O, Beloin C (2013) From in vitro to in vivo models of bacterial biofilm-related infections. Pathogens (Basel, Switzerland) 2(2):288–356. https://doi.org/10.3390/pathogens2020288
13. Lüneberg K, Prado B, Broszat M, Dalkmann P, Díaz D, Huebner J, Amelung W, López-Vidal Y, Siemens J, Grohmann E, Siebe C (2018) Water flow paths are hotspots for the dissemination of antibiotic resistance in soil. Chemosphere 193:1198–1206. https://doi.org/10.1016/j.chemosphere.2017.11.143
14. Mah TF, O'Toole GA (2001) Mechanisms of biofilm resistance to antimicrobial agents. Trends Microbiol 9(1):34–39
15. Maillard J-Y (2005) Antimicrobial biocides in the healthcare environment: efficacy, usage, policies, and perceived problems. Ther Clin Risk Manag 1(4):307–320
16. McBain AJ, Gilbert P (2001) Biocide tolerance and the harbingers of doom. Int Biodeter Biodegr 47(1):55–61
17. McDonnell G, Russell AD (1999) Antiseptics and disinfectants: activity, action, resistance. Clin Microbiol Rev 12(1):147–179
18. McNamara PJ, Levy SB (2016) Triclosan: an instructive tale. Antimicrob Agents Chemother 60(12):7015–7016. https://doi.org/10.1128/aac.02105-16
19. Miquel S, Lagrafeuille R, Souweine B, Forestier C (2016) Anti-biofilm activity as a health issue. Front Microbiol 7:592. https://doi.org/10.3389/fmicb.2016.00592
20. Oliveira WF, Silva PMS, Silva RCS, Silva GMM, Machado G, Coelho L, Correia MTS (2018) Staphylococcus aureus and Staphylococcus epidermidis infections on implants. J Hosp Infect 98(2):111–117. https://doi.org/10.1016/j.jhin.2017.11.008
21. Ortega Morente E, Fernandez-Fuentes MA, Grande Burgos MJ, Abriouel H, Perez Pulido R, Galvez A (2013) Biocide tolerance in bacteria. Int J Food Microbiol 162(1):13–25. https://doi.org/10.1016/j.ijfoodmicro.2012.12.028
22. Russell AD (2001) Mechanisms of bacterial insusceptibility to biocides. Am J Infect Control 29(4):259–261. https://doi.org/10.1067/mic.2001.115671
23. Russell AD (2004) Bacterial adaptation and resistance to antiseptics, disinfectants and preservatives is not a new phenomenon. J Hosp Infect 57(2):97–104. https://doi.org/10.1016/j.jhin.2004.01.004
24. Shi X, Zhu X (2009) Biofilm formation and food safety in food industries. Trends Food Sci Technol 20(9):407–413
25. Venter H, Henningsen ML, Begg SL (2017) Antimicrobial resistance in healthcare, agriculture and the environment: the biochemistry behind the headlines. Essays Biochem 61(1):1–10. https://doi.org/10.1042/ebc20160053
26. Wales AD, Davies RH (2015) Co-selection of resistance to antibiotics, biocides and heavy metals, and its relevance to foodborne pathogens. Antibiotics (Basel, Switzerland) 4(4):567–604. https://doi.org/10.3390/antibiotics4040567

27. Wang J, Foxman B, Mody L, Snitkin ES (2017) Network of microbial and antibiotic interactions drive colonization and infection with multidrug-resistant organisms. Proc Natl Acad Sci USA 114(39):10467–10472. https://doi.org/10.1073/pnas.1710235114
28. WHO (2015) Global action plan on antimicrobial resistance. WHO. http://www.who.int/medicines/publications/essentialmedicines/EML2015_8-May-15.pdf

Ethanol

2

2.1 Chemical Characterization

Ethanol is a simple alcohol. It is a volatile, flammable, colourless liquid with a slight characteristic odour. Ethanol is naturally produced by the fermentation of sugars by yeasts or via petrochemical processes. The basic chemical information on ethanol is summarized in Table 2.1.

2.2 Types of Application

According to the information provided by the European Chemicals Agency (ECHA), ethanol is used by consumers, in articles, by professional workers (widespread uses), in formulation or repacking, at industrial sites and in manufacturing. The use by consumers includes fuels, inks and toners [33]. Use by professional workers includes hand disinfection by healthcare workers and in veterinary medicine, skin antisepsis prior to surgery, disinfection of inanimate surfaces and rinsing of endoscope channels after manual processing. In 2009, the World Health Organization (WHO) has recommended to use alcohol-based hand rubs, e.g. based on ethanol, in specific situations during patient care for prevention of healthcare-associated infections [113]. Alcohol-based hand rubs, e.g. based on ethanol, are also recommended for the preoperative decontamination of hands for the prevention of surgical site infections [115].

Since 2015, the WHO has classified denatured ethanol at 70% as an antiseptic and at 80% (v/v) as a disinfectant for alcohol-based hand rubbing as an "essential medicine" [114], both for adults and children up to 12 years of age [116]. Alcohols including ethanol have been used effectively to disinfect oral and rectal thermometers [36, 100], hospital pagers [99], scissors [29] and stethoscopes [119]. Alcohols have been used to disinfect fibre-optic endoscopes [7, 37]. Alcohol towelettes are still used to disinfect small surfaces such as rubber stoppers of

G. Kampf, *Antiseptic Stewardship*, https://doi.org/10.1007/978-3-319-98785-9_2

Table 2.1 Basic chemical information on ethanol [33, 76]

CAS number	64-17-5
IUPAC name	Ethanol
Synonyms	Alcohol, absolute alcohol, ethyl alcohol, grain alcohol
Molecular formula	C_2H_6O
Molecular weight (g/mol)	46.042

multiple-dose medication vials or vaccine bottles [17, 71]. Furthermore, alcohol may be used to disinfect external surfaces of equipment (e.g. ventilators, manual ventilation bags) [112], cardiopulmonary resuscitation manikins [20], and ultrasound instruments or medication preparation areas [81]. In China, ethanol is used for hand hygiene, skin disinfection, surface disinfection and medical instrument disinfection at 75% [63].

2.2.1 European Chemicals Agency (European Union)

Ethanol is under review (June 2018) as an active biocidal agent for product types 1 (human hygiene), 2 (disinfectants and algaecides not intended for direct application to humans or animals), 4 (food and feed area) and 6 (preservatives for products during storage) [34].

2.2.2 Environmental Protection Agency (USA)

Ethanol has last been reregistered by the EPA in 1995. It is used as a component in a variety of commercial and household products including a sterilant, medical disinfectants, virucides, sanitizers, fungicides and plant regulators (ripener). Ethanol is used with quaternary ammonium compounds for swimming pool water systems [106].

2.2.3 Food and Drug Administration (USA)

In 1994, the tentative final monograph for healthcare antiseptic products classified ethanol between 60 and 95% as "generally recognized as safe and effective" for patient preoperative skin preparation, surgical hand scrubbing and for healthcare personnel hand wash [27]. In 2015, the classification was changed. Ethanol at 60 to 95% was now eligible for three types of application: patient preoperative skin preparation, healthcare personnel hand rub and surgical hand rub [28]. It is now classified in category IIISE indicating that available data are insufficient to classify ethanol as safe and effective, and further testing is required [28]. The main aspect is the safety under maximal use conditions [28].

2.2.4 Overall Environmental Impact

Ethanol is manufactured and/or imported in the European Economic Area in 1 to 10 million t per year [33]. Other release to the environment of this substance is likely to occur from outdoor use, indoor use (e.g. machine wash liquids/detergents, automotive care products, paints and coating or adhesives, fragrances and air fresheners), outdoor use in close systems with minimal release (e.g. hydraulic liquids in automotive suspension, lubricants in motor oil and break fluids) and indoor use in close systems with minimal release (e.g. cooling liquids in refrigerators, oil-based electric heaters) [33].

2.3 Spectrum of Antimicrobial Activity

2.3.1 Bactericidal Activity

2.3.1.1 Bacteriostatic Activity (MIC Values)

The bacteriostatic activity of ethanol begins at 3.1% with *Staphylococcus* spp. and has its highest MIC value against isolates of *E. coli* at 32% (Table 2.2). In addition, some authors reported that 70% ethanol is ineffective using a disc diffusion test, e.g. against 31 of 32 *S. aureus* isolates from insects [82], against 60% of 35 MRSA isolates and 66.7% of 60 MSSA isolates [19]. Due to the volatility of ethanol, however, the results of disc diffusion tests with ethanol should be evaluated with caution as they are likely to be false negative.

2.3.1.2 Bactericidal Activity (Suspension Tests)

Ethanol at 78% or more has comprehensive bactericidal activity within 30 s, at 85% even in 15 s (Table 2.3). It covers both culture collection strains and various clinical isolates. Lower concentrations of ethanol may require longer exposure times to reach an equivalent efficacy. These findings are supported by MBC values reported for 64 MDR clinical *A. baumannii* isolates (30% in 10 min), 56 QAC-tolerant *S. aureus* isolates (40–60% in 5 min) and 42 clinical MRSA isolates (30–60% in 5 min) [23, 63, 75].

2.3.1.3 Activity Against Bacteria in Biofilms

The efficacy of ethanol against bacteria in artificially grown biofilms is variable. Most studies with ethanol at 70% indicate a rather poor bactericidal efficacy within 60 min against *A. baumannii*, *P. aeruginosa*, *S.* Typhimurium and *S. aureus* with log reductions $\leq$ 2.0. Against *A. baumannii*, however, ethanol at 70% revealed in one study a very good bactericidal efficacy within 10 min. The efficacy of 35–70% ethanol against *E. coli* in biofilm was mostly good (Table 2.4).

The overall low bactericidal efficacy of ethanol at around 70% is confirmed by another study. Ethanol at 70% applied for up to 1 h to biofilm of *S.* Typhimurium, *E. coli*, *S. mutans* or *B. fragilis* on glass or rubber carrier was not effective enough to

Table 2.2 MIC values of various bacterial species to ethanol

Species	Strains/isolates	MIC value	References
A. baumannii	47 clinical isolates	7.5–22.5%	[61]
A. calcoaceticus	ATCC 19606	4.4%	[88]
B. stearothermophilus	ATCC 7953	8.8%	[88]
B. subtilis var. globigii	ATCC 9372	8.8%	[88]
E. cloacae	Strain IAL 1976	8.8%	[88]
E. faecalis	ATCC 29212	25%	[15]
E. hirae	Strain CIP 5855	8.8%	[70]
Enterocococcus spp.	6 glycopeptide-susceptible isolates	6.2–25%	[15]
Enterocococcus spp.	8 glycopeptide-resistant isolates	12.5–25%	[15]
E. coli	ATCC 25922	6.6%	[88]
E. coli	Reference strain and clinical isolate	10–32%	[59]
E. coli	ATCC 25922	17.5%	[70]
L. monocytogenes	Strain Scott A	5%	[80]
L. monocytogenes	10 isolates from food	6.3–12.5%	[1]
Micrococcus spp.	1 isolates from a clean room	6.3%	[22]
M. morganii	ATCC 25830	17.5%	[70]
P. aeruginosa	ATCC 27853	17.5%	[70]
S. marcescens	Strain IAL 1478	4.4%	[88]
S. aureus	MTCC 737	6.3%	[22]
S. aureus	ATCC 25923	8.8%	[88]
S. aureus	Strain CIP 53154	8.8%	[70]
Staphylococcus spp.	3 isolates from a clean room	3.1%	[22]

Table 2.3 Bactericidal activity of ethanol in suspension tests

Species	Strains/isolates	Exposure time	Concentration	$\log_{10}$ reduction	References
A. baumannii	ATCC 19606 and 2 clinical isolates incl. MDR	15 s	85% (P)	≥5.3	[51]
A. baumannii	81 clinical and environmental isolates	24 h	70% (S)	>5.0	[58]
A. calcoaceticus–baumannii	Clinical isolate	30 s	62% (P)	4.1	[45]
		60 s		5.1	
A. lwoffi	ATCC 15309 and 1 clinical isolate	15 s	85% (P)	≥5.3	[51]
B. fragilis	ATCC 25285 and 1 clinical isolate	15 s	85% (P)	≥6.6	[51]
B. cenocepacia	Strains LMG 16656 and LMG 18828	2 min, 5 min, 10 min	70% (S)	≥5.0	[87]
B. cepacia	ATCC 25416 and 1 clinical isolate	15 s	85% (P)	≥5.5	[51]

(continued)

Table 2.3 (continued)

Species	Strains/isolates	Exposure time	Concentration	$\log_{10}$ reduction	References
C. jejuni	ATCC BAA-1062, ATCC 33560 and 2 field strains	1 min	70% (S)	>6.0	[44]
C. difficile[a]	ATCC 9689 and 1 clinical isolate	15 s	85% (P)	≥5.3	[51]
C. jeikeium	ATCC 43216	5 s	75% (S)	>5.0	[118]
E. aerogenes	ATCC 13048 and 1 clinical isolate	15 s	85% (P)	≥5.9	[51]
E. cloacae	ATCC 13047 and 1 clinical isolate	15 s	85% (P)	≥6.5	[51]
E. faecalis	ATCC 29212 and 2 clinical isolates incl. VRE	15 s	85% (P)	≥7.1	[51]
E. faecium	ATCC 19434 and 2 clinical isolates incl. VRE	15 s	85% (P)	≥6.7	[51]
E. faecium and *E. faecalis*	NCTC 775 and 8 clinical isolates (4 of them vancomycin-resistant)	30 s	45% (S)	0.6–6.1	[16]
		1 min		1.7–7.7	
		5 min		3.8–7.7	
E. hirae	ATCC 10541	30 s	85% (P)	>5.1	[54]
E. hirae	ATCC 10541	30 s	78.2% (P)	≥4.9	[53]
E. coli	ATCC 11229 and 25922 and 3 clinical isolates incl. MDR	15 s	85% (P)	≥6.5	[51]
E. coli	NCTC 10538	30 s	85% (P)	>5.3	[54]
E. coli	NCTC 10538	30 s	78.2% (P)	≥5.1	[53]
E. coli	ATCC 25922	24 h	70% (S)	>5.0	[58]
H. influenzae	ATCC 19418 and 1 clinical isolate	15 s	85% (P)	≥5.3	[51]
H. parasuis	2 strains (serovars 1 and 5)	1 min	70% (S)	>6.0	[92]
				4.4–4.6[b]	
H. pylori	NCTC 11637, NCTC 11916 and 7 clinical isolates	15 s	80% (S)	>5.0	[2]
		15–30 s		>5.0[b]	
K. pneumoniae	ATCC 11296 and 1 clinical isolate	15 s	85% (P)	≥6.5	[51]
K. oxytoca	ATCC 43165 and 1 clinical isolate	15 s	85% (P)	≥6.6	[51]
L. innocua	Strain LCDC 86-417	1 min	70% (P)	>5.0	[11]
L. monocytogenes	ATCC 7644 and 1 clinical isolate	15 s	85% (P)	≥6.2	[51]
L. monocytogenes	Strain LCDC 88-702	1 min	70% (P)	>5.0	[11]
M. luteus	ATCC 7468 and 1 clinical isolate	15 s	85% (P)	≥5.4	[51]

(continued)

Table 2.3 (continued)

Species	Strains/isolates	Exposure time	Concentration	log_{10} reduction	References
P. mirabilis	ATCC 7002 and 1 clinical isolate	15 s	85% (P)	≥6.7	[51]
P. aeruginosa	ATCC 15442 and 27853 and 2 clinical isolates incl. MDR	15 s	85% (P)	≥6.6	[51]
P. aeruginosa	ATCC 15442	30 s	85% (P)	>5.3	[54]
P. aeruginosa	ATCC 15442	30 s	78.2% (P)	≥5.2	[53]
P. aeruginosa	ATCC 15442	2 min	40% (S)	>5.0	[107]
S. Enteritidis	ATCC 13076 and 1 clinical isolate	15 s	85% (P)	≥6.8	[51]
S. Typhimurium	ATCC 13311 and 1 clinical isolate	15 s	85% (P)	≥6.7	[51]
S. Typhimurium	ATCC 14028 and 3 *Salmonella* spp. isolates	5 min	70% (P)	≥5.0	[73]
S. marcescens	ATCC 14756 and 1 clinical isolate	15 s	85% (P)	≥5.6	[51]
S. sonnei	ATCC 11060 and 1 clinical isolate	15 s	85% (P)	≥6.5	[51]
S. aureus	ATCC 6538 and 29213 and 2 clinical isolates incl. MRSA/VISA	15 s	85% (P)	≥6.3	[51]
S. aureus	ATCC 6538	30 s	85% (P)	>5.3	[54]
S. aureus	ATCC 6538	30 s	78.2% (P)	≥4.9	[53]
S. aureus	Clinical MRSA isolate	30 s	62% (P)	3.5	[45]
		60 s		4.2	
S. aureus	ATCC 6538	2 min	40% (S)	>5.0	[107]
S. epidermidis	ATCC 12228 and 2 clinical isolates incl. VISE	15 s	85% (P)	≥5.6	[51]
S. epidermidis	ATCC 14990	5 s	75% (S)	>5.0	[118]
S. haemolyticus	ATCC 29970 and 1 clinical isolate	15 s	85% (P)	≥5.3	[51]
S. hominis	ATCC 27844 and 1 clinical isolate	15 s	85% (P)	≥5.4	[51]
S. saprophyticus	ATCC 15305 and 1 clinical isolate	15 s	85% (P)	≥5.4	[51]
S. pneumoniae	ATCC 6304 and 2 clinical isolate incl. PRSP	15 s	85% (P)	≥5.3	[51]
S. pyogenes	ATCC 19615 and 1 clinical isolate	15 s	85% (P)	≥5.5	[51]

P commercial product; *S* solution; [a]vegetative cell form; [b]with organic load

Table 2.4 Efficacy of ethanol-based formulations against bacteria in biofilms

Species	Strains/isolates	Type of biofilm	Exposure time	Concentration	log_{10} reduction	References
A. baumannii	ATCC 17978, ATCC 190451	24-h incubation on polystyrene tissue culture plates	10 min	70% (S)	No reduction	[74]
A. baumannii	ATCC 17978, ATCC 190451	5-d incubation on Foley catheter pieces	60 min	70% (S)	No reduction	[74]
A. baumannii	64 MDR clinical isolates	5-d incubation in polystyrene plates	10 min	70% (S)[a]	>7.8	[23]
				60% (S)	3.1–5.2	
				50% (S)	2.0–4.1	
				40% (S)	1.3–2.1	
				30% (S)	0.5–1.6	
				20% (S)	<1.0	
E. hirae	Strain CIP 5855	48-h incubation on polypropylene, PVC and silicone	30 min	35% (S)	>5.0	[70]
				17.5% (S)	3.1–5.0	
				8.75% (S)	0.0-1.0	
E. coli	ATCC 35218	48-h incubation on glass, polypropylene, polycarbonate, silicone and PVC	30 min	70% (S)	Complete inactivation	[69]
E. coli	Strain O157:H7	5-d incubation on stainless steel	1 min	58.8% (P)	>4.0	[8]
			5 min		>4.0	
E. coli	ATCC 25922	48-h incubation on polypropylene, PVC and silicone	30 min	35% (S)	≥ 5.0	[70]
				17.5% (S)	1.0–2.0	
				8.75% (S)	0.0	
L. plantarum	JCM 1149	24-h incubation on glass cover slips	30 min	40% (S)	5.8	[57]
				30% (S)	4.0	
				20% (S)	0.1	

(continued)

Table 2.4 (continued)

Species	Strains/isolates	Type of biofilm	Exposure time	Concentration	log_{10} reduction	References
M. morganii	ATCC 25830	48-h incubation on polypropylene, PVC and silicone	30 min	35% (S)	≥ 5.0	[70]
				17.5% (S)	0.3–2.6	
				8.75% (S)	0.0	
P. aeruginosa	ATCC 700928	24-h incubation in microplates	1 min	70% (S)	1.0	[104]
			5 min		1.0	
			60 min		1.4	
P. aeruginosa	ATCC 27853	48-h incubation on polypropylene, PVC and silicone	30 min	35% (S)	>5.0	[70]
				17.5% (S)	3.9–5.0	
				8.75% (S)	0.0	
S. Typhimurium	ATCC 14028	3-d incubation on a 96-peg lid	1 min	70% (S)	2.0	[117]
			5 min		1.9	
S. liquefaciens	Isolate from a raw-chicken processing plant	3-d incubation on stainless steel	6 min	75% (S)	3.6	[62]
S. putrefaciens	Isolate from a raw-chicken processing plant	3-d incubation on stainless steel	6 min	75% (S)	3.0	[62]
S. aureus	ATCC 35556, ATCC 29213 (both MSSA), ATCC 43300, strain L32 (both MRSA)	24-h incubation on polystyrene plates	24 h	95% (S)	No significant reduction	[65]
				80% (S)		
				60% (S)		
				40% (S)		
S. aureus	ATCC 6538	72-h incubation in microplates	1 min	70% (S)	1.0	[104]
			5 min		1.4	
			60 min		2.0	
S. aureus	ATCC 12600, 12692 and 49444	5-d incubation on stainless steel	1 min	58.8% (P)	>4.0	[8]

(continued)

Table 2.4 (continued)

Species	Strains/isolates	Type of biofilm	Exposure time	Concentration	log_{10} reduction	References
			5 min		>4.0	
S. aureus	CIP 53154	48-h incubation on polypropylene, PVC and silicone	30 min	35% (S)	≥ 5.0	[70]
				17.5% (S)	4.2–5.0	
				8.75% (S)	0.0–3.0	
S. aureus	Strain AH 2547	Overnight incubation on porcine skin	4 lateral wipes with soaked pads	10% (S)	0.4	[111]
S. epidermidis	ATCC 35934 and a biofilm deficient mutant M7	24-h incubation on polystyrene plates	24 h	95% (S)	Significant reduction	[65]
				80% (S)		
				60% (S)		
				40% (S)		
S. maltophilia	Clinical strain	24-h incubation in silicone catheter segment	1 h	40% (S)	>5.0	[86]
				25% (S)		
S. maltophilia	14 clinical strains	24- and 48-h incubation on polystyrene microtiter plates	1 h	40% (S)	>5.0	[86]
				25% (S)		
Mixed oral biofilm	*S. oralis* ATCC 10557, *S. gordonii* ATCC 10558 and *A. naeslundii* ATCC 19039	20-h incubation in biofilm reactor	1 h	40% (S)	1.2	[25]

[a]when 70% ethanol is combined with 2% chlorhexidine, the same effect can be achieved in 1 min

prevent survival of the *S. Typhimurium* on rubber and glass (10 min), *E. coli* on rubber (30 min) and glass (15 min) and *S. mutans* on rubber (1 h) [109]. Nevertheless, a solution of 80% ethanol was applied to cells from a *B. cepacia* biofilm grown with six isolates from disinfectants and aerosol solution for 5 d on silicone discs. A > 5.0 log reduction was found after only 15 s exposure indicating a strong bactericidal activity [72]. The efficacy in naturally grown mixed biofilm is likely to be lower. Mixed biofilm (e.g. *S. liquefaciens* and *S. putrefaciens*) was found to be more difficult to inactivate by ethanol at 75% compared to single-species biofilms [62].

B. subtilis biofilm colonies and pellicles are extremely liquid and gas repellent, greatly surpassing the properties of known repellent surfaces such as Teflon and lotus leaves. One study showed that the biofilm surface is persistently non-wetting against up to 80% ethanol as well as other organic solvents and commercial biocides. The biofilm non-wetting properties arose from both the polysaccharide and protein components of the extracellular matrix and were a synergistic result of surface chemistry, multiscale surface roughness and re-entrant topography. Moreover, gas impenetrability of the biofilm surface was reported, implying defence capability against vapour-phase antimicrobials as well [30].

2.3.1.4 Bactericidal Activity for Hygienic Hand Disinfection

The efficacy of ethanol-based hand rubs has been mostly evaluated on hands artificially contaminated with *E. coli* according to EN 1500 with an application of 3 ml for 30 s. A summary of published data has recently been published [50]. Preparations with up to 70% ethanol (w/w) mostly fail to meet the EN 1500 efficacy requirements, whereas solutions or gels with 80% (w/w) or more are mostly effective enough. The application of larger volumes (e.g. 6 ml) or smaller volumes (e.g. 2 ml) will yield different results [40, 42, 52, 66]. Application volumes of 1.5–2.0 ml are quite likely in clinical practice [52, 66].

According to ASTM E 2755, commercial preparations with volumes between 1.1 and 2 ml often reveal a log reduction between 2.0 and 3.3 on hands artificially contaminated with *S. marcescens* [50].

2.3.1.5 Bactericidal Activity for Surgical Hand Disinfection

The efficacy for surgical hand disinfection is often determined against the resident hand flora according to EN 12791, mostly with application times of 1.5 or 3 min. Formulations with ethanol of less than 80% (w/w) typically fail to meet the efficacy requirements even when applied for 5 min. Preparations with 80% or 85% (both w/w) are usually effective enough [50].

According to ASTM E 1115, the efficacy of a preparation with 61% ethanol against the resident hand flora is poor (immediate efficacy day 1: mean log reduction of 1.1). It is better with formulations based on 70% or 80% ethanol (both w/w) with 2.1 log and 3.1 log [50].

2.3.1.6 Bactericidal Activity in Carrier Tests

The bactericidal efficacy of ethanol in carrier tests depends on both the concentration and the exposure time. In most studies, ethanol at 70% was used. Ethanol at 70% was able to kill 10 bacterial species (*S. aureus*, *S. pyogenes*, *S. viridians*, *S. faecalis*, *E. coli*, *K. pneumoniae*, *P. vulgaris*, *P. pyocyanea*, *C. diphteriae*, *M. phlei*) in 30 s on dried films. Against *L. innocua* and *L. monocytogenes,* ethanol at 70% was effective within 1 min in a carrier test with 3.0–5.0 log; in the presence of serum, however, the efficacy was substantially lower with 1.0–2.0 log [11]. When *S. aureus* is placed on a glass cup carrier and exposed to 70% ethanol, a log reduction of 4.3 is found after 1 min and >6 after 10 min [14]. When *M. pneumoniae*, *M. gallisepticum* and *A. laidlawii* were exposed to 70% ethanol for 5 min on stainless steel, a sufficient efficacy was found with a log reduction >4.5 for all tested species [32]. An early study provides similar data with a strong effect of 70% ethanol against *S. aureus* and *P. aeruginosa* in 5 min which was lower at an ethanol concentration of 50%, especially against *S. aureus* [107]. In one study, ethanol (70%) with 0.3% of a phenolic compound was not effective against *S. aureus* and *P. aeruginosa* with a single application probably because of the fast drying within 1 min [83].

With ethanol at 90%, an exposure time of up to 5 min was necessary, similar to the concentration of 50% requiring up to 2.5 min [46]. Ethanol at 50% has been described to reduce *E. faecium* DSM 2146 on frosted glass strips by >6 log in 20 min [10].

2.3.2 Fungicidal Activity

2.3.2.1 Fungistatic Activity (MIC Values)

Ethanol between 4.6 and 18.4% inhibits the multiplication of different types of fungi. Higher MIC values were described with fungal cells obtained from biofilms, e.g. 24% for *C. glabrata* (Table 2.5).

2.3.2.2 Fungicidal Activity (Suspension Tests)

Ethanol is effective at 70% against healthcare-associated yeasts, at least within 5 min. At 85%, a sufficient activity was described in 30 s. Some food-associated spore-forming fungi such as *E. repens*, *M. ruber*, *P. caseifulvum*, *P. nalgiovense*, *P. roqueforti*, *P. solitum* and *P. verrucosum* are not sufficiently killed by 70% ethanol within 10 min (Table 2.6). In mixed suspensions of environmental isolates (*R. rubra*, *C. albicans*, *C. uniguttulatus*) and clinical isolates (*R. rubra*. *C. albicans*, *C. neoformans*), 70% ethanol is still fungicidal (log reduction > 6.0) in 5 min although the effect was somewhat smaller against the environmental mix [103].

2.3.2.3 Activity Against Fungi in Biofilms

Fungi in biofilms are more difficult to eradicate by ethanol. A *C. albicans* strain ATCC MYA-273, grown for 24 h on polystyrene plates for 24 h, was reduced by exposure to 70% ethanol by 1.5 log (5 min exposure), 2.8 log (7 min exposure) and >3.0 log (10 min exposure) [68]. Seventy per cent ethanol was even ineffective in 5 min to reduce *R. rubra*, *C. albicans*, *C. uniguttulatus* or *C. neoformans* in 24-h biofilms [103]. Four *Candida* strains grown in biofilm (2 *C. albicans*, *C. parapsilosis*, *C. glabrata*) were described to be 2–6 times less susceptible to

Table 2.5 MIC values of various fungal species to ethanol

Species	Strains/isolates	MIC value	References
C. albicans	1 strain	6–16%[a]	[59]
C. glabrata	1 strain	6–24%[a]	[59]
C. krusei	1 strain	4–20%[a]	[59]
C. tropicalis	1 strain	5–15%[a]	[59]
C. utilis	Strain IFO 0396	6.3%	[4]
H. anomala	Strain IFO 0118	10.5%	[4]
H. valbyensis	Strain IFO 011S	12.6%	[4]
S. arboriculus	Strain CBS 10644	8.2%	[6]
S. bayanus	4 strains from natural and fermentative habitats	7.7–8.5%	[6]
S. bayanus	Strain EC 1118	18.4%	[4]
S. cariocanus	Strain CBS 8841	7.1%	[6]
S. cerevisiae	10 strains from natural and fermentative habitats	9.6–14.1%	[6]
S. cerevisiae	Strain IFO 2363	11.3%	[4]
S. cerevisiae	Strain 9302	12.1%	[4]
S. cerevisiae	Strain IFO 2347	13.3%	[4]
S. cerevisiae	Strain Hakken No. 1	13.6%	[4]
S. kudriavzevii	5 strains from natural and fermentative habitats	4.6–7.2%	[6]
S. mikatae	Strain IFO 1815	8.1%	[6]
S. paradoxus	4 strains from natural and fermentative habitats	8.3–9.3%	[6]
S. pombe	Unknown	12.2%	[4]

[a]highest value obtained with biofilm cells

ethanol compared to planktonic cells of the same strain [77]. The susceptibility of *T. asahii* collected from biofilm to ethanol is also lower. The median MIC value is 25% compared to planktonic cells with 8% [60].

2.3.2.4 Fungicidal Activity for Hygienic Hand Disinfection

Ethanol at 70% was found to be very effective to reduce an artificial *C. albicans* contamination of fingertips within 20 s with a mean log reduction of 4.3. A hand gel based on 60% ethanol reached a similar reduction with 4.5 log [105].

2.3.2.5 Fungicidal Activity in Carrier Tests

Against *C. albicans*, *C. parapsilosis* and *C. tropicalis*, a log reduction >4.0 was found both on glass and steel carriers within 1 min exposure time for ethanol at 70% [105]. Spore-forming fungi are, however, more resistant. When spores of *T. mentagrophytes* are placed on a glass cup carrier and exposed to 70% ethanol, a log reduction <1.0 is found after 1 min and >5.0 after 10 min [14]. On a glass strip contaminated with spores of *A. niger* ATCC 16404, ethanol at 50% had basically no fungicidal activity within 20 min (<1.0 log) [10].

Table 2.6 Fungicidal activity of ethanol in suspension tests

Species	Strains/isolates	Exposure time	Concentration	$\log_{10}$ reduction	References
A. flavus	Bread isolate	10 min	70% (P)	4.0	[18]
A. niger	ATCC 16404	30 s	85% (P)	>4.4	[54]
A. niger	Bread isolate	10 min	70% (P)	>5.2	[18]
A. versicolor	2 cheese isolates	10 min	70% (P)	3.3–5.2	[18]
C. albicans	ATCC 10231	30 s	85% (P)	>4.4	[54]
C. albicans	1 human and 1 environmental isolate	5 min	70% (S)	>7.0	[103]
Cladosporium spp.	Bread isolate	10 min	70% (P)	>4.1	[18]
C. neoformans	1 clinical isolate	5 min	70% (S)	>7.0	[103]
C. uniguttulatus	1 clinical isolate	5 min	70% (S)	>7.0	[103]
E. repens	Bread factory isolate	10 min	70% (P)	3.0	[18]
D. hansenii	Cheese isolate	10 min	70% (P)	>4.5	[18]
H. burtonii	Bread isolate	10 min	70% (P)	>5.2	[18]
M. ruber	Bread isolate	10 min	70% (P)	1.3	[18]
M. suaveolens	Bread isolate	10 min	70% (P)	>4.5	[18]
N. pseudofischeri	Cherry filling isolate	10 min	70% (P)	4.0	[18]
P. anomala	Bread isolate	10 min	70% (P)	>5.9	[18]
P. caseifulvum	Cheese isolate	10 min	70% (P)	2.7	[18]
P. chrysogenum	Cheese isolate	10 min	70% (P)	>5.2	[18]
P. commune	2 cheese and 1 bread isolates	10 min	70% (P)	2.9–4.0	[18]
P. corylophilum	Bread isolate	10 min	70% (P)	4.6	[18]
P. crustosum	Cheese isolate	10 min	70% (P)	4.0	[18]
P. discolor	Cheese isolate	10 min	70% (P)	3.0	[18]
P. nalgiovense	2 cheese isolates	10 min	70% (P)	2.3–3.0	[18]
P. norvegensis	Cheese isolate	10 min	70% (P)	>5.2	[18]
P. roqueforti	2 bread isolates	10 min	70% (P)	3.4–3.6	[18]
P. solitum	Cheese isolate	10 min	70% (P)	3.2	[18]
P. verrucosum	Cheese isolate	10 min	70% (P)	3.1	[18]
R. rubra	1 clinical isolate	5 min	70% (S)	>7.0	[103]
S. brevicaulis	Cheese isolate	10 min	70% (P)	>4.2	[18]
T. delbrueckii	Cheese isolate	10 min	70% (P)	>4.8	[18]
Yeasts	25 strains isolated from food or food processing	5 min	70% (P)	$\geq$4.0	[94]

S solution; *P* commercial product

2.3.3 Mycobactericidal Activity

2.3.3.1 Mycobactericidal Activity (Suspension Tests)

Ethanol of 70% or more has sufficient efficacy against selected mycobacterial species within 1–5 min (Table 2.7). *M. bovis*, however, may not be susceptible enough to be completely killed by 70% ethanol in 20 min [93]. Few data were found against aquatic non-tuberculous mycobacteria. *M. marinum* was effectively reduced by ethanol at 50 and 70% in 1 min [67].

2.3.3.2 Activity Against Mycobacteria in Biofilms

One study from Japan indicates that ethanol at 80% has reduced activity against non-tuberculous mycobacteria in biofilm. In the first step, it was described to be effective against all 13 tested mycobacterial isolates. For the decontamination of one drain, however, it was not effective enough. Brushing the drain with 80% ethanol resulted in no further detection for 3 months indicating the presence of a biofilm on the inner drain surface [79].

2.3.3.3 Mycobactericidal Activity in Carrier Tests

In carrier tests, the mycobactericidal activity of ethanol is lower compared to suspension tests. The data summarized in Table 2.8 show that 70% ethanol has poor activity in 1 min and sufficient activity in at least 10 min.

2.3.3.4 Mycobactericidal Activity in Flexible Endoscopes

Ethanol at 70% may be used for flushing channels of flexible bronchoscopes after reprocessing with the aim of further reducing the final microbial burden [38]. One report from Japan describes that the use of an ethanol-based rinse as an additional procedure resulted in a significant reduction of isolating non-tuberculous mycobacteria from the fluid phase of colonic contents [56].

Table 2.7 Mycobactericidal activity of ethanol in suspension tests

Species	Strains/isolates	Exposure time	Concentration	$\log_{10}$ reduction	References
M. chelonae	2 clinical isolates	30 s	75% (S)	>5.0	[118]
M. nonchromogenicum	2 clinical isolates	30 s	75% (S)	>5.0	[118]
M. smegmatis	ATCC 14469	5 s	75% (S)	>5.0	[118]
M. smegmatis	Strain TMC 1515	1 min	70% (S)	>6.0 3.0–4.0[a]	[12]
M. terrae	ATCC 15755	30 s	85% (P)	>6.3	[54]
M. terrae	Isolate 232	5 min	70% (S)	>4.4	[108]
M. terrae	Isolate 373	5 min	70% (S)	>4.9	[108]
M. tuberculosis	Strain H37Rv	1 min	70% (S)	3.5–3.6	[13]
M. tuberculosis	ATCC 25618	5 min	70% (S)	>3.8	[108]

S solution; *P* commercial product; [a]with sputum

Table 2.8 Mycobactericidal efficacy of ethanol in carrier tests

Species	Strains/isolates	Exposure time	Concentration	$\log_{10}$ reduction	References
M. avium	DSM 44156	20 min	50% (S)	>6.0	[10]
M. bovis	ATCC 35743	1 min	70% (S)	2.7	[14]
		10 min		>5.0	
M. smegmatis	Strain TMC 1515	1 min	70% (S)	<1.0	[12]
M. tuberculosis	Strain H37Rv	1 min	70% (S)	1.9–2.0	[13]

S solution

2.4 Effect of Low-Level Exposure

2.4.1 Bacteria

Ethanol at 1.25–2.5% has been shown to significantly enhance *S. aureus* biofilm formation (Fig. 2.1) by up-regulation of some proteins with adhesive functions and others with cell maintenance functions and virulence factor EsxA [24]. Ethanol at 1 and 2% was also able to increase biofilm formation in a non-adherent *S. epidermidis* strain [21].

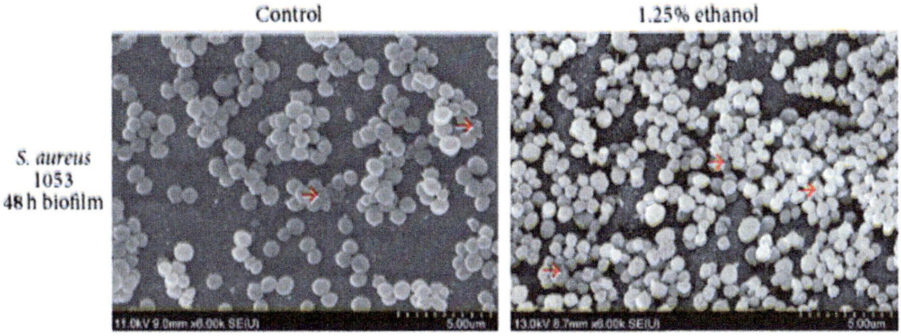

Fig. 2.1 SEM images of 48 h biofilm formed by an *S. aureus* isolate in medium (control) or 1.25% ethanol. Arrows: extracellular matrix [24]. Reproduced in parts without change from Cincarova L, Polansky O, Babak V, Kulich P, Kralik P. Changes in the Expression of Biofilm-Associated Surface Proteins in *Staphylococcus aureus* Food-Environmental Isolates Subjected to Sublethal Concentrations of Disinfectants. BioMed Research International 2016:4034517. https://doi.org/10.1155/2016/4034517. This is an open-access article distributed under the Creative Commons Attribution License

When a *Pseudomonas* spp. strain DJ-12 was exposed to 5% ethanol for 10 min, the cells were significantly more difficult to kill by 20% ethanol [85]. Cells treated with ethanol displayed irregular rod shapes with wrinkled surfaces [85]. Exposure of a *P. putida* strain S 12 to toluene increased to cellular tolerance to ethanol which was explained by an inhibitory effect of ethanol on the biosynthesis of saturated fatty acids [48]. Similar findings were reported with *L. monocytogenes*. When cells were exposed to 5% ethanol for 60 min, they were significantly more difficult to kill by ethanol at 17.5% [64]. The adapted cells were in addition more difficult to kill by 0.1% hydrogen peroxide [64]. Attachment of cells can be significantly increased in some *L. monocytogenes* strains when exposed to 2.5% ethanol, mostly at 10 °C [41].

The biofilm formation of 37 clinical, icaADBC-positive *S. epidermidis* isolates was investigated after exposure to ethanol at 1, 2, 4 or 6%. In 18 of the 37 strains, biofilm formation was inducible by ethanol exposure [55]. Ethanol at 0.2 and 0.5% could increase attachment of marine *P. aeruginosa* to polystyrene dishes and tissue culture dishes [35].

In *B. subtilis* cells, the transfer of the mobile genetic element Tn916, a conjugative transposon and the prototype of a large family of related elements, was increased 5-fold by exposure to 4% ethanol for up to 2 h. This may also result in a transfer of Tn916-like elements and any resistance genes they contain [97].

2.4.2 Yeasts

The maximum ethanol tolerance of *S. cerevisiae* has been described to be at 25% [110]. Ethanol at 2.5% can already increase *S. cerevisiae* colony growth by 20%, whereas ethanol at 5% or more inhibits yeast colony growth [98]. When *S. cerevisiae* cells are exposed for 30 min to sublethal concentrations of ethanol (8%), they become less susceptible to ethanol at a previously lethal concentration (14%) [26]. Specific activities of the glycolytic and alcohologenic enzymes within intact living cells remain high by the presence of sublethal ethanol [110]. Trehalose plays a role in ethanol tolerance at lethal ethanol concentrations but not at sublethal ethanol concentrations [9]. Using atomic force microscopy, it was shown that challenge of *S. cerevisiae* with 9% ethanol for 5 h reduces stiffness of glucose-grown yeast cells suggesting that the cell membrane contributes to the biophysical properties of yeast cells [96].

2.5 Resistance to Ethanol

No micro-organisms from bacteria, fungi and mycobacteria with a resistance to ethanol have been reported so far. There are also currently no MIC values available to describe an ethanol resistance.

It should be kept in mind that for more than 100 years ethanol has basically no effect on bacterial spores including spores of *C. difficile* indicating an intrinsic resistance [31, 39, 47, 49, 78, 91]. Bacterial spores are, however, not addressed in this book.

2.5.1 Resistance Mechanisms

No specific resistance mechanisms such as plasmids, efflux pumps or resistance genes have ever been described to explain an acquired bacterial or fungal resistance to ethanol. One study shows in 27 carbapenem-resistant *K. pneumoniae* isolates that the presence of single of multiple disinfectant resistance genes (qacA, qacΔE, qacE, acrA) is correlated with a higher ethanol MIC value (no resistance genes: 4 mg/l; all four resistance genes: 64 mg/l) [43]. The inducible Mar phenotype is associated with increased tolerance to multiple hydrophobic antibiotics as well as some highly hydrophobic organic solvents such as cyclohexane, mediated mainly through the AcrAB/TolC efflux system. AcrAB was found not to contribute to an increased ethanol tolerance [3]. In a preformed *S. aureus* biofilm, however, transcription of selected antibiotic resistance genes can be increased after 24 h exposure to ethanol at 100%, specifically some putative multidrug efflux pump genes [90].

2.6 Cross-Tolerance to Other Biocidal Agents

L. monocytogenes was more tolerant to hydrogen peroxide after low-level exposure to 5% ethanol for 60 min [64]. No further cross-resistance to other biocidal agents has so far been described.

2.7 Cross-Tolerance to Antibiotics

So far no cross-resistance between ethanol and antibiotics has been described.

2.8 Role of Biofilm

2.8.1 Effect on Biofilm Development

Most studies with Gram-positive bacteria show that ethanol can increase biofilm formation. Treatment of a preformed *S. aureus* (2 MSSA, 2 MRSA) and a *S. epidermidis* biofilm (1 of 2 strains) for 24 h with ethanol between 40 and 95% increased biofilm formation significantly. A higher ethanol concentration resulted in a higher biofilm formation [65]. A preformed *S. aureus* biofilm exposed for 24 h to ethanol at 20, 40, 60, 80 or 100% shows significantly higher biofilm levels compared to the untreated control biofilms [90]. Ethanol treatment also resulted in significantly greater transcript levels of both icaA and icaD. Both genes assist in the production of the polysaccharide intercellular adhesin and are thus vitally important for *S. aureus* biofilm formation [90].

At the same time, treatment of a preformed MRSA biofilm with 70% ethanol for 30 min has been shown to reduce biofilm cell growth by 20% [5]. Exposure of *S. aureus*, *C. albicans* or mixed biofilm to ethanol (range: 5–50%) for up to 24 h showed that metabolic biofilm activity is largely suppressed in all three types of biofilm by 30% ethanol or more [89]. *S. aureus*, however, may survive in low cell numbers regrown to 10^6 CFU per ml. With 50% ethanol, however, no regrowth was observed [89]. Only ethanol at 30% plus 4% trisodium citrate was capable to prevent biofilm formation (*E. coli*, *P. aeruginosa*, MRSA, MSSA, MRSE) for at least 72 h [102].

Ethanol at 11.6% in a solution based on 0.12% chlorhexidine used as a mouth rinse for 4 days had no additional preventive effect on subgingival biofilm formation [95]. Finally, biofilm formation was inhibited by 99% or more in *T. asahii* biofilm by exposure to ethanol at 25% for at least 8 h or ethanol at 50% for at least 4 h [60].

2.8.2 Effect on Biofilm Removal

Ethanol has variable but overall poor biofilm removal capacity. The removal rates are often <50% as shown with single-species biofilms obtained with *B. cenocepacia*, *P. aeruginosa*, *S. liquefaciens*, *S. putrefaciens* and *S. aureus*. It is also marginal with two types of triple species biofilms. Ethanol at 25–40% removed somewhat more biofilm obtained with *S. maltophilia* or *Y. enterocolitica* with removal rates between 30% and 88%. Higher and lower ethanol concentrations appear less effective for biofilm removal (Table 2.9). These findings are in line with other data showing that ethanol at 70% removes only a small proportion (approximately 12%) of a MRSA biofilm in 30 min grown on polystyrene microtiter plates [5].

2.8.3 Effect on Biofilm Fixation

No studies were found to evaluate the fixation potential of biofilms by exposure to ethanol.

Table 2.9 Biofilm removal by exposure to ethanol measured quantitatively as change of biofilm matrix

Type of biofilm	Concentration	Exposure time	Biofilm removal rate	References
B. cenocepacia LMG 18828, 4-h adhesion and 20-h incubation in polystyrene microtitre plates	70% (S)	2–10 min	20–30%	[87]
P. aeruginosa ATCC 700928, 24-h incubation in microplates	70% (S)	60 min	0%	[104]
S. liquefaciens raw-chicken plant isolate, 3-d incubation on stainless steel	75% (S)	6 min	27%	[62]
S. putrefaciens raw-chicken plant isolate, 3-d incubation on stainless steel	75% (S)	6 min	43%	[62]
S. aureus ATCC 6538, 72-h incubation in microplates	70% (S)	60 min	0%	[104]
S. maltophilia (14 clinical strains), 24-h incubation in polystyrene microtiter plates	40% (S)	1 h	30–75%	[86]
	25% (S)			
Y. enterocolitica ATCC 23715, 24-h incubation in PVC microtiter plates	100% (S)	15 min	53%	[84]
	75% (S)		56%	
	50% (S)		78%	
	40% (S)		88%	
	25% (S)		39%	
	20% (S)		37%	
	10% (S)		12%	
	5% (S)		0%	
Mixed-species biofilm (*S. oralis* ATCC 10557, *S. gordonii* ATCC 10558 and *A. naeslundii* ATCC 19039), 20-h incubation in a biofilm capillary reactor	40% (S)	1 h	No removal	[25]
Mixed-species biofilm (*S. oralis* ATCC 10557, *S. gordonii* ATCC 10558, *A. naeslundii* ATCC 19039), 20-h incubation in a biofilm capillary reactor	11.6% (S)	20 min	No evidence for removal or detachment	[101]

S solution

2.9 Summary

The principal antimicrobial activity of ethanol is summarized in Table 2.10.

The key findings on acquired resistance and cross-resistance including the role of biofilm for selecting resistant isolates are summarized in Table 2.11.

Table 2.10 Overview on the typical exposure times required for ethanol to achieve sufficient biocidal activity in suspension tests against the different target micro-organisms

Target micro-organisms	Species	Concentration	Exposure time
Bacteria	Most clinically relevant species including antibiotic-resistant isolates	≥ 78%	30 s[a]
		85%	15 s
Fungi	Healthcare-associated fungi such as *C. albicans* and *A. niger*	85%	30 s
	Various food-associated fungi	70%	≥ 10 min
Mycobacteria	*M. smegmatis, M. chelonae, M. nonchromogenicum, M. smegmatis*	70–75%	≥ 1 min
	M. terrae, M. tuberculosis	70%	5 min
	M. terrae	85%	30 s

[a]In biofilm, the efficacy is substantially lower depending on the species and the type of biofilm

Table 2.11 Key findings on ethanol resistance, the effect of low-level exposure, cross-tolerance to other biocides and antibiotics, and its effect on biofilm

Parameter	Species	Findings
Elevated MIC values	So far not reported.	
MIC value to determine resistance	Not proposed yet for bacteria, fungi or mycobacteria	
Cross-tolerance biocides	*L. monocytogenes*	Hydrogen peroxide after adaptation to ethanol
Cross-tolerance antibiotics	So far not reported	
Effect of low-level exposure	None	No MIC increase
	None	Weak MIC increase (≤ 4-fold)
	None	Strong MIC increase (>4-fold)
	L. monocytogenes, Pseudomonas spp., *S. cerevisiae*	Reduced susceptibility to lethal ethanol concentrations
	S. aureus, S. epidermidis	Increase of biofilm formation
	L. monocytogenes	Increase of surface attachment
	B. subtilis	Five-fold increase of mobile genetic element transfer (resistance genes)
Specific resistance mechanism	So far not reported	
Biofilm	Development	Enhancement in *S. aureus* and *S. epidermidis*
		Inhibition in *T. asahii* and MRSA
	Removal	Mostly moderate
	Fixation	Unknown

References

1. Aarnisalo K, Lundén J, Korkeala H, Wirtanen G (2007) Susceptibility of Listeria monocytogenes strains to disinfectants and chlorinated alkaline cleaners at cold temperatures. LWT Food Sci Technol 40(6):1041–1048
2. Akamatsu T, Tabata K, Hironga M, Kawakami H, Uyeda M (1996) Transmission of Helicobacter pylori infection via flexible fiberoptic endoscopy. Am J Infect Control 24(5): 396–401
3. Ankarloo J, Wikman S, Nicholls IA (2010) Escherichia coli mar and acrAB mutants display no tolerance to simple alcohols. Int J Mol Sci 11(4):1403–1412. https://doi.org/10.3390/ijms11041403
4. Antoce A, Takahashi K, Namolosanu I (1996) Characterization of ethanol tolerance of yeasts using a calorimetric technique. Vitis 35(2):105–106
5. Aparecida Guimaraes M, Rocchetto Coelho L, Rodrigues Souza R, Ferreira-Carvalho BT, Marie Sa Figueiredo A (2012) Impact of biocides on biofilm formation by methicillin-resistant Staphylococcus aureus (ST239-SCCmecIII) isolates. Microbiol Immunol 56(3):203–207. https://doi.org/10.1111/j.1348-0421.2011.00423.x
6. Arroyo-Lopez FN, Salvado Z, Tronchoni J, Guillamon JM, Barrio E, Querol A (2010) Susceptibility and resistance to ethanol in Saccharomyces strains isolated from wild and fermentative environments. Yeast (Chichester, England) 27(12):1005–1015. https://doi.org/10.1002/yea.1809
7. Babb JR, Bradley CR, Deverill CE, Ayliffe GA, Melikian V (1981) Recent advances in the cleaning and disinfection of fibrescopes. J Hosp Infect 2(4):329–340
8. Bae YM, Baek SY, Lee SY (2012) Resistance of pathogenic bacteria on the surface of stainless steel depending on attachment form and efficacy of chemical sanitizers. Int J Food Microbiol 153(3):465–473. https://doi.org/10.1016/j.ijfoodmicro.2011.12.017
9. Bandara A, Fraser S, Chambers PJ, Stanley GA (2009) Trehalose promotes the survival of Saccharomyces cerevisiae during lethal ethanol stress, but does not influence growth under sublethal ethanol stress. FEMS Yeast Res 9(8):1208–1216. https://doi.org/10.1111/j.1567-1364.2009.00569.x
10. Beekes M, Lemmer K, Thomzig A, Joncic M, Tintelnot K, Mielke M (2010) Fast, broad-range disinfection of bacteria, fungi, viruses and prions. J Gener Virol 91(Pt 2):580–589. https://doi.org/10.1099/vir.0.016337-0
11. Best M, Kennedy ME, Coates F (1990) Efficacy of a variety of disinfectants against Listeria spp. Appl Environ Microbiol 56(2):377–380
12. Best M, Sattar SA, Springthorpe VS, Kennedy ME (1988) Comparative mycobactericidal efficacy of chemical disinfectants in suspension and carrier tests. Appl Environ Microbiol 54:2856–2858
13. Best M, Sattar SA, Springthorpe VS, Kennedy ME (1990) Efficacies of selected disinfectants against *Mycobacterium tuberculosis*. J Clin Microbiol 28(10):2234–2239
14. Best M, Springthorpe VS, Sattar SA (1994) Feasibility of a combined carrier test for disinfectants: studies with a mixture of five types of microorganisms. Am J Infect Control 22(3): 152–162
15. Bhatia M, Mishra B, Thakur A, Dogra V, Loomba PS (2017) Evaluation of Susceptibility of glycopeptide-resistant and glycopeptide-sensitive enterococci to commonly used biocides in a super-speciality hospital: a pilot study. J Nat Sci Biol Med 8(2):199–202. https://doi.org/10.4103/0976-9668.210010
16. Bradley CR, Fraise AP (1996) Heat and chemical resistance of enterococci. J Hosp Infect 34:191–196
17. Buckley T, Dudley SM, Donowitz LG (1994) Defining unnecessary disinfection procedures for single-dose and multiple-dose vials. Am J Critical Care: An Official Publication, Am Assoc Critical-Care Nurses 3(6):448–451

18. Bundgaard-Nielsen K, Nielsen PV (1996) Fungicidal effect of 15 disinfectants against 25 fungal contaminants commonly found in bread and cheese manufacturing. J Food Prot 59 (3):268–275
19. Campos GB, Souza SG, Lob OT, Da Silva DC, Sousa DS, Oliveira PS, Santos VM, Amorim AT, Farias SV, Cruz MP, Yatsuda R, Marques LM (2012) Isolation, molecular characteristics and disinfection of methicillin-resistant Staphylococcus aureus from ICU units in Brazil. New Microbiol 35(2):183–190
20. Cavagnolo RZ (1985) Inactivation of herpesvirus on CPR manikins utilizing a currently recommended disinfecting procedure. Infection Control: IC 6(11):456–458
21. Chaieb K, Zmantar T, Souiden Y, Mahdouani K, Bakhrouf A (2011) XTT assay for evaluating the effect of alcohols, hydrogen peroxide and benzalkonium chloride on biofilm formation of Staphylococcus epidermidis. Microb Pathog 50(1):1–5. https://doi.org/10.1016/j.micpath.2010.11.004
22. Chand S, Saha K, Singh PK, Sri S, Malik N (2016) Determination of minimum inhibitory concentration (MIC) of routinely used disinfectants against microflora Isolated from clean rooms. Int J Curr Microbiol Appl Sci 5(1):334–341
23. Chiang SR, Jung F, Tang HJ, Chen CH, Chen CC, Chou HY, Chuang YC (2017) Desiccation and ethanol resistances of multidrug resistant Acinetobacter baumannii embedded in biofilm: the favorable antiseptic efficacy of combination chlorhexidine gluconate and ethanol. J Microbiol Immunol Infection = Wei mian yu gan ran za zhi. https://doi.org/10.1016/j.jmii.2017.02.003
24. Cincarova L, Polansky O, Babak V, Kulich P, Kralik P (2016) Changes in the expression of biofilm-associated surface proteins in Staphylococcus aureus food-environmental isolates subjected to sublethal concentrations of disinfectants. Biomed Res Int 2016:4034517. https://doi.org/10.1155/2016/4034517
25. Corbin A, Pitts B, Parker A, Stewart PS (2011) Antimicrobial penetration and efficacy in an in vitro oral biofilm model. Antimicrob Agents Chemother 55(7):3338–3344. https://doi.org/10.1128/aac.00206-11
26. Costa V, Reis E, Quintanilha A, Moradas-Ferreira P (1993) Acquisition of ethanol tolerance in Saccharomyces cerevisiae: the key role of the mitochondrial superoxide dismutase. Arch Biochem Biophys 300(2):608–614
27. Department of Health and Human Services; Food and Drug Administration (1994) Tentative final monograph for health care antiseptic products; proposed rule. Fed Reg 59(116): 31401–31452
28. Department of Health and Human Services; Food and Drug Administration (2015) Safety and effectiveness of healthcare antiseptics. Topical antimicrobial drug products for over-the-counter human use; proposed amendment of the tentative final monograph; reopening of administrative record; proposed rule. Fed Reg 80(84):25166–25205
29. Embil JM, Zhanel GG, Plourde PJ, Hoban D (2002) Scissors: a potential source of nosocomial infection. Infect Control Hosp Epidemiol 23(3):147–151. https://doi.org/10.1086/502026
30. Epstein AK, Pokroy B, Seminara A, Aizenberg J (2011) Bacterial biofilm shows persistent resistance to liquid wetting and gas penetration. Proc Natl Acad Sci USA 108(3):995–1000. https://doi.org/10.1073/pnas.1011033108
31. Epstein F (1896) Zur Frage der Alkoholdesinfektion. Z Hyg 24:1–21
32. Eterpi M, McDonnell G, Thomas V (2011) Decontamination efficacy against mycoplasma. Lett Appl Microbiol 52(2):150–155. https://doi.org/10.1111/j.1472-765X.2010.02979.x
33. European Chemicals Agency (ECHA) Ethanol. Substance information. https://echa.europa.eu/substance-information/-/substanceinfo/100.000.526. Accessed 30 Aug 2017
34. European Chemicals Agency (ECHA) Ethanol. Biocidal active substances https://echa.europa.eu/information-on-chemicals/biocidal-active-substances?p_p_id = echarevbiocides_WAR_echarevbiocidesportlet&p_p_lifecycle = 1&p_p_state = normal&p_p_mode = view&p_p_col_id = column-1&p_p_col_pos = 1&p_p_col_count = 2&_echarevbiocides_

WAR_echarevbiocidesportlet_javax.portlet.action = searchBiocidesAction. Accessed 30 Aug 2017
35. Fletcher M (1983) The effects of methanol, ethanol, propanol and butanol on bacterial attachment to surfaces. J Gen Microbiol 129(3):633–641
36. Frobisher M Jr, Sommermeyer L, Blackwell MJ (1953) Studies on disinfection of clinical thermometers I. Oral thermometers. Appl Microbiol 1(4):187–194
37. Garcia de Cabo A, Martinez Larriba PL, Checa Pinilla J, Guerra Sanz F (1978) A new method of disinfection of the flexible fibrebronchoscope. Thorax 33(2):270–272
38. Gavalda L, Olmo AR, Hernandez R, Dominguez MA, Salamonsen MR, Ayats J, Alcaide F, Soriano A, Rosell A (2015) Microbiological monitoring of flexible bronchoscopes after high-level disinfection and flushing channels with alcohol: results and costs. Respir Med 109(8):1079–1085. https://doi.org/10.1016/j.rmed.2015.04.015
39. Gershenfeld L (1938) The sterility of alcohol. Am J Med Sci 195(3):358–360
40. Goroncy-Bermes P, Koburger T, Meyer B (2010) Impact of the amount of hand rub applied in hygienic hand disinfection on the reduction of microbial counts on hands. J Hosp Infect 74(3):212–218
41. Gravesen A, Lekkas C, Knochel S (2005) Surface attachment of Listeria monocytogenes is induced by sublethal concentrations of alcohol at low temperatures. Appl Environ Microbiol 71(9):5601–5603. https://doi.org/10.1128/aem.71.9.5601-5603.2005
42. Guilhermetti M, Marques Wiirzler LA, Castanheira Facio B, da Silva Furlan M, Campo Meschial W, Bronharo Tognim MC, Botelho Garcia L, Luiz Cardoso C (2010) Antimicrobial efficacy of alcohol-based hand gels. J Hosp Infect 74(3):219–224. https://doi.org/10.1016/S0195-6701(09)00424-1, https://doi.org/10.1016/j.jhin.2009.09.019
43. Guo W, Shan K, Xu B, Li J (2015) Determining the resistance of carbapenem-resistant Klebsiella pneumoniae to common disinfectants and elucidating the underlying resistance mechanisms. Pathogens Global Health 109(4):184–192. https://doi.org/10.1179/2047773215y.0000000022
44. Gutierrez-Martin CB, Yubero S, Martinez S, Frandoloso R, Rodriguez-Ferri EF (2011) Evaluation of efficacy of several disinfectants against Campylobacter jejuni strains by a suspension test. Res Vet Sci 91(3):e44–47. https://doi.org/10.1016/j.rvsc.2011.01.020
45. Hall TJ, Wren MW, Jeanes A, Gant VA (2009) A comparison of the antibacterial efficacy and cytotoxicity to cultured human skin cells of 7 commercial hand rubs and Xgel, a new copper-based biocidal hand rub. Am J Infect Control 37(4):322–326
46. Hare R, Raik E, Gash S (1963) Efficiency of antiseptics when acting on dried organisms. BMJ 1(5329):496–500
47. Harrington C, Walker H (1903) The germicidal action of alcohol. Boston Med Surg J 148(21):548–552
48. Heipieper HJ, de Bont JA (1994) Adaptation of Pseudomonas putida S12 to ethanol and toluene at the level of fatty acid composition of membranes. Appl Environ Microbiol 60(12):4440–4444
49. Jabbar U, Leischner J, Kasper D, Gerber R, Sambol SP, Parada JP, Johnson S, Gerding DN (2010) Effectiveness of alcohol-based hand rubs for removal of Clostridium difficile spores from hands. Infect Control Hosp Epidemiol 31(6):565–570. https://doi.org/10.1086/652772
50. Kampf G (2017) Ethanol. In: Kampf G (ed) Kompendium Händehygiene. mhp-Verlag, Wiesbaden, pp 325–351
51. Kampf G, Hollingsworth A (2008) Comprehensive bactericidal activity of an ethanol-based hand gel in 15 seconds. Ann Clin Microbiol Antimicrob 7:2
52. Kampf G, Marschall S, Eggerstedt S, Ostermeyer C (2010) Efficacy of ethanol-based hand foams using clinically relevant amounts: a cross-over controlled study among healthy volunteers. BMC Infect Dis 10:78
53. Kampf G, Meyer B, Goroncy-Bermes P (2003) Comparison of two test methods for the determination of sufficient antimicrobial efficacy of three different alcohol-based hand rubs for hygienic hand disinfection. J Hosp Infect 55(3):220–225

54. Kampf G, Rudolf M, Labadie J-C, Barrett SP (2002) Spectrum of antimicrobial activity and user acceptability of the hand disinfectant agent Sterillium Gel. J Hosp Infect 52(2):141–147
55. Knobloch JK, Horstkotte MA, Rohde H, Kaulfers PM, Mack D (2002) Alcoholic ingredients in skin disinfectants increase biofilm expression of Staphylococcus epidermidis. J Antimicrob Chemother 49(4):683–687
56. Kobayashi Y, Takano T, Hirayama N, Sato N, Shimoide H (1995) Isolation of nontuberculous mycobacteria during colonoscopy. Kekkaku: [Tuberculosis] 70(11):629–634
57. Kubota H, Senda S, Tokuda H, Uchiyama H, Nomura N (2009) Stress resistance of biofilm and planktonic Lactobacillus plantarum subsp. plantarum JCM 1149. Food Microbiol 26(6):592–597. https://doi.org/10.1016/j.fm.2009.04.001
58. Lanjri S, Uwingabiye J, Frikh M, Abdellatifi L, Kasouati J, Maleb A, Bait A, Lemnouer A, Elouennass M (2017) In vitro evaluation of the susceptibility of Acinetobacter baumannii isolates to antiseptics and disinfectants: comparison between clinical and environmental isolates. Antimicrob Resist Infect Control 6:36. https://doi.org/10.1186/s13756-017-0195-y
59. Leung CY, Chan YC, Samaranayake LP, Seneviratne CJ (2012) Biocide resistance of Candida and Escherichia coli biofilms is associated with higher antioxidative capacities. J Hosp Infect 81(2):79–86. https://doi.org/10.1016/j.jhin.2011.09.014
60. Liao Y, Zhao H, Lu X, Yang S, Zhou J, Yang R (2015) Efficacy of ethanol against Trichosporon asahii biofilm in vitro. Med Mycol 53(4):396–404. https://doi.org/10.1093/mmy/myv006
61. Lin F, Xu Y, Chang Y, Liu C, Jia X, Ling B (2017) Molecular characterization of reduced susceptibility to biocides in clinical isolates of Acinetobacter baumannii. Front Microbiol 8:1836. https://doi.org/10.3389/fmicb.2017.01836
62. Liu J, Yu S, Han B, Chen J (2017) Effects of benzalkonium chloride and ethanol on dual-species biofilms of Serratia liquefaciens S1 and Shewanella putrefaciens S4. Food Control 78(Supplement C):196–202. https://doi.org/10.1016/j.foodcont.2017.02.063
63. Liu Q, Liu M, Wu Q, Li C, Zhou T, Ni Y (2009) Sensitivities to biocides and distribution of biocide resistance genes in quaternary ammonium compound tolerant Staphylococcus aureus isolated in a teaching hospital. Scand J Infect Dis 41(6–7):403–409. https://doi.org/10.1080/00365540902856545
64. Lou Y, Yousef AE (1997) Adaptation to sublethal environmental stresses protects Listeria monocytogenes against lethal preservation factors. Appl Environ Microbiol 63(4):1252–1255
65. Luther MK, Bilida S, Mermel LA, LaPlante KL (2015) Ethanol and isopropyl alcohol exposure increases biofilm formation in Staphylococcus aureus and Staphylococcus epidermidis. Infect Dis Ther 4(2):219–226. https://doi.org/10.1007/s40121-015-0065-y
66. Macinga DR, Shumaker DJ, Werner HP, Edmonds SL, Leslie RA, Parker AE, Arbogast JW (2014) The relative influences of product volume, delivery format and alcohol concentration on dry-time and efficacy of alcohol-based hand rubs. BMC Infect Dis 14:511. https://doi.org/10.1186/1471-2334-14-511
67. Mainous ME, Smith SA (2005) Efficacy of common disinfectants against mycobacterium marinum. J Aquat Anim Health 17(3):284–288. https://doi.org/10.1577/H04-051.1
68. Maisch T, Shimizu T, Isbary G, Heinlin J, Karrer S, Klampfl TG, Li YF, Morfill G, Zimmermann JL (2012) Contact-free inactivation of Candida albicans biofilms by cold atmospheric air plasma. Appl Environ Microbiol 78(12):4242–4247. https://doi.org/10.1128/aem.07235-11
69. Mariscal A, Carnero-Varo M, Gutierrez-Bedmar M, Garcia-Rodriguez A, Fernandez-Crehuet J (2007) A fluorescent method for assessing the antimicrobial efficacy of disinfectant against Escherichia coli ATCC 35218 biofilm. Appl Microbiol Biotechnol 77(1):233–240. https://doi.org/10.1007/s00253-007-1137-z
70. Mariscal A, Lopez-Gigosos RM, Carnero-Varo M, Fernandez-Crehuet J (2009) Fluorescent assay based on resazurin for detection of activity of disinfectants against bacterial biofilm. Appl Microbiol Biotechnol 82(4):773–783. https://doi.org/10.1007/s00253-009-1879-x

71. Mattner F, Gastmeier P (2004) Bacterial contamination of multiple-dose vials: a prevalence study. Am J Infect Control 32(1):12–16. https://doi.org/10.1016/j.ajic.2003.06.004
72. Miyano N, Oie S, Kamiya A (2003) Efficacy of disinfectants and hot water against biofilm cells of Burkholderia cepacia. Biol Pharm Bull 26(5):671–674
73. Moretro T, Vestby LK, Nesse LL, Storheim SE, Kotlarz K, Langsrud S (2009) Evaluation of efficacy of disinfectants against Salmonella from the feed industry. J Appl Microbiol 106 (3):1005–1012. https://doi.org/10.1111/j.1365-2672.2008.04067.x
74. Narayanan A, Nair MS, Karumathil DP, Baskaran SA, Venkitanarayanan K, Amalaradjou MA (2016) Inactivation of Acinetobacter baumannii biofilms on polystyrene, stainless steel, and urinary catheters by octenidine dihydrochloride. Front Microbiol 7:847. https://doi.org/10.3389/fmicb.2016.00847
75. Narui K, Takano M, Noguchi N, Sasatsu M (2007) Susceptibilities of methicillin-resistant Staphylococcus aureus isolates to seven biocides. Biol & Pharm Bull 30(3):585–587
76. National Center for Biotechnology Information Ethanol. PubChem Compound Database; CID = 702. https://pubchem.ncbi.nlm.nih.gov/compound/702. Accessed 30 Aug 2017
77. Nett JE, Guite KM, Ringeisen A, Holoyda KA, Andes DR (2008) Reduced biocide susceptibility in Candida albicans biofilms. Antimicrob Agents Chemother 52(9):3411–3413. https://doi.org/10.1128/aac.01656-07
78. Neufeld F, Schiemann O (1939) Über die Wirkung des Alkohols bei der Händedesinfektion. Z Hyg 121:312–333
79. Ogawa M, Nomoto M, Fukuda K, Miyamoto H, Taniguchi H (2011) Nontuberculous mycobacteria in wet areas of a hospital and standard residences. J UOEH 33(4):319–329
80. Oh DH, Marshall DL (1993) Antimicrobial activity of ethanol, glycerol monolaurate or lactic acid against Listeria monocytogenes. Int J Food Microbiol 20(4):239–246
81. Ohara T, Itoh Y, Itoh K (1998) Ultrasound instruments as possible vectors of staphylococcal infection. J Hosp Infect 40(1):73–77
82. Oliveira PS, Souza SG, Campos GB, da Silva DC, Sousa DS, Araujo SP, Ferreira LP, Santos VM, Amorim AT, Santos AM, Timenetsky J, Cruz MP, Yatsuda R, Marques LM (2014) Isolation, pathogenicity and disinfection of Staphylococcus aureus carried by insects in two public hospitals of Vitoria da Conquista, Bahia, Brazil. Brazilian J Infect Dis: An Off Public Brazilian Soc Infect Dis 18(2):129–136. https://doi.org/10.1016/j.bjid.2013.06.008
83. Omidbakhsh N (2010) Theoretical and experimental aspects of microbicidal activities of hard surface disinfectants: are their label claims based on testing under field conditions? J AOAC Int 93(6):1944–1951
84. Park HS, Ham Y, Shin K, Kim YS, Kim TJ (2015) Sanitizing effect of ethanol against biofilms formed by three gram-negative pathogenic bacteria. Curr Microbiol 71(1):70–75. https://doi.org/10.1007/s00284-015-0828-4
85. Park SH, Oh KH, Kim CK (2001) Adaptive and cross-protective responses of pseudomonas sp. DJ-12 to several aromatics and other stress shocks. Curr Microbiol 43(3):176–181. https://doi.org/10.1007/s002840010283
86. Passerini de Rossi B, Feldman L, Pineda MS, Vay C, Franco M (2012) Comparative in vitro efficacies of ethanol-, EDTA- and levofloxacin-based catheter lock solutions on eradication of Stenotrophomonas maltophilia biofilms. J Med Microbiol 61(Pt 9):1248–1253. https://doi.org/10.1099/jmm.0.039743-0
87. Peeters E, Nelis HJ, Coenye T (2008) Evaluation of the efficacy of disinfection procedures against Burkholderia cenocepacia biofilms. J Hosp Infect 70(4):361–368. https://doi.org/10.1016/j.jhin.2008.08.015
88. Penna TC, Mazzola PG, Silva Martins AM (2001) The efficacy of chemical agents in cleaning and disinfection programs. BMC Infect Dis 1:16
89. Peters BM, Ward RM, Rane HS, Lee SA, Noverr MC (2013) Efficacy of ethanol against Candida albicans and Staphylococcus aureus polymicrobial biofilms. Antimicrob Agents Chemother 57(1):74–82. https://doi.org/10.1128/aac.01599-12

90. Redelman CV, Maduakolam C, Anderson GG (2012) Alcohol treatment enhances Staphylococcus aureus biofilm development. FEMS Immunol Med Microbiol 66(3):411–418. https://doi.org/10.1111/1574-695x.12005
91. Reinicke EA (1894) Bakteriologische Untersuchungen über die Desinfektion der Hände. Zentralbl Gynäkol 47:1189–1199
92. Rodriguez Ferri EF, Martinez S, Frandoloso R, Yubero S, Gutierrez Martin CB (2010) Comparative efficacy of several disinfectants in suspension and carrier tests against Haemophilus parasuis serovars 1 and 5. Res Vet Sci 88(3):385–389. https://doi.org/10.1016/j.rvsc.2009.12.001
93. Rutala WA, Cole EC, Wannamaker NS, Weber DJ (1991) Inactivation of Mycobacterium tuberculosis and mycobacterium bovis by 14 hospital disinfectants. Am J Med 91(3b):267s–271s
94. Salo S, Wirtanen G (2005) Disinfectant efficacy on foodborne spoilage yeast strains. Food Bioprod Process 83(4):288–296
95. Santos GOD, Milanesi FC, Greggianin BF, Fernandes MI, Oppermann RV, Weidlich P (2017) Chlorhexidine with or without alcohol against biofilm formation: efficacy, adverse events and taste preference. Brazilian Oral Res 31:e32. https://doi.org/10.1590/1807-3107BOR-2017.vol31.0032
96. Schiavone M, Formosa-Dague C, Elsztein C, Teste MA, Martin-Yken H, De Morais MA, Jr., Dague E, Francois JM (2016) Evidence for a role for the plasma membrane in the nanomechanical properties of the cell wall as revealed by an atomic force microscopy study of the response of Saccharomyces cerevisiae to ethanol stress. Appl Environ Microbiol 82(15):4789–4801. https://doi.org/10.1128/aem.01213-16
97. Seier-Petersen MA, Jasni A, Aarestrup FM, Vigre H, Mullany P, Roberts AP, Agerso Y (2014) Effect of subinhibitory concentrations of four commonly used biocides on the conjugative transfer of Tn916 in Bacillus subtilis. J Antimicrob Chemother 69(2):343–348. https://doi.org/10.1093/jac/dkt370
98. Semchyshyn HM (2014) Hormetic concentrations of hydrogen peroxide but not ethanol induce cross-adaptation to different stresses in budding yeast. Int J Microbiol 2014:485792. https://doi.org/10.1155/2014/485792
99. Singh D, Kaur H, Gardner WG, Treen LB (2002) Bacterial contamination of hospital pagers. Infect Control Hosp Epidemiol 23(5):274–276. https://doi.org/10.1086/502048
100. Sommermeyer L, Frobisher M Jr (1953) Laboratory studies on disinfection of rectal thermometers. Nurs Res 2(2):85–89
101. Takenaka S, Trivedi HM, Corbin A, Pitts B, Stewart PS (2008) Direct visualization of spatial and temporal patterns of antimicrobial action within model oral biofilms. Appl Environ Microbiol 74(6):1869–1875. https://doi.org/10.1128/aem.02218-07
102. Takla TA, Zelenitsky SA, Vercaigne LM (2008) Effectiveness of a 30% ethanol/4% trisodium citrate locking solution in preventing biofilm formation by organisms causing haemodialysis catheter-related infections. J Antimicrob Chemother 62(5):1024–1026. https://doi.org/10.1093/jac/dkn291
103. Theraud M, Bedouin Y, Guiguen C, Gangneux JP (2004) Efficacy of antiseptics and disinfectants on clinical and environmental yeast isolates in planktonic and biofilm conditions. J Med Microbiol 53(Pt 10):1013–1018. https://doi.org/10.1099/jmm.0.05474-0
104. Tote K, Horemans T, Vanden Berghe D, Maes L, Cos P (2010) Inhibitory effect of biocides on the viable masses and matrices of Staphylococcus aureus and Pseudomonas aeruginosa biofilms. Appl Environ Microbiol 76(10):3135–3142. https://doi.org/10.1128/aem.02095-09
105. Traore O, Springthorpe VS, Sattar SA (2002) Testing chemical germicides against Candida species using quantitative carrier and fingerpad methods. J Hosp Infect 50(1):66–75. https://doi.org/10.1053/jhin.2001.1133
106. United States Environmental Protection Agency (1995) Reregistration Eligibility Decision (RED) aliphatic alcohols. https://www3.epa.gov/pesticides/chem_search/reg_actions/reregistration/red_G-4_1-Mar-95.pdf

107. van Klingeren B (1995) Disinfectant testing on surfaces. J Hosp Infect 30(Suppl):397–408
108. van Klingeren B, Pullen W (1987) Comparative testing of disinfectants against mycobacterium tuberculosis and mycobacterium terrae in a quantitative suspension test. J Hosp Infect 10(3):292–298. http://dx.doi.org/10.1016/0195-6701(87)90012-0
109. Vieira CD, Farias Lde M, Diniz CG, Alvarez-Leite ME, Camargo ER, Carvalho MA (2005) New methods in the evaluation of chemical disinfectants used in health care services. Am J Infect Control 33(3):162–169. https://doi.org/10.1016/j.ajic.2004.10.007
110. Wang M, Zhao J, Yang Z, Du Z, Yang Z (2007) Electrochemical insights into the ethanol tolerance of Saccharomyces cerevisiae. Bioelectrochemistry (Amsterdam, Netherlands) 71(2):107–112. https://doi.org/10.1016/j.bioelechem.2007.04.003
111. Wang Y, Leng V, Patel V, Phillips KS (2017) Injections through skin colonized with staphylococcus aureus biofilm introduce contamination despite standard antimicrobial preparation procedures. Sci Rep 7:45070. https://doi.org/10.1038/srep45070
112. Weber DJ, Wilson MB, Rutala WA, Thomann CA (1990) Manual ventilation bags as a source for bacterial colonization of intubated patients. Am Rev Respir Dis 142(4):892–894. https://doi.org/10.1164/ajrccm/142.4.892
113. WHO (2009) WHO guidelines on hand hygiene in health care. First Global Patient Safety Challenge Clean Care is Safer Care. WHO, Geneva
114. WHO (2015) WHO model list of essential medicines. WHO. http://www.who.int/medicines/publications/essentialmedicines/EML2015_8-May-15.pdf
115. WHO (2016) Global guidelines for the prevention of surgical site infections. WHO, Geneva
116. WHO (2017) WHO model list of essential medicines for children. WHO. Accessed 30 Aug 2017
117. Wong HS, Townsend KM, Fenwick SG, Trengove RD, O'Handley RM (2010) Comparative susceptibility of planktonic and 3-day-old Salmonella Typhimurium biofilms to disinfectants. J Appl Microbiol 108(6):2222–2228. https://doi.org/10.1111/j.1365-2672.2009.04630.x
118. Woo PC, Leung KW, Wong SS, Chong KT, Cheung EY, Yuen KY (2002) Relatively alcohol-resistant mycobacteria are emerging pathogens in patients receiving acupuncture treatment. J Clin Microbiol 40(4):1219–1224
119. Zachary KC, Bayne PS, Morrison VJ, Ford DS, Silver LC, Hooper DC (2001) Contamination of gowns, gloves, and stethoscopes with vancomycin-resistant enterococci. Infect Control Hosp Epidemiol 22(9):560–564. https://doi.org/10.1086/501952

Propan-1-ol

3

3.1 Chemical Characterization

Propan-1-ol is a primary alcohol and a colourless liquid. It is an isomer of propan-2-ol and formed naturally in small amounts during many fermentation processes. It was discovered in 1853 by Chancel [43]. The basic chemical information on propan-1-ol is summarized in Table 3.1.

3.2 Types of Application

The European Chemicals Agency (ECHA) describes that propan-1-ol is used by consumers in articles and by professional workers (widespread uses), in formulation or re-packing, at industrial sites and in manufacturing. Consumer use includes lubricants and greases, coating products, anti-freeze products, perfumes and fragrances, finger paints, adhesives and sealants, non-metal-surface treatment products, leather treatment products, polishes and waxes, washing and cleaning products, cosmetics and personal care products [9].

Use by professional workers includes laboratory chemicals, coating products, lubricants and greases, washing and cleaning products, metal working fluids and plant protection products [9] as well as hand disinfection by healthcare workers and in veterinary medicine, disinfection of inanimate surfaces and skin antisepsis prior to surgery [4]. In contrast to ethanol and propan-2-ol, the WHO does not specifically recommend to use alcohol-based hand rubs based on propan-1-ol in selected situations during patient care for prevention of healthcare-associated infections [39] or for the pre-operative decontamination of hands for the prevention of surgical site infections [41]. In contrast to ethanol or propan-2-ol, propan-1-ol is also not

G. Kampf, *Antiseptic Stewardship*, https://doi.org/10.1007/978-3-319-98785-9_3

Table 3.1 Basic chemical information on propan-1-ol [9, 27]

CAS number	71-23-8
IUPAC name	Propan-1-ol
Synonyms	n-propylalcohol, 1-propanol
Molecular formula	C_3H_8O
Molecular weight (g/mol)	60.096

mentioned in the list of "essential medicines" for adults or children up to 12 years of age [40, 42]. In EN 12791, propan-1-ol at 60% (v/v) is described as the reference alcohol for determination of the bactericidal efficacy of products for surgical hand disinfection or surgical scrubbing [7].

3.2.1 European Chemicals Agency (European Union)

Propan-1-ol has been approved in 2017 as an active biocidal agent for product types 1 (human hygiene), 2 (disinfectants and algaecides not intended for direct application to humans or animals) and 4 (food and feed area) [14].

3.2.2 Environmental Protection Agency (USA)

Propan-1-ol is not a registered active ingredient for pesticide products [38].

3.2.3 Food and Drug Administration (USA)

Propan-1-ol is not among those active ingredients currently evaluated for the monograph for healthcare antiseptics [6].

3.2.4 Overall Environmental Impact

Propan-1-ol is manufactured and/or imported in the European Economic Area in 10.000–100.000 t per year [9]. Other release to the environment of this substance is likely to occur from indoor use (e.g. machine wash liquids/detergents, automotive care products, paints and coating or adhesives, fragrances and air fresheners), outdoor use, indoor use in close systems with minimal release (e.g. cooling liquids in refrigerators, oil-based electric heaters) and outdoor use in close systems with minimal release (e.g. hydraulic liquids in automotive suspension, lubricants in motor oil and break fluids) [9].

3.3 Spectrum of Antimicrobial Activity

3.3.1 Bactericidal Activity

3.3.1.1 Bacteriostatic Activity (MIC Values)

A disinfectant based on >30% propan-1-ol in combination with 15–30% propan-2-ol has been described to have MIC values for 10 *L. monocytogenes* food isolates between 3.1 and 6.25% [1]. No additional data were found to describe concentrations of propan-1-ol with a bacteriostatic effect.

3.3.1.2 Bactericidal Activity (Suspension Tests)

Some data have been published on the efficacy of propan-1-ol against bacteria in suspension tests. They indicate a strong bactericidal activity of a 60% solution beginning after 15 s (Table 3.2). Additional data with mixed propanols (e.g. 30% propan-1-ol plus 45% propan-2-ol) indicate comprehensive bactericidal activity within 30 s [17, 19]. The bactericidal activity of 60% propan-1-ol is considered to be equal to propan-2-ol at 70% whereas lower concentrations such as 50% or 40% have a lower bactericidal activity [33]. For reducing the resident skin flora, propan-1-ol has been described as the most effective mono-alcohol [31].

3.3.1.3 Activity Against Bacteria in Biofilms

Propan-1-ol at 60% was reported to be effective to kill bacterial cells in a *S. epidermidis* biofilm by 5.0 log within 1 min [29].

3.3.1.4 Bactericidal Activity for Hygienic Hand Disinfection

Propan-1-ol at 40% was described to be effective in 1 min with a mean log reduction of 4.3 which was equivalent to the reference procedure with 4.2 log [34]. At a concentration of 50%, propan-1-ol is very effective in 30 s (5.0 log) and 1 min

Table 3.2 Bactericidal activity of propan-1-ol solutions in suspension tests

Species	Strains/isolates	Exposure time	Concentration	$\log_{10}$ reduction	References
A. baumannii	20 strains	15 s	60% (S)	>5.0	[44]
Enterococcus spp.	11 strains (8 *E. faecium*, 2 *E. faecalis*, 1 *E. gallinarum*)	15 s	60% (S)	>7.0	[16]
S. aureus	ATCC 6538, ATCC 43300, 2 clinical MSSA strains, 2 clinical MRSA strains	15 s	60% (S)	>5.0	[18]
		30 s	40% (S)	>5.0	
		60 s	30% (S)	>5.0[a]	

S Solution; [a]Only MRSA (MSSA was reduced in 15 s)

(4.9 log) [34, 36]. At 60%, it was reported to reach 5.5 log [34]. The efficacy of propan-1-ol against *E. coli* on artificially contaminated hands is considered to be at least as good as of propan-2-ol [36].

3.3.1.5 Bactericidal Activity for Surgical Hand Disinfection

In EN 12791 propan-1-ol at 60% (v/v) is described as the reference alcohol for determination of the efficacy of products for surgical hand disinfection or surgical scrubbing [7]. Numerous data sets according to EN 12791 have been published. The EN 12791 reference alcohol has a rather weak bactericidal efficacy for surgical hand disinfection when applied for only 1 min. Within the standard application time of 3 min, however, a mean log reduction between 2.0 and 3.0 is typically found immediately after application. The 3-h-value under the surgical glove is somewhat lower with 0.7–2.5 log [15]. Additional data with mixed propanols (e.g. 30% propan-1-ol plus 45% propan-2-ol) indicate sufficient bactericidal efficacy for surgical hand disinfection within 1.5 min [20–22, 35].

3.3.1.6 Bactericidal Activity in Carrier Tests

Propan-1-ol at 20% has been described to reduce *E. faecium* DSM 2146 on frosted glass strips by >6.0 log in 20 min [3]. No other data were found to describe the efficacy of propan-1-ol against bacteria in carrier tests.

3.3.2 Fungicidal Activity

At 14%, propan-1-ol has been described to inhibit multiplication of *C. albicans* suggesting a levurostatic activity at this concentration [24]. At 89.5%, propan-1-ol is effective against *C. albicans* [30]. A commercial product based on propan-1-ol (>30%) and propan-2-ol (15–30%) has been investigated for yeasticidal activity against 25 strains isolated from food or food processing. Within 5 min, the product reduced the number of yeast cells by at least 4.0 log steps of 19 species, some species were less susceptible [37]. On a glass strip contaminated with spores of *A. niger* ATCC 16404, propan-1-ol at 20% had basically no fungicidal activity within 20 min (<1.0 log) [3].

3.3.3 Mycobactericidal Activity

Propan-1-ol at 20% has been described to reduce *M. avium* DSM 44156 on frosted glass strips by >6.0 log in 20 min [3]. No other data were found to describe the efficacy of propan-1-ol against mycobacteria.

3.4 Effect of Low-Level Exposure

Propan-1-ol at 0.2, 1.5 and 2% can increase the attachment of marine *P. aeruginosa* to polystyrene dishes and tissue culture dishes [10]. The biofilm formation of 37 clinical, icaADBC-positive *S. epidermidis* isolates was investigated after exposure to propan-1-ol at 0.5, 1, 2 and 4%. In 15 of the 37 strains, biofilm formation was inducible by propan-1-ol exposure [23]. With *C. albicans* , it was described that propan-1-ol at 2% inhibited to some extent biofilm development [5]. In 2 food strains of *L. monocytogenes,* the propan-1-ol MIC values remained unchanged after exposure to sublethal concentration of propan-1-ol with MIC_{max} values of 6.25% [1].

3.5 Resistance to Propan-1-ol

No microorganisms with a resistance to propan-1-ol or propanol-based hand rubs have been reported so far [2, 26, 44]. It should be kept in mind that for more than 100 years alcohols such as propan-1-ol have basically no effect on bacterial spores including spores of *C. difficile* indicating an intrinsic resistance [8, 11–13, 28, 32]. Bacterial spores are, however, not addressed in this book.

3.5.1 Resistance Mechanisms

No specific resistance mechanism has ever been described to explain an acquired bacterial or fungal resistance to propan-1-ol.

3.5.2 Resistance Genes

The inducible Mar phenotype is associated with increased tolerance to multiple hydrophobic antibiotics as well as some highly hydrophobic organic solvents such as cyclohexane, mediated mainly through the AcrAB/TolC efflux system. AcrAB, however, was found not to contribute to an increased propan-1-ol tolerance [2].

3.6 Cross-Tolerance to Other Biocidal Agents

No cross-resistance to other biocidal agents has so far been described.

3.7 Cross-Tolerance to Antibiotics

So far, no cross-resistance between propan-1-ol and antibiotics has been described.

3.8 Role of Biofilm

3.8.1 Effect on Biofilm Development

No studies were found to evaluate the effect of propan-1-ol on biofilm development. Some biofilms may be able to produce propan-1-ol. An *E. coli* biofilm was able to produce propan-1-ol under hypoxic conditions within 48 h in a concentration of 0.125%. When the growth medium was supplemented with 0.4% of the amino acid threonine, the measured concentration of propan-1-ol was 0.45%. Other enterobacteriaceae species, e.g. *S. flexneri*, *S. enterica sv. Enteritidis* and *C. rodentium*, also produce propan-1-ol in anaerobic but not in aerobic planktonic cultures [25]. It is, however, not clear if the produced propan-1-ol has any effect on the biofilm itself.

3.8.2 Effect on Biofilm Removal

Based on the limited evidence obtained with *S. epidermidis,* the ability of 60% propan-1-ol to remove biofilm is overall poor with 0–40% (Table 3.3).

3.8.3 Effect on Biofilm Fixation

No studies were found to evaluate the fixation potential of biofilms by exposure to propan-1-ol.

Table 3.3 Biofilm removal by exposure to propan-1-ol solutions measured quantitatively as change of biofilm matrix

Type of biofilm	Concentration	Exposure time	Biofilm removal rate	References
S. epidermidis DSM 3269, 24-h incubation in polystyrene microtiter plates	60% (S)	1–60 min	0–40%	[29]
S. epidermidis (30 clinical isolates), 24-h incubation in polystyrene microtiter plates	60% (S)	1–60 min	0%	[29]

3.9 Summary

The principal antimicrobial activity of propan-1-ol is summarized in Table 3.4.

The key findings on acquired resistance and cross-resistance including the role of biofilm for selecting resistant isolates are summarized in Table 3.5.

Table 3.4 Overview on the typical exposure times required for propan-1-ol to achieve sufficient biocidal activity in suspension tests against the different target microorganisms

Target microorganisms	Species	Concentration	Exposure time
Bacteria	Some clinically relevant species	$\geq 60\%$	15 s[a]
Fungi	*C. albicans*	85%	30 s
	Food-associated yeasts	>30%[b]	≥ 5 min
Mycobacteria	Unknown		

[a]In biofilm, it may require 1 min or more depending on the species; [b]In combination with 15–30% propan-2-ol

Table 3.5 Key findings on propan-1-ol resistance, the effect of low-level exposure, cross-tolerance to other biocides and antibiotics, and its effect on biofilm

Parameter	Species	Findings
Elevated MIC values	So far not reported	
MIC value to determine resistance	Not proposed yet for bacteria, fungi or mycobacteria	
Cross-tolerance biocides	So far not reported	
Cross-tolerance antibiotics	So far not reported	
Effect of low-level exposure	*L. monocytogenes*	No MIC increase
	None	Weak MIC increase (≤ 4-fold)
	None	Strong MIC increase (>4-fold)
	S. epidermidis	Increase of biofilm formation
	C. albicans	Inhibition of biofilm development
	P. aeruginosa	Increase of surface attachment
Specific resistance mechanism	So far not reported	
Biofilm	Development	Unknown
	Removal	Poor
	Fixation	Unknown

References

1. Aarnisalo K, Lundén J, Korkeala H, Wirtanen G (2007) Susceptibility of Listeria monocytogenes strains to disinfectants and chlorinated alkaline cleaners at cold temperatures. LWT Food Sci Technol 40(6):1041–1048
2. Ankarloo J, Wikman S, Nicholls IA (2010) Escherichia coli mar and acrAB mutants display no tolerance to simple alcohols. Int J Mol Sci 11(4):1403–1412. https://doi.org/10.3390/ijms11041403
3. Beekes M, Lemmer K, Thomzig A, Joncic M, Tintelnot K, Mielke M (2010) Fast, broad-range disinfection of bacteria, fungi, viruses and prions. J Gen Virol 91(Pt 2):580–589. https://doi.org/10.1099/vir.0.016337-0
4. Bloß R, Meyer S, Kampf G (2010) Adsorption of active ingredients from surface disinfectants to different types of fabrics. J Hosp Infect 75:56–61
5. Chauhan NM, Shinde RB, Karuppayil SM (2013) Effect of alcohols on filamentation, growth, viability and biofilm development in Candida albicans. Brazilian J Microbiol: [Publication of the Brazilian Society for Microbiology] 44(4):1315–1320. https://doi.org/10.1590/s1517-83822014005000012
6. Department of Health and Human Services; Food and Drug Administration (2015) Safety and effectiveness of healthcare antiseptics. Topical antimicrobial drug products for over-the-counter human use; proposed amendment of the tentative final monograph; reopening of administrative record; proposed rule. Fed Reg 80(84):25166–25205
7. EN 12791:2015 (2015) Chemical disinfectants and antiseptics. Surgical hand disinfection. Test method and requirement (phase 2, step 2). In: CEN—Comité Européen de Normalisation, Brussels
8. Epstein F (1896) Zur Frage der Alkoholdesinfektion. Z Hyg 24:1–21
9. European Chemicals Agency (ECHA) Propan-1-ol. Substance information. https://echa.europa.eu/substance-information/-/substanceinfo/100.000.679. Accessed 2 Oct 2017
10. Fletcher M (1983) the effects of methanol, ethanol, propanol and butanol on bacterial attachment to surfaces. J Gen Microbiol 129(3):633–641
11. Gershenfeld L (1938) The sterility of alcohol. Am J Med Sci 195(3):358–360
12. Harrington C, Walker H (1903) The germicidal action of alcohol. Boston Med Surg J 148 (21):548–552
13. Jabbar U, Leischner J, Kasper D, Gerber R, Sambol SP, Parada JP, Johnson S, Gerding DN (2010) Effectiveness of alcohol-based hand rubs for removal of clostridium difficile spores from hands. Infect Control Hosp Epidemiol 31(6):565–570. https://doi.org/10.1086/652772
14. Juncker JC (2017) COMMISSION IMPLEMENTING REGULATION (EU) 2017/2001 of 8 November 2017 approving propan-1-ol as an existing active substance for use in biocidal products of product-type 1, 2 and 4. Off J Eur Union 60(L 290):1–3
15. Kampf G (2017) n-Propanol. In: Kampf G (ed) Kompendium Händehygiene. mhp-Verlag, Wiesbaden, pp 352–361
16. Kampf G, Höfer M, Wendt C (1999) Efficacy of hand disinfectants against vancomycin-resistant enterococci in vitro. J Hosp Infect 42(2):143–150
17. Kampf G, Hollingsworth A (2003) Validity of the four European test strains of prEN 12054 for the determination of comprehensive bactericidal activity of an alcohol-based hand rub. J Hosp Infect 55(3):226–231
18. Kampf G, Jarosch R, Rüden H (1997) Wirksamkeit alkoholischer Händedesinfektionsmittel gegenüber Methicillin-resistenten *Staphylococcus aureus* (MRSA). Der Chirurg; Zeitschrift fur alle Gebiete der operativen Medizen 68(3):264–270
19. Kampf G, Meyer B, Goroncy-Bermes P (2003) Comparison of two test methods for the determination of sufficient antimicrobial efficacy of three different alcohol-based hand rubs for hygienic hand disinfection. J Hosp Infect 55(3):220–225

20. Kampf G, Ostermeyer C (2009) A 1-minute hand wash does not impair the efficacy of a propanol-based hand rub in two consecutive surgical hand disinfection procedures. Eur J Clin Microbiol Infect Dis 28(11):1357–1362
21. Kampf G, Ostermeyer C, Heeg P (2005) Surgical hand disinfection with a propanol-based hand rub: equivalence of shorter application times. J Hosp Infect 59(4):304–310
22. Kampf G, Ostermeyer C, Kohlmann T (2008) Bacterial population kinetics on hands during 2 consecutive surgical hand disinfection procedures. Am J Infect Control 36(5):369–374
23. Knobloch JK, Horstkotte MA, Rohde H, Kaulfers PM, Mack D (2002) Alcoholic ingredients in skin disinfectants increase biofilm expression of Staphylococcus epidermidis. J Antimicrob Chemother 49(4):683–687
24. Lacroix J, Lacroix R, Reynouard F, Combescot C (1979) In vitro anti-yeast activity of 1- and 2-propanols. Effect of the addition of polyethylene glycol 400. C R Seances Soc Biol Fil 173 (3):547–552
25. Letoffe S, Chalabaev S, Dugay J, Stressmann F, Audrain B, Portais JC, Letisse F, Ghigo JM (2017) Biofilm microenvironment induces a widespread adaptive amino-acid fermentation pathway conferring strong fitness advantage in Escherichia coli. PLoS Genet 13(5):e1006800. https://doi.org/10.1371/journal.pgen.1006800
26. Martro E, Hernandez A, Ariza J, Dominguez MA, Matas L, Argerich MJ, Martin R, Ausina V (2003) Assessment of Acinetobacter baumannii susceptibility to antiseptics and disinfectants. J Hosp Infect 55(1):39–46
27. National Center for Biotechnology Information 1-propanol. PubChem Compound Database; CID = 1031. https://pubchem.ncbi.nlm.nih.gov/compound/1031. Accessed 2 Oct 2017
28. Neufeld F, Schiemann O (1939) Über die Wirkung des Alkohols bei der Händedesinfektion. Z Hyg 121:312–333
29. Presterl E, Suchomel M, Eder M, Reichmann S, Lassnigg A, Graninger W, Rotter M (2007) Effects of alcohols, povidone-iodine and hydrogen peroxide on biofilms of Staphylococcus epidermidis. J Antimicrob Chemother 60(2):417–420. https://doi.org/10.1093/jac/dkm221
30. Reichel M, Heisig P, Kampf G (2008) Pitfalls in efficacy testing—how important is the validation of neutralization of chlorhexidine digluconate? Ann Clin Microbiol Antimicrob 7:20
31. Reichel M, Heisig P, Kohlmann T, Kampf G (2009) Alcohols for skin antisepsis at clinically relevant skin sites. Antimicrob Agents Chemother 53(11):4778–4782
32. Reinicke EA (1894) Bakteriologische Untersuchungen über die Desinfektion der Hände. Zentralbl Gynäkol 47:1189–1199
33. Rotter M, Koller W, Kundi M (1977) Eignung dreier Alkohole für eine Standard-Desinfektionsmethode in der Wertbestimmung von Verfahren für die hygienische Händedesinfektion. Zentralbl Bakteriol Hyg I Abt Orig B 164:428–438
34. Rotter ML (1984) Hygienic hand disinfection. Infection Control: IC 5:18–22
35. Rotter ML, Kampf G, Suchomel M, Kundi M (2007) Long-term effect of a 1.5 minute surgical hand rub with a propanol-based product on the resident hand flora. J Hosp Infect 66 (1):84–85
36. Rotter ML, Koller W, Wewalka G, Werner HP, Ayliffe GAJ, Babb JR (1986) Evaluation of procedures for hygienic hand disinfection: controlled parallel experiments on the Vienna test model. J Hygiene 96:27–37
37. Salo S, Wirtanen G (2005) Disinfectant efficacy on foodborne spoilage yeast strains. Food Bioprod Process 83(4):288–296
38. United States Environmental Protection Agency (2005) Action memorandum. Inert reassessment n-propanol. https://www.epa.gov/sites/production/files/2015-04/documents/propanol.pdf
39. WHO (2009) WHO guidelines on hand hygiene in health care. First Global Patient Safety Challenge Clean Care is Safer Care. WHO, Geneva
40. WHO (2015) WHO model list of essential medicines. WHO. http://www.who.int/medicines/publications/essentialmedicines/EML2015_8-May-15.pdf

41. WHO (2016) Global guidelines for the prevention of surgical site infections. WHO, Geneva
42. WHO (2017) WHO model list of essential medicines for children. WHO. Accessed 30 Aug 2017
43. Wisniak J (2013) Gustav Charles Bonaventure Chancel. Educación Química 24(1):23–30. https://doi.org/10.1016/S0187-893X(13)73191-4
44. Wisplinghoff H, Schmitt R, Wohrmann A, Stefanik D, Seifert H (2007) Resistance to disinfectants in epidemiologically defined clinical isolates of Acinetobacter baumannii. J Hosp Infect 66(2):174–181. https://doi.org/10.1016/j.jhin.2007.02.016

4 Propan-2-ol

4.1 Chemical Characterization

Propan-2-ol is the simplest example of a secondary alcohol, where the alcohol carbon atom is attached to two other carbon atoms. It is a structural isomer of propan-1-ol and a colourless, flammable chemical compound with a strong odour. The basic chemical information on propan-2-ol is summarized in Table 4.1.

4.2 Types of Application

According to the information provided by the European Chemicals Agency (ECHA), propan-2-ol is used by consumers, in articles and by professional workers (widespread uses), in formulation or repacking, at industrial sites and in manufacturing [14]. Consumer use includes lubricants and greases, antifreeze products, coating products, adhesives and sealants, fillers, putties, plasters, modelling clay, finger paints, biocides (e.g. disinfectants, pest control products), polishes, waxes, fuels and toners [14].

Use by professional workers includes coating products, antifreeze products, fuels, lubricants and greases, inks and toners, polymers, water treatment chemicals, laboratory chemicals [14] as well as hand disinfection by healthcare workers and in veterinary medicine, skin antisepsis prior to surgery and disinfection of inanimate surfaces. In 2009, the World Health Organization (WHO) has recommended to use alcohol-based hand rubs, e.g. based on propan-2-ol, in specific situations during patient care for prevention of healthcare-associated infections [57]. Alcohol-based hand rubs, e.g. based on propan-2-ol, are also recommended for the preoperative decontamination of hands for the prevention of surgical site infections [59]. Since 2015, the WHO has classified propan-2-ol at 75% (v/v) as a disinfectant for alcohol-based hand rubbing as an "essential medicine" [58], both for adults and

G. Kampf, *Antiseptic Stewardship*, https://doi.org/10.1007/978-3-319-98785-9_4

Table 4.1 Basic chemical information on propan-2-ol [14, 41]

CAS number	67-63-0
IUPAC name	Propan-2-ol
Synonyms	Iso-propylalcohol, iso-propanol, 2-propanol
Molecular formula	C_3H_8O
Molecular weight (g/mol)	60.096

children up to 12 years of age [60]. Propan-2-ol is also a widely used biocidal ingredient in surface disinfectants [11, 44], e.g. for treatment of mobile phones or small surfaces on intensive care units [2, 51].

4.2.1 European Chemicals Agency (European Union)

Propan-2-ol has been approved in 2015 as an active biocidal agent for product types 1 (human hygiene), 2 (disinfectants and algaecides not intended for direct application to humans or animals) and 4 (food and feed area) [22].

4.2.2 Environmental Protection Agency (USA)

Propan-2-ol has last been reregistered by the EPA in 1995. It is used as a component of a variety of commercial and household products including a sterilant, medical disinfectants, virucides, sanitizers, fungicides and plant regulators (ripener). Propan-2-ol is used in conjunction with quaternary ammonium compounds, phenolic compounds, glycols, methyl salicylate and essential oils [54].

4.2.3 Food and Drug Administration (USA)

In 1994, the tentative final monograph for healthcare antiseptic products classified propan-2-ol between 70 and 91.3% as "generally recognized as safe and effective" for patient preoperative skin preparation, for surgical hand scrubbing and for healthcare personnel hand wash [9]. In 2015, the classification was changed. Propan-2-ol at 70–91.3% was now eligible for three types of application: patient preoperative skin preparation, healthcare personnel hand rub and surgical hand rub. It is now classified in category IIISE indicating that available data are insufficient to classify propan-2-ol as safe and effective, and further testing is required. The main aspect is the safety under maximal use conditions [10].

4.2.4 Overall Environmental Impact

Propan-2-ol is manufactured and/or imported in the European Economic Area in 100,000–1 million t per year [14]. Other release to the environment of this substance is likely to occur from outdoor use, indoor use (e.g. machine wash liquids/detergents, automotive care products, paints and coating or adhesives, fragrances and air fresheners), outdoor use in close systems with minimal release (e.g. hydraulic liquids in automotive suspension, lubricants in motor oil and break fluids) and indoor use in close systems with minimal release (e.g. cooling liquids in refrigerators, oil-based electric heaters) [14].

4.3 Spectrum of Antimicrobial Activity

4.3.1 Bactericidal Activity

4.3.1.1 Bacteriostatic Activity (MIC Values)

The MIC values obtained with different bacterial species are summarized in Table 4.2. Propan-2-ol has bacteriostatic activity between 3.1% in *Staphylococcus* spp. and 10% in *E. faecalis* and *S. aureus*.

4.3.1.2 Bactericidal Activity (Suspension Tests)

Some data have been published on the efficacy of propan-2-ol against bacteria in suspension tests. They indicate a strong bactericidal activity of a 70% solution beginning after 15 s. A concentration of 10% has mostly insufficient bactericidal activity within 5 min (Table 4.3). Additional data with mixed propanols (e.g. 45% propan-2-ol plus 30% propan-1-ol) indicate a comprehensive bactericidal activity within 30 s [25, 27].

Table 4.2 MIC values of various bacterial species to propan-2-ol

Species	Strains/isolates	MIC value	References
E. faecalis	9 isolates from swine meat production	8–10%	[46]
E. faecium	12 isolates from swine meat production	8%	[46]
E. coli	ATCC 8739	~5%	[37]
Micrococcus spp.	1 isolate from a clean room	6.3%	[5]
P. aeruginosa	ATCC 9027	≤5%	[37]
S. aureus	ATCC 6538	10%	[37]
S. aureus	MTCC 737	6.3%	[5]
Staphylococcus spp.	3 isolates from clean room	3.1–6.3%	[5]

Table 4.3 Bactericidal activity of propan-2-ol in suspension tests

Species	Strains/isolates	Exposure time	Concentration	$\log_{10}$ reduction	References
C. jejuni	ATCC BAA-1062, ATCC 33560 and 2 field strains	1 min	70% (S)	>6.0	[18]
E. faecalis	Strain Q33	5 min	70% (S)	5.0	[39]
E. faecium	VRE strain Z31901	5 min	70% (S)	5.0	[39]
Enterococcus spp.	11 strains (8 *E. faecium*, 2 *E. faecalis*, 1 *E. gallinarum*)	15 s	70% (P)[a]	>7.0	[24]
E. coli	NCTC 10538	5 min	70% (S)	5.0	[39]
F. nucleatum	NCTC 10562	5 min	10% (S)	0.7	[33]
H. parasuis	2 strains (serovars 1 and 5)	1 min	70% (S)	>6.0	[47]
				5.3–5.5[b]	
K. pneumoniae	NCIMB 13291	5 min	80% (S)	4.4	[33]
			50% (S)	4.4	
			10% (S)	1.3	
P. gingivalis	ATCC 53978	5 min	10% (S)	5.6	[33]
P. aeruginosa	NCIMB 10421	5 min	70% (S)	5.1	[39]
S. aureus	ATCC 6538, ATCC 43300, 4 clinical strains (2 MRSA, 2 MSSA)	15 s	70% (P)[a]	>8.0	[26]
S. aureus	NCTC 6571	5 min	70% (S)	4.8	[39]
S. aureus	MRSA strain 9543	5 min	70% (S)	5.0	[39]
S. epidermidis	Strain RP62A	30 s	70% (S)	6.5	[1]
S. epidermidis	Strain P69	5 min	70% (S)	5.2	[39]
S. mutans	NCTC 10449	5 min	80% (S)	5.4	[33]
			50% (S)	5.4	
			10% (S)	1.0	

S Solution; *P* Commercial product; [a]With 0.5% chlorhexidine; [b]With serum

4.3.1.3 Activity Against Bacteria in Biofilms

The efficacy of propan-2-ol against bacteria in artificially grown biofilms has been addressed in some studies; the results are summarized in Table 4.4. Propan-2-ol even at 70% has often poor bactericidal activity against bacterial cells grown in biofilms with log reductions <5.0 in up to 60 min which is described in the majority of studies (Table 4.4). It was also shown with *P. aeruginosa* and *P. fragi* biofilms that a substantial proportion of bacterial cells (5-15%) remains viable in the biofilm after exposure to a propan-2-ol-based disinfectant, whereas conventional cultivation yields negative results indicating that some bacterial cells remain viable but not culturable after disinfectant exposure [61].

4.3.1.4 Bactericidal Activity for Hygienic Hand Disinfection

Propan-2-ol at 60% (v/v) was first proposed as a reference treatment for hygienic hand disinfection in 1977 [56] and is since 1997 the reference alcohol to determine the efficacy of alcohol-based hand rubs for hygienic hand disinfection in EN 1500 [12]. Numerous data exist for the reference treatment (2 × 3 ml for 2 × 30 s) which usually achieves a mean log reduction on hands artificially contaminated with *E. coli* of 4.6 [28, 29].

The efficacy of hand rubs based on propan-2-ol has been mostly evaluated on hands artificially contaminated with *E. coli* according to EN 1500 with an application of 3 ml for 30 s. A summary of published data has recently been published [23]. Hand rubs with a propan-2-ol concentration <75% may fail to meet the EN 1500 efficacy requirements depending on the entire product composition. Longer application times or larger volumes mostly yield better results.

4.3.1.5 Bactericidal Activity for Surgical Hand Disinfection

The efficacy for surgical hand disinfection is determined in many countries without an artificial contamination of hands against the resident hand flora according to EN 12791, mostly with application times of 1.5 or 3 min. Formulations with propan-2-ol of at least 70% (w/w) usually meet the efficacy requirements when applied for 1.5 or 3 min [49]. Additional data with mixed propanols (e.g. 45% propan-2-ol plus 30% propan-1-ol) indicate sufficient bactericidal efficacy for surgical hand disinfection within 1.5 min [30–32, 48].

4.3.1.6 Bactericidal Activity in Carrier Tests

Propan-2-ol at 70% has been described to be effective with a 5.4 log reduction in 30 s against *S. epidermidis* in a carrier test even when the cells were grown in a biofilm [1]. The effect is lower with 2.8 log when an organic load (10% serum) is added [1]. Against strains from six bacterial species (*E. faecalis*, *E. faecium* VRE, *E. coli*, *P. aeruginosa*, *S. aureus*, MRSA, *S. epidermidis*), the efficacy of 70% propan-2-ol on glass carriers was good with log reductions between 3.8 and 5.0 in 1 min [39].

Table 4.4 Efficacy of formulations based in propan-2-ol against bacteria in biofilms

Species	Strains/isolates	Type of biofilm	Exposure time	Concentration	$\log_{10}$ reduction	References
A. baumannii	64 MDR clinical isolates	5-d incubation in polystyrene plates	1 min	70% (P)[a]	3.5	[7]
			3 min		>8.2	
P. aeruginosa	ATCC 700928	24-h incubation in microplates	1 min	70% (S)	0.7	[53]
			5 min		1.4	
			60 min		>5.0	
P. aeruginosa	ATCC 15442	24-h incubation in microplates	30 min	80% (S)	2.8	[35]
				70% (S)	2.6	
				50% (S)	2.5	
				25% (S)	1.6	
S. aureus	ATCC 35556, ATCC 29213 (both MSSA), ATCC 43300, strain L32 (both MRSA)	24-h incubation on polystyrene plates	24 h	95% (S)	No significant reduction	[38]
				80% (S)		
				60% (S)		
				40% (S)		
S. aureus	ATCC 6538	72-h incubation in microplates	1 min	70% (S)	1.2	[53]
			5 min		1.4	
			60 min		1.7	
S. capitis	Strain CBS 517	24-h incubation in microtitre plates	30 s	70% (S)	0.2[b]	[52]
S. epidermidis	ATCC 35984 and a biofilm-deficient mutant M7	24-h incubation on polystyrene plates	24 h	95% (S)	"significant reduction"	[38]
				80% (S)		
				60% (S)		
				40% (S)		

(continued)

Table 4.4 (continued)

Species	Strains/isolates	Type of biofilm	Exposure time	Concentration	log_{10} reduction	References
S. epidermidis	Strain 9142	24-h incubation in microtitre plates	30 s	70% (S)	0.3^b	[52]

S Solution; *P* Commercial product; [a]Plus 0.5% chlorhexidine; [b]Significantly lower compared to planktonic cells

Table 4.5 MIC values of various fungal species to propan-2-ol

Species	Strains/isolates	MIC value	References
C. albicans	ATCC 10231	≤5%	[37]
T. rubrum	IP 1464.83	≤5%	[37]

4.3.2 Fungicidal Activity

4.3.2.1 Fungistatic Activity (MIC Values)

The MIC values obtained with different fungal species are summarized in Table 4.5. Propan-2-ol has fungistatic activity at a concentration ≤5%.

4.3.2.2 Fungicidal Activity (Suspension Tests)

Published data on the fungicidal activity obtained with propan-2-ol are summarized in Table 4.6. Propan-2-ol at 60% or 70% was mostly effective against various types of yeasts within 5–10 min. The efficacy of 70% propan-2-ol against food-associated fungi such as *E. repens*, *M. ruber*, *N. pseudofischeri*, *P. caseifulvum*, *P. discolor*, *P. nalgiovense* and *P. verrucosum* is low within 10 min (Table 4.6).

Table 4.6 Fungicidal activity of propan-2-ol in suspension tests

Species	Strains/isolates	Exposure time	Concentration	$\log_{10}$ reduction	References
A. flavus	Bread isolate	10 min	70% (P)	4.0	[3]
A. niger	Bread isolate	10 min	70% (P)	>5.2	[3]
A. versicolor	2 cheese isolates	10 min	70% (P)	3.3–5.2	[3]
C. albicans	NCPF 3179	5 min	80% (S)	4.4	[33]
			50% (S)	4.4	
			10% (S)	1.1	
Cladosporium spp.	Bread isolate	10 min	70% (P)	>4.1	[3]
D. hansenii	Cheese isolate	10 min	70% (P)	>4.5	[3]
E. repens	Bread factory isolate	10 min	70% (P)	2.3	[3]
H. burtonii	Bread isolate	10 min	70% (P)	>5.2	[3]
M. ruber	Bread isolate	10 min	70% (P)	0.9	[3]
M. suaveolens	Bread isolate	10 min	70% (P)	>4.5	[3]
N. pseudofischeri	Cherry filling isolate	10 min	70% (P)	3.3	[3]
P. anomala	Bread isolate	10 min	70% (P)	>5.9	[3]
P. caseifulvum	Cheese isolate	10 min	70% (P)	2.7	[3]
P. chrysogenum	Cheese isolate	10 min	70% (P)	>5.2	[3]
P. commune	2 cheese and 1 bread isolates	10 min	70% (P)	2.7–4.0	[3]
P. corylophilum	Bread isolate	10 min	70% (P)	>4.8	[3]
P. crustosum	Cheese isolate	10 min	70% (P)	4.0	[3]
P. discolor	Cheese isolate	10 min	70% (P)	3.0	[3]

(continued)

Table 4.6 (continued)

Species	Strains/isolates	Exposure time	Concentration	log_{10} reduction	References
P. nalgiovense	2 cheese isolates	10 min	70% (P)	2.3–3.0	[3]
P. norvegensis	Cheese isolate	10 min	70% (P)	>5.2	[3]
P. roqueforti	2 bread isolates	10 min	70% (P)	3.5–5.2	[3]
P. solitum	Cheese isolate	10 min	70% (P)	3.7	[3]
P. verrucosum	Cheese isolate	10 min	70% (P)	3.1	[3]
S. brevicaulis	Cheese isolate	10 min	70% (P)	>4.2	[3]
T. delbrueckii	Cheese isolate	10 min	70% (P)	>4.8	[3]
Yeasts	25 strains isolated from food or food processing	5 min	60% (P)[a]	$\geq$ 4.0	[50]

S Solution; *P* Commercial product; [a]With additional QAC (<5%)

4.3.3 Mycobactericidal Activity

Early data from 1953 indicate a tuberculocidal activity of propan-2-ol at 50–70% [15]. In suspension tests, only few data were published on the mycobactericidal activity of propan-2-ol. At 60% it was mostly effective against both *M. tuberculosis* and *M. terrae* within 5 min (Table 4.7). Whether the tuberculocidal efficacy covers other mycobacterial species is not yet described in the literature.

4.4 Effect of Low-Level Exposure

Only very few data exist in the literature on the effect of low-level exposure of micro-organisms against propan-2-ol. One study indicates that attachment of cells can be significantly increased in some *L. monocytogenes* strains when exposed to 2.5% propan-2-ol, mostly at 10 °C [17]. The biofilm formation of 37 clinical, icaADBC-positive *S. epidermidis* isolates was investigated after exposure to propan-2-ol at 1, 2, 4 or 6%. In 14 of the 37 strains, biofilm formation was inducible by propan-2-ol exposure [34]. Propan-2-ol at 1 and 2% was able to increase biofilm formation in a non adherent *S. epidermidis* strain [4]. With *C. albicans*, it was described that propan-2-ol at 2% inhibited biofilm development to some extent [6].

Table 4.7 Mycobactericidal activity of propan-2-ol in suspension tests

Species	Strains/isolates	Exposure time	Concentration	log_{10} reduction	References
M. terrae	Isolate 232	5 min	60% (S)	>4.6	[55]
	Isolate 373			>4.9	
M. tuberculosis	ATCC 25618	5 min	60% (S)	3.8	[55]

S Solution

In *E. coli,* it was shown that low-level exposure to variable propan-2-ol concentrations up to 2.7% for up to 24 d reduced the susceptibility of the six tested strains to propan-2-ol substantially. But no MIC_{max} values were described after adaptation, and the stability of the lower susceptibility is also unknown [20].

4.5 Resistance to Propan-2-ol

A propan-2-ol-tolerant *S. mizutae* has been recently isolated from an oil–soil mixture. It was able to multiply in propan-2-ol solutions without further C supplementation at concentrations between 0.2 and 3.8% indicating the potential for tolerance to propan-2-ol to specific environmental bacterial species [40]. In 2018, recent *E. faecium* isolates from Australia (2011–2015) were reported to be more tolerant to 23% and 70% (v/v) propan-2-ol compared to previously isolated *E. faecium* isolates (1997–2010) [43]. No other micro-organisms with a resistance to propan-2-ol have been reported so far. It should be kept in mind that for more than 100 years alcohols such as propan-2-ol have basically no effect on bacterial spores including spores of *C. difficile* indicating an intrinsic resistance [13, 16, 19, 21, 42, 45]. Bacterial spores are, however, not addressed in this book.

4.5.1 Resistance Mechanisms

In *E. faecium* it was shown that propan-2-ol-tolerant isolates accumulated mutations in genes involved in carbohydrate uptake and metabolism [43]. No other specific resistance mechanism has so far been described to explain an acquired bacterial or fungal resistance to propan-2-ol.

4.5.2 Resistance Genes

In *E. coli* strains, five mutations (relA, marC, proQ, yfgO and rraA) provided the increase of tolerance to propan-2-ol. Expression levels of genes related to biosynthetic pathways of amino acids, iron ion homoeostasis and energy metabolisms were changed in the tolerant strains [20].

4.6 Cross-Tolerance to Other Biocidal Agents

No cross-tolerance to other biocidal agents has so far been described.

4.7 Cross-Tolerance to Antibiotics

So far no cross-tolerance between propan-2-ol and antibiotics has been described.

Table 4.8 Biofilm removal by exposure to propan-2-ol solutions measured quantitatively as change of biofilm matrix

Type of biofilm	Concentration	Exposure time	Biofilm removal rate	References
S. aureus ATCC 6538, 72-h incubation in microplates	70% (S)	60 min	0%	[53]
P. aeruginosa ATCC 700928, 24-h incubation in microplates	70% (S)	60 min	0%	[53]

S Solution

4.8 Role of Biofilm

4.8.1 Effect on Biofilm Development

Treatment of a preformed *S. aureus* (2 MSSA, 2 MRSA) and a *S. epidermidis* biofilm (2 strains) for 24 h with propan-2-ol between 40 and 95% increased biofilm formation significantly. A higher propan-2-ol concentration resulted in a higher biofilm formation [38].

4.8.2 Effect on Biofilm Removal

The ability of propan-2-ol to remove biofilm is overall very poor (Table 4.8). This finding is supported by data showing that protein removal rates from different types of surfaces by propan-2-ol are overall low [36].

4.8.3 Effect on Biofilm Fixation

No studies were found to evaluate the fixation potential of biofilms by exposure to propan-2-ol. It is, however, likely that propan-2-ol induces fixation of an existing biofilm to some extent because the substance is known for its fixative properties (e.g. bacteria and blood) [8].

4.9 Summary

The principal antimicrobial activity of propan-2-ol is summarized in Table 4.9.

The key findings on acquired resistance and cross-resistance including the role of biofilm for selecting resistant isolates are summarized in Table 4.10.

Table 4.9 Overview on the typical exposure times required for propan-2-ol to achieve sufficient biocidal activity in suspension tests against the different target micro-organisms

Target micro-organisms	Species	Concentration	Exposure time
Bacteria	Most clinically relevant species including some antibiotic-resistant isolates	$\geq$70%	15 s[a]
Fungi	*C. albicans*	50%	5 min
	Food-associated yeasts	70%	10 min
Mycobacteria	*M. terrae, M. tuberculosis*	60%	5 min

[a]In biofilm there may be no sufficient efficacy in 60 min depending on the species and the type of biofilm

Table 4.10 Key findings on propan-2-ol resistance, the effect of low-level exposure, cross-tolerance to other biocides and antibiotics and its effect on biofilm

Parameter	Species	Findings
Elevated MIC values	*E. faecium*	Increased tolerance possible
MIC value to determine resistance	Not proposed yet for bacteria, fungi or mycobacteria	
Cross-tolerance biocides	So far not reported	
Cross-tolerance antibiotics	So far not reported	
Effect of low-level exposure	None	No MIC increase
	None	Weak MIC increase ($\leq$4-fold)
	None	Strong MIC increase (>4-fold)
	E. coli	Reduced susceptibility to lethal propan-2-ol concentrations (adaptation)
	S. epidermidis	Increase of biofilm formation in some isolates
	C. albicans	Inhibition of biofilm development
	L. monocytogenes	Increase of surface attachment
Specific resistance mechanism	So far not reported	
Biofilm	Development	Enhancement in *S. epidermidis* and *S. aureus*
	Removal	None in *P. aeruginosa* and *S. aureus*
	Fixation	Unknown

References

1. Adams D, Quayum M, Worthington T, Lambert P, Elliott T (2005) Evaluation of a 2% chlorhexidine gluconate in 70% isopropyl alcohol skin disinfectant. J Hosp Infect 61(4): 287–290
2. Boyce JM (2018) Alcohols as surface disinfectants in healthcare settings. Infect Control Hosp Epidemiol 39(3):323–328. https://doi.org/10.1017/ice.2017.301

3. Bundgaard-Nielsen K, Nielsen PV (1996) Fungicidal effect of 15 disinfectants against 25 fungal contaminants commonly found in bread and cheese manufacturing. J Food Prot 59(3):268–275
4. Chaieb K, Zmantar T, Souiden Y, Mahdouani K, Bakhrouf A (2011) XTT assay for evaluating the effect of alcohols, hydrogen peroxide and benzalkonium chloride on biofilm formation of Staphylococcus epidermidis. Microb Pathog 50(1):1–5. https://doi.org/10.1016/j.micpath.2010.11.004
5. Chand S, Saha K, Singh PK, Sri S, Malik N (2016) Determination of minimum inhibitory concentration (MIC) of routinely used disinfectants against microflora isolated from clean rooms. Int J Curr Microbiol Appl Sci 5(1):334–341
6. Chauhan NM, Shinde RB, Karuppayil SM (2013) Effect of alcohols on filamentation, growth, viability and biofilm development in Candida albicans. Brazilian J Microbiol [Publication of the Brazilian Society for Microbiology] 44(4):1315–1320. https://doi.org/10.1590/s1517-83822014005000012
7. Chiang SR, Jung F, Tang HJ, Chen CH, Chen CC, Chou HY, Chuang YC (2017) Desiccation and ethanol resistances of multidrug resistant Acinetobacter baumannii embedded in biofilm: the favorable antiseptic efficacy of combination chlorhexidine gluconate and ethanol. J Microbiol Immunol infection = Wei mian yu gan ran za zhi. https://doi.org/10.1016/j.jmii.2017.02.003
8. Costa DM, Lopes LKO, Hu H, Tipple AFV, Vickery K (2017) Alcohol fixation of bacteria to surgical instruments increases cleaning difficulty and may contribute to sterilization inefficacy. Am J Infect Control 45(8):e81–e86. https://doi.org/10.1016/j.ajic.2017.04.286
9. Department of Health and Human Services; Food and Drug Administration (1994) Tentative final monograph for health care antiseptic products; proposed rule. Fed Reg 59(116):31401–31452
10. Department of Health and Human Services; Food and Drug Administration (2015) Safety and effectiveness of healthcare antiseptics. Topical antimicrobial drug products for over-the-counter human use; proposed amendment of the tentative final monograph; reopening of administrative record; proposed rule. Fed Reg 80(84):25166–25205
11. Eaton T (2009) Cleanroom airborne particulate limits and 70% isopropyl alcohol: a lingering problem for pharmaceutical manufacturing? PDA J Pharm Sci Technol 63(6):559–567
12. EN 1500:2013 (2013) Chemical disinfectants and antiseptics. Hygienic hand disinfection. Test method and requirement (phase 2, step 2). In: CEN—Comité Européen de Normalisation, Brussels
13. Epstein F (1896) Zur Frage der Alkoholdesinfektion. Z Hyg 24:1–21
14. European Chemicals Agency (ECHA) Propan-2-ol. Substance information. https://echa.europa.eu/substance-information/-/substanceinfo/100.000.601. Accessed 27 Sept 2017
15. Frobisher M (1953) A study of the effect of alcohols on tubercle bacilli and other bacteria in sputum. Am Rev Tuberc 68:419–424
16. Gershenfeld L (1938) The sterility of alcohol. Am J Med Sci 195(3):358–360
17. Gravesen A, Lekkas C, Knochel S (2005) Surface attachment of Listeria monocytogenes is induced by sublethal concentrations of alcohol at low temperatures. Appl Environ Microbiol 71(9):5601–5603. https://doi.org/10.1128/aem.71.9.5601-5603.2005
18. Gutierrez-Martin CB, Yubero S, Martinez S, Frandoloso R, Rodriguez-Ferri EF (2011) Evaluation of efficacy of several disinfectants against Campylobacter jejuni strains by a suspension test. Res Vet Sci 91(3):e44–47. https://doi.org/10.1016/j.rvsc.2011.01.020
19. Harrington C, Walker H (1903) The germicidal action of alcohol. Boston Med Surg J 148(21):548–552
20. Horinouchi T, Sakai A, Kotani H, Tanabe K, Furusawa C (2017) Improvement of isopropanol tolerance of Escherichia coli using adaptive laboratory evolution and omics technologies. J Biotechnol 255:47–56. https://doi.org/10.1016/j.jbiotec.2017.06.408
21. Jabbar U, Leischner J, Kasper D, Gerber R, Sambol SP, Parada JP, Johnson S, Gerding DN (2010) Effectiveness of alcohol-based hand rubs for removal of Clostridium difficile spores from hands. Infect Control Hosp Epidemiol 31(6):565–570. https://doi.org/10.1086/652772
22. Juncker JC (2015) COMMISSION IMPLEMENTING REGULATION (EU) 2015/407 of 11 March 2015 approving propan-2-ol as an active substance for use in biocidal products for product-types 1, 2 and 4. Off J Eur Union 58(L 67):15–17

23. Kampf G (2017) Iso-propanol. In: Kampf G (ed) Kompendium Händehygiene. mhp-Verlag, Wiesbaden, pp 362–375
24. Kampf G, Höfer M, Wendt C (1999) Efficacy of hand disinfectants against vancomycin-resistant enterococci in vitro. J Hosp Infect 42(2):143–150
25. Kampf G, Hollingsworth A (2003) Validity of the four European test strains of prEN 12054 for the determination of comprehensive bactericidal activity of an alcohol-based hand rub. J Hosp Infect 55(3):226–231
26. Kampf G, Jarosch R, Rüden H (1998) Limited effectiveness of chlorhexidine based hand disinfectants against methicillin-resistant *Staphylococcus aureus* (MRSA). J Hosp Infect 38 (4):297–303
27. Kampf G, Meyer B, Goroncy-Bermes P (2003) Comparison of two test methods for the determination of sufficient antimicrobial efficacy of three different alcohol-based hand rubs for hygienic hand disinfection. J Hosp Infect 55(3):220–225
28. Kampf G, Ostermeyer C (2002) Intra-laboratory reproducibility of the hand hygiene reference procedures of EN 1499 (hygienic hand wash) and EN 1500 (hygienic hand disinfection). J Hosp Infect 52(3):219–224
29. Kampf G, Ostermeyer C (2003) Inter-laboratory reproducibility of the EN 1500 reference hand disinfection. J Hosp Infect 53(4):304–306
30. Kampf G, Ostermeyer C (2009) A 1-minute hand wash does not impair the efficacy of a propanol-based hand rub in two consecutive surgical hand disinfection procedures. Eur J Clin Microbiol Infect Dis 28(11):1357–1362
31. Kampf G, Ostermeyer C, Heeg P (2005) Surgical hand disinfection with a propanol-based hand rub: equivalence of shorter application times. J Hosp Infect 59(4):304–310
32. Kampf G, Ostermeyer C, Kohlmann T (2008) Bacterial population kinetics on hands during 2 consecutive surgical hand disinfection procedures. Am J Infect Control 36(5):369–374
33. Kiesow A, Sarembe S, Pizzey RL, Axe AS, Bradshaw DJ (2016) Material compatibility and antimicrobial activity of consumer products commonly used to clean dentures. J Prosthet Dent 115 (2):189–198.e188. https://doi.org/10.1016/j.prosdent.2015.08.010
34. Knobloch JK, Horstkotte MA, Rohde H, Kaulfers PM, Mack D (2002) Alcoholic ingredients in skin disinfectants increase biofilm expression of staphylococcus epidermidis. J Antimicrob Chemother 49(4):683–687
35. Konrat K, Schwebke I, Laue M, Dittmann C, Levin K, Andrich R, Arvand M, Schaudinn C (2016) The bead assay for biofilms: a quick, easy and robust method for testing disinfectants. PLoS ONE 11(6):e0157663. https://doi.org/10.1371/journal.pone.0157663
36. Kratz F, Grass S, Umanskaya N, Scheibe C, Muller-Renno C, Davoudi N, Hannig M, Ziegler C (2015) Cleaning of biomaterial surfaces: protein removal by different solvents. Colloids Surf, B 128:28–35. https://doi.org/10.1016/j.colsurfb.2015.02.016
37. Lens C, Malet G, Cupferman S (2016) Antimicrobial activity of butyl acetate, ethyl acetate and Isopropyl alcohol on undesirable microorganisms in cosmetic products. Int J Cosmet Sci 38(5):476–480. https://doi.org/10.1111/ics.12314
38. Luther MK, Bilida S, Mermel LA, LaPlante KL (2015) Ethanol and Isopropyl alcohol exposure increases biofilm formation in staphylococcus aureus and staphylococcus epidermidis. Infect Dis Ther 4(2):219–226. https://doi.org/10.1007/s40121-015-0065-y
39. Messager S, Goddard PA, Dettmar PW, Maillard JY (2001) Determination of the antibacterial efficacy of several antiseptics tested on skin by an 'ex-vivo' test. J Med Microbiol 50(3):284–292. https://doi.org/10.1099/0022-1317-50-3-284
40. Mohammad BT, Wright PC, Bustard MT (2006) Bioconversion of isopropanol by a solvent tolerant Sphingobacterium mizutae strain. J Ind Microbiol Biotechnol 33(12):975–983. https://doi.org/10.1007/s10295-006-0143-y
41. National Center for Biotechnology Information Isopropanol. PubChem Compound Database; CID = 3776. https://pubchem.ncbi.nlm.nih.gov/compound/3776. Accessed 27 Sept 2017
42. Neufeld F, Schiemann O (1939) Über die Wirkung des Alkohols bei der Händedesinfektion. Z Hyg 121:312–333

43. Pidot SJ, Gao W, Buultjens AH, Monk IR, Guerillot R, Carter GP et al. (2018) Increasing tolerance of hospital Enterococcus faecium to handwash alcohols. Sci Transl Med. 10(452) eaar6115. https://doi.org/10.1126/scitranslmed.aar6115
44. Rabenau HF, Steinmann J, Rapp I, Schwebke I, Eggers M (2014) Evaluation of a virucidal quantitative carrier test for surface disinfectants. PLoS ONE 9(1):e86128. https://doi.org/10.1371/journal.pone.0086128
45. Reinicke EA (1894) Bakteriologische Untersuchungen über die Desinfektion der Hände. Zentralbl Gynäkol 47:1189–1199
46. Rizzotti L, Rossi F, Torriani S (2016) Biocide and antibiotic resistance of Enterococcus faecalis and Enterococcus faecium isolated from the swine meat chain. Food Microbiol 60:160–164. https://doi.org/10.1016/j.fm.2016.07.009
47. Rodriguez Ferri EF, Martinez S, Frandoloso R, Yubero S, Gutierrez Martin CB (2010) Comparative efficacy of several disinfectants in suspension and carrier tests against Haemophilus parasuis serovars 1 and 5. Res Vet Sci 88(3):385–389. https://doi.org/10.1016/j.rvsc.2009.12.001
48. Rotter ML, Kampf G, Suchomel M, Kundi M (2007) Long-term effect of a 1.5 minute surgical hand rub with a propanol-based product on the resident hand flora. J Hosp Infect 66(1):84–85
49. Rotter ML, Kampf G, Suchomel M, Kundi M (2007) Population kinetics of the skin flora on gloved hands following surgical hand disinfection with 3 propanol-based hand rubs: a prospective, randomized, double-blind trial. Infect Control Hosp Epidemiol 28(3):346–350
50. Salo S, Wirtanen G (2005) Disinfectant efficacy on foodborne spoilage yeast strains. Food Bioprod Process 83(4):288–296
51. Singh S, Acharya S, Bhat M, Rao SK, Pentapati KC (2010) Mobile phone hygiene: potential risks posed by use in the clinics of an Indian dental school. J Dent Educ 74(10):1153–1158
52. Taha M, Kalab M, Yi QL, Landry C, Greco-Stewart V, Brassinga AK, Sifri CD, Ramirez-Arcos S (2014) Biofilm-forming skin microflora bacteria are resistant to the bactericidal action of disinfectants used during blood donation. Transfusion 54(11):2974–2982. https://doi.org/10.1111/trf.12728
53. Tote K, Horemans T, Vanden Berghe D, Maes L, Cos P (2010) Inhibitory effect of biocides on the viable masses and matrices of Staphylococcus aureus and Pseudomonas aeruginosa biofilms. Appl Environ Microbiol 76(10):3135–3142. https://doi.org/10.1128/aem.02095-09
54. United States Environmental Protection Agency (1995) Reregistration eligibility decision (RED) aliphatic alcohols. https://www3.epa.gov/pesticides/chem_search/reg_actions/reregistration/red_G-4_1-Mar-95.pdf
55. van Klingeren B, Pullen W (1987) Comparative testing of disinfectants against Mycobacterium tuberculosis and Mycobacterium terrae in a quantitative suspension test. J Hosp Infect 10(3):292–298. https://doi.org/10.1016/0195-6701(87)90012-0
56. Wewalka G, Rotter M, Koller W, Stanek G (1977) Wirkungsvergleich von 14 Verfahren zur hygienischen Händedesinfektion. Zentralblatt für Bakteriologie und Hygiene, I Abt Orig B 165:242–249
57. WHO (2009) WHO guidelines on hand hygiene in health care. First Global Patient Safety Challenge Clean Care is Safer Care. WHO, Geneva
58. WHO (2015) WHO model list of essential medicines. WHO. http://www.who.int/medicines/publications/essentialmedicines/EML2015_8-May-15.pdf
59. WHO (2016) Global guidelines for the prevention of surgical site infections. WHO, Geneva
60. WHO (2017) WHO model list of essential medicines for children. WHO. Accessed 30 Aug 30 2017
61. Wirtanen G, Salo S, Helander IM, Mattila-Sandholm T (2001) Microbiological methods for testing disinfectant efficiency on Pseudomonas biofilm. Colloids Surf, B 20(1):37–50

Peracetic Acid

5

5.1 Chemical Characterization

Peracetic acid is an organic peroxide and a colourless liquid with a characteristic acrid odour reminiscent of acetic acid. The basic chemical information on peracetic acid is summarized in Table 5.1. The stability of peracetic acid depends on the formulation. The concentration may go down from 250 mg/l to undetectable levels within 4 days or may go down from 500 mg/l on day 1 to 400 mg/l on days 2–24 [24]. Lack of stability of a two-component peracetic acid-based surface disinfectant has been associated with an increase of *C. difficile* infections. The label concentration of 1,500 mg/l peracetic acid was neither achieved in newly activated product (mean: 400 mg/l) nor in in-use product solutions (mean: 180 mg/l) [18].

It is in the meantime possible to deliver peracetic acid in combination with hydrogen peroxide with localised potent, non-toxic bactericidal activity (1.5-h exposure time against MRSA and carbapenem-resistant *E. coli*) using the pre-cursor compounds tetraacetylethylenediamine and sodium percarbonate loaded into thermally induced phase separation microparticles [109].

5.2 Types of Application

Peracetic acid is used in the European Union by consumers and professional workers (widespread uses), in formulation or re-packing, at industrial sites and in manufacturing. It is used in washing and cleaning products. Professional workers use peracetic acid in washing and cleaning products, biocides (e.g. disinfectants, pest control products) and laboratory chemicals. It is used in health services, scientific research and development, and manufacturing of textile, leather or fur. It is also used indoors (e.g. machine wash liquids or detergents, automotive care products, paints and coating or adhesives, fragrances and air fresheners) and in closed systems (e.g. cooling liquids in refrigerators, oil-based electric heaters) [41].

G. Kampf, *Antiseptic Stewardship*, https://doi.org/10.1007/978-3-319-98785-9_5

Table 5.1 Basic chemical information on peracetic acid [5, 46]

CAS number	79-21-0
IUPAC name	Ethaneperoxoic acid
Synonyms	Acetyl hydroperoxide, ethaneperoxoic acid, peroxyacetic acid
Molecular formula	$C_2H_4O_3$
Molecular weight (g/mol)	76.05

Based on the evaluation of the EPA in 1993, it was and is used in the USA as a disinfectant for dialyzers and dialysis equipment, anaesthesia equipment, aseptic packaging and related surfaces in food processing plants, respiratory equipment, endoscopes, endotracheal tubes, dental hand instruments and burs, and surgical instruments. It is also used for disinfection of reverse osmosis membranes and their associated distribution systems, and for disinfection of hospital non-critical items made of plastic or stainless steel. It is used for disinfection of various types of hard non-food contact surfaces, as a sanitizer of food contact and non-food contact surfaces and equipment in food production, for disinfection of animal life science laboratories, livestock premises, dairy cattle and goat premises, poultry premises, transportation vehicle treatment, feeding and watering appliance treatment, and for disinfection of farm buildings and premises [118].

5.2.1 European Chemicals Agency (European Union)

In 2016, the European Commission has approved peracetic acid as an active biocidal agent for use in product type 1 (human hygiene), product type 2 (disinfectants and algaecides not intended for direct application to humans or animals), product type 3 (veterinary hygiene), product type 4 (food and feed area), product type 5 (drinking water) and product type 6 (preservatives for products during storage) [65]. Peracetic acid has also been approved in 2016 for uses in product type 11 (preservatives for liquid-cooling and processing systems) and product type 12 (slimicides) [66].

5.2.2 Environmental Protection Agency (USA)

The EPA has reregistered peracetic acid in the group of peroxy compounds 1993 as an active ingredient in pesticides [118].

5.2.3 Overall Environmental Impact

In the European Union, peracetic acid is manufactured and/or imported in 1,000–10,000 t per year [41]. The evaluation of peracetic acid by the ECHA revealed that the substance is not persistent because of its high reactivity and rapid degradation. Therefore, no residues appear either in food or in the environment [42]. Peracetic acid decomposes rapidly in all environmental compartments, i.e. in surface water, soil, air and active sludge. The degradation products of peracetic acid are oxygen, acetic acid and hydrogen peroxide (see also Chap. 6 on hydrogen peroxide). Acetic acid and hydrogen peroxide are further degraded to water, carbon dioxide and oxygen. In addition, peracetic acid decomposes already in sewage before reaching the sewage treatment plants [42].

5.3 Spectrum of Antimicrobial Activity

5.3.1 Bactericidal Activity

5.3.1.1 Bacteriostatic Activity (MIC Values)

The reported MIC values for peracetic acid are variable (Table 5.2), even within the same bacterial species such as *E. coli* (16–2,310 mg/l) or *S. aureus* (160–4,620 mg/l). The highest MIC values were reported with *B. subtilis* (18,500 mg/l), *A. calcoaceticus*, *E. cloacae* and *S. marcescens* (all 9,250 mg/l). When bacterial cells from food contact surfaces were obtained from a single-species biofilm, the MIC values were in a similar range to those obtained with planktonic cells as shown with *E. coli* (625 mg/l), *Klebsiella* spp. (1,250 mg/l), *S. aureus* (625–1,250 mg/l) and *S. epidermidis* (625–1,250 mg/l) [62].

5.3.1.2 Bactericidal Activity (Suspension Tests)

The bactericidal activity expressed as log reductions obtained with peracetic acid against various bacterial species is summarized in Table 5.3. Formulations based on peracetic acid at 0.03% are mostly bactericidal ($\geq$ 5.0 log reduction) within 30 min and at 0.32–0.5% within 5 min. One study, however, found a lower effect against *E. coli*, *Streptococcus* spp. and *S. aureus* with 0.5% peracetic acid in 30 min. At 1.6%, peracetic acid was also bactericidal with 3 min (most tested species) or 5 min (*E. faecalis*). Only *T. whipplei* was reduced by two products based on peracetic acid after 5–60 min by <3.0 log [77]. The bactericidal effect includes antibiotic-resistant isolates as shown with *E. coli*. The presence of organic load reduces the bactericidal effect, as shown with 12.5% skimmed milk [68]. The bactericidal activity can be enhanced by silver [79]. False-positive results due to insufficient neutralization are unlikely because peracetic acid at 0.26% was easy to neutralize [39].

The minimum bactericidal concentrations of peracetic acid, expressed as MBC values with a 5-min exposure time, were variable with 0.0005–0.45% which is equivalent to 4.8–4,050 mg/l (Table 5.4). For *E. coli* and *S. aureus*, it is noteworthy that the MBC values are in a similar range as the MIC values (see Table 5.2).

Peracetic acid at 0.0002% applied for 45 min in wastewater was equally effective against *E. coli* and coliforms and against antibiotic-resistant *E. coli* and coliforms [132]. Peracetic acid between 0.00009 and 0.0002% used for wastewater disinfection reduced uropathogenic *E. coli* on average by 52%. It also reduced the proportion of antimicrobial resistance gene-carrying uropathogenic *E. coli* pathotypes in municipal wastewaters [10].

Table 5.2 MIC values of various bacterial species to peracetic acid

Species	Strains/isolates	MIC value (mg/l)	References
A. actinomycetemcomitans	ATCC 29523	<1,000	[125]
A. calcoaceticus	ATCC 19606	9,250	[98]
B. fragilis	ATCC 25285	<1,000	[125]
B. stearothermophilus	ATCC 7953	4,620	[98]
B. subtilis var. globigii	ATCC 9372	18,500	[98]
E. cloacae	Strain IAL 1976	9,250	[98]
E. hirae	CIP 5855	160	[86]
E. coli	ATCC 25922	16	[86]
E. coli	ATCC 25922	312	[33]
E. coli	74 isolates from food contact surfaces	625–1,250	[62]
E. coli	ATCC 25922	<1,000	[125]
E. coli	ATCC 25922	2,310	[98]
F. psychrophilum	5 fresh trout isolates	31.2–125	[51]
Klebsiella spp.	30 isolates from food contact surfaces	625	[62]
L. monocytogenes	1 poultry isolate and 1 sheep spinal cord isolate	100–110	[4]
L. monocytogenes	12 strains from animals, food or contact surfaces	115–2,713	[101]
L. monocytogenes	10 isolates from food	125–250[a]	[1]
L. monocytogenes	ATCC 7644	156	[33]
L. monocytogenes	6 strains from a cheese processing facility	250	[103]
M. morganii	ATCC 25830	16	[86]
P. aeruginosa	NCTC 6749 and 3 extensively resistant clinical isolates	12–23	[130]
P. aeruginosa	ATCC 27853	16	[86]
P. aeruginosa	ATCC 10145	<1,000	[125]
P. intermedia	ATCC 25611	<1,000	[125]
S. enterica	2 poultry isolates	70–80	[4]
S. Typhimurium	ATCC 14028	<1,000	[125]
S. marcescens	Strain IAL 1478	9,250	[98]
S. aureus	CIP 53154	160	[86]
S. aureus	ATCC 25923	312	[33]
S. aureus	4 strains (CECT976, RN4220, SA1199b, XU212)	600–750	[50]
S. aureus	22 isolates from food contact surfaces	625	[62]
S. aureus	ATCC 33591	<1,000	[125]
S. aureus	ATCC 25923	4,620	[98]
S. epidermidis	65 isolates from food contact surfaces	312.5–625	[62]
S. mutans	ATCC 25175	<1,000	[125]

[a]In combination with hydrogen peroxide and acetic acid

Table 5.3 Bactericidal activity of peracetic acid in suspension tests

Species	Strains/isolates	Exposure time	Concentration	log_{10} reduction	References
A. anitratus	3 clinical isolates	3 min	1.6% (P)	>6.0	[126]
A. salmonicida	ATCC 14174	30 min	0.005% (P)	≥ 5.1	[123]
			0.0025% (P)	<4.3	
B. abortus	B19 vaccine strain	5 min	0.46% (P)	>7.0	[19]
			0.32% (P)	>7.0	
B. cereus[a]	ATCC 14579	30 min	0.03% (S)	>5.0	[105]
B. mallei	ATCC 23344	5 min	0.46% (P)	>7.0	[19]
			0.32% (P)	>7.0	
B. melitensis	Strain RKI 16M	5 min	0.46% (P)	>7.0	[19]
			0.32% (P)	>7.0	
B. pseudomallei	Strain A101-10	5 min	0.46% (P)	>7.0	[19]
			0.32% (P)	>7.0	
B. pseudomallei	10 patient isolates	30 min	0.26% (S)	>5.0	[131]
C. perfringens[a]	No information	30 min	1% (P)	≥ 4.0	[68]
			0.5% (P)	≥ 4.0	
C. perfringens[a]	Strain CDC 1861	30 min	0.03% (S)	4.1	[105]
C. piscicola	ATCC 35586	30 min	0.01% (P)	≥ 5.2	[123]
			0.005% (P)	4.4–5.4	
E. cloacae	3 clinical isolates	3 min	1.6% (P)	>6.0	[126]
E. faecalis	3 clinical isolates	5 min	1.6% (P)	>6.0	[126]
Enterococcus spp.	1 VRE blood culture isolate	30 min	0.2% (P)	2.2–8.0[b]	[89]
			0.1% (P)	2.0–4.5[b]	
			0.01% (P)	1.8–2.9[b]	
E. coli	No information	30 min	1% (P)	3.7	[68]
			0.5% (P)	3.7	
E. coli	Food isolate 0157:H7	30 min	0.03% (S)	>6.9	[105]
E. coli	ATCC 25922	30 min	0.01% (P)	≥ 6.5	[89]
F. tularensis	Strain SCHU S4	5 min	0.46% (P)	>7.0	[19]
			0.32% (P)	>7.0	
K. pneumoniae	3 clinical isolates	3 min	1.6% (P)	>6.0	[126]
L. garvieae	NCIMB 702927	30 min	0.01% (P)	4.0–5.8	[123]
			0.005% (P)	<3.8–5.7	
L. monocytogenes	Food isolate	30 min	0.03% (S)	>6.1	[105]
L. monocytogenes	20 environmental and food isolates	5 min	0.002–0.008% (P)	≥ 5.0	[27]
L. monocytogenes	Strain LO28	5 min	0.0005% (P)	4.7	[91]
P. aeruginosa	3 clinical isolates	3 min	1.6% (P)	>6.0	[126]
P. aeruginosa	NCTC 6749	30 s	0.2% (P)	≥ 7.3	[12]
P. aeruginosa	ATCC 27853	30 min	0.03% (S)	5.0	[105]

(continued)

Table 5.3 (continued)

Species	Strains/isolates	Exposure time	Concentration	$\log_{10}$ reduction	References
P. aeruginosa	Clinical isolate	5 min	0.0045% (S)	$\geq$5.0	[127]
S. Typhimurium	ATCC 14028	30 min	0.03% (S)	$\geq$6.4	[105]
S. enteritidis	No information	30 min	1% (P)	$\geq$4.0	[68]
			0.5% (P)	$\geq$4.0	
S. marcescens	14 strains from contaminated alkylamine disinfectant footbaths (dairy)	5 min	0.2% (P)	>5.0	[80]
S. sonnei	Food isolate	30 min	0.03% (S)	>6.3	[105]
S. aureus	ATCC 25923 and 2 clinical isolates	3 min	1.6% (P)	>6.0	[126]
S. aureus	No information	30 min	1% (P)	<3.0	[68]
			0.5% (P)	<3.0	
S. aureus	NCTC 4163	1 min	0.2% (P)	$\geq$7.3	[12]
S. aureus	ATCC 25923 and a MRSA blood culture isolate	30 min	0.1% (P)	$\geq$5.0	[89]
			0.01% (P)	2.8–5.0[b]	
S. aureus	ATCC 25923	30 min	0.03% (S)	6.6	[105]
S. epidermidis	ATCC 12228	30 min	0.03% (S)	>6.3	[105]
Streptococcus spp.	No information	30 min	1% (P)	<3.0	[68]
			0.5% (P)	<3.0	
V. cholerae	Strain C6706	30 min	0.03% (S)	>6.4	[105]
V. parahaemolyticus	Strain NY477	30 min	0.03% (S)	>6.2	[105]
V. vulnificus	Strain LA M624	30 min	0.03% (S)	>6.3	[105]
Y. enterocolitica	Strain 8081	30 min	0.03% (S)	>6.8	[105]
Y. pestis	NCTC 2028	5 min	0.46% (P)	>7.0	[19]
			0.32% (P)	>7.0	
Y. ruckeri	ATCC 29473	30 min	0.01% (P)	$\geq$5.0	[123]
			0.005% (P)	$\leq$4.7	

P commercial product; *S* solution; [a]vegetative cell form; [b]depending on the presence of organic load

5.3.1.3 Activity Against Bacteria in Biofilms

Effectively killing the bacterial cells grown in biofilm is more difficult [37]. *E. coli* cells in biofilms were largely killed within 10 min by products with a concentration of at least 0.016% peracetic acid (Table 5.5). In 74 isolates from food contact surfaces, however, a peracetic acid concentration >4% was necessary to achieve a bactericidal effect in 5 min [62]. Other studies show that the susceptibility of *E. coli* cells grown in biofilm for 48 h in microplates is 50 times or even 100 times lower compared to planktonic cells [48, 114]. Some authors report a 25–33 times lower susceptibility with *E. coli* in biofilm [28]. The reduction in sensitivity in *E. coli* CIP

Table 5.4 MBC values of various bacterial species to peracetic acid (5-min exposure time)

Species	Strains/isolates	MBC value (mg/l)	References
A. baumannii	Clinical isolate	1,792	[37]
B. subtilis	ATCC 6633	4.8	[13]
B. cepacia	Clinical isolate	384	[37]
E. coli	Strain PHL 628	7.4	[13]
E. coli	ATCC 25922	256	[37]
E. coli	74 isolates from food contact surfaces	625–1,250	[62]
E. faecalis	ATCC 19433	8.5	[13]
E. faecalis	Clinical isolate	384	[37]
E. faecium	Clinical isolate	384	[37]
Klebsiella spp.	30 isolates from food contact surfaces	625–1,250	[62]
L. monocytogenes	Strain EGDe	9.1	[13]
P. aeruginosa	ATCC 15442	10.3	[13]
P. aeruginosa	Clinical isolate	384	[37]
S. enterica	Strain S24	8.2	[13]
Salmonella spp.	11 strains (untreated wastewater)	11	[38]
	10 strains (treated wastewater)	13	
S. aureus	ATCC 6538	10.8	[13]
S. aureus	54 MRSA strains isolated in Canary black pigs	253–4,050	[40]
S. aureus	ATCC 6538 and 12 isolates from fishery products	300–450	[122]
S. aureus	22 isolates from food contact surfaces	625–1,250	[62]
S. aureus	Clinical MRSA isolate	768	[37]
S. epidermidis	65 isolates from food contact surfaces	625–1,250	[62]
S. epidermidis	Clinical isolate	2,048	[37]
S. maltophilia	Clinical isolate	1,792	[37]

54127 was attributed to a reduced accessibility of the bacterial cells to the disinfectants, due to the fact that the former adhered to a support [97].

L. monocytogenes in biofilms were quite effectively reduced by products based on peracetic acid, for example when used at 2% for at least 6 min (Table 5.5). One study even showed that cells of *L. monocytogenes* grown in biofilm (4 or 11 d on stainless steel or polypropylene) did not show a reduction of susceptibility to peracetic acid (10-min exposure time) compared to planktonic cells [104]. Peracetic acid was also able to inactivate *L. monocytogenes* biofilms on stainless steel but it was not able to remove adherent cells of *L. monocytogenes* from polystyrene microplates [81].

Exposure to peracetic acid may enhance persistence of micro-organisms in biofilms. For example, a *L. monocytogenes* biofilm (static or continuous flow)

Table 5.5 Bactericidal activity of peracetic acid against bacterial cells in biofilms

Species	Strains/isolates	Type of biofilm	Exposure time	Concentration	$\log_{10}$ reduction	References
C. jejuni	30 strains from chicken carcasses	48-h incubation in 96-well plates	24 h	0.8% (S)	≥ 4.3	[90]
C. jejuni	2 isolates from chicken	48-h incubation on PVC coupons	45 s and 180 s	0.02%[a] (P)	>3.6	[116]
				0.005%[b] (P)	2.7	
E. hirae	CIP 5855	48-h incubation on polypropylene, PVC and silicone	10 min	0.016% (P)	>5.0	[86]
				0.0016% (P)	2.1–3.9	
				0.00016% (P)	0.0	
E. coli	ATCC 25922	48-h incubation on polypropylene, PVC and silicone	10 min	0.016% (P)	4.7–5.0	[86]
				0.0016% (P)	0.0–3.0	
				0.00016% (P)	0.0	
E. coli	ATCC 43895	14-d incubation on stainless steel sheets	30 s	0.015% (P)	0.5–3.2	[113]
			2 min		1.0–4.2	
			5 min		1.0–4.2	
L. monocytogenes	Strain Scott A	12-d incubation on stainless steel at 20 °C	1 min	2% (P)	1.5–2.0	[9]
			2 min		1.5–4.5	
			3 min		2.0–5.0	
			6 min		≥ 4.0	
L. monocytogenes	Strain Scott A	19-d incubation on stainless steel at 5 °C	1 min	2% (P)	2.5–3.5	[9]
			2 min		3.5–4.4	
			3 min		≥ 3.5	
			6 min		≥ 3.5	

(continued)

Table 5.5 (continued)

Species	Strains/isolates	Type of biofilm	Exposure time	Concentration	$\log_{10}$ reduction	References
L. monocytogenes	12 strains from food controls	72-h incubation on polystyrene and stainless steel	5 min	0.2% (P)	2.8–3.5	[101]
L. monocytogenes	20 environmental and food isolates	48-h incubation in microtiter plates	5 min	0.015–0.035% (P)	≥ 5.0	[27]
L. monocytogenes	11 strains from different origins	48-h incubation in polystyrene microtiter plates and on stainless steel	6 min	0.001% (S)	3.0	[76]
M. morganii	ATCC 25830	48-h incubation on polypropylene, PVC and silicone	10 min	0.016% (P)	4.2–5.0	[86]
				0.0016% (P)	0.0–2.0	
				0.00016% (P)	0.0	
P. aeruginosa	ATCC 700928	24-h incubation in microplates	1 min	0.3% (S)	2.0	[115]
			5 min		2.0	
			60 min		2.0	
P. aeruginosa	Strain PA01	24-h incubation in microplates	1, 5, 15, 30 and 60 min	0.3% (S)	2.0–2.7	[75]
P. aeruginosa	ATCC 15442	24-h incubation in glass and PTFE beads	10 min	0.3% (S)	6.9	[73]
				0.2% (S)	4.3	
				0.1% (S)	3.4	
				0.05% (S)	2.1	
P. aeruginosa	ATCC 27853	48-h incubation on polypropylene, PVC and silicone	10 min	0.016% (P)	≥ 5.0	[86]
				0.0016% (P)	0.0–4.3	
				0.00016% (P)	0.0	

(continued)

Table 5.5 (continued)

Species	Strains/isolates	Type of biofilm	Exposure time	Concentration	$\log_{10}$ reduction	References
S. enterica	8 strains from different origins	48-h incubation in polystyrene microtiter plates and on stainless steel	6 min	0.001% (S)	5.2	[76]
S. Typhimurium	3 strains (FMCC B-137, FMCC B-193, FMCC B-415)	6-d incubation on stainless steel	6 min	0.001% (S)	1.4–2.0	[49]
S. aureus	AH478	24-h incubation in microplates	5 min	0.35% (S)	>7.3	[15]
S. aureus	ATCC 6538	72-h incubation in microplates	1 min	0.3% (S)	1.7	[115]
			5 min		2.0	
			60 min		2.0	
S. aureus	ATCC 6538	24-h incubation in microplates	1, 5, 15, 30 and 60 min	0.3% (S)	1.8–2.5	[75]
S. aureus	ATCC 6538 and 12 isolates from fishery products	48-h incubation on stainless steel coupons	30 min	0.15–0.4% (S)	≥ 5.0	[122]
S. aureus	CIP 53154	48-h incubation on polypropylene, PVC and silicone	10 min	0.016% (P)	≥ 5.0	[86]
				0.0016% (P)	4.8–5.0	
				0.00016% (P)	0.0–1.8	
S. aureus	3 strains (FMCC B-134, FMCC B-135, FMCC B-410)	6-d incubation on stainless steel	6 min	0.001% (S)	0.9–1.6	[49]
S. aureus	Strain S3	15-d incubation on polypropylene and stainless steel	30 s	0.003% (S)	2.6–3.7	[29]

(continued)

Table 5.5 (continued)

Species	Strains/isolates	Type of biofilm	Exposure time	Concentration	$\log_{10}$ reduction	References
Various species	*L. monocytogenes* strain Scott A and *Pseudomonas* spp. strain M-21, a meat processing plant isolate	48-h incubation on stainless steel coupons	1 min	0.008% (P)	≥ 7.0	[45]
			5 min			
Various species	*S. aureus* strain RN 4220 and *B. subtilis* (WD isolate)	24-h incubation in microplates	5 min	0.35% (S)	5.8[c]	[15]
Various species	*S. aureus* strain RN 4220 and *B. subtilis* strain 168	24-h incubation in microplates	5 min	0.35% (S)	>6.2[c]	[15]

S solution; *P* commercial product; [a]plus 0.095% hydrogen peroxide; [b]plus 0.024% hydrogen peroxide; [c]reduction of *S. aureus* showed increased resistance after exposure to 0.002% peracetic acid in a wild-type strain both in static and continuous flow biofilm [119]. HrcA and DnaK play an important role in the resistance of *L. monocytogenes* planktonic and biofilm cells against disinfectants [119]. In single-species biofilms, *L. monocytogenes* developed higher tolerance to cleaning and disinfection over time for the peracetic acid disinfectant, indicating that a broad-spectrum mechanism was involved [44].

P. aeruginosa cells in biofilms showed a variable susceptibility to peracetic acid. The majority of studies indicate that biofilm treatment with 0.3% peracetic acid for 60 min resulted in some bactericidal effect with 2.0–2.7 log while in one study a 6.9 log reduction was described in 10 min (Table 5.5). The susceptibility of *P. aeruginosa* in biofilm is lower in older biofilms (192 h versus 48 h or 24 h) [2]. This correlation was also described with a *P. marginalis* biofilm grown for 24 h at 30 °C (1.2 times less susceptible to peracetic acid compared to planktonic cells) or 48 h (4.8 times less susceptible) [78]. In order to kill *P. aeruginosa* in a 96-h biofilm within 5 min, peracetic acid of at least 2.5% was necessary, whereas *P. aeruginosa* survives at 2.0% for 5 min [2]. Other authors have reported that *P. aeruginosa* cells grown in biofilm for 24 h in microtiterplates were 15–20 times less susceptible to peracetic acid (5-min exposure time) compared to planktonic cells [14]. A biofilm of *P. aeruginosa* on stainless steel required 80× concentration of a formulation with peracetic acid, hydrogen peroxide and silver to achieve a 5.0 log reduction [114]. Only one study describes that *P. aeruginosa* cells in a 24-h biofilm can be reduced by exposure for 15 min at 37 °C to a formulation based on only 0.0042% peracetic acid by 5.2 log steps [87]. The effect in biofilm cells in endoscope channels has also been described. A formulation based on peracetic acid at 0.15% was effective in original channels of an endoscope as part of manual processing (10-min disinfection) to yield negative cultures after disinfection when the channels were allowed to build *P. aeruginosa* (ATCC 27853) biofilm over 5 d. However, 0.06% of cells in residual biofilm were still viable after disinfection [96].

Peracetic acid is able to diffuse inside the clusters of a *P. aeruginosa* biofilm; the biocidal compounds may partly have been consumed through quenching reactions with exopolymeric substances, leading to the greater biofilm resistance observed (27). In line with this, it was observed that disruption of the biofilm and the washing of cells enabled the recovery of the same susceptibility as that observed for planktonic cells; this finding was consistent with the fact that biofilm resistance appeared mainly to be due to the presence of the exopolymeric matrix. The efficacy of oxidizing agents is indeed well known to be profoundly affected by the presence of organic materials such as the constituents of the biofilm matrix (polysaccharides, proteins and nucleic acids). In addition, the presence of protective enzymes such as catalases in the extracellular matrix has also been reported to be involved in the resistance of *P. aeruginosa* biofilms to oxidizing agents (27).

S. aureus cells in biofilms grown for 24 h were mostly susceptible to $\geq$0.3% peracetic acid but the susceptibility was substantially lower when the biofilm was grown for 72 h (Table 5.5). Similar results were described by other authors. For example, peracetic acid at 0.5% removed all *S. aureus* cells within 15 s in biofilm grown in polystyrene microtiter plates [81]. Peracetic acid was also able to inactivate *S. aureus* biofilms on stainless steel and to remove adherent cells of *S. aureus* from polystyrene microplates [81]. Lower concentrations of peracetic acid are less effective. Data from Brazil indicate that 0.003% peracetic acid was not sufficient to remove *S. aureus* from a 15-day biofilm from stainless steel and polypropylene [29]. A formulation based on peracetic acid at 0.15%, however, was effective in original channels of an endoscope as part of manual processing (10-min disinfection) to yield negative cultures after disinfection when the channels were allowed to build *S. aureus* (ATCC 29213) biofilm over 5 d. However, 0.06% of cells in the residual biofilm were still viable after disinfection [96]. Finally, a biofilm of *S. aureus* on stainless steel required 100× concentration of a formulation with peracetic acid, hydrogen peroxide and silver to achieve a 5.0 log reduction [114]. In 22 *S. aureus* isolates from food contact surfaces, a peracetic acid concentration >4% was necessary to achieve a bactericidal effect against biofilm cells within 5 min [62].

For *S. epidermidis,* a 3.4× decrease of susceptibility was reported for biofilm-grown cells (48-h incubation in microplates) compared to planktonic cells [48]. Other authors report a two times lower susceptibility with *S. epidermidis* in biofilm [28]. An *E. faecalis* biofilm attached to dentin (5 days incubation) and irrigated for 3 min with 2% peracetic acid had an increase of dead cells in the biofilm from 13.8% to 50.5% indicating a rather poor efficacy [6]. In 65 *S. epidermidis* isolates from food contact surfaces, a peracetic acid concentration >4% was necessary to achieve a bactericidal effect against biofilm cells within 5 min [62]. A biofilm of *E. hirae* on stainless steel required 10× concentration of a product with peracetic acid, hydrogen peroxide and silver to achieve a 5.0 log reduction [114]. On various plastic materials, the effect of 0.016% peracetic acid was good within 10 min (>5.0 log; Table 5.5).

Other species such as *C. jejuni* (at least 0.02% peracetic acid in 3 min) or *M. morganii* (at least 0.016% peracetic acid in 10 min) are rather easily inactivated in biofilms. With 30 strains of *C. jejuni* in biofilm from chicken carcasses, it was shown that exposure to 0.8% peracetic acid for 24 h resulted in survival of seven strains (23.3%) in the presence of peracetic acid with 1.3–2.2 log; the persistence was probably strain dependent [90]. The results with *Salmonella* spp. are conflicting for 0.001% peracetic acid in 6 min (Table 5.5). In 30 *Klebsiella* spp. isolates from food contact surfaces, a peracetic acid concentration >4% was necessary to achieve a bactericidal effect against biofilm cells within 5 min [62]. Peracetic acid at 0.2% may also prevent survival of *S. typhimurium*, *E. coli*, *S. mutans* or *B. fragilis* in biofilm on glass or rubber carriers within 60 min [125].

Some studies have looked at the efficacy of peracetic acid in mixed biofilms. Micro-organisms in a waterborne mixed biofilm grown over 50 days on silicone tubes were killed by peracetic acid at 0.5% in 30 min by >5.0 log [43]. Mixed biofilm (*L. monocytogenes* and *L. plantarum*) had a similar susceptibility to the bactericidal activity of 0.01% peracetic acid in 15 min compared to single-species biofilm of *L. monocytogenes* and *L. plantarum* (all 3.5–5.0 log) [120].

5.3.1.4 Bactericidal Activity in Carrier Tests

Formulations based on 0.14–0.18% peracetic acid are mostly effective against different bacterial species in carrier tests within 10 min. Only one study suggests that a VRE isolate may not be reduced by 0.2% peracetic acid in 30 min (Table 5.6).

Table 5.6 Bactericidal activity of commercial products (P) based on peracetic acid in carrier tests

Species	Strains/isolates	Exposure time	Concentration	$\log_{10}$ reduction	References
E. cloacae	11 clinical isolates	10 min	0.18% (P)	>4.0[a]	[57]
E. cloacae	17 MDR clinical isolates	10 min	0.18% (P)	>4.0	[57]
Enterococcus spp.	1 VRE blood culture isolate	30 min	0.2% (P)	None	[89]
Enterococcus spp.	3 VRE strains (2 vanA, 1 vanB)	3 min	0.14% (P)	>5.4	[30]
E. coli	ATCC 25922	30 min	0.2% (P)	$\geq$5.0	[89]
E. coli	6 clinical isolates	10 min	0.18% (P)	>4.0	[57]
K. pneumoniae	3 clinical isolates	10 min	0.18% (P)	>4.0	[57]
L. monocytogenes	5 food strains	10 min	0.1%[b] (P)	>4.0	[1]
L. monocytogenes	Strain LO28	5 min	0.0005% (P)	3.3	[91]
P. aeruginosa	8 clinical isolates	10 min	0.18% (P)	>4.0	[57]
P. mirabilis	5 clinical isolates	10 min	0.18% (P)	>4.0	[57]
S. marcescens	3 clinical isolates	10 min	0.18% (P)	>4.0	[57]
S. aureus	ATCC 25923 and a MRSA blood culture isolate	30 min	0.2% (P)	2.0–3.5[c]	[89]
S. aureus	6 clinical isolates	10 min	0.18% (P)	>4.0	[57]
S. aureus	ATCC 43300 and 2 MRSA clinical isolates	3 min	0.14% (P)	>4.8	[30]

[a]One isolate with a log reduction <4.0; [b]contains in addition hydrogen peroxide and acetic acid; [c]depending on the type of organic load

5.3.1.5 Bactericidal Activity in Endoscopes or Test Tubes

Peracetic acid was described to be effective at a concentration of 0.05% applied for 45 min for manual disinfection of different types of flexible endoscopes. In 9 of 10 gastroscopes and colonoscopes, no bioburden was detected in the samplings of the suction channel after peracetic acid treatment [124]. Various studies describe its efficacy in automated processing. When endoscopes were artificially contaminated with *P. aeruginosa* and treated with 0.2% peracetic acid for 12 min at 53 °C, the bacterial counts were reduced by at least 6.0 log [12, 36]. Similar results were found with the same type of treatment in test tubes artificially contaminated with *E. faecium* [3]. An automatic processing without a specific description of peracetic acid application parameter (Steris 20) was described to eliminate *Enterococcus* spp. from artificially contaminated colonoscopes [26]. A similar result was described with four common types of endoscopes contaminated with *P. aeruginosa*, VRE and MRSA, also without a specific description of peracetic acid application parameter [106]. Another automated process without a peracetic acid treatment specification was effective to kill *H. pylori* on artificially contaminated endoscopes [25]. In an experimental model, peracetic acid at 0.085% was effective in reducing MRSA, VRE and *C. difficile* in a flexible gastrointestinal endoscope [22].

5.3.2 Fungicidal Activity

5.3.2.1 Fungicidal Activity (Suspension Tests)

Peracetic acid is effective against yeasts such as *C. albicans* at 0.25% within 1 min or at 0.1% in 15–30 min. The efficacy of 0.3% peracetic acid is poor in 10 min against selected fungi obtained from food. Many *Aspergillus* spp. are killed by 0.01% peracetic acid in 10 min but *A. brasiliensis* was somewhat more resistant even to 0.225% peracetic acid (Table 5.7). Some authors describe that the effect of peracetic acid at 0.1 and 0.3% is also rather weak against *Aspergillus* and *Penicillum* spp. [74]. The environmental saprophytic fungi *C. globosum* and *C. funicola* also showed a high resistance to peracetic acid [94]. In drinking water, an effect of at least 5.0 log is achieved against *C. albicans* by 0.001% peracetic acid within 24 h [108].

5.3.2.2 Activity Against Fungi in Biofilms

In one study, biofilms were grown with *C. albicans* (strain SC 5314), *C. orthopsilosis* (5 strains) and *C. parapsilosis* sensu *strictu* (5 strains). The *Candida* cells in the biofilms were reduced to <10 CFU/ml in up to 48 h with a product based on 0.083% peracetic acid and 0.26% hydrogen peroxide [100].

Table 5.7 Fungicidal activity of peracetic acid in suspension tests

Species	Strains/isolates	Exposure time	Concentration	$\log_{10}$ reduction	References
A. brasiliensis	ATCC 16404	15 min	0.225% (P)	3.3	[60]
A. flavipes	3 clinical or environmental isolates	10 min	0.01% (S)	>4.0	[108]
		30 min	0.002% (S)	>4.0	
A. flavus	4 clinical or environmental isolates	5 min	0.01% (S)	>4.0	[108]
		30 min	0.002% (S)	>4.0	
A. fumigatus	4 clinical or environmental isolates	10 min	0.01% (S)	>4.0	[108]
		4 h	0.005% (S)	>4.0	
A. nidulans	4 clinical or environmental isolates	30 min	0.005% (S)	>4.0	[108]
		4 h	0.002% (S)	>4.0	
A. terreus	2 clinical or environmental isolates	10 min	0.005% (S)	>4.0	[108]
		4 h	0.002% (S)	>4.0	
A. ustus	2 clinical or environmental isolates	10 min	0.005% (S)	>4.0	[108]
		1 h	0.002% (S)	>4.0	
A. versicolor	2 clinical or environmental isolates	10 min	0.005% (S)	>4.0	[108]
		1 h	0.002% (S)	>4.0	
C. albicans	3 clinical isolates	3 min	1.6% (P)	>6.0	[126]
C. albicans	ATCC 10231	1 min	0.25% (P)	>4.0	[102]
		3 min	0.025% (P)	>4.0	
C. albicans	ATCC 10231	30 min	0.2% (P)	≥ 7.0	[89]
			0.1% (P)	4.0–8.0[a]	
			0.01% (P)	2.2–8.0[a]	
C. albicans	ATCC 10231	15 min	0.1% (P)	>4.0	[60]
C. albicans	ATCC 10231	10 min	0.025%[b] (P)	>4.0	[71]
C. krusei	ATCC 14243	30 min	0.01% (P)	≥ 4.0	[89]
			0.1% (P)	≥ 5.0	
E. repens	Isolate from bread factory	10 min	0.3% (P)	0.0	[16]
P. anomala	Isolate from bread	10 min	0.3% (P)	0.0	[16]
P. roqueforti	Isolate from bread	10 min	0.3% (P)	0.3	[16]

S solution; *P* commercial product; [a]depending on the type of organic load; [b]contains also hydrogen peroxide (approximately 0.1%)

5.3.2.3 Fungicidal Activity in Carrier Tests

A product based on 0.18% peracetic acid was very effective in carrier tests against different types of yeasts (four clinical isolates of *C. albicans*, one clinical isolate of *C. krusei*, one clinical isolate of *C. parapsilosis* and one clinical isolate of *C. tropicalis*) with log reductions >4.0 in a 10-min exposure time [57]. Against four clinical isolates of *C. auris*, *C. albicans* ATCC 10231 and *C. glabrata* ATCC 2001, peracetic acid (0.2% for 5 min) led to a significant reduction on a contaminated cellulose matrix (3.1–6.6 log), on stainless steel (2.2–3.0 log) and on polyester coverslips (4.4–6.8 log) [72]. On stainless steel squares, peracetic acid at 0.2% reduced *C. albicans* ATCC 10231 within 30 min by 1.5–2.1 log, *C. krusei* ATCC 14243 was more susceptible with a 3.2–5.4 log reduction [89].

5.3.2.4 Fungicidal Activity in Endoscopes or Test Tubes

One study described the effect of peracetic acid in endoscopes artificially contaminated with *A. niger* undergoing automated processing with 0.2% peracetic acid for 12 min at 53 °C. The fungal cell load was reduced from at least 1.1×10^6 to 0 per endoscope [36].

5.3.3 Mycobactericidal Activity

5.3.3.1 Mycobactericidal Activity (Suspension Tests)

Products or solutions based on 0.35% peracetic acid were effective against *M. chelonae* and *M. tuberculosis* in 1 min, *M. avium*, *M. smegmatis* and *M. xenopi* in 2 min and *M. bovis* in 5 min. A lower concentration such as 0.2% required between 5 and 15 min to be mycobactericidal. The minimum bactericidal concentration for a *M. chelonae* strain (647P-Mc) was with 0.294% in a similar range [13]. Most studies with glutaraldehyde-resistant isolates of *M. chelonae* indicate that 0.35% peracetic acid is effective against them in 1–4 min. Only one study with 0.035% peracetic acid described a poor activity in 60 min (up to 1.9 log), whereas an ATCC strain of *M. chelonae* was effectively reduced by >5.6 log (Table 5.8).

In addition, three commercial solutions based on an unknown concentration of peracetic acid were described to be effective in 15 min against glutaraldehyde-resistant *M. massiliense* isolates [82]. One recent study described a broad mycobactericidal efficacy (*M. avium*, *M. abscessus*, *M. bovis*, *M. chelonae* and *M. terrae*) of commercial peracetic acid-based products (Reliance DG and S40) with >5.0 log reduction in 5 min but did not mention the concentration of the active agent so that it cannot be included in Table 5.8 [17, 67].

5.3.3.2 Activity Against Mycobacteria in Biofilms

A *M. abscessus* (INCQS 594) biofilm was produced in original channels of an endoscope over 15 d. A product based on 0.15% peracetic acid applied for 10 min as part of manual processing was effective into yield negative cultures after disinfection. However, 0.06% of cells in residual biofilm were still viable after disinfection [96].

Table 5.8 Mycobactericidal activity of products or solutions based on peracetic acid in suspension tests

Species	Strains/isolates	Exposure time	Concentration	log_{10} reduction	References
M. abscessus	ATCC 19977	5 min	0.08%[a] (P)	>6.2	[111]
M. abscessus	ATCC 19977	5 min	0.07% (S)	5.3	[111]
M. avium	NCTC 10437	2 min	0.35% (P)	>6.0	[107]
M. avium-intracellulare	Clinical isolate	4 min	0.35% (P)	5.2	[85]
M. avium-intracellulare	Clinical isolate	4 min	0.35% (P)	>5.2	[53]
M. avium	NCTC 10437	5 min	0.35% (P)	≥6.0	[59]
M. avium	Clinical isolate (strain 3051)	5 min	0.35% (P)	≥6.5	[59]
M. avium-intracellulare	Clinical strain 104	5 min	0.26% (P)	>5.0	[55]
		20 min[b]		>5.0	
M. avium-intracellulare	6 fresh clinical isolates	15 min	0.2% (P)	>5.0	[58]
		50 min		>5.0	
M. bovis	NCTC 10772	5 min	0.35% (P)	≥4.2	[59]
M. bovis	ATCC 35743	10 min	0.08%[a] (P)	4.6	[111]
M. bovis	ATCC 35743	10 min	0.07% (S)	4.9	[111]
M. chelonae	NCTC 946	1 min	0.35% (P)	>5.8	[85]
M. chelonae	NCTC 946	1 min	0.35% (P)	>5.5	[53]
M. chelonae	Glutaraldehyde-resistant isolate WD 1	1 min	0.35% (P)	4.1	[85]
M. chelonae	Glutaraldehyde-resistant isolate WD 2	1 min	0.35% (P)	4.0	[85]
M. chelonae	NCTC 946	1 min	0.35% (P)	>5.0	[52]
M. chelonae	2 glutaraldehyde-resistant isolates from WD from different hospitals	1 min	0.35% (P)	>4.0	[52]
		4 min		>5.0	
M. chelonae	Clinical isolate	2 min	0.35% (P)	>5.0	[107]
M. chelonae	Strain Epping	4 min	0.35% (P)	>6.1	[53]
M. chelonae	ATCC 35752	5 min	0.26% (P)	>5.0	[55]
M. chelonae subsp. abscessus	CMCC 93326	5 min	0.2% (S)	≥6.0	[128]
M. chelonae	5 glutaraldehyde-resistant isolates	10 min	0.08%[a] (P)	>6.2	[111]
M. chelonae	5 glutaraldehyde -resistant isolates	60 min	0.07% (S)	0.6–6.3	[111]
M. chelonae	3 glutaraldehyde -resistant strains from WD from different hospitals	10, 30 and 60 min	0.035% (P)	0.1–1.9	[121]
M. chelonae	ATCC 14998	10, 30 and 60 min	0.035% (P)	>5.6	[121]

(continued)

Table 5.8 (continued)

Species	Strains/isolates	Exposure time	Concentration	log_{10} reduction	References
M. fortuitum	ATCC 609	10 min	1.6% (P)	>6.0	[126]
M. fortuitum	NCTC 10394	4 min	0.35% (P)	>6.0	[53]
M. fortuitum	Clinical strain	10 min	0.26% (P)	>5.0	[55]
M. kansasii	WD isolate	1 min	0.35% (P)	>5.4	[85]
M. smegmatis	NCTC 8159	2 min	0.35% (P)	>9.0	[107]
M. tuberculosis	NCTC 7416	1 min	0.35% (P)	>5.1	[85]
M. tuberculosis	Strain H37Rv	4 min	0.35% (P)	>5.1	[53]
M. tuberculosis	Strain H37Rv	5 min	0.35% (P)	≥ 5.2	[59]
M. tuberculosis	MDR clinical isolate (strain 98)	5 min	0.35% (P)	≥ 5.5	[59]
M. tuberculosis	Strain H37Rv	10 min	0.26% (P)	>5.0	[55]
M. tuberculosis	CMCC 93020	5 min	0.2% (S)	4.5	[128]
		20 min		≥ 6.0	
M. tuberculosis	6 fresh clinical isolates	15 min	0.2% (P)	>5.0	[58]
		50 min		>5.0	
M. xenopi	NCTC 10042	2 min	0.35% (P)	>5.0	[107]

S solution; *P* commercial product; [a]with 1% hydrogen peroxide; [b]with organic load

5.3.3.3 Mycobactericidal Activity in Carrier Tests

Commercial products were mostly effective in carrier tests against the different mycobacterial species including *M. avium*, *M. bovis*, *M. chelonae*, *M. fortuitum* and *M. tuberculosis* with peracetic acid at 0.26% in 30 min or 0.35% in 5 min (Table 5.9).

5.3.3.4 Mycobactericidal Activity in Endoscopes or Test Tubes

Most data were published with bronchoscopes. In one study, bronchoscope was artificially contaminated with *M. tuberculosis*, *M. avium-intracellulare* and *M. chelonae*, followed by 10 automated processing with wash cycles per organism. A commercial product based on 0.35% peracetic acid was used for disinfection over 5 or 10 min. Without physical pre-cleaning, *M. tuberculosis* and *M. chelonae* were not recovered after 5-min disinfection but *M. avium-intracellulare* was recovered after 1 of 10 washes (5 min: 310 CFU/ml; 10 min: 2,800 CFU/ml). With physical pre-cleaning, *M. avium-intracellulare* was never recovered after 5-min disinfection [92]. In another study, a bronchoscope was artificially contaminated with *M. gordonae* (10^5 or 10^8 CFU per ml). Processing consisted of manual cleaning including use of a brush, followed by automated processing with a formulation based on 0.2% peracetic acid for 10 or 20 min at 25 °C. *M. gordonae* was not recovered after 10 min for any type of contamination [61]. In a third study, a bronchoscope was artificially contaminated with *M. tuberculosis* (n = 5) and *M. avium-intracellulare* (n = 5). Processing consisted of a cleaning step and

Table 5.9 Mycobactericidal activity of commercial products (P) based on peracetic acid in carrier tests

Species	Strains/isolates	Exposure time	Concentration	$\log_{10}$ reduction	References
M. avium	NCTC 10437	5 min	0.35% (P)	≥4.5	[59]
M. avium	Clinical isolate (strain 3051)	5 min	0.35% (P)	≥5.5	[59]
M. avium-intracellulare	Clinical strain 104	5 min	0.26% (P)	>5.0	[55]
		10 min		>5.0[a]	
M. avium	ATCC 25291	10 min	0.18% (P)	>4.0	[57]
M. bovis	NCTC 10772	5 min	0.35% (P)	≥3.2	[59]
M. chelonae	ATCC 35752	5 min	0.26% (P)	>5.0	[55]
M. fortuitum	Clinical strain	20 min	0.26% (P)	>5.0	[55]
		30 min		4.8[a]	
M. fortuitum	ATCC 609	10 min	0.18% (P)	>4.0	[57]
M. tuberculosis	Strain H37Rv	5 min	0.35% (P)	≥4.2	[59]
M. tuberculosis	MDR clinical isolate (strain 98)	5 min	0.35% (P)	≥5.0	[59]
M. tuberculosis	Strain H37Rv	5 min	0.26% (P)	>5.0	[55]
		10 min		>5.0[a]	

[a]In the presence of organic load

manual disinfection by immersion in a formulation based on 0.26% peracetic acid for 10 or 20 min. After disinfection for 10 min, *M. avium-intracellulare* was not recovered in 4 of 5 samples (log > 5.0), and one sample reached a log 4.2. After disinfection for 20 min, all 5 *M. avium-intracellulare* samples were without growth (log > 5.0). *M. tuberculosis* was never recovered after processing, and all samples were without growth (log > 5.0) [56]. Finally, five colonoscopes and five duodenoscopes were artificially contaminated with *M. chelonae* and processed automatically with a formulation based on 0.2% peracetic acid for 12 min at 50–56 °C. All scope cultures were negative after processing, and high-level disinfection was achieved [47]. Similar results were described in test tubes artificially contaminated with *M. chelonae* after automated processing with a product based on 0.2% peracetic acid (12 min at 53 °C). Colony counts were reduced from 6.9×10^5 to 0 per lumen [3]. A sufficient efficacy result was described for automated peracetic acid processing with four common types of endoscopes contaminated with a glutaraldehyde-resistant *M. chelonae* but without a specific description of peracetic acid application parameter [106]. Overall, the use of products based on 0.35 or 0.26% peracetic acid for 10 min was very effective against mycobacteria in channels of flexible endoscopes, similar to 0.2% peracetic acid for 12 min at 50–56 °C.

5.4 Effect of Low-Level Exposure

Exposure of *S. enterica* and *L. monocytogenes* to sublethal concentrations of peracetic acid changed the MIC values in comparison to unexposed cells only marginally ($\leq$1.1-fold increase) [4]. Similar findings were reported with *L. monocytogenes* strain EGD and a commercial disinfectant based on peracetic acid and hydrogen peroxide [69]. In *E. coli* O157:H7, however, cultures exhibited increased tolerance to peroxidative stress when acutely exposed to a sublethal concentration of 0.1% peracetic acid (1.0 log reduction in <2 h compared to 3.0 log reduction in 45 min) suggesting that acute sublethal contact with peracetic acid may cause resistance to lethal peracetic acid treatments of unknown stability [134]. When *P. aeruginosa* was exposed to sublethal concentrations of peracetic acid (0.0076%), many genes associated with cellular protective processes were induced, the transcription of genes involved in primary metabolic pathways was repressed, and the transcription of genes encoding membrane proteins and small molecule transporters was altered [20]. Water treatment with 0.0005% peracetic acid (corresponding to the MIC value) for 1 h was able to reduce the viable cell number of *S. Typhimurium* strain LT2 in sewage effluent by 5.0 log. The cells, however, retained their ability to adhere and to invade HeLa cells indicating a potential risk of pathogenic bacteria disseminating in natural and bathing water [63]. When *S. Typhimurium* in sterilized sewage water was exposed to peracetic acid at 0.0001–0.0007%, a log reduction > 5.0 was found but 500 cells per ml were still viable but not culturable. With a higher concentration of 0.0015%, a log reduction > 7.0 was found but 500 cells per ml were still viable but not culturable. Only with a concentration of 0.002%, a log reduction > 7.0 was found and no viable cells were found anymore [64].

Exposure of *S. aureus* to sublethal concentrations of peracetic acid (0.0076%) significantly altered the regulation of membrane transport genes, selectively induced DNA repair and replication genes, and differently repressed primary metabolism-related genes between the two growth states. Most intriguingly, many virulence factor genes were induced upon the exposure, which proposes a possibility that the pathogenesis of *S. aureus* may be stimulated in response to peracetic acid [21]. In *L. monocytogenes*, however, a disinfectant based on peracetic acid reduced expression of virulence genes [70]. Low-level exposure of *E. faecium* to peracetic acid (<0.001%) did not select for antibiotic resistance genes (ermB) [117].

5.5 Resistance to Peracetic Acid

The risk of the development of resistance is regarded to be very low due to the low specificity of reactions of peracetic acid [42]. The primary mode of action of peracetic acid is oxidation. It denatures proteins, disrupts cell wall permeability, and oxidizes sulfhydral and sulphur bonds in proteins, enzymes and other metabolites [42]. The bactericidal effect is explained by hydroxyl- and carbon-centred radicals which are produced in the bacterial cell. Hydroxyl radicals are the lethal species [23].

5.5.1 Insufficient Efficacy in Suspension Tests

A *R. erythropolis* isolate from a dairy production facility was described to resist the efficacy of 0.2% peracetic acid in 5 min with only 0.5 log reduction [11]. The susceptibility of micro-organisms to peracetic acid can be reduced, e.g. by exposure of *S. Typhimurium* to sublethal concentrations of terpenes which reduces the efficacy of 0.0003% peracetic acid (5 min) on average by 1.3 log [35]. After adaptation to nalidixic acid, a lower susceptibility of *E. coli* 0157:H7 (1.6 versus 2.3 log reduction), *L. monocytogenes* (0.1 versus 1.8) and *Salmonella* spp. (1.4 versus 2.1) to 0.007% peracetic acid for 3 min was described on mung bean sprouts [95].

5.5.2 Persistence Despite Disinfection with Peracetic Acid as Recommended

Use of peracetic acid as recommended does not always ensure sufficient antimicrobial activity so that some species may survive the treatment and may persist, e.g. on flexible endoscopes. Persistence of various bacteria, mainly Gram-negative incl. *P. aeruginosa*, has been reported after automated processing of flexible endoscopes with 0.2% peracetic acid for 12 min at 50–56 °C indicating that the formulation was not effective enough [31].

One pseudo-outbreak and one outbreak were suspected to be caused by technical malfunction indicating that the results may have been the same when another biocide would have been used for disinfection. One pseudo-outbreak was assumed to be caused by biofilm formation enabling persistence despite best practice decontamination processes (Table 5.10).

5.5.3 Resistance Mechanisms

So far there are no reports on the cellular mechanisms of reduced bacterial susceptibility to peracetic acid, probably because reports of resistance of the cell itself to peracetic acid are so rare [129]. *C. globosum* and *C. funicola* showed high resistance to peracetic acid which was probably not acquired. They had thick cell walls as ascospores that can impede the action mechanism of peracetic acid [94].

5.5.4 Resistance Genes

5.5.4.1 Peracetic Acid Resistance Genes

No peracetic resistance genes have been identified so far.

Table 5.10 Pseudo-outbreaks and outbreaks associated with suspected insufficient microbiocidal activity of peracetic acid

Species	Place of persistence	Treatments	Clinical impact	Suspected reason for resistance	References
S. maltophilia	Ultrasound endoscopes (air channel, water channel, suction channel, channel separators, balloon, elevator)	Automatic processing with peracetic acid-based product (Neodisher Septo PAC 1%) for 10 min during disinfection	Pseudo-outbreak with 3 patients (bronchial aspirates)	Formation of niches in the ultrasound endoscopes caused by differences in temperature and pressure; cross-contamination to bronchoscopes via connecting tubes (drying cabinet)	[112]
P. aeruginosa (imipenem-resistant)	None	Automated processing with peracetic acid for disinfection	Outbreak with 3 cases of infection and 15 patients with transient colonization after bronchoscopy	Incorrect connectors for suction channel obstructing peracetic acid flow; no malfunction warning	[110]
F. oxysporum	Internal lumen of bronchoscope	Automated processing with peracetic acid for disinfection (Gigasept Autoscope)	Pseudo-outbreak with 2 patients	Presumably biofilm formation enabling persistence despite best practice decontamination processes	[8]

5.5.4.2 Effect of Peracetic Acid on Antibiotic Resistance Genes

The effect of peracetic acid on antibiotic resistance genes is poor. One study shows that peracetic acid is effective to remove bacterial cells in a wastewater treatment plant. However, the stress imposed by peracetic acid selected for bacterial aggregates and stimulated the selection of antibiotic resistance genes during the incubation experiment [32]. In addition, it was found that nine antibiotic resistance genes (ampC, mecA, ermB, sul1, sul2, tetA, tetO, tetW, vanA) were not reduced by peracetic acid disinfection in wastewater [84]. Other authors also postulated that the effect of peracetic acid used in wastewater on the resistance genes from uropathogenic *E. coli* is unclear [10].

5.6 Cross-Tolerance to Other Biocidal Agents

Vegetative cells of a rinse water isolate of *B. subtilis* with reduced susceptibility to chlorine dioxide have shown cross-resistance to other oxidising agents such as peracetic acid [88].

5.7 Cross-Tolerance to Antibiotics

Cross-resistance to antibiotics has not been reported yet. In wastewater, peracetic acid at concentrations of 0.0005–0.002% was able to transform different beta-lactam antibiotics which may be an advantage to reduce antibiotic selection pressure in wastewater [133].

5.8 Role of Biofilm

Peracetic acid is able to diffuse inside biofilm clusters. The biocidal compound may partly be consumed through quenching reactions with exopolymeric substances, leading to the greater biofilm resistance [14]. Biofilm resistance appears mainly to be due to the presence of the exopolymeric matrix. The efficacy of oxidizing agents is indeed well known to be profoundly affected by the presence of organic materials such as the constituents of the biofilm matrix (polysaccharides, proteins and nucleic acids) [14]. In addition, the presence of protective enzymes such as catalases in the extracellular matrix has also been reported to be involved in the resistance of *P. aeruginosa* biofilms to oxidizing agents [14].

5.8.1 Effect on Biofilm Development

Formation of a biofilm of *C. sakazakii*, an emerging opportunistic food-borne pathogen, is impaired to some extent in the presence of peracetic acid (range: 6% lower with 0.005% peracetic acid, 41% lower with 0.02% peracetic acid) [7]. Biofilm formation in microtiter wells by *C. albicans* (strain SC 5314), *C. orthopsilosis* (5 strains) and *C. parapsilosis sensu strictu* (5 strains) is inhibited by peracetic acid at 0.021% (*C. albicans* and *C. orthopsilosis*) or 0.041% (*C. parapsilosis*) [100]. Peracetic acid was also capable to significantly reduce biofilm formation of the *S. aureus* strain 9213 [79]. Dental unit waterlines from five new units were exposed for 30 days to a product based on 0.26% peracetic acid for 5 cycles à 5 min per day. No biofilm at all was found indicating an effective prevention of biofilm formation [93].

5.8.2 Effect on Biofilm Removal

As described in Table 5.11 biofilm of *E. coli*, *P. aeruginosa* and *S. aureus* are partially removed by exposure to peracetic acid between 0 and 63% although most data indicate a poor biofilm removal rate <33%. An *E. faecalis* biofilm attached to dentin (5 days incubation) and irrigated for 3 min with 2% peracetic acid had a reduction of biovolume from 63.5 to 14.9 mm^3 indicating also only partial biofilm removal [6]. Silicone tubes with mixed biofilm after 50-d perfusion with tap water (water systems in hospitals) were exposed for 1 h to a 0.5% or 1% solution of a product based on peracetic acid and tenside. No clear changes of biofilm thickness were observed by SEM [43].

5.8.3 Effect on Biofilm Fixation

The biofilm fixation potential of peracetic acid clearly depends on the composition of the product and may vary between 0 and 54% (Table 5.12). A possible fixation of *P. aeruginosa* biofilm (relative residual protein quantity) was investigated in PTFE tubes by a peracetic acid solution, a peracetic acid product (both 0.15%) and sterile water. After nine treatment cycles, protein was lower with the peracetic acid solution (24.9 μg/cm^2) or the peracetic acid product (24.7 μg/cm^2), compared to distilled water (57.3 μg/cm^2) also indicating a low fixation potential [99].

5.9 Summary

The principal antimicrobial activity of peracetic acid is summarized in Table 5.13.

The key findings on resistance and cross-resistance including the role of biofilm for selecting resistant isolates are summarized in Table 5.14.

Table 5.11 Biofilm removal rate (quantitative determination of biofilm matrix) by exposure to products or solutions based on peracetic acid

Type of biofilm	Concentration	Exposure time	Biofilm removal rate	References
A. acidoterrestris biofilm on stainless steel, nylon and PVC surfaces	0.05% (S)	10 min	"small numbers of cells left"	[34]
E. coli 54127 biofilm on haemolysis glass tubes	0.35% (P)	5 min	0%	[54]
	0.09–0.15% (P)	5 min	0%	
	0.1–0.25%[a] (P)	5 min	8%	
	0.087%[a] (P)	15 min	14%	
E. coli 54127 biofilm on haemolysis glass tubes	0.11% (P)	15 min	0%	[83]
	0.11% (P)[b]	15 min	16%	
	0.11% (P)[c]	15 min	16%	
P. aeruginosa biofilm on 96-well plates	0.3% (S)	1, 5, 15, 30 and 60 min	41–63%	[75]
S. aureus biofilm on stainless steel and polypropylene coupons	3% (S)	30 s	"partial removal"	[29]
S. aureus biofilm on 96-well plates	0.3% (S)	1, 5, 15, 30 and 60 min	14–32%	[75]

P commercial product; *S* solution; [a]plus QAC; [b]plus surfactant; [c]plus surfactants

Table 5.12 Biofilm fixation rate (quantitative determination of biofilm matrix) by exposure to peracetic acid commercial products (P)

Type of biofilm	Concentration	Exposure time	Biofilm fixation rate	References
E. coli 54127 biofilm on haemolysis glass tubes	0.35% (P)	5 min	34%	[54]
	0.09–0.15% (P)	5 min	54%	
	0.1–0.25% (P)[a]	5 min	0%	
	0.087% (P)[a]	15 min	0%	
E. coli 54127 biofilm on haemolysis glass tubes	0.11% (P)	15 min	3%	[83]
	0.11% (P)[b]	15 min	0%	
	0.11% (P)[c]	15 min	0%	

[a]Plus QAC; [b]plus surfactant; [c]plus surfactants

Table 5.13 Overview on the typical exposure times required for peracetic acid to achieve sufficient biocidal activity against the different target micro-organisms

Target micro-organisms	Species	Concentration	Exposure time
Bacteria	Most clinically relevant species except *T. whipplei*, selected *Streptococcus* spp. and *S. aureus* isolates	1.6%	3–5 min[a]
		0.32%	5 min[a]
		0.03%	30 min[a]
Fungi	*Candida* spp.	0.25%	1 min
		0.1%	15–30 min
	Many *Aspergillus* spp. except *A. brasiliensis*	0.01%	10 min
	No sufficient activity against various fungi from food	0.3%	10 min
Mycobacteria	*M. chelonae*, *M. tuberculosis*	0.35%	1 min
	M. avium-intracellulare, *M. smegmatis*, *M. xenopi*	0.35%	2 min
	M. fortuitum	0.35%	4 min
	M. bovis	0.35%	5 min

[a]In biofilm, the efficacy will be lower

Table 5.14 Key findings on peracetic acid resistance, the effect of low-level exposure, cross-tolerance to other biocides and antibiotics, and its effect on biofilm

Parameter	Species	Findings
Elevated MIC values	So far not reported.	
Low efficacy in suspension tests	*R. erythropolis*	Possibly natural resistance (dairy farm isolate)
MIC value to determine the resistance	Not proposed yet for bacteria, fungi or mycobacteria	
Cross-tolerance biocides	*B. subtilis* (vegetative cells)	Other oxidising agents
Cross-tolerance antibiotics	So far not reported	
	Peracetic acid (0.0005–0.002%) can transform different beta-lactam antibiotics in wastewater (reduction of selection pressure)	
Resistance mechanisms	Selected fungi	Thick cell walls of ascospores
	Selected bacterial isolates	Unknown

(continued)

Table 5.14 (continued)

Parameter	Species	Findings
Effect of low-level exposure	*S. Enterica, L. monocytogenes*	No MIC increase
	None	Weak MIC increase (>4-fold)
	E. coli	Strong (>4-fold) decrease of lethal effect
	S. Typhimurium	Survivors may be viable but not culturable
	E. faecium	No selection for antibiotic resistance genes
	L. monocytogenes	Reduced expression of virulence genes
	S. aureus	Induction of virulence factor genes
	P. aeruginosa	Induction of genes responsible for cellular protective processes
Biofilm	Development	Inhibition in *C. sakazakii, Candida* spp. and *S. aureus*
		Prevention of biofilm formation in new dental unit waterlines
	Removal	Mostly poor
	Fixation	Mostly low

References

1. Aarnisalo K, Lundén J, Korkeala H, Wirtanen G (2007) Susceptibility of Listeria monocytogenes strains to disinfectants and chlorinated alkaline cleaners at cold temperatures. LWT Food Sci Technol 40(6):1041–1048
2. Akinbobola AB, Sherry L, McKay WG, Ramage G, Williams C (2017) Tolerance of Pseudomonas aeruginosa in in-vitro biofilms to high-level peracetic acid disinfection. J Hosp Infect 97(2):162–168. https://doi.org/10.1016/j.jhin.2017.06.024
3. Alfa MJ, DeGagne P, Olson N, Hizon R (1998) Comparison of liquid chemical sterilization with peracetic acid and ethylene oxide sterilization for long narrow lumens. Am J Infect Control 26(5):469–477. https://doi.org/10.1016/S0196-6553(98)70018-5
4. Alonso-Hernando A, Alonso-Calleja C, Capita R (2010) Effects of exposure to poultry chemical decontaminants on the membrane fluidity of Listeria monocytogenes and Salmonella enterica strains. Int J Food Microbiol 137(2–3):130–136. https://doi.org/10.1016/j.ijfoodmicro.2009.11.022
5. Anonymous (2010) Peracetic acid. In: Committee on Acute Exposure Guideline Levels (ed) Acute exposure guideline levels for selected airborne chemicals (vol 8). The National Academic Press, Washington, pp 327–367
6. Arias-Moliz MT, Ordinola-Zapata R, Baca P, Ruiz-Linares M, Garcia Garcia E, Hungaro Duarte MA, Monteiro Bramante C, Ferrer-Luque CM (2015) Antimicrobial activity of Chlorhexidine, Peracetic acid and Sodium hypochlorite/etidronate irrigant solutions against Enterococcus faecalis biofilms. Int Endod J 48(12):1188–1193. https://doi.org/10.1111/iej.12424
7. Bang HJ, Park SY, Kim SE, Rahaman MDF, Ha SD (2017) Synergistic effects of combined ultrasound and peroxyacetic acid treatments against Cronobacter sakazakii biofilms on fresh

cucumber. LWT Food Sci Technol 84(Supplement C):91–98. https://doi.org/10.1016/j.lwt.2017.05.037
8. Barton E, Borman A, Johnson E, Sherlock J, Giles A (2016) Pseudo-outbreak of Fusarium oxysporum associated with bronchoscopy. J Hosp Infect 94(2):197–198. https://doi.org/10.1016/j.jhin.2016.06.016
9. Belessi CE, Gounadaki AS, Psomas AN, Skandamis PN (2011) Efficiency of different sanitation methods on Listeria monocytogenes biofilms formed under various environmental conditions. Int J Food Microbiol 145(Suppl 1):S46–52. https://doi.org/10.1016/j.ijfoodmicro.2010.10.020
10. Biswal BK, Khairallah R, Bibi K, Mazza A, Gehr R, Masson L, Frigon D (2014) Impact of UV and peracetic acid disinfection on the prevalence of virulence and antimicrobial resistance genes in uropathogenic Escherichia coli in wastewater effluents. Appl Environ Microbiol 80(12):3656–3666. https://doi.org/10.1128/aem.00418-14
11. Bore E, Langsrud S (2005) Characterization of micro-organisms isolated from dairy industry after cleaning and fogging disinfection with alkyl amine and peracetic acid. J Appl Microbiol 98(1):96–105. https://doi.org/10.1111/j.1365-2672.2004.02436.x
12. Bradley CR, Babb JR, Ayliffe GA (1995) Evaluation of the steris system 1 peracetic acid endoscope processor. J Hosp Infect 29(2):143–151
13. Bridier A, Briandet R, Thomas V, Dubois-Brissonnet F (2011) Comparative biocidal activity of peracetic acid, benzalkonium chloride and ortho-phthalaldehyde on 77 bacterial strains. J Hosp Infect 78(3):208–213. https://doi.org/10.1016/j.jhin.2011.03.014
14. Bridier A, Dubois-Brissonnet F, Greub G, Thomas V, Briandet R (2011) Dynamics of the action of biocides in Pseudomonas aeruginosa biofilms. Antimicrob Agents Chemother 55 (6):2648–2654. https://doi.org/10.1128/aac.01760-10
15. Bridier A, Sanchez-Vizuete Mdel P, Le Coq D, Aymerich S, Meylheuc T, Maillard JY, Thomas V, Dubois-Brissonnet F, Briandet R (2012) Biofilms of a Bacillus subtilis hospital isolate protect Staphylococcus aureus from biocide action. PLoS ONE 7(9):e44506. https://doi.org/10.1371/journal.pone.0044506
16. Bundgaard-Nielsen K, Nielsen PV (1996) Fungicidal effect of 15 disinfectants against 25 fungal contaminants commonly found in bread and cheese manufacturing. J Food Prot 59 (3):268–275
17. Burgess W, Margolis A, Gibbs S, Duarte RS, Jackson M (2017) Disinfectant susceptibility profiling of glutaraldehyde-resistant nontuberculous mycobacteria. Infect Control Hosp Epidemiol 38(7):784–791. https://doi.org/10.1017/ice.2017.75
18. Cadnum JL, Jencson AL, O'Donnell MC, Flannery ER, Nerandzic MM, Donskey CJ (2017) An increase in healthcare-associated clostridium difficile infection associated with use of a defective peracetic acid-based surface disinfectant. Infect Control Hosp Epidemiol 38 (3):300–305. https://doi.org/10.1017/ice.2016.275
19. Candeliere A, Campese E, Donatiello A, Pagano S, Iatarola M, Tolve F, Antonino L, Fasanella A (2016) Biocidal and sporicidal efficacy of pathoster ((R)) 0.35% and pathoster ((R)) 0.50% against bacterial agents in potential bioterrorism use. Health Security 14 (4):250–257. https://doi.org/10.1089/hs.2016.0003
20. Chang W, Small DA, Toghrol F, Bentley WE (2005) Microarray analysis of toxicogenomic effects of peracetic acid on Pseudomonas aeruginosa. Environ Sci Technol 39(15):5893–5899
21. Chang W, Toghrol F, Bentley WE (2006) Toxicogenomic response of Staphylococcus aureus to peracetic acid. Environ Sci Technol 40(16):5124–5131
22. Chenjiao W, Hongyan Z, Qing G, Xiaoqi Z, Liying G, Ying F (2016) In-use evaluation of peracetic acid for high-level disinfection of endoscopes. Gastroenterol Nursing: Off J Soc Gastroenterol Nurses Assoc 39(2):116–120. https://doi.org/10.1097/sga.0000000000000192
23. Clapp PA, Davies MJ, French MS, Gilbert BC (1994) The bactericidal action of peroxides; an E.P.R. spin-trapping study. Free Radical Res 21(3):147–167

24. Costa SA, Paula OF, Silva CR, Leao MV, Santos SS (2015) Stability of antimicrobial activity of peracetic acid solutions used in the final disinfection process. Brazilian Oral Res 29. https://doi.org/10.1590/1807-3107bor-2015.vol29.0038
25. Cronmiller JR, Nelson DK, Jackson DK, Kim CH (1999) Efficacy of conventional endoscopic disinfection and sterilization methods against Helicobacter pylori contamination. Helicobacter 4(3):198–203
26. Cronmiller JR, Nelson DK, Salman G, Jackson DK, Dean RS, Hsu JJ, Kim CH (1999) Antimicrobial efficacy of endoscopic disinfection procedures: a controlled, multifactorial investigation. Gastrointest Endosc 50(2):152–158
27. Cruz CD, Fletcher GC (2012) Assessing manufacturers' recommended concentrations of commercial sanitizers on inactivation of Listeria monocytogenes. Food Control 26(1):194–199. https://doi.org/10.1016/j.foodcont.2012.01.041
28. Das JR, Bhakoo M, Jones MV, Gilbert P (1998) Changes in the biocide susceptibility of Staphylococcus epidermidis and Escherichia coli cells associated with rapid attachment to plastic surfaces. J Appl Microbiol 84(5):852–858
29. de Souza EL, Meira QG, de Medeiros Barbosa I, Athayde AJ, da Conceicao ML, de Siqueira Junior JP (2014) Biofilm formation by Staphylococcus aureus from food contact surfaces in a meat-based broth and sensitivity to sanitizers. Brazilian J Microbiol: [Publication of the Brazilian Society for Microbiology] 45(1):67–75
30. Deshpande A, Mana TS, Cadnum JL, Jencson AC, Sitzlar B, Fertelli D, Hurless K, Kundrapu S, Sunkesula VC, Donskey CJ (2014) Evaluation of a sporicidal peracetic acid/hydrogen peroxide-based daily disinfectant cleaner. Infect Control Hosp Epidemiol 35 (11):1414–1416. https://doi.org/10.1086/678416
31. Deva AK, Vickery K, Zou J, West RH, Selby W, Benn RA, Harris JP, Cossart YE (1998) Detection of persistent vegetative bacteria and amplified viral nucleic acid from in-use testing of gastrointestinal endoscopes. J Hosp Infect 39(2):149–157
32. Di Cesare A, Fontaneto D, Doppelbauer J, Corno G (2016) Fitness and recovery of bacterial communities and antibiotic resistance genes in urban wastewaters exposed to classical disinfection treatments. Environ Sci Technol 50(18):10153–10161. https://doi.org/10.1021/acs.est.6b02268
33. Dominciano LCC, Oliveira CAF, Lee SH, Corassin CH (2016) Individual and combined antimicrobial activity of oleuropein and chemical sanitizers. J Food Chem Nanotechnol 2 (3):124–127
34. dos Anjos MM, Ruiz SP, Nakamura CV, de Abreu Filho BA (2013) Resistance of Alicyclobacillus acidoterrestris spores and biofilm to industrial sanitizers. J Food Prot 76 (8):1408–1413. https://doi.org/10.4315/0362-028x.jfp-13-020
35. Dubois-Brissonnet F, Naitali M, Mafu AA, Briandet R (2011) Induction of fatty acid composition modifications and tolerance to biocides in Salmonella enterica serovar Typhimurium by plant-derived terpenes. Appl Environ Microbiol 77(3):906–910. https://doi.org/10.1128/aem.01480-10
36. Duc DL, Ribiollet A, Dode X, Ducel G, Marchetti B, Calop J (2001) Evaluation of the microbicidal efficacy of Steris System I for digestive endoscopes using GERMANDE and ASTM validation protocols. J Hosp Infect 48(2):135–141. https://doi.org/10.1053/jhin.2001.0900
37. El-Azizi M, Farag N, Khardori N (2016) Efficacy of selected biocides in the decontamination of common nosocomial bacterial pathogens in biofilm and planktonic forms. Comp Immunol Microbiol Infect Dis 47:60–71. https://doi.org/10.1016/j.cimid.2016.06.002
38. Espigares E, Bueno A, Espigares M, Galvez R (2006) Isolation of Salmonella serotypes in wastewater and effluent: effect of treatment and potential risk. Int J Hyg Environ Health 209 (1):103–107. https://doi.org/10.1016/j.ijheh.2005.08.006
39. Espigares E, Bueno A, Fernandez-Crehuet M, Espigares M (2003) Efficacy of some neutralizers in suspension tests determining the activity of disinfectants. J Hosp Infect 55 (2):137–140

40. Espigares E, Moreno Roldan E, Espigares M, Abreu R, Castro B, Dib AL, Arias A (2017) Phenotypic resistance to disinfectants and antibiotics in methicillin-resistant Staphylococcus aureus strains isolated from pigs. Zoonoses Public Health 64(4):272–280. https://doi.org/10.1111/zph.12308
41. European Chemicals Agency (ECHA) Peracetic acid. Substance information. https://echa.europa.eu/substance-information/-/substanceinfo/100.001.079. Accessed 25 Oct 2017
42. European Chemicals Agency (ECHA) (2015) Opinion on the application for approval of the active substance: peracetic acid. Product-type: 2. ECHA/BPC/068/2015. https://echa.europa.eu/documents/10162/e10165ca10148f10165-10168c10158-10164baf-10168ced-10162ece65470ffa
43. Exner M, Tuschewitzki GJ, Scharnagel J (1987) Influence of biofilms by chemical disinfectants and mechanical cleaning. Zentralbl Bakteriol Mikrobiol Hyg B 183(5–6):549–563
44. Fagerlund A, Moretro T, Heir E, Briandet R, Langsrud S (2017) Cleaning and disinfection of biofilms composed of Listeria monocytogenes and background microbiota from meat processing surfaces. Appl Environ Microbiol. https://doi.org/10.1128/aem.01046-17
45. Fatemi P, Frank JF (1999) Inactivation of Listeria monocytogenes/Pseudomonas biofilms by peracid sanitizers. J Food Prot 62(7):761–765
46. Finland (2015) Assessment report. Peracetic acid. Product-types 1–6
47. Foliente RL, Kovacs BJ, Aprecio RM, Bains HJ, Kettering JD, Chen YK (2001) Efficacy of high-level disinfectants for reprocessing GI endoscopes in simulated-use testing. Gastrointest Endosc 53(4):456–462. https://doi.org/10.1067/mge.2001.113380
48. Gilbert P, Das JR, Jones MV, Allison DG (2001) Assessment of resistance towards biocides following the attachment of micro-organisms to, and growth on, surfaces. J Appl Microbiol 91(2):248–254
49. Gkana EN, Giaouris ED, Doulgeraki AI, Kathariou S, Nychas GJE (2017) Biofilm formation by Salmonella Typhimurium and Staphylococcus aureus on stainless steel under either mono- or dual-species multi-strain conditions and resistance of sessile communities to sub-lethal chemical disinfection. Food Control 73 (Part B):838–846. https://doi.org/10.1016/j.foodcont.2016.09.038
50. Gomes IB, Malheiro J, Mergulhao F, Maillard JY, Simoes M (2016) Comparison of the efficacy of natural-based and synthetic biocides to disinfect silicone and stainless steel surfaces. Pathog Dis 74(4):ftw014. https://doi.org/10.1093/femspd/ftw014
51. Grasteau A, Guiraud T, Daniel P, Calvez S, Chesneau V, Le Hénaff M (2015) Evaluation of glutaraldehyde, chloramine-t, bronopol, incimaxx aquatic® and hydrogen peroxide as biocides against flavobacterium psychrophilum for sanitization of rainbow trout eyed eggs. J Aquac Res Develop 6(12):382
52. Griffiths PA, Babb JR, Bradley CR, Fraise AP (1997) Glutaraldehyde-resistant Mycobacterium chelonae from endoscope washer disinfectors. J Appl Microbiol 82(4):519–526
53. Griffiths PA, Babb JR, Fraise AP (1999) Mycobactericidal activity of selected disinfectants using a quantitative suspension test. J Hosp Infect 41(2):111–121
54. Henoun Loukili N, Becker H, Harno J, Bientz M, Meunier O (2004) Effect of peracetic acid and aldehyde disinfectants on biofilm. J Hosp Infect 58(2):151–154
55. Hernández A, Martró E, Matas L, Ausina V (2003) In-vitro evaluation of Perasafe compared with 2% alkaline glutaraldehyde against Mycobacterium spp. J Hosp Infect 54(1):52–56
56. Hernández A, Martró E, Puzo C, Matas L, Burgués C, Vázquez N, Castella J, Ausina V (2003) In-use evaluation of Perasafe compared with Cidex in fibreoptic bronchoscope disinfection. J Hosp Infect 54(1):46–51
57. Herruzo R, Vizcaino MJ, Herruzo I (2010) Efficacy of a new peracetic acid-based disinfectant agent ('Adaspor ready to use'). J Hosp Infect 74(2):192–193. https://doi.org/10.1016/j.jhin.2009.10.019
58. Holton J, Nye P, McDonald V (1994) Efficacy of selected disinfectants against mycobacteria and cryptosporidia. J Hosp Infect 27(2):105–115

59. Holton J, Shetty N, McDonald V (1995) Efficacy of 'Nu-Cidex' (0.35% peracetic acid) against mycobacteria and cryptosporidia. J Hosp Infect 31(3):235–237
60. Humphreys PN, Finan P, Rout S, Hewitt J, Thistlethwaite P, Barnes S, Pilling S (2013) A systematic evaluation of a peracetic-acid-based high performance disinfectant. J Infect Prevent 14(4):126–131. https://doi.org/10.1177/1757177413476125
61. Jackson J, Leggett JE, Wilson DA, Gilbert DN (1996) Mycobacterium gordonae in fiberoptic bronchoscopes. Am J Infect Control 24(1):19–23
62. Jaglic Z, Červinková D, Vlková H, Michu E, Kunová G, Babák V (2012) Bacterial biofilms resist oxidising agents due to the presence of organic matter. Czech J Food Sci 30(2):178–187
63. Jolivet-Gougeon A, Sauvager F, Arturo-Schaan M, Bonnaure-Mallet M, Cormier M (2003) Influence of peracetic acid on adhesion/invasion of Salmonella enterica serotype typhimurium LT2. Cell Biol Toxicol 19(2):83–93
64. Jolivet-Gougeon A, Sauvager F, Bonnaure-Mallet M, Colwell RR, Cormier M (2006) Virulence of viable but nonculturable S. Typhimurium LT2 after peracetic acid treatment. Int J Food Microbiol 112(2):147–152. https://doi.org/10.1016/j.ijfoodmicro.2006.06.019
65. Juncker JC (2016) COMMISSION IMPLEMENTING REGULATION (EU) 2016/672 of 29 April 2016 approving peracetic acid as an existing active substance for use in biocidal products for product-types 1, 2, 3, 4, 5 and 6. Off J Eur Union 59(L 116):3–7
66. Juncker JC (2016) COMMISSION IMPLEMENTING REGULATION (EU) 2016/2290 of 16 December 2016 approving peracetic acid as an existing active substance for use in biocidal products of product-types 11 and 12. Off J Eur Union 59(L 344):71–73
67. Kampf G (2017) Black box oxidizers. Infect Control Hosp Epidemiol 38(11):1387–1388. https://doi.org/10.1017/ice.2017.199
68. Kassaify ZG, El Hakim RG, Rayya EG, Shaib HA, Barbour EK (2007) Preliminary study on the efficacy and safety of eight individual and blended disinfectants against poultry and dairy indicator organisms. Veterinaria Italiana 43(4):821–830
69. Kastbjerg VG, Gram L (2012) Industrial disinfectants do not select for resistance in Listeria monocytogenes following long term exposure. Int J Food Microbiol 160(1):11–15. https://doi.org/10.1016/j.ijfoodmicro.2012.09.009
70. Kastbjerg VG, Larsen MH, Gram L, Ingmer H (2010) Influence of sublethal concentrations of common disinfectants on expression of virulence genes in Listeria monocytogenes. Appl Environ Microbiol 76(1):303–309. https://doi.org/10.1128/aem.00925-09
71. Katara G, Hemvani N, Chitnis S, Chitnis V, Chitnis D (2016) Efficacy studies on peracetic acid against pathogenic microorganisms. J Patient Saf Infect Control 4(1):17–21. https://doi.org/10.4103/2214-207x.203545
72. Kean R, Sherry L, Townsend E, McKloud E, Short B, Akinbobola A, Mackay WG, Williams C, Jones BL, Ramage G (2018) Surface disinfection challenges for Candida auris: an in-vitro study. J Hosp Infect 98(4):433–436. https://doi.org/10.1016/j.jhin.2017.11.015
73. Konrat K, Schwebke I, Laue M, Dittmann C, Levin K, Andrich R, Arvand M, Schaudinn C (2016) The bead assay for biofilms: a quick, easy and robust method for testing disinfectants. PLoS ONE 11(6):e0157663. https://doi.org/10.1371/journal.pone.0157663
74. Korukluoglu M, Sahan Y, Yigit A (2006) The fungicidal efficacy of various commercial disinfectants used in the food industry. Ann Microbiol 56(4):325–330
75. Köse H, Yapar N (2017) The comparison of various disinfectants' efficacy on Staphylococcus aureus and Pseudomonas aeruginosa biofilm layers. Turkish J Med Sci 47(4):1287–1294
76. Kostaki M, Chorianopoulos N, Braxou E, Nychas GJ, Giaouris E (2012) Differential biofilm formation and chemical disinfection resistance of sessile cells of Listeria monocytogenes strains under monospecies and dual-species (with Salmonella enterica) conditions. Appl Environ Microbiol 78(8):2586–2595. https://doi.org/10.1128/aem.07099-11
77. La Scola B, Rolain J-M, Maurin M, Raoult D (2003) Can Whipple's disease be transmitted by gastroscopes? Infect Control Hosp Epidemiol 24(3):191–194

78. Lagace L, Jacques M, Mafu AA, Roy D (2006) Biofilm formation and biocides sensitivity of Pseudomonas marginalis isolated from a maple sap collection system. J Food Prot 69 (10):2411–2416
79. Lalueza P, Carmona D, Monzón M, Arruebo M, Santamaría J (2012) Strong bactericidal synergy between peracetic acid and silver-exchanged zeolites. Microporous Mesoporous Mater 156(Supplement C):171–175. https://doi.org/10.1016/j.micromeso.2012.02.035
80. Langsrud S, Moretro T, Sundheim G (2003) Characterization of Serratia marcescens surviving in disinfecting footbaths. J Appl Microbiol 95(1):186–195
81. Lee SH, Cappato LP, Corassin CH, Cruz AG, Oliveira CA (2016) Effect of peracetic acid on biofilms formed by Staphylococcus aureus and Listeria monocytogenes isolated from dairy plants. J Dairy Sci 99(3):2384–2390. https://doi.org/10.3168/jds.2015-10007
82. Lorena NS, Pitombo MB, Cortes PB, Maya MC, Silva MG, Carvalho AC, Coelho FS, Miyazaki NH, Marques EA, Chebabo A, Freitas AD, Lupi O, Duarte RS (2010) Mycobacterium massiliense BRA100 strain recovered from postsurgical infections: resistance to high concentrations of glutaraldehyde and alternative solutions for high level disinfection. Acta cirurgica brasileira 25(5):455–459
83. Loukili NH, Granbastien B, Faure K, Guery B, Beaucaire G (2006) Effect of different stabilized preparations of peracetic acid on biofilm. J Hosp Infect 63(1):70–72. https://doi.org/10.1016/j.jhin.2005.11.015
84. Luprano ML, De Sanctis M, Del Moro G, Di Iaconi C, Lopez A, Levantesi C (2016) Antibiotic resistance genes fate and removal by a technological treatment solution for water reuse in agriculture. Sci Total Environ 571:809–818. https://doi.org/10.1016/j.scitotenv.2016.07.055
85. Lynam PA, Babb JR, Fraise AP (1995) Comparison of the mycobactericidal activity of 2% alkaline glutaraldehyde and 'Nu-Cidex' (0.35% peracetic acid). J Hosp Infect 30(3):237–240
86. Mariscal A, Lopez-Gigosos RM, Carnero-Varo M, Fernandez-Crehuet J (2009) Fluorescent assay based on resazurin for detection of activity of disinfectants against bacterial biofilm. Appl Microbiol Biotechnol 82(4):773–783. https://doi.org/10.1007/s00253-009-1879-x
87. Martín-Espada MC, D'ors A, Bartolomé MC, Pereira M, Sánchez-Fortún S (2014) Peracetic acid disinfectant efficacy against Pseudomonas aeruginosa biofilms on polystyrene surfaces and comparison between methods to measure it. LWT Food Sci Technol 56(1):58–61
88. Martin DJ, Denyer SP, McDonnell G, Maillard JY (2008) Resistance and cross-resistance to oxidising agents of bacterial isolates from endoscope washer disinfectors. J Hosp Infect 69 (4):377–383. https://doi.org/10.1016/j.jhin.2008.04.010
89. Meade E, Garvey M (2018) Efficacy testing of novel chemical disinfectants on clinically relevant microbial pathogens. Am J Infect Control 46(1):44–49. https://doi.org/10.1016/j.ajic.2017.07.001
90. Melo RT, Mendonca EP, Monteiro GP, Siqueira MC, Pereira CB, Peres P, Fernandez H, Rossi DA (2017) Intrinsic and extrinsic aspects on Campylobacter jejuni biofilms. Front Microbiol 8:1332. https://doi.org/10.3389/fmicb.2017.01332
91. Meylheuc T, Renault M, Bellon-Fontaine MN (2006) Adsorption of a biosurfactant on surfaces to enhance the disinfection of surfaces contaminated with Listeria monocytogenes. Int J Food Microbiol 109(1–2):71–78. https://doi.org/10.1016/j.ijfoodmicro.2006.01.013
92. Middleton AM, Chadwick MV, Gaya H (1997) Disinfection of bronchoscopes, contaminated in vitro with Mycobacterium tuberculosis, Mycobacterium avium-intracellulare and Mycobacterium chelonae in sputum, using stabilized, buffered peracetic acid solution ('Nu-Cidex'). J Hosp Infect 37(2):137–143
93. Montebugnoli L, Chersoni S, Prati C, Dolci G (2004) A between-patient disinfection method to control water line contamination and biofilm inside dental units. J Hosp Infect 56(4):297–304. https://doi.org/10.1016/j.jhin.2004.01.015
94. Nakayama M, Hosoya K, Tomiyama D, Tsugukuni T, Matsuzawa T, Imanishi Y, Yaguchi T (2013) Method for rapid detection and identification of chaetomium and evaluation of

resistance to peracetic acid. J Food Prot 76(6):999–1005. https://doi.org/10.4315/0362-028x.jfp-12-543
95. Neo SY, Lim PY, Phua LK, Khoo GH, Kim SJ, Lee SC, Yuk HG (2013) Efficacy of chlorine and peroxyacetic acid on reduction of natural microflora, Escherichia coli O157:H7, Listeria monocyotgenes and Salmonella spp. on mung bean sprouts. Food Microbiol 36(2):475–480. https://doi.org/10.1016/j.fm.2013.05.001
96. Neves MS, da Silva MG, Ventura GM, Cortes PB, Duarte RS, de Souza HS (2016) Effectiveness of current disinfection procedures against biofilm on contaminated GI endoscopes. Gastrointest Endosc 83(5):944–953. https://doi.org/10.1016/j.gie.2015.09.016
97. Ntsama-Essomba C, Bouttier S, Ramaldes M, Dubois-Brissonnet F, Fourniat J (1997) Resistance of Escherichia coli growing as biofilms to disinfectants. Vet Res 28(4):353–363
98. Penna TC, Mazzola PG, Silva Martins AM (2001) The efficacy of chemical agents in cleaning and disinfection programs. BMC Infect Dis 1:16
99. Pineau L, Desbuquois C, Marchetti B, Luu Duc D (2008) Comparison of the fixative properties of five disinfectant solutions. J Hosp Infect 68(2):171–177. https://doi.org/10.1016/j.jhin.2007.10.021
100. Pires RH, da Silva Jde F, Gomes Martins CH, Fusco Almeida AM, Pienna Soares C, Soares Mendes-Giannini MJ (2013) Effectiveness of disinfectants used in hemodialysis against both Candida orthopsilosis and C. parapsilosis sensu stricto biofilms. Antimicrob Agents Chemother 57(5):2417–2421. https://doi.org/10.1128/aac.01308-12
101. Poimenidou SV, Chrysadakou M, Tzakoniati A, Bikouli VC, Nychas GJ, Skandamis PN (2016) Variability of Listeria monocytogenes strains in biofilm formation on stainless steel and polystyrene materials and resistance to peracetic acid and quaternary ammonium compounds. Int J Food Microbiol 237:164–171. https://doi.org/10.1016/j.ijfoodmicro.2016.08.029
102. Reis L, Zanetti AL, Castro Junior OV, Martinez EF (2012) Use of 0.25% and 0.025% peracetic acid as disinfectant agent for chemically activated acrylic resin: an in vitro study. Rev Gaúcha Odontol 60(3):315–320
103. Ruckerl I, Muhterem-Uyar M, Muri-Klinger S, Wagner KH, Wagner M, Stessl B (2014) L. monocytogenes in a cheese processing facility: Learning from contamination scenarios over three years of sampling. Int J Food Microbiol 189:98–105. https://doi.org/10.1016/j.ijfoodmicro.2014.08.001
104. Saa Ibusquiza P, Herrera JJ, Cabo ML (2011) Resistance to benzalkonium chloride, peracetic acid and nisin during formation of mature biofilms by Listeria monocytogenes. Food Microbiol 28(3):418–425. https://doi.org/10.1016/j.fm.2010.09.014
105. Sagripanti J-L, Eklund CA, Trost PA, Jinneman KC, Abeyta C, Kaysner CA, Hill WE (1997) Comparative sensitivity of 13 species of pathogenic bacteria to seven chemical germicides. Am J Infect Control 25(4):335–339
106. Sattar SA, Kibbee RJ, Tetro JA, Rook TA (2006) Experimental evaluation of an automated endoscope reprocessor with in situ generation of peracetic acid for disinfection of semicritical devices. Infect Control Hosp Epidemiol 27(11):1193–1199. https://doi.org/10.1086/508830
107. Shetty N, Srinivasan S, Holton J, Ridgway GL (1999) Evaluation of microbicidal activity of a new disinfectant: Sterilox 2500 against *Clostridium difficile* spores, *Helicobacter pylori*, vancomycin resistant Enterococcus species, *Candida albicans* and several Mycobacterium species. J Hosp Infect 41:101–105
108. Sisti M, Brandi G, De Santi M, Rinaldi L, Schiavano GF (2012) Disinfection efficacy of chlorine and peracetic acid alone or in combination against Aspergillus spp. and Candida albicans in drinking water. J Water Health 10(1):11–19. https://doi.org/10.2166/wh.2011.150
109. Sofokleous P, Ali S, Wilson P, Buanz A, Gaisford S, Mistry D, Fellows A, Day RM (2017) Sustained antimicrobial activity and reduced toxicity of oxidative biocides through biodegradable microparticles. Acta Biomater 64:301–312. https://doi.org/10.1016/j.actbio.2017.10.001

110. Sorin M, Segal-Maurer S, Mariano N, Urban C, Combest A, Rahal JJ (2001) Nosocomial transmission of imipenem-resistant Pseudomonas aeruginosa following bronchoscopy associated with improper connection to the Steris System 1 processor. Infect Control Hosp Epidemiol 22(7):409–413. https://doi.org/10.1086/501925
111. Stanley PM (1999) Efficacy of peroxygen compounds against glutaraldehyde-resistant mycobacteria. Am J Infect Control 27(4):339–343
112. Stigt JA, Wolfhagen MJ, Smulders P, Lammers V (2015) The Identification of Stenotrophomonas maltophilia contamination in ultrasound endoscopes and reproduction of decontamination failure by deliberate soiling tests. Respiration; Int Rev Thoracic Dis 89 (6):565–571. https://doi.org/10.1159/000381725
113. Stopforth JD, Samelis J, Sofos JN, Kendall PA, Smith GC (2003) Influence of extended acid stressing in fresh beef decontamination runoff fluids on sanitizer resistance of acid-adapted Escherichia coli O157:H7 in biofilms. J Food Prot 66(12):2258–2266
114. Surdeau N, Laurent-Maquin D, Bouthors S, Gelle MP (2006) Sensitivity of bacterial biofilms and planktonic cells to a new antimicrobial agent, Oxsil 320N. J Hosp Infect 62 (4):487–493. https://doi.org/10.1016/j.jhin.2005.09.003
115. Tote K, Horemans T, Vanden Berghe D, Maes L, Cos P (2010) Inhibitory effect of biocides on the viable masses and matrices of Staphylococcus aureus and Pseudomonas aeruginosa biofilms. Appl Environ Microbiol 76(10):3135–3142. https://doi.org/10.1128/aem.02095-09
116. Trachoo N, Frank JF (2002) Effectiveness of chemical sanitizers against Campylobacter jejuni-containing biofilms. J Food Prot 65(7):1117–1121
117. Turolla A, Sabatino R, Fontaneto D, Eckert EM, Colinas N, Corno G, Citterio B, Biavasco F, Antonelli M, Mauro A, Mangiaterra G, Di Cesare A (2017) Defence strategies and antibiotic resistance gene abundance in enterococci under stress by exposure to low doses of peracetic acid. Chemosphere 185:480–488. https://doi.org/10.1016/j.chemosphere.2017.07.032
118. United States Environmental Protection Agency (1993) EPA R.E.D. Facts. Peroxy Compounds. https://www3.epa.gov/pesticides/chem_search/reg_actions/reregistration/red_G-67_1-Dec-93.pdf
119. van der Veen S, Abee T (2010) HrcA and DnaK are important for static and continuous-flow biofilm formation and disinfectant resistance in Listeria monocytogenes. Microbiology (Reading, England) 156(Pt 12):3782–3790. https://doi.org/10.1099/mic.0.043000-0
120. van der Veen S, Abee T (2011) Mixed species biofilms of Listeria monocytogenes and Lactobacillus plantarum show enhanced resistance to benzalkonium chloride and peracetic acid. Int J Food Microbiol 144(3):421–431. https://doi.org/10.1016/j.ijfoodmicro.2010.10.029
121. van Klingeren B, Pullen W (1993) Glutaraldehyde resistant mycobacteria from endoscope washers. J Hosp Infect 25(2):147–149
122. Vázquez-Sánchez D, Cabo ML, Ibusquiza PS, Rodríguez-Herrera JJ (2014) Biofilm-forming ability and resistance to industrial disinfectants of Staphylococcus aureus isolated from fishery products. Food Control 39(Supplement C):8–16. https://doi.org/10.1016/j.foodcont.2013.09.029
123. Verner-Jeffreys DW, Joiner CL, Bagwell NJ, Reese RA, Husby A, Dixon PF (2009) Development of bactericidal and virucidal testing standards for aquaculture disinfectants. Aquaculture 286(3):190–197. https://doi.org/10.1016/j.aquaculture.2008.10.001
124. Vesley D, Melson J, Stanley P (1999) Microbial bioburden in endoscope reprocessing and an in-use evaluation of the high-level disinfection capabilities of Cidex PA. Gastroenterol Nursing: Off J Soc Gastroenterol Nurses Assoc 22(2):63–68
125. Vieira CD, Farias Lde M, Diniz CG, Alvarez-Leite ME, Camargo ER, Carvalho MA (2005) New methods in the evaluation of chemical disinfectants used in health care services. Am J Infect Control 33(3):162–169. https://doi.org/10.1016/j.ajic.2004.10.007
126. Vizcaino-Alcaide MJ, Herruzo-Cabrera R, Fernandez-Acenero MJ (2003) Comparison of the disinfectant efficacy of Perasafe and 2% glutaraldehyde in in vitro tests. J Hosp Infect 53:124–128

127. Walsh SE, Maillard JY, Russell AD (1999) Ortho-phthalaldehyde: a possible alternative to glutaraldehyde for high level disinfection. J Appl Microbiol 86(6):1039–1046
128. Wang GQ, Zhang CW, Liu HC, Chen ZB (2005) Comparison of susceptibilities of M. tuberculosis H37Ra and M. chelonei subsp. abscessus to disinfectants. Biomed Environ Sci: BES 18(2):124–127
129. Wessels S, Ingmer H (2013) Modes of action of three disinfectant active substances: a review. Regul Toxicol Pharmacol: RTP 67(3):456–467. https://doi.org/10.1016/j.yrtph.2013.09.006
130. Witney AA, Gould KA, Pope CF, Bolt F, Stoker NG, Cubbon MD, Bradley CR, Fraise A, Breathnach AS, Butcher PD, Planche TD, Hinds J (2014) Genome sequencing and characterization of an extensively drug-resistant sequence type 111 serotype O12 hospital outbreak strain of Pseudomonas aeruginosa. Clin Microbiol Infect 20(10):O609–618. https://doi.org/10.1111/1469-0691.12528
131. Wuthiekanun V, Wongsuwan G, Pangmee S, Teerawattanasook N, Day NP, Peacock SJ (2011) Perasafe, Virkon and bleach are bactericidal for Burkholderia pseudomallei, a select agent and the cause of melioidosis. J Hosp Infect 77(2):183–184. https://doi.org/10.1016/j.jhin.2010.06.026
132. Zanotto C, Bissa M, Illiano E, Mezzanotte V, Marazzi F, Turolla A, Antonelli M, De Giuli Morghen C, Radaelli A (2016) Identification of antibiotic-resistant Escherichia coli isolated from a municipal wastewater treatment plant. Chemosphere 164:627–633. https://doi.org/10.1016/j.chemosphere.2016.08.040
133. Zhang K, Zhou X, Du P, Zhang T, Cai M, Sun P, Huang CH (2017) Oxidation of beta-lactam antibiotics by peracetic acid: reaction kinetics, product and pathway evaluation. Water Res 123:153–161. https://doi.org/10.1016/j.watres.2017.06.057
134. Zook CD, Busta FF, Brady LJ (2001) Sublethal sanitizer stress and adaptive response of Escherichia coli O157:H7. J Food Prot 64(6):767–769

Hydrogen Peroxide

6

6.1 Chemical Characterization

Hydrogen peroxide is the simplest peroxide. Its chemistry is dominated by the nature of its unstable peroxide bond. The basic chemical information on hydrogen peroxide is summarized in Table 6.1.

Pure hydrogen peroxide does not exist commercially. Hydrogen peroxide is always directly produced as an aqueous solution which contains 35–70% of hydrogen peroxide (w/w). Aqueous solutions of hydrogen peroxide are used as biocidal products. Commercial hydrogen peroxide grades are stabilized to prevent or slow down its decomposition. The stabilizers are of several types. It may be mineral acids to keep the solution acidic (stability is at a maximum at pH 3.5–4.5), it may be complexing or chelating agents to inhibit metal-catalysed decomposition, or it may be colloidal to neutralize small amounts of colloidal catalysts or absorb impurities [33].

6.2 Types of Application

The use of hydrogen peroxide includes food production, processing and handling, disinfection of hard surfaces in health care, veterinary medicine and institutions, and use for critical, semicritical and non-critical hospital items including flexible endoscopes [81, 120]. Based on the Finnish assessment report on hydrogen peroxide, uses include disinfection of human skin with 7.4 or 4.9% (w/w) hydrogen peroxide by private and professional users (product type 1), surface disinfection in private or public hygiene disinfection of rooms using the vaporized hydrogen peroxide process (250–400 ppm in air, equivalent to 0.025–0.04%) (product type 2), disinfection of animal housing by spraying aqueous solutions of 7.4% (w/w) of hydrogen peroxide (product type 3), disinfection of packaging for food products by immersion into 35% (w/w) aqueous hydrogen peroxide solutions (product type 4),

G. Kampf, *Antiseptic Stewardship*, https://doi.org/10.1007/978-3-319-98785-9_6

Table 6.1 Basic chemical information on hydrogen peroxide [33, 77]

CAS number	7722-84-1
IUPAC name	Hydrogen peroxide
Synonyms	Dihydrogen dioxide, hydrogen dioxide
Molecular formula	H_2O_2
Molecular weight (g/mol)	34.01

surface disinfection by vaporized hydrogen peroxide process in food processing facilities (product type 4), disinfection of distribution systems for drinking water at 4% (w/w) (product type 4), disinfection of drinking water for humans and animals (product type 5), and preservation of paper additives with up to 1.0% (w/w) hydrogen peroxide (product type 6) [33].

6.2.1 European Chemicals Agency (European Union)

In 2015, it has been approved by the European Commission as an active substance for use in biocidal products for product types 1 (human hygiene), 2 (disinfectants and algaecides not intended for direct application to humans or animals), 3 (veterinary hygiene), 4 (food and feed area), 5 (drinking water) and 6 (preservatives for products during storage) [52].

6.2.2 Environmental Protection Agency (USA)

Hydrogen peroxide was first registered as a pesticide in the USA in 1977. The overall assessment revealed that the use of products containing hydrogen peroxide will not pose unreasonable risks or adverse effects to humans or the environment [120].

6.2.3 Overall Environmental Impact

Hydrogen peroxide is manufactured or imported in the European Economic Area in 1–10 million t per year [28]. It decomposes rapidly in different environmental compartments. The following processes are involved in the decomposition or degradation of hydrogen peroxide in the environment: biotic degradation catalysed by microbial catalase and peroxidase enzymes, abiotic degradation by transition metal (Fe, Mn, Cu) and heavy metal catalysed decomposition or oxidation or reduction reactions with organic compounds or formation of addition compounds with organic or inorganic substances. Hydrogen peroxide decomposes into water and oxygen ($2\ H_2O_2 \rightarrow 2\ H_2O + O_2$). The rate of this reaction depends on the contact with catalytic materials and other factors such as heat and sunlight. Hydrogen peroxide shows a very rapid biodegradation in sewage sludge with a 50% dissipation time (DT50) of 2 min at 20 °C. Ready biodegradability has not been unequivocally demonstrated as the standard ready biodegradability tests are not suitable for inorganic substances. Rapid degradation of hydrogen peroxide has also been observed in surface water and soil compartments. This degradation has been proposed to be mainly microbially derived based on the difference in degradation rates between the natural and filtered or sterilized samples [33].

6.3 Spectrum of Antimicrobial Activity

6.3.1 Bactericidal Activity

6.3.1.1 Bacteriostatic Activity (MIC Values)

The MIC values obtained with different bacterial species are summarized in Table 6.2; they were between 0.5 and 12,784 mg/l (equivalent to 0.00005 and 1.28%) indicating a broad range of susceptibility to hydrogen peroxide. Some bacterial species such as *S. sanguis* have been described to be able to produce hydrogen peroxide at bacteriostatic concentrations [44].

Table 6.2 MIC values of various bacterial species to hydrogen peroxide

Species	Strains/isolates	MIC value (mg/l)	References
A. baumannii	47 clinical isolates	1,598–12,784	[65]
A. calcoaceticus	ATCC 19606	469	[85]
Acinetobacter spp.	5 clinical strains, NCTC 13424 and ATCC 17978	238–476	[86]
B. stearothermophilus	ATCC 7953	1,875	[85]
B. subtilis var. globigii	ATCC 9372	1,875	[85]
E. cloacae	Strain IAL 1976	1,250	[85]
E. faecalis	52 isolates from livestock	80–160	[1]
E. faecalis	9 isolates from swine meat production	120	[95]
E. faecium	78 isolates from livestock	80–160	[1]
E. faecium	12 isolates from swine meat production	120–140	[95]
E. coli	Reference strain and clinical isolate	0.5–1	[62]
E. coli	Strain JM 101	3.4	[48]
E. coli	202 isolates from livestock	40–160	[1]
E. coli	ATCC 11229	72	[49]
E. coli	ATCC 25922	234	[22]

(continued)

Table 6.2 (continued)

Species	Strains/isolates	MIC value (mg/l)	References
E. coli	ATCC 25922	2,505	[85]
F. psychrophilum	5 fresh trout isolates	3.1–62.5	[39]
K. pneumoniae	2 clinical strains, NCTC 13439, NCTC 13443, NCTC 13368, MGH 78578, NCTC 9633	238–476	[86]
L. monocytogenes	ATCC 19111	100	[49]
L. monocytogenes	6 strains from a cheese processing facility	125	[99]
L. monocytogenes	ATCC 7644	469	[22]
P. aeruginosa	ATCC 15442	100	[49]
P. aeruginosa	6 clinical strains and NCTC 13359	476	[86]
S. choleraesuis	ATCC 10708	128	[49]
S. marcescens	Strain IAL 1478	625	[85]
Salmonella spp.	156 isolates from livestock	20–80	[1]
S. aureus	43 isolates from livestock	20–40	[1]
S. aureus	ATCC 6538	72	[49]
S. aureus	ATCC 25923	100	[49]
S. aureus	ATCC 25923	117	[22]
S. aureus	ATCC 25923	938	[85]
S. hyicus	38 isolates from livestock	20	[1]

6.3.1.2 Bactericidal Activity (Suspension Tests)

Hydrogen peroxide between 0.5 and 10% was found to be mostly bactericidal within 30 min, and lower concentrations such as 0.3% may require longer exposure times or additional substances to enhance the bactericidal activity. *E. faecium* was rather resistant to 3% hydrogen peroxide with ≤ 2.1 log in 10 min (Table 6.3). Other studies indicate that the bactericidal activity of hydrogen peroxide at 0.85–3.4% against *S. aureus* exposed for 1 min can be significantly improved by photolysis at 365–400 nm [116]. In swimming pool water, hydrogen peroxide at 0.015% did not show any relevant efficacy within 30 min against *P. aeruginosa* (0.2 log), *E. coli* (0.1 log), *S. aureus* (0.3 log) and *L. pneumophila* (0.4 log). A formulation with the same concentration of hydrogen peroxide and additional silver ions at 15 ppb did not improve the bactericidal activity [10].

In order to kill bacterial cells with hydrogen peroxide in 5 min, higher concentrations may be necessary depending on the species (Table 6.4). The MBC values were between 0.173–12.5% (equivalent to 1,730 and 125.000 mg/l). *Enterococcus* and *Listeria* seem to be the least susceptible genus.

The combination with formic acid can significantly increase the bactericidal activity of 0.039–0.39% hydrogen peroxide against bacterial species (*E. hirae*,

Table 6.3 Bactericidal activity of hydrogen peroxide in suspension tests

Species	Strains/isolates	Exposure time	Concentration	log_{10} reduction	References
B. cenocepacia	LMG 16656, LMG 18828	30 min	3% (S)	≥ 5.0	[84]
			1% (S)	≥ 5.0	
			0.5% (S)	≥ 5.0	
			0.3% (S)	4.0	
B. cereus[a]	ATCC 14579	30 min	10% (S)	>5.0	[101]
C. jejuni	ATCC BAA-1062, ATCC 33560 and 2 field strains	1 min	3% (S)	4.4–6.0	[42]
C. perfringens[a]	Strain CDC 1861	30 min	10% (S)	>6.3	[101]
E. faecium	ATCC 6057	30 s	3% (S)	0.2–0.5[b]	[90]
		1 min		0.3–0.9[b]	
		10 min		0.1–2.1[b]	
E. coli	Food isolate 0157:H7	30 min	10% (S)	>6.9	[101]
E. coli	NCTC 10536	30 s	3% (S)	≥ 6.3	[90]
E. coli	CCUG 44857, ATCC 10536	10 min	0.01375%[c] (P)	>5.0	[11]
H. parasuis	2 strains (serovars 1 and 5)	1 min	3% (S)	>6.0	[96]
				5.6[d]	
L. monocytogenes	Food isolate	30 min	10% (S)	>6.1	[101]
L. monocytogenes	Strain Scott A	10 min	0.01375%[c] (P)	>5.0	[11]
L. innocua	ATCC 33090	10 min	0.01375%[c] (P)	>5.0	[11]
P. aeruginosa	ATCC 27853	30 min	10% (S)	>6.1	[101]
P. aeruginosa	ATCC 700928	1 min	5% (S)	3.7	[118]
		5 min		≥ 5.2	
P. aeruginosa	ATCC 15442	30 s	3% (S)	3.9–6.1[b]	[90]
		1 min		4.7–7.3[b]	
		10 min		4.8–7.1[b]	
P. aeruginosa	ATCC 9027	2–6 h	3% (P)	4.9	[98]
S. Typhimurium	ATCC 14028	30 min	10% (S)	>6.4	[101]
S. sonnei	Food isolate	30 min	10% (S)	>6.3	[101]
S. aureus	ATCC 25923	30 min	10% (S)	5.6	[101]
S. aureus	Newman laboratory strain	5 min	6% (S)	≥ 5.0	[60]
S. aureus	ATCC 6538	1 min	5% (S)	0.6	[118]
		5 min		4.7	
		15 min		≥ 5.4	
S. aureus	ATCC 6538	30 s	3% (S)	0.2–0.6[b]	[90]
		1 min		0.2–0.3[b]	
		10 min		1.3–4.7[b]	
S. aureus	IFO 13276	1 h	3% (S)	≥ 5.0	[126]
S. aureus	ATCC 6538	2–6 h	3% (P)	4.2–5.3	[98]
S. aureus	ATCC 6538	10 min	0.0275%[e] (P)	>5.0	[11]
S. epidermidis	ATCC 12228	30 min	10% (S)	>6.3	[101]

(continued)

Table 6.3 (continued)

Species	Strains/isolates	Exposure time	Concentration	$\log_{10}$ reduction	References
S. epidermidis	ATCC 17917	2–6 h	3% (P)	4.2–4.7	[98]
S. marcescens	ATCC 13880	2–6 h	3% (P)	4.6–5.0	[98]
V. cholerae	Strain C6706	30 min	10% (S)	>6.4	[101]
V. parahaemolyticus	Strain NY477	30 min	10% (S)	>6.2	[101]
V. vulnificus	Strain LA M624	30 min	10% (S)	>6.3	[101]
Y. enterocolitica	Strain 8081	30 min	10% (S)	>6.8	[101]

S solution; *P* commercial product; [a]vegetative cell form; [b]depending on the type of organic load; [c]plus 0.0029% peracetic acid; [d]plus organic load; [e]plus 0.0058% peracetic acid

Table 6.4 MBC values (5 min exposure) of various bacterial species to hydrogen peroxide

Species	Strains/isolates	MBC value (%)	References
A. baumannii	Clinical isolate	0.17	[26]
B. cepacia	Clinical isolate	0.17	[26]
E. faecalis	Clinical isolate	0.41	[26]
E. faecium	Clinical isolate	0.41	[26]
E. hirae	ATCC 10541	12.5	[72]
E. coli	ATCC 25922	0.14	[26]
E. coli	ATCC 11229	1.56	[72]
L. monocytogenes	ATCC 19115	6.25	[72]
P. aeruginosa	Clinical isolate	0.41	[26]
P. aeruginosa	CIP A 22	1.56	[72]
Salmonella spp.	Strain 276	3.12	[72]
S. aureus	Clinical MRSA isolate	0.41	[26]
S. aureus	ATCC 9144	3.12	[72]
S. epidermidis	Clinical isolate	0.41	[26]
S. maltophilia	Clinical isolate	0.41	[26]

S. aureus, *E. coli*, *P. aeruginosa*, *L. monocytogenes*, *S. Typhimurium*). The combination with acetic acid can also significantly increase the bactericidal activity against most of the bacterial species except *S. Typhimurium* [72].

6.3.1.3 Activity Against Bacteria in Biofilms

The majority of studies show that hydrogen peroxide is less effective against bacterial cells in biofilms. While a bactericidal activity against planktonic cells is mostly achieved with 0.5% hydrogen peroxide in 30 min, a 4.0 log reduction in biofilm is mostly not achieved using 2 or 3% hydrogen peroxide for up to 30 min or 5% hydrogen peroxide for up to 60 min (Table 6.5). This finding has been reported also in other experimental settings [26, 86], also with *F. noatunensis subsp. orientalis*, an emergent fish pathogen [107]. The duration of biofilm incubation

Table 6.5 Efficacy of hydrogen peroxide against bacteria in biofilms

Species	Strains/isolates	Type of biofilm	Exposure time	Concentration	log_{10} reduction	References
E. faecalis	ATCC 29212	8-d incubation in polystyrene pegs	5 min	2% (P)	3.8	[19]
E. hirae	CIP 58.55	24-h incubation on stainless steel	5 min	Not described but with silver and peracetic acid (P)	1.9	[113]
E. coli	ATCC 35218	48-h incubation on glass, polypropylene, polycarbonate, silicone and PVC	30 min	3% (S)	"Complete inactivation"	[70]
E. coli O157:H7	ATCC 35150, ATCC 43889, ATCC 43890	24-h incubation on stainless steel	30 s	2% (P)	2.0	[6]
				1% (P)	1.1	
				0.5% (P)	0.7	
E. coli	3 avian pathogenic strains	24-h incubation on polystyrene	30 min	1% (S)	≥ 3.8	[83]
				0.5% (S)	3.4–3.8	
		24-h incubation on PVC		1% (S)	≥ 4.3	
				0.5% (S)	3.5–4.3	
E. coli	CIP 54.127	24-h incubation on stainless steel	5 min	Not described but with silver and peracetic acid (P)	1.7	[113]
L. innocua	Food contact surface isolate	48-h incubation on stainless steel coupons	10 min	0.25–0.3%[c] (P)	2.5	[58]
					7.3[b]	
				0.125–0.15%[a] (P)	2.0	
					6.1[b]	
L. monocytogenes	ATCC 15313, ATCC 19114, ATCC 19115	24-h incubation on stainless steel	30 s	2% (P)	1.2	[6]
				1% (P)	0.7	
				0.5% (P)	0.7	
M. luteus	Food contact surface isolate	48-h incubation on stainless steel coupons	10 min	0.25–0.3%[c] (P)	2.7	[58]
					7.4[b]	
				0.125–0.15%[a] (P)	2.5	
					6.0[b]	

(continued)

Table 6.5 (continued)

Species	Strains/isolates	Type of biofilm	Exposure time	Concentration	log_{10} reduction	References
P. aeruginosa	ATCC 700928	24-h incubation in microplates	1 min	5% (S)	0.3	[118]
			5 min		0.7	
			60 min		2.0	
P. aeruginosa	Strain PA01	24-h incubation in microtitre plates	1, 5, 15, 30 and 60 min	5% (S)	0.8–2.0	[57]
P. aeruginosa	ATCC 15442	8-d incubation in polystyrene pegs	5 min	2% (P)	5.8	[19]
P. aeruginosa	CIP A22	24-h incubation on stainless steel	5 min	Not described but with silver and peracetic acid (P)	1.3	[113]
P. fluorescens	JCM 2779	2 nights incubation on glass slides	10 s	1.1% (S)	0.6	[114]
P. putida	Food contact surface isolate	48-h incubation on stainless steel coupons	10 min	0.25–0.3%[c] (P)	3.7	[58]
				0.125–0.15%[a] (P)	1.4	
					7.4[b]	
S. Typhimurium	ATCC 19585, ATCC 43971, DT 104	24-h incubation on stainless steel	30 s	2% (P)	1.8	[6]
				1% (P)	1.4	
				0.5% (P)	0.9	
S. aureus	ATCC 6538	72-h incubation in microplates	1 min	5% (S)	0.8	[118]
			5 min		0.8	
			60 min		0.7	
S. aureus	ATCC 6538	24-h incubation in microtitre plates	1, 5, 15, 30 and 60 min	5% (S)	0.7–2.0	[57]
S. aureus	CIP 53.154	24-h incubation on stainless steel	5 min	Not described but with silver and peracetic acid (P)	2.6	[113]
S. epidermidis	30 clinical isolates	24-h incubation in polystyrene microtiter plates	5 min	5% (S)	≥ 5.0	[92]
				3% (S)	≥ 5.0	
				0.5% (S)	<5.0	

(continued)

Table 6.5 (continued)

Species	Strains/isolates	Type of biofilm	Exposure time	Concentration	log_{10} reduction	References
S. hominis	Food contact surface isolate	48-h incubation on stainless steel coupons	10 min	0.25–0.3%[c] (P)	2.6	[58]
					7.5[b]	
				0.125–0.15%[a] (P)	1.9	
					6.3[b]	
S. mutans	Strain C180-2	24-h incubation on titanium discs	5 min	10% (P)	3.0	[79]
S. mutans	Strain JCM 5705	24-h incubation on a hydroxyapatite disk	1 min	3% (S)	3.9[d]	[75]
					5.3[e]	
S. mutans	Strain JCM 5705	24-h incubation on a hydroxyapatite disk	4 min	3% (S)	4.5–4.9	[106]
S. mutans	DSM 20523	72-h incubation on titanium discs	30 min	1.5% (S)	6.9	[56]
Various species	Dental unit waterlines	Natural biofilm from dental unit waterlines	2 d	35% (S)	0.2–0.5	[64]
Various species	Polymicrobial biofilm from saliva	48-h incubation on titanium discs	5 min	10% (P)	1.1	[79]
Various species	Mixed oral biofilm	12-h incubation in the oral cavity on titanium surfaces	1 min	3% (S)	"significant reduction"	[36]
Various species	Human saliva bacteria	72-h incubation on titanium discs	30 min	1.5% (S)	1.5	[56]
Various species	Subgingival plaque bacteria	Overnight incubation on titanium discs	30 min	1.5% (S)	1.9	[56]

P commercial product; *S* solution; [a]plus peracetic acid at 0.01–0.025%; [b]planktonic cells; [c]plus peracetic acid at 0.02–0.05%; [d]with photolysis at 400 nm; [e]with photolysis at 365 nm seems to have no relevance for the magnitude of bactericidal efficacy of hydrogen peroxide on bacterial cells in biofilm. No correlation was described with a *P. marginalis* biofilm grown for 24 at 30 °C (41 times less susceptible to hydrogen peroxide compared to planktonic cells) or 48 h (33 times less susceptible) [59].

Various biofilm forms were found in *P. aeruginosa* biofilms. Susceptibility to hydrogen peroxide was significantly lower when the wild-type form was found (5.0 log reduction in 4 h with 0.85% hydrogen peroxide) compared to wrinkly variant biofilms (5.0 log reduction in 4 h with 6.8% hydrogen peroxide), also described as hyperbiofilm formation with increased initial attachment, cell clusters formed earlier and much bigger and a 9-fold lower detachment rate [9].

In waterborne biofilms, the efficacy of hydrogen peroxide has been described to be strong. Micro-organisms in waterborne mixed biofilm (50 days) on silicone tubes were killed by hydrogen peroxide at 1.5% in 30 min by >5.0 log, whereas 1% hydrogen peroxide revealed <5.0 log in 60 min [29]. Treatment of drinking water biofilm with 3% hydrogen peroxide resulted in an immense population shift. It implies that half of biofilm members disappeared after treatment and were replaced by other micro-organisms, which were better adapted to these conditions [97]. The combination of hydrogen peroxide with other antimicrobial agents such as peracetic acid, silver or formaldehyde may reveal different results suggesting that these treatments have different effects on the biofilm community depending on the composition and concentration of the disinfectants [97].

The biofilm can act as a catalyst in the oxidation–reduction process resulting in degradation of peroxides without necessarily damaging the micro-organisms within the biofilm [71]. Studies on the penetration of hydrogen peroxide into *P. aeruginosa* biofilms suggest that hydrogen peroxide is neutralized in the surface layers of the biofilm at a faster rate than it can diffuse into the biofilm interior [84]. This may be an explanation for the lower bactericidal efficacy of hydrogen peroxide on bacterial cells in biofilms. Surviving bacterial cells may regrow. Regrowth within 24 h has been described of a *B. cenocepacia* biofilm after treatments with 0.3–3% for 30 min indicating that *B. cenocepacia* biofilms are highly resistant to hydrogen peroxide [84].

6.3.1.4 Bactericidal Activity in Carrier Tests

The published data indicate an overall good and quick bactericidal activity of hydrogen peroxide in carrier tests at concentrations between 0.5 and 7% within short exposure time of 1–5 min (Table 6.6). According to ASTM E 2967, the efficacy of disinfectant wipes soaked with 0.5% accelerated hydrogen peroxide was determined on stainless steel carriers contaminated with *S. aureus* (ATCC 6538) or *A. baumannii* (ATCC 19568). With a 10 s wipe, the bacterial load was reduced by at least 7.0 log, and the control wipe without the disinfectant yielded a 3.0 log reduction. From none of the disinfected surfaces, a transfer of the test organism to another sterile surface was observed [103].

Table 6.6 Bactericidal activity of hydrogen peroxide in carrier tests

Species	Strains/isolates	Exposure time	Concentration	$\log_{10}$ reduction	References
Enterococcus spp.	3 VRE strains (2 vanA, 1 vanB)	3 min	0.64%[a] (P)	>5.4	[21]
P. aeruginosa	ATCC 15442	5 min	7% (S)	≥6.0	[102]
P. aeruginosa	ATCC 15442	5 min	2% (P)	7.2	[81]
S. choleraesuis	ATCC 10708	5 min	7% (S)	≥6.0	[102]
S. choleraesuis	ATCC 10708	5 min	2% (P)	6.9	[81]
S. aureus	ATCC 6538	5 min	7% (S)	≥6.0	[102]
S. aureus	ATCC 6538	5 min	2% (P)	6.7	[81]
S. aureus	ATCC 43300 and 2 clinical MRSA isolates	3 min	0.64%[a] (P)	>4.8	[21]
S. aureus	ATCC 6538	1 min	0.5% (P)	≥6.0	[82]

P commercial product; *S* solution; [a]plus peracetic acid at 0.14%

6.3.1.5 Bacterial Activity of Fumigation

In recent years, some manufacturers have offered vaporized hydrogen peroxide for disinfection of surfaces and rooms [100]. It has not replaced surface disinfection by wiping but may be useful in specific clinical situations such as terminal disinfection of a hospital room when the previous patient had been colonized or infected with MRSA, VRE, *Acinetobacter* spp. or *C. difficile* [100]. Based on a review from 2013, the overall bactericidal efficacy is good with a reduction of surfaces contaminated with MRSA, *Serratia* spp., *C. difficile* or Gram-negative bacterial species between 88 and 100% [100]. Some studies have in addition addressed the reduction of viable bacterial cells on surfaces by hydrogen peroxide vapour. Although the concentration of hydrogen peroxide and the exposure time are not described in all studies, the overall bactericidal effect is good in open rooms without any barriers (Table 6.7). One study described a bactericidal efficacy of commercially available hydrogen peroxide fumigation systems (Bioquell Q10 and Deprox) against a clinical ESBL *K. pneumoniae* isolate and a clinical EMRSA-15 isolate with 6.3 log reduction but did not mention the aerial concentration of the active agent or the exposure time so that it could not be included in Table 5.7 [2].

6.3.1.6 Bactericidal Activity in Other Applications

One study looked at the efficacy of a formulation based on 1% hydrogen peroxide on seven different dental instruments. *S. aureus*, *P. aeruginosa* and *S. marcescens* were reduced during immersion by 5.0 log within 1–60 min [3]. Hydrogen peroxide was somewhat effective with 1.1 log (1%) and 1.5 log (2%) against *E. coli* 0157:H7 on baby spinach when used for 5 min [47]. When hydrogen peroxide at 0.35% was applied with an atomizer to cement floor surfaces contaminated with isolates of *Salmonella* spp. for an exposure up to 60 min, only 1.2% of all *Salmonella* strains were eliminated [69]. Wipes based on 0.5% hydrogen peroxide (10 min application time) were effective to reduce *Y. pseudotuberculosis* (ATCC 6902) and

Table 6.7 Bactericidal activity of fumigated hydrogen peroxide on inanimate surfaces

Species	Strains/isolates	Exposure time	Aerial concentration	$\log_{10}$ reduction	References
A. baumannii	Multidrug-resistant clinical isolate	2.5 h	5% (P)	4.4–4.7	[89]
				2.9–3.8[a]	
A. baumannii	Multidrug-resistant clinical isolate	50–52 min	0.05–0.06% (P)	4.7–5.1	[61]
E. faecium	DSM 17050 (VRE)	50–52 min	0.05–0.06% (P)	4.0–4.1	[61]
E. coli O157:H7	ATCC 35150, ATCC 43889, ATCC 43890	60 min	0.5% (P)	≥2.0	[17]
		30 min	0.25% (P)	≥2.7	
L. monocytogenes	ATCC 7644, ATCC 19114, ATCC 19115	60 min	0.5% (P)	≥3.0	[17]
		30 min	0.25% (P)	≥2.0	
S. Typhimurium	ATCC 19586, ATCC 43174, DT104 Killercow	60 min	0.5% (P)	≥2.6	[17]
		30 min	0.25% (P)	≥2.8	
S. aureus	MRSA strain NCTC 8325	2.5 h	5% (P)	4.5–4.7	[89]
				1.5–3.5[a]	
S. aureus	ATCC 43300 (MRSA)	50–52 min	0.05–0.06% (P)	4.4–4.7	[61]

P Commercial product; [a]with a barrier such as a drawer or a covered petri dish

B. thailandensis (ATCC 700388) cells from a pulse oximeter sensor, and the efficacy against *S. aureus* (ATCC 6538) was somewhat lower [76]. Hydrogen peroxide at 3% was not effective enough within 1 min for disinfection of titanium implants contaminated with *S. sanguinis* or *S. epidermidis* [14].

6.3.2 Fungicidal Activity

6.3.2.1 Fungistatic Activity (MIC Values)

MIC values for *C. albicans*, *C. glabrata*, *C. krusei* and *C. tropicalis* were found between 0.1 and 4.5 mg/l (equivalent to 0.00001–0.00045%) [62]. With *C. albicans*, a MIC value of 0.0234% was described [31]. The MIC value for *P. expansum*, an apple isolate, was 0.05% [121].

6.3.2.2 Fungicidal Activity (Suspension Tests)

The fungicidal activity of 3% hydrogen peroxide is overall poor at exposure times up to 10 min. Even within 2–6 h, a log reduction ≥4.0 is not commonly found, not even against *C. albicans* (Table 6.8).

6.3.2.3 Activity Against Fungi in Biofilms

The activity of hydrogen peroxide against fungi in biofilms is lower compared to planktonic cells. One study shows that four *Candida* strains in biofilm (two

Table 6.8 Fungicidal activity of hydrogen peroxide in suspension tests

Species	Strains/isolates	Exposure time	Concentration	log_{10} reduction	References
A. fumigatus	ATCC 10894	2–6 h	3% (P)	0.3–2.1	[98]
C. albicans	ATCC 10231	30 s	3% (S)	0.1–0.4[a]	[90]
		1 min		0.1–0.2[a]	
		10 min		0.1–0.3[a]	
C. albicans	1 human and 1 environmental isolate	5 min	3% (S)	1.0	[115]
C. albicans	IFO 1594	30 min	3% (S)	≥4.0	[126]
C. albicans	ATCC 10231	2–6 h	3% (P)	3.3–4.3	[98]
C. neoformans	1 clinical isolate	5 min	3% (S)	0.7	[115]
C. uniguttulatus	1 clinical isolate	5 min	3% (S)	0.3	[115]
E. repens	Isolate from bread factory	10 min	3% (P)	0.0	[13]
F. solani	ATCC 36031	2–6 h	3% (P)	2.3–3.7	[98]
P. roqueforti	Isolate from bread	10 min	3% (P)	0.3	[13]
P. anomala	Isolate from bread	10 min	3% (P)	0.0	[13]
R. rubra	1 clinical isolate	5 min	3% (S)	0.6	[115]

P commercial product; *S* solution; [a]depending on the type of organic load

C. albicans, *C. parapsilosis*, *C. glabrata*) were 2–8 times less susceptible to hydrogen peroxide compared to planktonic cells of the same strain [78]. Killing *C. albicans* (strain SC 5314) and *C. parapsilosis* sensu *strictu* (5 strains) biofilm cells to a level below 10 CFU/ml was possible with hydrogen peroxide at 1.87% in up to 48 h. With five strains of *C. orthopsilosis*, it required a hydrogen peroxide concentration of 3.75% [88].

6.3.2.4 Fungicidal Activity in Carrier Tests

The fungicidal activity of hydrogen peroxide in carrier tests depends on the concentration, the fungal species and exposure time. With 0.5% hydrogen peroxide, a significant reduction was achieved in 5 min as shown with *A. fumigatus* and *T. mentagrophytes* (Table 6.9). When spores of *T. mentagrophytes*, however, were placed on a glass cup carrier and exposed to 3% hydrogen peroxide, a log reduction <1.0 is found after 1 and 10 min [8]. Against *C. albicans*, a log reduction >4.0 was found both on stainless steel carriers within a 1 min exposure time for hydrogen peroxide at 7.5%. *C. parapsilosis* and *C. tropicalis* were more resistant to hydrogen peroxide and required 10 min to yield a similar effect [119].

6.3.2.5 Bactericidal Activity in Other Applications

In swimming pool water, hydrogen peroxide at 0.015% did not show any efficacy within 30 min against *C. albicans* (0.0 log). A formulation with the same concentration of hydrogen peroxide and additional silver ions at 15 ppb did not improve the yeasticidal activity [10]. When *C. albicans* is allowed to adhere to soft

Table 6.9 Fungicidal activity of hydrogen peroxide in carrier tests

Species	Strains/isolates	Exposure time	Concentration	$\log_{10}$ reduction	References
A. fumigatus	ATCC 16404	1–20 min	0.27%[a] (P)	≤ 1.0	[25]
			0.54%[b] (P)	≥ 4.0[c]	
T. mentagrophytes	ATCC 9533	20 min	7% (S)	≥ 5.9	[21]
T. mentagrophytes	ATCC 9533	5 min	2% (P)	6.1	[81]
T. mentagrophytes	ATCC 9533	5 min	0.5% (P)	5.5	[82]

P commercial product; *S* solution [a]with additional peracetic acid at 0.045%; [b]with additional peracetic acid at 0.09%; [c]after 5 min

denture lining material for 2.5 h, immersion of the contaminated and carefully washed material in a solution of 3% hydrogen peroxide does not reduce the number of adherent cells significantly [12]. On seven different dental instruments, a formulation with 1% hydrogen peroxide reduced *C. albicans* during immersion by 5.0 log within 1–15 min after cleaning and within 1–40 min without cleaning [3]. Hydrogen peroxide at 3% was effective within 1 min for disinfection of titanium implants contaminated with *C. albicans* [14].

6.3.3 Mycobactericidal Activity

6.3.3.1 Mycobactericidal Activity (Suspension Tests)

Hydrogen peroxide at 0.5% showed sufficient activity (≥ 4.0 log) within 5 min against *M. bovis* and *M. terrae*. Against other mycobacteria, hydrogen peroxide at 3% was not sufficiently active within 60 min such as *M. avium-intracellulare*, *M. fortuitum* and *M. tuberculosis*. Even at 10%, the activity of hydrogen peroxide was not sufficient against glutaraldehyde-resistant *M. chelonae* isolates within 60 min (Table 6.10). One recent study described a broad mycobactericidal efficacy (*M. avium*, *M. abscessus*, *M. bovis*, *M. chelonae* and *M. terrae*) of a commercial hydrogen peroxide-based product (Resert XL HLD) with >5.0 log reduction in 5 min but did not mention the concentration of the active agent so that it cannot be included in Table 6.10 [15, 53].

6.3.3.2 Activity Against Mycobacteria in Biofilms

One study with *M. phlei* indicates that the susceptibility of cells grown in biofilm is lower to hydrogen peroxide (MBEC: > 0.25% in 30 min) compared to planktonic cells (MBC: 0.2% in 30 min) [7].

6.3.3.3 Mycobactericidal Activity in Carrier Tests

M. terrae was found to be rather susceptible against hydrogen peroxide, whereas a glutaraldehyde-resistant *M. chelonae* isolate required a concentration of 7% hydrogen peroxide to achieve ≥ 4.0 log in 4 min (Table 6.11).

Table 6.10 Mycobactericidal activity of hydrogen peroxide in suspension tests

Species	Strains/isolates	Exposure time	Concentration	log_{10} reduction	References
M. abscessus	ATCC 19977	60 min	10% (P)	5.5	[109]
M. avium-intracellulare	Clinical isolate	60 min	3% (P)	0.0–0.1	[41]
			1% (P)	0.0	
M. avium-intracellulare	6 fresh clinical isolates	15 min	1% (P)	1.0–1.5	[45]
		50 min		2.0	
M. bovis	ATCC 35743	60 min	10% (P)	>6.2	[109]
M. bovis	OT 451C150	5 min	0.5% (P)	≥6.8	[82]
M. chelonae	5 glutaraldehyde-resistant isolates	60 min	10% (P)	0.8–3.4	[109]
M. chelonae	NCTC 946	60 min	3% (P)	3.0–3.2	[41]
		20 min	1% (P)	>4.8	
M. chelonae	Strain Epping	60 min	3% (P)	0.0–0.2	[41]
			1% (P)	0.0–2.3	
M. chelonae	NCTC 946	1, 4, 10, 20 and 60 min	1% (P)	>4 after 10 min	[40]
M. chelonae	2 isolates from WD from different hospitals, UK	1, 4, 10, 20 and 60 min	1% (P)	0.0–0.2	[40]
M. fortuitum	NCTC 10394	60 min	3% (P)	0.1–0.4	[41]
			1% (P)	0.0–0.2	
M. terrae	ATCC 15755	5 min	0.5% (P)	≥6.4	[82]
M. tuberculosis	Strain H37Rv	60 min	3% (P)	0.6–1.7	[41]
			1% (P)	0.4–0.5	
M. tuberculosis	6 fresh clinical isolates	15 min	1% (P)	1.5–2.5	[45]
		50 min		2.5	
M. tuberculosis	15 isoniazid-sensitive catalase-positive clinical strains	90 min	0.02% (S)	0.0–3.0	[110]
	24 isoniazid-negative clinical strains			0.7–6.3[a]	

P commercial product; *S* solution; [a]depending on the catalase activity

Table 6.11 Mycobactericidal activity of hydrogen peroxide in carrier tests

Species	Strains/isolates	Exposure time (min)	Concentration	log_{10} reduction	References
M. chelonae	Glutaraldehyde-resistant isolate	4	7% (P)	≥4.0	[46]
M. terrae	ATCC 15755	25	7% (S)	≥6.4	[21]
M. terrae	ATCC 15755	5	2% (P)	6.5	[81]

P commercial product; *S* solution

6.3.3.4 Mycobactericidal Activity in Flexible Endoscopes

Only few data were found describing the mycobactericidal activity of hydrogen peroxide for disinfection of flexible endoscopes. In one study, five colonoscopes and five duodenoscopes were artificially contaminated with *M. chelonae* and immersed after cleaning in a formulation based on 7.5% hydrogen peroxide for 30 min. On average, 40 CFU were found per scope after processing which was described as sufficient to achieve high-level disinfection [34].

6.4 Effect of Low-Level Exposure

6.4.1 Bacteria

Some authors have looked at the effect of hydrogen peroxide low-level exposure on various bacterial species (Table 6.12).

The data by Soumet et al. suggest that most species do not respond with a lower susceptibility to hydrogen peroxide [108]. Most other authors described an induction of resistance to hydrogen peroxide (*E. coli*, *S. typhimurium*) or stannous acid (*E. coli*), and a reduction of virulence gene expression (*L. monocytogenes*). Another finding was described in *B. subtilis* cells. The transfer of the mobile genetic element Tn916, a conjugative transposon and the prototype of a large family of related elements, was not increased by exposure to 0.002% hydrogen peroxide for up to 2 h [104].

6.4.2 Yeasts

The yeast *S. cerevisiae* (strain CY 4) has been described to react to a 1 h exposure with 0.007% hydrogen peroxide with a reduced susceptibility against 0.07% hydrogen peroxide (10 h exposure) [38]. A cross-resistance to 20% ethanol was found when *S. cerevisiae* was exposed to hydrogen peroxide at hormetic concentrations (0.00017–0.0017%). The regulatory protein Yap1 played an important role in the hormetic effects by low concentrations of hydrogen peroxide [105]. Pretreatment of *S. cerevisiae* with 0.0007% hydrogen peroxide promoted an increase in catalase activity [32].

6.4.3 Mycobacteria

M. smegmatis was used to look at changes on a cellular level caused to low-level exposure to hydrogen peroxide. When exposed to 0.00068% hydrogen peroxide, expression of approximately 10% of the genes in the *M. smegmatis* genome was significantly changed. In contrast, 29.3% of *M. smegmatis* genes were significantly changed in response to 0.0238% hydrogen peroxide. Transcriptional analysis suggested that a metabolic switch in glycolysis/gluconeogenesis and fatty acid metabolism was potentially involved in the response to the 0.00068% hydrogen

Table 6.12 Effect of hydrogen peroxide low-level-exposure on various bacterial species

Species	Strains/isolates	Concentration and exposure time	Increase in MIC	MIC_{max} (mg/l)	Stability of MIC change	Associated changes	References
C. coli	16 strains from pig faeces or pork meat	"sublethal" for 7 d	None	No data	Not applicable	None described	[108]
E. faecalis	Strain 155600A	0.068% for 4 h	No data	No data	Not applicable	No change of susceptibility to 0.17% hydrogen peroxide in the presence of 100 mg/l indoor or outdoor dust	[112]
E. coli	54 strains from pig faeces or pork meat	"sublethal" for 7 d	None	No data	Not applicable	None described	[108]
E. coli	AB 1157	0.0002% for 20 min	No data	No data	Not applicable	Induction of resistance to stannous chloride at 0.166 mM	[5]
E. coli	K12	0.0001% for 1 h	Induction of resistance to 0.034% hydrogen peroxide	No data	No data	Induction of *oxyR*	[24]
E. coli	K12 (strain 155065A)	0.068% for 4 h	Increase of susceptibility to 0.17% hydrogen peroxide in the presence of 100 mg/l indoor or outdoor dust	No data	Not applicable	None described	[112]
E. coli	NCIMB 8545	0.001% for 30 s, 5 min and 24 h	≤ 2-fold	>0.3%	No data	Unstable resistance[a] to ampicillin	[123]
L. monocytogenes	Strain Scott A	0.05% for 1 h	Induction of resistance to 0.1% hydrogen peroxide	No data	No data	None described	[67]
L. monocytogenes	31 strains from pig faeces or pork meat	"sublethal" for 7 d	None	No data	Not applicable	None described	[108]

(continued)

Table 6.12 (continued)

Species	Strains/isolates	Concentration and exposure time	Increase in MIC	MIC_{max} (mg/l)	Stability of MIC change	Associated changes	References
L. monocytogenes	Strain EGD	"sublethal" for 48 h	None	No data	Not applicable	Reduction of virulence gene expression	[54]
P. aeruginosa	Strain 155250A	0.068% for 4 h	No change of susceptibility to 0.17% hydrogen peroxide in the presence of 100 mg/l indoor or outdoor dust	No data	Not applicable	None described	[112]
S. enterica	35 strains from pig faeces or pork meat	"sublethal" for 7 d	None	No data	Not applicable	None described	[108]
S. Typhimurium	Strain LT2	0.0002% for 60 min	Resistant to killing by 0.03% hydrogen peroxide[b]	No data	No data	Cells pretreated with 60 μM hydrogen peroxide in the presence of chloramphenicol did not acquire the resistance indicating a requirement for de novo protein synthesis for the adaptation	[18]
S. aureus	NCIMB 9518	0.001% for 30 s, 5 min and 24 h	None	0.13%	No data	Unstable resistance[a] to ciprofloxacin	[123]

[a]Disk diffusion test; [b]4-fold to 5-fold increase of catalyse activity

peroxide treatment but not to the 0.0238% hydrogen peroxide treatment. It was also observed that transcriptional levels of genes encoding ribosomes decreased when bacterial cells were treated with 0.0238% hydrogen peroxide. This result suggests that 0.0238% hydrogen peroxide treatment affected the protein synthesis apparatus and thus reduced protein synthesis, resulting in reduced bacterial growth [63].

6.5 Resistance to Hydrogen Peroxide

6.5.1 Species with Resistance to Hydrogen Peroxide

Some strains of the oral cavity bacterial species *A. actinomycetemcomitans* were described to have a low susceptibility to 0.0034% hydrogen peroxide (1 h exposure) which was not explained by catalase activity [73]. An spacecraft-associated *Acinetobacter* spp. named gyllenbergii 2P01AA was isolated with very high catalase-specific activities resulting in no viable cell loss in 0.34% hydrogen peroxide for 1 h [20].

6.5.2 Resistance Mechanisms

Hydrogen peroxide is degraded by peroxidases and catalases, the latter being able both to reduce hydrogen peroxide to water and to oxidize it to molecular oxygen. The catalase–peroxidase family of enzymes is involved in removing hydrogen peroxide. They are bifunctional enzymes; capable of either reducing hydrogen peroxide with an external reductant (peroxidase activity) or disproportionating it to water and oxygen (catalase activity). Nature has evolved three protein families that are able to catalyse this dismutation at reasonable rates. Two of the protein families are heme enzymes: typical catalases and catalase–peroxidizes [127]. It has been proposed that a catalase–peroxidase gene was originally transferred from an archaeon to a pathogenic bacterium, either directly or through an intermediate with more frequent physical contact with Archaea. The presence of two dissimilar catalase–peroxidases in *E. coli* and *L. pneumophila* strongly suggest they were on the receiving end of a lateral transfer [30]. Typical catalases comprise the most abundant group found in Eubacteria, Archaebacteria, Protista, Fungi, Plantae, and Animalia, whereas catalase–peroxidizes are not found in plants and animals and exhibit both catalatic and peroxidatic activities. The third group is a minor bacterial protein family with a dimanganese active site called manganese catalases. Although catalysing the same reaction, the three groups differ significantly in their overall and active-site architecture and the mechanism of reaction [127]. In *S. aureus*, hydrogen peroxide resistance can be partly explained by the high catalase activity in the dead cell fraction (at least 90%) of a decline phase cell suspension compared to rather susceptible cells obtained in the stationary phase [68].

KatG has raised considerable interest, because it represents the only peroxidase with a reasonably high catalatic activity around neutral pH, besides a usual peroxidase activity [127]. katG genes are distributed in approximately 40% of bacterial genomes [127].

6.5.3 Resistance Genes

The relevance of selected resistance genes for tolerance to hydrogen peroxide is summarized in Table 6.13. Most of the resistance genes are directly involved in hydrogen peroxide tolerance. Some are expressed 2-fold to 3.8-fold higher in biofilm cells which is a possible explanation for the reduced bactericidal efficacy of hydrogen peroxide against bacteria in biofilms.

Table 6.13 Examples of resistance genes and their impact on tolerance to hydrogen peroxide

Resistance gene	Species	Relevance	References
katA	*Serratia* sp. LCN16	Directly involved in the high tolerance to hydrogen peroxide	[122]
	B. subtilis	Directly involved in the high tolerance to hydrogen peroxide	[27]
	P. aeruginosa	Directly involved in the high tolerance to hydrogen peroxide, both in planktonic and biofilm cells; involved in UVA tolerance	[55, 87]
katE	*A. baumannii*, *A. nosocomialis*	Primary catalase responsible for hydrogen peroxide degradation in stationary-phase bacteria	[111]
	B. longum	Improvement of tolerance to hydrogen peroxide	[43]
katG	*A. baumannii*, *A. nosocomialis*, *V. cholerae*, *X. citri subsp. citri*	Predominant role in the resistance to hydrogen peroxide	[37, 111, 117, 124]
	X. citri subsp. citri	Impaired development of biofilm structures	[117]
kat1	*C. albicans*	Expression 3.8-fold higher in biofilm cells	[62]
oxyR	*Serratia* sp. LCN16	Directly involved in the high tolerance to hydrogen peroxide	[122]
	L. monocytogenes	Directly involved in tolerance to hydrogen peroxide	[18]
	E. coli	Protects cells against endogenous hydrogen peroxide; but virtually all of the oxidase-generated hydrogen peroxide will diffuse across the outer membrane and be lost to the external world, rather than enter the cytoplasm where hydrogen peroxide sensitive enzymes are located.	[93]
	P. chlororaphis	Regulates multiple pathways to enhance the survival of *P. chlororaphis* GP72 exposed to different oxidative stresses	[125]
sod1	*C. albicans*	Expression 2-fold higher in biofilm cells	[62]

6.6 Cross-Tolerance to Other Biocidal Agents

In *E. coli*, a cross-tolerance to hypochlorous acid has been reported after low-level exposure to hydrogen peroxide [24]. Hydrogen peroxide (2 mg/l for 30 min) has also the capacity to induce a function which reduces the killing effects of aldehydes (formaldehyde at 6 mM and glutaraldehyde at 0.1 mM) in *E. coli* WP2 cells (cross-adaptive response). The function is controlled by the recA gene without involvement of an SOS response [80]. And in *S. Typhimurium*, a cross-resistance to other agents (N-ethylmaleimide, 1-chloro-2,4-dinitrobenzene and menadione) and heat (50 °C) has been reported after low-level exposure to hydrogen peroxide [18]. No other cross-resistance to other biocidal agents has so far been described.

6.7 Cross-Resistances to Antibiotics

So far, no cross-tolerance between hydrogen peroxide and antibiotics has been described.

6.8 Role of Biofilm

6.8.1 Effect on Biofilm Development

Some studies indicate that hydrogen peroxide inhibits biofilm formation. For example, *C. albicans* (strain SC 5314) biofilm formation in microtiter wells was inhibited by 470 mg/l hydrogen peroxide, *C. orthopsilosis* (five strains) biofilm by 930 mg/l and *C. parapsilosis* sensu *strictu* (five strains) by 470 mg/l [88]. In *S. epidermidis*, exposure to 0.034, 0.017, 0.125 and 0.25% hydrogen peroxide reduced biofilm formation significantly [35], whereas exposure to 1% hydrogen peroxide increased biofilm formation in a non-adherent *S. epidermidis* strain [16].

Other studies indicate that hydrogen peroxide promotes biofilm formation. Exogenous addition of hydrogen peroxide promoted biofilm formation in *A. oleivorans* wild-type cells, which suggested that biofilm development is linked to defence against hydrogen peroxide [51]. In *P. aeruginosa,* sublethal concentrations of hydrogen peroxide stimulated biofilm formation [91]. Hydrogen peroxide produced at low levels by the periodontal pathogen *A. actinomycetemcomitans* enhances biofilm formation by *S. parasanguinis* [23]. When 35% hydrogen peroxide was applied to enamel specimen (total exposure for 3 × 8 min), *S. sanguinis* biofilm formation was promoted, but was reduced when 25% hydrogen peroxide was used for bleaching. No difference in *S. mutans* biofilm formation was observed [50].

Biofilm formation was not prevented in potable water distribution systems by hydrogen peroxide at 16.5 mg/l. Biofilm formation was related to the depletion of residual disinfectant concentration [74].

6.8.2 Effect on Biofilm Removal

Biofilm can partially be removed by hydrogen peroxide (Table 6.14). In order to remove at least half of the biofilm mass, a concentration of 3% hydrogen peroxide seems necessary. With *S. aureus*, it was shown that biofilm removal in 5 min is higher with 5% hydrogen peroxide compared to 2.5 or 1.25% (all higher than control) [118]. No biofilm disruption was found with *X. citri subsp. citri*, causing citrus bacterial canker, on borosilicate plates or lemon leaves, by 30 min treatment with products based on unknown concentrations of hydrogen peroxide plus silver and hydrogen peroxide plus peracetic acid [94].

Table 6.14 Biofilm removal rate by exposure to hydrogen peroxide

Type of biofilm	Concentration	Exposure time	Biofilm removal rate	References
B. cenocepacia LMG 18828, 4 h adhesion and 20-h incubation in polystyrene microtitre plates	3% (S)	2–10 min	55%	[84]
	1% (S)		37%	
	0.5% (S)		45%	
	0.3% (S)		<10%	
P. aeruginosa ATCC 700928, 24-h incubation in microplates	5% (S)	1 min	68%	[118]
		5 min	75%	
		60 min	85%	
P. aeruginosa strain PA01, 24-h incubation on 96 well plates	5% (S)	1 min	51%	[57]
		5 min	33%	
		15 min	28%	
		30 min	35%	
		60 min	31%	
P. aeruginosa ATCC 700928, 24-h incubation in microplates	5% (S)	1 min	68%	[118]
		60 min	85%	
S. aureus (MRSA) isolate, 18-h incubation in polystyrene plates	7%[a] (P)	30 min	>95%	[4]
S. aureus ATCC 6538, 72-h incubation in microplates	5% (S)	1 min	89%	[118]
		5 min	85%	
		60 min	84%	
S. aureus ATCC 6538, 24-h incubation on 96 well plates	5% (S)	1 min	70%	[57]
		5 min	77%	
		15 min	79%	
		30 min	79%	
		60 min	80%	

(continued)

Table 6.14 (continued)

Type of biofilm	Concentration	Exposure time	Biofilm removal rate	References
S. aureus ATCC 6538, 72-h incubation in microplates	5% (S)	1 min	89%	[118]
		60 min	84%	
S. epidermidis (30 clinical isolates), 24-h incubation in polystyrene microtiter plates	5% (S)	1 min	69%	[92]
	3% (S)		63%	
	0.5% (S)		22%	
S. mutans C180-2, 24-h incubation on titanium discs	10% (P)	5 min	38%	[79]
Mixed biofilm with *E. faecalis* ATCC 29212 and *P. aeruginosa* ATCC 15442, 8-d incubation in polystyrene pegs	2% (P)	5 min	Protein removal: 48–69%[b]	[19]
			Carbohydrate removal: 88–99%[b]	
Various species in a natural biofilm from dental unit waterlines	35% (S)	2 d	"no biofilm removal"	[64]
Natural mature biofilm from dental unit waterlines (10–14 years old)	7% (S)	24 h	"noticable biofilm removal"	[66]
Natural mature biofilm from dental unit waterlines (10–14 years old)	3% (S)	24 h	"noticable biofilm removal"	[66]
Natural mature biofilm from dental unit waterlines (10–14 years old)	2% (S)	24 h	"minimal biofilm disruption"	[66]

S solution; *P* commercial product; [a]plus 0.2% peracetic acid; [b]depending on the type of cleaner used before hydrogen peroxide exposure

6.8.3 Effect on Biofilm Fixation

No studies were found to evaluate a fixation potential of hydrogen peroxide on biofilms.

6.9 Summary

The principal antimicrobial activity of hydrogen peroxide is summarized in Table 6.15.

The key findings on acquired resistance and cross-resistance including the role of biofilm in selecting resistant isolates are summarized in Table 6.16.

Table 6.15 Overview on the typical exposure times required for hydrogen peroxide to achieve sufficient biocidal activity in suspension tests against the different target micro-organisms

Target micro-organisms	Species	Concentration (%)	Exposure time
Bacteria	Most clinically relevant species including antibiotic-resistant isolates	0.5	30 min[a]
Fungi	*C. albicans*	3	30 min
	Other fungi	>3	>6 h
Mycobacteria	*M. bovis*, *M. terrae*	0.5	5 min
	M. avium, *M. fortuitum*, *M. tuberculosis*	>3	>60 min
	M. chelonae (some glutaraldehyde-resistant isolates)	>10	>60 min

[a]In biofilm there may be no sufficient efficacy in 60 min depending on the species and the type of biofilm

Table 6.16 Key findings on hydrogen peroxide resistance, the effect of low-level exposure, cross-tolerance to other biocides and antibiotics, and its effect on biofilm

Parameter	Species	Findings
Elevated MIC values	So far not reported.	
MIC value to determine resistance	Not proposed yet for bacteria, fungi or mycobacteria	
Cross-tolerance biocides	*E. coli*	Hypochlorous acid and aldehydes (after low-level exposure)
	S. Typhimurium	N-ethylmaleimide, 1-chloro-2,4-dinitrobenzene and menadione
Cross-tolerance antibiotics	So far not reported.	
Effect of low-level exposure	*C. coli*, *E. faecalis*, *P. aeruginosa*, *S. enterica*, *S. aureus*	No MIC increase
	E. coli, *L. monocytogenes*, *S. cerevisiae*	Weak MIC increase (≤ 4-fold)
	None	Strong (>4-fold) MIC increase
	S. Typhimuriun, *S. cerevisiae*	Increase of catalase activity
	E. coli	Cross-tolerance to stannous chloride
	S. cerevisiae	Cross-tolerance to ethanol
	L. monocytogenes	Reduction of virulence gene expression
	B. subtilis	No increase of transposon transfer
Specific resistance mechanism	Peroxidases and catalases encoded by various genes	
Biofilm	Development	Enhancement in *A. oleivorans*, *P. aeruginosa*, *S. epidermidis* and *S. parasanguinis*
		No effect in *S. mutans*
		Inhibition in *Candida* spp. and *S. epidermidis*
	Removal	Variable between 28 and 89%
	Fixation	Unknown

References

1. Aarestrup FM, Hasman H (2004) Susceptibility of different bacterial species isolated from food animals to copper sulphate, zinc chloride and antimicrobial substances used for disinfection. Vet Microbiol 100(1–2):83–89. https://doi.org/10.1016/j.vetmic.2004.01.013
2. Ali S, Muzslay M, Bruce M, Jeanes A, Moore G, Wilson AP (2016) Efficacy of two hydrogen peroxide vapour aerial decontamination systems for enhanced disinfection of meticillin-resistant Staphylococcus aureus, Klebsiella pneumoniae and Clostridium difficile in single isolation rooms. J Hosp Infect 93(1):70–77. https://doi.org/10.1016/j.jhin.2016.01.016
3. Angelillo IF, Bianco A, Nobile CG, Pavia M (1998) Evaluation of the efficacy of glutaraldehyde and peroxygen for disinfection of dental instruments. Lett Appl Microbiol 27 (5):292–296
4. Aparecida Guimaraes M, Rocchetto Coelho L, Rodrigues Souza R, Ferreira-Carvalho BT, Marie Sa Figueiredo A (2012) Impact of biocides on biofilm formation by methicillin-resistant Staphylococcus aureus (ST239-SCCmecIII) isolates. Microbiol Immunol 56(3):203–207. https://doi.org/10.1111/j.1348-0421.2011.00423.x
5. Assis ML, De Mattos JC, Caceres MR, Dantas FJ, Asad LM, Asad NR, Bezerra RJ, Caldeira-de-Araujo A, Bernardo-Filho M (2002) Adaptive response to H(2)O(2) protects against SnCl(2) damage: the OxyR system involvement. Biochimie 84(4):291–294
6. Ban GH, Kang DH (2016) Effect of sanitizer combined with steam heating on the inactivation of foodborne pathogens in a biofilm on stainless steel. Food Microbiol 55:47–54. https://doi.org/10.1016/j.fm.2015.11.003
7. Bardouniotis E, Huddleston W, Ceri H, Olson ME (2001) Characterization of biofilm growth and biocide susceptibility testing of Mycobacterium phlei using the MBEC assay system. FEMS Microbiol Lett 203(2):263–267
8. Best M, Springthorpe VS, Sattar SA (1994) Feasibility of a combined carrier test for disinfectants: studies with a mixture of five types of microorganisms. Am J Infect Control 22 (3):152–162
9. Boles BR, Thoendel M, Singh PK (2004) Self-generated diversity produces "insurance effects" in biofilm communities. Proc Natl Acad Sci USA 101(47):16630–16635. https://doi.org/10.1073/pnas.0407460101
10. Borgmann-Strahsen R (2003) Comparative assessment of di#erent biocides in swimming pool water. Int Biodeter Biodegr 51(4):291–297
11. Brinez WJ, Roig-Sagués AX, Hernández Herrero MM, López-Pedemonte T, Guamis B (2006) Bactericidal efficacy of peracetic acid in combination with hydrogen peroxide against pathogenic and non pathogenic strains of Staphylococcus spp., Listeria spp. and Escherichia coli. Food Control 17(7):516–521
12. Buergers R, Rosentritt M, Schneider-Brachert W, Behr M, Handel G, Hahnel S (2008) Efficacy of denture disinfection methods in controlling Candida albicans colonization in vitro. Acta Odontol Scand 66(3):174–180. https://doi.org/10.1080/00016350802165614
13. Bundgaard-Nielsen K, Nielsen PV (1996) Fungicidal effect of 15 disinfectants against 25 fungal contaminants commonly found in bread and cheese manufacturing. J Food Prot 59 (3):268–275
14. Burgers R, Witecy C, Hahnel S, Gosau M (2012) The effect of various topical peri-implantitis antiseptics on Staphylococcus epidermidis, Candida albicans, and Streptococcus sanguinis. Arch Oral Biol 57(7):940–947. https://doi.org/10.1016/j.archoralbio.2012.01.015
15. Burgess W, Margolis A, Gibbs S, Duarte RS, Jackson M (2017) Disinfectant Susceptibility Profiling of Glutaraldehyde-Resistant Nontuberculous Mycobacteria. Infect Control Hosp Epidemiol 38(7):784–791. https://doi.org/10.1017/ice.2017.75
16. Chaieb K, Zmantar T, Souiden Y, Mahdouani K, Bakhrouf A (2011) XTT assay for evaluating the effect of alcohols, hydrogen peroxide and benzalkonium chloride on biofilm

formation of Staphylococcus epidermidis. Microb Pathog 50(1):1–5. https://doi.org/10.1016/j.micpath.2010.11.004
17. Choi NY, Baek SY, Yoon JH, Choi MR, Kang DH, Lee SY (2012) Efficacy of aerosolized hydrogen peroxide-based sanitizer on the reduction of pathogenic bacteria on a stainless steel surface. Food Control 27(1):57–63
18. Christman MF, Morgan RW, Jacobson FS, Ames BN (1985) Positive control of a regulon for defenses against oxidative stress and some heat-shock proteins in Salmonella typhimurium. Cell 41(3):753–762
19. da Costa Luciano C, Olson N, Tipple AF, Alfa M (2016) Evaluation of the ability of different detergents and disinfectants to remove and kill organisms in traditional biofilm. Am J Infect Control 44(11):e243–e249. https://doi.org/10.1016/j.ajic.2016.03.040
20. Derecho I, McCoy KB, Vaishampayan P, Venkateswaran K, Mogul R (2014) Characterization of hydrogen peroxide-resistant Acinetobacter species isolated during the Mars Phoenix spacecraft assembly. Astrobiology 14(10):837–847. https://doi.org/10.1089/ast.2014.1193
21. Deshpande A, Mana TS, Cadnum JL, Jencson AC, Sitzlar B, Fertelli D, Hurless K, Kundrapu S, Sunkesula VC, Donskey CJ (2014) Evaluation of a sporicidal peracetic acid/hydrogen peroxide-based daily disinfectant cleaner. Infect Control Hosp Epidemiol 35 (11):1414–1416. https://doi.org/10.1086/678416
22. Dominciano LCC, Oliveira CAF, Lee SH, Corassin CH (2016) Individual and combined antimicrobial activity of Oleuropein and chemical sanitizers. J Food Chem Nanotechnol 2 (3):124–127
23. Duan D, Scoffield JA, Zhou X, Wu H (2016) Fine-tuned production of hydrogen peroxide promotes biofilm formation of Streptococcus parasanguinis by a pathogenic cohabitant Aggregatibacter actinomycetemcomitans. Environ Microbiol 18(11):4023–4036. https://doi.org/10.1111/1462-2920.13425
24. Dukan S, Touati D (1996) Hypochlorous acid stress in Escherichia coli: resistance, DNA damage, and comparison with hydrogen peroxide stress. J Bacteriol 178(21):6145–6150
25. Eissa ME, Abd El Naby M, Beshir MM (2014) Bacterial versus fungal spore resistance to peroxygen biocide on inanimate surfaces. Bull Facult Pharm 52 (2):219–224. doi:https://doi.org/10.1016/j.bfopcu.2014.06.003
26. El-Azizi M, Farag N, Khardori N (2016) Efficacy of selected biocides in the decontamination of common nosocomial bacterial pathogens in biofilm and planktonic forms. Comp Immunol Microbiol Infect Dis 47:60–71. https://doi.org/10.1016/j.cimid.2016.06.002
27. Engelmann S, Hecker M (1996) Impaired oxidative stress resistance of Bacillus subtilis sigB mutants and the role of katA and katE. FEMS Microbiol Lett 145(1):63–69
28. European Chemicals Agency (ECHA) Hydrogen peroxide. Substance information. https://echa.europa.eu/substance-information/-/substanceinfo/100.028.878. Accessed 12 Oct 2017
29. Exner M, Tuschewitzki GJ, Scharnagel J (1987) Influence of biofilms by chemical disinfectants and mechanical cleaning. Zentralbl Bakteriol Mikrobiol Hyg B 183(5–6):549–563
30. Faguy DM, Doolittle WF (2000) Horizontal transfer of catalase-peroxidase genes between archaea and pathogenic bacteria. Trends Genet: TIG 16(5):196–197
31. Ferguson JW, Hatton JF, Gillespie MJ (2002) Effectiveness of intracanal irrigants and medications against the yeast Candida albicans. J Endod 28(2):68–71. https://doi.org/10.1097/00004770-200202000-00004
32. Fernandes PN, Mannarino SC, Silva CG, Pereira MD, Panek AD, Eleutherio EC (2007) Oxidative stress response in eukaryotes: effect of glutathione, superoxide dismutase and catalase on adaptation to peroxide and menadione stresses in Saccharomyces cerevisiae. Redox Rep: Commun Free Radical Res 12(5):236–244. https://doi.org/10.1179/135100007x200344
33. Finland (2015) Assessment report. Hydrogen peroxide. Product-types 1–6

34. Foliente RL, Kovacs BJ, Aprecio RM, Bains HJ, Kettering JD, Chen YK (2001) Efficacy of high-level disinfectants for reprocessing GI endoscopes in simulated-use testing. Gastrointest Endosc 53(4):456–462. https://doi.org/10.1067/mge.2001.113380
35. Glynn AA, O'Donnell ST, Molony DC, Sheehan E, McCormack DJ, O'Gara JP (2009) Hydrogen peroxide induced repression of icaADBC transcription and biofilm development in Staphylococcus epidermidis. J Orthop Res: Off Publ Orthop Res Soc 27(5):627–630. https://doi.org/10.1002/jor.20758
36. Gosau M, Hahnel S, Schwarz F, Gerlach T, Reichert TE, Burgers R (2010) Effect of six different peri-implantitis disinfection methods on in vivo human oral biofilm. Clin Oral Implant Res 21(8):866–872. https://doi.org/10.1111/j.1600-0501.2009.01908.x
37. Goulart CL, Barbosa LC, Bisch PM, von Kruger WM (2016) Catalases and PhoB/PhoR system independently contribute to oxidative stress resistance in Vibrio cholerae O1. Microbiology (Reading, England) 162(11):1955–1962. https://doi.org/10.1099/mic.0.000364
38. Grant CM, MacIver FH, Dawes IW (1997) Mitochondrial function is required for resistance to oxidative stress in the yeast Saccharomyces cerevisiae. FEBS Lett 410(2–3):219–222
39. Grasteau A, Guiraud T, Daniel P, Calvez S, Chesneau V, Le Hénaff M (2015) Evaluation of Glutaraldehyde, Chloramine-T, Bronopol, Incimaxx Aquatic® and Hydrogen Peroxide as Biocides against Flavobacterium psychrophilum for Sanitization of Rainbow Trout Eyed Eggs. J Aquac Res Dev 6(12):382
40. Griffiths PA, Babb JR, Bradley CR, Fraise AP (1997) Glutaraldehyde-resistant Mycobacterium chelonae from endoscope washer disinfectors. J Appl Microbiol 82(4):519–526
41. Griffiths PA, Babb JR, Fraise AP (1999) Mycobactericidal activity of selected disinfectants using a quantitative suspension test. J Hosp Infect 41(2):111–121
42. Gutierrez-Martin CB, Yubero S, Martinez S, Frandoloso R, Rodriguez-Ferri EF (2011) Evaluation of efficacy of several disinfectants against Campylobacter jejuni strains by a suspension test. Res Vet Sci 91(3):e44–e47. https://doi.org/10.1016/j.rvsc.2011.01.020
43. He J, Sakaguchi K, Suzuki T (2012) Acquired tolerance to oxidative stress in Bifidobacterium longum 105-A via expression of a catalase gene. Appl Environ Microbiol 78 (8):2988–2990. https://doi.org/10.1128/aem.07093-11
44. Holmberg K, Hallander HO (1973) Production of bactericidal concentrations of hydrogen peroxide by Streptococcus sanguis. Arch Oral Biol 18(3):423–434
45. Holton J, Nye P, McDonald V (1994) Efficacy of selected disinfectants against mycobacteria and cryptosporidia. J Hosp Infect 27(2):105–115
46. Howie R, Alfa MJ, Coombs K (2008) Survival of enveloped and non-enveloped viruses on surfaces compared with other micro-organisms and impact of suboptimal disinfectant exposure. J Hosp Infect 69(4):368–376. https://doi.org/10.1016/j.jhin.2008.04.024
47. Huang Y, Chen H (2011) Effect of organic acids, hydrogen peroxide and mild heat on inactivation of Escherichia coli O157:H7 on baby spinach. Food Control 22(8):1178–1183
48. Hyslop PA, Hinshaw DB, Scraufstatter IU, Cochrane CG, Kunz S, Vosbeck K (1995) Hydrogen peroxide as a potent bacteriostatic antibiotic: implications for host defense. Free Radic Biol Med 19(1):31–37
49. Iniguez-Moreno M, Avila-Novoa MG, Iniguez-Moreno E, Guerrero-Medina PJ, Gutierrez Lomeli M (2017) Antimicrobial activity of disinfectants commonly used in the food industry in Mexico. J Glob Antimicrob Res 10:143–147. https://doi.org/10.1016/j.jgar.2017.05.013
50. Ittatirut S, Matangkasombut O, Thanyasrisung P (2014) In-office bleaching gel with 35% hydrogen peroxide enhanced biofilm formation of early colonizing streptococci on human enamel. J Dent 42(11):1480–1486. https://doi.org/10.1016/j.jdent.2014.08.003
51. Jang IA, Kim J, Park W (2016) Endogenous hydrogen peroxide increases biofilm formation by inducing exopolysaccharide production in Acinetobacter oleivorans DR1. Sci Rep 6:21121. https://doi.org/10.1038/srep21121

52. Juncker JC (2015) COMMISSION IMPLEMENTING REGULATION (EU) 2015/1730 of 28 September 2015 approving hydrogen peroxide as an existing active substance for use in biocidal products for product-types 1, 2, 3, 4, 5 and 6. Off J Eur Union 58(L 252):27–32
53. Kampf G (2017) Black Box Oxidizers. Infect Control Hosp Epidemiol 38(11):1387–1388. https://doi.org/10.1017/ice.2017.199
54. Kastbjerg VG, Larsen MH, Gram L, Ingmer H (2010) Influence of sublethal concentrations of common disinfectants on expression of virulence genes in Listeria monocytogenes. Appl Environ Microbiol 76(1):303–309. https://doi.org/10.1128/aem.00925-09
55. Khakimova M, Ahlgren HG, Harrison JJ, English AM, Nguyen D (2013) The stringent response controls catalases in Pseudomonas aeruginosa and is required for hydrogen peroxide and antibiotic tolerance. J Bacteriol 195(9):2011–2020. https://doi.org/10.1128/jb.02061-12
56. Koban I, Geisel MH, Holtfreter B, Jablonowski L, Hubner NO, Matthes R, Masur K, Weltmann KD, Kramer A, Kocher T (2013) Synergistic effects of nonthermal plasma and disinfecting agents against dental biofilms in vitro. ISRN Dent 2013:573262. https://doi.org/10.1155/2013/573262
57. Köse H, Yapar N (2017) The comparison of various disinfectants' efficacy on Staphylococcus aureus and Pseudomonas aeruginosa biofilm layers. Turk J Med Sci 47(4):1287–1294
58. Krolasik J, Zakowska Z, Krepska M, Klimek L (2010) Resistance of bacterial biofilms formed on stainless steel surface to disinfecting agent. Pol J Microbiol 59(4):281–287
59. Lagace L, Jacques M, Mafu AA, Roy D (2006) Biofilm formation and biocides sensitivity of Pseudomonas marginalis isolated from a maple sap collection system. J Food Prot 69 (10):2411–2416
60. Leitch CS, Leitch AE, Tidman MJ (2015) Quantitative evaluation of dermatological antiseptics. Clin Exp Dermatol 40(8):912–915. https://doi.org/10.1111/ced.12745
61. Lemmen S, Scheithauer S, Hafner H, Yezli S, Mohr M, Otter JA (2015) Evaluation of hydrogen peroxide vapor for the inactivation of nosocomial pathogens on porous and nonporous surfaces. Am J Infect Control 43(1):82–85. https://doi.org/10.1016/j.ajic.2014.10.007
62. Leung CY, Chan YC, Samaranayake LP, Seneviratne CJ (2012) Biocide resistance of Candida and Escherichia coli biofilms is associated with higher antioxidative capacities. J Hosp Infect 81(2):79–86. https://doi.org/10.1016/j.jhin.2011.09.014
63. Li X, Wu J, Han J, Hu Y, Mi K (2015) Distinct Responses of Mycobacterium smegmatis to Exposure to Low and High Levels of Hydrogen Peroxide. PLoS ONE 10(7):e0134595. https://doi.org/10.1371/journal.pone.0134595
64. Liaqat I, Sabri AN (2008) Effect of biocides on biofilm bacteria from dental unit water lines. Curr Microbiol 56(6):619–624. https://doi.org/10.1007/s00284-008-9136-6
65. Lin F, Xu Y, Chang Y, Liu C, Jia X, Ling B (2017) Molecular characterization of reduced susceptibility to biocides in clinical isolates of Acinetobacter baumannii. Front Microbiol 8:1836. https://doi.org/10.3389/fmicb.2017.01836
66. Lin SM, Svoboda KK, Giletto A, Seibert J, Puttaiah R (2011) Effects of hydrogen peroxide on dental unit biofilms and treatment water contamination. Eur J Dent 5(1):47–59
67. Lou Y, Yousef AE (1997) Adaptation to sublethal environmental stresses protects Listeria monocytogenes against lethal preservation factors. Appl Environ Microbiol 63(4):1252–1255
68. Luppens SB, Rombouts FM, Abee T (2002) The effect of the growth phase of Staphylococcus aureus on resistance to disinfectants in a suspension test. J Food Prot 65 (1):124–129
69. Marin C, Hernandiz A, Lainez M (2009) Biofilm development capacity of Salmonella strains isolated in poultry risk factors and their resistance against disinfectants. Poult Sci 88(2):424–431. https://doi.org/10.3382/ps.2008-00241

70. Mariscal A, Carnero-Varo M, Gutierrez-Bedmar M, Garcia-Rodriguez A, Fernandez-Crehuet J (2007) A fluorescent method for assessing the antimicrobial efficacy of disinfectant against Escherichia coli ATCC 35218 biofilm. Appl Microbiol Biotechnol 77 (1):233–240. https://doi.org/10.1007/s00253-007-1137-z
71. Martienssen M (2000) Simultaneous catalytic detoxification and biodegradation of organic peroxides during the biofilm process. Water Res 34(16):3917–3926
72. Martin H, Maris P (2012) Synergism between hydrogen peroxide and seventeen acids against six bacterial strains. J Appl Microbiol 113(3):578–590. https://doi.org/10.1111/j.1365-2672.2012.05364.x
73. Miyasaki KT, Wilson ME, Zambon JJ, Genco RJ (1985) Influence of endogenous catalase activity on the sensitivity of the oral bacterium Actinobacillus actinomycetemcomitans and the oral haemophili to the bactericidal properties of hydrogen peroxide. Arch Oral Biol 30 (11–12):843–848
74. Momba MNB, Cloete TE, Venter SN, Kfir R (1998) Evaluation of the impact of disinfection processes on the formation of biofilms in potable surface water distribution systems. Water Sci Technol 38(8):283–289. https://doi.org/10.1016/S0273-1223(98)00703-3
75. Nakamura K, Shirato M, Kanno T, Ortengren U, Lingstrom P, Niwano Y (2016) Antimicrobial activity of hydroxyl radicals generated by hydrogen peroxide photolysis against Streptococcus mutans biofilm. Int J Antimicrob Agents 48(4):373–380. https://doi.org/10.1016/j.ijantimicag.2016.06.007
76. Nandy P, Lucas AD, Gonzalez EA, Hitchins VM (2016) Efficacy of commercially available wipes for disinfection of pulse oximeter sensors. Am J Infect Control 44(3):304–310. https://doi.org/10.1016/j.ajic.2015.09.028
77. National Center for Biotechnology Information Hydrogen peroxide. PubChem Compound Database; CID=784. https://pubchem.ncbi.nlm.nih.gov/compound/784. Accessed 7 Oct 2017
78. Nett JE, Guite KM, Ringeisen A, Holoyda KA, Andes DR (2008) Reduced biocide susceptibility in Candida albicans biofilms. Antimicrob Agents Chemother 52(9):3411–3413. https://doi.org/10.1128/aac.01656-07
79. Ntrouka V, Hoogenkamp M, Zaura E, van der Weijden F (2011) The effect of chemotherapeutic agents on titanium-adherent biofilms. Clin Oral Implant Res 22 (11):1227–1234. https://doi.org/10.1111/j.1600-0501.2010.02085.x
80. Nunoshiba T, Hashimoto M, Nishioka H (1991) Cross-adaptive response in Escherichia coli caused by pretreatment with H2O2 against formaldehyde and other aldehyde compounds. Mutat Res 255(3):265–271
81. Omidbakhsh N (2006) A new peroxide-based flexible endoscope-compatible high-level disinfectant. Am J Infect Control 34(9):571–577. https://doi.org/10.1016/j.ajic.2006.02.003
82. Omidbakhsh N, Sattar SA (2006) Broad-spectrum microbicidal activity, toxicologic assessment, and materials compatibility of a new generation of accelerated hydrogen peroxide-based environmental surface disinfectant. Am J Infect Control 34(5):251–257. https://doi.org/10.1016/j.ajic.2005.06.002
83. Oosterik LH, Tuntufye HN, Butaye P, Goddeeris BM (2014) Effect of serogroup, surface material and disinfectant on biofilm formation by avian pathogenic Escherichia coli. Vet J (London, England: 1997) 202(3):561–565. https://doi.org/10.1016/j.tvjl.2014.10.001
84. Peeters E, Nelis HJ, Coenye T (2008) Evaluation of the efficacy of disinfection procedures against Burkholderia cenocepacia biofilms. J Hosp Infect 70(4):361–368. https://doi.org/10.1016/j.jhin.2008.08.015
85. Penna TC, Mazzola PG, Silva Martins AM (2001) The efficacy of chemical agents in cleaning and disinfection programs. BMC Infect Dis 1:16
86. Perumal PK, Wand ME, Sutton JM, Bock LJ (2014) Evaluation of the effectiveness of hydrogen-peroxide-based disinfectants on biofilms formed by Gram-negative pathogens. J Hosp Infect 87(4):227–233. https://doi.org/10.1016/j.jhin.2014.05.004

87. Pezzoni M, Pizarro RA, Costa CS (2014) Protective role of extracellular catalase (KatA) against UVA radiation in Pseudomonas aeruginosa biofilms. J Photochem Photobiol, B 131:53–64. https://doi.org/10.1016/j.jphotobiol.2014.01.005
88. Pires RH, da Silva Jde F, Gomes Martins CH, Fusco Almeida AM, Pienna Soares C, Soares Mendes-Giannini MJ (2013) Effectiveness of disinfectants used in hemodialysis against both Candida orthopsilosis and C. parapsilosis sensu stricto biofilms. Antimicrob Agents Chemother 57(5):2417–2421. https://doi.org/10.1128/aac.01308-12
89. Piskin N, Celebi G, Kulah C, Mengeloglu Z, Yumusak M (2011) Activity of a dry mist-generated hydrogen peroxide disinfection system against methicillin-resistant Staphylococcus aureus and Acinetobacter baumannii. Am J Infect Control 39(9):757–762. https://doi.org/10.1016/j.ajic.2010.12.003
90. Pitten F-A, Werner H-P, Kramer A (2003) A standardized test to assess the impact of different organic challenges on the antimicrobial activity of antiseptics. J Hosp Infect 55 (2):108–115
91. Pliuta VA, Andreenko IuV, Kuznetsov AE, Khmel IA (2013) Formation of the Pseudomonas aeruginosa PAO1 biofilms in the presence of hydrogen peroxide; the effect of the AiiA gene. Molekuliarnaia genetika, mikrobiologiia i virusologiia 4:10–14
92. Presterl E, Suchomel M, Eder M, Reichmann S, Lassnigg A, Graninger W, Rotter M (2007) Effects of alcohols, povidone-iodine and hydrogen peroxide on biofilms of Staphylococcus epidermidis. J Antimicrob Chemother 60(2):417–420. https://doi.org/10.1093/jac/dkm221
93. Ravindra Kumar S, Imlay JA (2013) How Escherichia coli tolerates profuse hydrogen peroxide formation by a catabolic pathway. J Bacteriol 195(20):4569–4579. https://doi.org/10.1128/jb.00737-13
94. Redondo C, Sena-Véleza M, Gell I, Ferragud E, Sabuquillo P, Graham JH, Cubero J (2015) Influence of selected bactericides on biofilm formation and viability of Xanthomonas citri subsp. citri. Crop Protect 78:204–213
95. Rizzotti L, Rossi F, Torriani S (2016) Biocide and antibiotic resistance of Enterococcus faecalis and Enterococcus faecium isolated from the swine meat chain. Food Microbiol 60:160–164. https://doi.org/10.1016/j.fm.2016.07.009
96. Rodriguez Ferri EF, Martinez S, Frandoloso R, Yubero S, Gutierrez Martin CB (2010) Comparative efficacy of several disinfectants in suspension and carrier tests against Haemophilus parasuis serovars 1 and 5. Res Vet Sci 88(3):385–389. https://doi.org/10.1016/j.rvsc.2009.12.001
97. Roeder RS, Lenz J, Tarne P, Gebel J, Exner M, Szewzyk U (2010) Long-term effects of disinfectants on the community composition of drinking water biofilms. Int J Hyg Environ Health 213(3):183–189. https://doi.org/10.1016/j.ijheh.2010.04.007
98. Rosenthal RA, Bell WM, Abshire R (1999) Disinfecting action of a new multi-purpose disinfection solution for contact lenses. Cont Lens Anterior Eye 22(4):104–109
99. Ruckerl I, Muhterem-Uyar M, Muri-Klinger S, Wagner KH, Wagner M, Stessl B (2014) L. monocytogenes in a cheese processing facility: learning from contamination scenarios over three years of sampling. Int J Food Microbiol 189:98–105. https://doi.org/10.1016/j.ijfoodmicro.2014.08.001
100. Rutala WA, Weber DJ (2013) Disinfectants used for environmental disinfection and new room decontamination technology. Am J Infect Control 41(5 Suppl):S36–S41. https://doi.org/10.1016/j.ajic.2012.11.006
101. Sagripanti J-L, Eklund CA, Trost PA, Jinneman KC, Abeyta C, Kaysner CA, Hill WE (1997) Comparative sensitivity of 13 species of pathogenic bacteria to seven chemical germicides. Am J Infect Control 25(4):335–339
102. Sattar SA, Adegbunrin O, Ramirez J (2002) Combined application of simulated reuse and quantitative carrier tests to assess high-level disinfection: experiments with an accelerated hydrogen peroxide-based formulation. Am J Infect Control 30(8):449–457
103. Sattar SA, Bradley C, Kibbee R, Wesgate R, Wilkinson MA, Sharpe T, Maillard JY (2015) Disinfectant wipes are appropriate to control microbial bioburden from surfaces: use of a

new ASTM standard test protocol to demonstrate efficacy. J Hosp Infect 91(4):319–325. https://doi.org/10.1016/j.jhin.2015.08.026
104. Seier-Petersen MA, Jasni A, Aarestrup FM, Vigre H, Mullany P, Roberts AP, Agerso Y (2014) Effect of subinhibitory concentrations of four commonly used biocides on the conjugative transfer of Tn916 in Bacillus subtilis. J Antimicrob Chemother 69(2):343–348. https://doi.org/10.1093/jac/dkt370
105. Semchyshyn HM (2014) Hormetic concentrations of hydrogen peroxide but not ethanol induce cross-adaptation to different stresses in budding yeast. Int J Microbiol 2014:485792. https://doi.org/10.1155/2014/485792
106. Shirato M, Nakamura K, Kanno T, Lingstrom P, Niwano Y, Ortengren U (2017) Time-kill kinetic analysis of antimicrobial chemotherapy based on hydrogen peroxide photolysis against Streptococcus mutans biofilm. J Photochem Photobiol, B 173:434–440. https://doi.org/10.1016/j.jphotobiol.2017.06.023
107. Soto E, Halliday-Simmonds I, Francis S, Kearney MT, Hansen JD (2015) Biofilm formation of Francisella noatunensis subsp. orientalis. Vet Microbiol 181(3–4):313–317. https://doi.org/10.1016/j.vetmic.2015.10.007
108. Soumet C, Meheust D, Pissavin C, Le Grandois P, Fremaux B, Feurer C, Le Roux A, Denis M, Maris P (2016) Reduced susceptibilities to biocides and resistance to antibiotics in food-associated bacteria following exposure to quaternary ammonium compounds. J Appl Microbiol 121(5):1275–1281. https://doi.org/10.1111/jam.13247
109. Stanley PM (1999) Efficacy of peroxygen compounds against glutaraldehyde-resistant mycobacteria. Am J Infect Control 27(4):339–343
110. Subbaiah TV, Mitchison DA, Selkon JD (1960) The Susceptibility to Hydrogen Peroxide of Indian and British Isoniazid-Sensitive and lsoniazid-Resistant Tubercle Bacilli. Tubercle 41 (5):323–333
111. Sun D, Crowell SA, Harding CM, De Silva PM, Harrison A, Fernando DM, Mason KM, Santana E, Loewen PC, Kumar A, Liu Y (2016) KatG and KatE confer Acinetobacter resistance to hydrogen peroxide but sensitize bacteria to killing by phagocytic respiratory burst. Life Sci 148:31–40. https://doi.org/10.1016/j.lfs.2016.02.015
112. Suraju MO, Lalinde-Barnes S, Sanamvenkata S, Esmaeili M, Shishodia S, Rosenzweig JA (2015) The effects of indoor and outdoor dust exposure on the growth, sensitivity to oxidative-stress, and biofilm production of three opportunistic bacterial pathogens. Sci Total Environ 538:949–958. https://doi.org/10.1016/j.scitotenv.2015.08.063
113. Surdeau N, Laurent-Maquin D, Bouthors S, Gelle MP (2006) Sensitivity of bacterial biofilms and planktonic cells to a new antimicrobial agent, Oxsil 320N. J Hosp Infect 62 (4):487–493. https://doi.org/10.1016/j.jhin.2005.09.003
114. Tachikawa M, Yamanaka K (2014) Synergistic disinfection and removal of biofilms by a sequential two-step treatment with ozone followed by hydrogen peroxide. Water Res 64:94–101. https://doi.org/10.1016/j.watres.2014.06.047
115. Theraud M, Bedouin Y, Guiguen C, Gangneux JP (2004) Efficacy of antiseptics and disinfectants on clinical and environmental yeast isolates in planktonic and biofilm conditions. J Med Microbiol 53(Pt 10):1013–1018. https://doi.org/10.1099/jmm.0.05474-0
116. Toki T, Nakamura K, Kurauchi M, Kanno T, Katsuda Y, Ikai H, Hayashi E, Egusa H, Sasaki K, Niwano Y (2015) Synergistic interaction between wavelength of light and concentration of H(2)O(2) in bactericidal activity of photolysis of H(2)O(2). J Biosci Bioeng 119(3):358–362. https://doi.org/10.1016/j.jbiosc.2014.08.015
117. Tondo ML, Delprato ML, Kraiselburd I, Fernandez Zenoff MV, Farias ME, Orellano EG (2016) KatG, the Bifunctional Catalase of Xanthomonas citri subsp. citri, Responds to Hydrogen Peroxide and Contributes to Epiphytic Survival on Citrus Leaves. PLoS ONE 11 (3):e0151657. https://doi.org/10.1371/journal.pone.0151657
118. Tote K, Horemans T, Vanden Berghe D, Maes L, Cos P (2010) Inhibitory effect of biocides on the viable masses and matrices of Staphylococcus aureus and Pseudomonas aeruginosa biofilms. Appl Environ Microbiol 76(10):3135–3142. https://doi.org/10.1128/aem.02095-09

119. Traore O, Springthorpe VS, Sattar SA (2002) Testing chemical germicides against Candida species using quantitative carrier and fingerpad methods. J Hosp Infect 50(1):66–75. https://doi.org/10.1053/jhin.2001.1133
120. United States Environmental Protection Agency (1993) EPA R.E.D. Facts. Peroxy Compounds. https://www3.epa.gov/pesticides/chem_search/reg_actions/reregistration/red_G-67_1-Dec-93.pdf
121. Venturini ME, Blanco D, Oria R (2002) In vitro antifungal activity of several antimicrobial compounds against Penicillium expansum. J Food Prot 65(5):834–839
122. Vicente CS, Nascimento FX, Ikuyo Y, Cock PJ, Mota M, Hasegawa K (2016) The genome and genetics of a high oxidative stress tolerant Serratia sp. LCN16 isolated from the plant parasitic nematode Bursaphelenchus xylophilus. BMC Genom 17:301. https://doi.org/10.1186/s12864-016-2626-1
123. Wesgate R, Grasha P, Maillard JY (2016) Use of a predictive protocol to measure the antimicrobial resistance risks associated with biocidal product usage. Am J Infect Control 44(4):458–464. https://doi.org/10.1016/j.ajic.2015.11.009
124. Wright MS, Mountain S, Beeri K, Adams MD (2017) Assessment of insertion sequence mobilization as an adaptive response to oxidative stress in Acinetobacter baumannii using IS-seq. J Bacteriol 199(9). https://doi.org/10.1128/jb.00833-16
125. Xie K, Peng H, Hu H, Wang W, Zhang X (2013) OxyR, an important oxidative stress regulator to phenazines production and hydrogen peroxide resistance in Pseudomonas chlororaphis GP72. Microbiol Res 168(10):646–653. https://doi.org/10.1016/j.micres.2013.05.001
126. Yanai R, Yamada N, Ueda K, Tajiri M, Matsumoto T, Kido K, Nakamura S, Saito F, Nishida T (2006) Evaluation of povidone-iodine as a disinfectant solution for contact lenses: antimicrobial activity and cytotoxicity for corneal epithelial cells. Cont Lens Anterior Eye 29(2):85–91. https://doi.org/10.1016/j.clae.2006.02.006
127. Zamocky M, Furtmuller PG, Obinger C (2008) Evolution of catalases from bacteria to humans. Antioxid Redox Signal 10(9):1527–1548. https://doi.org/10.1089/ars.2008.2046

Glutaraldehyde

7

7.1 Chemical Characterization

Glutaraldehyde is a colourless liquid with a pungent odour. It is an oily liquid at room temperature and miscible with water, alcohol and benzene. The basic chemical information on glutaraldehyde is summarized in Table 7.1.

7.2 Types of Application

In the European Union, glutaraldehyde is mainly used in human medicine for disinfection of inanimate surfaces (variable concentrations depending on the composition of the formula, e.g. 1.4–3 g/l) [34], for reprocessing flexible endoscopes (usually at 20 g/l) [16, 45, 127] or for disinfection of medical instruments (usually at 20 g/l with 30 min exposure time) [76, 83]. Glutaraldehyde at 2% can be found as a disinfectant in the WHO model list of essential medicines [125]. In the veterinary field, glutaraldehyde is used at 0.625–1.25 g/l (120–240 min exposure time) for disinfection of the environment [25]. For poultry farm disinfection, the typical concentration is 1 g/l by spraying [34]. For pig farm disinfection, the typical concentration is 20 g/l by fogging [34]. For machinery and food processing surface disinfection, the typical concentration of glutaraldehyde is 1 g/l [34]. When used as a preservative, the concentrations are typically 1 g/l for detergents and 0.025–0.2 g/l for most applications except oilfield applications [34].

In the USA, it is used in the agricultural setting for egg sanitation, in hatcheries, setters and chick processing facilities; in animal housing buildings; on farm equipment, trays, racks, carts, chick boxes, cages, trucks, vehicles and other hard surfaces. In commercial, institutional and industrial settings, it is used in laboratories, biomedical research facilities, nursing homes, veterinary hospitals and facilities, on cages, urinals and hard surfaces, and in the treatment of medical waste, human waste and animal waste. It is also used to disinfect hospital, medical, and dental office equipment/premises/surfaces and solid and liquid medical waste.

G. Kampf, *Antiseptic Stewardship*, https://doi.org/10.1007/978-3-319-98785-9_7

Table 7.1 Basic chemical information on glutaraldehyde [34]

CAS number	111-30-8
IUPAC name	1,5-pentanedial
Synonyms	Glutaral, glutardialdehyde, glutaric dialdehyde
Molecular formula	$C_5H_8O_2$
Molecular weight (g/mol)	100.11

Glutaraldehyde is used in oil storage tanks; water floods; drilling muds, drilling, completion, and workover fluids; packer fluids; gas production and transmission pipe systems; gas storage wells and systems; hydrotesting; pipeline pigging and scraping operations; paper mills and paper mill process water systems; pigments, filler slurries and water-based coatings for paper and paperboard; metalworking fluids; water-based conveyor lubricants; air washer and industrial scrubbing; systems/recirculating cooling and process water systems; service water and auxiliary systems; heat transfer systems; industrial wastewater systems; and sugar beet mills and process water systems [112].

7.2.1 European Chemicals Agency (European Union)

The European Commission has approved glutaraldehyde in 2015 as an active biocidal agent for various types of disinfectants [66]: hard surface disinfection in hospitals and industrial areas (PT 2), poultry farm and pig farm disinfection (PT 3) and food vessel disinfection, machinery disinfection and food processing surface disinfection (PT 4). In addition, it has been assessed as a preservative for detergents (e.g. laundry softeners, liquid detergent, wax emulsion or car polish) and paper wet-end additives preservation and paper coatings preservation (PT 6), closed and open recirculating cooling systems (PT 11) and slimicides for paper pulp (e.g., wet-end or paper de-inking slimicides) (PT 12). It meets the criteria for classification as respiratory sensitizer and as skin sensitizer subcategory 1A. That is why it was considered a candidate for substitution [66]. For product types 1 (human hygiene) and 13 (working or cutting fluid preservatives), it has not been approved [11].

7.2.2 Environmental Protection Agency (USA)

The EPA has reregistered glutaraldehyde in 2007 as an active ingredient in pesticides [112].

7.2.3 Overall Environmental Impact

In the European Union, glutaraldehyde is manufactured and/or imported in at least 1000 t per year [39]. Australia is one of very few countries that has published its sources of emission. The primary sources of glutaraldehyde are the industries that use it. Some of them are crude oil and natural gas extraction, beverage manufacturers, hospitals and x-ray processing. These emissions are mainly to the air and water. Other possible emitters of glutaraldehyde are medical offices, veterinary clinics, water in cooling systems, food processing facilities, tanneries, household disinfectants and agriculture sanitising. It may also be emitted from agricultural chemicals, disinfecting, sterilizing, sanitizing, household disinfectants and furniture polish. There is no known source of natural glutaraldehyde [6].

Aquatic exposure of microorganisms may enhance tolerance or resistance as an adaptive response. Results from environmental partitioning studies indicate that glutaraldehyde tends to remain in the aquatic compartment and has little tendency to bioaccumulate [75]. Aqueous solutions of glutaraldehyde are stable at room temperature under acidic to neutral conditions, and to sunlight, but unstable at elevated temperatures and under alkaline conditions. Glutaraldehyde is readily biodegradable in the freshwater environment and has the potential to biodegrade in the marine environment [75]. Half-life catabolism based on the loss of glutaraldehyde from the water phase of a river water–sediment system was described as 10.6 h aerobically and 7.7 h anaerobically [74]. The extrapolated half-life of abiotic degradation was 508 days at pH 5, 102 days at pH 7 and 46 days at pH 9 [74].

7.3 Spectrum of Antimicrobial Activity

7.3.1 Bactericidal Activity

7.3.1.1 Bacteriostatic Activity (MIC Values)

The MIC values obtained with different bacterial species are summarized in Table 7.2.

MIC values for *S. aureus* are between 500 and 10.000 mg/l, for *E. coli* between 150 and 10,000 mg/l, and for other Gram-negative bacterial species between 0.66 and 15.000 mg/l (Table 7.2).

7.3.1.2 Bactericidal Activity (Suspension Tests)

Glutaraldehyde at 2% achieves a 5.0 log reduction against most bacterial species within 3 min including *A. anitratus*, *E. cloacae*, *E. faecalis*, *E. faecium*, *K. pneumoniae*, *P. aeruginosa* and *S. aureus* (Table 7.3). It should, however, be considered that 2% glutaraldehyde is considered quite difficult to neutralize [37]

The findings are supported by a data showing that 131 clinical isolates of MDR *A. baumannii* were killed by a formulation based on 2% glutaraldehyde in 5 or 10 min [69] and that 20 clinical strains from seven different bacterial species (*A. anitratus*, *A. xylosoxidans*, *E. coli*, *P. aeruginosa*, *P. cepacia*, *S. aureus*, *S. maltophilia*) are killed by 2% glutaraldehyde within 10 min [82]. Only *T. whipplei* is not susceptible enough to 2% glutaraldehyde in 60 min [72].

In addition, the MBC values obtained with different bacterial species are summarized in Table 7.4. They are between 313 and 2,018 mg/l glutaraldehyde within 5 min and are lower with a contact time of 30 min.

Table 7.2 MIC values of various bacterial species to glutaraldehyde

Species	Strains/isolates	MIC value (mg/l)	References
A. calcoaceticus	ATCC 19606	3,250	[89]
A. actinomycetemcomitans	ATCC 29523	<5,000	[116]
B. subtilis var. globigii	ATCC 9372	3,250	[89]
B. stearothermophilus	ATCC 7953	1,875	[89]
B. fragilis	ATCC 25285	5,000	[116]
B. melitensis	Epidemic bovine strain	1,250	[123]
E. cloacae	Strain IAL 1976	3,250	[89]
E. coli	1 clinical strain (VU3695)	150	[7]
E. coli	NCTC 10418	800	[107]
E. coli	NCTC 8196	500	[61]
E. coli	ATCC 25922	3,250	[89]
E. coli	ATCC 25922	10,000	[116]
F. psychrophilum	5 fresh trout isolates	160–800	[47]
Halomonas spp.	DSM 7328 (strain MAC)	100	[7]
H. pylori	4 strains	0.66–4.5	[22]
K. pneumoniae	27 carbapenem-resistant clinical isolates	4–32	[52]
P. intermedia	ATCC 25611	<5,000	[116]
P. aeruginosa	ATCC 19582	1,000	[61]
P. aeruginosa	NCTC6749 and 3 extensively resistant clinical isolates	1,250–2,500	[126]
P. aeruginosa	91 clinical isolates, 37 hospital environmental isolates	$\leq$ 5,000	[88]
P. aeruginosa	ATCC 10145	15,000	[116]
P. mirabilis	11 clinical strains	1,600	[106]
P. vulgaris	NCTC 4635	250	[61]
S. Typhimurium	ATCC 14028	15,000	[116]
S. marcescens	Strain IAL 1478	1,375	[89]
S. aureus	NCTC 4613	500	[61]
S. aureus	4 strains (CECT976, RN4220, SA1199b, XU212)	750–800	[45]
S. aureus	ATCC 25923	1,875	[89]
S. aureus	ATCC 33591	10,000	[116]
S. mutans	ATCC 25175	10,000	[116]
5 Gram-negative species	35 isolates	1,400–8,000	[107]

Table 7.3 Bactericidal activity of glutaraldehyde in suspension tests

Species	Strains/isolates	Exposure time	Concentration	Log_{10} reduction	References
A. anitratus	3 clinical isolates	3 min	2% (P)	>6.0	[119]
Acinetobacter spp.	KN 93-25	30 s	3.34% (P)	>5.0	[1]
			2.34% (P)		
			2% (P)		
			1.67% (P)		
B. cereus[a]	ATCC 14579	30 min	2% (S)	>5.0	[94]
C. perfringens[a]	Strain CDC 1861	30 min	2% (S)	>6.3	[94]
E. cloacae	3 clinical isolates	3 min	2% (P)	>6.0	[119]
E. faecalis	KN 93-35	30 s	3.34% (P)	>5.0	[1]
			2.34% (P)		
			2% (P)		
			1.67% (P)		
E. faecalis	3 clinical isolates	3 min	2% (P)	>6.0	[119]
E. faecalis and *E. faecium*	NCTC 775 and 8 clinical isolates (4 of them vancomycin-resistant)	30 s	0.2% (S)	0.2–6.1	[17]
		1 min		0.9–6.1	
		5 min		3.7–7.4	
E. coli	ATCC 25922 and KN 93-152	30 s	3.34% (P)	>5.0	[1]
			2.34% (P)		
			2% (P)		
			1.67% (P)		
E. coli	Food isolate 0157:H7	30 min	2% (S)	>6.9	[94]
H. pylori	NCTC 11637, NCTC 11916 and 7 clinical isolates	15 s	0.5% (P)	>5.0	[2]
		15–30 s		>5.0[b]	
K. pneumoniae	3 clinical isolates	3 min	2% (P)	>6.0	[119]
L. monocytogenes	Food isolate	30 min	2% (S)	>6.1	[94]
P. aeruginosa	ATCC 27853	30 s	3.34% (P)	>5.0	[1]
			2.34% (P)		
			2% (P)		
			1.67% (P)		
P. aeruginosa	3 clinical isolates	3 min	2% (P)	>6.0	[119]
P. aeruginosa	ATCC 27853	30 min	2% (S)	3.8	[94]
P. aeruginosa	Clinical isolate	5 min	0.045% (S)	≥5.0	[120]
S. Typhimurium	ATCC 14028	30 min	2% (S)	>6.4	[94]
S. sonnei	Food isolate	30 min	2% (S)	>6.3	[94]
S. aureus	ATCC 25923 and KN 93-256	30 s	3.34% (P)	>5.0	[1]
			2.34% (P)		
			2% (P)		
			1.67% (P)		

(continued)

Table 7.3 (continued)

Species	Strains/isolates	Exposure time	Concentration	Log_{10} reduction	References
S. aureus	ATCC 25923 and 2 clinical isolates	3 min	2% (P)	>6.0	[119]
S. aureus	ATCC 25923	30 min	2% (S)	>6.5	[94]
S. epidermidis	KN 93-188	30 s	3.34% (P)	>5.0	[1]
			2.34% (P)		
			2% (P)		
			1.67% (P)		
S. epidermidis	ATCC 12228	30 min	2% (S)	>6.3	[94]
T. whipplei	Strain Twist-Marseille	60 min	2% (P)	<3.0	[72]
V. cholerae	Strain C6706	30 min	2% (S)	>6.4	[94]
V. parahaemolyticus	Strain NY477	30 min	2% (S)	>6.2	[94]
V. vulnificus	Strain LA M624	30 min	2% (S)	>6.3	[94]
X. maltophilia	KN 93-17	30 s	3.34% (P)	>5.0	[1]
			2.34% (P)		
			2% (P)		
			1.67% (P)		
Y. enterocolitica	Strain 8081	30 min	2% (S)	>6.8	[94]

P commercial product; *S* solution; [a]vegetative cell form; [b]with organic load

Table 7.4 MBC values (5 min) of various bacterial species to glutaraldehyde

Species	Strains/isolates	MBC value (mg/l)	References
A. baumannii	Clinical isolate	1,024	[35]
B. cepacia	Clinical isolate	512	[35]
E. faecalis	Clinical isolate	2,048	[35]
E. faecium	Clinical isolate	2,048	[35]
E. coli	ATCC 25922	512	[35]
P. aeruginosa	Clinical isolate	512	[35]
Salmonella spp.	11 strains (untreated wastewater)	665 ± 228[b]	[36]
	10 strains (treated wastewater)	619 ± 178[b]	
S. aureus	54 MRSA strains isolated in Canary black pigs	313–1,250	[38]
S. aureus	Clinical MRSA isolate	512	[35]
S. aureus	42 MRSA clinical isolates	256–2,048	[83]
		64–256[a]	
S. aureus	56 isolates (QAC tolerant)	8–512[a]	[76]
S. epidermidis	Clinical isolate	1,792	[35]
S. maltophilia	Clinical isolate	512	[35]

[a]30-min exposure time; [b]mean with stdev

7.3.1.3 Activity Against Bacteria in Biofilms

The efficacy of glutaraldehyde is impaired when bacteria are present in biofilms. Glutaraldehyde at 2% did not achieve a 5.0 log reduction in 3 min anymore. It seems necessary to exposure a biofilm for more than 30 min to achieve at least 2.0 log by $\geq$2% glutaraldehyde (Table 7.5).

These findings are supported by other reports. The eradication or reduction of biofilm cells of various bacterial species by glutaraldehyde, for example, required much longer time than that of planktonic cells in suspensions [35, 108]. Another study described that glutaraldehyde at 0.0025% was 47 times less effective against *P. aeruginosa* in biofilm compared to planktonic cells; the resistance factor was lower at 0.005% (36 times less effective) and 0.01% (20 times less effective) [50]. For preventing survival of bacteria in biofilms, glutaraldehyde was not completely effective. Glutaraldehyde at 2% applied for up to 1 h to biofilm of *S. Typhimurium*, *E. coli*, *S. mutans* or *B. fragilis* on glass or rubber carrier was not effective enough to prevent survival of the *S. Typhimurium* on rubber (45 min), *S. mutans* on glass (1 h) and *B. fragilis* on glass (30 min) [116]. Similar findings were reported from endoscope channels. A product based on 2% glutaraldehyde was effective in 20 min in original channels of an endoscope as part of manual processing and yielded negative cultures after disinfection when the channels were allowed to build *S. aureus* (ATCC 29213) or *P. aeruginosa* (ATCC 27853) biofilm over 5 d. However, 0.68% of cells in residual biofilm were still viable after disinfection [84]. The reduced efficacy of glutaraldehyde against bacteria in biofilms is partly explained by a transport limitation of the biocide into the biofilm as shown with *E. aerogenes* [105].

Table 7.5 Efficacy of glutaraldehyde against bacteria in biofilms

Species	Strains/isolates	Type of biofilm	Exposure time (min)	Concentration	$\log_{10}$ reduction	References
E. faecalis	ATCC 29212	8-d incubation in polystyrene pegs	20	2.6% (P)	3.9	[28]
P. aeruginosa	ATCC 15442	24-h incubation in microplates	30	5% (S)	6.6	[70]
				1% (S)	3.5	
				0.5% (S)	2.9	
				0.1% (S)	1.3	
				0.1% (S)	6.0[a]	
P. aeruginosa	ATCC 15442	8-d incubation in polystyrene pegs	20	2.6% (P)	5.3	[28]
S. aureus	9 ST239 isolates (MRSA)	18-h incubation in polystyrene microtiter plates	30	2% (P)	1.8	[5]

P commercial product; *S* solution; [a]planktonic cells

Biofilms can become acclimated to glutaraldehyde and eventually can degrade it. Acclimation to the biocide took longer at the higher biocide concentrations. The degree of biocide degradation and chemical oxygen demand (COD) removal depended on acclimation period, the presence of other organic matters and the amount of mineral salts available. Glutaraldehyde at up to 80 mg/l had no effect on treatment efficiency and populations of biofilms and planktonic phase of the system, whereas glutaraldehyde at 180 mg/l caused a progressive decline in all measured values. The presence of biofilm provided additional resistance to glutaraldehyde to bacteria because the biocide had to penetrate through biofilm to reach bacteria [73].

7.3.1.4 Bactericidal Activity in Carrier Tests

In carrier tests, 2% glutaraldehyde was able to reduce *E. faecalis* and *P. aeruginosa* by >4.0 log on a PVC carrier surface in 1 min [60]. When *S. aureus* is placed on a glass cup carrier and exposed to 2% glutaraldehyde, a log reduction >6.0 is found after 1 min [15]. Against *L. innocua* and *L. monocytogenes*, 2% glutaraldehyde was effective within 1 min in a carrier test with 3.0–6.0 log; in the presence of serum, however, the effect was lower with 3.0–4.0 log [12].

7.3.1.5 Bactericidal Activity in Endoscopes or Test Tubes

Glutaraldehyde at 2% was described to be effective for disinfection in manual processing in 20 min to eliminate *Enterococcus* spp. from artificially contaminated colonoscopes [27]. The same concentration was effective in 10 min to kill *H. pylori* on artificially contaminated endoscopes during manual disinfection [26]. When used during automated processing for disinfection at 55 °C, the entire process was effective to reduce *E. faecium* by at least 9.0 log in artificially contaminated test tubes [128]. When used at 1.5% for 45 min for manual disinfection, glutaraldehyde was still effective. When no bioburden after treatment was considered to be effective and the suction channel was sampled, 1.5% glutaraldehyde was effective only in 4 of 10 gastroscopes and colonoscopes [115].

In endoscope channels, treatment of a *P. aeruginosa* or *E. faecalis* or *C. albicans* biofilm with a formulation containing 2.6% glutaraldehyde for 20 min showed the micro-organisms can outgrow after 6–15 days after disinfection treatment indicating biofilm as a reservoir for microbial persistence despite negative cultures soon after reprocessing endoscopes [3].

7.3.1.6 Bactericidal Activity in Other Applications

A product based on 2% glutaraldehyde reduced *S. aureus*, *P. aeruginosa* and *S. marcescens* on seven different dental instruments by 5.0 log within 1 min during immersion [4]. Impregnation of polyurethane with glutaraldehyde (i.e., incorporation into polyurethane) has some bactericidal effect as shown with *E. coli* and *S. aureus* but is not maintained for more than 2 weeks. Coating of polyurethane with glutaraldehyde (i.e., applied to the polymer surface) has no substantial bactericidal efficacy [96]. When glutaraldehyde at 0.5% was applied with an atomizer to cement floor surfaces contaminated with isolates of *Salmonella* spp. for an exposure up to 60 min, 30% of all *Salmonella* strains were eliminated [80].

7.3.2 Fungicidal Activity

7.3.2.1 Fungicidal Activity (Suspension Tests)

Glutaraldehyde at 2% has yeasticidal activity within 3 min (Table 7.6). Against other types of fungi such as *A. niger*, *A. terreus*, *M. racemosus* or *R. nigricans*, longer exposure times up to 30 min are necessary to achieve a 4.0 log reduction.

7.3.2.2 Fungicidal Activity in Carrier Tests

On a PVC carrier surface, 2% glutaraldehyde results in a >4.0 log reduction of *C. albicans* in 1 min showing strong yeasticidal activity [60]. When *C. albicans* is allowed to adhere to soft denture lining material for 2.5 h, immersion of the contaminated and carefully washed material in a solution of 2% glutaraldehyde did not reduce the number of adherent cells significantly [18]. When spores of

Table 7.6 Fungicidal activity of glutaraldehyde in suspension tests

Species	Strains/isolates	Exposure time	Concentration	$\log_{10}$ reduction	References
A. fumigatus	15 clinical isolates	5 min	1.6% (P)	≥4.0	[110]
A. niger	ATCC 6275	3 min	3.34% (P)	>5.0	[1]
		10–15 min	2.34% (P)	>5.0	
		15–30 min	2% (P)	>5.0	
		30–45 min	1.67% (P)	>5.0	
A. terreus	KN 93-11	3 min	3.34% (P)	>5.0	[1]
		10–15 min	2.34% (P)	>5.0	
		15–30 min	2% (P)	>5.0	
		30–45 min	1.67% (P)	>5.0	
C. albicans	3 clinical isolates	3 min	2% (P)	>6.0	[119]
C. albicans	ATCC 10231	30 s	3.34% (P)	>5.0	[1]
			2.34% (P)	>5.0	
			2% (P)	>5.0	
			1.67% (P)	>5.0	
M. racemosus	KN 93-5	3 min	3.34% (P)	>5.0	[1]
		5 min	2.34% (P)	>5.0	
		5–10 min	2% (P)	>5.0	
		10-15 min	1.67% (P)	>5.0	
R. nigricans	SN 32	3 min	3.34% (P)	>5.0	[1]
		10-15 min	2.34% (P)	>5.0	
		15–30 min	2% (P)	>5.0	
		15–30 min	1.67% (P)	>5.0	
T. mentagrophytes	ATCC 26323	30 min	0.1% (S)	≥4.0	[54]

P commercial product; *S* solution

T. mentagrophytes are placed on a glass cup carrier and exposed to 2% glutaraldehyde, a log reduction >5.0 is found after 1 min [15].

A product based on 2% glutaraldehyde reduced *C. albicans* on seven different dental instruments by 5.0 log within 1–3 min during immersion [4].

7.3.3 Mycobactericidal Activity

7.3.3.1 Mycobactericidal Activity (Suspension Tests)

Table 7.7 illustrates that 2% glutaraldehyde is effective with at least a 4.0 log reduction against *M. smegmatis* (2 min), *M. chelonae*, *M. fortuitum* and *M. terrae* (5 min), *M. bovis* and *M. tuberculosis* (30 min) including MDR *M. tuberculosis* strains [91], *M. avium-intracellulare* and *M. xenopi* (60 min). *M. smegmatis* is known to be rather susceptible to glutaraldehyde with a MIC value <0.5% [116]. The overall exposure times for *M. tuberculosis*, *M. kansasii* and *M. avium* are supported by one more study using suspension tests but without a description of specific log reductions [92]. An acidic solution of 2% glutaraldehyde may be somewhat more effective against *M. bovis* compared to an alkaline solution [93].

Table 7.7 Mycobactericidal activity of glutaraldehyde in suspension tests

Species	Strains/isolates	Exposure time	Concentration	$\log_{10}$ reduction	References
M. abscessus	ATCC 19977	5 min	2.4% (P)	>6.2	[104]
M. abscessus	Strain CRM-0270	5 min	1.8% (P)	>5.0	[19]
			1.5% (P)		
M. abscessus	Strain CIP 108297	5 min	1.8% (P)	>5.0	[19]
		10 min	1.5% (P)		
M. abscessus	NCTC 10882	30 min	0.5% (S)	5.1	[44]
M. avium	KN 93-13	10–15 min[a] 10–30 min[b]	3.34% (P)	>5.0	[1]
			2.34% (P)		
			2% (P)		
			1.67% (P)		
M. avium-intracellulare	Clinical strain 104	5 min[a]	2% (P)	>5.0	[56]
		30 min[b]			
M. avium	NCTC 10437	10 min[b]	2% (S)	>5.0	[99]
M. avium	Clinical isolate (strain 3051)	10 min	2% (P)	4.4	[59]
		30 min		≥ 6.5	
M. avium	NCTC 10437	30 min	2% (P)	≥ 6.0	[59]
M. avium-intracellulare	Clinical isolate	60 min	2% (P)	>6.0	[78]
M. avium-intracellulare	Clinical isolate	60 min	2% (P)	>6.9	[49]

(continued)

Table 7.7 (continued)

Species	Strains/isolates	Exposure time	Concentration	log_{10} reduction	References
M. avium-intracellulare	1 fresh clinical isolate	60 min	2% (P)	>5.0	[58]
M. avium	Strain 104	5 min	1.8% (P)	>5.0	[19]
		30 min	1.5% (P)		
M. avium	TMC 724	2 min	1% (P)	≥ 4.0	[24]
M. intracellulare	Strain 637	5 min	1% (P)	3.2	[24]
M. bovis	ATCC 35743	10 min	2.4% (P)	4.1	[104]
M. bovis	ATCC 35743	10 min	2% (P)	≥ 3.0	[62]
M. bovis	NCTC 10772	30 min	2% (P)	≥ 4.2	[59]
M. bovis	Pasteur strain 1173 P2	5 min	1.8% (P)	>5.0	[19]
		15 min	1.5% (P)		
M. bovis	TMC 412	5 min	1% (P)	3.7	[24]
	TMC 1012	5 min	1% (P)	≥ 4.0	
M. chelonae	NCTC 946	1 min	2% (P)	>5.6	[49]
M. chelonae	NCTC 946	1 min	2% (P)	>5.6	[78]
M. chelonae	ATCC 35752	5 min[a]	2% (P)	>5.0	[56]
		30 min[b]			
M. chelonae	Clinical isolate	10 min[b]	2% (S)	>5.0	[99]
M. chelonae	ATCC 35752	5 min	1.8% (P)	>5.0	[19]
			1.5% (P)		
M. chelonae subsp. abscessus	CMCC 93326	5 min	1% (P)	4.5	[121]
		10 min		≥ 6.0	
M. chelonae	NCTC 946	5 min	0.5% (S)	>5.0	[44]
M. fortuitum	NCTC 10394	1 min	2% (P)	>6.1	[49]
M. fortuitum	ATCC 609	3 min	2% (P)	>6.0	[119]
M. fortuitum	Clinical strain	5 min	2% (P)	>5.0	[56]
M. kansasii	KN 93-21	10–30 min[a] 10–45 min[b]	3.34% (P)	>5.0	[1]
			2.34% (P)		
			2% (P)		
			1.67% (P)		
M. kansasii	WD isolate	10 min	2% (P)	>5.6	[78]
M. kansasii	TMC 1201	3 min	1% (P)	≥ 4.0	[24]
M. marinum	TMC 1219	1 min	1% (P)	≥ 4.0	[24]
M. scrofulaceum	TMC 1316	3 min	1% (P)	≥ 4.0	[24]
M. smegmatis	Strain TMC 1515	1 min	2% (S)	>6.0	[13]
M. smegmatis	TMC 1515	1 min	2% (P)[c]	≥ 6.2	[14]
M. smegmatis	NCTC 8159	2 min	2% (S)	>9.0	[99]
M. smegmatis	TMC 1515	1 min	1% (P)	≥ 4.0	[24]
M. terrae	Strain JCM12143	1 min	3.5% (P)	>5.0	[63]
		5 min	2.25% (P)	>5.0	
		5 min	2% (S)	>5.0	

(continued)

Table 7.7 (continued)

Species	Strains/isolates	Exposure time	Concentration	$\log_{10}$ reduction	References
M. terrae	ATCC 15755	5 min	1.8% (P)	>5.0	[19]
		15 min	1.5% (P)	>4.0	
M. terrae	NCTC 10856	10 min	0.5% (S)	>5.0	[44]
M. tuberculosis	KN 93-7	1–5 min[a] 3–15 min[b]	3.34% (P)	>5.0	[1]
			2.34% (P)		
			2% (P)		
			1.67% (P)		
M. tuberculosis	Strain H37Rv	1–3 min	2% (P)	≥4.0	[24]
M. tuberculosis	Strain H37Rv	1 min	2% (P)[c]	≥5.2	[14]
		30 min	2% (P)	≥5.3	
M. tuberculosis	4 strains of H37Rv, 8 MDR clinical isolates, 7 drug-resistant isolates	10 min	2% (S)	>4.0	[91]
M. tuberculosis	NCTC 7416	10 min	2% (P)	4.6	[78]
M. tuberculosis	Strain H37Rv	10 min	2% (P)	>4.6	[49]
M. tuberculosis	MDR clinical isolate (strain 98)	10 min	2% (P)	4.3	[59]
		30 min		≥5.5	
M. tuberculosis	Strain H37Rv	10 min[a]	2% (P)	>5.0	[56]
		30 min[b]			
M. tuberculosis	1 fresh clinical isolate	15 min	2% (P)	4.0	[58]
M. tuberculosis	Strain H37Rv	30 min	2% (P)	≥5.2	[59]
M. tuberculosis	Strain H37Rv (TMC 102)	5 min	1% (P)	3.8	[24]
M. tuberculosis	CMCC 93020	5 min	1% (P)	4.3	[121]
		10 min		5.2	
M. xenopi	NCTC 10042	10 min[b]	2% (S)	>5.0	[99]
M. xenopi	Strain CIP 104035T	15 min	2% (P)	>5.0	[29]
M. xenopi	One environmental isolate from soil and one clinical isolate	60 min	2% (P)	2.5	[29]
M. xenopi	2 clinical isolates	60 min	2% (P)	>5.0	[29]

P commercial product; *S* solution; [a]without organic load; [b]with organic load; [c]glutaraldehyde-phenate

The mycobactericidal efficacy of glutaraldehyde can be explained by significant protein coagulation as demonstrated in *M. chelonae* spheroplasts although concentrations <0.5% caused no protein coagulation. Glutaraldehyde is an effective cross-linking agent, and its own uptake may be decreased by virtue of its extensive cross-linking at the bacterial cell surface [43].

7.3.3.2 Activity Against Mycobacteria in Biofilms

A product based on 2% glutaraldehyde was effective in original channels of an endoscope as part of manual processing to yield negative cultures after 20 min disinfection when the channels were allowed to build *M. abscessus* (INCQS 594) biofilm over 15 d. However, 0.75% of cells in residual biofilm were still viable after disinfection [84]. One study with *M. phlei* indicates that the susceptibility of biofilm-grown cells to glutaraldehyde is lower (MBEC: 0.125% in 30 min) compared to planktonic cells (MBC: 0.0156% in 30 min) [10].

7.3.3.3 Mycobactericidal Activity in Carrier Tests

In carrier tests, 2% glutaraldehyde was mostly effective against different mycobacterial species including *M. bovis*, *M. chelonae*, *M. fortuitum*, *M. terrae*, *M. smegmatis* and *M. tuberculosis* with at least a 4.0 log reduction in 10 min (Table 7.8). Only *M. avium* was somewhat more resistant requiring mostly a 30 min exposure to achieve the same effect.

Table 7.8 Mycobactericidal activity of glutaraldehyde in carrier tests

Species	Strains/isolates	Exposure time	Concentration	$\log_{10}$ reduction	References
M. avium-intracellulare	Clinical strain 104	5 min[a]	2% (P)	>5.0	[56]
		10 min[b]			
M. avium	NCTC 10437	30 min	2% (P)	≥4.5	[59]
M. avium	Clinical isolate (strain 3051)	30 min	2% (P)	≥5.5	[59]
M. bovis	ATCC 35743	1 min	2% (S)	1.6	[15]
		10 min		>5.0	
M. bovis	NCTC 10772	10 min	2% (P)	≥3.2	[59]
M. chelonae	NCTC 946	1 min	2% (S)	≥4.0	[120]
M. chelonae var. abscessus	NCTC 10882	5 min	2% (S)	≥4.0	[120]
M. chelonae	ATCC 35752	5 min[a]	2% (P)	>5.0	[56]
		10 min[b]			
M. fortuitum	Clinical strain	5 min	2% (P)	>5.0	[56]
M. terrae	NCTC 10856	1 min	2% (S)	≥4.0	[120]
M. smegmatis	Strain TMC 1515	1 min	2% (S)	>6.0	[13]
M. smegmatis	TMC 1515	10 min	2% (P)	≥4.0	[14]
M. tuberculosis	Strain H37Rv	5 min	2% (P)	>5.0	[56]
M. tuberculosis	Strain H37Rv	10 min	2% (P)	≥4.0	[14]
M. tuberculosis	Strain H37Rv	30 min	2% (P)	≥4.2	[59]
M. tuberculosis	MDR clinical isolate (strain 98)	30 min	2% (P)	≥5.0	[59]

P commercial product; *S* solution; [a]without organic load; [b]with organic load

7.3.3.4 Mycobactericidal Activity in Flexible Endoscopes

Various studies have been published on the efficacy of glutaraldehyde against mycobacteria used for contamination of bronchoscopes, colonoscopes and duodenoscopes. When bronchoscopes were contaminated with *M. tuberculosis*, disinfection with 2% glutaraldehyde for 10 or 15 min was usually sufficient to achieve negative samplings [9, 30, 53, 57] or log reductions $\geq$4.0 [81]. Similar results were obtained when bronchoscopes were contaminated with *M. gordonae* [30, 64] or *M. avium-intracellulare* [57], or when colonoscopes and duodenoscopes were artificially contaminated with *M. chelonae* (20 min immersion time) [41]. Only a high inoculum of 10^8 CFU per ml *M. gordonae* required either a 20 min immersion time with 2% glutaraldehyde or a concentration of 3.2% for the 10 min immersion time [64].

One study with bronchoscopes artificially contaminated at the suction and biopsy channel with *M. tuberculosis*, however, indicates that even ten automatic processings using 2% activated glutaraldehyde for 15 min results in detection of *M. tuberculosis* after processing [85]. After prolongation of the processing to 60 min, *M. tuberculosis* was still detected after five processings [85]. The log reduction observed after artificial contamination with *M. avium-intracellulare* was even lower with 2.2 after 15 min and 2.4 after 60 min [85]. These results support the lower susceptibility of *M. avium-intracellulare* to glutaraldehyde compared to other mycobacterial species (see also Tables 7.7 and 7.8).

7.4 Effect of Low-Level Exposure

Adaptation experiments with dilutions of a product based on 23% glutaraldehyde and 5% BAC revealed that exposure to *Salmonella* spp. isolates obtained mainly from broiler farms did not change the MIC to the product in a relevant proportion ($\leq$2-fold) [46]. No studies were found that have systematically addressed possible cellular changes or changes of susceptibility to biocidal agents due to low-level exposure to glutaraldehyde.

7.5 Resistance to Glutaraldehyde

Commonly accepted break points to determine resistance to glutaraldehyde do not exist yet. So far, MIC values for *S. aureus* have been reported to be between 500 and 10,000 mg/l, for *E. coli* between 150 and 10,000 mg/l, and for the Gram-negative bacterial species between 4 and 15,000 mg/l (see also Table 7.2). For *Bacillus* strains, a MIC value >4,000 mg/l (0.4%) has been proposed to describe resistance [97]. It may also be suitable for other bacterial species when looking at the published MIC values.

7.5.1 Bacteria

7.5.1.1 Persistence Despite Disinfection with Glutaraldehyde as Recommended

One pseudo-outbreak involving six patients with two *P. aeruginosa* strains was reported after use of flexible endoscopes. The various types of endoscopes were processed automatically with a product based on 20% glutaraldehyde used at 1% for disinfection. The strains were isolated in the rinsing water and the drain. Insufficient killing by the bactericidal concentration of the disinfectant was described and assessed as resistance to glutaraldehyde [111]. Another report describes persistence of various mainly Gram-negative bacteria incl. *P. aeruginosa* after automated or manual processing of flexible endoscopes with 2% glutaraldehyde for 20 min. The persistence was mainly explained by insufficient cleaning prior to disinfection [32].

7.5.1.2 Insufficient Efficacy in Suspension Tests

The two *P. aeruginosa* strains described in Sect. 7.5.1.1 were evaluated in suspension tests with the aim to verify if they have a reduced susceptibility to glutaraldehyde [68]. The manufacturer of the product claimed a bactericidal activity at 50–55 °C within 5 min [111]. The two *P. aeruginosa* strains were reduced by the product (0.2% glutaraldehyde final concentration) within 5 min by 3.9 log at 50 °C and by 4.7 log at 55 °C. At 20 °C, the log reduction was only 0.9. It is noteworthy that a solution of 0.2% glutaraldehyde was somewhat more effective with 5.1 log at 50 °C and 6.3 log at 55 °C. The reason for the reduced susceptibility of the *P. aeruginosa* strains to a formulation based on glutaraldehyde is unknown. Another *Pseudomonas* spp. (*P. fluorescens* ATCC 13525) was also found to be less susceptible to glutaraldehyde. Exposure to 1.43% glutaraldehyde for 30 min did not completely inactivate *P. fluorescens* [101]. In 2014, surfaces and air were sampled in a hospital for bacterial contamination. A total of 104 bacterial isolates were obtained. The efficacy of a disinfectant based on 2% glutaraldehyde was tested as recommended by the manufacturer claiming a bactericidal activity. In 7.7% of the samples, bacterial growth was observed in the presence of the biocidal agent with the highest detection rates among enterobacteriaceae (22.2%) [65].

B. megaterium is another species with a report on a reduced susceptibility to glutaraldehyde. It was isolated from a washer disinfector despite the use of glutaraldehyde for disinfection. In suspension tests, it was reduced only by 2.3 log by a 20 min exposure to 2.5% glutaraldehyde. The reason for the reduced susceptibility to glutaraldehyde is unknown [40].

7.5.2 Mycobacteria

7.5.2.1 Persistence Despite Disinfection with Glutaraldehyde as Recommended

Use of glutaraldehyde as recommended does not always ensure sufficient antimicrobial activity so that specific species may survive the treatment and may persist

Table 7.9 Pseudo-outbreaks or infections associated with suspected insufficient mycobactericidal activity of glutaraldehyde

Species	Place of persistence	Treatments	Clinical impact	Suspected reason for pseudo-outbreaks and infections	References
M. abscessus	Bronchoscopes, digestive endoscopes, and disinfection machines	Automatic processing with 2% glutaraldehyde for disinfection	Pseudo-outbreak involving five patients after bronchoscopy	No evidence for resistance to glutaraldehyde	[51]
M. chelonae	Glutaraldehyde solutions of three different washer disinfectors; endoscopes; automated washers	Manual processing with 2% glutaraldehyde for 40 min during disinfection; or automatic processing with 2% glutaraldehyde for 28 min during disinfection	Pseudo-outbreak involving 22 patients	Lack of routine disinfection cycles (monthly 8-h disinfection of the machine with 2% glutaraldehyde) and biofilm formation may have contributed to the initial contamination of the automated washers	[71]
M. chelonae	Rinse water from the bronchoscope disinfecting machine	2% glutaraldehyde for disinfection	Pseudo-outbreak involving 14 patients	Presence of a biofilm inside the machine	[42]
M. chelonae	Suction channel of four different bronchoscopes	Automatic processing with 2.3% glutaraldehyde for a 10 min disinfection; replacement of the disinfectant every 4 weeks; no disinfection of the machine itself	Pseudo-outbreak involving 18 patients	Presumably the tap water used to rinse the bronchoscopes	[122]
M. chelonae	An autocleaner and a bronchoscope	Not described in abstract	Pseudo-outbreak involving 12 patients	Presumably the autocleaner	[20]
M. fortuitum	Bronchoscope	Manual processing with 2% glutaraldehyde for 30 min or automatic processing	Recurrent episodes of mycobacterial cross-contamination of bronchoscopy specimens	Suction valve continued to be contaminated	[124]
M. massiliense	No devices investigated	Commercial 2% glutaraldehyde solution was used for the disinfection of the surgical instruments (15 to 30 min of exposure) in all institutions that had confirmed cases	Epidemic with 172 confirmed cases of postsurgery infections	Possibly selective pressure of 2% glutaraldehyde use and the inadequate mechanical cleaning of surgical instruments have facilitated the occurrence of outbreaks	[33]

on flexible endoscopes or instruments. Some pseudo-outbreaks have been reported after bronchoscopy caused by *M. chelonae*, *M. abscessus* or *M. fortuitum*. An epidemic with surgical site infections caused by *M. massiliense* has also been published (Table 7.9).

Most pseudo-outbreaks were described with *M. chelonae* after bronchoscopy (Table 7.9). The presumed reasons were suspected presence of biofilm inside the machine or suspected contamination of tap water. These findings are supported by other authors. In one study, five gastrointestinal endoscopes were contaminated with *M. chelonae*, followed by cleaning and disinfection with 2% alkaline glutaraldehyde. One out of five scopes showed consistent growth in channels after 10 min disinfection. With membrane filtration, one colony was still detected in one scope after 45 min disinfection [113]. *M. chelonae* could be eliminated by increasing glutaraldehyde to 3%, changing the glutaraldehyde solution once per week, recirculating used disinfectant and an additional disinfection procedure before automatic bronchoscope processing using 70% alcohol [109].

Selection pressure of 2% glutaraldehyde use and inadequate mechanical cleaning of surgical instruments has been suspected to have facilitated the occurrence of the outbreak of 172 confirmed cases of postsurgery infections in Brazil (Table 7.9).

7.5.2.2 Insufficient Efficacy in Suspension Tests

A few studies describe environmental or clinical isolates with a reduced susceptibility to glutaraldehyde compared to culture collection strains typically used for biocidal efficacy testing (Table 7.10).

Some strains and isolates of *M. chelonae* from washer disinfectors were resistant to 2% glutaraldehyde; the measured log reduction was mostly <1.0 in 60 min which is insufficient to be considered as mycobactericidal. Similar results were observed in a glass carrier test with two isolates of *M. chelonae* suspected to be resistant to glutaraldehyde. A 30 min, exposure to 2% glutaraldehyde reduced the cell count by 0.3 and 0.5 log, whereas NCTC strains were killed in 1 or 5 min [120]. On a PVC carrier surface, a *M. chelonae* strain suspected to be resistant to glutaraldehyde was reduced by 2% glutaraldehyde by 2.0 log in 20 min, whereas a strain susceptible to glutaraldehyde (ATCC 19977) was reduced by >4.0 log in 1 min [60]. The lower susceptibility of *M. chelonae* isolates was explained by a possible biofilm formation [48], a possible selection of glutaraldehyde resistance due to reduction of glutaraldehyde level by 50% in one week [114] or by a contaminated water tank [86].

The two clinical isolates of *M. massiliense* associated with an epidemic of surgical site infections also revealed a lower susceptibility to glutaraldehyde; they were able to survive in 1.5–7% glutaraldehyde showing growth after exposure for 30 min. The reason for the reduced susceptibility was unknown [77].

The frequency of mycobacterial isolates with a reduced susceptibility to glutaraldehyde is difficult to determine. An analysis of 117 clinical isolates of rapid growing non-tuberculous mycobacteria showed a reduced susceptibility to 0.5% glutaraldehyde in six clinical isolates of *M. abscessus* compared to ATCC control strains [31].

Table 7.10 Results obtained from suspension tests with isolates of mycobacteria suspected to be resistant to glutaraldehyde

Species	Strains/isolates	Exposure time	Concentration	$\log_{10}$ reduction	References
M. chelonae	Strain Epping	60 min	0.5% (S)	0.2	[44]
	Strain Harefield	60 min		0.2	
	NCTC 946[a]	5 min		>5.0	
M. chelonae	Strain Epping	60 min	2% (P)	0.3	[49]
	NCTC 946[a]	1 min		>5.6	
M. chelonae	Strain WD 1	60 min	2% (P)	0.6	[78]
	Strain WD 2	60 min		0.3	
	NCTC 946[a]	1 min		>5.6	
M. chelonae	2 isolates from WD	$\leq$60 min	2% (P)	0.0–0.6	[48]
	NCTC 946[a]	1 min		>5.8	
M. chelonae	3 strains from WD	$\leq$60 min	2% (P)	0.0–0.6	[114]
	ATCC 14998[a]	10 min		>5.3	
M. chelonae	5 isolates	$\leq$60 min	2.4% (P)	1.3–2.1	[104]
M. chelonae	1 isolate from WD	60 min	2% (S)	3.5	[86]
M. chelonae	Strain Epping	30 min	1.5% (P)	0.0	[19]
	Strain Harefield	30 min		0.1	
	Strain 9917	30 min		0.0	
	ATCC 35752[a]	5 min		>5.0	
M. chelonae	Strain Epping	30 min	1.8% (P)	1.0	[19]
	Strain Harefield	30 min		0.4	
	Strain 9917	30 min		0.9	
	ATCC 35752[a]	5 min		>5.0	
M. gordonae	1 isolate from WD	10 min	2.5% (P)	3.3	[40]
	ATCC 14470[a]	10 min		6.2	

P commercial product; *S* solution; [a]comparison to standard culture collection strains

7.5.3 Resistance Mechanisms

The mechanisms of resistance to glutaraldehyde have mainly been studied in *Pseudomonas*. In *P. aeruginosa* and *P. fluorescens* biofilms, it can be explained by efflux pumps. Induction of known modulators of biofilm formation, including phosphonate degradation, lipid biosynthesis, and polyamine biosynthesis, may in addition contribute to biofilm resistance and resilience [117]. Produced water induced genes in *P. fluorescens* involved in osmotic stress, energy production and conversion, membrane integrity and protein transport following produced water exposure, which facilitates bacterial survival and alters biocide tolerance [118]. And a class I integron was detected in 22 of 36 MDR *P. aeruginosa* isolates. Integron

I-positive isolates showed reduced susceptibility to tested biocides including glutaraldehyde. Class I integron may also be responsible for generating MDR *P. aeruginosa* isolates with reduced susceptibility to biocides [67].

In *E. coli* and *Halomonas* spp., resistance to glutaraldehyde depends on the composition and structure of the outer membrane [7]. In *H. pylori*, an Imp/OstA protein was identified that was associated with glutaraldehyde resistance in a clinical strain. Disruption of this protein results in altering membrane permeability, sensitivity to organic solvent and susceptibility to antibiotics [22]. The resistance mechanism in *H. pylori* is described in more detail by Chiu et al. [23].

A plasmid pTZ22 was detected in *S. aureus* exhibiting resistance to glutaraldehyde resulting in MIC values of up to 1,600 mg/l [95].

The mechanism of glutaraldehyde resistance in *M. chelonae* is not yet understood. No changes were identified in the extractable fatty acids or the mycolic acid components of the cell wall, but a reduction in each of the resistant strains in the arabinogalactan/arabinomannan portion of the cell wall was detected [79]. Resistance is not explained by efflux pumps [86].

7.5.4 Resistance Genes

No specific genes have been identified to explain resistance to glutaraldehyde. However, a correlation was described in 27 carbapenem-resistant clinical *K. pneumoniae* isolates between the presence of drug resistance genes (qacA, qacΔE, qacE and acrA) and a higher tolerance to killing or growth inhibition by disinfectants including glutaraldehyde [52].

7.6 Cross-Tolerance to Other Biocidal Agents

A cross-adaptive response was demonstrated when *E. coli* WP2 cells were pretreated with hydrogen peroxide (60 μM for 30 min) followed by challenging treatment with aldehyde compounds including glutaraldehyde. These results suggest that hydrogen peroxide has the capacity to induce a function which reduces the killing effects of aldehydes, and the function is controlled by the recA gene without involvement of SOS response [87].

Cross-resistance may be found to other aldehydes. For example, formaldehyde-tolerant *E. coli* and *Halomonas* spp. strains were also tolerant to high concentrations of glutaraldehyde (1,000 mg/l) and acetaldehyde (500 mg/l) [7]. A *B. cepacia* isolate that was originally isolated from a contaminated matrix (used as a preservative) was selected with glutaraldehyde as glutaraldehyde-resistant and exhibited cross-resistance to formaldehyde [21].

7.7 Cross-Tolerance to Antibiotics

Increased tolerance to glutaraldehyde of the glutaraldehyde-resistant mutants of *M. chelonae* was matched by increased tolerance to rifampicin and ethambutol but not isoniazid [79]. Another study shows that all of nine glutaraldehyde-tolerant *M. chelonae* isolates were either resistant or intermediately resistant to two or three classes of antibiotics (mostly rifampicin and isoniazid) but only one of nine glutaraldehyde-susceptible *M. chelonae* isolates [86].

7.8 Role of Biofilm

7.8.1 Effect on Biofilm Development

Data on biofilm development were not found.

7.8.2 Effect on Biofilm Removal

Removal of biofilm by glutaraldehyde is mostly poor with $\leq 10\%$ as shown with *B. cereus*, *P. fluorescens* and dual species biofilms (Table 7.11). In addition, removal of a mixed biofilm or *Acinetobacter* biofilm from brass coupons by glutaraldehyde solutions was also described to be low [98].

Table 7.11 Biofilm removal rate (quantitative determination of biofilm matrix) by exposure to products or solutions based on glutaraldehyde

Type of biofilm	Concentration	Exposure time	Biofilm removal rate	References
B. cereus biofilm on stainless steel	Aldehyde-based product ("GLUT") at 200 mg/l	No data in abstract	<10%	[100]
P. fluorescens ATCC 13525 biofilm on stainless steel	200 mg/l (S)	1–2 h	2–18%	[102]
P. fluorescens biofilm on stainless steel	Aldehyde-based product ("GLUT") at 200 mg/l	No data in abstract	<10%	[100]
S. aureus biofilm (9 MRSA isolates of strain ST239 and isolate MBM 9393) on polystyrene	2% (S)	30 min	Approximately 72%	[5]
Dual species biofilm (*P. fluorescens, B. cereus*) on stainless steel	Aldehyde-based product ("GLUT") at 200 mg/l	No data in abstract	<10%	[100]

P commercial product; *S* solution

Data from a study on protein removal from an *E. faecalis* ATCC 29212 and *P. aeruginosa* ATCC 15442 biofilm (8-d incubation in polystyrene pegs) with exposure to a formulation based on 2.6% glutaraldehyde for 20 min revealed protein removal rates between 20.8 and 56.0% depending on the type of cleaner used before. The removal rates of carbohydrate were higher with 91.5–98.9% [28].

7.8.3 Effect on Biofilm Fixation

Glutaraldehyde is described as a protective agent against cell lysis [8]. A *P. aeruginosa* biofilm on PTFE tubes showed that protein was higher after nine treatment cycles with 2% glutaraldehyde solution (89.0 μg/cm^2) compared to distiled water (57.3 μg/cm^2) indicating a substantial fixation of the biofilm [90]. Glutaraldehyde at concentrations between 100, 200, 500 and 1000 mg/l showed variable fixation of a *P. fluorescens* biofilm with biofilm mass reduction rates from 77% (no exposure to glutaraldehyde) to 64% (100 mg/l) and 38% (1000 mg/l) indicating an increasingly difficult removal of biofilm after exposure to a higher concentration of glutaraldehyde [103]. Two products based on 2% glutaraldehyde revealed fixation rates of an *E. coli* 54127 biofilm on haemolysis glass tubes between 62 and 97% [55].

7.9 Summary

The principal antimicrobial activity of 2% glutaraldehyde is summarized in Table 7.12.

The key findings on acquired resistance and cross-resistance including the role of biofilm in selecting resistant isolates are summarized in Table 7.13.

Table 7.12 Overview on the typical exposure times required for 2% glutaraldehyde to achieve sufficient biocidal activity against the different target microorganisms

Target microorganisms	Species	Exposure time
Bacteria	Most clinically relevant species except *T. whipplei*	3 min[a]
Fungi	*Candida* spp.	3 min
	Other types of fungi	≤ 30 min
Mycobacteria	*M. smegmatis*	2 min
	M. chelonae, M. fortuitum, M. terrae	5 min
	M. bovis, M. tuberculosis	30 min
	M. avium-intracellulare, M. xenopi	60 min

[a]In biofilm often no sufficient efficacy in 30 min

Table 7.13 Key findings on acquired glutaraldehyde resistance, the effect of low-level exposure, cross-tolerance to other biocides and antibiotics, and its effect on biofilm

Parameter	Species	Findings
Elevated MIC values	*S. Typhimurium*	$\leq$ 15,000 mg/l
	P. aeruginosa	$\leq$ 15,000 mg/l
	S. aureus	$\leq$ 10,000 mg/l
	S. mutans	$\leq$ 10,000 mg/l
	E. coli	$\leq$ 10,000 mg/l
	B. fragilis	$\leq$ 5,000 mg/l
Infections suggestive of tolerance	*M. chelonae*, *M. abscessus*, *M. fortuitum*	Few pseudo-outbreaks associated with bronchoscopies; mycobacteria persisted despite processing with glutaraldehyde correlating with insufficient efficacy in suspension tests
	M. massiliense	Epidemic with 172 confirmed cases of postsurgery infections; inadequate mechanical cleaning of surgical instruments and possibly selection pressure by glutaraldehyde
	P. aeruginosa	One pseudo-outbreak with contaminated washer-disinfector correlating with lower susceptibility of strain to glutaraldehyde in suspension tests
	M. chelonae, *M. gordonae*	Various isolates with little or no mycobactericidal efficacy of 2% glutaraldehyde in suspension tests (60 min)
	B. megaterium	WD isolate with reduced susceptibility to 2.6% glutaraldehyde in suspension tests (20 min)
MIC value to determine resistance	Most species	Not available
	Bacillus species	>4,000 mg/l (proposal in the literature)
Cross-tolerance biocides	*E. coli*	Hydrogen peroxide has the capacity to induce a function which reduces the killing effects of aldehydes
	E. coli, *Halomonas* spp., *B. cepacia*	Other aldehydes
Cross-tolerance antibiotics	*M. chelonae*	Often rifampicin, sometimes isoniazid
Specific resistance mechanism	*M. chelonae*	Unknown but not by efflux pumps.
	Pseudomonas spp.	Efflux pumps, class I integron
	E. coli, *Halomonas* spp.	Composition and structure of the outer membrane
	S. aureus	Plasmid
Effect of low-level exposure	None	No MIC increase
	Salmonella spp.	Weak MIC increase ($\leq$ 4-fold)
	None	Strong MIC increase (>4-fold)
Biofilm	Development	Unknown
	Removal	Poor; mostly <10%
	Fixation	Strong: mostly >60%

References

1. Akamatsu T, Tabata K, Hironaga M, Uyeda M (1997) Evaluation of the efficacy of a 3.2% glutaraldehyde product for disinfection of fibreoptic endoscopes with an automatic machine. J Hosp Infect 35(1):47–57
2. Akamatsu T, Tabata K, Hironga M, Kawakami H, Uyeda M (1996) Transmission of Helicobacter pylori infection via flexible fiberoptic endoscopy. Am J Infect Control 24 (5):396–401
3. Alfa MJ, Howie R (2009) Modeling microbial survival in buildup biofilm for complex medical devices. BMC Infect Dis 9:56. https://doi.org/10.1186/1471-2334-9-56
4. Angelillo IF, Bianco A, Nobile CG, Pavia M (1998) Evaluation of the efficacy of glutaraldehyde and peroxygen for disinfection of dental instruments. Lett Appl Microbiol 27 (5):292–296
5. Aparecida Guimaraes M, Rocchetto Coelho L, Rodrigues Souza R, Ferreira-Carvalho BT, Marie Sa Figueiredo A (2012) Impact of biocides on biofilm formation by methicillin-resistant Staphylococcus aureus (ST239-SCCmecIII) isolates. Microbiol Immunol 56(3):203–207. https://doi.org/10.1111/j.1348-0421.2011.00423.x
6. Australian Government (2014) Glutaraldehyde: Sources of emissions. http://www.npi.gov.au/resource/glutaraldehyde-sources-emissions
7. Azachi M, Henis Y, Shapira R, Oren A (1996) The role of the outer membrane in formaldehyde tolerance in Escherichia coli VU3695 and Halomonas sp. MAC. Microbiology (Reading, England) 142(Pt 5):1249–1254. https://doi.org/10.1099/13500872-142-5-1249
8. Azeredo J, Henriques M, Sillankorva S, Oliveira R (2003) Extraction of exopolymers from biofilms: the protective effect of glutaraldehyde. Water Sci Technol 47(5):175–179
9. Bar W, Marquez de Bar G, Naumann A, Rusch-Gerdes S (2001) Contamination of bronchoscopes with Mycobacterium tuberculosis and successful sterilization by low-temperature hydrogen peroxide plasma sterilization. Am J Infect Control 29(5):306–311
10. Bardouniotis E, Huddleston W, Ceri H, Olson ME (2001) Characterization of biofilm growth and biocide susceptibility testing of Mycobacterium phlei using the MBEC assay system. FEMS Microbiol Lett 203(2):263–267
11. Barroso JM (2014) COMMISSION IMPLEMENTING DECISION of 24 April 2014 on the non-approval of certain biocidal active substances pursuant to Regulation (EU) No 528/2012 of the European Parliament and of the Council. Off J Eur Union 57(L 124):27–29
12. Best M, Kennedy ME, Coates F (1990) Efficacy of a variety of disinfectants against Listeria spp. Appl Environ Microbiol 56(2):377–380
13. Best M, Sattar SA, Springthorpe VS, Kennedy ME (1988) Comparative mycobactericidal efficacy of chemical disinfectants in suspension and carrier tests. Appl Environ Microbiol 54:2856–2858
14. Best M, Sattar SA, Springthorpe VS, Kennedy ME (1990) Efficacies of selected disinfectants against *Mycobacterium tuberculosis*. J Clin Microbiol 28(10):2234–2239
15. Best M, Springthorpe VS, Sattar SA (1994) Feasibility of a combined carrier test for disinfectants: studies with a mixture of five types of microorganisms. Am J Infect Control 22 (3):152–162
16. Bordas JM, Marcos-Maeso MA, Perez MJ, Llach J, Gines A, Pique JM (2005) GI flexible endoscope disinfection: "in use" test comparative study. Hepatogastroenterology 52 (63):800–807
17. Bradley CR, Fraise AP (1996) Heat and chemical resistance of enterococci. J Hosp Infect 34:191–196
18. Buergers R, Rosentritt M, Schneider-Brachert W, Behr M, Handel G, Hahnel S (2008) Efficacy of denture disinfection methods in controlling Candida albicans colonization in vitro. Acta Odontol Scand 66(3):174–180. https://doi.org/10.1080/00016350802165614

19. Burgess W, Margolis A, Gibbs S, Duarte RS, Jackson M (2017) Disinfectant susceptibility profiling of glutaraldehyde-resistant Nontuberculous Mycobacteria. Infect Control Hosp Epidemiol 38(7):784–791. https://doi.org/10.1017/ice.2017.75
20. Campagnaro RL, Teichtahl H, Dwyer B (1994) A pseudoepidemic of Mycobacterium chelonae: contamination of a bronchoscope and autocleaner. Aust N Z J Med 24(6):693–695
21. Chapman JS, Diehl MA, Fearnside KB (1998) Preservative tolerance and resistance. Int J Cosmet Sci 20(1):31–39. https://doi.org/10.1046/j.1467-2494.1998.171733.x
22. Chiu HC, Lin TL, Wang JT (2007) Identification and characterization of an organic solvent tolerance gene in Helicobacter pylori. Helicobacter 12(1):74–81. https://doi.org/10.1111/j.1523-5378.2007.00473.x
23. Chiu HC, Lin TL, Yang JC, Wang JT (2009) Synergistic effect of imp/ostA and msbA in hydrophobic drug resistance of Helicobacter pylori. BMC Microbiol 9:136. https://doi.org/10.1186/1471-2180-9-136
24. Collins FM, Montalbine V (1976) Mycobactericidal activity of glutaraldehyde solutions. J Clin Microbiol 4(5):408–412
25. Couto N, Belas A, Tilley P, Couto I, Gama LT, Kadlec K, Schwarz S, Pomba C (2013) Biocide and antimicrobial susceptibility of methicillin-resistant staphylococcal isolates from horses. Vet Microbiol 166(1–2):299–303. https://doi.org/10.1016/j.vetmic.2013.05.011
26. Cronmiller JR, Nelson DK, Jackson DK, Kim CH (1999) Efficacy of conventional endoscopic disinfection and sterilization methods against Helicobacter pylori contamination. Helicobacter 4(3):198–203
27. Cronmiller JR, Nelson DK, Salman G, Jackson DK, Dean RS, Hsu JJ, Kim CH (1999) Antimicrobial efficacy of endoscopic disinfection procedures: a controlled, multifactorial investigation. Gastrointest Endosc 50(2):152–158
28. da Costa Luciano C, Olson N, Tipple AF, Alfa M (2016) Evaluation of the ability of different detergents and disinfectants to remove and kill organisms in traditional biofilm. Am J Infect Control 44(11):e243–e249. https://doi.org/10.1016/j.ajic.2016.03.040
29. Dauendorffer JN, Laurain C, Weber M, Dailloux M (2000) Evaluation of the bactericidal efficiency of a 2% alkaline glutaraldehyde solution on Mycobacterium xenopi. J Hosp Infect 46(1):73–76. https://doi.org/10.1053/jhin.2000.0793
30. Davis D, Bonekat HW, Andrews D, Shigeoka JW (1984) Disinfection of the flexible fibreoptic bronchoscope against Mycobacterium tuberculosis and M gordonae. Thorax 39 (10):785–788
31. De Groote MA, Gibbs S, de Moura VC, Burgess W, Richardson K, Kasperbauer S, Madinger N, Jackson M (2014) Analysis of a panel of rapidly growing mycobacteria for resistance to aldehyde-based disinfectants. Am J Infect Control 42(8):932–934. https://doi.org/10.1016/j.ajic.2014.05.014
32. Deva AK, Vickery K, Zou J, West RH, Selby W, Benn RA, Harris JP, Cossart YE (1998) Detection of persistent vegetative bacteria and amplified viral nucleic acid from in-use testing of gastrointestinal endoscopes. J Hosp Infect 39(2):149–157
33. Duarte RS, Lourenco MC, Fonseca Lde S, Leao SC, Amorim Ede L, Rocha IL, Coelho FS, Viana-Niero C, Gomes KM, da Silva MG, Lorena NS, Pitombo MB, Ferreira RM, Garcia MH, de Oliveira GP, Lupi O, Vilaca BR, Serradas LR, Chebabo A, Marques EA, Teixeira LM, Dalcolmo M, Senna SG, Sampaio JL (2009) Epidemic of postsurgical infections caused by Mycobacterium massiliense. J Clin Microbiol 47(7):2149–2155. https://doi.org/10.1128/jcm.00027-09
34. eCA Finland (2014) Assessment report. Glutaraldehyde Product-type 2, 3, 4, 6, 11, 12.
35. El-Azizi M, Farag N, Khardori N (2016) Efficacy of selected biocides in the decontamination of common nosocomial bacterial pathogens in biofilm and planktonic forms. Comp Immunol Microbiol Infect Dis 47:60–71. https://doi.org/10.1016/j.cimid.2016.06.002
36. Espigares E, Bueno A, Espigares M, Galvez R (2006) Isolation of Salmonella serotypes in wastewater and effluent: Effect of treatment and potential risk. Int J Hyg Environ Health 209 (1):103–107. https://doi.org/10.1016/j.ijheh.2005.08.006

37. Espigares E, Bueno A, Fernandez-Crehuet M, Espigares M (2003) Efficacy of some neutralizers in suspension tests determining the activity of disinfectants. J Hosp Infect 55 (2):137–140
38. Espigares E, Moreno Roldan E, Espigares M, Abreu R, Castro B, Dib AL, Arias A (2017) Phenotypic Resistance to Disinfectants and Antibiotics in Methicillin-Resistant Staphylococcus aureus Strains Isolated from Pigs. Zoonoses Public Health 64(4):272–280. https://doi.org/10.1111/zph.12308
39. European Chemicals Agency (ECHA) Glutaral. Substance information. https://echa.europa.eu/substance-information/-/substanceinfo/100.003.506. Accessed 16 Nov 2017
40. Fisher CW, Fiorello A, Shaffer D, Jackson M, McDonnell GE (2012) Aldehyde-resistant mycobacteria bacteria associated with the use of endoscope reprocessing systems. Am J Infect Control 40(9):880–882. https://doi.org/10.1016/j.ajic.2011.11.004
41. Foliente RL, Kovacs BJ, Aprecio RM, Bains HJ, Kettering JD, Chen YK (2001) Efficacy of high-level disinfectants for reprocessing GI endoscopes in simulated-use testing. Gastrointest Endosc 53(4):456–462. https://doi.org/10.1067/mge.2001.113380
42. Fraser VJ, Jones M, Murray PR, Medoff G, Zhang Y, Wallace RJ Jr (1992) Contamination of flexible fiberoptic bronchoscopes with Mycobacterium chelonae linked to an automated bronchoscope disinfection machine. Am Rev Respir Dis 145(4 Pt 1):853–855. https://doi.org/10.1164/ajrccm/145.4_Pt_1.853
43. Fraud S, Hann AC, Maillard JY, Russell AD (2003) Effects of ortho-phthalaldehyde, glutaraldehyde and chlorhexidine diacetate on Mycobacterium chelonae and Mycobacterium abscessus strains with modified permeability. J Antimicrob Chemother 51(3):575–584
44. Fraud S, Maillard JY, Russell AD (2001) Comparison of the mycobactericidal activity of ortho- phthalaldehyde, glutaraldehyde and other dialdehydes by a quantitative suspension test. J Hosp Infect 48(3):214–221. https://doi.org/10.1053/jhin.2001.1009
45. Gomes IB, Malheiro J, Mergulhao F, Maillard JY, Simoes M (2016) Comparison of the efficacy of natural-based and synthetic biocides to disinfect silicone and stainless steel surfaces. Pathog Dis 74(4):ftw014. https://doi.org/10.1093/femspd/ftw014
46. Gradel KO, Randall L, Sayers AR, Davies RH (2005) Possible associations between Salmonella persistence in poultry houses and resistance to commonly used disinfectants and a putative role of mar. Vet Microbiol 107 (1–2):127–138. https://doi.org/10.1016/j.vetmic.2005.01.013
47. Grasteau A, Guiraud T, Daniel P, Calvez S, Chesneau V, Le Hénaff M (2015) Evaluation of Glutaraldehyde, Chloramine-T, Bronopol, Incimaxx Aquatic® and Hydrogen Peroxide as Biocides against Flavobacterium psychrophilum for Sanitization of Rainbow Trout Eyed Eggs. J Aquac Res Development 6(12):382
48. Griffiths PA, Babb JR, Bradley CR, Fraise AP (1997) Glutaraldehyde-resistant Mycobacterium chelonae from endoscope washer disinfectors. J Appl Microbiol 82(4):519–526
49. Griffiths PA, Babb JR, Fraise AP (1999) Mycobactericidal activity of selected disinfectants using a quantitative suspension test. J Hosp Infect 41(2):111–121
50. Grobe KJ, Zahller J, Stewart PS (2002) Role of dose concentration in biocide efficacy against Pseudomonas aeruginosa biofilms. J Ind Microbiol Biotechnol 29(1):10–15. https://doi.org/10.1038/sj.jim.7000256
51. Guimarães T, Chimara E, do Prado GVB, Ferrazoli L, Carvalho NGF, Simeão FCdS, de Souza AR, Costa CAR, Viana Niero C, Brianesi UA, di Gioia TR, Gomes LMB, Spadão FdS, Silva MdG, de Moura EGH, Levin AS (2016) Pseudooutbreak of rapidly growing mycobacteria due to Mycobacterium abscessus subsp bolletii in a digestive and respiratory endoscopy unit caused by the same clone as that of a countrywide outbreak. Am J Infect Control 44(11):e221-e226. http://dx.doi.org/10.1016/j.ajic.2016.06.019
52. Guo W, Shan K, Xu B, Li J (2015) Determining the resistance of carbapenem-resistant Klebsiella pneumoniae to common disinfectants and elucidating the underlying resistance mechanisms. Pathog Glob Health 109(4):184–192. https://doi.org/10.1179/2047773215y.0000000022

53. Hanson PJ, Chadwick MV, Gaya H, Collins JV (1992) A study of glutaraldehyde disinfection of fibreoptic bronchoscopes experimentally contaminated with Mycobacterium tuberculosis. J Hosp Infect 22(2):137–142
54. Hashimoto T, Blumenthal HJ (1978) Survival and resistance of Trichophyton mentagrophytes arthrospores. Appl Environ Microbiol 35(2):274–277
55. Henoun Loukili N, Becker H, Harno J, Bientz M, Meunier O (2004) Effect of peracetic acid and aldehyde disinfectants on biofilm. J Hosp Infect 58(2):151–154
56. Hernández A, Martró E, Matas L, Ausina V (2003) In-vitro evaluation of Perasafe compared with 2% alkaline glutaraldehyde against Mycobacterium spp. J Hosp Infect 54(1):52–56
57. Hernández A, Martró E, Puzo C, Matas L, Burgués C, Vázquez N, Castella J, Ausina V (2003) In-use evaluation of Perasafe compared with Cidex in fibreoptic bronchoscope disinfection. J Hosp Infect 54(1):46–51
58. Holton J, Nye P, McDonald V (1994) Efficacy of selected disinfectants against mycobacteria and cryptosporidia. J Hosp Infect 27(2):105–115
59. Holton J, Shetty N, McDonald V (1995) Efficacy of 'Nu-Cidex' (0.35% peracetic acid) against mycobacteria and cryptosporidia. J Hosp Infect 31(3):235–237
60. Howie R, Alfa MJ, Coombs K (2008) Survival of enveloped and non-enveloped viruses on surfaces compared with other micro-organisms and impact of suboptimal disinfectant exposure. J Hosp Infect 69(4):368–376. https://doi.org/10.1016/j.jhin.2008.04.024
61. Isenberg HD (1985) Clinical laboratory studies of disinfection with Sporicidin. J Clin Microbiol 22(5):735–739
62. Isenberg HD, Giugliano ER, France K, Alperstein P (1988) Evaluation of three disinfectants after in-use stress. J Hosp Infect 11(3):278–285
63. Iwasawa A, Niwano Y, Kohno M, Ayaki M (2011) Bactericidal effects and cytotoxicity of new aromatic dialdehyde disinfectants (ortho-phthalaldehyde). Biocontrol Sci 16(4):165–170
64. Jackson J, Leggett JE, Wilson DA, Gilbert DN (1996) Mycobacterium gordonae in fiberoptic bronchoscopes. Am J Infect Control 24(1):19–23
65. Jomha MY, Yusef H, Holail H (2014) Antimicrobial and biocide resistance of bacteria in a Lebanese tertiary care hospital. J Glob Antimicrob Res 2(4):299–305. https://doi.org/10.1016/j.jgar.2014.09.001
66. Juncker JC (2015) COMMISSION IMPLEMENTING REGULATION (EU) 2015/1759 of 28 September 2015 approving glutaraldehyde as an existing active substance for use in biocidal products for product- types 2, 3, 4, 6, 11 and 12. Off J Eur Union 58(L 257):19–26
67. Kadry AA, Serry FM, El-Ganiny AM, El-Baz AM (2017) Integron occurrence is linked to reduced biocide susceptibility in multidrug resistant Pseudomonas aeruginosa. Br J Biomed Sci 74(2):78–84. https://doi.org/10.1080/09674845.2017.1278884
68. Kampf G, Ostermeyer C, Tschudin-Sutter S, Widmer AF (2013) Resistance or adaptation? How susceptible is a 'glutaraldehyde-resistant' Pseudomonas aeruginosa isolate in the absence of selection pressure? J Hosp Infect 84(4):316–318. https://doi.org/10.1016/j.jhin.2013.05.010
69. Khalilzadegan S, Sade M, Godarzi H, Eslami G, Hallajzade M, Fallah F, Yadegarnia D (2016) Beta-Lactamase Encoded Genes blaTEM and blaCTX Among Acinetobacter baumannii Species Isolated From Medical Devices of Intensive Care Units in Tehran Hospitals. Jundishapur J Microbiol 9(5):e14990. https://doi.org/10.5812/jjm.14990
70. Konrat K, Schwebke I, Laue M, Dittmann C, Levin K, Andrich R, Arvand M, Schaudinn C (2016) The bead assay for biofilms: a quick, easy and robust method for testing disinfectants. PLoS ONE 11(6):e0157663. https://doi.org/10.1371/journal.pone.0157663
71. Kressel AB, Kidd F (2001) Pseudo-outbreak of Mycobacterium chelonae and Methylobacterium mesophilicum caused by contamination of an automated endoscopy washer. Infect Control Hosp Epidemiol 22(7):414–418. https://doi.org/10.1086/501926
72. La Scola B, Rolain J-M, Maurin M, Raoult D (2003) Can Whipple's disease be transmitted by gastroscopes? Infect Control Hosp Epidemiol 24(3):191–194

73. Laopaiboon L, Phukoetphim N, Laopaiboon P (2006) Effect of glutaraldehyde biocide on laboratory-scale rotating biological contactors and biocide efficacy. Electron J Biotechnol 9 (4):10
74. Leung HW (2001) Aerobic and anaerobic metabolism of glutaraldehyde in a river water-sediment system. Arch Environ Contam Toxicol 41(3):267–273. https://doi.org/10.1007/s002440010248
75. Leung HW (2001) Ecotoxicology of glutaraldehyde: review of environmental fate and effects studies. Ecotoxicol Environ Saf 49(1):26–39. https://doi.org/10.1006/eesa.2000.2031
76. Liu Q, Liu M, Wu Q, Li C, Zhou T, Ni Y (2009) Sensitivities to biocides and distribution of biocide resistance genes in quaternary ammonium compound tolerant Staphylococcus aureus isolated in a teaching hospital. Scand J Infect Dis 41(6–7):403–409. https://doi.org/10.1080/00365540902856545
77. Lorena NS, Pitombo MB, Cortes PB, Maya MC, Silva MG, Carvalho AC, Coelho FS, Miyazaki NH, Marques EA, Chebabo A, Freitas AD, Lupi O, Duarte RS (2010) Mycobacterium massiliense BRA100 strain recovered from postsurgical infections: resistance to high concentrations of glutaraldehyde and alternative solutions for high level disinfection. Acta cirurgica brasileira 25(5):455–459
78. Lynam PA, Babb JR, Fraise AP (1995) Comparison of the mycobactericidal activity of 2% alkaline glutaraldehyde and 'Nu-Cidex' (0.35% peracetic acid). J Hosp Infect 30(3):237–240
79. Manzoor SE, Lambert PA, Griffiths PA, Gill MJ, Fraise AP (1999) Reduced glutaraldehyde susceptibility in Mycobacterium chelonae associated with altered cell wall polysaccharides. J Antimicrob Chemother 43(6):759–765
80. Marin C, Hernandiz A, Lainez M (2009) Biofilm development capacity of Salmonella strains isolated in poultry risk factors and their resistance against disinfectants. Poult Sci 88(2):424–431. https://doi.org/10.3382/ps.2008-00241
81. Middleton AM, Chadwick MV, Sanderson JL, Gaya H (2000) Comparison of a solution of super-oxidized water (Sterilox) with glutaraldehyde for the disinfection of bronchoscopes, contaminated. J Hosp Infect 45(4):278–282. https://doi.org/10.1053/jhin.2000.0772
82. Namba Y, Suzuki A, Takeshima N, Kato N (1985) Comparative study of bactericidal activities of six different disinfectants. Nagoya J Med Sci 47(3–4):101–112
83. Narui K, Takano M, Noguchi N, Sasatsu M (2007) Susceptibilities of methicillin-resistant Staphylococcus aureus isolates to seven biocides. Biol Pharm Bull 30(3):585–587
84. Neves MS, da Silva MG, Ventura GM, Cortes PB, Duarte RS, de Souza HS (2016) Effectiveness of current disinfection procedures against biofilm on contaminated GI endoscopes. Gastrointest Endosc 83(5):944–953. https://doi.org/10.1016/j.gie.2015.09.016
85. Nicholson G, Hudson RA, Chadwick MV, Gaya H (1995) The efficacy of the disinfection of bronchoscopes contaminated in vitro with Mycobacterium tuberculosis and Mycobacterium avium-intracellulare in sputum: a comparison of Sactimed-I-Sinald and glutaraldehyde. J Hosp Infect 29(4):257–264
86. Nomura K, Ogawa M, Miyamoto H, Muratani T, Taniguchi H (2004) Antibiotic susceptibility of glutaraldehyde-tolerant Mycobacterium chelonae from bronchoscope washing machines. Am J Infect Control 32(4):185–188. https://doi.org/10.1016/j.ajic.2003.07.007
87. Nunoshiba T, Hashimoto M, Nishioka H (1991) Cross-adaptive response in Escherichia coli caused by pretreatment with H_2O_2 against formaldehyde and other aldehyde compounds. Mutat Res 255(3):265–271
88. Orsi GB, Tomao P, Visca P (1995) In vitro activity of commercially manufactured disinfectants against Pseudomonas aeruginosa. Eur J Epidemiol 11(4):453–457
89. Penna TC, Mazzola PG, Silva Martins AM (2001) The efficacy of chemical agents in cleaning and disinfection programs. BMC Infect Dis 1:16
90. Pineau L, Desbuquois C, Marchetti B, Luu Duc D (2008) Comparison of the fixative properties of five disinfectant solutions. J Hosp Infect 68(2):171–177. https://doi.org/10.1016/j.jhin.2007.10.021

91. Rikimaru T, Kondo M, Kajimura K, Hashimoto K, Oyamada K, Sagawa K, Tanoue S, Oizumi K (2002) Bactericidal activities of commonly used antiseptics against multidrug-resistant Mycobacterium tuberculosis. Dermatology (Basel, Switzerland) 204 Suppl 1:15–20. https://doi.org/10.1159/000057719
92. Rikimaru T, Kondo M, Kondo S, Oizumi K (2000) Efficacy of common antiseptics against mycobacteria. Int J Tuberc Lung Dis: Off J Int Union Against Tuberc Lung Dis 4(6):570–576
93. Rutala WA, Cole EC, Wannamaker NS, Weber DJ (1991) Inactivation of Mycobacterium tuberculosis and Mycobacterium bovis by 14 hospital disinfectants. Am J Med 91(3b):267s–271s
94. Sagripanti J-L, Eklund CA, Trost PA, Jinneman KC, Abeyta C, Kaysner CA, Hill WE (1997) Comparative sensitivity of 13 species of pathogenic bacteria to seven chemical germicides. Am J Infect Control 25(4):335–339
95. Sasatsu M, Shibata Y, Noguchi N, Kono M (1992) High-level resistance to ethidium bromide and antiseptics in Staphylococcus aureus. FEMS Microbiol Lett 72(2):109–113
96. Sehmi SK, Allan E, MacRobert AJ, Parkin I (2016) The bactericidal activity of glutaraldehyde-impregnated polyurethane. MicrobiologyOpen 5(5):891–897. https://doi.org/10.1002/mbo3.378
97. Serry FM, Kadry AA, Abdelrahman AA (2003) Potential biological indicators for glutaraldehyde and formaldehyde sterilization processes. J Ind Microbiol Biotechnol 30 (3):135–140. https://doi.org/10.1007/s10295-002-0007-z
98. Shakeri S, Kermanshahi RK, Moghaddam MM, Emtiazi G (2007) Assessment of biofilm cell removal and killing and biocide efficacy using the microtiter plate test. Biofouling 23(1–2):79–86. https://doi.org/10.1080/08927010701190011
99. Shetty N, Srinivasan S, Holton J, Ridgway GL (1999) Evaluation of microbicidal activity of a new disinfectant: Sterilox 2500 against *Clostridium difficile* spores, *Helicobacter pylori*, vancomycin resistant Enterococcus species, *Candida albicans* and several Mycobacterium species. J Hosp Infect 41:101–105
100. Simoes LC, Lemos M, Araujo P, Pereira AM, Simoes M (2011) The effects of glutaraldehyde on the control of single and dual biofilms of Bacillus cereus and Pseudomonas fluorescens. Biofouling 27(3):337–346. https://doi.org/10.1080/08927014.2011.575935
101. Simoes M, Pereira MO, Machado I, Simoes LC, Vieira MJ (2006) Comparative antibacterial potential of selected aldehyde-based biocides and surfactants against planktonic Pseudomonas fluorescens. J Ind Microbiol Biotechnol 33(9):741–749. https://doi.org/10.1007/s10295-006-0120-5
102. Simoes M, Pereira MO, Vieira MJ (2003) Monitoring the effects of biocide treatment of Pseudomonas fluorescens biofilms formed under different flow regimes. Water Sci Technol 47(5):217–223
103. Simoes M, Pereira MO, Vieira MJ (2005) Effect of mechanical stress on biofilms challenged by different chemicals. Water Res 39(20):5142–5152. https://doi.org/10.1016/j.watres.2005.09.028
104. Stanley PM (1999) Efficacy of peroxygen compounds against glutaraldehyde-resistant mycobacteria. Am J Infect Control 27(4):339–343
105. Stewart PS, Grab L, Diemer JA (1998) Analysis of biocide transport limitation in an artificial biofilm system. J Appl Microbiol 85(3):495–500
106. Stickler DJ (1974) Chlorhexidine resistance in *Proteus mirabilis*. J Clin Pathol 27(4):284–287
107. Stickler DJ, Thomas B, Chawla JC (1981) Antiseptic and antibiotic resistance in gram-negative bacteria causing urinary tract infection in spinal cord injured patients. Paraplegia 19:50–58

108. Takeo Y, Oie S, Kamiya A, Konishi H, Nakazawa T (1994) Efficacy of disinfectants against biofilm cells of Pseudomonas aeruginosa. Microbios 79(318):19–26
109. Takigawa K, Fujita J, Negayama K, Terada S, Yamaji S, Kawanishi K, Takahara J (1995) Eradication of contaminating Mycobacterium chelonae from bronchofibrescopes and an automated bronchoscope disinfection machine. Respir Med 89(6):423–427
110. Tortorano AM, Viviani MA, Biraghi E, Rigoni AL, Prigitano A, Grillot R (2005) In vitro testing of fungicidal activity of biocides against Aspergillus fumigatus. J Med Microbiol 54 (Pt 10):955–957. https://doi.org/10.1099/jmm.0.45997-0
111. Tschudin-Sutter S, Frei R, Kampf G, Tamm M, Pflimlin E, Battegay M, Widmer AF (2011) Emergence of glutaraldehyde-resistant *Pseudomonas aeruginosa*. Infect Control Hosp Epidemiol 32(12):1173–1178
112. United States Environmental Protection Agency (2007) Reregistration eligibility decision for glutaraldehyde https://www3.epa.gov/pesticides/chem_search/reg_actions/reregistration/red_PC-043901_28-Sep-07.pdf
113. Urayama S, Kozarek RA, Sumida S, Raltz S, Merriam L, Pethigal P (1996) Mycobacteria and glutaraldehyde: is high-level disinfection of endoscopes possible? Gastrointest Endosc 43(5):451–456
114. van Klingeren B, Pullen W (1993) Glutaraldehyde resistant mycobacteria from endoscope washers. J Hosp Infect 25(2):147–149
115. Vesley D, Melson J, Stanley P (1999) Microbial bioburden in endoscope reprocessing and an in-use evaluation of the high-level disinfection capabilities of Cidex PA. Gastroenterol Nurs: Off J Soc Gastroenterol Nurs Associates 22(2):63–68
116. Vieira CD, Farias Lde M, Diniz CG, Alvarez-Leite ME, Camargo ER, Carvalho MA (2005) New methods in the evaluation of chemical disinfectants used in health care services. Am J Infect Control 33(3):162–169. https://doi.org/10.1016/j.ajic.2004.10.007
117. Vikram A, Bomberger JM, Bibby KJ (2015) Efflux as a glutaraldehyde resistance mechanism in Pseudomonas fluorescens and Pseudomonas aeruginosa biofilms. Antimicrob Agents Chemother 59(6):3433–3440. https://doi.org/10.1128/aac.05152-14
118. Vikram A, Lipus D, Bibby K (2014) Produced water exposure alters bacterial response to biocides. Environ Sci Technol 48(21):13001–13009. https://doi.org/10.1021/es5036915
119. Vizcaino-Alcaide MJ, Herruzo-Cabrera R, Fernandez-Acenero MJ (2003) Comparison of the disinfectant efficacy of Perasafe and 2% glutaraldehyde in in vitro tests. J Hosp Infect 53:124–128
120. Walsh SE, Maillard JY, Russell AD (1999) Ortho-phthalaldehyde: a possible alternative to glutaraldehyde for high level disinfection. J Appl Microbiol 86(6):1039–1046
121. Wang GQ, Zhang CW, Liu HC, Chen ZB (2005) Comparison of susceptibilities of M. tuberculosis H37Ra and M. chelonei subsp. abscessus to disinfectants. Biomed Environ Sci: BES 18(2):124–127
122. Wang HC, Liaw YS, Yang PC, Kuo SH, Luh KT (1995) A pseudoepidemic of Mycobacterium chelonae infection caused by contamination of a fibreoptic bronchoscope suction channel. Eur Res J 8(8):1259–1262
123. Wang Z, Bie P, Cheng J, Wu Q, Lu L (2015) In vitro evaluation of six chemical agents on smooth Brucella melitensis strain. Ann Clin Microbiol Antimicrob 14:16. https://doi.org/10.1186/s12941 015 0077 1
124. Wheeler PW, Lancaster D, Kaiser AB (1989) Bronchopulmonary cross-colonization and infection related to mycobacterial contamination of suction valves of bronchoscopes. J Infect Dis 159(5):954–958
125. WHO (2017) WHO model list of essential medicines. WHO. http://www.who.int/medicines/publications/essentialmedicines/20th_EML2017_FINAL_amendedAug2017.pdf
126. Witney AA, Gould KA, Pope CF, Bolt F, Stoker NG, Cubbon MD, Bradley CR, Fraise A, Breathnach AS, Butcher PD, Planche TD, Hinds J (2014) Genome sequencing and characterization of an extensively drug-resistant sequence type 111 serotype O12 hospital

outbreak strain of Pseudomonas aeruginosa. Clin Microbiol Infect 20(10):O609–O618. https://doi.org/10.1111/1469-0691.12528
127. Zhang X, Kong J, Tang P, Wang S, Hyder Q, Sun G, Zhang R, Yang Y (2011) Current status of cleaning and disinfection for gastrointestinal endoscopy in China: a survey of 122 endoscopy units. Dig Liver Dis: Off J Ital Soc Gastroenterol Ital Assoc Study Liver 43 (4):305–308. https://doi.org/10.1016/j.dld.2010.12.010
128. Zühlsdorf B, Kampf G (2006) Evaluation of the effectiveness of an enzymatic cleaner and a glutaraldehyde-based disinfectant for chemothermal processing of flexible endoscopes in washer-disinfectors in accordance with prEN ISO 15883. Endoscopy 38(6):586–591

Sodium Hypochlorite 8

8.1 Chemical Characterization

The active substance released from sodium hypochlorite in aqueous solutions is active chlorine. The hypochlorite ion is in equilibrium with hypochlorous acid and chlorine. The equilibrium depends on the pH value: chlorine is available only below pH 4; in the neutral pH range, hypochlorous acid is the predominant species, and at pH values higher than 10, the only species present is the hypochlorite ion. The disinfecting efficiency of hypochlorite aqueous solution is dependent on the active chlorine concentration and decreases with an increase in pH and vice versa, which is parallel to the concentration of un-dissociated hypochlorous acid. The activity is strongly reduced by the presence of organic load and in general by the presence of particles. The chlorination and the oxidation reaction of hypochlorite are unspecific [75]. Hypochlorous acid is naturally generated in neutrophils leading to non-specific oxidation in phagocytized bacteria [45]. The basic chemical information on sodium hypochlorite is summarized in Table 8.1.

The stability of sodium hypochlorite solution used as disinfectants can be maintained at 4 °C for 2 years, but after 2 years at 24 °C the concentration of available chlorine was less than 50% of the original. A more rapid deterioration occurred in solutions containing approximately 10% chlorine than in those containing 5% or 1% chlorine. At pH 5–6, decomposition was rapid, whereas increasing the pH increased the stability. In general, hypochlorites are stable if kept in a cool place, out of direct sunlight and in purpose made containers [73], but are dependent on the formulation and may be less than 200 days [138] or only some weeks [36].

8.2 Types of Application

Sodium hypochlorite is used by consumers and by professional workers (widespread uses), in formulation or re-packing, at industrial sites and in manufacturing. Consumer use includes washing and cleaning products, textile treatment products

G. Kampf, *Antiseptic Stewardship*, https://doi.org/10.1007/978-3-319-98785-9_8

Table 8.1 Basic chemical information on sodium hypochlorite [55, 124]

CAS-number	7681-52-9
IUPAC name	Sodium hypochlorite
Synonyms	Bleach, sodium salt from hypochlorous acid
Molecular formula	NaClO
Molecular weight (g/mol)	74.439

and dyes, water treatment chemicals, perfumes, fragrances, cosmetics and personal care products. Professional use includes washing and cleaning products, formulation of mixtures and/or re-packaging, manufacturing of food products, chemicals, textile, leather or fur and wood and wood products. It is also used in laundries, swimming pools, ponds, drinking water, and other water and wastewater systems; on food and non-food contact surfaces; and as a postharvest, seed or soil treatment on various fruit and vegetable crops [164].

Sodium hypochlorite is used in biocides (e.g. wiping disinfectants for surfaces with 0.05–0.5%, spraying disinfectants with up to 3%, or skin disinfection with 0.1%), pH regulators and water treatment products, paper chemicals and dyes [55, 136]. In health care in Japan, for example, the substance has been used as a hand scrub (0.01–0.05%), surgical site antiseptic (0.01–0.05%), mucosa and wound antiseptic (0.005–0.01%), surface disinfectant (0.0125–0.05%) and instrument disinfectant (0.0125–0.05%) [123]. In China, sodium hypochlorite is used for disinfection of surfaces and medical instruments at 0.05% available chlorine, typically at 10–30 min exposure times [103]. Hypochlorite is used for wound antisepsis and is superior to povidone iodine for the treatment of contaminated acute and chronic wounds [93]. It is also used for antiseptic treatment of burns [126].

8.2.1 European Chemicals Agency (European Union)

"Active chlorine released from sodium hypochlorite" has been approved in 2017 as an active biocidal agent for product types 1 (human hygiene), 2 (disinfectants and algaecides not intended for direct application to humans or animals), 3 (veterinary hygiene), 4 (food and feed area) and 5 (drinking water) [81]. The substance is still under review (June 2018) for product types 11 (preservatives for liquid-cooling and processing systems) and 12 (slimicides). Ready-to-use products typically contain 0.05% or 5% sodium hypochlorite [75].

8.2.2 Environmental Protection Agency (USA)

Sodium hypochlorite was first registered for use as pesticide in 1957. It was re-registered in 1991 based on the 1986 registration standard [164].

8.2.3 Overall Environmental Impact

Sodium hypochlorite is manufactured and/or imported in the European Economic Area in 1–10 million t per year [55]. Active chlorine reacts rapidly with organic matter in the sewer, sewage treatment plant, surface water and soil. Where organic and nitrogenous materials are present, it acts as a highly reactive oxidizing agent. It reacts rapidly with organic matter, and most of the active chlorine (≈99%) is converted to inorganic chloride [75]. In seawater, chlorine levels decline rapidly [164]. Contamination of soils due to direct application of chlorinated water will not be of permanent origin. The high content of organic matter in a soil will allow a quick (order of seconds) reduction of HClO, too. Hypochlorite reacts rapidly in soil with soil organics. The ultimate fate of hypochlorite in soil is a reduction in chloride [75].

At environmental pH values (6.5–8.5), half of the active chlorine is present in the un-dissociated form of hypochlorous acid and half is dissociated to the hypochlorite anion. Only the hypochlorous acid fraction is volatile, but the amount of hypochlorous acid that could volatilize from water into air is expected to be very low. Active chlorine does not bioaccumulate or bioconcentrate due to its high water solubility and high reactivity [75].

8.3 Spectrum of Antimicrobial Activity

8.3.1 Bactericidal Activity

8.3.1.1 Bacteriostatic Activity (MIC Values)

The MIC values for sodium hypochlorite obtained with different bacterial species are summarized in Table 8.2.

The range of MIC values to sodium hypochlorite is broad. In *E. faecalis*, it varies between 125 and 32,000 mg/l, in *E. coli* between 0.05 and 12,000 mg/l, in *Lactobacillus* spp. between 64 and 4,096 mg/l, in *L. monocytogenes* between 512 and 7,800 mg/, in *P. aeruginosa* between 1 and 8,192 mg/l, and in *S. aureus* between 10 and 16,384 mg/l (Table 8.2).

8.3.1.2 Bactericidal Activity (Suspension Tests)

The spectrum of bactericidal activity expressed as log reductions obtained with sodium hypochlorite against various bacterial species is summarized in Table 8.3.

Sodium hypochlorite at 100 mg/l does not have sufficient bactericidal activity within 15 min as demonstrated with *A. baumannii*, *A. lwoffii*, *E. cloacae*, *E. coli*, *K. oxytoca*, *K. pneumoniae*, *P. aeruginosa* and *S. maltophilia*. At 500 mg/l, it reduces bacterial counts by at least 5.0 log within 30 min against the majority of bacterial species except *C. perfringens*, *P. aeruginosa*, *S.* Typhimurium and *S. aureus*. At 1,000 mg/l, it is bactericidal within 5 or 10 min against most bacterial species with

Table 8.2 MIC values of various bacterial species to sodium hypochlorite

Species	Strains/isolates	MIC value (mg/l)	References
A. actinomycetemcomitans	ATCC 29523	<1,250	[169]
A. actinomycetemcomitans	ATCC 33384	5,000	[82]
A. baumannii	47 clinical isolates	160–640	[100]
A. calcoaceticus	Drinking water isolate	125	[67]
A. naeslundii	ATCC 12104	1,250	[82]
Acinetobacter spp.	Strain SH-94B from fairy shrimps	20	[147]
Aeromonas spp.	4 isolates from fairy shrimps	5–20	[147]
Arcobacter spp.	32 isolates from chicken slaughterhouse	200–500[a]	[140]
B. fragilis	ATCC 25285	<1,250	[169]
B. adolescentis	4 isolates from faeces of healthy humans	64–2,048	[52]
B. animalis subsp. lactis	8 isolates from faeces of healthy humans	512–2,048	[52]
B. bifidum	31 isolates from faeces of healthy humans	64–2,048	[52]
B. breve	5 isolates from faeces of healthy humans	1,024–2,048	[52]
B. catenulatum	1 isolate from faeces of a healthy human	1,024	[52]
B. infantis	2 isolates from faeces of healthy humans	1,024–2,048	[52]
B. longum	25 isolates from faeces of healthy humans	16–2,048	[52]
B. pseudocatenulatum	15 isolates from faeces of healthy humans	128–1,024	[52]
B. pseudolongum	1 isolate from faeces of a healthy human	1,024	[52]
B. thermoacidophilum	6 isolates from faeces of healthy humans	512–2,048	[52]
B. suis	1 isolate from faeces of a healthy human	1,024	[52]
C. gingivalis	ATCC 33624	2,500	[82]
C. rectus	ATCC 33238	2,500	[82]
E. corrodens	ATCC 23834	5,000	[82]
Enterobacter spp.	54 worldwide strains from hospital- and community-acquired infections	2,000–16,000[a]	[121]
E. faecalis	1 clinical isolate	125	[104]
E. faecalis	ATCC 29212	390	[163]
E. faecalis	ATCC 29212	781	[80]
E. faecalis	ATCC 29212	1,000	[54]
E. faecalis	ATCC 29212	1,250	[20]
E. faecalis	ATCC 29212	1,400	[176]
E. faecalis	56 worldwide strains from hospital- and community-acquired infections	2,000–32,000[a]	[121]
E. faecalis	9 isolates from swine meat production	2,200–5,200	[142]
E. faecalis	ATCC 29212	6,250	[11]
E. faecium	53 worldwide strains from hospital- and community-acquired infections	2,000–16,000[a]	[121]
E. faecium	12 isolates from swine meat production	2,200–5,200	[142]
E. hirae	CIP 5855	1	[109]
Enterocococcus spp.	6 glycopeptide-susceptible isolates and 8 glycopeptide-resistant isolates	1,250–2,500	[20]
E. rhusiopathiae	60 isolates from various sources	160–300	[60]

(continued)

Table 8.2 (continued)

Species	Strains/isolates	MIC value (mg/l)	References
E. coli	Reference strain and clinical isolate	0.05–0.12	[98]
E. coli	ATCC 25922	1	[109]
E. coli	NCTC 10418	125	[104]
E. coli	ATCC 12806	239	[32]
E. coli	306 worldwide strains from hospital- and community-acquired infections	500–16,000[a]	[121]
E. coli	74 isolates from food contact surfaces	2,031–4,063	[77]
E. coli	ATCC 25922	2,500	[47]
E. coli	ATCC 25922	5,000	[169]
E. coli	ATCC 11229	12,000	[74]
E. nodatum	ATCC 33270	5,000	[82]
F. alocis	ATCC 33099	2,500	[82]
F. nucleatum	ATCC 25586	5,000	[82]
K. pneumoniae	60 worldwide strains from hospital- and community-acquired infections	2,000–16,000[a]	[121]
Klebsiella spp.	30 isolates from food contact surfaces	1,016–2,031	[77]
L. acidophilus	4 strains from different origins	512–1,024	[9]
L. amylovorus	7 strains from different origins	128–512	[9]
L. brevis	13 strains from different origins	256–4,096	[9]
L. bulgaricus	6 strains from different origins	512–1,024	[9]
L. coryniformis	3 strains from different origins	2,048–4,096	[9]
L. fermentum	4 strains from different origins	512–1,024	[9]
L. helveticus	39 strains from different origins	256–2,048	[9]
L. paracasei	75 strains from different origins	256–4,096	[9]
L. plantarum	43 strains from different origins	256–4,096	[9]
L. reuteri	42 strains from different origins	64–512	[9]
L. rhamnosus	9 strains from different origins	512–4,096	[9]
L. garvieae	42 isolates from different origins	200–3,200	[9]
L. monocytogenes	ATCC 19112, ATCC 19113, ATCC 19114, ATCC 19115, ATCC 19116, ATCC 19117, ATCC 19118, ATCC 7644, ATCC 13992	512	[63]
L. monocytogenes	ATCC 7644	1,250	[47]
L. monocytogenes	10 isolates from food	1,560–7,800	[1]
L. monocytogenes	4 isolates from food (ice cream, poultry)	2,500–5,000	[105]
L. monocytogenes	ATCC 19111	16,384	[74]
M. morganii	ATCC 25830	10	[109]
P. gingivalis	ATCC 33277 and 3 clinical isolates	390–6,250	[82]
P. micra	ATCC 33270	2,500	[82]
P. intermedia	ATCC 25611	5,000	[82]
P. aeruginosa	ATCC 27853	1	[109]
P. aeruginosa	NCIMB 12469	250	[104]

(continued)

Table 8.2 (continued)

Species	Strains/isolates	MIC value (mg/l)	References
P. aeruginosa	ATCC 27853	1,000	[54]
P. aeruginosa	ATCC 10145	5,000	[169]
P. aeruginosa	ATCC 15442	8,192	[74]
P. intermedia	ATCC 25611	<1,250	[169]
S. choleraesuis	ATCC 10708	8,192	[74]
S. enterica	10 multidrug-resistant strains isolated from poultry	390–440	[119]
S. Typhimurium	ATCC 14028	2,500	[169]
S. Typhimurium	1 poultry isolate	6,000	[31]
Salmonella spp.	901 worldwide strains from hospital- and community-acquired infections	500–4,000[a]	[121]
S. aureus	CIP 53154	10	[109]
S. aureus	NCTC 6571	250	[104]
S. aureus	1,635 worldwide strains from hospital- and community-acquired infections	250–4,000[a]	[121]
S. aureus	ATCC 25923	312	[47]
S. aureus	ATCC 25923	1,000	[4]
S. aureus	ATCC 6538	1,000	[54]
S. aureus	22 isolates from food contact surfaces	2,031–4,063	[77]
S. aureus	ATCC 25923	9,000	[74]
S. aureus	ATCC 33591	10,000	[169]
S. aureus	ATCC 6538	16,384	[74]
S. epidermidis	65 isolates from food contact surfaces	2,031–4,063	[77]
S. maltophilia	Drinking water isolate	175	[67]
S. gordonii	ATCC 10558	5,000	[82]
S. mutans	ATCC 25175	5,000	[169]
S. salivarus	44 strains from different origins	400–1,600	[9]
S. thermophilus	135 strains from different origins	50–400	[9]
T. forsythia	ATCC 43037 and 3 clinical isolates	390–3,130	[82]

[a]Active chlorine

the exception of *A. baumannii*, *K. pneumoniae* and *S. maltophilia*. At 5,000 mg/l or more, sodium hypochlorite was found to be bactericidal within 30 min (Table 8.3).

The minimum bactericidal concentration depends on the species and begins at 4 mg/l sodium hypochlorite in *S. aureus* but can also be as high as 25,000 mg/l sodium hypochlorite in *E. faecalis* (Table 8.4). It is noteworthy that the bactericidal concentration is for some species at the same level as the bacteriostatic concentration of sodium hypochlorite (see also Table 8.2).

The bactericidal activity can be impaired. In the presence of 20% serum or more, 2,500 mg/l sodium hypochlorite was basically not bactericidal anymore in 5 or 10 min. The negative effect of 10% serum on the bactericidal activity was much lower [21].

Table 8.3 Bactericidal activity of sodium hypochlorite in suspension tests

Species	Strains/isolates	Exposure time	Concentration (mg/l)	log_{10} reduction	References
A. baumannii	13 clinical strains	5 min	1,000 (P)	4.0	[50]
		15 min	100 (P)	0.4	
A. lwoffii	2 clinical strains	5 min	1,000 (P)	5.6	[50]
		15 min	100 (P)	0.2	
B. cereus[a]	ATCC 14579	30 min	500 (S)	>5.0	[148]
B. cenocepacia	LMG 16656, LMG 18828	5 min	3,000 (S)	≥ 5.0	[135]
			1,000 (S)		
			500 (S)		
B. pseudomallei	10 clinical isolates	30 min	5,000 (P)	≥ 5.0	[175]
C. jejuni	ATCC BAA-1062, ATCC 33560 and 2 field strains	1 min	5,000 (S)	2.7–6.1	[71]
Campylobacter spp.	46 strains from broilers and pigs	5 min	6,300 (S)	≥ 5.0	[13]
C. perfringens[a]	Strain CDC 1861	30 min	500 (S)	0.1	[148]
E. cloacae	3 clinical strains	5 min	1,000 (P)	5.9	[50]
		15 min	100 (P)	0.2	
E. faecalis	ATCC 29212	20 min	25,000 (S)	≥ 5.0	[160]
E. faecalis	ATCC 35550	10 min	25,000 (S)	≥ 5.0	[78]
		1 min	52,500 (S)		
E. coli	NCTC 8196	5 min	2,500 (S)	≥ 5.0	[21]
		10 min	1,000 (S)		
E. coli	17 clinical strains	5 min	1,000 (P)	5.9	[50]
		15 min	100 (P)	3.2	
E. coli	ATCC 43895 (0157:H7)	5 min	512 (S)[a]	≥ 8.0	[141]
			256 (S)[a]		
			128 (S)[a]		
E. coli	Food isolate strain 0157: H7	30 min	500 (S)	6.2	[148]
E. coli	K12 strain	10 min	16 (S)[b]	>5.0	[33]
		2 h	4 (S)[b]		
E. coli	6 Shiga toxigenic strains	2.5 min	10 (S)[b]	≥ 5.0	[3]
		5 min	1 (S)[b]		
F. nucleatum	NCTC 10562	5 min	6,000 (S)	5.4	[88]
H. parasuis	2 strains (serovars 1 and 5)	1 min	500 (S)	4.3–5.4	[143]
				1.8–2.5[b]	
H. pylori	NCTC 11637, NCTC 11916 and 7 clinical isolates	30 s	150 (P)	>5.0	[2]
		1–30 min[c]			
K. oxytoca	5 clinical strains	5 min	1,000 (P)	6.1	[50]
		15 min	100 (P)	1.6	

(continued)

Table 8.3 (continued)

Species	Strains/isolates	Exposure time	Concentration (mg/l)	log_{10} reduction	References
K. pneumoniae	NCIMB 13291	5 min	48,000 (S)	4.4	[88]
			30,000 (S)		
			6,000 (S)		
K. pneumoniae	15 clinical strains	5 min	1,000 (P)	4.7	[50]
		15 min	100 (P)	0.6	
L. monocytogenes	Strain Scott A	5 min	512 (S)[a]	≥ 8.0	[141]
			256 (S)[a]		
			128 (S)[a]		
L. monocytogenes	Food isolate	30 min	500 (S)	>6.1	[148]
L. monocytogenes	20 environmental and food isolates	5 min	30–60 (P)	≥ 5.0	[40]
L. monocytogenes	LO28	5 min	0.6[b] (S)	≥ 5.0	[116]
P. gingivalis	ATCC 53978	5 min	6,000 (S)	8.6	[88]
P. aeruginosa	ATCC 15442	5 min	6,300 (S)	≥ 5.0	[13]
P. aeruginosa	NCTC 6570	5 min	2,500 (S)	≥ 5.0	[21]
		10 min	1,000 (S)	≥ 5.0	
P. aeruginosa	20 clinical strains	5 min	1,000 (P)	6.2	[50]
		15 min	100 (P)	0.4	
P. aeruginosa	ATCC 15442	5 min	512 (S)[a]	≥ 8.0	[141]
			256 (S)[a]		
			128 (S)[a]		
P. aeruginosa	ATCC 27853	30 min	500 (S)	1.3	[148]
P. aeruginosa	NCTC 6749	2 min	5 (P)[b]	≥ 5.0	[36]
		10 min	5 (P)[b]		
S. Typhimurium	ATCC 14028	30 min	500 (S)	4.1	[148]
S. enteritidis	Chicken isolate	10, 30 and 60 min	400 (P)	4.0–5.0	[95]
			300 (P)	2.9–5.0	
			200 (P)	2.2–4.0	
S. enteritidis	ATCC 13076	5 min	512 (S)[a]	≥ 8.0	[141]
			256 (S)[a]		
			128 (S)[a]		
S. sonnei	Food isolate	30 min	500 (S)	>6.3	[148]
S. aureus	ATCC 29213	30 min	25,000 (S)	≥ 5.0	[160]
S. aureus	ATCC 6538	5 min	6,300 (S)	≥ 5.0	[13]
S. aureus	NCTC 4163	5 min	2,500 (S)	≥ 5.0	[21]
		10 min	1,000 (S)		
S. aureus	ATCC 6538	5 min	512 (S)[a]	≥ 8.0	[141]
			256 (S)[a]		
			128 (S)[a]		
S. aureus	ATCC 25923	30 min	500 (S)	4.8	[148]

(continued)

Table 8.3 (continued)

Species	Strains/isolates	Exposure time	Concentration (mg/l)	$\log_{10}$ reduction	References
S. aureus	Strain DFSN_B26 (cheese derived)	6 min	450 (S)	3.5	[168]
			350 (S)	2.8	
			250 (S)	1.7	
S. aureus	Human isolate	10, 30 and 60 min	400 (P)	1.8–5.0	[95]
			300 (P)	2.1–3.2	
			200 (P)	0.7–2.3	
S. aureus	NCTC 4163	2 min	12.5 (P)[b]	≥ 5.0	[36]
		10 min	10 (P)[b]		
S. epidermidis	ATCC 12228	30 min	500 (S)	6.3	[148]
S. maltophilia	2 clinical strains	5 min	1,000 (P)	0.0	[50]
		15 min	100 (P)	0.2	
S. mutans	NCTC 10449	5 min	48,000 (S)	5.4	[88]
			30,000 (S)		
			6,000 (S)		
V. cholerae	Strain C6706	30 min	500 (S)	>6.4	[148]
V. parahaemolyticus	Strain NY477	30 min	500 (S)	>6.2	[148]
V. parahaemolyticus	ATCC 2210001	30 s	35[b] (S)	≥ 5.0	[139]
V. vulnificus	Strain LA M624	30 min	500 (S)	>6.3	[148]
V. vulnificus	Strain KCTC 2962	30 s	35[c] (S)	≥ 5.0	[139]
Y. enterocolitica	Strain 8081	30 min	500 (S)	>6.8	[148]
Mixed anaerobic species	*A. actinomycetemcomitans* ATCC 43718, *A. viscosus* DSMZ 43798, *F. nucleatum* ATCC 10953, *P. gingivalis* ATCC 33277, *V. atypica* ATCC 17744 and *S. gordonii* ATCC 33399	30 s	500 (S)	7.5	[44]

P commercial product; *S* solution; [a]vegetative cell form; [b]free chlorine; [c]with organic load

8.3.1.3 Activity Against Bacteria in Biofilms

The bactericidal activity of sodium hypochlorite against bacteria in biofilms is summarized in Table 8.5.

At 10 mg/l, sodium hypochlorite has only poor activity against bacterial cells in biofilms. At 100 mg/l, it reduces bacterial cells against the majority of selected species by 4.0 log within 30 min (*B. cepacia*, *E. hirae*, *E. coli*, *M. morganii*, *P. aeruginosa*, *S. aureus*), whereas the effect is low within a 30 s exposure time (*E. coli*, *L. monocytogenes*, *S. Typhimurium*) or a 5 min exposure time (*S. Enteritidis*, *S. aureus*) or against MRSA in biofilm [131]. At 10,000 mg/l, a good bactericidal activity was found within 30 min against *E. coli* and *E. faecalis* but not against

Table 8.4 MBC values of various bacterial species to sodium hypochlorite (5 min exposure time)

Species	Strains/isolates	MBC value (mg/l)	References
E. faecalis	ATCC 29212	1,562–25,000	[163]
E. coli	74 isolates from food contact surfaces	2,031–4,063	[77]
Klebsiella spp.	30 isolates from food contact surfaces	1,016–2,031	[77]
L. monocytogenes	ATCC 19112, ATCC 19113, ATCC 19114, ATCC 19115, ATCC 19116, ATCC 19117, ATCC 19118, ATCC 7644, ATCC 13992	512	[63]
P. aeruginosa	31 isolates from burns	15–30	[37]
Salmonella spp.	11 strains (untreated wastewater) 10 strains (treated wastewater)	34 ± 9^a 41 ± 14^a	[53]
S. aureus	56 isolates (QAC tolerant)	$4–32^b$	[103]
S. aureus	12 isolates from burns	15–60	[37]
S. aureus	42 clinical MRSA isolates	16–128	[123]
S. aureus	ATCC 6538 and 12 isolates from fishery products	$600–900^c$	[166]
S. aureus	22 isolates from food contact surfaces	2,031–4,063	[77]
S. epidermidis	65 isolates from food contact surfaces	2,031–4,063	[77]
S. pseudintermedius	12 methicillin-resistant isolates from canine skin	$1,922^c$	[134]
S. pyogenes	5 isolates from burns	4–15	[37]

[a]Mean with stdev; [b]available chlorine; [c]30-min exposure time

P. aeruginosa and *S. aureus*. Sodium hypochlorite at 52,500 mg/l was bactericidal against *E. faecalis* biofilm bacteria within 30 min unless the biofilm was mature (3 w) and prepared in dental root canals, or it was used from in dental unit waterlines (mixed biofilm) (Table 8.5).

Overall, the susceptibility of bacteria in biofilms to sodium hypochlorite seems to be variable, especially when compared to the data obtained with planktonic cells (Table 8.3). Some studies suggest a higher resistance of biofilm cells. *P. marginalis* cells grown for 24 at 30 °C in a biofilm were described to be 9.2 times less susceptible to sodium hypochlorite compared to planktonic cells. When the cells were grown in biofilm for 48 h, they were even 13.5 times less susceptible [96]. These findings are supported by data showing that the eradication of biofilm cells of *P. aeruginosa* by sodium hypochlorite required much longer time than that of planktonic cells in suspensions [157]. A 24 h biofilm on polystyrene microtiter plates grown by eight strains of *P. aeruginosa* was quite susceptible to sodium

Table 8.5 Efficacy of sodium hypochlorite against bacteria in biofilms

Species	Strains/ solates	Type of biofilm	Exposure time	Concentration (mg/l)	$\log_{10}$ reduction	References
A. calcoaceticus	Drinking water isolate	24-h incubation on PVC	30 min	125 (S)	0.9	[67]
				0.5 (S)	0.1	
B. cepacia	6 isolates from disinfectants and aerosol solution	5-d incubation on silicone discs	15 s	100 (S)	≥ 5.0	[117]
C. jejuni	30 strains from chicken carcasses	48-h incubation in 96-well plates	24 h	10,000 (S)	≥ 5.1	[114]
E. hirae	CIP 5855	48-h incubation on polypropylene, PVC and silicone	30 min	100 (S)	>5.0	[109]
				10 (S)	4.0–5.0	
E. faecalis	ATCC 29212	3-w incubation on pieces of cellulose nitrate membranes	10 s	52,500 (P)	"complete elimination"	[65]
E. faecalis	ATCC 29212	3-w incubation in single-rooted teeth canals	30 s	52,500 (P)	1.2–1.3	[159]
			1 min		1.4	
			5 min		1.7–2.1	
E. faecalis	ATCC 29212	24-h incubation in 48-well plates	1 min	52,500 (S)	5.2	[107]
				10,000 (S)	2.3	
E. faecalis	Strain A197A	3-w incubation in root samples	3 min	52,500 (P)	>7.0	[69]
				25,000 (P)		
E. faecalis	ATCC 29212	3-w incubation on dentin discs	10 min	52,500 (S)	1.2	[26]
E. faecalis	ATCC 29212	6-w incubation on teeth	30 min	52,500 (S)	≥ 5.0	[177]
				25,000 (P)	5.0	
				25,000 (S)	≥ 5.0	
E. faecalis	Not described.	4-, 6- or 10-w incubation on human teeth	10 min	50,000 (S)	"complete inactivation"	[61]
				25,000 (S)	"complete inactivation"	
				10,000 (S)	0.7–0.8	

(continued)

Table 8.5 (continued)

Species	Strains/isolates	Type of biofilm	Exposure time	Concentration (mg/l)	log_{10} reduction	References
E. faecalis	ATCC 700802	4-w incubation on human teeth	2 min	40,000 (S)	6.2	[34]
				10,000 (S)	4.2	
E. faecalis	Not described.	5-d incubation on dentin	3 min	25,000 (S)	0.7	[8]
E. faecalis	ATCC 29212	48-h incubation in canals of single-rooted teeth	5 min	25,000 (S)	5.5	[165]
E. coli	ATCC 35218	48-h incubation on glass, polypropylene, polycarbonate, silicone and PVC	30 min	10,000 (S)	"Complete inactivation"	[108]
E. coli	B6-914 strain 0157:H7	2-h incubation on cantaloupe rind surfaces	5 min	2,000 (S)	≥ 5.5	[62]
				200 (S)	1.3	
E. coli	B6-914 strain 0157:H7	12-h incubation on cantaloupe rind surfaces	5 min	2,000 (S)	2.0	[62]
				200 (S)	0.9	
E. coli	B6-914 strain 0157:H7	24-h incubation on cantaloupe rind surfaces	5 min	2,000 (S)	1.5	[62]
				200 (S)	0.7	
E. coli	B6-914 strain 0157:H7	2-h incubation on cover glass	5 min	320 (S)	≥ 4.7	[62]
				160 (S)	≥ 4.7	
				80 (S)	≥ 4.7	
				40 (S)	1.2	
E. coli	B6-914 strain 0157:H7	12-h incubation on cover glass	5 min	320 (S)	≥ 6.0	[62]
				160 (S)	1.7	
				80 (S)	0.9	
				40 (S)	0.6	

(continued)

Table 8.5 (continued)

Species	Strains/isolates	Type of biofilm	Exposure time	Concentration (mg/l)	$\log_{10}$ reduction	References
E. coli	B6-914 strain 0157:H7	24-h incubation on cover glass	5 min	320 (S)	≥ 6.5	[62]
				160 (S)	1.4	
				80 (S)	1.1	
				40 (S)	0.8	
E. coli	Strain O157, isolate from food poisoning outbreak	8-d incubation on stainless steel	5 min	200 (S)[a]	5.5	[162]
				100 (S)[a]	4.3	
				50 (S)[a]	2.7	
				25 (S)[a]	0.7	
E. coli	ATCC 35150, ATCC 43889, ATCC 43890	24-h incubation on stainless steel	30 s	100 (P)	1.1	[14]
				50 (P)	1.0	
				20 (P)	0.6	
E. coli	ATCC 25922	48-h incubation on polypropylene, PVC and silicone	30 min	100 (S)	>5.0	[109]
				10 (S)	2.0 – >5.0	
E. coli	K-12 MG1655	4-d incubation on polycarbonate	10 min	10 (P)	1.8	[153]
F. nucleatum	ATCC 25586	4-d incubation on glass slides	1 min	50,000 (P)	0.3	[12]
L. plantarum	JCM 1149	24-h incubation on glass cover slips	30 min	12.5–275 (S)	0.1–1.1	[94]
L. monocytogenes	20 environmental and food isolates	48-h incubation in microtiter plates	5 min	1,800–4,600 (P)	≥ 5.0	[40]
L. monocytogenes	ATCC 15313, ATCC 19114, ATCC 19115	24-h incubation on stainless steel	30 s	100 (P)	1.3	[14]
				50 (P)	1.1	
				20 (P)	0.4	

(continued)

Table 8.5 (continued)

Species	Strains/isolates	Type of biofilm	Exposure time	Concentration (mg/l)	$\log_{10}$ reduction	References
L. monocytogenes	11 strains from different origins	48-h incubation in polystyrene microtiter plates and on stainless steel	6 min	10 (S)	1.5–1.8	[92]
M. morganii	ATCC 25830	48-h incubation on polypropylene, PVC and silicone	30 min	100 (S)	≥5.0	[109]
				10 (S)	2.1–4.6	
P. aeruginosa	ATCC 19142	6-d incubation on aluminium	1 min	25,000 (S)	3.0	[46]
			5 min		4.0	
			20 min		7.0	
P. aeruginosa	ATCC 19142	6-d incubation on stainless steel	1 min	25,000 (S)	4.0	[46]
			5 min		6.0	
			20 min		7.0	
P. aeruginosa	Strain PA01	24-h incubation in microtiter plates	1, 5, 15, 30 and 60 min	10,000 (S)	1.3–2.7	[91]
P. aeruginosa	ATCC 700928	24-h incubation in microplates	1 min	10,000 (S)	1.1	[161]
			5 min		1.2	
			60 min		1.2	
P. aeruginosa	ATCC 27853	48-h incubation on polypropylene, PVC and silicone	30 min	100 (S)	>5.0	[109]
				10 (S)	4.0–5.0	
S. Enteritidis	Isolate from food poisoning outbreak	8-d incubation on stainless steel	5 min	200 (S)[a]	3.9	[162]
				100 (S)[a]	2.5	
				50 (S)[a]	1.6	
				25 (S)[a]	0.9	

(continued)

Table 8.5 (continued)

Species	Strains/isolates	Type of biofilm	Exposure time	Concentration (mg/l)	log_{10} reduction	References
S. enterica	2 strains	2-d incubation in biofilm reactor	10 min	200 (S)	0.1–0.2	[39]
			45 min		0.2–0.3	
			90 min		0.8–1.0	
S. enterica	4 strains	7-d incubation in biofilm reactor	10 min	200 (S)	0.2–0.4	[39]
			45 min		0.3–0.7	
			90 min		0.3–1.0	
S. enterica	8 strains from different origins	48-h incubation in polystyrene microtiter plates and on stainless steel	6 min	10 (S)	2.2–2.8	[92]
S. Typhimurium	ATCC 14028	3-d incubation on a 96-peg lid	1 min	5,250 (S)	2.1	[174]
				2,625 (S)	4.0	
				1,310 (S)	≥ 7.0	
			5 min	5,250 (S)	≥ 6.0	
				2,625 (S)	≥ 6.0	
				1,310 (S)	≥ 6.0	
S. Typhimurium	ATCC 19585, ATCC 43971, DT 104	24-h incubation on stainless steel	30 s	100 (P)	2.2	[14]
				50 (P)	1.5	
				20 (P)	1.2	
S. Typhimurium	ATCC 14028	24-h incubation on acrylic and stainless steel coupons	5 min	50 (P)	≥ 7.0	[125]
S. Typhimurium	3 strains (FMCC B-137, FMCC B-193, FMCC B-415)	6-d incubation on stainless steel	6 min	10 (S)	0.3–0.8	[66]
S. aureus	ATCC 25923	3-w incubation on pieces of cellulose nitrate membranes	10 s	52,500 (P)	"complete elimination"	[65]

(continued)

Table 8.5 (continued)

Species	Strains/isolates	Type of biofilm	Exposure time	Concentration (mg/l)	log_{10} reduction	References
S. aureus	ATCC 6538	24-h incubation on glass coupons	5 min	30,000 (S)	≥ 4.0	[106]
				3,000 (S)	2.3–2.8	
S. aureus	Strain DFSN_B26 (cheese derived)	96-h incubation in polystyrene 96-well plates	6 min	25,000 (S)	1.4	[168]
				15,000 (S)	0.8	
				7,500 (S)	1.2	
S. aureus	ATCC 25923	12-d incubation on polycarbonate coupons (dry surface biofilm)	10 min	20,000 (S)	≥ 7.0	[4]
				1,000 (S)		
S. aureus	ATCC 6538 and 12 isolates from fishery products	48-h incubation on stainless steel coupons	30 min	18,000 (S)	≥ 5.0	[166]
				5,000 (S)		
S. aureus	ATCC 6538	72-h incubation in microplates	1 min	10,000 (S)	2.0	[161]
			5 min			
			60 min			
S. aureus	ATCC 6538	24 incubation in microtiter plates	1, 5, 15, 30 and 60 min	10,000 (S)	2.3–2.5	[91]
S. aureus	Strain S3	15-d incubation on polypropylene and stainless steel	30 s	250 (S)	1.9–2.7	[43]
S. aureus	Isolate from food poisoning outbreak	8-d incubation on stainless steel	5 min	200 (S)[a]	2.0	[162]
				100 (S)[a]	1.8	
				50 (S)[a]	1.5	
				25 (S)[a]	1.2	
S. aureus	CIP 53154	48-h incubation on polypropylene, PVC and silicone	30 min	100 (S)	>5.0	[109]
				10 (S)	≥ 5.0	

(continued)

Table 8.5 (continued)

Species	Strains/isolates	Type of biofilm	Exposure time	Concentration (mg/l)	log_{10} reduction	References
S. aureus	3 strains (FMCC B-134, FMCC B-135, FMCC B-410)	6-d incubation on stainless steel	6 min	10 (S)	0.4–0.6	[66]
S. maltophilia	Drinking water isolate	24-h incubation on PVC	30 min	175 (S)	0.5	[67]
				0.5 (S)	0.0	
S. mutans	DSM 20523	72-h incubation on titanium discs	30 min	6,000 (S)	6.9	[89]
Y. enterocolitica	16 food isolates	1–5-d incubation on stainless steel	1 min	50 (P)	3.0–5.0	[172]
Mixed species	Species from dental unit waterlines	Natural biofilm	2 d	52,500 (S)	0.8–1.0	[99]
Mixed species	Polymicrobial samples from infected root canals	3-w incubation on teeth	3 min	25,000 (S)	1.9	[145]
Mixed species	Mixed oral biofilm	12-h incubation in the oral cavity on titanium surfaces	1 min	10,000 (S)	"significant reduction"	[68]
Mixed species	*S. gordonii* ATCC 10558, *P. gingivalis* ATCC 33277, *T. forsythia* ATCC 43037, *F. nucleatum* ATCC 25586, *A. naeslundii* ATCC 12104, and *P. micra* ATCC 33270	4-d incubation in 96-well plates	1 h	9,500 (S)	6.0–7.0	[82]

(continued)

Table 8.5 (continued)

Species	Strains/isolates	Type of biofilm	Exposure time	Concentration (mg/l)	$\log_{10}$ reduction	References
Mixed species	Human saliva bacteria	72-h incubation on titanium discs	30 min	6,000 (S)	2.4	[89]
Mixed species	Subgingival plaque bacteria	Overnight incubation on titanium discs	30 min	6,000 (S)	1.8	[89]
Mixed species	*L. monocytogenes* strain Scott A and *Pseudomonas* spp. strain M-21, a meat processing plant isolate	48-h incubation on stainless steel coupons	1 min	80 (P)	≥ 7.0	[56]
			5 min			

P commercial product; *S* solution; [a]free chlorine

hypochlorite requiring concentrations between 350 and 500 mg/l to achieve a bactericidal effect in 60 min [133].

The reduced efficacy of hypochlorite against bacteria in biofilms is partly explained by a transport limitation of the biocide into the biofilm as shown with *E. aerogenes* [155]. In a non-typable *H. influenzae* biofilm, it was shown that resistance to sodium hypochlorite is mediated to a large part by the cohesive and protective properties of the biofilm matrix [76]. Tests with *E. coli* CIP 54127 obtained from culture on tryptic soy agar or in the form of biofilms showed a great impairment of bactericidal activity of sodium hypochlorite against biofilm cells. The reduction in sensitivity was attributed to a reduced accessibility of the bacterial cells to the disinfectants, due to the fact that the former adhered to a support [128]. Finally, biofilm treated with sodium hypochlorite may still serve as a bacterial reservoir. Sodium hypochlorite at 10,000 mg/l applied for up to 1 h to biofilm of *S.* Typhimurium, *E. coli*, *S. mutans* or *B. fragilis* on glass or rubber carrier was not effective enough to prevent survival of the *S.* Typhimurium on rubber (1 h) and *S. mutans* on glass (30 min) [169].

Compared to MBC values obtained with planktonic cells (Table 8.4), the minimum bactericidal concentration of sodium hypochlorite against selected biofilm cells was much higher with >65,000 mg/l (Table 8.6).

Microbial persistence has been described for various species despite treatment of biofilms with sodium hypochlorite. *S. aureus*, for example, was reduced by sodium hypochlorite in biofilm by 7.0 log. Staining of residual biofilm showed that live *S. aureus* cells remained with approximately 0.8% of the initial biofilm bacteria [4]. A polymicrobial biofilm from infected root canals (3 w incubation on teeth) was treated with a solution of 25,000 mg/l sodium hypochlorite for 3 min. A proportion of 4.3% viable cells remained in the biofilm [145]. A low number of survivors (1.3 log) was also described with *C. jejuni* in biofilm for 6 of 30 strains from chicken carcasses after exposure to 10,000 mg/l sodium hypochlorite for 24 h [114]. *L. pneumophila* may also survive in low numbers for 28 d in the presence of chlorine at up to 0.4 mg/l. Immediately after exposure to 50 mg/l chlorine for 1 h, the biofilms yielded no recoverable colonies, but colonies did reappear in low numbers over the following days. Despite chlorination at 50 mg/l for 1 h, both one- and two-month-old *L. pneumophila* biofilms were able to survive this treatment and to continue to grow, ultimately exceeding 10^6 cfu per disc [38].

Bacterial persistence after sodium hypochlorite exposure followed by outgrowth of the survivors from the biofilm may increase the level of antibiotic resistance

Table 8.6 MBC values for sodium hypochlorite solutions (5 min exposure time) obtained with bacterial cells from biofilms

Species	Strains/isolates	MBC value (mg/l)	References
E. coli	74 isolates from food contact surfaces	>65,000[a]	[77]
Klebsiella spp.	30 isolates from food contact surfaces	>65,000[a]	[77]
S. aureus	22 isolates from food contact surfaces	>65,000[a]	[77]
S. epidermidis	65 isolates from food contact surfaces	>65,000[a]	[77]

[a]Free chlorine

genes in water. When ciprofloxacin was exposed to 1 mg/l sodium hypochlorite in drinking water distribution systems, the piperazine ring was destroyed by chlorination. Correspondingly, specific antibiotic resistance genes such as mexA and qnrS increased in effluents, while qnrA and qnrB increased in biofilms indicating growth of these bacterial genera by transformation of ciprofloxacin chlorination products in drinking water distribution systems [171].

8.3.1.4 Bactericidal Activity in Carrier Tests

A low concentration of sodium hypochlorite (0.6 mg/l free chlorine) was able to reduce *L. monocytogenes* (strain LO28) within 5 min by 3.5–4.0 log, depending on the type of carrier [116]. When *S. aureus* was placed on a glass cup carrier and exposed to 5,500 mg/l sodium hypochlorite, a log reduction of >6.0 was found after 1 min [19]. In a quantitative carrier test, sodium hypochlorite (500 mg/l) was effective against *S. aureus* and *P. aeruginosa* with a single application and within the drying time of 3 min [132]. On surfaces, it was found with five bacterial species that a concentration of 512 mg/l is necessary to achieve a log reduction between 2.0 (*P. aeruginosa*) or 5.0 (*L. monocytogenes*) within 5 min [141].

Wipes based on 0.55% sodium hypochlorite and 0.94% sodium hypochlorite (both 10 min application time) were effective to reduce *Y. pseudotuberculosis* (ATCC 6902), *S. aureus* (ATCC 6538) and *B. thailandensis* (ATCC 700388) cells from a pulse oximeter sensor [122]. According to ASTM E 2967, the efficacy of disinfectant wipes soaked with sodium hypochlorite (1000 mg/l active chlorine) was determined on stainless steel carriers contaminated with *S. aureus* (ATCC 6538) or *A. baumannii* (ATCC 19568). With a 10 s wipe the bacterial load was reduced by at least 7.0 log, the control wipe without the disinfectant yielded a 3.0 log reduction. From none of the disinfected surfaces, a transfer of the test organism to another sterile surface was observed [149]. Against *L. innocua* and *L. monocytogenes*, sodium hypochlorite at 60 mg/l was very effective within 1 min in a carrier test with >5.0 log; in the presence of serum, however, the effect was marginal with 2.0 log [16].

8.3.1.5 Bactericidal Activity in Other Applications

Sodium hypochlorite at 10,000 mg/l was quite effective within 1 min for disinfection of titanium implants contaminated with *S. sanguinis* or *S. epidermidis* [28]. The substance was also very effective at 52,500 mg/l against *S. aureus*, *P. aeruginosa*, group D *Streptococcus* and *B. subtilis* within 5 min for disinfection of dentures [144]. Infected root canals from teeth with apical periodontitis were irrigated with 25,000 mg/l sodium hypochlorite. The mean bacterial cells count was reduced by 2.5 log with 6 of 16 canals yielding negative cultures [152]. The efficacy of 42,000 mg/l sodium hypochlorite against *E. faecalis* in root canals (5 min exposure time) was best at a pH value of 6.5 compared to equivalent solutions at pH values of 7.5 and 12 [115]. In swimming pool water, sodium hypochlorite with 1 mg/l active chlorine showed good bactericidal activity within 10 min against *P. aeruginosa* ($\geq$3.9 log), *E. coli* ($\geq$4.2 log), *S. aureus* ($\geq$3.9 log) and *L. pneumophila* ($\geq$3.9 log) [23].

8.3.2 Fungicidal Activity

8.3.2.1 Fungistatic Activity (MIC Values)

The majority of MIC values for *Candida* spp. *Aspergillus* spp., *Penicillum* spp., *Mucor* spp., *Rhizopus* spp. and *Trichoderma* spp. is 2,048 mg/l sodium hypochlorite or lower (Table 8.7). For *C. albicans*, an epidemiologic cut-off value of 8,200 mg/l active chlorine has been proposed to determine resistance to sodium hypochlorite [121]. Most *C. albicans* isolates described in Table 8.7 would have to be regarded as susceptible to sodium hypochlorite. For other fungal species, no such cut-off value is currently available.

Table 8.7 MIC values for different fungal species obtained with sodium hypochlorite

Species	Strains/isolates	MIC value (mg/l)	References
A. flavus	7 isolates from surfaces in a veterinary hospital	40–160	[110]
A. flavus	3 clinical, 3 airborne and 2 food isolates	512–2,048	[83]
A. fumigatus	9 isolates from surfaces in a veterinary hospital	40–160	[110]
A. fumigatus	6 clinical and 14 airborne isolates	128–2,048	[83]
A. niger	2 isolates from surfaces in a veterinary hospital	40–160	[110]
A. niger	2 airborne and 2 food isolates	256–512	[83]
A. ochraceus	2 food isolates	1,024–2,048	[83]
C. albicans	Not described	0.025–0.05	[98]
C. albicans	Not described	<10	[57]
C. albicans	Strain USP 562	1,000	[54]
C. albicans	Not described	1,620	[176]
C. albicans	200 worldwide strains from hospital- and community-acquired infections	2,000–16,000[a]	[121]
C. albicans	ATCC 90028	3,125	[11]
C. glabrata	Not described	0.0125–0.05	[98]
C. krusei	Not described	0.0125–0.025	[98]
C. tropicalis	Not described	0.025–0.05	[98]
Candida spp.	9 strains	312–1,250	[58]
Mucor spp.	2 clinical and 1 food isolates	512–2,048	[83]
P. aurantiogriseum	Food isolate	256	[83]
P. citrinum	15 airborne isolates	128–2,048	[83]
P. crysogenum	14 airborne isolates	128–2,048	[83]
P. expansum	Apple isolate	1,000[a]	[167]
P. paneum	2 food isolates	512	[83]
P. roquefortii	4 food isolates	256	[83]
Rhizopus spp.	2 clinical and 1 food isolate	512–2,048	[83]
Trichoderma spp.	Food isolate	1,024	[83]

[a]Active chlorine

8.3.2.2 Fungicidal Activity (Suspension Tests)

C. albicans was reduced by at least 4.0 log with sodium hypochlorite at 1,000 mg/l in 5 min. A higher concentration was necessary against *Cryptococcus* spp. (3,800 mg/l), conidia, food-related yeasts or ascospores (30,000 mg/l) although few species of the latter group were not killed in a sufficient level such as *M. ruber* and *P. commune* (Table 8.8). Against food-associated fungi, the efficacy is variable. Some authors confirm that the effect of sodium hypochlorite is increasing with higher concentrations (5,000–20,000 mg/l) and that it has a quite strong activity against moulds in up to 36 min [90].

Table 8.8 Fungicidal activity of sodium hypochlorite in suspension tests

Species	Strains/isolates	Exposure time	Concentration (mg/l)	$\log_{10}$ reduction	References
A. flavus	Bread isolate	10 min	30,000 (P)	4.3	[27]
A. niger	Bread isolate	10 min	30,000 (P)	>5.2	[27]
A. ochraceus	2 clinical isolates	15 min	10,000 (S)[a]	≥4.0	[70]
A. versicolor	2 cheese isolates	10 min	30,000 (P)	>4.5	[27]
C. albicans	NCPF 3179	5 min	48,000 (S)	4.4	[88]
			30,000 (S)		
			6,000 (S)		
C. albicans	ATCC 10231	5 min	25,000 (S)	≥5.0	[160]
C. albicans	1 human and 1 environmental isolate	5 min	3,800 (S)	>7.0	[158]
C. albicans	ATCC 10231 and one clinical isolate	5 min	1,000 (P)	≥4.6	[120]
C. auris	NCPF 8971, NCPF 8977, NCPF 8984, NCPF 8985	5 min	1,000 (P)	≥4.7	[120]
Candida spp.	3 clinical isolates (*C. albicans*, *C. krusei*, *C. parapsilosis*)	15 min	10,000 (S)[a]	≥4.0	[70]
C. neoformans	1 clinical isolate	5 min	3,800 (S)	>7.0	[158]
C. uniguttulatus	1 clinical isolate	5 min	3,800 (S)	>7.0	[158]
Cladosporium spp.	Bread isolate	10 min	30,000 (P)	>4.1	[27]
D. hansenii	Cheese isolate	10 min	30,000 (P)	>4.5	[27]
E. repens	Bread factory isolate	10 min	30,000 (P)	>4.5	[27]
H. burtonii	Bread isolate	10 min	30,000 (P)	>5.2	[27]
M. ruber	Bread isolate	10 min	30,000 (P)	2.9	[27]
M. suaveolens	Bread isolate	10 min	30,000 (P)	>4.5	[27]
N. pseudofischeri	Cherry filling isolate	10 min	30,000 (P)	4.0	[27]
P. anomala	Bread isolate	10 min	30,000 (P)	>5.2	[27]
P. caseifulvum	Cheese isolate	10 min	30,000 (P)	>4.9	[27]
P. chrysogenum	Cheese isolate	10 min	30,000 (P)	>5.2	[27]

(continued)

Table 8.8 (continued)

Species	Strains/isolates	Exposure time	Concentration (mg/l)	log_{10} reduction	References
P. commune	2 cheese and 1 bread isolates	10 min	30,000 (P)	1.7–5.2	[27]
P. corylophilum	Bread isolate	10 min	30,000 (P)	>4.8	[27]
P. crustosum	Cheese isolate	10 min	30,000 (P)	>5.2	[27]
P. discolor	Cheese isolate	10 min	30,000 (P)	>5.2	[27]
P. nalgiovense	2 cheese isolates	10 min	30,000 (P)	>4.2	[27]
P. norvegensis	Cheese isolate	10 min	30,000 (P)	>5.9	[27]
P. roqueforti	2 bread isolates	10 min	30,000 (P)	>5.2	[27]
P. solitum	Cheese isolate	10 min	30,000 (P)	>4.8	[27]
P. verrucosum	Cheese isolate	10 min	30,000 (P)	>4.2	[27]
R. rubra	1 clinical isolate	5 min	3,800 (S)	>7.0	[158]
S. brevicaulis	Cheese isolate	10 min	30,000 (P)	>4.2	[27]
T. delbrueckii	Cheese isolate	10 min	30,000 (P)	>4.8	[27]
Mixed species	Environmental isolates (*R. rubra*, *C. albicans*, *C. uniguttulatus*)	5 min	3,800 (S)	≥6.0	[158]
Mixed species	Clinical isolates (*R. rubra*, *C. albicans*, *C. neoformans*)	5 min	3,800 (S)	≥6.0	[158]

P commercial product; *S* solution; [a]free chlorine

8.3.2.3 Activity Against Fungi in Biofilms

The overall fungicidal activity of sodium hypochlorite at 8 to 3,800 mg/l against fungal cells in biofilms is poor (<1.0 log) as shown with *C. albicans* and other yeasts. On selected dental materials, the efficacy of 5,000 or 20,000 mg/l sodium hypochlorite is better with log reductions between 1.5 and 3.3. Some studies suggest a good effect at 52,500 mg/l against *C. albicans* in biofilms (Table 8.9).

The biofilm may also serve as a viral reservoir despite treatment with sodium hypochlorite. In a *C. albicans* biofilm, it was shown that viruses such as the herpes simplex virus type 1 and the coxsackievirus B5 can be embedded in biofilm, and infectious virus can be released again from biofilm without damaging it. In addition, the fungal biofilm reduces virus sensitivity to sodium hypochlorite (1:400 for 30 min) [112].

8.3.2.4 Fungicidal Activity in Carrier Tests

When spores of *T. mentagrophytes* are placed on a glass cup carrier and exposed to 5,500 mg/l sodium hypochlorite, a log reduction of at least 5.0 is found after 1 min [19]. Against 4 clinical isolates of *C. auris*, *C. albicans* ATCC 10231 and

Table 8.9 Efficacy of sodium hypochlorite against fungal cells in biofilms

Species	Strains/isolates	Type of biofilm	Exposure time	Concentration (mg/l)	$\log_{10}$ reduction	References
C. albicans	ATCC 10231	3-w incubation in single-rooted teeth canals	30 s	52,500 (P)	0.8	[159]
			1 min		1.2–1.3	
			5 min		1.4–1.5	
C. albicans	ATCC 10231D-5	3-w incubation on pieces of cellulose nitrate membranes	60 s	52,500 (P)	"complete elimination"	[65]
C. albicans	ATCC 90028	14-d incubation in canals of single-rooted human teeth	3 min	52,500 (P)	4.0	[51]
				25,000 (P)		
C. albicans	Not described	4-w incubation in roots of sterile teeth	10 min	30,000 (S)	0.3	[176]
C. albicans	ATCC 10231	2.5-h incubation on silicone-based soft denture liners	15 min	20,000 (S)	1.5–2.0	[72]
C. albicans	ATCC 10231	24-h incubation on silicone-based soft denture liners	15 min	20,000 (S)	1.5–2.2	[72]
C. albicans	ATCC 10231	7- or 16-d incubation on polymethylmethacrylate	30 min	6,000 (S)	1.6–1.9	[111]
C. albicans	ATCC 90028	72-h incubation on saliva-coated polymethylmethacrylate and polyamide resin discs	10 min	5,000 (S)	3.0–3.3	[42]
C. albicans	1 clinical and 1 environmental isolate	24-h incubation in microtitre plates	5 min	3,800 (S)	0.0	[158]
C. albicans	Strain SC 5314	24-h incubation in microtitre plates	5–60 min	500 (S)	0.0	[137]
C. albicans	ATCC 60193	48-h incubation in microtitre plates	3×1 min	8.4 (S)	≤ 0.3	[58]
			6×1 min		≤ 0.1	
			8 h		0.1–0.6	
			2×8 h		0.1–0.3	

(continued)

Table 8.9 (continued)

Species	Strains/isolates	Type of biofilm	Exposure time	Concentration (mg/l)	$\log_{10}$ reduction	References
C. glabrata	ATCC 2001	72-h incubation on saliva-coated polymethylmethacrylate and polyamide resin discs	10 min	5,000 (S)	3.1–3.3	[42]
C. orthopsilosis	5 strains (WCO139, WCO147, WCO154, HMCOB, HMCOC)	24-h incubation in microtitre plates	5–60 min	500 (S)	0.0	[137]
C. parapsilosis	5 strains (WCP14, WCP16, WCP17, WCP24, WCP82)	24-h incubation in microtitre plates	5–60 min	500 (S)	0.0	[137]
C. neoformans	1 clinical isolate	24-h incubation in microtitre plates	5 min	3,800 (S)	0.0	[158]
C. uniguttulatus	1 patient room isolate	24-h incubation in microtitre plates	5 min	3,800 (S)	0.0	[158]
R. rubra	1 clinical isolate	24-h incubation in microtitre plates	5 min	3,800 (S)	0.0	[158]

P commercial product; *S* solution

C. glabrata ATCC 2001, sodium hypochlorite (1,000 mg/l for 5 min) lead to a significant reduction on contaminated cellulose matrix (3.2–6.0 log), stainless steel (1.9–2.7 log) and polyester coverslips (1.0–2.0 log). The yeasticidal activity was only better on polyester coverslips (3.2–4.1 log) either when it was applied for 10 min or when the concentration was increased to 10,000 mg/l [86].

8.3.2.5 Fungicidal Activity in Other Applications

Sodium hypochlorite at 10,000 mg/l was quite effective within 1 min for disinfection of titanium implants contaminated with *C. albicans* [28]. When *C. albicans* is allowed to adhere to soft denture lining material for 2.5 h, immersion of the contaminated and carefully washed material in a solution of 10,000 mg/l sodium hypochlorite reduces the number of adherent cells significantly [25]. In swimming pool water, sodium hypochlorite with 1 mg/l active chlorine showed good efficacy within 10 min against *C. albicans* ($\geq$ 3.0 log) [23].

8.3.3 Mycobactericidal Activity

8.3.3.1 Mycobactericidal Activity (Suspension Tests)

Sodium hypochlorite is effective against *M. smegmatis* at 6 mg/l in 1 min. *M. tuberculosis* and *M. bovis*, however, are less susceptible to sodium hypochlorite so that a concentration of 1,000 mg/l is needed against the two species with a 20 min exposure time. But even higher concentrations such as 6,000 or 10,000 mg/l are not sufficiently effective within 1 min against *M. tuberculosis* (Table 8.10). In the presence of organic load such as sputum, sodium hypochlorite at 50,000 and 30,000 mg/l was able to completely inactivate *M. tuberculosis* within 5 min, 10,000 mg/l sodium hypochlorite required longer exposure times [6].

8.3.3.2 Activity Against Mycobacteria in Biofilms

One study with *M. phlei* indicates that the susceptibility of biofilm-grown cells to sodium hypochlorite is lower (MBEC: 125 mg/l in 30 min) compared to planktonic cells (MBC: 63 mg/l in 30 min) [15]. When *M. avium* biofilm is allowed to form on copper or iron water pipes, exposure to 0.3 or 0.6 mg/l free chlorine from sodium hypochlorite lowered *M. avium* levels on copper pipes to 0.2–0.4 CFU per cm^2 but not on iron pipes (5.5–5.9 CFU per cm^2) [127]. The presence of biofilms in heater-cooler units was associated with persistence of *M. chimaera* leading to invasive infections after open cardiac surgery [170]. Biofilms appear to support mycobacterial growth and protect the organism, which makes reliable disinfection of colonized water systems difficult to achieve. Sodium hypochlorite was only able to eliminate *M. chimaera* during full system decontamination when used weekly in addition to other measures [64].

Table 8.10 Mycobactericidal activity of products or solutions based on sodium hypochlorite in suspension tests

Species	Strains/isolates	Exposure time	Concentration (mg/l)	$\log_{10}$ reduction	References
M. bovis	ATCC 35743	20 min	1,000[a] (P)	≥6.0[b]	[146]
			100[a] (P)	<6.0	
M. smegmatis	Strain TMC 1515	1 min	100[a] (S)	>6.0	[17]
			60[a] (S)	>6.0	
			6[a] (S)	>5.0	
M. tuberculosis	Strain H37Rv	1 min	10,000[a] (S)	3.0–3.1	[18]
M. tuberculosis	Strain H37Rv	1 min	6,000[a] (S)	2.4–2.7	[18]
M. tuberculosis complex	10 multidrug-resistant and 10 sensitive strains	10 min	5,000[a] (S)	>7.9	[113]
			1,000[a] (S)	4.4	
M. tuberculosis	1 clinical isolate	20 min	1,000[a] (P)	≥5.0	[146]
			100[a] (P)	<5.0	

S solution; *P* commercial product; [a]available chlorine; [b]1 of 10 experiments <6.0

8.3.3.3 Mycobactericidal Activity in Carrier Tests

M. smegmatis is susceptible to 100 mg/l sodium hypochlorite in carrier tests within 1 min. *M. bovis* requires a higher concentration (5,000 mg/l) for a 5.0 log reduction. Sodium hypochlorite at >10,000 mg/l was not sufficiently effective against *M. tuberculosis* within 1 min (Table 8.11).

Table 8.11 Mycobactericidal activity of solutions (S) based on sodium hypochlorite in carrier tests

Species	Strain/isolate	Exposure time	Concentration (mg/l)	$\log_{10}$ reduction	References
M. bovis	ATCC 35743	1 min	5,000[a] (S)	>5.0	[19]
M. smegmatis	Strain TMC 1515	1 min	100[a] (S)	5.0–6.0	[17]
			60[a] (S)	4.0–6.0	
			6[a] (S)	2.0–3.0	
M. tuberculosis	Strain H37Rv	1 min	10,000[a] (S)	3.2	[18]
M. tuberculosis	Strain H37Rv	1 min	6,000[a] (S)	2.1	[18]

[a]available chlorine per ml

8.4 Effect of Low-Level Exposure

The effect of exposing micro-organisms to low levels of sodium hypochlorite has been studied extensively. The results are summarized in Table 8.12.

In *C. coli* and some strains of *E. coli*, *L. monocytogenes* and *S. enterica*, no change of susceptibility was found. But other strains of the same species showed changes.

When *E. coli* is exposed to sublethal concentrations of sodium hypochlorite, various effects can be observed. The susceptibility to sodium hypochlorite may be weakly reduced (1.7-fold stable increase in MIC), the VBNC cellular state may be strongly induced including an enhanced persistence of the VBNC cells in the presence of nine typical antibiotics at 16 to 256 x MIC, and a cross-tolerance to sodium nitrite and hydrogen peroxide may be found. At 0.3–0.5 mg/l, a decrease in conjugative plasmid transfer has been described (Table 8.12). Adapted *E. coli* cells change shape from rod-shaped to coccoid-shaped; outer cell layer changed from undulating and rough to smooth similar to viable but not culturable cells [32].

In some *L. monocytogenes* strains, the tolerance to sodium hypochlorite can weakly increase (2-fold stable increase in MIC) after low-level exposure. In adapted strains, a cross-tolerance to benzalkonium chloride, another quaternary ammonium compound and alkylamine can be detected. Virulence gene expression has been reduced by low-level exposure (Table 8.12). The changes of adapted *L. monocytogenes* cell are illustrated in Fig. 8.1.

The results described with different *Salmonella* spp. are conflicting. Two studies indicated an increase in the MIC up to 3.5-fold with an associated strain-dependant increase in antibiotic resistance or an increase in biofilm formation. Another study found that previous exposure to sodium hypochlorite makes the surviving cells more susceptible to the biocidal agents explained by an increase in cell permeability (Table 8.12).

Biofilm production was enhanced in strains of *E. coli*, *S. Typhimurium* and MRSA and impaired in *E. faecalis* (Table 8.12). In addition, the transfer of the mobile genetic element Tn916, a conjugative transposon and the prototype of a large family of related elements, was not increased in *B. subtilis* cells by exposure to 1,250 mg/l sodium hypochlorite for up to 2 h [150]. Finally, the catheter exit sites of patients with continuous ambulatory peritoneal dialysis were sampled over at least 6 months. Thirteen CNS isolates were sampled from patients using sodium hypochlorite as disinfectant. No development of tolerance was found [97].

Table 8.12 Effects observed after low-level exposure of various bacterial species to sodium hypochlorite

Species	Strains/isolates	Type of exposure	Increase in MIC	MIC_{max} (mg/l)	Stability of MIC change	Associated changes	References
C. coli	16 strains from pig faeces or pork meat	7 d at various concentrations	None	No data	Not applicable	None reported	[154]
E. faecalis	5 human isolates	48 h at 590 mg/l	No data	No data	Not applicable	Lower biofilm production in 4 isolates, higher biofilm production in 1 isolate	[173]
E. coli	54 strains from pig faeces or pork meat	7 d at various concentrations	None	No data	Not applicable	None reported	[154]
E. coli	ATCC 12806	Several passages with gradually higher concentrations	1.7-fold	403	Stable for 7 d	Marked ability to form biofilm in the presence of hypochlorite; resistance[a] to spectinomycin, ampicillin-sulbactam, nalidixic acid	[32]
E. coli	ATCC 12806	Several passages with gradually higher concentrations	1.7-fold	403	Stable for 7 d	Cross-adaptation[b] with sodium nitrite	[5]
E. coli	CMCC44103	Up to 24 h at 0.5 mg/l	No data	No data	Not applicable	Induction of the VBNC state for 10^5 CFU per ml after 6 h; enhanced persistence[c] to ampicillin, gentamicin, polymyxin, ciprofloxacin, rifampicin, clarithromycin, chloromycetin, tetracycline and terramycin	[101]
E. coli	Strain HB 101	6 h at 0.3–0.5 mg/l	No data	No data	Not applicable	Decrease in conjugative plasmid transfer below detection limit; no change of conjugative plasmid transfer with 0.05–0.2 mg/l for 6 h	[102]

(continued)

Table 8.12 (continued)

Species	Strains/isolates	Type of exposure	Increase in MIC	MIC_{max} (mg/l)	Stability of MIC change	Associated changes	References
E. coli	Strain K12	0.3 mg/l for 1 h	No data	No data	Not applicable	Induction of protection against hydrogen peroxide	[49]
L. monocytogenes	31 strains from pig faeces or pork meat	7 d at various concentrations	None	No data	Not applicable	None reported	[154]
L. monocytogenes	Strain EGD	300–420 bacterial generations	None	No data	Not applicable	None reported	[84]
L. monocytogenes	2 isolates from a freezer at a meat plant	Several passages with gradually higher concentrations	None	3,130	Not applicable	None reported	[1]
L. monocytogenes	2 food isolates (ice cream, poultry)	2 h at sublethal concentration	No increase in MIC	5,000	Not applicable	None reported	[105]
		2 h, followed by 24 h at sublethal concentration	Up to 2-fold increase in MIC	5,000		Cross-adaptation[b] to BAC, another quaternary ammonium compound and alkylamine	
L. monocytogens	ATCC 19112, ATCC 19113, ATCC 19114, ATCC 19115, ATCC 19116, ATCC 19117, ATCC 19118, ATCC 7644, ATCC 13992	10 d at sublethal concentrations	2-fold	512	Unstable over 5 d (5 strains), stable for 5 d (4 strains)	Cell changes (see Fig. 8.1)	[63]
L. monocytogenes	Strain EGD	48 h at inhibitory concentrations	No data	No data	Not applicable	Reduction of virulence gene expression	[85]
S. enterica	35 strains from pig faeces or pork meat	7 d at various concentrations	None	No data	Not applicable	None reported	[154]

(continued)

Table 8.12 (continued)

Species	Strains/isolates	Type of exposure	Increase in MIC	MIC_{max} (mg/l)	Stability of MIC change	Associated changes	References
S. enterica	10 multidrug-resistant strains from poultry	Several passages with gradually higher concentrations	$\leq$3.5-fold (strain dependent)	680	No data	Strain-dependant increase in antibiotic multiresistance (1 to 6 more antibiotics classified as resistant[a], mostly gentamicin and amikacin [4 of 10] and ceftazidime and tobramycin [3 of 10])	[119]
S. enteritidis	Chicken product isolate	10–60 min on contaminated cloths (500 mg/l), followed by the same treatment on the same cloths one day later	Increase in susceptibility from <1.0 log (first treatment) to 4.0 log (second treatment)	No data	No data	Membrane damage in the majority of bacterial cells after the first treatment, resulting in increased membrane permeability	[95]
S. Typhimurium	1 poultry isolate	Not described	1.7-fold	10.1	"stable"	Previous adaptation enhanced biofilm formation 2.6-fold	[31]
S. aureus	Strain 48a (MRSA) isolated from a poultry hamburger	Several passages with gradually higher concentrations	1.7-fold	8,400	Stable for 10 d	Marked enhancement of biofilm formation	[29]
S. aureus	Human isolate	10–60 min on contaminated cloths (500 mg/l), followed by the same treatment on the same cloths one day later	Increase in susceptibility from <1.0 log (first treatment) to 4.0 log (second treatment)	No data	No data	Membrane damage in the majority of bacterial cells after the first treatment, resulting in increased membrane permeability	[95]

[a]Disk diffusion test; [b]microdilution broth method; [c]VBNC cells remained viable at antibiotic concentrations of 16 to 256 x MIC

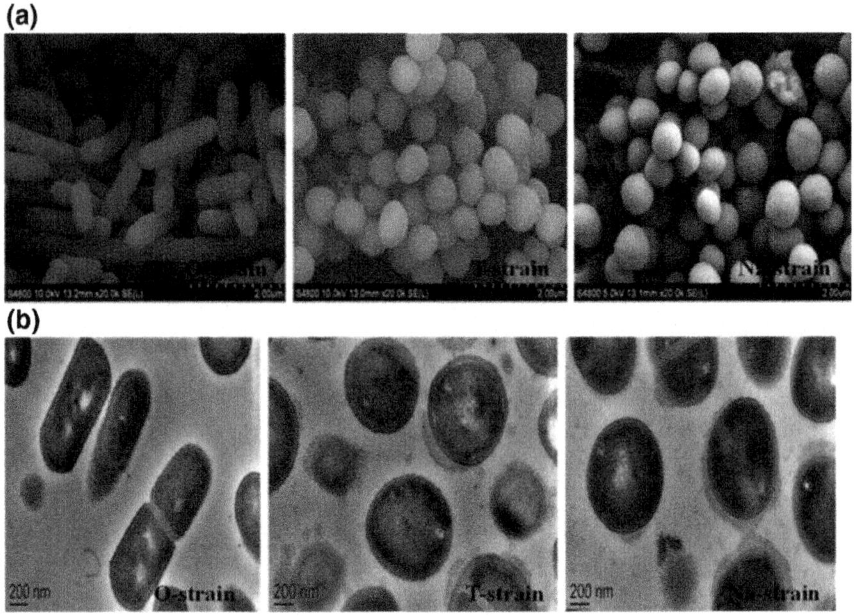

Fig. 8.1 Scanning electron micrographs (**a**) and transmission electron micrographs (**b**) of *L. monocytogenes* strains (ATCC 19112). O-strain represents original strains grown in TSB without disinfectant. T-strain represents strains adapted to chloramines-T. Na-strain represents strains adapted to sodium hypochlorite [63]; Reprinted from Food Control, Volume number 46, Authors Gao H and Liu C, Biochemical and morphological alteration of *Listeria monocytogenes* under environmental stress caused by chloramine-T and sodium hypochlorite, pp. 455–461, Copyright 2014, with permission from Elsevier

8.5 Resistance to Sodium Hypochlorite

Resistance to sodium hypochlorite is very uncommon. A comparison with the proposed epidemiologic cut-off values based on an analysis of MIC values from 3,319 clinical isolates indicates that isolates from clinical specimen and from food processing and animals are usually susceptible to sodium hypochlorite with MIC values below 4,100 mg/l active chlorine (*Enterobacter* spp., *Salmonella* spp., *S. aureus*) or 8,200 mg/l active chlorine (*E. faecium*, *E. faecalis*, *E. coli*, *K. pneumoniae*) [121]. The majority of MIC values summarized in Table 8.2 indicate susceptibility of the bacterial species to sodium hypochlorite based on the proposed break points.

A *Methylobacterium* spp. isolate was described to resist exposure to 10,000 mg/l sodium hypochlorite (5 min) with a log reduction of 0.0 [22]. And a *R. erythropolis* isolate was insufficiently reduced by the same type of exposure with 4.5 log [22].

From Turkey, a VRE strain and an ESBL *K. pneumoniae* strain were found to be "resistant" to 5,500 mg/l sodium hypochlorite [59]. The proposed epidemiological break point, however, is for both species 8,200 mg/l active chlorine suggesting that a bactericidal effect should still be expected [121].

Some species may survive treatment with a rather low concentration of sodium hypochlorite. In 2014, surfaces and air were sampled in a hospital in Beirut for bacterial contamination. A total of 104 bacterial isolates were obtained. The efficacy of a disinfectant based on sodium hypochlorite was tested in a concentration as recommended by the manufacturer claiming a bactericidal activity (500 mg/l active chlorine). In 5.8% of the samples, bacterial growth was observed in the presence of the biocidal agent with the highest rates among bacilli (25.0%) and *S. aureus* (20.1%) [79].

8.5.1 Resistance Mechanisms

It is assumed that resistance to hypochlorite, as shown in *E. coli*, is largely mediated by genes involved in resistance to hydrogen peroxide or is induced by hydrogen peroxide stress, supporting the idea that similar reactive oxygen species are generated in vivo by both reactants [49].

8.5.2 Resistance Genes

8.5.2.1 Sodium Hypochlorite Resistance Genes

No specific sodium hypochlorite resistance genes have been identified so far.

8.5.2.2 Effect of Sodium Hypochlorite on Antibiotic Resistance Genes

A weak reduction of antibiotic resistance genes or plasmids (mostly ≤ 1.0 log) by exposure of various species to sodium hypochlorite has been described. When wastewater was treated with 10 mg/l free chlorine, *E. coli* could be reduced by 4.9 log in 3 min. Three antibiotic resistance genes (sul1, blaTEM, blaCTX-M) were reduced significantly but only by 0.8–0.9 log. The antibiotic resistance plasmid pB10 from an *E. coli* strain was also reduced by 1.0 log in artificial wastewater by exposure to 10 mg/l free chlorine for 15 min [130]. The effect of the same treatment on a somatic coliphage in wastewater showed a reduction by at least 1.0 log in 30 min. The antibiotic resistance genes were not significantly reduced (0.2–0.6 log) indicating that the prevalence of antibiotic resistance genes, particularly in the bacteriophage fraction, poses the threat of the spread of antibiotic resistance genes and their incorporation into a new bacterial background that could lead to the emergence of new resistant clones [30]. Similar findings were reported with tetracycline-resistant bacterial species such as *Acinetobacter*, *Aeromonas*, *Chryseobacterium*, *E. coli*, *Pseudomonas* and *Serratia*. The bacterial cells were reduced by at least 5.0 log when exposed for 10 min to 0.5 mg/l. The tet(W) gene, however, was mostly reduced by 0.0–0.9 log immediately after exposure to the disinfectant except for *Acinetobacter* (1.8 log) and *Chryseobacterium* (4.0 log). In *Chryseobacterium*, the tet(W) concentration increased again after 24 h [156]. In wastewater treatment, a higher concentration of active chlorine (range: 2–32 mg/l) decreases the abundance of antibiotic resistance genes linearly [178].

8.6 Cross-Tolerance to Other Biocidal Agents

A cross-adaptation to benzalkonium chloride, another quaternary ammonium compound and alkylamine has been described in two *L. monocytogenes* isolates from a freezer at a meat plant exposed to sodium hypochlorite for 2 h followed by another 24 h resulting in a lower susceptibility to sodium hypochlorite (up to 2-fold increase in MIC) [105]. *E. coli* ATCC 12806 showed cross-adaptation with sodium nitrite in addition to a reduced susceptibility to sodium hypochlorite (1.7-fold increase in MIC) after several passages with gradually higher concentrations of sodium hypochlorite [5]. And induction of protection against hydrogen peroxide was described in *E. coli* K12 after low-level sodium hypochlorite exposure. An increased tolerance to sodium hypochlorite, however, was not described [49].

8.7 Cross-Tolerance to Antibiotics

In 1,632 clinical *S. aureus* isolates, no correlation of susceptibility profiles was found to sodium hypochlorite and any clinically relevant antibiotic [129]. Some adapted Gram-negative bacterial species, however, have occasionally developed an associated resistance to specific antibiotics. It is particularly interesting that *E. coli* was found to be viable but non-culturable after low-level exposure to sodium hypochlorite and that the same adapted cells were able to better persist in the presence of various antibiotics (Table 8.13).

When surviving bacteria after drinking water chlorination with sodium hypochlorite were evaluated, 22 genera including members of *Paenibacillus*, *Burkholderia*, *Escherichia*, *Sphingomonas* and *Dermacoccus* species were isolated with weak but significant cross-tolerance between chlorine and tetracycline, sulphamethoxazole and amoxicillin but not ciprofloxacin suggesting that chlorine-tolerant bacteria are more likely to also be antibiotic resistant [87].

Table 8.13 Associated tolerance or resistance to antibiotics in bacterial species with a reduced susceptibility to sodium hypochlorite

Species	Strains/isolates	Associated tolerance or resistance	References
E. coli	VBNC cell state after low-level exposure to sodium hypochlorite	Improved persistence[a] to ampicillin, gentamicin, polymyxin, ciprofloxacin, terramycin, tetracycline, rifampicin, clarithromycin and chloromycetin	[101]
S. Anatum	Adapted poultry isolate	Resistance[b] to gentamicin	[119]
S. Enteritidis	Adapted poultry isolate	Resistance[b] to ceftazidime	[119]
S. Hadar	Adapted poultry isolate	Resistance[b] to amikacin, ampicillin, chloramphenicol and nitrofurantoin	[119]
S. Infantis	Adapted poultry isolate	Resistance[b] to gentamicin, ceftazidime, amikacin, tobramycin, cefoxitin and tetracycline	[119]
S. Kentucky	Adapted poultry isolate	Resistance[b] to amikacin and ampicillin/sulbactam	[119]
S. Thompson	Adapted poultry isolate	Resistance[b] to gentamicin, ceftazidime, tobramycin, cefoxitin, cefazolin and nalidixic acid	[119]
S. Thyphimurium	Adapted poultry isolate	Resistance[b] to amikacin, tobramycin, cefazolin, cefotaxime	[119]
S. Virchow	Adapted poultry isolate	Resistance[b] to teicoplanin	[119]
Salmonella spp.	Adapted poultry isolate (strain 1,4, [5],12:i-)	Resistance[b] to gentamicin, nitrofurantoin, cephalothin, cefepime and enrofloxacin	[119]

[a]VBNC cells remained viable at antibiotic concentrations of 16 to 256 x MIC; [b]disk diffusion test

8.8 Role of Biofilm

8.8.1 Effect on Biofilm Development

In the majority of species, sodium hypochlorite is capable to stimulate biofilm production. In MRSA, it was found that sublethal concentrations of sodium hypochlorite can markedly enhance biofilm formation [29]. In *E. coli* ATCC 12806, exposure to gradually higher concentrations of sodium hypochlorite increased the ability to form biofilm [32]. And in a *S. Typhimurium* poultry isolate, previous adaptation to sodium hypochlorite enhanced biofilm formation 2.6-fold [31].

In five human isolates of *E. faecalis*, however, biofilm production was reduced by low-level sodium hypochlorite exposure in four isolates and enhanced in one isolate [173]. Against various yeasts, sodium hypochlorite inhibited biofilm formation at 1,250 mg/l (5 strains of *C. orthopsilosis*) or 2,500 mg/l (*C. albicans*, strain SC 5314; 5 strains of *C. parapsilosis* sensu *strictu*) [137].

8.8.2 Effect on Biofilm Removal

Sodium hypochlorite has some biofilm removal capacity in single-species biofilms (Table 8.14). The removal rate is higher in young biofilms (e.g. 57–62%) and lower in mature biofilms (e.g. ≤ 30%) as shown with *C. krusei* and *P. aeruginosa*. A stronger biofilm removal is seen with longer exposure times as demonstrated with *S. aureus* or *P. aeruginosa* (Table 8.14).

Table 8.14 Biofilm removal rate (quantitative determination of biofilm matrix) by exposure to solutions based on sodium hypochlorite

Type of biofilm	Concentration (mg/l)	Exposure time	Biofilm removal rate	References
A. acidoterrestris DSMZ 3922 biofilm on stainless steel, nylon and PVC surfaces	1,000 (S)	10 min	Partial removal	[48]
B. cenocepacia LMG 18828, 4-h adhesion and 20-h incubation in polystyrene microtitre plates	3,000 (S)	5 min	82%	[135]
	1,000 (S)		65%	
	500 (S)		58%	
C. albicans ATCC 90028, 24-h incubation on acrylic resin specimens	20,000 (P)	5 min	Significant reduction (removal of the majority of biofilm cells)	[41]
	10,000 (P)	10 min		
C. krusei (apple juice processing isolate), incubation for 1–4 d	500[a] (S)	5 min	57–62% (1 d)	[24]
			57–62% (2 d)	
			39–46% (3 d)	
			≤ 30% (4 d)	
E. faecalis ATCC 19433 biofilm, 10-d incubation of apical canal models	25,000 (S)	30 s[b]	44%	[118]
P. aeruginosa ATCC 19142 biofilm, 6-d incubation on stainless steel or aluminium	25,000[c] (S)	5 min	"complete removal"	[46]
P. aeruginosa ATCC 19142 biofilm, 12-d incubation on stainless steel or aluminium	25,000[c] (S)	5 min	"complete removal"	[46]
P. aeruginosa ATCC 19142 biofilm, 18-d incubation on stainless steel or aluminium	25,000[c] (S)	5 min	"traces of carbohydrates remained on steel"	[46]
P. aeruginosa ATCC 700928, 24-h incubation in microplates	10,000 (S)	1 min	66%	[161]
		5 min	76%	
		15 min	91%	
		30 min	85%	
		60 min	92%	
P. aeruginosa biofilm strain PA01, 24-h on 96-well plates	10,000 (S)	1 min	9%	[91]
		5 min	38%	
		15 min	61%	
		30 min	84%	
		60 min	83%	

(continued)

Table 8.14 (continued)

Type of biofilm	Concentration (mg/l)	Exposure time	Biofilm removal rate	References
P. aeruginosa (8 dairy isolates exhibiting high biofilm formation, 24-h incubation in microtiter plates)	650–800 (S)	5 min	"eradication"	[133]
	500–750 (S)	15 min		
	450–600 (S)	30 min		
	350–500 (S)	60 min		
S. aureus ATCC 6538, 72-h incubation in microplates	10,000 (S)	1 min	0%	[161]
		5 min	21%	
		15 min	45%	
		30 min	54%	
		60 min	55%	
S. aureus biofilm ATCC 6538, 24-h incubation on 96-well plates	10,000 (S)	1 min	0%	[91]
		5 min	14%	
		15 min	51%	
		30 min	76%	
		60 min	50%	
S. aureus (MRSA) strain ST 239, 18-h incubation in polystyrene microtiter plates	10,000[a] (S)	30 min	"complete removal"	[7]
S. aureus ATCC 25923 dry surface biofilm, 12-d incubation on polycarbonate coupons	1,000–20,000 (S)	10 min	$\geq 90\%$	[4]
Mixed species in a natural biofilm from dental unit waterlines	52,500 (S)	2 d	"no effective biofilm removal"	[99]
Mixed species in root canals from human mandibular premolars	25,500 (P)	1 min[b]	No biofilm removal	[35]
Mixed species (*S. gordonii* ATCC 10558, *P. gingivalis* ATCC 33277, *T. forsythia* ATCC 43037, *F. nucleatum* ATCC 25586, *A. naeslundii* ATCC 12104, and *P. micra* ATCC 33270), 4-d incubation in 96-well plates	9,500 (S)	1 h	No biofilm removal	[82]

[a]Active chlorine; [b]irrigation; [c]plus 0.25% hydrogen peroxide

Multispecies natural biofilms with multiple species seem to be almost impossible to remove by exposure to sodium hypochlorite, even at 52,500 mg/l (Table 8.14). When, however, dentures were repeatedly treated with 1,000 or 2,000 mg/l sodium hypochlorite for 20 min per day for 14 days, significantly lower biofilm coverage was found compared to saline treatment indicating partial biofilm removal [10]. Based on the data of Table 8.14, it seems impossible to define a minimum sodium hypochlorite concentration and exposure time resulting in a consistent biofilm removal rate >90%.

8.8.3 Effect on Biofilm Fixation

No studies were found describing a fixation potential by sodium hypochlorite. On the contrary, sodium hypochlorite weakened the biofilm mechanical stability to some extent in *P. fluorescens* ATCC 13525 grown for 7 d on stainless steel, e.g. by 35% at 50 mg/l, by 55% at 200 mg/l, by 63% at 300 mg/l and by 65% at 500 mg/l [151].

8.9 Summary

The principal antimicrobial activity of sodium hypochlorite is summarized in Table 8.15.

The key findings on acquired resistance and cross-resistance including the role of biofilm in selecting resistant isolates are summarized in Table 8.16.

Table 8.15 Overview on the typical exposure times required for sodium hypochlorite to achieve sufficient biocidal activity against the different target micro-organisms

Target micro-organisms	Species	Concentration (mg/l)	Exposure time (min)
Bacteria	Most clinically relevant species	5,000	30[a]
Fungi	*C. albicans*	1,000	5[a]
	Other mainly food-associated fungi (except *M. ruber* and *P. commune*)	30,000	10
Mycobacteria	*M. smegmatis*	6	1
	M. tuberculosis	1,000	10
	M. bovis	1,000	20

[a]In biofilm the efficacy will be lower

Table 8.16 Key findings on acquired sodium hypochlorite resistance, the effect of low-level exposure, cross-tolerance to other biocides and antibiotics, and its effect on biofilm

Parameter	Species	Findings
Elevated MIC values	None	
Insufficient efficacy in suspension tests	*Methylobacterium* spp.	Possibly intrinsic resistance
	R. erythropolis	Possibly intrinsic resistance
Proposed MIC value to determine resistance	*C. albicans*	8,200 mg/l[a]
	Enterobacter spp.	4,100 mg/l[a]
	E. faecium	8,200 mg/l[a]
	E. faecalis	8,200 mg/l[a]
	E. coli	8,200 mg/l[a]
	K. pneumoniae	8,200 mg/l[a]
	Salmonella spp.	4,100 mg/l[a]
	S. aureus	4,100 mg/l[a]

(continued)

Table 8.16 (continued)

Parameter	Species	Findings
Cross-tolerance biocides	*E. coli*	Sodium nitrite, hydrogen peroxide
	L. monocytogenes	Benzalkonium chloride, another quaternary ammonium compound, alkylamine
Cross-tolerance antibiotics	*S. aureus*	No correlation
	Salmonella spp. (9 species)	Resistance in some adapted isolates to various antibiotics incl. gentamicin, ceftazidime and tobramycin
Resistance mechanisms	*E. coli*	Genes involved in resistance to hydrogen peroxide
Effect of low-level exposure	*C. coli*, *E. coli*, *L. monocytogenes*, *S. enterica*	No MIC increase
	E. coli, *L. monocytogenes*, *S. enterica*	Weak MIC increase ($\leq$4-fold)
	None	Strong MIC increase (>4-fold)
	E. coli, *S. Typhimurium*, MRSA	Increase in biofilm formation
	E. faecalis	Decrease in biofilm formation
	E. coli	VBNC state with enhanced antibiotic tolerance
	L. monocytogenes	Reduced virulence gene expression
	E. coli	Reduced plasmid transfer
	B. subtilis	No increase in transposon Tn916 transfer
	E. coli	Cross-tolerance to sodium nitrite and hydrogen peroxide
	L. monocytogenes	Cross-tolerance to benzalkonium chloride, another quaternary ammonium compound and alkylamine
Biofilm	Development	Increase in *E. coli*, MRSA, *S. Typhimurium*
		Inhibition in *E. faecalis*, *Candida* spp.
	Removal	Variable (0–100% removal)
		No removal in mixed natural biofilms
	Fixation	Unknown

[a]Active chlorine

References

1. Aarnisalo K, Lundén J, Korkeala H, Wirtanen G (2007) Susceptibility of Listeria monocytogenes strains to disinfectants and chlorinated alkaline cleaners at cold temperatures. LWT Food Sci Technol 40(6):1041–1048
2. Akamatsu T, Tabata K, Hironga M, Kawakami H, Uyeda M (1996) Transmission of Helicobacter pylori infection via flexible fiberoptic endoscopy. Am J Infect Control 24 (5):396–401
3. Akhtar M, Maserati A, Diez-Gonzalez F, Sampedro F (2016) Does antibiotic resistance influence shiga-toxigenic Escherichia coli O26 and O103 survival to stress environments? Food Control 68:330–336. https://doi.org/10.1016/j.foodcont.2016.04.011
4. Almatroudi A, Gosbell IB, Hu H, Jensen SO, Espedido BA, Tahir S, Glasbey TO, Legge P, Whiteley G, Deva A, Vickery K (2016) Staphylococcus aureus dry-surface biofilms are not killed by sodium hypochlorite: implications for infection control. J Hosp Infect 93(3):263–270. https://doi.org/10.1016/j.jhin.2016.03.020
5. Alonso-Calleja C, Guerrero-Ramos E, Alonso-Hernando A, Capita R (2015) Adaptation and cross-adaptation of Escherichia coli ATCC 12806 to several food-grade biocides. Food Control 56(Supplement C):86–94. https://doi.org/10.1016/j.foodcont.2015.03.012
6. Antony T, Srikanth P, Edwin B (2014) Mycobactericidal activity of various concentrations of bleach. Indian J Tuberc 61(3):257–260
7. Aparecida Guimaraes M, Rocchetto Coelho L, Rodrigues Souza R, Ferreira-Carvalho BT, Marie Sa Figueiredo A (2012) Impact of biocides on biofilm formation by methicillin-resistant Staphylococcus aureus (ST239-SCCmecIII) isolates. Microbiol Immunol 56(3):203–207. https://doi.org/10.1111/j.1348-0421.2011.00423.x
8. Arias-Moliz MT, Ordinola-Zapata R, Baca P, Ruiz-Linares M, Garcia Garcia E, Hungaro Duarte MA, Monteiro Bramante C, Ferrer-Luque CM (2015) Antimicrobial activity of Chlorhexidine, Peracetic acid and Sodium hypochlorite/etidronate irrigant solutions against Enterococcus faecalis biofilms. Int Endod J 48(12):1188–1193. https://doi.org/10.1111/iej.12424
9. Arioli S, Elli M, Ricci G, Mora D (2013) Assessment of the susceptibility of lactic acid bacteria to biocides. Int J Food Microbiol 163(1):1–5. https://doi.org/10.1016/j.ijfoodmicro.2013.02.002
10. Arruda CNF, Salles MM, Badaro MM, de Cassia Oliveira V, Macedo AP, Silva-Lovato CH, de Freitas Oliveira Paranhos H (2017) Effect of sodium hypochlorite and Ricinus communis solutions on control of denture biofilm: a randomized crossover clinical trial. J Prosthet Dent 117(6):729–734. https://doi.org/10.1016/j.prosdent.2016.08.035
11. Arslan S, Ozbilge H, Kaya EG, Er O (2011) In vitro antimicrobial activity of propolis, BioPure MTAD, sodium hypochlorite, and chlorhexidine on Enterococcus faecalis and Candida albicans. Saudi Med J 32(5):479–483
12. Ashok R, Ganesh A, Deivanayagam K (2017) Bactericidal effect of different anti-microbial agents on Fusobacterium nucleatum biofilm. Cureus 9(6):e1335. https://doi.org/10.7759/cureus.1335
13. Avrain L, Allain L, Vernozy-Rozand C, Kempf I (2003) Disinfectant susceptibility testing of avian and swine Campylobacter isolates by a filtration method. Vet Microbiol 96(1):35–40
14. Ban GH, Kang DH (2016) Effect of sanitizer combined with steam heating on the inactivation of foodborne pathogens in a biofilm on stainless steel. Food Microbiol 55:47–54. https://doi.org/10.1016/j.fm.2015.11.003
15. Bardouniotis E, Huddleston W, Ceri H, Olson ME (2001) Characterization of biofilm growth and biocide susceptibility testing of Mycobacterium phlei using the MBEC assay system. FEMS Microbiol Lett 203(2):263–267
16. Best M, Kennedy ME, Coates F (1990) Efficacy of a variety of disinfectants against Listeria spp. Appl Environ Microbiol 56(2):377–380

17. Best M, Sattar SA, Springthorpe VS, Kennedy ME (1988) Comparative mycobactericidal efficacy of chemical disinfectants in suspension and carrier tests. Appl Environ Microbiol 54:2856–2858
18. Best M, Sattar SA, Springthorpe VS, Kennedy ME (1990) Efficacies of selected disinfectants against *Mycobacterium tuberculosis*. J Clin Microbiol 28(10):2234–2239
19. Best M, Springthorpe VS, Sattar SA (1994) Feasibility of a combined carrier test for disinfectants: studies with a mixture of five types of microorganisms. Am J Infect Control 22 (3):152–162
20. Bhatia M, Mishra B, Thakur A, Dogra V, Loomba PS (2017) Evaluation of susceptibility of glycopeptide-resistant and glycopeptide-sensitive enterococci to commonly used biocides in a super-speciality hospital: a pilot study. J Nat Sci, Biol, Med 8(2):199–202. https://doi.org/10.4103/0976-9668.210010
21. Bloomfield SF, Miller EA (1989) A comparison of hypochlorite and phenolic disinfectants for disinfection of clean and soiled surfaces and blood spillages. J Hosp Infect 13(3):231–239
22. Bore E, Langsrud S (2005) Characterization of micro-organisms isolated from dairy industry after cleaning and fogging disinfection with alkyl amine and peracetic acid. J Appl Microbiol 98(1):96–105. https://doi.org/10.1111/j.1365-2672.2004.02436.x
23. Borgmann-Strahsen R (2003) Comparative assessment of deferent biocides in swimming pool water. Int Biodeter Biodegr 51(4):291–297
24. Brugnoni LI, Cubitto MA, Lozano JE (2012) Candida krusei development on turbulent flow regimes: Biofilm formation and efficiency of cleaning and disinfection program. J Food Eng 111(4):546–552. https://doi.org/10.1016/j.jfoodeng.2012.03.023
25. Buergers R, Rosentritt M, Schneider-Brachert W, Behr M, Handel G, Hahnel S (2008) Efficacy of denture disinfection methods in controlling Candida albicans colonization in vitro. Acta Odontol Scand 66(3):174–180. https://doi.org/10.1080/00016350802165614
26. Bukhary S, Balto H (2017) Antibacterial efficacy of Octenisept, Alexidine, Chlorhexidine, and Sodium Hypochlorite against Enterococcus faecalis Biofilms. J Endod 43(4):643–647. https://doi.org/10.1016/j.joen.2016.09.013
27. Bundgaard-Nielsen K, Nielsen PV (1996) Fungicidal effect of 15 disinfectants against 25 fungal contaminants commonly found in bread and cheese manufacturing. J Food Prot 59 (3):268–275
28. Burgers R, Witecy C, Hahnel S, Gosau M (2012) The effect of various topical peri-implantitis antiseptics on Staphylococcus epidermidis, Candida albicans, and Streptococcus sanguinis. Arch Oral Biol 57(7):940–947. https://doi.org/10.1016/j.archoralbio.2012.01.015
29. Buzón-Durán L, Alonso-Calleja C, Riesco-Peláez F, Capita R (2017) Effect of sub-inhibitory concentrations of biocides on the architecture and viability of MRSA biofilms. Food Microbiol 65(Supplement C):294–301. https://doi.org/10.1016/j.fm.2017.01.003
30. Calero-Caceres W, Muniesa M (2016) Persistence of naturally occurring antibiotic resistance genes in the bacteria and bacteriophage fractions of wastewater. Water Res 95:11–18. https://doi.org/10.1016/j.watres.2016.03.006
31. Capita R, Buzon-Duran L, Riesco-Pelaez F, Alonso-Calleja C (2017) Effect of Sub-lethal concentrations of biocides on the structural parameters and viability of the biofilms formed by Salmonella Typhimurium. Foodborne Pathog Dis 14(6):350–356. https://doi.org/10.1089/fpd.2016.2241
32. Capita R, Riesco-Pelaez F, Alonso-Hernando A, Alonso-Calleja C (2014) Exposure of Escherichia coli ATCC 12806 to sublethal concentrations of food-grade biocides influences its ability to form biofilm, resistance to antimicrobials, and ultrastructure. Appl Environ Microbiol 80(4):1268–1280. https://doi.org/10.1128/aem.02283-13

33. Chamakura K, Perez-Ballestero R, Luo Z, Bashir S, Liu J (2011) Comparison of bactericidal activities of silver nanoparticles with common chemical disinfectants. Colloids Surf B 84 (1):88–96. https://doi.org/10.1016/j.colsurfb.2010.12.020
34. Christo JE, Zilm PS, Sullivan T, Cathro PR (2016) Efficacy of low concentrations of sodium hypochlorite and low-powered Er, Cr:YSGG laser activated irrigation against an Enterococcus faecalis biofilm. Int Endod J 49(3):279–286. https://doi.org/10.1111/iej.12447
35. Coaguila-Llerena H, Stefanini da Silva V, Tanomaru-Filho M, Guerreiro Tanomaru JM, Faria G (2018) Cleaning capacity of octenidine as root canal irrigant: a scanning electron microscopy study. Microsc Res Tech. https://doi.org/10.1002/jemt.23007
36. Coates D (1985) A comparison of sodium hypochlorite and sodium dichloroisocyanurate products. J Hosp Infect 6(1):31–40
37. Coetzee E, Whitelaw A, Kahn D, Rode H (2012) The use of topical, un-buffered sodium hypochlorite in the management of burn wound infection. Burns: J Int Soc Burn Injuries 38 (4):529–533. https://doi.org/10.1016/j.burns.2011.10.008
38. Cooper IR, Hanlon GW (2010) Resistance of Legionella pneumophila serotype 1 biofilms to chlorine-based disinfection. J Hosp Infect 74(2):152–159. https://doi.org/10.1016/j.jhin.2009.07.005
39. Corcoran M, Morris D, De Lappe N, O'Connor J, Lalor P, Dockery P, Cormican M (2014) Commonly used disinfectants fail to eradicate Salmonella enterica biofilms from food contact surface materials. Appl Environ Microbiol 80(4):1507–1514. https://doi.org/10.1128/aem.03109-13
40. Cruz CD, Fletcher GC (2012) Assessing manufacturers' recommended concentrations of commercial sanitizers on inactivation of Listeria monocytogenes. Food Control 26(1):194–199. https://doi.org/10.1016/j.foodcont.2012.01.041
41. da Silva PM, Acosta EJ, Pinto Lde R, Graeff M, Spolidorio DM, Almeida RS, Porto VC (2011) Microscopical analysis of Candida albicans biofilms on heat-polymerised acrylic resin after chlorhexidine gluconate and sodium hypochlorite treatments. Mycoses 54(6): e712–e717. https://doi.org/10.1111/j.1439-0507.2010.02005.x [doi]
42. de Freitas Fernandes FS, Pereira-Cenci T, da Silva WJ, Filho AP, Straioto FG, Del Bel Cury AA (2011) Efficacy of denture cleansers on Candida spp. biofilm formed on polyamide and polymethyl methacrylate resins. J Prosthet Dent 105(1):51–58
43. de Souza EL, Meira QG, de Medeiros Barbosa I, Athayde AJ, da Conceicao ML, de Siqueira Junior JP (2014) Biofilm formation by Staphylococcus aureus from food contact surfaces in a meat-based broth and sensitivity to sanitizers. Braz J Microbiol: [Publication of the Brazilian Society for Microbiology] 45(1):67–75
44. Decker EM, Bartha V, Kopunic A, von Ohle C (2017) Antimicrobial efficiency of mouthrinses versus and in combination with different photodynamic therapies on periodontal pathogens in an experimental study. J Periodontal Res 52(2):162–175. https://doi.org/10.1111/jre.12379
45. Degrossoli A, Muller A, Xie K, Schneider JF, Bader V, Winklhofer KF, Meyer AJ, Leichert LI (2018) Neutrophil-generated HOCl leads to non-specific thiol oxidation in phagocytized bacteria. eLife 7. https://doi.org/10.7554/elife.32288
46. DeQueiroz GA, Day DF (2007) Antimicrobial activity and effectiveness of a combination of sodium hypochlorite and hydrogen peroxide in killing and removing Pseudomonas aeruginosa biofilms from surfaces. J Appl Microbiol 103(4):794–802. https://doi.org/10.1111/j.1365-2672.2007.03299.x
47. Dominciano LCC, Oliveira CAF, Lee SH, Corassin CH (2016) Individual and combined antimicrobial activity of Oleuropein and chemical sanitizers. J Food Chem Nanotechnol 2 (3):124–127
48. dos Anjos MM, Ruiz SP, Nakamura CV, de Abreu Filho BA (2013) Resistance of Alicyclobacillus acidoterrestris spores and biofilm to industrial sanitizers. J Food Prot 76 (8):1408–1413. https://doi.org/10.4315/0362-028x.jfp-13-020

49. Dukan S, Touati D (1996) Hypochlorous acid stress in Escherichia coli: resistance, DNA damage, and comparison with hydrogen peroxide stress. J Bacteriol 178(21):6145–6150
50. Ekizoglu MT, Özalp M, Sultan N, Gür D (2003) An investigation of the bactericidal effect of certain antiseptics and disinfectants on some hospital isolates of gram-negative bacteria. Infect Control Hosp Epidemiol 24(3):225–227
51. Eldeniz AU, Guneser MB, Akbulut MB (2015) Comparative antifungal efficacy of light-activated disinfection and octenidine hydrochloride with contemporary endodontic irrigants. Lasers Med Sci 30(2):669–675. https://doi.org/10.1007/s10103-013-1387-1
52. Elli M, Arioli S, Guglielmetti S, Mora D (2013) Biocide susceptibility in bifidobacteria of human origin. J Glob Antimicrob Res 1(2):97–101. https://doi.org/10.1016/j.jgar.2013.03.007
53. Espigares E, Bueno A, Espigares M, Galvez R (2006) Isolation of Salmonella serotypes in wastewater and effluent: effect of treatment and potential risk. Int J Hyg Environ Health 209 (1):103–107. https://doi.org/10.1016/j.ijheh.2005.08.006
54. Estrela CR, Estrela C, Reis C, Bammann LL, Pecora JD (2003) Control of microorganisms in vitro by endodontic irrigants. Braz Dent J 14(3):187–192
55. European Chemicals Agency (ECHA) Sodium hypochlorite. Substance information. https://echa.europa.eu/substance-information/-/substanceinfo/100.028.790. Accessed 27 Oct 2017
56. Fatemi P, Frank JF (1999) Inactivation of Listeria monocytogenes/Pseudomonas biofilms by peracid sanitizers. J Food Prot 62(7):761–765
57. Ferguson JW, Hatton JF, Gillespie MJ (2002) Effectiveness of intracanal irrigants and medications against the yeast Candida albicans. J Endod 28(2):68–71. https://doi.org/10.1097/00004770-200202000-00004
58. Ferreira GLS, Rosalen PL, Peixoto LR, Perez A, Carlo FGC, Castellano LRC, Lima JM, Freires IA, Lima EO, Castro RD (2017) Antibiofilm activity and mechanism of action of the disinfectant chloramine T on Candida spp., and Its toxicity against human cells. Molecules (Basel, Switzerland) 22(9). https://doi.org/10.3390/molecules22091527
59. Ficici SE, Durmaz G, Ilhan S, Akgun Y, Kosgeroglu N (2002) Bactericidal effects of commonly used antiseptics/disinfectants on nosocomial bacterial pathogens and the relationship between antibacterial and biocide resistance. Mikrobiyoloji Bul 36(3–4): 259–269
60. Fidalgo SG, Longbottom CJ, Rjley TV (2002) Susceptibility of Erysipelothrix rhusiopathiae to antimicrobial agents and home disinfectants. Pathology 34(5):462–465
61. Frough-Reyhani M, Ghasemi N, Soroush-Barhaghi M, Amini M, Gholizadeh Y (2016) Antimicrobial efficacy of different concentration of sodium hypochlorite on the biofilm of Enterococcus faecalis at different stages of development. J Clin Exp Dent 8(5):e480–e484. https://doi.org/10.4317/jced.53158
62. Fu Y, Deering AJ, Bhunia AK, Yao Y (2017) Biofilm of Escherichia coli O157:H7 on cantaloupe surface is resistant to lauroyl arginate ethyl and sodium hypochlorite. Int J Food Microbiol 260:11–16. https://doi.org/10.1016/j.ijfoodmicro.2017.08.008
63. Gao H, Liu C (2014) Biochemical and morphological alteration of Listeria monocytogenes under environmental stress caused by chloramine-T and sodium hypochlorite. Food Control 46(Supplement C):455–461. https://doi.org/10.1016/j.foodcont.2014.05.016
64. Garvey MI, Ashford R, Bradley CW, Bradley CR, Martin TA, Walker J, Jumaa P (2016) Decontamination of heater-cooler units associated with contamination by atypical mycobacteria. J Hosp Infect 93(3):229–234. https://doi.org/10.1016/j.jhin.2016.02.007
65. Ghivari SB, Bhattacharya H, Bhat KG, Pujar MA (2017) Antimicrobial activity of root canal irrigants against biofilm forming pathogens- An in vitro study. J Conservative Dent: JCD 20 (3):147–151. https://doi.org/10.4103/jcd.jcd_38_16
66. Gkana EN, Giaouris ED, Doulgeraki AI, Kathariou S, Nychas GJE (2017) Biofilm formation by Salmonella Typhimurium and Staphylococcus aureus on stainless steel under either mono- or dual-species multi-strain conditions and resistance of sessile communities to

sub-lethal chemical disinfection. Food Control 73(Part B):838–846. https://doi.org/10.1016/j.foodcont.2016.09.038

67. Gomes IB, Simoes M, Simoes LC (2016) The effects of sodium hypochlorite against selected drinking water-isolated bacteria in planktonic and sessile states. Sci Total Environ 565:40–48. https://doi.org/10.1016/j.scitotenv.2016.04.136
68. Gosau M, Hahnel S, Schwarz F, Gerlach T, Reichert TE, Burgers R (2010) Effect of six different peri-implantitis disinfection methods on in vivo human oral biofilm. Clin Oral Implant Res 21(8):866–872. https://doi.org/10.1111/j.1600-0501.2009.01908.x
69. Guneser MB, Akbulut MB, Eldeniz AU (2016) Antibacterial effect of chlorhexidine-cetrimide combination, Salvia officinalis plant extract and octenidine in comparison with conventional endodontic irrigants. Dent Mater J 35(5):736–741. https://doi.org/10.4012/dmj.2015-159
70. Gupta AK, Ahmad I, Summerbell RC (2002) Fungicidal activities of commonly used disinfectants and antifungal pharmaceutical spray preparations against clinical strains of Aspergillus and Candida species. Med Mycol 40(2):201–208
71. Gutierrez-Martin CB, Yubero S, Martinez S, Frandoloso R, Rodriguez-Ferri EF (2011) Evaluation of efficacy of several disinfectants against Campylobacter jejuni strains by a suspension test. Res Vet Sci 91(3):e44–e47. https://doi.org/10.1016/j.rvsc.2011.01.020
72. Hahnel S, Rosentritt M, Burgers R, Handel G, Lang R (2012) Candida albicans biofilm formation on soft denture liners and efficacy of cleaning protocols. Gerodontology 29(2): e383–e391. https://doi.org/10.1111/j.1741-2358.2011.00485.x
73. Hoffman PN, Death JE, DC (1981) The stability of sodium hypochlorite solutions. In: Collins CH, Allwood MC, Bloomfield SF, Fox A (eds) Disinfectants: their use and evaluation of effectiveness. Academic Press, London, pp 77–83
74. Iniguez-Moreno M, Avila-Novoa MG, Iniguez-Moreno E, Guerrero-Medina PJ, Gutierrez-Lomeli M (2017) Antimicrobial activity of disinfectants commonly used in the food industry in Mexico. J Glob Antimicrob Res 10:143–147. https://doi.org/10.1016/j.jgar.2017.05.013
75. Italy (2017) Assessment report. Active chlorine released from sodium hypochlorite. Product-type 2.
76. Izano EA, Shah SM, Kaplan JB (2009) Intercellular adhesion and biocide resistance in nontypeable Haemophilus influenzae biofilms. Microb Pathog 46(4):207–213. https://doi.org/10.1016/j.micpath.2009.01.004
77. Jaglic Z, Červinková D, Vlková H, Michu E, Kunová G, Babák V (2012) Bacterial biofilms resist oxidising agents due to the presence of organic matter. Czech J Food Sci 30(2): 178–187
78. Jayahari NK, Niranjan NT, Kanaparthy A (2014) The efficacy of passion fruit juice as an endodontic irrigant compared with sodium hypochlorite solution: an in vitro study. J Invest Clin Dent 5(2):154–160. https://doi.org/10.1111/jicd.12023
79. Jomha MY, Yusef H, Holail H (2014) Antimicrobial and biocide resistance of bacteria in a Lebanese tertiary care hospital. J Glob Ant Res 2(4):299–305. https://doi.org/10.1016/j.jgar.2014.09.001
80. Joy Sinha D, K DSN, Jaiswal N, Vasudeva A, Prabha Tyagi S, Pratap Singh U (2017) Antibacterial effect of Azadirachta indica (Neem) or Curcuma longa (Turmeric) against Enterococcus faecalis compared with that of 5% sodium hypochlorite or 2% chlorhexidine in vitro. Bull Tokyo Dent Coll 58(2):103–109
81. Juncker JC (2017) COMMISSION IMPLEMENTING REGULATION (EU) 2017/1273 of 14 July 2017 approving active chlorine released from sodium hypochlorite as an existing active substance for use in biocidal products of product-types 1, 2, 3, 4 and 5. Off J Eur Union 60(L 184):13–16
82. Jurczyk K, Nietzsche S, Ender C, Sculean A, Eick S (2016) In-vitro activity of sodium-hypochlorite gel on bacteria associated with periodontitis. Clin Oral Invest 20 (8):2165–2173. https://doi.org/10.1007/s00784-016-1711-9

83. Kalkanci A, Elli M, Adil Fouad A, Yesilyurt E, Jabban Khalil I (2015) Assessment of susceptibility of mould isolates towards biocides. Journal de mycologie medicale 25(4):280–286. https://doi.org/10.1016/j.mycmed.2015.08.001
84. Kastbjerg VG, Gram L (2012) Industrial disinfectants do not select for resistance in Listeria monocytogenes following long term exposure. Int J Food Microbiol 160(1):11–15. https://doi.org/10.1016/j.ijfoodmicro.2012.09.009
85. Kastbjerg VG, Larsen MH, Gram L, Ingmer H (2010) Influence of sublethal concentrations of common disinfectants on expression of virulence genes in Listeria monocytogenes. Appl Environ Microbiol 76(1):303–309. https://doi.org/10.1128/aem.00925-09
86. Kean R, Sherry L, Townsend E, McKloud E, Short B, Akinbobola A, Mackay WG, Williams C, Jones BL, Ramage G (2018) Surface disinfection challenges for Candida auris: an in-vitro study. J Hosp Infect 98(4):433–436. https://doi.org/10.1016/j.jhin.2017.11.015
87. Khan S, Beattie TK, Knapp CW (2016) Relationship between antibiotic- and disinfectant-resistance profiles in bacteria harvested from tap water. Chemosphere 152:132–141. https://doi.org/10.1016/j.chemosphere.2016.02.086
88. Kiesow A, Sarembe S, Pizzey RL, Axe AS, Bradshaw DJ (2016) Material compatibility and antimicrobial activity of consumer products commonly used to clean dentures. J Prosthet Dent 115 (2):189–198.e188. https://doi.org/10.1016/j.prosdent.2015.08.010
89. Koban I, Geisel MH, Holtfreter B, Jablonowski L, Hubner NO, Matthes R, Masur K, Weltmann KD, Kramer A, Kocher T (2013) Synergistic effects of nonthermal plasma and disinfecting agents against dental biofilms in vitro. ISRN Dent 2013:573262. https://doi.org/10.1155/2013/573262
90. Korukluoglu M, Sahan Y, Yigit A (2006) The fungicidal efficacy of various commercial disinfectants used in the food industry. Ann Microbiol 56(4):325–330
91. Köse H, Yapar N (2017) The comparison of various disinfectants' efficacy on Staphylococcus aureus and Pseudomonas aeruginosa biofilm layers. Turkish J Med Sci 47(4):1287–1294
92. Kostaki M, Chorianopoulos N, Braxou E, Nychas GJ, Giaouris E (2012) Differential biofilm formation and chemical disinfection resistance of sessile cells of Listeria monocytogenes strains under monospecies and dual-species (with Salmonella enterica) conditions. Appl Environ Microbiol 78(8):2586–2595. https://doi.org/10.1128/aem.07099-11
93. Kramer A, Dissemond J, Kim S, Willy C, Mayer D, Papke R, Tuchmann F, Assadian O (2017) Consensus on wound antisepsis: update 2018. Skin Pharmacol Physiol 31(1):28–58. https://doi.org/10.1159/000481545
94. Kubota H, Senda S, Tokuda H, Uchiyama H, Nomura N (2009) Stress resistance of biofilm and planktonic Lactobacillus plantarum subsp. plantarum JCM 1149. Food Microbiol 26 (6):592–597. https://doi.org/10.1016/j.fm.2009.04.001
95. Kusumaningrum HD, Paltinaite R, Koomen AJ, Hazeleger WC, Rombouts FM, Beumer RR (2003) Tolerance of Salmonella Enteritidis and Staphylococcus aureus to surface cleaning and household bleach. J Food Prot 66(12):2289–2295
96. Lagace L, Jacques M, Mafu AA, Roy D (2006) Biofilm formation and biocides sensitivity of Pseudomonas marginalis isolated from a maple sap collection system. J Food Prot 69 (10):2411–2416
97. Lanker Klossner B, Widmer HR, Frey F (1997) Nondevelopment of resistance by bacteria during hospital use of povidone-iodine. Dermatology (Basel, Switzerland) 195(Suppl 2):10–13. https://doi.org/10.1159/000246024
98. Leung CY, Chan YC, Samaranayake LP, Seneviratne CJ (2012) Biocide resistance of Candida and Escherichia coli biofilms is associated with higher antioxidative capacities. J Hosp Infect 81(2):79–86. https://doi.org/10.1016/j.jhin.2011.09.014
99. Liaqat I, Sabri AN (2008) Effect of biocides on biofilm bacteria from dental unit water lines. Curr Microbiol 56(6):619–624. https://doi.org/10.1007/s00284-008-9136-6
100. Lin F, Xu Y, Chang Y, Liu C, Jia X, Ling B (2017) Molecular characterization of reduced susceptibility to biocides in clinical isolates of Acinetobacter baumannii. Front Microbiol 8:1836. https://doi.org/10.3389/fmicb.2017.01836

101. Lin H, Ye C, Chen S, Zhang S, Yu X (2017) Viable but non-culturable E. coli induced by low level chlorination have higher persistence to antibiotics than their culturable counterparts. Environ Poll (Barking, Essex: 1987) 230:242–249. https://doi.org/10.1016/j.envpol.2017.06.047
102. Lin W, Li S, Zhang S, Yu X (2016) Reduction in horizontal transfer of conjugative plasmid by UV irradiation and low-level chlorination. Water Res 91(Supplement C):331–338. https://doi.org/10.1016/j.watres.2016.01.020
103. Liu Q, Liu M, Wu Q, Li C, Zhou T, Ni Y (2009) Sensitivities to biocides and distribution of biocide resistance genes in quaternary ammonium compound tolerant Staphylococcus aureus isolated in a teaching hospital. Scand J Infect Dis 41(6–7):403–409. https://doi.org/10.1080/00365540902856545
104. Locker J, Fitzgerald P, Sharp D (2014) Antibacterial validation of electrogenerated hypochlorite using carbon-based electrodes. Lett Appl Microbiol 59(6):636–641. https://doi.org/10.1111/lam.12324
105. Lunden J, Autio T, Markkula A, Hellstrom S, Korkeala H (2003) Adaptive and cross-adaptive responses of persistent and non-persistent Listeria monocytogenes strains to disinfectants. Int J Food Microbiol 82(3):265–272
106. Luppens SB, Reij MW, van der Heijden RW, Rombouts FM, Abee T (2002) Development of a standard test to assess the resistance of Staphylococcus aureus biofilm cells to disinfectants. Appl Environ Microbiol 68(9):4194–4200
107. Ma J, Tong Z, Ling J, Liu H, Wei X (2015) The effects of sodium hypochlorite and chlorhexidine irrigants on the antibacterial activities of alkaline media against Enterococcus faecalis. Arch Oral Biol 60(7):1075–1081. https://doi.org/10.1016/j.archoralbio.2015.04.008
108. Mariscal A, Carnero-Varo M, Gutierrez-Bedmar M, Garcia-Rodriguez A, Fernandez-Crehuet J (2007) A fluorescent method for assessing the antimicrobial efficacy of disinfectant against Escherichia coli ATCC 35218 biofilm. Appl Microbiol Biotechnol 77 (1):233–240. https://doi.org/10.1007/s00253-007-1137-z
109. Mariscal A, Lopez-Gigosos RM, Carnero-Varo M, Fernandez-Crehuet J (2009) Fluorescent assay based on resazurin for detection of activity of disinfectants against bacterial biofilm. Appl Microbiol Biotechnol 82(4):773–783. https://doi.org/10.1007/s00253-009-1879-x
110. Mattei AS, Madrid IM, Santin R, Schuch LF, Meireles MC (2013) In vitro activity of disinfectants against Aspergillus spp. Braz J Microbiol: [Publication of the Brazilian Society for Microbiology] 44(2):481–484. https://doi.org/10.1590/s1517-83822013000200024
111. Matthes R, Jablonowski L, Koban I, Quade A, Hubner NO, Schlueter R, Weltmann KD, von Woedtke T, Kramer A, Kocher T (2015) In vitro treatment of Candida albicans biofilms on denture base material with volume dielectric barrier discharge plasma (VDBD) compared with common chemical antiseptics. Clin Oral Invest 19(9):2319–2326. https://doi.org/10.1007/s00784-015-1463-y
112. Mazaheritehrani E, Sala A, Orsi CF, Neglia RG, Morace G, Blasi E, Cermelli C (2014) Human pathogenic viruses are retained in and released by Candida albicans biofilm in vitro. Virus Res 179:153–160. https://doi.org/10.1016/j.virusres.2013.10.018
113. Mekonnen D, Admassu A, Wassie B, Biadglegne F (2015) Evaluation of the efficacy of bleach routinely used in health facilities against Mycobacterium tuberculosis isolates in Ethiopia. Pan Afr Med J 21:317. https://doi.org/10.11604/pamj.2015.21.317.5456
114. Melo RT, Mendonca EP, Monteiro GP, Siqueira MC, Pereira CB, Peres P, Fernandez H, Rossi DA (2017) Intrinsic and extrinsic aspects on Campylobacter jejuni Biofilms. Front Microbiol 8:1332. https://doi.org/10.3389/fmicb.2017.01332
115. Mercade M, Duran-Sindreu F, Kuttler S, Roig M, Durany N (2009) Antimicrobial efficacy of 4.2% sodium hypochlorite adjusted to pH 12, 7.5, and 6.5 in infected human root canals. Oral Surg Oral Med Oral Pathol Oral Radiol Endod 107(2):295–298. https://doi.org/10.1016/j.tripleo.2008.05.006

116. Meylheuc T, Renault M, Bellon-Fontaine MN (2006) Adsorption of a biosurfactant on surfaces to enhance the disinfection of surfaces contaminated with Listeria monocytogenes. Int J Food Microbiol 109(1–2):71–78. https://doi.org/10.1016/j.ijfoodmicro.2006.01.013
117. Miyano N, Oie S, Kamiya A (2003) Efficacy of disinfectants and hot water against biofilm cells of Burkholderia cepacia. Biol Pharm Bull 26(5):671–674
118. Mohmmed SA, Vianna ME, Penny MR, Hilton ST, Mordan N, Knowles JC (2016) A novel experimental approach to investigate the effect of different agitation methods using sodium hypochlorite as an irrigant on the rate of bacterial biofilm removal from the wall of a simulated root canal model. Dent Mater: Off Publ Acad Den Mater 32(10):1289–1300. https://doi.org/10.1016/j.dental.2016.07.013
119. Molina-González D, Alonso-Calleja C, Alonso-Hernando A, Capita R (2014) Effect of sub-lethal concentrations of biocides on the susceptibility to antibiotics of multi-drug resistant Salmonella enterica strains. Food Control 40(Supplement C):329–334. https://doi.org/10.1016/j.foodcont.2013.11.046
120. Moore G, Schelenz S, Borman AM, Johnson EM, Brown CS (2017) Yeasticidal activity of chemical disinfectants and antiseptics against Candida auris. J Hosp Infect 97(4):371–375. https://doi.org/10.1016/j.jhin.2017.08.019
121. Morrissey I, Oggioni MR, Knight D, Curiao T, Coque T, Kalkanci A, Martinez JL (2014) Evaluation of epidemiological cut-off values indicates that biocide resistant subpopulations are uncommon in natural isolates of clinically-relevant microorganisms. PLoS ONE 9(1): e86669. https://doi.org/10.1371/journal.pone.0086669
122. Nandy P, Lucas AD, Gonzalez EA, Hitchins VM (2016) Efficacy of commercially available wipes for disinfection of pulse oximeter sensors. Am J Infect Control 44(3):304–310. https://doi.org/10.1016/j.ajic.2015.09.028
123. Narui K, Takano M, Noguchi N, Sasatsu M (2007) Susceptibilities of methicillin-resistant Staphylococcus aureus isolates to seven biocides. Biol Pharma Bull 30(3):585–587
124. National Center for Biotechnology Information Sodium hypochlorite. PubChem Compound Database; CID=23665760. https://pubchem.ncbi.nlm.nih.gov/compound/23665760. Accessed 27 Oct 2017
125. Nguyen HDN, Yuk HG (2013) Changes in resistance of Salmonella Typhimurium biofilms formed under various conditions to industrial sanitizers. Food Control 29(1):236–240. https://doi.org/10.1016/j.foodcont.2012.06.006
126. Norman G, Christie J, Liu Z, Westby MJ, Jefferies JM, Hudson T, Edwards J, Mohapatra DP, Hassan IA, Dumville JC (2017) Antiseptics for burns. Cochrane database Syst Rev 7: Cd011821. https://doi.org/10.1002/14651858.cd011821.pub2
127. Norton CD, LeChevallier MW, Falkinham JO 3rd (2004) Survival of Mycobacterium avium in a model distribution system. Water Res 38(6):1457–1466. https://doi.org/10.1016/j.watres.2003.07.008
128. Ntsama-Essomba C, Bouttier S, Ramaldes M, Dubois-Brissonnet F, Fourniat J (1997) Resistance of Escherichia coli growing as biofilms to disinfectants. Vet Res 28(4):353–363
129. Oggioni MR, Coelho JR, Furi L, Knight DR, Viti C, Orefici G, Martinez JL, Freitas AT, Coque TM, Morrissey I (2015) Significant differences characterise the correlation coefficients between biocide and antibiotic susceptibility profiles in Staphylococcus aureus. Curr Pharm Des 21(16):2054–2057
130. Oh J, Salcedo DE, Medriano CA, Kim S (2014) Comparison of different disinfection processes in the effective removal of antibiotic-resistant bacteria and genes. J Environ Sci (China) 26(6):1238–1242. https://doi.org/10.1016/s1001-0742(13)60594-x
131. Oie S, Huang Y, Kamiya A, Konishi H, Nakazawa T (1996) Efficacy of disinfectants against biofilm cells of methicillin-resistant *Staphylococcus aureus*. Microbios 85:223–230
132. Omidbakhsh N (2010) Theoretical and experimental aspects of microbicidal activities of hard surface disinfectants: are their label claims based on testing under field conditions? J AOAC Int 93(6):1944–1951

133. Pagedar A, Singh J (2015) Evaluation of antibiofilm effect of benzalkonium chloride, iodophore and sodium hypochlorite against biofilm of Pseudomonas aeruginosa of dairy origin. J Food Sci Technol 52(8):5317–5322. https://doi.org/10.1007/s13197-014-1575-4
134. Pariser M, Gard S, Gram D, Schmeitzel L (2013) An in vitro study to determine the minimal bactericidal concentration of sodium hypochlorite (bleach) required to inhibit meticillin-resistant Staphylococcus pseudintermedius strains isolated from canine skin. Vet Dermatol 24(6):632–634, e156–e637. https://doi.org/10.1111/vde.12079
135. Peeters E, Nelis HJ, Coenye T (2008) Evaluation of the efficacy of disinfection procedures against Burkholderia cenocepacia biofilms. J Hosp Infect 70(4):361–368. https://doi.org/10.1016/j.jhin.2008.08.015
136. Petti S, Polimeni A, Dancer SJ (2013) Effect of disposable barriers, disinfection, and cleaning on controlling methicillin-resistant Staphylococcus aureus environmental contamination. Am J Infect Control 41(9):836–840. https://doi.org/10.1016/j.ajic.2012.09.021
137. Pires RH, da Silva Jde F, Gomes Martins CH, Fusco Almeida AM, Pienna Soares C, Soares Mendes-Giannini MJ (2013) Effectiveness of disinfectants used in hemodialysis against both Candida orthopsilosis and C. parapsilosis sensu stricto biofilms. Antimicrob Agents Chemother 57(5):2417–2421. https://doi.org/10.1128/aac.01308-12
138. Piskin B, Turkun M (1995) Stability of various sodium hypochlorite solutions. J Endod 21 (5):253–255
139. Quan Y, Choi KD, Chung D, Shin IS (2010) Evaluation of bactericidal activity of weakly acidic electrolyzed water (WAEW) against Vibrio vulnificus and Vibrio parahaemolyticus. Int J Food Microbiol 136(3):255–260. https://doi.org/10.1016/j.ijfoodmicro.2009.11.005
140. Rasmussen LH, Kjeldgaard J, Christensen JP, Ingmer H (2013) Multilocus sequence typing and biocide tolerance of Arcobacter butzleri from Danish broiler carcasses. BMC Res Notes 6:322. https://doi.org/10.1186/1756-0500-6-322
141. Riazi S, Matthews KR (2011) Failure of foodborne pathogens to develop resistance to sanitizers following repeated exposure to common sanitizers. Int Biodeter Biodegr 65 (2):374–378. https://doi.org/10.1016/j.ibiod.2010.12.001
142. Rizzotti L, Rossi F, Torriani S (2016) Biocide and antibiotic resistance of Enterococcus faecalis and Enterococcus faecium isolated from the swine meat chain. Food Microbiol 60:160–164. https://doi.org/10.1016/j.fm.2016.07.009
143. Rodriguez Ferri EF, Martinez S, Frandoloso R, Yubero S, Gutierrez Martin CB (2010) Comparative efficacy of several disinfectants in suspension and carrier tests against Haemophilus parasuis serovars 1 and 5. Res Vet Sci 88(3):385–389. https://doi.org/10.1016/j.rvsc.2009.12.001
144. Rudd RW, Senia ES, McCleskey FK, Adams ED (1984) Sterilization of complete dentures with sodium hypochlorite. J Prosth Dent 51(3):318–321. https://doi.org/10.1016/0022-3913(84)90212-9
145. Ruiz-Linares M, Aguado-Perez B, Baca P, Arias-Moliz MT, Ferrer-Luque CM (2017) Efficacy of antimicrobial solutions against polymicrobial root canal biofilm. Int Endod J 50 (1):77–83. https://doi.org/10.1111/iej.12598
146. Rutala WA, Cole EC, Wannamaker NS, Weber DJ (1991) Inactivation of Mycobacterium tuberculosis and Mycobacterium bovis by 14 hospital disinfectants. Am J Med 91(3b):267s–271s
147. Saejung C, Hatai K, Sanoamuang L (2014) The in-vitro antibacterial effects of organic salts, chemical disinfectants and antibiotics against pathogens of black disease in fairy shrimp of Thailand. J Fish Dis 37(1):33–41. https://doi.org/10.1111/j.1365-2761.2012.01452.x
148. Sagripanti J-L, Eklund CA, Trost PA, Jinneman KC, Abeyta C, Kaysner CA, Hill WE (1997) Comparative sensitivity of 13 species of pathogenic bacteria to seven chemical germicides. Am J Infect Control 25(4):335–339
149. Sattar SA, Bradley C, Kibbee R, Wesgate R, Wilkinson MA, Sharpe T, Maillard JY (2015) Disinfectant wipes are appropriate to control microbial bioburden from surfaces: use of a

new ASTM standard test protocol to demonstrate efficacy. J Hosp Infect 91(4):319–325. https://doi.org/10.1016/j.jhin.2015.08.026

150. Seier-Petersen MA, Jasni A, Aarestrup FM, Vigre H, Mullany P, Roberts AP, Agerso Y (2014) Effect of subinhibitory concentrations of four commonly used biocides on the conjugative transfer of Tn916 in Bacillus subtilis. J Antimicrob Chemother 69(2):343–348. https://doi.org/10.1093/jac/dkt370
151. Simoes M, Pereira MO, Vieira MJ (2005) Effect of mechanical stress on biofilms challenged by different chemicals. Water Res 39(20):5142–5152. https://doi.org/10.1016/j.watres.2005.09.028
152. Siqueira JF Jr, Rocas IN, Paiva SS, Guimaraes-Pinto T, Magalhaes KM, Lima KC (2007) Bacteriologic investigation of the effects of sodium hypochlorite and chlorhexidine during the endodontic treatment of teeth with apical periodontitis. Oral Surg Oral Med Oral Pathol Oral Radiol Endod 104(1):122–130. https://doi.org/10.1016/j.tripleo.2007.01.027
153. Song L, Wu J, Xi C (2012) Biofilms on environmental surfaces: evaluation of the disinfection efficacy of a novel steam vapor system. Am J Infect Control 40(10):926–930. https://doi.org/10.1016/j.ajic.2011.11.013
154. Soumet C, Meheust D, Pissavin C, Le Grandois P, Fremaux B, Feurer C, Le Roux A, Denis M, Maris P (2016) Reduced susceptibilities to biocides and resistance to antibiotics in food-associated bacteria following exposure to quaternary ammonium compounds. J Appl Microbiol 121(5):1275–1281. https://doi.org/10.1111/jam.13247
155. Stewart PS, Grab L, Diemer JA (1998) Analysis of biocide transport limitation in an artificial biofilm system. J Appl Microbiol 85(3):495–500
156. Sullivan BA, Vance CC, Gentry TJ, Karthikeyan R (2017) Effects of chlorination and ultraviolet light on environmental tetracycline-resistant bacteria and tet(W) in water. J Environ Chem Eng 5(1):777–784. https://doi.org/10.1016/j.jece.2016.12.052
157. Takeo Y, Oie S, Kamiya A, Konishi H, Nakazawa T (1994) Efficacy of disinfectants against biofilm cells of Pseudomonas aeruginosa. Microbios 79(318):19–26
158. Theraud M, Bedouin Y, Guiguen C, Gangneux JP (2004) Efficacy of antiseptics and disinfectants on clinical and environmental yeast isolates in planktonic and biofilm conditions. J Med Microbiol 53(Pt 10):1013–1018. https://doi.org/10.1099/jmm.0.05474-0
159. Tirali RE, Bodur H, Ece G (2012) In vitro antimicrobial activity of sodium hypochlorite, chlorhexidine gluconate and octenidine dihydrochloride in elimination of microorganisms within dentinal tubules of primary and permanent teeth. Medicina oral, patologia oral y cirugia bucal 17(3):e517–e522
160. Tirali RE, Turan Y, Akal N, Karahan ZC (2009) In vitro antimicrobial activity of several concentrations of NaOCl and Octenisept in elimination of endodontic pathogens. Oral Surg Oral Med Oral Pathol Oral Radiol Endod 108(5):e117–e120. https://doi.org/10.1016/j.tripleo.2009.07.012
161. Tote K, Horemans T, Vanden Berghe D, Maes L, Cos P (2010) Inhibitory effect of biocides on the viable masses and matrices of Staphylococcus aureus and Pseudomonas aeruginosa biofilms. Appl Environ Microbiol 76(10):3135–3142. https://doi.org/10.1128/aem.02095-09
162. Ueda S, Kuwabara Y (2007) Susceptibility of biofilm Escherichia coli, Salmonella Enteritidis and Staphylococcus aureus to detergents and sanitizers. Biocontrol Sci 12(4):149–153
163. Ulusoy AT, Kalyoncuoglu E, Reis A, Cehreli ZC (2016) Antibacterial effect of N-acetylcysteine and taurolidine on planktonic and biofilm forms of Enterococcus faecalis. Dent Traumatol: Off Publ Int Assoc Dent Traumatol 32(3):212–218. https://doi.org/10.1111/edt.12237
164. United States Environmental Protection Agency (1991) EPA R.E.D. Facts. Sodium and calcium hypochlorite salts. https://www3.epa.gov/pesticides/chem_search/reg_actions/reregistration/fs_G-77_1-Sep-91.pdf
165. Vaziri S, Kangarlou A, Shahbazi R, Nazari Nasab A, Naseri M (2012) Comparison of the bactericidal efficacy of photodynamic therapy, 2.5% sodium hypochlorite, and 2%

chlorhexidine against Enterococcous faecalis in root canals; an in vitro study. Dent Res J 9 (5):613–618
166. Vázquez-Sánchez D, Cabo ML, Ibusquiza PS, Rodríguez-Herrera JJ (2014) Biofilm-forming ability and resistance to industrial disinfectants of Staphylococcus aureus isolated from fishery products. Food Control 39(Supplement C):8–16. https://doi.org/10.1016/j.foodcont.2013.09.029
167. Venturini ME, Blanco D, Oria R (2002) In vitro antifungal activity of several antimicrobial compounds against Penicillium expansum. J Food Prot 65(5):834–839
168. Vetas D, Dimitropoulou E, Mitropoulou G, Kourkoutas Y, Giaouris E (2017) Disinfection efficiencies of sage and spearmint essential oils against planktonic and biofilm Staphylococcus aureus cells in comparison with sodium hypochlorite. Int J Food Microbiol 257:19–25. https://doi.org/10.1016/j.ijfoodmicro.2017.06.003
169. Vieira CD, Farias Lde M, Diniz CG, Alvarez-Leite ME, Camargo ER, Carvalho MA (2005) New methods in the evaluation of chemical disinfectants used in health care services. Am J Infect Control 33(3):162–169. https://doi.org/10.1016/j.ajic.2004.10.007
170. Walker J, Moore G, Collins S, Parks S, Garvey MI, Lamagni T, Smith G, Dawkin L, Goldenberg S, Chand M (2017) Microbiological problems and biofilms associated with Mycobacterium chimaera in heater-cooler units used for cardiopulmonary bypass. J Hosp Infect 96(3):209–220. https://doi.org/10.1016/j.jhin.2017.04.014
171. Wang H, Hu C, Liu L, Xing X (2017) Interaction of ciprofloxacin chlorination products with bacteria in drinking water distribution systems. J Hazard Mater 339:174–181. https://doi.org/10.1016/j.jhazmat.2017.06.033
172. Wang H, Tay M, Palmer J, Flint S (2017) Biofilm formation of Yersinia enterocolitica and its persistence following treatment with different sanitation agents. Food Control 73(Part B):433–437. https://doi.org/10.1016/j.foodcont.2016.08.033
173. Wilson CE, Cathro PC, Rogers AH, Briggs N, Zilm PS (2015) Clonal diversity in biofilm formation by Enterococcus faecalis in response to environmental stress associated with endodontic irrigants and medicaments. Int Endod J 48(3):210–219. https://doi.org/10.1111/iej.12301
174. Wong HS, Townsend KM, Fenwick SG, Trengove RD, O'Handley RM (2010) Comparative susceptibility of planktonic and 3-day-old Salmonella Typhimurium biofilms to disinfectants. J Appl Microbiol 108(6):2222–2228. https://doi.org/10.1111/j.1365-2672.2009.04630.x
175. Wuthiekanun V, Wongsuwan G, Pangmee S, Teerawattanasook N, Day NP, Peacock SJ (2011) Perasafe, Virkon and bleach are bactericidal for Burkholderia pseudomallei, a select agent and the cause of melioidosis. J Hosp Infect 77(2):183–184. https://doi.org/10.1016/j.jhin.2010.06.026
176. Yadav P, Chaudhary S, Saxena RK, Talwar S, Yadav S (2017) Evaluation of Antimicrobial and Antifungal efficacy of Chitosan as endodontic irrigant against Enterococcus Faecalis and Candida Albicans Biofilm formed on tooth substrate. J Clin Exp Dent 9(3):e361–e367. https://doi.org/10.4317/jced.53210
177. Zand V, Lotfi M, Soroush MH, Abdollahi AA, Sadeghi M, Mojadadi A (2016) Antibacterial efficacy of different concentrations of sodium hypochlorite gel and solution on Enterococcus faecalis biofilm. Iran Endod J 11(4):315–319. https://doi.org/10.22037/iej.2016.11
178. Zheng J, Su C, Zhou J, Xu L, Qian Y, Chen H (2017) Effects and mechanisms of ultraviolet, chlorination, and ozone disinfection on antibiotic resistance genes in secondary effluents of municipal wastewater treatment plants. Chem Eng J 317(Supplement C):309–316. https://doi.org/10.1016/j.cej.2017.02.076

Triclosan

9

9.1 Chemical Characterization

Triclosan was developed in the 1960s and patented in 1964 [62]. It is a white powdered solid with a slight aromatic phenolic odour. Categorized as a polychloro phenoxy phenol, triclosan is a chlorinated aromatic compound that has functional groups representative of both ethers and phenols. It is poorly soluble in water but dissolves well in alcohols [142]. The basic chemical information on triclosan is summarized in Table 9.1.

9.2 Types of Application

Triclosan is typically used in cosmetic and personal care consumer products, and for preservation [41, 145]. It is also used by professionals, e.g. in detergents (0.4–1%) and in alcohols (0.2–0.5%) used for hygienic and surgical hand antisepsis or preoperative skin disinfection [116]. It has also been used for antiseptic body baths to control MRSA [147]. This agent is incorporated into some soaps at 1% (w/v) as well as being integrated into various dressings and bandages for release over time onto the skin [142] or on self-disinfecting surfaces [140].

9.2.1 European Chemicals Agency (European Union)

In 2010, the Scientific Committee on Consumer Safety has evaluated the risk of antimicrobial resistance and recommended only the prudent use of triclosan, for example in applications where a health benefit can be demonstrated [120]. Triclosan has also been evaluated as an active biocidal substance. In 2014, it was not approved for product types 2 (disinfectants and algaecides not intended for direct application to humans or animals), 7 (film preservatives) and 9 (fibre, leather,

G. Kampf, *Antiseptic Stewardship*, https://doi.org/10.1007/978-3-319-98785-9_9

Table 9.1 Basic chemical information on triclosan [102]

CAS-number	3380-34-5
IUPAC-name	5-chloro-2-(2,4-dichlorophenoxy)phenol
Synonyms	Irgasan DP-300
Molecular formula	$C_{12}H_7Cl_3O_2$
Molecular weight (g/mol)	289.536

rubber and polymerised materials preservatives), followed by non-approval in 2016 for product type 1 (human hygiene) [5, 69].

9.2.2 Environmental Protection Agency (USA)

Triclosan was first registered by the EPA in 1969. The Agency has found in 2008 that currently registered uses of triclosan are eligible for reregistration provided the conditions and requirements for reregistration identified in the reregistration eligibility decision are implemented [136]. Use as a materials preservative in paint has been excluded but has also been requested to be voluntarily cancelled by the registrants. Appendix A contains a list of uses such as material preservatives in personal care products, textiles and fibres, home furnishing, carpets and rugs, PVC, plastics, sponges, polymer compounds, construction and building materials, adhesives, ice making equipment and sporting goods [136].

9.2.3 Food and Drug Administration (USA)

One group of antiseptic products is intended for use by healthcare professionals in a hospital setting or other healthcare situations outside the hospital. Based on the proposed amendment of the tentative final monograph for healthcare antiseptic products in 2015, triclosan is eligible as an active ingredient in patient preoperative skin preparations, healthcare personnel hand wash products and surgical hand scrub products [37]. In the 1994 tentative final monograph for healthcare antiseptic products, triclosan was classified IIIE indicating that additional effectiveness data are needed [36]. The proposed rule from 2015 classifies triclosan as IIISE indicating that additional safety and effectiveness data are needed [37]. The main aspects of safety are human pharmacokinetics, animal pharmacokinetics, potential hormonal effects and resistance potential [37].

Another group of antiseptic products is intended for use by consumers and classified as over-the-counter antiseptic products. In 2016, the FDA misbranded triclosan and 18 other active ingredients used in consumer antiseptic wash products intended for use with water and are rinsed off after use, including hand washes and body washes [38]. The submitted data were not sufficient to classify triclosan as generally safe and effective. A key aspect was that the risk from the use of a consumer antiseptic wash drug product must be balanced by a demonstration—

through studies that demonstrate a direct clinical benefit (i.e. a reduction of infection)—that the product is superior to washing with non-antibacterial soap and water in reducing infection. If the active ingredient in a drug product (in this case: triclosan) carries the potential risk associated with the drug (e.g. reproductive toxicity or carcinogenicity or resistance), but does not provide a clinical benefit, then the benefit-to-risk calculation shifts towards a not generally safe and effective status for that drug. The decision by the FDA was welcomed by the scientific community [63, 95].

9.2.4 Overall Environmental Impact

Triclosan is manufactured and/or imported in the European Economic Area in 10–100 t per year [41], globally more than 1,500 t are produced per year [54]. Triclosan is often detected in water samples in effluent-dominated urban streams [133] leading to exposure of aquatic micro-organisms [34]. It has been estimated that in France alone a total of 11.2–23.5 t per year are added to the wastewater, mainly by personal care products [53]. An environmental risk has mainly been found for aquatic and sediment-dwelling organisms exposed to triclosan in the surface water and sediment compartments, indicating the environmental risk to be concerned due to the high levels of triclosan [61]. The potential environmental risk of triclosan is considered to be high especially in rivers where water scarcity results in low dilution capacity [113]. Triclosan-resistant faecal coliforms were isolated in 79–94% from surfaces waters located near wastewater treatment plants. Environmental faecal coliforms isolates resistant to high-level triclosan included species of *Escherichia*, *Enterobacter*, *Serratia* and *Citrobacter*. A significant relationship between triclosan resistance and multiple antibiotic resistances was described [99].

9.3 Spectrum of Antimicrobial Activity

9.3.1 Bactericidal Activity

9.3.1.1 Bacteriostatic Activity (MIC Values)

The MIC values for triclosan obtained with different bacterial species are summarized in Table 9.2. The highest MIC value (2,500 mg/l) was described in an extensively resistant clinical *P. aeruginosa* isolate. Resistant isolates were also found among Gram-negative species such as *P. aeruginosa* (up to 2,500 mg/l), followed by *E. coli* (up to 1,000 mg/l), *B. cepacia* (up to 500 mg/l), *A. baumannii* and selected *Lactobacillus* spp. and *Salmonella* spp. (up to 256 mg/l), *S. marcescens* and *B. cepacia* (up to 232 mg/l) and *C. freundii* (>100 mg/l). Taking into account the proposed epidemiological cut-off value for *E. coli* with 2 mg/l some isolates can be classified as resistant [100]. Most *Enterobacter* spp. isolates were below the proposed epidemiological cut-off value of 1 mg/l similar to *Klebsiella* spp. with

2 mg/l and *Salmonella* spp. with 8 mg/l [100]. Gram-positive bacterial species tend to be more susceptible to triclosan. The highest MIC values were described in *Bifidobacterium* spp. (up to 512 mg/l), *C. perfringens* and selected *Lactobacillus* spp. (256 mg/l) and *Enterococcus* spp. (128 mg/l). The majority of MIC values of the tested isolates of *E. faecalis* (ECOFF: 16 mg/l), *E. faecium* (ECOFF: 32 mg/l) and *S. aureus* (ECOFF: 0.5 mg/l) can be classified as susceptible to triclosan [100]. It is noteworthy that in 34 *S. epidermidis* isolates from the 1960s, the MIC values were lower (range: 0.0156–0.125 mg/l) compared to 64 isolates from 2010 to 2011 (0.0156–4.0 mg/l) [126].

Table 9.2 MIC values of various bacterial species to triclosan

Species	Number of strains/isolates	MIC value (mg/l)	References
A. baumannii	3 isolates from domestic surfaces	2	[22]
A. baumannii	47 clinical isolates	2–256	[84]
A. johnsonii	NCIMB 12460	0.094	[28]
	"triclosan-tolerant strain"	21.9	
A. johnsonii	NCIMB 12460	4–5	[82]
	triclosan-tolerant industrial strain	>100	
A. proteolyticus	Environmental strain M9.12	19.5	[83]
A. xylosoxidans	Environmental strain M4.31	1	[83]
B. cereus	Environmental strain M7.15	1	[83]
B. cereus	MRBG 4.21 (kitchen drain biofilm isolate)	7.3	[46]
B. adolescentis	4 isolates from faeces of healthy humans	8–64	[39]
B. animalis subsp. lactis	8 isolates from faeces of healthy humans	64–512	[39]
B. bifidum	31 isolates from faeces of healthy humans	4–512	[39]
B. breve	5 isolates from faeces of healthy humans	8–128	[39]
B. catenulatum	1 isolate from faeces of a healthy human	256	[39]
B. infantis	2 isolates from faeces of healthy humans	32–256	[39]
B. longum	25 isolates from faeces of healthy humans	8–256	[39]
B. pseudocatenulatum	15 isolates from faeces of healthy humans	32–256	[39]
B. pseudolongum	1 isolate from faeces of a healthy human	256	[39]
B. thermoacidophilum	6 isolates from faeces of healthy humans	16–512	[39]
B. suis	1 isolate from faeces of a healthy human	64	[39]
B. cepacia complex	38 clinical, non-clinical and environmental strains	50–500	[117]
B. cepacia	ATCC BAA-245	232	[46]
C. coli	8 strains from poultry	16–64	[89]
	6 strains from humans	32–64	
	4 strains from pigs	32	
	1 strain from water	64	

(continued)

Table 9.2 (continued)

Species	Number of strains/isolates	MIC value (mg/l)	References
C. jejuni	5 strains from humans	8–32	[89]
	5 strains from water	8–32	
	3 strains from poultry	16–32	
C. indologenes	Environmental strain M9.15	1	[83]
Chryseobacterium spp.	Environmental strain FR2 9.17	15.6	[83]
C. freundii	3 isolates from domestic surfaces	2	[22]
C. freundii	NCIMB 11490	>100	[82]
	triclosan-tolerant industrial strain		
C. perfringens	ATCC 13124	256	[77]
C. indologenes	MRBG 4.29 (kitchen drain biofilm isolate)	0.9	[46]
C. xerosis	WIBG 1.2 (wound isolate)	7.3	[46]
E. asburiae	Enteric strain M21.2	1	[83]
E. cloacae	5 isolates from domestic surfaces	0.5–1	[22]
Enterobacter spp.	54 worldwide strains from hospital- and community-acquired infections	0.03–8	[100]
E. faecalis	56 worldwide strains from hospital- and community-acquired infections	0.5–16	[100]
E. faecalis	9 isolates from swine meat production	2–30	[114]
E. faecalis	WIBG 1.1 (wound isolate)	3.3	[46]
E. faecium	12 isolates from swine meat production	2–16	[114]
E. faecium	53 worldwide strains from hospital- and community-acquired infections	2–64	[100]
E. faecalis	ATCC 29212	16	[77]
Enterococcus spp.	122 strains (*E. faecalis*, *E. faecium*) from different traditional fermented foods	0.1–0.25	[81]
Enterococcus spp.	3 isolates from domestic surfaces	>2	[22]
Enterococcus spp.	4 VRE strains	3–4	[129]
Enterococcus spp.	Clinical VRE isolate	128	[77]
E. coli	306 worldwide strains from hospital- and community-acquired infections	0.015–2	[100]
E. coli	Strain HEC30	0.06	[32]
E. coli	3 strains	0.06–0.25	[139]
E. coli	ATCC 8739	0.2	[28]
	"triclosan-tolerant strain"	20–1,000	
E. coli	5 isolates from domestic surfaces	0.3–0.5	[22]
E. coli	ATCC 25922	0.5	[46]
E. coli	ATCC 25922	0.5	[40]
E. coli	ATCC 25922 and 4 clinical isolates	0.5–64	[4]
E. coli	ATCC 25922 and enteric strain M20.1	1.3–2	[83]
E. coli	ATCC 35218	2–4	[77]

(continued)

Table 9.2 (continued)

Species	Number of strains/isolates	MIC value (mg/l)	References
E. coli	13 bovine and 7 equine strains	3.1–12.5	[123]
E. coli	27 isolates from hen eggshells	5	[59]
Eubacterium spp.	Environmental strain M4.14	15.6	[83]
F. nucleatum	Dental strains M20.2 and M20.3	1–3.3	[83]
H. gallinarum	Environmental strain M4.27	31.3	[83]
H. influenzae	ATCC 49247	0.125–32	[77]
K. oxytoca	Enteric strain M21.3	1	[83]
K. oxytoca	2 isolates from domestic surfaces	≥2	[22]
K. planticola	Enteric strain M21.1	1	[83]
K. pneumoniae	60 worldwide strains from hospital- and community-acquired infections	0.015–4	[100]
K. pneumoniae	Strain 39.11	0.5	[32]
K. pneumoniae	ATCC 13883	0.9	[46]
Klebsiella spp.	37 isolates predominately from a variety of human infections pre-1949 ("Murray isolates") and 39 "modern strains" (2007–2012)	0.007–0.5 (old isolates) 0.125–2 (modern isolates)	[138]
L. acidophilus	4 strains from different origins	16	[3]
L. amylovorus	7 strains from different origins	64–256	[3]
L. brevis	13 strains from different origins	16–64	[3]
L. bulgaricus	6 strains from different origins	8–16	[3]
L. coryniformis	3 strains from different origins	64	[3]
L. fermentum	4 strains from different origins	16–64	[3]
L. garvieae	42 isolates from different origins	1–4	[3]
L. helveticus	39 strains from different origins	2–8	[3]
L. paracasei	75 strains from different origins	2–256	[3]
L. pentosus	60 strains from naturally fermented Aloreña green table olives	0.01–5.0	[15]
L. plantarum	43 strains from different origins	16–256	[3]
L. reuteri	42 strains from different origins	8–256	[3]
L. rhamnosus	Dental strain M6.1	2	[83]
L. rhamnosus	9 strains from different origins	8–64	[3]
L. lactis	Dental strain M6.3	2	[83]
L. pseudomesenteroides	13 strains from naturally fermented Aloreña green table olives	0.1–5.0	[15]
Megasphaera spp.	Dental strain M20.9	2	[83]
M. luteus	Environmental strain M9.25	1	[83]
M. luteus	MRBG 9.25 (skin isolate)	7.3	[46]
M. phyllosphaeriae	Environmental strain M4.30	2	[83]
P. multocida	ATCC 11039 and 2 strains	0.06–0.25	[40]

(continued)

Table 9.2 (continued)

Species	Number of strains/isolates	MIC value (mg/l)	References
P. aeruginosa	111 clinical isolates	1–500	[78]
P. aeruginosa	8 isolates from domestic surfaces	≥2	[22]
P. aeruginosa	PA01	>32	[40]
P. aeruginosa	ATCC 15442	>512	[77]
P. aeruginosa	ATCC 9027	>1,000	[46]
P. aeruginosa	NCTC6749 and 3 extensively resistant clinical isolates	2,500	[144]
P. alkylphenolia	2 isolates from meat chain production	>10	[80]
P. fluorescens	3 isolates from meat chain production	0.0025–>10	[80]
P. fragi	4 isolates from meat chain production	0.0025–10	[80]
P. lundensis	34 isolates from meat chain production	0.0025–>10	[80]
P. putida	9 isolates from meat chain production	0.0025–>10	[80]
S. enterica	368 animal isolates and 60 human isolates	0.25–4	[25]
S. Enteritidis	NCTC 8513	2.6	[83]
S. Infantis	NCIMB 13036	1.6	[83]
S. Typhimurium	3 strains	0.06–0.25	[139]
S. Typhimurium	ATCC 23564 and NCTC 74	2	[83]
Salmonella spp.	901 worldwide strains from hospital- and community-acquired infections	0.03–8	[100]
Salmonella spp.	375 avian isolates	0.0625–0.5	[111]
Salmonella spp.	465 isolates from 6 different slaughterhouses	0.25–8	[51]
Salmonella spp.	112 isolates from meat	2.5–250	[52]
S. marcescens	ATCC 13880	232	[46]
S. aureus	7 isolates from domestic surfaces, 3 isolates from household individuals	<0.004–0.128	[22]
S. aureus	256 clinical isolates (87 MRSA, 169 MSSA)	0.005–0.32	[78]
S. aureus	1,388 clinical isolates	≤0.015–32	[20]
S. aureus	1,635 worldwide strains from hospital- and community-acquired infections	0.015–0.5	[100]
S. aureus	NCTC 6571, NCTC 83254 and 17 clinical MRSA isolates	0.025–1	[129]
S. aureus	NCTC 6571, 17 clinical isolates and 15 MRSA strains	0.025–1	[130]
S. aureus	NCIMB 9518	0.053	[28]
	"triclosan-tolerant strain"	3.2	
S. aureus	198 clinical isolates (161 MRSA, 37 MSSA)	<0.06–16	[68]
S. aureus	ATCC 6538	0.125	[77]
S. aureus	ATCC 29213	0.125	[4]
	1 clinical MRSA strain	64	

(continued)

Table 9.2 (continued)

Species	Number of strains/isolates	MIC value (mg/l)	References
S. aureus	ATCC 6538	0.2	[46]
S. aureus	Clinical MRSA isolate	64	[77]
S. capitis	MRBG 9.34 (skin isolate)	24.2	[46]
S. caprae	MRBG 9.3 (skin isolate)	12.3	[46]
S. epidermidis	34 blood culture isolates from the 1960s and 64 blood culture isolates from 2010 to 2011	0.0156–0.125 ("old isolates")	[126]
		0.0156–4.0 ("current isolates")	
S. epidermidis	MRBG 9.33 (skin isolate)	13.3	[46]
S. haemolyticus	21 clinical isolates	0.008–0.25	[27]
S. haemolyticus	MRBG 9.35 (skin isolate)	0.4	[46]
S. lugdunensis	MRBG 9.36 (skin isolate)	0.9	[46]
S. pseudointermedius	20 MRSP and 20 MSSP from dogs (35) and cats (5)	<0.003–4	[29]
S. pseudointermedius	25 MSSP and 25 MRSP from dogs with skin and soft tissue infections	≤ 0.5	[137]
S. warneri	MRBG 9.27 (skin isolate)	0.9	[46]
Staphylococcus spp.	14 methicillin-resistant isolates from horses (*S. cohnii, S. lentus, S. fleurettii, S. sciuri, S. haemolyticus, S. aureus*)	0.003–4	[30]
Staphylococcus spp.	11 species (skin isolates)	0.5–1	[83]
S. maltophilia	Environmental strain M9.13	9.8	[83]
S. maltophilia	MRBG 4.17 (kitchen drain biofilm isolate)	14.5	[46]
S. anginosus	Dental strain M5.2	3.9	[83]
S. multivorum	Environmental strain M9.19	7.8	[83]
S. pneumoniae	ATCC 49619	8	[77]
S. proteomaculans	Environmental strain M4.8	31.3	[83]
S. salivarus	44 strains from different origins	2–4	[3]
S. thermophilus	135 strains from different origins	1–8	[3]
V. dispar	Dental strain M20.6	2	[83]
Veillonella spp.	Dental strains M20.4 and M20.7	2–3.9	[83]
Various species[a]	40 Gram-negative isolates from cow milk and goat cheese production	2.5–25	[44]
Various species[a]	80 Gram-positive isolates from cow milk and goat cheese production	2.5–250	[44]
Various species[a]	378 isolates from organic food	10–2,000	[43]

[a]No MIC values per species

9.3.1.2 Bactericidal Activity (Suspension Tests)

The spectrum of bactericidal activity expressed as log reductions obtained with triclosan against various bacterial species is summarized in Table 9.3. Triclosan at 1% has bactericidal activity within 3 min. At 0.6%, an exposure time of 5 min is not consistently sufficient to reach a 5.0 log reduction, and MRSA may be less susceptible. Lower triclosan concentrations yield an inconsistent picture of the bactericidal activity.

9.3.1.3 Activity Against Bacteria in Biofilms

The bactericidal activity of triclosan against bacteria in biofilms is summarized in Table 9.4. The available data indicate that triclosan at 1% has only poor bactericidal activity within 24 h with a log reduction mostly <1.0 against species grown in biofilms except for *L. monocytogenes* in a 24-h biofilm.

Triclosan may accumulate in biofilms as shown in river water downstream of a wastewater treatment plant. Biofilm was able to uptake triclosan present at very low concentrations in water below the detection limit and bioaccumulate them [67].

9.3.1.4 Bactericidal Activity in Other Applications

In surgical scrubbing, liquid soaps based on triclosan may be used [42]. In one study, 3 ml of a soap based on 1% triclosan had basically no bactericidal effect when applied for 3 min for surgical hand disinfection [86]. When used at 1% in combination with 70% iso-propanol, it has only poor persistent bactericidal efficacy on the resident skin flora measured for up to 24 h [85]. In antibacterial soaps, triclosan at 0.3% was found to have no superior efficacy to plain soap on hands artificially contaminated with *S. marcescens* [76]. Similar data obtained with *E. coli* indicate that soaps based on 0.3%, 0.5% or 2% triclosan have only limited bactericidal efficacy on artificially contaminated hands [7]. In sutures, triclosan has some bactericidal activity, especially against *S. aureus* [55, 88]. The use of triclosan sutures has been shown to prevent on average one surgical site infection in 36 children undergoing elective or daytime emergency surgery [110].

Against 7 strains from 6 bacterial species (*E. faecalis*, VRE, *E. coli*, *P. aeruginosa*, *S. aureus*, MRSA, *S. epidermidis*), the efficacy of 0.5% triclosan on glass carriers was good with log reduction between 4.0 and 5.2 in 1 min [98].

9.3.2 Fungicidal Activity

9.3.2.1 Fungistatic Activity (MIC Values)

An overview on published MIC values obtained with different fungal species can be found in Table 9.5. For *C. albicans,* MIC values between 0.06 and 32 mg/l were described. Considering the proposed epidemiological cut-off value of 16 mg/l, most *C. albicans* isolates can be considered to be susceptible to triclosan [100]. The MIC values for various fungal strains from hospital air, clinical specimen or food were all 8 mg/l or less.

Table 9.3 Bactericidal activity of triclosan in suspension tests

Species	Strains/isolates	Exposure time	Concentration	$\log_{10}$ reduction	References
A. baumannii	20 clinical strains	15 s	1% (P)	>5.0	[143]
E. faecalis	Strain Q33	5 min	0.5% (S)	5.2	[98]
E. faecium	VRE strain Z31901	5 min	0.5% (S)	4.5	[98]
E. hirae	ATCC 10541	3 min	1%[a] (P)	≥ 5.0	[86]
Enterococcus spp.	ATCC 13820 plus a gentamicin- and a vancomycin-resistant strain	30 s	0.001% (S)	0.1–0.2	[129]
			0.0015% (S)	3.3–6.0	
E. coli	NCTC 10536	3 min	1%[a] (P)	≥ 5.0	[86]
E. coli	ATCC 25922 and 1 clinical isolate	5 min	0.6% (S)	>5.0	[4]
			0.2% (S)	1.9–2.1	
			0.02% (S)	0.3–2.1	
E. coli	NCTC 10538	5 min	0.5% (S)	4.7	[98]
E. coli	ATCC 11229	30 min	0.05% (S)	≥ 3.0	[101]
P. aeruginosa	ATCC 15442	1 min	2% (S)	5.2	[77]
		60 min	1% (S)	5.2	
		6 h	0.25% (S)	5.1	
P. aeruginosa	ATCC 15442	3 min	1%[a] (P)	≥ 5.0	[86]
P. aeruginosa	NCIMB 10421	5 min	0.5% (S)	4.5	[98]
S. aureus	ATCC 6538	3 min	1%[a] (P)	≥ 5.0	[86]
S. aureus	ATCC 29213 and 2 clinical MRSA strains	5 min	0.6% (S)	2.7–5.0	[4]
			0.2% (S)	0.1–1.2	
			0.02% (S)	0.0–0.7	
S. aureus	NCTC 6571	5 min	0.5% (S)	4.8	[98]
S. aureus	MRSA strain 9543	5 min	0.5% (S)	5.0	[98]
S. aureus	ATCC 6538	1 min	0.25% (S)	5.2	[77]
		5 min	0.1% (S)	5.2	
		6 h	0.025%	5.0	
S. aureus	ATCC 6538	5 min	0.06% (S)	5.5	[20]
			0.01% (S)	0.3	
S. aureus	ATCC 6538	30 min	0.025% (S)	≥ 3.0	[101]
S. aureus	NCTC 6571 plus 2 MRSA strains	30 s	0.001% (S)	0.5–0.8	[129]
			0.0005% (S)	0.0–0.2	
		20 min	0.001% (S)	3.0–6.0	
			0.0005% (S)	0.3–0.4	
		60 min	0.001% (S)	>6.0	
			0.0005% (S)	0.4–0.7	
S. epidermidis	Strain P69	5 min	0.5% (S)	5.0	[98]
S. haemolyticus	Triclosan-tolerant strain (fabI)	5 min	0.06% (S)	4.7	[20]
			0.01% (S)	0.4	

P product; *S* solution; [a]diluted to 0.55%

Table 9.4 Efficacy of triclosan against bacteria in biofilms

Species	Strains/isolates	Type of biofilm	Exposure time	Concentration	$\log_{10}$ reduction	References
L. monocytogenes	6 strains from various sources	24-h incubation in polystyrene microtiter plates	60 min	0.25% (S)	0.9	[13]
				0.5% (S)	1.6	
				1% (S)	≥ 6.1	
P. aeruginosa	8 clinical isolates	24-h incubation on stainless steel, Teflon and polyethylene	24 h	1% (P)	0.8	[127]
S. aureus	8 clinical MRSA isolates	24-h incubation on stainless steel, Teflon and polyethylene	24 h	1% (P)	1.0	[127]
Mixed oral biofilm	*S. oralis* ATCC 10557, *S. gordonii* ATCC 10558, and *A. naeslundii* ATCC 19039	20-h incubation in biofilm reactor	1 h	0.03% (S)	0.9	[26]
Mixed species	Oral biofilm bacteria	12-h incubation in the oral cavity on titanium surfaces	1 min	0.3% (P)	"significant reduction"	[57]

P product; *S* solution

Table 9.5 MIC values of various fungal species to triclosan

Species	Strains/isolates	MIC value (mg/l)	References
A. flavus	3 clinical, 3 airborne and 2 food isolates	0.12–8	[73]
A. fumigatus	6 clinical and 14 airborne isolates	0.12–2	[73]
A. niger	2 airborne and 2 food isolates	0.12–2	[73]
A. ochraceus	2 food isolates	8	[73]
C. albicans	20 clinical isolates from oropharyngeal candidiasis cases	6.3–25	[109]
C. albicans	ATCC 10231	8–32	[77]
C. albicans	10 strains	16	[66]
C. dubliniensis	20 fluconazole-susceptible clinical isolates from oropharyngeal candidiasis cases	0.8–25	[109]
	20 fluconazole-resistant clinical isolates from oropharyngeal candidiasis cases	6.3–25	
Mucor spp.	2 clinical and 1 food isolates	0.5–4	[73]
P. aurantiogriseum	Food isolate	2	[73]
P. citrinum	15 airborne isolates	0.12–1	[73]
P. crysogenum	14 airborne isolates	0.12–1	[73]
P. paneum	2 food isolates	4	[73]
P. roquefortii	4 food isolates	2	[73]
Rhizopus spp.	2 clinical and 1 food isolates	2–4	[73]
Trichoderma spp.	Food isolate	8	[73]

9.3.2.2 Fungicidal Activity (Suspension Tests)

Data from suspension tests to describe the fungicidal activity of products based on triclosan are summarized in Table 9.6. Triclosan at 1% has yeasticidal activity within 1 min, at 0.1% it requires 60 min, and at 0.005% even 6 h.

9.3.2.3 Activity Against Fungi in Biofilms

No studies were found to evaluate the fungicidal activity of triclosan against fungal cells in mature biofilms. Only an early stage biofilm has been evaluated. When

Table 9.6 Fungicidal activity of solutions (S) of triclosan in suspension tests

Species	Strains/isolates	Exposure time	Concentration	$\log_{10}$ reduction	References
C. albicans	ATCC 10231	1 min	1% (S)	4.0	[77]
		60 min	0.1% (S)	4.2	
		6 h	0.005% (S)	4.2	
C. dubliniesensis	1 strain	24 h	0.0004% (S)	≥ 5.0	[66]
C. glabrata	1 strain	24 h	0.0016% (S)	≥ 5.0	[66]
C. parapsilosis	1 strain	24 h	0.0008% (S)	≥ 5.0	[66]
C. tropicalis	1 strain	24 h	0.0008% (S)	≥ 5.0	[66]

C. albicans is allowed to adhere to soft denture lining material for 2.5 h, immersion of the contaminated and carefully washed material in a commercial product based on 0.3% triclosan does not reduce the number of adherent cells significantly [12].

9.3.3 Mycobactericidal Activity

Shortly after describing that triclosan inhibits lipid synthesis in *E. coli* via the target enoyl reductase (FabI) [94], it was found that a similar triclosan target also exists in *M. smegmatis* [92] indicating that triclosan is likely to have mycobactericidal activity [87]. The *M. tuberculosis* H37Rv strain was described with MIC values between 21.7 and 40 mg/l triclosan [48, 56], and four *M. abscessus* isolates were more susceptible with a common MIC value of 8 mg/l [74]. Data obtained with suspension tests were not found.

Two outbreak descriptions indicate that triclosan is effective against waterborne mycobacteria which transiently colonize human skin. One outbreak involved 11 proven and 4 presumptive cases of surgical site infection after breast implants caused by *M. jacuzzii*. The source was a surgeon who carried the strain on his eyebrows, scalp, face, nose, ears and groin. In addition, the strain was detected in the surgeon's outdoor whirlpool. Stopping the use of the whirlpool bath and using a triclosan soap and shampoo for daily body washing controlled the outbreak [107]. A similar outbreak involving 10 patients with surgical site infections after breast implant surgery caused by *M. jacuzzii* was also traced to a hot tub of a surgeon. Various measures were implemented including using a triclosan shampoo by the surgeon for five years. The outbreak was finally controlled [118].

9.4 Effect of Low-Level Exposure

The effects of low-level triclosan exposure have been studied extensively. The results are summarized in Table 9.7.

No change of susceptibility was described in *A. baumannii*, *A. naeslundii*, *A. xylosoxidans*, *B. cereus*, *B. cepacia*, *C. coli*, *C. indologenes*, *Chryseobacterium* spp., *E. faecalis*, *E. faecium*, *E. coli*, *Eubacterium* spp., *F. nucleatum*, *H. gallinarum*, *K. oxytoca*, *K. planticola*, *L. rhamnosus*, *L. lactis*, *M. luteus*, *Megasphaera* spp., *N. subflava*, *P. gingivalis*, *P. aeruginosa*, *S.* Enteritidis, *S.* Infantis, *S.* Typhimurium, *S. marcescens*, *S. capitis*, *Staphylococcus* spp., *S. anginosus*, *S. multivorum*, *S. sanguis*, *S. mutans* and *V. dispar*.

A weak MIC change ($\leq$ 4-fold) was found *B. cereus*, *B. licheniformis*, *Bacillus* spp., *C. jejuni*, *C. indologenes*, *Chryseobacterium* spp., *E. casseliflavus*, *E. faecium*, *Enterococcus* spp., *K. oxytoca*, *M. luteus*, *M. phyllosphaeriae*, *P. ananatis*, *Pantoea* spp., *P. nigrescens*, *P. putida*, *Salmonella* spp., *S. aureus*, *S. caprae*, *S. epidermidis*, *S. saprophyticus*, *S. maltophilia*, *S. proteomaculans*, *S. oralis* and *Veillonella* spp..

Table 9.7 Effects observed after low-level exposure of various bacterial species to triclosan

Species	Strains/isolates	Type of exposure	Increase in MIC	MIC_{max} (mg/l)	Stability of MIC change	Associated changes	References
A. baumannii	Strain MBRG15.1 from a domestic kitchen drain biofilm	14 passages at various concentrations	None	125	Not applicable	None reported	[31]
A. baumannii	MDR strain ZJ06	10 d at increasing concentrations	16-fold	8	"stable"	Several general protective mechanisms were enhanced	[105]
A. naeslundii	Strain WVU627	10 passages à 4 d at various concentrations	None	4.9	Not applicable	No MIC increase[a] to chlorhexidine, metronidazole and tetracycline	[91]
A. proteolyticus	Environmental strain M9.12	10 passages à 4 d at various concentrations	8-fold	156	No data	None described	[83]
A. xylosoxidans	Environmental strain M4.31	10 passages à 4 d at various concentrations	None	1.2	Not applicable	None described	[83]
B. cereus	Environmental strain M7.15	10 passages à 4 d at various concentrations	None	1	Not applicable	None described	[83]
B. cereus	2 isolates from organic food	Several passages with gradually higher concentrations	≤ 2.5-fold	0.25	No data	20-fold increased tolerance[a] to sodium nitrate (1 strain)	[50]
B. cereus	5 biocide-sensitive strains from organic foods	Several passages with gradually higher concentrations	3-fold–4-fold	40	Unstable for 20 d (3 strains), stable for 20 d (2 strains)	Cross-adaptation[a] to benzalkonium chloride (4-fold–100-fold), hexachlorophene (up to 20-fold), chlorhexidine (up to 16-fold) and DDAB[b] (up to 4-fold); cross-resistance[a] to sulfamethoxazol (3 strains), ampicillin, cefotaxime and ceftazidime (1 strain each)	[49]

(continued)

Table 9.7 (continued)

Species	Strains/isolates	Type of exposure	Increase in MIC	MIC_{max} (mg/l)	Stability of MIC change	Associated changes	References
B. cereus	MRBG 4.21 (kitchen drain biofilm isolate)	40 d at increasing concentrations	4-fold	29	Unstable for 14 d	None described	[46]
B. licheniformis	1 biocide-sensitive strains from organic foods	Several passages with gradually higher concentrations	3-fold	15	Unstable for 20 d	Cross-adaptation[a] to benzalkonium chloride (25-fold), hexachlorophene (5-fold) and chlorhexidine (2-fold); cross-resistance[a] to sulfamethoxazol	[49]
Bacillus spp.	4 biocide-sensitive strains from organic foods	Several passages with gradually higher concentrations	2-fold	10	Unstable for 20 d	Cross-adaptation[a] to benzalkonium chloride (10-fold–25-fold) and chlorhexidine (up to 18-fold); cross-resistance[a] to sulfamethoxazol (2 strains), cefotaxime and ceftazidime (1 strain each)	[49]
B. cepacia	ATCC BAA-245	40 d at increasing concentrations	None	116	Not applicable	None described	[46]
C. coli	ATCC 33559 and a poultry isolate	Up to 15 passages with gradually higher concentrations	None	1.0	Not applicable	None described	[90]
C. jejuni	NCTC 11168, ATCC 33560 and a poultry isolate	Up to 15 passages with gradually higher concentrations	2-fold (NCTC and ATCC strain)	2.0	Stable for 5 d	None described	[90]

(continued)

Table 9.7 (continued)

Species	Strains/isolates	Type of exposure	Increase in MIC	MIC_{max} (mg/l)	Stability of MIC change	Associated changes	References
C. indologenes	Environmental strain M9.15	10 passages à 4 d at various concentrations	None	1	Not applicable	None described	[83]
C. indologenes	MRBG 4.29 (kitchen drain biofilm isolate)	40 d at increasing concentrations	4-fold	3.6	Unstable for 14 d	None described	[46]
C. xerosis	WIBG 1.2 (wound isolate)	40 d at increasing concentrations	8-fold	58	Unstable for 14 d	None described	[46]
C. sakazakii	Strain MBRG15.5 from a domestic kitchen drain biofilm	14 passages at various concentrations	64-fold	500	Stable for 14 d	None reported	[31]
Chryseobacterium spp.	Environmental strain FR2 9.17	10 passages à 4 d at various concentrations	None	15.6	Not applicable	None described	[83]
Chrysobacterium spp.	1 biocide-sensitive strain from organic foods	Several passages with gradually higher concentrations	2-fold	20	Unstable for 20 d	Cross-adaptation[a] to hexachlorophen (5-fold)	[49]
E. gergoviae	ATCC 33028	Various passages with increasing concentrations	2-fold–8-fold	20	No data	None described	[104]
Enterobacter spp.	5 biocide-sensitive strains from organic foods	Several passages with gradually higher concentrations	2-fold–15-fold	20	Unstable for 20 d	Cross-adaptation[a] to benzalkonium chloride (up to 20-fold), hexachlorophene (up to 6-fold), chlorhexidine (up to 18-fold) and DDAB[b] (up to 6-fold); cross-resistance[a] to sulfamethoxazol, ampicillin and ceftazidime (2 strains) and cefotaxime (1 strain)	[49]

(continued)

Table 9.7 (continued)

Species	Strains/isolates	Type of exposure	Increase in MIC	MIC_{max} (mg/l)	Stability of MIC change	Associated changes	References
E. casseliflavus	2 biocide-sensitive strains from organic foods	Several passages with gradually higher concentrations	2-fold	10	Unstable for 20 d	Cross-adaptation[a] to benzalkonium chloride (50-fold), hexachlorophene (5-fold), chlorhexidine (4-fold); cross-resistance[a] to cefotaxime	[49]
E. faecalis	1 strain of unknown origin	14 passages at various concentrations	None	62.5	Not applicable	None reported	[31]
E. faecalis	WIBG 1.1 (wound isolate)	40 d at increasing concentrations	18-fold	58	Unstable for 14 d	None described	[46]
E. faecium	2 isolates from organic food	Several passages with gradually higher concentrations	None	0.1	Not applicable	None reported	[50]
E. faecium	5 biocide-sensitive strains from organic foods	Several passages with gradually higher concentrations	2-fold–4-fold	40	Unstable for 20 d (3 strains), stable for 20 d (2 strains)	Cross-adaptation[a] to benzalkonium chloride (10-fold–100-fold), hexachlorophene (up to 5-fold) and chlorhexidine (up to 2-fold); cross-resistance[a] to cefotaxime and ceftazidime (1 strain each)	[49]
Enterococcus spp.	2 biocide-sensitive strains from organic foods	Several passages with gradually higher concentrations	2-fold	20	Unstable for 20 d	Cross-adaptation[a] to benzalkonium chloride (10-fold–25-fold), hexachlorophene (up to 5-fold) and chlorhexidine (8-fold); cross-resistance[a] to ceftazidimee (1 strain)	[49]

(continued)

Table 9.7 (continued)

Species	Strains/isolates	Type of exposure	Increase in MIC	MIC_{max} (mg/l)	Stability of MIC change	Associated changes	References
E. coli	ATCC 25922 and enteric strain M20.1	10 passages à 4 d at various concentrations	None	2	Not applicable	None described	[83]
E. coli	NCTC 12900 strain O157	2 passages (P1 and P2) at variable concentrations	16-fold (P1) 8,192-fold (P2)	2,048	Stable for 30 d	Increased tolerance[b] to amoxicillin-clavulanic acid (0 mm), amoxicillin (0 mm), chloramphenicol (5 mm), imipenem (11 mm), tetracycline (14 mm), trimethoprim (0 mm), erythromycin and chlorhexidine (0 mm).	[9]
E. coli	ATCC 25922 and strain MBRG15.4 from a domestic kitchen drain biofilm	14 passages at various concentrations	31-fold–125-fold	125	Stable for 14 d	None reported	[31]
E. coli	NCIMB 8545	0.0004% for 30 s, 5 min and 24 h	32-fold–39-fold	78	Stable for 10 d	MBC increased 4-fold	[141]
E. coli	ATCC 25922	40 d at increasing concentrations	58-fold	29	Stable for 14 d	Increase of biofilm formation	[46]
E. coli	ATCC 8729	10 passages à 4 d at various concentrations	391-fold	39.1	No data	No MIC increase[a] to chlorhexidine, metronidazole and tetracycline	[91]
E. coli	Equine strains 4 and 48, bovine strain T3 5H5	Not described	640-fold	>8,000	Stable for 7 d	None described	[123]

(continued)

Table 9.7 (continued)

Species	Strains/isolates	Type of exposure	Increase in MIC	MIC_{max} (mg/l)	Stability of MIC change	Associated changes	References
E. coli	NCTC 12900 and NCTC 43888 (both O157 H7), 3 clinical strains (O55:H7, O55: H29, O111:H24), ATCC 27325 and NCIMB 10115 (both K-12)	6 passages at variable concentrations	2,048-fold–8,192-fold	2,048	No data	Strain O157:H7: increased tolerance[a] to amoxicillin-clavulanic acid (256 mg/l), amoxicillin (>256 mg/l), chloramphenicol (256 mg/l), tetracycline (>256 mg/l), trimethoprim (>256 mg/l), benzalkonium chloride (256 mg/l) and chlorhexidine (256 mg/l) Strain O55:H7: increased tolerance[a] to trimethoprim (256 mg/l) Strain K-12: increased tolerance[a] to chloramphenicol (256 mg/l)	[10]
E. coli	CV601	0.1 mg/l for 3 h	No data	No data	Not applicable	Induction of horizontal gene transfer (sulfonamide resistance by conjugation)	[71]
E. coli	Strain MG1655	10 d at 0.03 mg/l	No data	No data	Not applicable	No increase of quinolone-resistant[c] mutants; a Asp87Gly GyrA mutant demonstrated greatly increased fitness in the presence of triclosan	[139]

(continued)

Table 9.7 (continued)

Species	Strains/isolates	Type of exposure	Increase in MIC	MIC_{max} (mg/l)	Stability of MIC change	Associated changes	References
E. coli	Triclosan-resistant mutant of an O157: H19 isolate	6 mg/l for 30 min	No data	>8,000	Not applicable	Increase of biofilm formation; significant changes in protein expression levels	[122]
Eubacterium spp.	Environmental strain M4.14	10 passages à 4 d at various concentrations	None	15.6	Not applicable	None described	[83]
F. nucleatum	ATCC 10953	10 passages à 4 d at various concentrations	None	9.8	Not applicable	2-fold MIC increase[a] to metronidazole, no MIC increase[a] to chlorhexidine and tetracycline	[91]
F. nucleatum	Dental strains M20.2 and M20.3	10 passages à 4 d at various concentrations	None	3.3	Not applicable	None described	[83]
H. gallinarum	Environmental strain M4.27	10 passages à 4 d at various concentrations	None	31.3	Not applicable	None described	[83]
K. pneumoniae	ATCC 13883	40 d at increasing concentrations	129-fold	116	Stable for 14 d	None described	[46]
K. oxytoca	Enteric strain M21.3	10 passages à 4 d at various concentrations	None	1	Not applicable	None described	[83]
K. oxytoca	2 biocide-sensitive strains from organic foods	Several passages with gradually higher concentrations	2-fold–3-fold	20	Unstable for 20 d	Cross-adaptation[a] to benzalkonium chloride (up to 40-fold), chlorhexidine (up to 18-fold) and DDAB[b] (up to 4-fold)	[49]
K. planticola	Enteric strain M21.1	10 passages à 4 d at various concentrations	None	1	Not applicable	None described	[83]

(continued)

Table 9.7 (continued)

Species	Strains/isolates	Type of exposure	Increase in MIC	MIC_{max} (mg/l)	Stability of MIC change	Associated changes	References
L. pentosus	7 strains from naturally fermented Aloreña green table olives	48 h at 1 mg/l	No data	No data	Not applicable	Increased tolerance[a] to ampicillin (up to 100-fold), chloramphenicol (up to 200-fold), ciprofloxacin (up to 7-fold), teicoplanin (up to 340-fold), tetracycline (up to 80-fold) and trimethoprim (up to 15-fold); no increase of MIC[a] with clindamycin, erythromycin and streptomycin.	[15]
L. pentosus	Strain MP-10	48 h at 1 mg/l	No data	No data	Not applicable	Increase in growth rate, improved survival at pH 1.5 and in the presence of 2–3% bile	[14]
L. rhamnosus	Strain AC413	10 passages à 4 d at various concentrations	None	6.8	Not applicable	3-fold MIC increase[a] to chlorhexidine, no MIC increase[a] to metronidazole and tetracycline	[91]
L. rhamnosus	Dental strain M6.1	10 passages à 4 d at various concentrations	None	2	Not applicable	None described	[83]
L. lactis	Dental strain M6.3	10 passages à 4 d at various concentrations	None	2	Not applicable	None described	[83]

(continued)

Table 9.7 (continued)

Species	Strains/isolates	Type of exposure	Increase in MIC	MIC_{max} (mg/l)	Stability of MIC change	Associated changes	References
L. pseudomesenteroides	1 strain from naturally fermented Aloreña green table olives	48 h at 1 mg/l	No data	No data	Not applicable	Increased tolerance[a] to chloramphenicol (2-fold), ciprofloxacin (7-fold) and tetracycline (2-fold); no increase of MIC[a] with ampicillin, clindamycin, erythromycin, streptomycin, teicoplanin and trimethoprim.	[15]
L. monocytogenes	8 strains from food and animals	4 × 24 h (1 and 4 mg/l)	No data	16.0	Not applicable	Gentamicin resistance[c] frequency increased 10-fold to 10,000-fold	[17]
M. luteus	Environmental strain M9.25	10 passages à 4 d at various concentrations	None	1	Not applicable	None described	[83]
M. luteus	MRBG 9.25 (skin isolate)	40 d at increasing concentrations	1.7-fold	12.1	Unstable for 14 d	None described	[46]
M. osloensis	Strain MBRG15.3 from a domestic kitchen drain biofilm	14 passages at various concentrations	16-fold	15.6	Stable for 14 d	None reported	[31]
M. phyllosphaeriae	Environmental strain M4.30	10 passages à 4 d at various concentrations	4-fold	7.8	No data	None described	[83]
Megasphaera spp.	Dental strain M20.9	10 passages à 4 d at various concentrations	None	2	Not applicable	None described	[83]

(continued)

Table 9.7 (continued)

Species	Strains/isolates	Type of exposure	Increase in MIC	MIC_{max} (mg/l)	Stability of MIC change	Associated changes	References
N. subflava	Strain A1078	10 passages à 4 d at various concentrations	None	0.1	Not applicable	2-fold MIC increase[a] to tetracycline, no MIC increase[a] to chlorhexidine and metronidazole	[91]
P. agglomerans	1 biocide-sensitive strain from organic foods	Several passages with gradually higher concentrations	150-fold	15	Unstable for 20 d	Cross-adaptation[a] to benzalkonium chloride (10-fold), hexachlorophene (up to 5-fold), chlorhexidine (5-fold) and DDAB[b] (3-fold); cross-resistance[a] to sulfamethoxazol, ampicillin and ceftazidime	[49]
P. ananatis	1 isolate from organic food	Several passages with gradually higher concentrations	2.5-fold	0.25	No data	20-fold increased tolerance[a] to sodium nitrate	[50]
P. ananatis	2 biocide-sensitive strains from organic foods	Several passages with gradually higher concentrations	5-fold–200-fold	200	Unstable for 20 d (1 strain), stable for 20 d (1 strain)	Cross-adaptation[a] to benzalkonium chloride (20-fold–30-fold), hexachlorophene (5-fold–30-fold), chlorhexidine (4-fold–10-fold) and DDAB[b] (3-fold); cross-resistance[a] to sulfamethoxazol and trimethoprim/sulfamethoxazol (both strains), ampicillin and cefotaxime (1 strain)	[49]

(continued)

Table 9.7 (continued)

Species	Strains/isolates	Type of exposure	Increase in MIC	MIC_{max} (mg/l)	Stability of MIC change	Associated changes	References
Pantoea spp.	2 biocide-sensitive strains from organic foods	Several passages with gradually higher concentrations	2-fold–3-fold	20	Unstable for 20 d	Cross-adaptation[a] to benzalkonium chloride (up to 30-fold), chlorhexidine (up to 2-fold) and DDAB[b] (up to 4-fold); cross-resistance[a] to sulfamethoxazol, ceftazidime and cefotaxime (1 strain each)	[49]
P. gingivalis	Strain W50	10 passages à 4 d at various concentrations	None	3.9	Not applicable	2-fold MIC increase[a] to metronidazole, no MIC increase[a] to chlorhexidine and tetracycline	[91]
P. nigrescens	Strain T588	10 passages à 4 d at various concentrations	2-fold	7.8	No data	2.4-fold MIC increase[a] to chlorhexidine, no MIC increase[a] to metronidazole and tetracycline	[91]
P. aeruginosa	ATCC 9027	14 passages at various concentrations	None	>1,000	Not applicable	MIC was initially already >1,000 mg/l	[31]
P. putida	Strain MBRG15.2 from a domestic kitchen drain biofilm	14 passages at various concentrations	4-fold	62.5	Stable for 14 d	None reported	[31]
S. Enteritidis	NCTC 8513	10 passages à 4 d at various concentrations	None	2.6	Not applicable	None described	[83]
S. Enteritidis	Clinical isolate	Several passages with gradually higher concentrations	32-fold	512	Stable for 30 d	None described	[11]

(continued)

Table 9.7 (continued)

Species	Strains/isolates	Type of exposure	Increase in MIC	MIC_{max} (mg/l)	Stability of MIC change	Associated changes	References
S. *Enteritidis*	NCTC 13349	8 days at increasing concentrations	1,000-fold	100	"stable"	None described	[23]
S. *Infantis*	NCIMB 13036	10 passages à 4 d at various concentrations	None	1.6	Not applicable	None described	[83]
S. *Typhimurium*	ATCC 23564 and NCTC 74	10 passages à 4 d at various concentrations	None	2	Not applicable	None described	[83]
S. *Typhimurium*	NCTC 74	Several passages with gradually higher concentrations	64-fold	512	Stable for 30 d	None described	[11]
S. *Typhimurium*	Strain SL1344	8 days at increasing concentrations	1,500-fold	150	"stable"	None described	[23]
S. *Virchow*	Food isolate	Several passages with gradually higher concentrations	64-fold	1,014	Stable for 30 d	None described	[11]
Salmonella spp.	2 isolates from organic food	Several passages with gradually higher concentrations	$\leq$ 2.5-fold	0.25	No data	None reported	[50]
Salmonella spp.	3 biocide-sensitive strains from organic foods	Several passages with gradually higher concentrations	2-fold–200-fold	3,000	Stable for 20 d (2 strains), unstable for 20 d (1 strain)	Cross-adaptation[a] to benzalkonium chloride and hexachlorophene (up to 40-fold), chlorhexidine (up to 18-fold) and DDAB[b] (up to 3-fold); cross-resistance[a] to trimethoprim/sulfamethoxazol, cefotaxime and nalidixic acid (2 strains each), ampicillin, sulfamethoxazol and imipenem (1 strain each)	[49]

(continued)

Table 9.7 (continued)

Species	Strains/isolates	Type of exposure	Increase in MIC	MIC_{max} (mg/l)	Stability of MIC change	Associated changes	References
Salmonella spp.	7 broiler house isolates	Several passages with gradually higher concentrations	4-fold–515-fold in 6 isolates	>129	Stable for 6 d	None described	[58]
Salmonella spp.	6 strains with higher MICs to biocidal products	8 days at increasing concentrations	500-fold–10,000-fold in 3 strains	>1,000	"stable"	One strain displayed a decreased susceptibility[d] to piperacillin (16 mg/l), Ceftiofur (>8 mg/l), amikacin (16 mg/l), kanamycin (32 mg/l), chloramphenicol (16 mg/l), cefoxitin (32 mg/l) and sulfisoxazole (>256 mg/l)	[23]
S. marcescens	ATCC 13880	40 d at increasing concentrations	None	116	Not applicable	None described	[46]
S. aureus	NCIMB 9518	0.0004% for 30 s, 5 min and 24 h	≤ 5-fold	625	Unstable for 10 d	MBC increased 2-fold to 74-fold	[141]
S. aureus	NCTC 6571 and 2 MRSA strains	Several passages with gradually higher concentrations	5-fold–50-fold	5	"unstable"	None described	[129]
S. aureus	ATCC 6538	40 d at increasing concentrations	5-fold–69-fold	29	Stable for 14 d	Decrease of biofilm formation	[46]
S. aureus	3 EMRSA-15 strains	Up to 72 h on polymer impregnated with 0.2% triclosan	8-fold–67-fold	4	No data	Change to small colony variant	[8]
S. aureus	ATCC 6538	14 passages at various concentrations	313-fold	62.5	Stable for 14 d	None reported	[31]

(continued)

Table 9.7 (continued)

Species	Strains/isolates	Type of exposure	Increase in MIC	MIC_{max} (mg/l)	Stability of MIC change	Associated changes	References
S. capitis	MRBG 9.34 (skin isolate)	40 d at increasing concentrations	None	29	Not applicable	None described	[46]
S. caprae	MRBG 9.3 (skin isolate)	40 d at increasing concentrations	2.4-fold	29	Unstable for 14 d	None described	[46]
S. epidermidis	MRBG 9.33 (skin isolate)	40 d at increasing concentrations	2.9-fold	38.7	Unstable for 14 d	Increase of biofilm formation	[46]
S. epidermidis	ATCC 35983	20 passages at various concentrations	8-fold	20	Stable for 20 d	None reported	[132]
S. haemolyticus	MRBG 9.35 (skin isolate)	40 d at increasing concentrations	73-fold	29	Unstable for 14 d	None described	[46]
S. lugdunensis	MRBG 9.36 (skin isolate)	40 d at increasing concentrations	32-fold	29	Unstable for 14 d	Decrease of biofilm formation	[46]
S. saprophyticus	2 isolates from organic food	Several passages with gradually higher concentrations	≤2.5-fold	0.25	No data	None reported	[50]
S. saprophyticus	3 biocide-sensitive strains from organic foods	Several passages with gradually higher concentrations	3-fold–5-fold	40	Stable for 20 d (2 strains), unstable for 20 d (1 strain)	Cross-adaptation[a] to benzalkonium chloride (up to 300-fold), hexachlorophene (up to 30-fold) and chlorhexidine (up to 6-fold); cross-resistance[a] to sulfamethoxazol (2 strains), cefotaxime and ceftazidimee (1 strain each)	[49]
S. warneri	MRBG 9.27 (skin isolate)	40 d at increasing concentrations	27-fold	24.2	Unstable for 14 d	None described	[46]

(continued)

Table 9.7 (continued)

Species	Strains/isolates	Type of exposure	Increase in MIC	MIC_{max} (mg/l)	Stability of MIC change	Associated changes	References
S. xylosus	1 biocide-sensitive strain from organic foods	Several passages with gradually higher concentrations	5-fold	25	Unstable for 20 d	Cross-adaptation[a] to DDAB (5-fold); cross-resistance[a] to sulfamethoxazol and ceftazidime	[49]
Staphylococcus spp.	11 species (skin isolates)	10 passages à 4 d at various concentrations	None	1.0	Not applicable	None described	[83]
Staphylococcus spp.	2 biocide-sensitive strains from organic foods	Several passages with gradually higher concentrations	2-fold–150-fold	15	Unstable for 20 d	Cross-adaptation[a] to hexachlorophene (up to 5-fold) and DDAB (up to 5-fold); cross-resistance[a] to sulfamethoxazol and ceftazidime (both strains), cefotaxime and ampicillin (1 strain)	[49]
S. maltophilia	Environmental strain M9.13	10 passages à 4 d at various concentrations	4-fold	39.1	No data	None described	[83]
S. maltophilia	MRBG 4.17 (kitchen drain biofilm isolate)	40 d at increasing concentrations	16-fold	232	Unstable for 14 d	None described	[46]
S. anginosus	Dental strain M5.2	10 passages à 4 d at various concentrations	None	3.9	Not applicable	None described	[83]
S. multivorum	Environmental strain M9.19	10 passages à 4 d at various concentrations	None	2	Not applicable	None described	[83]
S. proteomaculans	Environmental strain M4.8	10 passages à 4 d at various concentrations	2-fold	62.5	No data	None described	[83]

(continued)

Table 9.7 (continued)

Species	Strains/isolates	Type of exposure	Increase in MIC	MIC_{max} (mg/l)	Stability of MIC change	Associated changes	References
S. oralis	NCTC 11427	10 passages à 4 d at various concentrations	1.7-fold	13.0	No data	2-fold MIC increase[a] to chlorhexidine and metronidazole, no MIC increase[a] to tetracycline	[91]
S. sanguis	NCTC 7863	10 passages à 4 d at various concentrations	None	3.9	Not applicable	2-fold MIC increase[a] to chlorhexidine and metronidazole, no MIC increase[a] to tetracycline	[91]
S. mutans	NCTC 10832	10 passages à 4 d at various concentrations	None	11.7	Not applicable	2.7-fold MIC increase[a] to chlorhexidine, 2-fold MIC increase to tetracycline, no MIC increase[a] to metronidazole	[91]
V. dispar	ATCC 17745	10 passages à 4 d at various concentrations	None	4.9	Not applicable	No MIC increase[a] to chlorhexidine, metronidazole and tetracycline	[91]
V. dispar	Dental strain M20.6	10 passages à 4 d at various concentrations	None	1	Not applicable	None described	[83]
Veillonella spp.	Dental strains M20.4 and M20.7	10 passages à 4 d at various concentrations	1.7-fold	3.3	No data	None described	[83]

[a]Broth microdilution method; [b]disc diffusion test; [c]agar dilution method; [d]NARMS plates

A strong (>4-fold) and unstable MIC change was observed with *C. xerosis*, *Enterobacter* spp., *E. faecalis*, *P. agglomerans*, *P. ananatis*, *Salmonella* spp., *S. aureus*, *S. haemolyticus*, *S. lugdunensis*, *S. saprophyticus*, *S. warneri*, *S. xylosus*, *Staphylococcus* spp. and *S. maltophilia*. In other species, a strong and stable MIC change was described such as *A. baumannii*, *C. sakazakii*, *E. coli*, *K. pneumoniae*, *M. osloensis*, *P. ananatis*, *S.* Enteritidis, *S.* Typhimurium, *S.* Virchow, *Salmonella* spp., *S. aureus*, *S. epidermidis* and *S. saprophyticus*. In isolates or strains of *A. proteolyticus*, *E. gergoviae*, *E. coli* and *S. aureus*, a strong adaptive response was described but its stability not investigated.

The strongest MIC increase was found in *Salmonella* spp. (up to 10,000-fold), *E. coli* (up to 8,192-fold), *S. aureus* (up to 313-fold), *P. ananatis* (up to 200-fold) and *P. agglomerans* and *Staphylococcus* spp. (up to 150-fold). The highest MIC values after low-level triclosan exposure were described in *E. coli* (>8,000 mg/l), *Salmonella* spp. (3,000 mg/l), *P. aeruginosa* (>1,000 mg/l), *S. aureus* (625 mg/l) and *C. sakazakii* (500 mg/l).

Cross-adaptation to chlorhexidine was described in various species such as *B. cereus*, *B. licheniformis*, *Enterobacter* spp., *E. casseliflavus*, *E. faecium*, *Enterococcus* spp., *E. coli*, *K. oxytoca*, *P. agglomerans*, *P. ananatis*, *Pantoea* spp., *P. nigrescens*, *Salmonella* spp., *S. saprophyticus*, *S. oralis*, *S. sanguis* and *S. mutans*. A cross-adaptive response to benzalkonium chloride was found in *B. cereus*, *B. licheniformis*, *Enterobacter* spp., *E. casseliflavus*, *E. faecium*, *Enterococcus* spp., *E. coli*, *K. oxytoca*, *P. agglomerans*, *P. ananatis*, *Pantoea* spp., *Salmonella* spp. and *S. saprophyticus*. MIC values to hexachlorophene increased after triclosan exposure in *B. cereus*, *B. licheniformis*, *Chrysobacterium* spp., *Enterobacter* spp., *E. casseliflavus*, *E. faecium*, *Enterococcus* spp., *P. agglomerans*, *P. ananatis*, *Salmonella* spp., *S. saprophyticus* and *Staphylococcus* spp.. Similar cross-reactive MIC changes to DDAB were noticed in *Enterobacter* spp., *K. oxytoca*, *P. agglomerans*, *P. ananatis*, *Pantoea* spp., *Salmonella* spp., *S. saprophyticus*, *S. xylosus* and *Staphylococcus* spp. Finally, *B. cereus* and *P. ananatis* were less susceptible to sodium nitrate after low-level triclosan exposure.

In various bacterial species, it was found that triclosan-adapted strains show increased tolerance or even resistance to selected antibiotics. A detailed description per species can be found in Table 9.7. Additional studies show that in *R. rubrum*, the degree of triclosan resistance depends on the initial exposure concentration and that similar resistance degrees can be the result of different defence mechanisms, which all have distinct antibiotic cross-resistance profiles [106]. In addition, exposure of seven species (*A. baumannii*, *C. sakazakii*, *E. faecalis*, *E. coli*, *P. aeruginosa*, *P. putida*, *S. aureus*) over 14 passages of 4 d each to increasing triclosan concentrations on agar was associated with both increases and decreases in antibiotic susceptibility but its effect was typically small relative to the differences observed among microbicides. Susceptibility changes resulting in resistance were not observed [47].

Biofilm formation was enhanced in *E. coli* and *S. epidermidis* but reduced in *S. aureus* and *S. lugdunensis* (Table 9.7). In *E. coli*, low-level triclosan exposure induced horizontal gene transfer (sulfonamide resistance by conjugation). In *A. baumannii*, several general protective mechanisms were enhanced. And in EMRSA strains, a change to the small colony variant was observed. In addition, sub-lethal concentrations of triclosan also induced discernible changes in the proteome of exposed *Salmonella* providing insights into mechanisms of response and tolerance [24].

9.5 Resistance to Triclosan

9.5.1 Resistance Mechanisms

Triclosan resistance mechanisms include target mutations, increased target expression, active efflux from the cell, and enzymatic inactivation and degradation [108, 119]. Efflux pumps were mostly described to explain triclosan resistance [18]. In *P. aeruginosa,* intrinsic resistance (MIC $\geq$ 1,000 mg/l) to triclosan was solely attributable to the expression of efflux pumps [19]. In *E. coli,* overexpression of the multidrug efflux pump locus acrAB, or of marA or soxS, both encoding positive regulators of acrAB, decreased susceptibility to triclosan 2-fold [93]. In *S. Typhimurium,* the multidrug efflux systems, EmrAB and AcrEF, play a role in the phenotypic susceptibility to triclosan, and overexpression of the genes emrAB or acrEF can partially compensate for a functional inactivity of the primary transporter AcrAB. As a consequence, the contribution of these efflux pumps should be a consideration when designing studies investigating cross-resistance between triclosan and antibiotic agents [112]. In addition, multidrug-resistant and triclosan-resistant strains of *S. enterica* showed increased efflux activity compared with strains with reduced susceptibility to triclosan alone [25]. In *K. pneumoniae,* the kpnGH efflux pump was described with a wide substrate specificity of the transporter including 14 antibiotics and triclosan. kpnGH mediates antimicrobial resistance by active extrusion in *K. pneumoniae* [128].

Gene expression is also involved. *A. baumannii* responds to triclosan by altering the expression of genes involved in fatty acid metabolism, antibiotic resistance and amino acid metabolism as shown with a triclosan-resistant mutant strain of *A. baumannii* ATCC 17978 [45]. In addition, the outer membrane exclusionary properties of *P. aeruginosa* for non-polar molecules confer intrinsic resistance to low concentrations of triclosan such as might be expected to occur in environmental residues. Moreover, a role for outer cell envelope impermeability is suggested for resistance to high triclosan concentrations in vitro [16]. In *S. aureus*, gene expression profiling demonstrated that an alteration in cell membrane structural and functional gene expression is likely responsible for triclosan and ciprofloxacin resistance [135]. In an *E. coli* strain, 47 genes were confirmed to enhance the resistance to triclosan. These genes, including the FabI target, were involved in inner or outer membrane synthesis, cell surface material synthesis, transcriptional activation, sugar phosphotransferase (PTS) systems, various transporter systems, cell division, and ATPase and reductase/dehydrogenase reactions. In particular, overexpression of pgsA, rcsA, or gapC conferred to *E. coli* cells a similar level of triclosan resistance induced by fabI overexpression. These results indicate that triclosan may have multiple targets other than well-known FabI and that there are several undefined novel mechanisms for the resistance development to triclosan, thus probably inducing cross-antibiotic resistance [146]. In *R. rubrum,* triclosan resistance was a result of a FabI1 (G98 V) mutation. This point mutation led to an even higher level of triclosan resistance (MIC > 16 mg/l) in combination with constitutive up-regulation of mexB and mexF efflux pump homologues [106].

The structural basis of triclosan resistance has been explored in *E. coli*. It was found that overall structural change of protein is minimal in triclosan resistance except that a flexible a-helical turn around triclosan is slightly pushed away due to the presence of the bulky valine group. However, triclosan shows substantial edge-to-face aromatic (p) interactions with both the flexible R192-F203 region and the residues in the close vicinity of G93. The weakening of some edge-to-face aromatic interactions around triclosan in the G93 V mutant results in serious resistance to triclosan [125].

In *S. aureus,* an additional sh-fabI allele derived from *S. haemolyticus* was detected. Detection of sh-fabI as a novel resistance mechanism with high potential for horizontal gene transfer demonstrates for the first time that a biocide could exert a selective pressure able to drive the spread of a resistance determinant in a human pathogen [20]. In addition, both the introduction of a plasmid expressing the saFabI gene or a missense mutation in the chromosomal saFabI gene led to triclosan resistance in *S. aureus* [65]. *S. aureus* is also able to form small colony variants which are characterized by impaired growth, down-regulation of genes for metabolism and virulence while sigB and genes important for persistence and biofilm formation are up-regulated. Small colony variants are resistant to various antibiotics and triclosan [72].

Some species such as *A. xylosoxidans* or *P. putida* are typically found in soil but can also cause infections in humans. These species are able to use triclosan as the sole carbon source resulting in an almost complete removal of triclosan within 2–8 d [97].

9.5.2 Resistance Genes

So far, no specific triclosan resistance gene has been identified. The antiseptic resistance genes cepA, qacΔE and qacE had no impact on the MICs of a soap based on 1% triclosan [1]. Among 120 isolates from cow milk and goat cheese production a correlation between biocide tolerance and the presence of beta-lactamase genes was observed [44]. In dust, a significant positive association between the ubiquitous antimicrobial triclosan and the relative abundance of the antibiotic resistance gene erm(X) was observed, a 23S rRNA methyltransferase implicated in resistance to several antibiotics [64].

9.5.3 Infections Associated with Resistance to Triclosan

Some reports of infections caused by contaminated triclosan soaps have been described (Table 9.8). The suspected mode of transmission was via the transiently contaminated hands of healthcare workers.

Table 9.8 Infections associated with resistance to triclosan

Bacterial species	Type and number of infections	Patient population	Source of infection and role of triclosan resistance	References
P. aeruginosa	5 cases; pneumonia (2) septicaemia (1) and asymptomatic patients (2)	Haematology unit	Contaminated triclosan (0.5%) soap dispenser acted as a continuous source of infections; MIC value of 2,125 mg/l	[33, 79]
S. marcescens	Sporadic cases with no identifiable source	Surgical intensive care unit	Contaminated triclosan (1%) soap but no infections could be attributed to the contaminated soap	[6]
S. marcescens	3 cases of conjunctivitis	Newborn nursery	Contaminated triclosan (0.5%) soap bottles, one in use and one unopened	[96]

9.6 Cross-Tolerance to Other Biocidal Agents

Cross-adaptation to chlorhexidine, benzalkonium chloride, hexachlorophene, DDAB and sodium nitrate has been for numerous bacterial species after low-level triclosan exposure. They are described in Table 9.7.

9.7 Cross-Tolerance to Antibiotics

The triclosan resistance mechanisms are the same types of mechanisms involved in antibiotic resistance and some of them account for the observed cross-tolerance with antibiotics in laboratory isolates. Therefore, there is a link between triclosan and antibiotics, and the widespread use of triclosan-containing antiseptics and disinfectants may indeed aid in the development of microbial resistance, in particular cross-resistance to antibiotics [119].

Low-level triclosan exposure can cause antibiotic resistance in various bacterial species (see Table 9.7). Other studies indicate a variable cross-tolerance (Table 9.9).

Cross-tolerance seems quite common in *Salmonella*. Repeated in vitro exposure of *S. Typhimurium* cells to triclosan selects for reduced susceptibility to several antibiotics (chloramphenicol, tetracycline, ampicillin, acriflavine). Resistance to disinfectants was observed only after exposure to gradually increasing concentrations of triclosan, accompanied by a 2,000-fold increase in its MIC. This is associated with overexpression of AcrAB efflux pump [75]. Another study shows that among 4% of 428, *S. enterica* isolates with a decreased triclosan susceptibility 56% were multidrug-resistant compared with 12% of triclosan-sensitive isolates [25]. Antibiotic-resistant *E. coli* and *Salmonella* spp. with efflux pumps isolated from

Table 9.9 Triclosan and associated antibiotic tolerance

Species	Strains/isolates	Associated tolerance or resistance	References
A. johnsonii	Triclosan-tolerant strain	None[a] (33 different antibiotics)	[28]
A. johnsonii	Triclosan-tolerant strain	Chloramphenicol[a]	[82]
E. coli	Triclosan-tolerant strain	None[a] (33 different antibiotics)[b]	[28]
Pseudomonas spp.	52 isolates from meat chain production	Ampicillin, amoxicillin, erythromycin, imipenem and trimethoprim[c]	[80]
S. aureus	Triclosan-tolerant strain	None[a] (33 different antibiotics)	[28]
S. aureus	1,632 clinical isolates	No cross-resistance[c] to any clinically relevant antibiotic	[103]

[a]Disc diffusion test; [b]significantly higher susceptibility to aminoglycoside antibiotics; [c]broth microdilution method

poultry and clinical samples have also been reported to be less susceptible to triclosan [134].

Use of a toothpaste twice daily with triclosan resulted in 3.6 mg/l triclosan in saliva immediately after tooth brushing. The concentration decreased gradually to 0.6 mg/l after 15 min. There were no differences of susceptibility between streptococcal strains collected at days 0 and 14 to triclosan or five specific antibiotics (benzylpenicillin, gentamicin, erythromycin, tetracycline, fusidic acid) [131]. In the domestic setting no cross-resistance to antibiotics and antibacterial agents was found in target bacteria from antibacterial product users and non-users [22].

9.8 Role of Biofilm

9.8.1 Effect on Biofilm Development

Some studies indicate that triclosan does not inhibit biofilm formation. Attachment of *S. mutans* and *P. gingivalis* to polymethylmethacrylate (PMMA) or titanium was not impaired by an 18 h exposure to triclosan between 0.01% [115]. When a plastic based on acrylonitrile–butadiene–styrene with and without 5% triclosan was exposed for 1–3 weeks to drinking water, no significant differences were observed between the biofilm populations attached to triclosan plate and control plate surfaces. These results call into question the long-term utility of triclosan incorporation into this type of plastic [70].

Other studies indicate that triclosan can inhibit biofilm formation. For example, triclosan was described to inhibit biofilm formation to some extent on coated polyglactin sutures 3–0 coating compared to non-coated sutures [121]. And a commercially available mouth rinse containing triclosan resulted in a significantly higher percentage of plaque-free surfaces compared to another triclosan-free mouth rinse, both at 24 h and at 72 h but not at 48 h and 96 h indicating some retardation of bacterial biofilm down growth from the supra- to the subgingival environment [2]. The inhibition of biofilm formation of triclosan on vascular catheters can be significantly increased by DispersinB as shown with *S. aureus*, *S. epidermidis* and *E. coli* [35]. When a *P. acnes* biofilm was cultured on 96-well plates for 24 h and then exposed for another 24 h to 0.1% triclosan, the biofilm mass was approximately 90% lower compared to the negative control without triclosan [21]. And triclosan at 2.5 mg/l inhibited biofilm formation in two outbreak *S. Enteritidis* strains [60].

In combination with xylitol and polyhexamethylene biguanide, it was used for coating central venous catheters. A biofilm disaggregation with significant reduction of micro-organism's adherence was observed in coated fragments. In vivo anti-adherence results demonstrated a reduction of early biofilm formation of *S. aureus* ATCC 25923, mainly in an external surface of the coated central venous catheter [124].

9.8.2 Effect on Biofilm Removal

S. oralis (ATCC 10557), *S. gordonii* (ATCC 10558) and *A. naeslundii* (ATCC 19039) were incubated for 20 h in a biofilm capillary reactor and exposed for 1 h with a solution of 0.03% triclosan. No removal of biofilm was observed [26].

9.8.3 Effect on Biofilm Fixation

No data were found to evaluate the potential biofilm fixation properties of triclosan.

9.9 Summary

The principal antimicrobial activity of triclosan is summarized in Table 9.10.

The key findings on acquired resistance and cross-resistance including the role of biofilm for selecting resistant isolates are summarized in Table 9.11.

Table 9.10 Overview on the typical exposure times required for triclosan to achieve sufficient biocidal activity against the different target micro-organisms

Target micro-organisms	Species	Concentration	Exposure time (min)
Bacteria	Most bacterial species	1%	3[a]
		0.6%	5[a]
Fungi	*C. albicans*	1%	1
		0.1%	60
Mycobacteria	Unknown		

[a]In biofilm, the efficacy will be lower

Table 9.11 Key findings on acquired triclosan resistance, the effect of low-level exposure, cross-tolerance to other biocides and antibiotics, and its effect on biofilm

Parameter	Species	Findings
Elevated MIC values	*P. aeruginosa*	≤2,500 mg/l
	E. coli	≤1,000 mg/l
	Bifidobacterium spp.	≤512 mg/l
	A. johnsonii, C. perfringens, Lactobacillus spp.	≤256 mg/l
	Salmonella spp.	≤250 mg/l
	S. marcescens	≤232 mg/l
	Enterococcus spp.	≤128 mg/l
	C. freundii	≤100 mg/l
	S. aureus	≤64 mg/l
Proposed MIC values to determine resistance	*C. albicans*	16 mg/l
	Enterobacter spp.	1 mg/l
	E. faecium	32 mg/l
	E. faecalis	16 mg/l
	E. coli	2 mg/l
	K. pneumoniae	2 mg/l
	Salmonella spp.	8 mg/l
	S. aureus	0.5 mg/l
Cross-tolerance biocides	*B. cereus, B. licheniformis, Enterobacter* spp., *E. casseliflavus, E. faecium, Enterococcus* spp., *E. coli, K. oxytoca, P. agglomerans, P. ananatis, Pantoea* spp., *P. nigrescens, Salmonella* spp., *S. saprophyticus, S. oralis, S. sanguis* and *S. mutans*	Chlorhexidine
	B. cereus, B. licheniformis, Enterobacter spp., *E. casseliflavus, E. faecium, Enterococcus* spp., *E. coli, K. oxytoca, P. agglomerans, P. ananatis, Pantoea* spp., *Salmonella* spp. and *S. saprophyticus*	Benzalkonium chloride
	B. cereus, B. licheniformis, Chrysobacterium spp., *Enterobacter* spp., *E. casseliflavus, E. faecium, Enterococcus* spp., *P. agglomerans, P. ananatis, Salmonella* spp., *S. saprophyticus* and *Staphylococcus* spp.	Hexachlorophen
	Enterobacter spp., *K. oxytoca, P. agglomerans, P. ananatis, Pantoea* spp., *Salmonella* spp., *S. saprophyticus, S. xylosus* and *Staphylococcus* spp.	DDAB
	B. cereus and *P. ananatis*	Sodium nitrate

(continued)

Table 9.11 (continued)

Parameter	Species	Findings
Cross-tolerance antibiotics	*B. cereus*, *B. licheniformis*, *Bacillus* spp., *Enterobacter* spp., *E. casseliflavus*, *E. faecium*, *Enterococcus* spp., *E. coli*, *L. pentosus*, *L. pseudomesenteroides*, *L. monocytogenes*, *P. agglomerans*, *P. ananatis*, *Pantoea* spp., *Salmonella* spp., *S. saprophyticus*, *S. xylosus* and *Staphylococcus* spp.	Possible in selected strains to various types of antibiotics
Resistance mechanisms	*P. aeruginosa*, *E. coli*, *S.* Typhimurium, *S. enterica*, *K. pneumoniae*	Efflux pumps
	P. aeruginosa	Outer membrane changes
	A. baumannii, *S. aureus*	Gene expression changes
	A. xylosoxidans and *P. putida*	Use of triclosan as sole carbon source
	R. rubrum	FabI point mutation
Effect of low-level exposure	*A. baumannii*, *A. naeslundii*, *A. xylosoxidans*, *B.* cereus, *B. cepacia*, *C. coli*, *C. indologenes*, *Chryseobacterium* spp., *E. faecalis*, *E. faecium*, *E. coli*, *Eubacterium* spp., *F. nucleatum*, *H. gallinarum*, *K. oxytoca*, *K. planticola*, *L. rhamnosus*, *L. lactis*, *M. luteus*, *Megasphaera* spp., *N. subflava*, *P. gingivalis*, *P. aeruginosa*, *S.* Enteritidis, *S.* Infantis, *S.* Typhimurium, *S. marcescens*, *S. capitis*, *Staphylococcus* spp., *S. anginosus*, *S. multivorum*, *S. sanguis*, *S. mutans* and *V. dispar*	No MIC increase
	B. cereus, *B. licheniformis*, *Bacillus* spp., *C. jejuni*, *C. indologenes*, *Chryseobacterium* spp., *E. casseliflavus*, *E. faecium*, *Enterococcus* spp., *K. oxytoca*, *M. luteus*, *M. phyllosphaeriae*, *P. ananatis*, *Pantoea* spp., *P. nigrescens*, *P. putida*, *Salmonella* spp., *S. aureus*, *S. caprae*, *S. epidermidis*, *S. saprophyticus*, *S. maltophilia*, *S. proteomaculans*, *S. oralis* and *Veillonella* spp.	Weak MIC increase ($\leq$ 4-fold)
	C. xerosis, *Enterobacter* spp., *E. faecalis*, *P. agglomerans*, *P. ananatis*, *Salmonella* spp., *S. aureus*, *S. haemolyticus*, *S. lugdunensis*, *S. saprophyticus*, *S. warneri*, *S. xylosus*, *Staphylococcus* spp. and *S. maltophilia*	Strong (>4-fold) but unstable MIC increase

(continued)

Table 9.11 (continued)

Parameter	Species	Findings
	A. baumannii, *C. sakazakii*, *E. coli*, *K. pneumoniae*, *M. osloensis*, *P. ananatis*, *S.* Enteritidis, *S.* Typhimurium, *S.* Virchow, *Salmonella* spp., *S. aureus*, *S. epidermidis* and *S. saprophyticus*	Strong and stable MIC increase
	A. proteolyticus, *E. gergoviae*, *E. coli* and *S. aureus*	Strong MIC increase (unknown stability)
	Salmonella spp. (up to 10,000-fold)	Strongest MIC change after low-level exposure
	E. coli (up to 8,192-fold)	
	S. aureus (up to 313-fold)	
	P. ananatis (up to 200-fold)	
	P. agglomerans (up to 150-fold)	
	Staphylococcus spp. (up to 150-fold)	
	E. coli (>8,000 mg/l)	Highest MIC values after low-level exposure
	Salmonella spp. (3,000 mg/l)	
	P. aeruginosa (>1,000 mg/l)	
	S. aureus (625 mg/l)	
	C. sakazakii (500 mg/l).	
	E. coli	Induction of horizontal gene transfer
	S. aureus	Change to small colony variant
	A. baumannii	Enhancement of several general protective mechanisms
Biofilm	Development	Enhancement in *E. coli* and *S. epidermidis*
		No effect in *S. mutans* and *P. gingivalis*
		Inhibition in *S. aureus*, *S. epidermidis*, *S. lugdunensis*, *E. coli*, *P. acnes* and *S.* Enteritidis
	Removal	None
	Fixation	Unknown

References

1. Abuzaid A, Hamouda A, Amyes SG (2012) Klebsiella pneumoniae susceptibility to biocides and its association with cepA, qacDeltaE and qacE efflux pump genes and antibiotic resistance. J Hosp Infect 81(2):87–91. https://doi.org/10.1016/j.jhin.2012.03.003
2. Andrade E, Weidlich P, Angst PD, Gomes SC, Oppermann RV (2015) Efficacy of a triclosan formula in controlling early subgingival biofilm formation: a randomized trial. Braz Oral Res 29. https://doi.org/10.1590/1807-3107bor-2015.vol29.0065

3. Arioli S, Elli M, Ricci G, Mora D (2013) Assessment of the susceptibility of lactic acid bacteria to biocides. Int J Food Microbiol 163(1):1–5. https://doi.org/10.1016/j.ijfoodmicro.2013.02.002
4. Assadian O, Wehse K, Hubner NO, Koburger T, Bagel S, Jethon F, Kramer A (2011) Minimum inhibitory (MIC) and minimum microbicidal concentration (MMC) of polihexanide and triclosan against antibiotic sensitive and resistant Staphylococcus aureus and Escherichia coli strains. GMS Krankenhaushygiene interdisziplinar 6(1):Doc06. https://doi.org/10.3205/dgkh000163
5. Barroso JM (2014) COMMISSION IMPLEMENTING DECISION of 24 April 2014 on the non-approval of certain biocidal active substances pursuant to Regulation (EU) No 528/2012 of the European Parliament and of the Council. Off J Eur Union 57(L 124):27–29
6. Barry MA, Craven DE, Goularte TA, Lichtenberg DA (1984) *Serratia marcescens* contamination of antiseptic soap containing triclosan: implications for nosocomial infections. Infect Control 5(9):427–430
7. Bartzokas CA, Gibson MF, Graham R, Pinder DC (1983) A comparison of triclosan and chlorhexidine preparations with 60 per cent isopropyl alcohol for hygienic hand disinfection. J Hosp Infect 4:245–255
8. Bayston R, Ashraf W, Smith T (2007) Triclosan resistance in methicillin-resistant Staphylococcus aureus expressed as small colony variants: a novel mode of evasion of susceptibility to antiseptics. J Antimicrob Chemother 59(5):848–853. https://doi.org/10.1093/jac/dkm031
9. Braoudaki M, Hilton AC (2004) Adaptive resistance to biocides in Salmonella enterica and Escherichia coli O157 and cross-resistance to antimicrobial agents. J Clin Microbiol 42(1):73–78
10. Braoudaki M, Hilton AC (2004) Low level of cross-resistance between triclosan and antibiotics in Escherichia coli K-12 and E. coli O55 compared to E. coli O157. FEMS Microbiol Lett 235(2):305–309. https://doi.org/10.1016/j.femsle.2004.04.049
11. Braoudaki M, Hilton AC (2005) Mechanisms of resistance in Salmonella enterica adapted to erythromycin, benzalkonium chloride and triclosan. Int J Antimicrob Agents 25(1):31–37. https://doi.org/10.1016/j.ijantimicag.2004.07.016
12. Buergers R, Rosentritt M, Schneider-Brachert W, Behr M, Handel G, Hahnel S (2008) Efficacy of denture disinfection methods in controlling Candida albicans colonization in vitro. Acta Odontol Scand 66(3):174–180. https://doi.org/10.1080/00016350802165614
13. Caballero Gómez N, Abriouel H, Grande MJ, Pérez Pulido R, Gálvez A (2012) Effect of enterocin AS-48 in combination with biocides on planktonic and sessile Listeria monocytogenes. Food Microbiol 30(1):51–58. https://doi.org/10.1016/j.fm.2011.12.013
14. Casado Munoz Mdel C, Benomar N, Ennahar S, Horvatovich P, Lavilla Lerma L, Knapp CW, Galvez A, Abriouel H (2016) Comparative proteomic analysis of a potentially probiotic Lactobacillus pentosus MP-10 for the identification of key proteins involved in antibiotic resistance and biocide tolerance. Int J Food Microbiol 222:8–15. https://doi.org/10.1016/j.ijfoodmicro.2016.01.012
15. Casado Munoz Mdel C, Benomar N, Lavilla Lerma L, Knapp CW, Galvez A, Abriouel H (2016) Biocide tolerance, phenotypic and molecular response of lactic acid bacteria isolated from naturally-fermented Alorena table to different physico-chemical stresses. Food Microbiol 60:1–12. https://doi.org/10.1016/j.fm.2016.06.013
16. Champlin FR, Ellison ML, Bullard JW, Conrad RS (2005) Effect of outer membrane permeabilisation on intrinsic resistance to low triclosan levels in Pseudomonas aeruginosa. Int J Antimicrob Agents 26(2):159–164. https://doi.org/10.1016/j.ijantimicag.2005.04.020
17. Christensen EG, Gram L, Kastbjerg VG (2011) Sublethal triclosan exposure decreases susceptibility to gentamicin and other aminoglycosides in Listeria monocytogenes. Antimicrob Agents Chemother 55(9):4064–4071. https://doi.org/10.1128/aac.00460-11
18. Chuanchuen R, Beinlich K, Hoang TT, Becher A, Karkhoff-Schweizer RR, Schweizer HP (2001) Cross-resistance between triclosan and antibiotics in *Pseudomonas aeruginosa* is

mediated by multi-drug efflux pumps: exposure of a susceptible mutant strain to triclosan selects *nfxB* mutants overexpressing MexCD-OprJ. Antimicrob Agents Chemother 45:428–432
19. Chuanchuen R, Karkhoff-Schweizer RR, Schweizer HP (2003) High-level triclosan resistance in Pseudomonas aeruginosa is solely a result of efflux. Am J Infect Control 31 (2):124–127
20. Ciusa ML, Furi L, Knight D, Decorosi F, Fondi M, Raggi C, Coelho JR, Aragones L, Moce L, Visa P, Freitas AT, Baldassarri L, Fani R, Viti C, Orefici G, Martinez JL, Morrissey I, Oggioni MR (2012) A novel resistance mechanism to triclosan that suggests horizontal gene transfer and demonstrates a potential selective pressure for reduced biocide susceptibility in clinical strains of Staphylococcus aureus. Int J Antimicrob Agents 40 (3):210–220. https://doi.org/10.1016/j.ijantimicag.2012.04.021
21. Coenye T, Peeters E, Nelis HJ (2007) Biofilm formation by Propionibacterium acnes is associated with increased resistance to antimicrobial agents and increased production of putative virulence factors. Res Microbiol 158(4):386–392. https://doi.org/10.1016/j.resmic.2007.02.001
22. Cole EC, Addison RM, Rubino JR, Leese KE, Dulaney PD, Newell MS, Wilkins J, Gaber DJ, Wineinger T, Criger DA (2003) Investigation of antibiotic and antibacterial agent cross-resistance in target bacteria from homes of antibacterial product users and nonusers. J Appl Microbiol 95(4):664–676
23. Condell O, Iversen C, Cooney S, Power KA, Walsh C, Burgess C, Fanning S (2012) Efficacy of biocides used in the modern food industry to control salmonella enterica, and links between biocide tolerance and resistance to clinically relevant antimicrobial compounds. Appl Environ Microbiol 78(9):3087–3097. https://doi.org/10.1128/aem.07534-11
24. Condell O, Sheridan A, Power KA, Bonilla-Santiago R, Sergeant K, Renaut J, Burgess C, Fanning S, Nally JE (2012) Comparative proteomic analysis of Salmonella tolerance to the biocide active agent triclosan. J Proteomics 75(14):4505–4519. https://doi.org/10.1016/j.jprot.2012.04.044
25. Copitch JL, Whitehead RN, Webber MA (2010) Prevalence of decreased susceptibility to triclosan in Salmonella enterica isolates from animals and humans and association with multiple drug resistance. Int J Antimicrob Agents 36(3):247–251. https://doi.org/10.1016/j.ijantimicag.2010.04.012
26. Corbin A, Pitts B, Parker A, Stewart PS (2011) Antimicrobial penetration and efficacy in an in vitro oral biofilm model. Antimicrob Agents Chemother 55(7):3338–3344. https://doi.org/10.1128/aac.00206-11
27. Correa JE, De Paulis A, Predari S, Sordelli DO, Jeric PE (2008) First report of qacG, qacH and qacJ genes in Staphylococcus haemolyticus human clinical isolates. J Antimicrob Chemother 62(5):956–960. https://doi.org/10.1093/jac/dkn327
28. Cottell A, Denyer SP, Hanlon GW, Ochs D, Maillard JY (2009) Triclosan-tolerant bacteria: changes in susceptibility to antibiotics. J Hosp Infect 72(1):71–76. https://doi.org/10.1016/j.jhin.2009.01.014
29. Couto N, Belas A, Couto I, Perreten V, Pomba C (2014) Genetic relatedness, antimicrobial and biocide susceptibility comparative analysis of methicillin-resistant and -susceptible Staphylococcus pseudintermedius from Portugal. Microb Drug Res (Larchmont, NY) 20 (4):364–371. https://doi.org/10.1089/mdr.2013.0043
30. Couto N, Belas A, Tilley P, Couto I, Gama LT, Kadlec K, Schwarz S, Pomba C (2013) Biocide and antimicrobial susceptibility of methicillin-resistant staphylococcal isolates from horses. Vet Microbiol 166(1–2):299–303. https://doi.org/10.1016/j.vetmic.2013.05.011
31. Cowley NL, Forbes S, Amezquita A, McClure P, Humphreys GJ, McBain AJ (2015) Effects of formulation on microbicide potency and mitigation of the development of bacterial insusceptibility. Appl Environ Microbiol 81(20):7330–7338. https://doi.org/10.1128/aem.01985-15
32. Curiao T, Marchi E, Viti C, Oggioni MR, Baquero F, Martinez JL, Coque TM (2015) Polymorphic variation in susceptibility and metabolism of triclosan-resistant mutants of

Escherichia coli and Klebsiella pneumoniae clinical strains obtained after exposure to biocides and antibiotics. Antimicrob Agents Chemother 59(6):3413–3423. https://doi.org/10.1128/aac.00187-15
33. D'Arezzo S, Lanini S, Puro V, Ippolito G, Visca P (2012) High-level tolerance to triclosan may play a role in Pseudomonas aeruginosa antibiotic resistance in immunocompromised hosts: evidence from outbreak investigation. BMC Res Notes 5:43. https://doi.org/10.1186/1756-0500-5-43
34. Dann AB, Hontela A (2011) Triclosan: environmental exposure, toxicity and mechanisms of action. J Appl Toxicol: JAT 31(4):285–311. https://doi.org/10.1002/jat.1660
35. Darouiche RO, Mansouri MD, Gawande PV, Madhyastha S (2009) Antimicrobial and antibiofilm efficacy of triclosan and DispersinB combination. J Antimicrob Chemother 64(1):88–93. https://doi.org/10.1093/jac/dkp158
36. Department of Health and Human Services; Food and Drug Administration (1994) Tentative final monograph for health care antiseptic products; proposed rule. Fed Reg 59(116):31401–31452
37. Department of Health and Human Services; Food and Drug Administration (2015) Safety and effectiveness of healthcare antiseptics. Topical antimicrobial drug products for over-the-counter human use; proposed amendment of the tentative final monograph; reopening of administrative record; proposed rule. Fed Reg 80(84):25166–25205
38. Department of Health and Human Services; Food and Drug Administration (2016) Safety and effectiveness of consumer antiseptics; topical antimicrobial drug products for over-the-counter human use. Fed Reg 81(172):61106–61130
39. Elli M, Arioli S, Guglielmetti S, Mora D (2013) Biocide susceptibility in bifidobacteria of human origin. J Glob Antimicrob Res 1(2):97–101. https://doi.org/10.1016/j.jgar.2013.03.007
40. Ellison ML, Champlin FR (2007) Outer membrane permeability for nonpolar antimicrobial agents underlies extreme susceptibility of Pasteurella multocida to the hydrophobic biocide triclosan. Vet Microbiol 124(3–4):310–318. https://doi.org/10.1016/j.vetmic.2007.04.038
41. European Chemicals Agency (ECHA) Triclosan. Substance information. https://echa.europa.eu/substance-information/-/substanceinfo/100.020.167. Accessed 16 Nov 2017
42. Faoagali J, Fong J, George N, Mahoney P, O'Rourke V (1995) Comparison of the immediate, residual, and cumulative antibacterial effects of Novaderm, Novascrub, Betadine Surgical Scrub, Hibiclens, and liquid soap. Am J Infect Control 23(6):337–343
43. Fernández-Fuentes MA, Ortega Morente E, Abriouel H, Pérez Pulido R, Gálvez A (2012) Isolation and identification of bacteria from organic foods: sensitivity to biocides and antibiotics. Food Control 26(1):73–78. https://doi.org/10.1016/j.foodcont.2012.01.017
44. Fernandez Marquez ML, Grande Burgos MJ, Lopez Aguayo MC, Perez Pulido R, Galvez A, Lucas R (2017) Characterization of biocide-tolerant bacteria isolated from cheese and dairy small-medium enterprises. Food Microbiol 62:77–81. https://doi.org/10.1016/j.fm.2016.10.008
45. Fernando DM, Chong P, Singh M, Spicer V, Unger M, Loewen PC, Westmacott G, Kumar A (2017) Multi-omics approach to study global changes in a triclosan-resistant mutant strain of Acinetobacter baumannii ATCC 17978. Int J Antimicrob Agents 49(1):74–80. https://doi.org/10.1016/j.ijantimicag.2016.10.014
46. Forbes S, Dobson CB, Humphreys GJ, McBain AJ (2014) Transient and sustained bacterial adaptation following repeated sublethal exposure to microbicides and a novel human antimicrobial peptide. Antimicrob Agents Chemother 58(10):5809–5817. https://doi.org/10.1128/aac.03364-14
47. Forbes S, Knight CG, Cowley NL, Amezquita A, McClure P, Humphreys G, McBain AJ (2016) Variable effects of exposure to formulated microbicides on antibiotic susceptibility in Firmicutes and Proteobacteria. Appl Environ Microbiol 82(12):3591–3598. https://doi.org/10.1128/aem.00701-16
48. Freundlich JS, Wang F, Vilcheze C, Gulten G, Langley R, Schiehser GA, Jacobus DP, Jacobs WR Jr, Sacchettini JC (2009) Triclosan derivatives: towards potent inhibitors of

drug-sensitive and drug-resistant Mycobacterium tuberculosis. ChemMedChem 4(2):241–248. https://doi.org/10.1002/cmdc.200800261

49. Gadea R, Fernández Fuentes MA, Pérez Pulido R, Gálvez A, Ortega E (2016) Adaptive tolerance to phenolic biocides in bacteria from organic foods: Effects on antimicrobial susceptibility and tolerance to physical stresses. Food Res Int 85(Supplement C):131–143. https://doi.org/10.1016/j.foodres.2016.04.033
50. Gadea R, Glibota N, Pérez Pulido R, Gálvez A, Ortega E (2017) Effects of exposure to biocides on susceptibility to essential oils and chemical preservatives in bacteria from organic foods. Food Control 80(Supplement C):176–182. https://doi.org/10.1016/j.foodcont.2017.05.002
51. Gantzhorn MR, Pedersen K, Olsen JE, Thomsen LE (2014) Biocide and antibiotic susceptibility of Salmonella isolates obtained before and after cleaning at six Danish pig slaughterhouses. Int J Food Microbiol 181:53–59. https://doi.org/10.1016/j.ijfoodmicro.2014.04.021
52. Garrido AM, Burgos MJ, Marquez ML, Aguayo MC, Pulido RP, del Arbol JT, Galvez A, Lopez RL (2015) Biocide tolerance in Salmonella from meats in Southern Spain. Braz J Microbiol: [Publication of the Brazilian Society for Microbiology] 46(4):1177–1181. https://doi.org/10.1590/s1517-838246420140396
53. Gasperi J, Geara D, Lorgeoux C, Bressy A, Zedek S, Rocher V, El Samrani A, Chebbo G, Moilleron R (2014) First assessment of triclosan, triclocarban and paraben mass loads at a very large regional scale: case of Paris conurbation (France). Sci Total Environ 493:854–861. https://doi.org/10.1016/j.scitotenv.2014.06.079
54. Giuliano CA, Rybak MJ (2015) Efficacy of triclosan as an antimicrobial hand soap and its potential impact on antimicrobial resistance: a focused review. Pharmacotherapy 35(3):328–336. https://doi.org/10.1002/phar.1553
55. Gomez-Alonso A, Garcia-Criado FJ, Parreno-Manchado FC, Garcia-Sanchez JE, Garcia-Sanchez E, Parreno-Manchado A, Zambrano-Cuadrado Y (2007) Study of the efficacy of Coated VICRYL Plus Antibacterial suture (coated Polyglactin 910 suture with Triclosan) in two animal models of general surgery. J Infect 54(1):82–88. https://doi.org/10.1016/j.jinf.2006.01.008
56. Gomez A, Andreu N, Ferrer-Navarro M, Yero D, Gibert I (2016) Triclosan-induced genes Rv1686c-Rv1687c and Rv3161c are not involved in triclosan resistance in Mycobacterium tuberculosis. Sci R 6:26221. https://doi.org/10.1038/srep26221
57. Gosau M, Hahnel S, Schwarz F, Gerlach T, Reichert TE, Burgers R (2010) Effect of six different peri-implantitis disinfection methods on in vivo human oral biofilm. Clin Oral Implant Res 21(8):866–872. https://doi.org/10.1111/j.1600-0501.2009.01908.x
58. Gradel KO, Randall L, Sayers AR, Davies RH (2005) Possible associations between Salmonella persistence in poultry houses and resistance to commonly used disinfectants and a putative role of mar. Vet Microbiol 107(1–2):127-138. S0378-1135(05)00036-2 [pii]. https://doi.org/10.1016/j.vetmic.2005.01.013 [doi]
59. Grande Burgos MJ, Fernandez Marquez ML, Perez Pulido R, Galvez A, Lucas Lopez R (2016) Virulence factors and antimicrobial resistance in Escherichia coli strains isolated from hen egg shells. Int J Food Microbiol 238:89–95. https://doi.org/10.1016/j.ijfoodmicro.2016.08.037
60. Grande Burgos MJ, Lucas López R, López Aguayo M, Pérez Pulido R, Gálvez A (2013) Inhibition of planktonic and sessile Salmonella enterica cells by combinations of enterocin AS-48, polymyxin B and biocides. Food Control 30(1):214–221. https://doi.org/10.1016/j.foodcont.2012.07.011
61. Guo J, Iwata H (2017) Risk assessment of triclosan in the global environment using a probabilistic approach. Ecotoxicol Environ Saf 143:111–119. https://doi.org/10.1016/j.ecoenv.2017.05.020
62. Halden RU (2014) On the need and speed of regulating triclosan and triclocarban in the United States. Environ Sci Technol 48(7):3603–3611. https://doi.org/10.1021/es500495p

63. Halden RU, Lindeman AE, Aiello AE, Andrews D, Arnold WA, Fair P, Fuoco RE, Geer LA, Johnson PI, Lohmann R, McNeill K, Sacks VP, Schettler T, Weber R, Zoeller RT, Blum A (2017) The florence statement on Triclosan and Triclocarban. Environ Health Perspect 125(6):064501. https://doi.org/10.1289/ehp1788
64. Hartmann EM, Hickey R, Hsu T, Betancourt Roman CM, Chen J, Schwager R, Kline J, Brown GZ, Halden RU, Huttenhower C, Green JL (2016) Antimicrobial chemicals are associated with elevated antibiotic resistance genes in the indoor dust microbiome. Environ Sci Technol 50(18):9807–9815. https://doi.org/10.1021/acs.est.6b00262
65. Heath RJ, Li J, Roland GE, Rock CO (2000) Inhibition of the Staphylococcus aureus NADPH-dependent enoyl-acyl carrier protein reductase by triclosan and hexachlorophene. J Biol Chem 275(7):4654–4659
66. Higgins J, Pinjon E, Oltean HN, White TC, Kelly SL, Martel CM, Sullivan DJ, Coleman DC, Moran GP (2012) Triclosan antagonizes fluconazole activity against Candida albicans. J Dent Res 91(1):65–70. https://doi.org/10.1177/0022034511425046
67. Huerta B, Rodriguez-Mozaz S, Nannou C, Nakis L, Ruhi A, Acuna V, Sabater S, Barcelo D (2016) Determination of a broad spectrum of pharmaceuticals and endocrine disruptors in biofilm from a waste water treatment plant-impacted river. Sci Total Environ 540:241–249. https://doi.org/10.1016/j.scitotenv.2015.05.049
68. Hughes C, Ferguson J (2017) Phenotypic chlorhexidine and triclosan susceptibility in clinical Staphylococcus aureus isolates in Australia. Pathology 49(6):633–637. https://doi.org/10.1016/j.pathol.2017.05.008
69. Juncker JC (2016) COMMISSION IMPLEMENTING REGULATION (EU) 2016/110 of 27 January 2016 not approving triclosan as an existing active substance for use in biocidal products for product- type 1. Off J Eur Union 59(L 21):86–87
70. Junker LM, Hay AG (2004) Effects of triclosan incorporation into ABS plastic on biofilm communities. J Antimicrob Chemother 53(6):989–996. https://doi.org/10.1093/jac/dkh196
71. Jutkina J, Marathe NP, Flach CF, Larsson DGJ (2017) Antibiotics and common antibacterial biocides stimulate horizontal transfer of resistance at low concentrations. Sci Total Environ 616–617:172–178. https://doi.org/10.1016/j.scitotenv.2017.10.312
72. Kahl BC (2014) Small colony variants (SCVs) of Staphylococcus aureus–a bacterial survival strategy. Infect, Genet Evol: J Mol Epidemiol Evol Genet Infect Dis 21:515–522. https://doi.org/10.1016/j.meegid.2013.05.016
73. Kalkanci A, Elli M, Adil Fouad A, Yesilyurt E, Jabban Khalil I (2015) Assessment of susceptibility of mould isolates towards biocides. Journal de mycologie medicale 25(4):280–286. https://doi.org/10.1016/j.mycmed.2015.08.001
74. Kapoor R, Yadav JS (2012) Expanding the mycobacterial diversity of metalworking fluids (MWFs): evidence showing MWF colonization by Mycobacterium abscessus. FEMS Microbiol Ecol 79(2):392–399. https://doi.org/10.1111/j.1574-6941.2011.01227.x
75. Karatzas KA, Webber MA, Jorgensen F, Woodward MJ, Piddock LJ, Humphrey TJ (2007) Prolonged treatment of Salmonella enterica serovar Typhimurium with commercial disinfectants selects for multiple antibiotic resistance, increased efflux and reduced invasiveness. J Antimicrob Chemother 60(5):947–955. https://doi.org/10.1093/jac/dkm314
76. Kim SA, Moon H, Lee K, Rhee MS (2015) Bactericidal effects of triclosan in soap both in vitro and in vivo. J Antimicrob Chemother 70(12):3345–3352. https://doi.org/10.1093/jac/dkv275
77. Koburger T, Hübner N-O, Braun M, Siebert J, Kramer A (2010) Standardized comparison of antiseptic efficacy of triclosan, PVP-iodine, octenidine dihydrochloride, polyhexanide and chlorhexidine digluconate. J Antimicrob Chemother 65(8):1712–1719
78. Lambert RJ (2004) Comparative analysis of antibiotic and antimicrobial biocide susceptibility data in clinical isolates of methicillin-sensitive Staphylococcus aureus, methicillin-resistant Staphylococcus aureus and Pseudomonas aeruginosa between 1989 and 2000. J Appl Microbiol 97(4):699–711. https://doi.org/10.1111/j.1365-2672.2004.02345.x

79. Lanini S, D'Arezzo S, Puro V, Martini L, Imperi F, Piselli P, Montanaro M, Paoletti S, Visca P, Ippolito G (2011) Molecular epidemiology of a Pseudomonas aeruginosa hospital outbreak driven by a contaminated disinfectant-soap dispenser. PLoS ONE 6(2):e17064. https://doi.org/10.1371/journal.pone.0017064
80. Lavilla Lerma L, Benomar N, Casado Munoz Mdel C, Galvez A, Abriouel H (2015) Correlation between antibiotic and biocide resistance in mesophilic and psychrotrophic Pseudomonas spp. isolated from slaughterhouse surfaces throughout meat chain production. Food Microbiol 51:33–44. https://doi.org/10.1016/j.fm.2015.04.010
81. Lavilla Lerma L, Benomar N, Valenzuela AS, Casado Munoz Mdel C, Galvez A, Abriouel H (2014) Role of EfrAB efflux pump in biocide tolerance and antibiotic resistance of Enterococcus faecalis and Enterococcus faecium isolated from traditional fermented foods and the effect of EDTA as EfrAB inhibitor. Food Microbiol 44:249–257. https://doi.org/10.1016/j.fm.2014.06.009
82. Lear JC, Maillard JY, Dettmar PW, Goddard PA, Russell AD (2006) Chloroxylenol- and triclosan-tolerant bacteria from industrial sources—susceptibility to antibiotics and other biocides. Int Biodeter Biodegr 57(1):51–56. https://doi.org/10.1016/j.ibiod.2005.11.002
83. Ledder RG, Gilbert P, Willis C, McBain AJ (2006) Effects of chronic triclosan exposure upon the antimicrobial susceptibility of 40 ex-situ environmental and human isolates. J Appl Microbiol 100(5):1132–1140. https://doi.org/10.1111/j.1365-2672.2006.02811.x
84. Lin F, Xu Y, Chang Y, Liu C, Jia X, Ling B (2017) Molecular characterization of reduced susceptibility to biocides in clinical isolates of acinetobacter baumannii. Front Microbiol 8:1836. https://doi.org/10.3389/fmicb.2017.01836
85. Macias JH, Alvarez MF, Arreguin V, Munoz JM, Macias AE, Alvarez JA (2016) Chlorhexidine avoids skin bacteria recolonization more than triclosan. Am J Infect Control 44(12):1530–1534. https://doi.org/10.1016/j.ajic.2016.04.235
86. Marchetti MG, Kampf G, Finzi G, Salvatorelli G (2003) Evaluation of the bactericidal effect of five products for surgical hand disinfection according to prEN 12054 and prEN 12791. J Hosp Infect 54(1):63–67
87. Massengo-Tiasse RP, Cronan JE (2009) Diversity in enoyl-acyl carrier protein reductases. Cell Mol Life Sci: CMLS 66(9):1507–1517. https://doi.org/10.1007/s00018-009-8704-7
88. Matalon S, Kozlovsky A, Kfir A, Levartovsky S, Mazor Y, Slutzky H (2013) The effect of commonly used sutures on inflammation inducing pathogens - an in vitro study. J Craniomaxillofac Surg: Off Publ Eur Assoc Craniomaxillofac Surg 41(7):593–597. https://doi.org/10.1016/j.jcms.2012.11.033
89. Mavri A, Mozina SS (2012) Involvement of efflux mechanisms in biocide resistance of Campylobacter jejuni and Campylobacter coli. J Med Microbiol 61(Pt 6):800–808. https://doi.org/10.1099/jmm.0.041467-0
90. Mavri A, Smole Mozina S (2013) Development of antimicrobial resistance in Campylobacter jejuni and Campylobacter coli adapted to biocides. Int J Food Microbiol 160 (3):304–312. https://doi.org/10.1016/j.ijfoodmicro.2012.11.006
91. McBain AJ, Ledder RG, Sreenivasan P, Gilbert P (2004) Selection for high-level resistance by chronic triclosan exposure is not universal. J Antimicrob Chemother 53(5):772–777. https://doi.org/10.1093/jac/dkh168
92. McMurry LM, McDermott PF, Levy SB (1999) Genetic evidence that InhA of *Mycobacterium smegmatis* is a target for triclosan. Antimicrob Agents Chemother 43 (3):711–713
93. McMurry LM, Oethinger M, Levy SB (1998) Overexpression of *marA*, *soxS*, or *acrAB* produces resistance to triclosan in laboratory and clinical strains of *Escherichia coli*. FEMS Microbiol Lett 166:305–309
94. McMurry LM, Oethinger M, Levy SB (1998) Triclosan targets lipid synthesis. Nature 394:531–532
95. McNamara PJ, Levy SB (2016) Triclosan: an Instructive Tale. Antimicrob Agents Chemother 60(12):7015–7016. https://doi.org/10.1128/aac.02105-16

96. McNaughton M, Mazinke N, Thomas E (1995) Newborn conjunctivitis associated with triclosan 0.5% antiseptic intrinsically contaminated with Serratia marcescens. The Canadian journal of infection control: the official journal of the Community & Hospital Infection Control Association-Canada = Revue canadienne de prevention des infections/Association pour la prevention des infections a l'ho 10(1):7–8
97. Meade MJ, Waddell RL, Callahan TM (2001) Soil bacteria *Pseudomonas putida* and *Alcaligenes xylosoxidans* subsp. denitrificans inactivate triclosan in liquid and solid substrates. Fed Eur Microbiol Soc, Microbiol Lett 204:45–48
98. Messager S, Goddard PA, Dettmar PW, Maillard JY (2001) Determination of the antibacterial efficacy of several antiseptics tested on skin by an 'ex-vivo' test. J Med Microbiol 50(3):284–292. https://doi.org/10.1099/0022-1317-50-3-284
99. Middleton JH, Salierno JD (2013) Antibiotic resistance in triclosan tolerant fecal coliforms isolated from surface waters near wastewater treatment plant outflows (Morris County, NJ, USA). Ecotoxicol Environ Saf 88:79–88. https://doi.org/10.1016/j.ecoenv.2012.10.025
100. Morrissey I, Oggioni MR, Knight D, Curiao T, Coque T, Kalkanci A, Martinez JL (2014) Evaluation of epidemiological cut-off values indicates that biocide resistant subpopulations are uncommon in natural isolates of clinically-relevant microorganisms. PLoS ONE 9(1): e86669. https://doi.org/10.1371/journal.pone.0086669
101. Muller G, Kramer A (2008) Biocompatibility index of antiseptic agents by parallel assessment of antimicrobial activity and cellular cytotoxicity. J Antimicrob Chemother 61 (6):1281–1287. https://doi.org/10.1093/jac/dkn125
102. National Center for Biotechnology Information Triclosan. PubChem Compound Database; CID=5564. https://pubchem.ncbi.nlm.nih.gov/compound/5564. Accessed 16 Nov 2017
103. Oggioni MR, Coelho JR, Furi L, Knight DR, Viti C, Orefici G, Martinez JL, Freitas AT, Coque TM, Morrissey I (2015) Significant differences characterise the correlation coefficients between biocide and antibiotic susceptibility profiles in Staphylococcus aureus. Curr Pharm Des 21(16):2054–2057
104. Periame M, Pages JM, Davin-Regli A (2015) Enterobacter gergoviae membrane modifications are involved in the adaptive response to preservatives used in cosmetic industry. J Appl Microbiol 118(1):49–61. https://doi.org/10.1111/jam.12676
105. Pi B, Yu D, Hua X, Ruan Z, Yu Y (2017) Genomic and transcriptome analysis of triclosan response of a multidrug-resistant Acinetobacter baumannii strain, MDR-ZJ06. Arch Microbiol 199(2):223–230. https://doi.org/10.1007/s00203-016-1295-4
106. Pycke BF, Crabbe A, Verstraete W, Leys N (2010) Characterization of triclosan-resistant mutants reveals multiple antimicrobial resistance mechanisms in Rhodospirillum rubrum S1H. Appl Environ Microbiol 76(10):3116–3123. https://doi.org/10.1128/aem.02757-09
107. Rahav G, Pitlik S, Amitai Z, Lavy A, Blech M, Keller N, Smollan G, Lewis M, Zlotkin A (2006) An outbreak of Mycobacterium jacuzzii infection following insertion of breast implants. Clin Infect Dis: Off Publ Infect Dis Soc Am 43(7):823–830. https://doi.org/10.1086/507535
108. Rawat R, Whitty A, Tonge PJ (2003) The isoniazid-NAD adduct is a slow, tight-binding inhibitor of InhA, the Mycobacterium tuberculosis enoyl reductase: adduct affinity and drug resistance. Proc Natl Acad Sci USA 100(24):13881–13886. https://doi.org/10.1073/pnas.2235848100
109. Reginato CF, Bandeira LA, Zanette RA, Santurio JM, Alves SH, Danesi CC (2017) Antifungal activity of synthetic antiseptics and natural compounds against Candida dubliniensis before and after in vitro fluconazole exposure. Rev Soc Bras Med Trop 50 (1):75–79. https://doi.org/10.1590/0037-8682-0461-2016
110. Renko M, Paalanne N, Tapiainen T, Hinkkainen M, Pokka T, Kinnula S, Sinikumpu JJ, Uhari M, Serlo W (2017) Triclosan-containing sutures versus ordinary sutures for reducing surgical site infections in children: a double-blind, randomised controlled trial. Lancet Infect Dis 17(1):50–57. https://doi.org/10.1016/s1473-3099(16)30373-5

111. Rensch U, Klein G, Schwarz S, Kaspar H, de Jong A, Kehrenberg C (2013) Comparative analysis of the susceptibility to triclosan and three other biocides of avian Salmonella enterica isolates collected 1979 through 1994 and 2004 through 2010. J Food Prot 76 (4):653–656. https://doi.org/10.4315/0362-028x.jfp-12-420
112. Rensch U, Nishino K, Klein G, Kehrenberg C (2014) Salmonella enterica serovar Typhimurium multidrug efflux pumps EmrAB and AcrEF support the major efflux system AcrAB in decreased susceptibility to triclosan. Int J Antimicrob Agents 44(2):179–180. https://doi.org/10.1016/j.ijantimicag.2014.04.015
113. Ricart M, Guasch H, Alberch M, Barcelo D, Bonnineau C, Geiszinger A, Farre M, Ferrer J, Ricciardi F, Romani AM, Morin S, Proia L, Sala L, Sureda D, Sabater S (2010) Triclosan persistence through wastewater treatment plants and its potential toxic effects on river biofilms. Aquat Toxicol (Amsterdam, Netherlands) 100(4):346–353. https://doi.org/10.1016/j.aquatox.2010.08.010
114. Rizzotti L, Rossi F, Torriani S (2016) Biocide and antibiotic resistance of Enterococcus faecalis and Enterococcus faecium isolated from the swine meat chain. Food Microbiol 60:160–164. https://doi.org/10.1016/j.fm.2016.07.009
115. Roberts JL, Khan S, Emanuel C, Powell LC, Pritchard MF, Onsoyen E, Myrvold R, Thomas DW, Hill KE (2013) An in vitro study of alginate oligomer therapies on oral biofilms. J Dent 41(10):892–899. https://doi.org/10.1016/j.jdent.2013.07.011
116. Rochon-Edouard S, Pons JL, Veber B, Larkin M, Vassal S, Lemeland JF (2004) Comparative in vitro and in vivo study of nine alcohol-based handrubs. Am J Infect Control 32(4):200–204. https://doi.org/10.1016/j.ajic.2003.08.003
117. Rose H, Baldwin A, Dowson CG, Mahenthiralingam E (2009) Biocide susceptibility of the Burkholderia cepacia complex. J Antimicrob Chemother 63(3):502–510. https://doi.org/10.1093/jac/dkn540
118. Scheflan M, Wixtrom RN (2016) Over troubled water: an outbreak of infection due to a new species of Mycobacterium following implant-based breast surgery. Plast Reconstr Surg 137 (1):97–105. https://doi.org/10.1097/prs.0000000000001854
119. Schweizer HP (2001) Triclosan: a widely used biocide and its links to antibiotics. Fed Eur Microbiol Soc, Microbiol Lett 202:1–7
120. Scientific Committee On Consumer Safety S (2010) Opinion on triclosan. Antimicrobial Resistance. https://ec.europa.eu/health/scientific_committees/consumer_safety/docs/sccs_o_023.pdf. Accessed 28 Nov 2017
121. Sethi KS, Karde PA, Joshi CP (2016) Comparative evaluation of sutures coated with triclosan and chlorhexidine for oral biofilm inhibition potential and antimicrobial activity against periodontal pathogens: an in vitro study. Indian J Dent Res: Off Publ Indian Soc Dent Res 27(5):535–539. https://doi.org/10.4103/0970-9290.195644
122. Sheridan A, Lenahan M, Condell O, Bonilla-Santiago R, Sergeant K, Renaut J, Duffy G, Fanning S, Nally JE, Burgess CM (2013) Proteomic and phenotypic analysis of triclosan tolerant verocytotoxigenic Escherichia coli O157:H19. J Proteomics 80:78–90. https://doi.org/10.1016/j.jprot.2012.12.025
123. Sheridan A, Lenahan M, Duffy G, Fanning S, Burgess C (2012) The potential for biocide tolerance in Escherichia coli and its impact on the response to food processing stresses. Food Control 26(1):98–106
124. Silva Paes Leme AF, Ferreira AS, Alves FA, de Azevedo BM, de Bretas LP, Farias RE, Oliveira MG, Raposo NR (2015) An effective and biocompatible antibiofilm coating for central venous catheter. Can J Microbiol 61(5):357–365. https://doi.org/10.1139/cjm-2014-0783
125. Singh NJ, Shin D, Lee HM, Kim HT, Chang HJ, Cho JM, Kim KS, Ro S (2011) Structural basis of triclosan resistance. J Struct Biol 174(1):173–179. https://doi.org/10.1016/j.jsb.2010.11.008

126. Skovgaard S, Nielsen LN, Larsen MH, Skov RL, Ingmer H, Westh H (2013) Staphylococcus epidermidis isolated in 1965 are more susceptible to triclosan than current isolates. PLoS ONE 8(4):e62197. https://doi.org/10.1371/journal.pone.0062197
127. Smith K, Hunter IS (2008) Efficacy of common hospital biocides with biofilms of multi-drug resistant clinical isolates. J Med Microbiol 57(Pt 8):966–973. https://doi.org/10.1099/jmm.0.47668-0
128. Srinivasan VB, Singh BB, Priyadarshi N, Chauhan NK, Rajamohan G (2014) Role of novel multidrug efflux pump involved in drug resistance in Klebsiella pneumoniae. PLoS ONE 9 (5):e96288. https://doi.org/10.1371/journal.pone.0096288
129. Suller MTE, Russell AD (1999) Antibiotic and biocide resistance in methicillin-resistant *Staphylococcus aureus* and vancomycin-resistant enterococcus. J Hosp Infect 43:281–291
130. Suller MTE, Russell AD (2000) Triclosan and antibiotic resistance in *Staphylococcus aureus*. J Antimicrob Chemother 46:11–18
131. Sullivan A, Wretlind B, Nord CE (2003) Will triclosan in toothpaste select for resistant oral streptococci? Clin Microbiol Infect 9(4):306–309
132. Tambe SM, Sampath L, Modak SM (2001) *In vitro* evaluation of the risk of developing bacterial resistance to antiseptics and antibiotics used in medical devices. J Antimicrob Chemother 47(5):589–598
133. Tamura I, Yasuda Y, Kagota KI, Yoneda S, Nakada N, Kumar V, Kameda Y, Kimura K, Tatarazako N, Yamamoto H (2017) Contribution of pharmaceuticals and personal care products (PPCPs) to whole toxicity of water samples collected in effluent-dominated urban streams. Ecotoxicol Environ Saf 144:338–350. https://doi.org/10.1016/j.ecoenv.2017.06.032
134. Thorrold CA, Letsoalo ME, Duse AG, Marais E (2007) Efflux pump activity in fluoroquinolone and tetracycline resistant Salmonella and E. coli implicated in reduced susceptibility to household antimicrobial cleaning agents. Int J Food Microbiol 113(3):315–320. https://doi.org/10.1016/j.ijfoodmicro.2006.08.008
135. Tkachenko O, Shepard J, Aris VM, Joy A, Bello A, Londono I, Marku J, Soteropoulos P, Peteroy-Kelly MA (2007) A triclosan-ciprofloxacin cross-resistant mutant strain of Staphylococcus aureus displays an alteration in the expression of several cell membrane structural and functional genes. Res Microbiol 158(8–9):651–658. https://doi.org/10.1016/j.resmic.2007.09.003
136. United States Environmental Protection Agency (2008) Reregistration Eligibility Decision for Triclosan. https://archive.epa.gov/pesticides/reregistration/web/pdf/2340red.pdf
137. Valentine BK, Dew W, Yu A, Weese JS (2012) In vitro evaluation of topical biocide and antimicrobial susceptibility of Staphylococcus pseudintermedius from dogs. Vet Dermatol 23(6):e493–e495. https://doi.org/10.1111/j.1365-3164.2012.01095.x
138. Wand ME, Baker KS, Benthall G, McGregor H, McCowen JW, Deheer-Graham A, Sutton JM (2015) Characterization of pre-antibiotic era Klebsiella pneumoniae isolates with respect to antibiotic/disinfectant susceptibility and virulence in Galleria mellonella. Antimicrob Agents Chemother 59(7):3966–3972. https://doi.org/10.1128/aac.05009-14
139. Webber MA, Buckner MMC, Redgrave LS, Ifill G, Mitchenall LA, Webb C, Iddles R, Maxwell A, Piddock LJV (2017) Quinolone-resistant gyrase mutants demonstrate decreased susceptibility to triclosan. J Antimicrob Chemother 72(10):2755–2763. https://doi.org/10.1093/jac/dkx201
140. Weber DJ, Rutala WA (2013) Self-disinfecting surfaces: review of current methodologies and future prospects. Am J Infect Control 41(5 Suppl):S31–S35. https://doi.org/10.1016/j.ajic.2012.12.005
141. Wesgate R, Grasha P, Maillard JY (2016) Use of a predictive protocol to measure the antimicrobial resistance risks associated with biocidal product usage. Am J Infect Control 44 (4):458–464. https://doi.org/10.1016/j.ajic.2015.11.009
142. WHO (2009) WHO guidelines on hand hygiene in health care. First Global Patient Safety Challenge Clean Care is Safer Care, WHO, Geneva

143. Wisplinghoff H, Schmitt R, Wohrmann A, Stefanik D, Seifert H (2007) Resistance to disinfectants in epidemiologically defined clinical isolates of Acinetobacter baumannii. J Hosp Infect 66(2):174–181. https://doi.org/10.1016/j.jhin.2007.02.016
144. Witney AA, Gould KA, Pope CF, Bolt F, Stoker NG, Cubbon MD, Bradley CR, Fraise A, Breathnach AS, Butcher PD, Planche TD, Hinds J (2014) Genome sequencing and characterization of an extensively drug-resistant sequence type 111 serotype O12 hospital outbreak strain of Pseudomonas aeruginosa. Clin Microbiol Infect 20(10):O609–O618. https://doi.org/10.1111/1469-0691.12528
145. Yigit N, Aktas E, Ayyildiz A (2008) Antifungal activity of toothpastes against oral Candida isolates. Journal de mycologie medicale 18(3):141–146. https://doi.org/10.1016/j.mycmed.2008.06.003
146. Yu BJ, Kim JA, Ju HM, Choi SK, Hwang SJ, Park S, Kim E, Pan JG (2012) Genome-wide enrichment screening reveals multiple targets and resistance genes for triclosan in Escherichia coli. J Microbiol (Seoul, Korea) 50(5):785–791. https://doi.org/10.1007/s12275-012-2439-0
147. Zastrow K-D, Kramer A, Bauch B (2001) Konzept zur Dekontamination von MRSA-Patienten. Hyg + Med 26(9):344–348

Benzalkonium Chloride 10

10.1 Chemical Characterization

Benzalkonium chloride (BAC) is a type of cationic surfactant [101]. It is a mixture of alkyl benzyl dimethyl ammonium chlorides, in which the alkyl group has various even-numbered alkyl chain lengths. BAC comprises of 24 compounds that are structurally similar quaternary ammonium compounds ("quats"). They are characterized by having a positively charged nitrogen covalently bonded to three alkyl group substituents and a benzyl substituent [388]. In finished form, these quats are salts with the positively charged nitrogen (cation) balanced by a negatively charged molecule (anion). The most common anion for the quats in this cluster is chloride.

The basic chemical information of three typical mixtures described as benzalkonium chloride is summarized in Table 10.1.

In the majority of studies, the CAS number is not mentioned when BAC was used as a biocidal agent. That is why the specific chemical identity of the substance under investigation is not always clear. Nevertheless, data on BAC were reviewed and summarized because it was considered unlikely that a specific mixture of alkyl benzyl dimethyl ammonium chlorides would yield results that are not typical for the entire group of mixtures. This possible limitation should be kept in mind for the entire chapter.

10.2 Types of Application

BAC is used for a variety of different applications. In China, BAC is used for hand scrubs, skin disinfection and mucosa and wound disinfection (500–1,000 mg/l; 3–5 min) and surface disinfection (1,000–2,000 mg/l; 30 min) [220]. In Japan, it is

G. Kampf, *Antiseptic Stewardship*, https://doi.org/10.1007/978-3-319-98785-9_10

Table 10.1 Basic chemical information on typical mixtures described as benzalkonium chloride (BAC) [153]

Type of BAC	C12–18 mixture	C12–16 mixture	C12–14 mixture
Components of mixture with examples of composition	C12 (61%) C14 (23%) C16 (11%) C18 (5%)	C12 (39–76%) C14 (20–52%) C16 (<12%)	C12 (70%) C14 (30%)
CAS number	68391-01-5	68424-85-1	85409-22-9
Synonyms	N-Alkyl dimethyl benzyl ammonium chloride (C12-C18)	Benzyl-C12-16-alkyldimethyl chlorides, alkyl (C12-16) dimethyl benzyl ammonium chloride	Benzyl-C12-14-alkyldimethyl ammonium chlorides, C12-14 ADBAC

used for hand scrubbing (500–1,000 mg/l), as a surgical site antiseptic (100–500 mg/l), as a mucosa and wound antiseptic (100–500 mg/l), as a surface disinfectant (500–2,000 mg/l) and as an instrument disinfectant (500–1,000 mg/l) [286]. In Europe, it can be found in many surface disinfectants and instrument disinfectants or disinfectant cleaners. It is occasionally used in alcohol-based skin or hand disinfectants. Its use in the veterinary field includes environmental treatment at 60–120 mg/l (120 min), treatment of surgical sites at 100–500 mg/l, hand scrubbing at 500–1,000 mg/l (1–3 min) and skin and wound treatment at 1,000–2,000 mg/l (5–10 min) [74]. As a wood preservative, it is used at a final concentration between 4,000 and 20,000 mg/l [153].

The use of BAC in the USA is broad. It is used in agricultural, food handling and medical settings. Examples of registered uses for BAC in these settings include application to indoor and outdoor hard surfaces (e.g. walls, floors, tables, toilets and fixtures), eating utensils, laundry, carpets, agricultural tools and vehicles, egg shells, hands and gloves, shoes, milking equipment, and udders, humidifiers, recreational vehicle tanks, medical instruments, human remains, ultrasonic tanks, reverse osmosis units and water storage tanks. There are also BAC end-user products that are used in residential and commercial swimming pools, in aquatic areas such as decorative ponds, decorative fountains, and agricultural watering lines, and in industrial process and water systems such as once-through and re-circulating cooling water systems, cooling towers, evaporative condensers, pasteurizers, drilling mud, packer fluids, oil well injection and wastewater systems, and in pulp and paper products, water and chemicals. Additionally, BAC end-user products are used for wood preservation [388].

10.2.1 European Chemicals Agency (European Union)

The C12-C18 BAC, the C12-C16 BAC and the C12-C14 BAC are still under review as active biocidal substances (June 2018) for product types 1 (human hygiene), 2 (disinfectants and algaecides not intended for direct application to humans or animals), 3 (veterinary hygiene), 4 (food and feed area), 10 (construction material preservatives), 11 (preservatives for liquid-cooling and processing systems), 12 (slimicides) and 22 (embalming and taxidermist fluids). So far, only the C12-C16 BAC has been approved for product type 8 (wood preservatives) [20].

10.2.2 Environmental Protection Agency (USA)

In the USA, the first product containing BAC was registered in 1947. The oldest active product containing BAC was registered in 1956 [388]. In 2006, the active agent was last reviewed. As a result of this review, EPA has determined that BAC-containing products are eligible for re-registration, provided that risk mitigation measures are adopted and labels are amended accordingly [388].

10.2.3 Food and Drug Administration (USA)

In 2015, BAC was eligible for five types of application in health care: patient preoperative skin preparation, healthcare personnel hand wash, healthcare personnel hand rub, surgical hand scrub and surgical hand rub [82]. It is classified in category IIISE indicating that available data are insufficient to classify BAC as safe and effective, and further testing is required [82]. The main aspects on safety are oral carcinogenicity and resistance potential [82].

10.2.4 Overall Environmental Impact

In 1997, it was described that hospitals alone emit between 4.5 and 362 g BAC per bed and year [195]. Recent data from Germany indicate that on average 0.042 g BAC per bed and day are emitted from hospitals (equivalent to 15.33 g per bed and year) [379]. In 2006, the EPA concluded that BAC is hydrolytically stable under abiotic and buffered conditions over the pH 5–9 range. The calculated half-lives for BAC were 379 days at pH 9 and 150–183 days at pH 5 and pH 7. BAC is also stable to photodegradation in pH 7 buffered aqueous solutions [388]. Based on a biodegradation study, BAC readily degrades into 60% carbon dioxide in 13 days. BAC is immobile in soil. The available soil mobility study shows that BAC has a strong tendency to bind to sediment and soil. Due to its strong adsorption to soils [176], BAC was not expected to contaminate surface and groundwaters [388]. Nevertheless, cationic surfactants are widely detected in the environment, and an antagonistic effect of interactions on biodegradation of two widely used types of benzalkonium chlorides has been described suggesting further investigation on the degradation of mixture of QACs in wastewater effluents and biosolids [177]. Samples from 11 Swedish sewage treatment plants revealed that quaternary ammonium compounds were the most abundant substances in the particulate phases with levels up to 370 μg per g. The majority of the QACs were mostly found in the particle phase of the incoming water, but all could be detected in the water phase as well [305]. Overall, wastewater, wastewater treatment plants, wastewater sludge, sludge-applied soil, surface waters and aquatic sediments are environments where BAC resistance evolves and proliferates. BAC-resistant bacteria may be transferred from outdoor environments to indoors such as homes and healthcare facilities [376].

10.3 Spectrum of Antimicrobial Activity

10.3.1 Bactericidal Activity

10.3.1.1 Bacteriostatic Activity (MIC Values)

The MIC values for benzalkonium chloride obtained with different bacterial species are summarized in Table 10.2. The most susceptible bacterial species were *Lactobacillus* spp. (MICs up to 5 mg/l), *Campylobacter* spp. (up to 32 mg/l) and coagulase-negative Staphylococcus spp. (up to 64 mg/l). *S. aureus* was mostly below the recommended epidemiological cut-off value (16 mg/l) [271] although few MIC values up to 1,250 mg/l have been reported. *Enterococcus* spp. were also often below the recommended epidemiological cut-off value of 8 mg/l [271] but were also described with MIC values up to 250 mg/l. *K. pneumoniae* was among the most susceptible Gram-negative species with MIC values up to 64 mg/l which is just above the recommended epidemiological cut-off value (32 mg/l). *E. coli* (up to 156 mg/l), *C. freundii* (up to 190 mg/l), *Acinetobacter* spp. (up to 200 mg/l), *Salmonella* spp. (up to 256 mg/l), *A. xylosoxidans* and *B. cepacia* (up to 500 mg/l), *Pseudomonas* spp. (up to 5,000 mg/l), *B. cereus* and *E. meningoseptica* (up to 7,800 mg/l) and *A. hydrophila* (up to 31,300 mg/l) were often susceptible although some isolates were above the recommended epidemiological cut-off value (64 mg/l for *E. coli*, 128 mg/l for *Salmonella* spp.). The highest MIC values were described with *A. hydrophila* (up to 31,300 mg/l), *B. cereus* and *E. meningoseptica* (up to 7,800 mg/l) *P. aeruginosa* (up to 5,000 mg/l), *L. monocytogenes* (up to 625 mg/l; proposed breakpoint: >7.5 mg/l) [361], *E. cloacae* (up to 512 mg/l but mostly below the recommended epidemiological cut-off value of 32 mg/l), *A. xylosoxidans* and *B. cepacia* (up to 500 mg/l) and *P. mirabilis* (up to 400 mg/l). Overall, it is important to know that with BAC the result of MIC testing depends to some extent on the media composition and plate material showing the need to standardize biocide susceptibility testing [31].

Only few data are available to describe the MIC values for bacterial species that were obtained from biofilms. They are summarized in Table 10.3. Overall, the MIC values were higher compared those obtained with planktonic cells (Table 10.2).

Table 10.2 MIC values of various bacterial species to benzalkonium chloride

Species	Strains/isolates	MIC value (mg/l)	References
A. baumannii	3 isolates from domestic surfaces	1.7–3.4	[65]
A. baumannii	47 clinical isolates	4–32	[215]
A. baumannii	51 carbapenem-resistant clinical isolates	4–64	[222]
A. baumannii	JCM 6841	21	[417]
A. baumannii	2 blood culture isolates from oncology patients	50–100	[145]
A. johnsonii	NCIMB 12460	30–40	[206]
	Triclosan-tolerant industrial strain	120–130	
Acinetobacter spp.[a]	283 clinical isolates (273 *A. calcoaceticus-A. baumannii complex*, 7 *A. lwoffii*, 3 *A. junii*)	5–50	[171]
Acinetobacter spp.[a]	283 clinical isolates (273 *A. calcoaceticus-A. baumannii complex*, 7 *A. lwoffii*, 3 *A. junii*)	10–200	[172]
A. xylosoxidans	Domestic drain biofilm isolate MBRG 4.31	31.2	[266]
A. xylosoxidans	2 clinical isolates	63–500	[279]
A. hydrophila	Domestic drain biofilm isolate MBRG 4.3	31.2	[266]
A. hydrophilia	Blood culture isolate from an oncology patient	100	[145]
A. hydrophila	Domestic drain biofilm isolate	31,300	[108]
A. jandaei	Domestic drain biofilm isolate MBRG 9.11	31.2	[266]
A. proteolyticus	Domestic drain biofilm isolate MBRG 9.12	3.9	[266]
Alcaligenes spp.	2 blood culture isolates from oncology patients	25–75	[145]
B. cereus	Domestic drain biofilm isolate MBRG 4.21	6.5	[266]
B. cereus	Domestic drain biofilm isolate	7,800	[108]
B. adolescentis	4 isolates from faeces of healthy humans	4–16	[95]
B. animalis subsp. lactis	8 isolates from faeces of healthy humans	4–64	[95]
B. bifidum	31 isolates from faeces of healthy humans	2–128	[95]
B. breve	5 isolates from faeces of healthy humans	4–16	[95]
B. catenulatum	1 isolate from faeces of a healthy human	4	[95]
B. infantis	2 isolates from faeces of healthy humans	4–64	[95]
B. longum	25 isolates from faeces of healthy humans	4–128	[95]
B. pseudocatenulatum	15 isolates from faeces of healthy humans	16–64	[95]
B. pseudolongum	1 isolate from faeces of a healthy human	128	[95]
B. thermoacidophilum	6 isolates from faeces of healthy humans	4–64	[95]
B. suis	1 isolate from faeces of a healthy human	8	[95]

(continued)

Table 10.2 (continued)

Species	Strains/isolates	MIC value (mg/l)	References
B. cepacia complex	38 clinical, non-clinical and environmental strains	50–400	[330]
B. cepacia	JCM 5964	213	[417]
B. cepacia complex	*B. lata* strain 383	500	[181]
B. cepacia	1 wash basin isolate	63–500	[279]
C. coli	8 strains from poultry	0.06–1	[245]
	6 strains from humans	0.25–1	
	4 strains from pigs	0.25–0.5	
	1 strain from water	0.5	
C. jejuni	5 strains from water	0.06–2	[245]
	3 strains from poultry	0.5–1	
	5 strains from humans	1–4	
C. jejuni	81 isolates from poultry slaughterhouses	<32	[312]
C. acidivorans	Blood culture isolate from an oncology patient	100	[145]
C. pseudogenitalum	Human skin isolate MBRG 9.24	15.6	[266]
C. renale group	Human skin isolate MBRG 9.13	7.8	[266]
C. indologenes	Domestic drain biofilm isolate MBRG 9.15	31.2	[266]
C. indologenes	Blood culture isolate from an oncology patient	100	[145]
C. meningosepticum	Blood culture isolate from an oncology patient	75	[145]
Chrysobacterium spp.	Domestic drain biofilm isolate MBRG 9.17	31.2	[266]
C. luteola	Blood culture isolate from an oncology patient	100	[145]
C. freundii	3 isolates from domestic surfaces	13.6	[65]
C. freundii	NCIMB 11490	120–130	[206]
	Triclosan-tolerant industrial strain	180–190	
Citrobacter spp.	Domestic drain biofilm isolate MBRG 9.18	26	[266]
E. cloacae	Strain 17/97 (clinical isolate)	0.012–0.024	[238]
E. cloacae	5 isolates from domestic surfaces	6.8–13.6	[65]
E. cloacae	43 ESBL patient isolates (haematology ward)	64–512	[57]
Enterobacter spp.[a]	54 worldwide strains from hospital- and community-acquired infections	4–64	[271]
E. meningoseptica	Domestic drain biofilm isolate	7,800	[108]
E. casseliflavus	5 isolates from dust samples collected in breeding pig facilities	2–4	[35]

(continued)

Table 10.2 (continued)

Species	Strains/isolates	MIC value (mg/l)	References
E. casseliflavus	1 dairy isolate	4	[34]
E. durans	5 isolates from clinical, veterinary and dairy sources	4	[34]
E. faecalis	56 worldwide strains from hospital- and community-acquired infections	0.5–16	[271]
E. faecalis	53 isolates from dust samples collected in breeding pig facilities	2–4	[35]
E. faecalis	46 isolates from clinical, veterinary and dairy sources	2–4	[34]
E. faecalis	9 isolates from swine meat production	2–4	[325]
E. faecalis	52 isolates from livestock	2–8	[1]
E. faecalis	ATCC 29212	8	[417]
E. faecalis	1 cariogenic strain isolated from children	>256	[188]
E. faecium	53 worldwide strains from hospital- and community-acquired infections	0.5–16	[271]
E. faecium	22 isolates from dust samples collected in breeding pig facilities	2–4	[35]
E. faecium	12 isolates from swine meat production	2–4	[325]
E. faecium	78 isolates from livestock	2–16	[1]
E. faecium	17 isolates from clinical, veterinary and dairy sources	4–8	[34]
E. gallinarum	2 isolates from dust samples collected in breeding pig facilities	2–4	[35]
E. hirae	39 isolates from dust samples collected in breeding pig facilities	1–4	[35]
E. hirae	3 isolates from dairy and clinical sources	2–4	[34]
E. hirae	ATCC 10541	8	[417]
E. raffinosus	2 isolates from dust samples collected in breeding pig facilities	1–2	[35]
E. saccharolyticus	Domestic drain biofilm isolate MBRG 9.16	31.2	[266]
E. solitarius	1 veterinary isolate	2	[34]
Enterococcus spp.[a]	122 strains (*E. faecalis*, *E. faecium*) from different traditional fermented foods	<0.1	[204]
Enterococcus spp.[a]	25 isolates from dust samples collected in breeding pig facilities	<0.25–4	[35]
Enterococcus spp.[a]	7 isolates from domestic surfaces, 3 isolates from household individuals	0.4–0.9	[65]
Enterococcus spp.[a]	4 VRE strains	5–6	[366]
Enterococcus spp.[a]	69 clinical isolates	8–16	[151]
Enterococcus spp.[a]	272 strains from various sources	25–250	[391]

(continued)

Table 10.2 (continued)

Species	Strains/isolates	MIC value (mg/l)	References
E. rhusiopathiae	60 isolates from various sources	0.15	[105]
E. coli	306 worldwide strains from hospital- and community-acquired infections	2–128	[271]
E. coli	12 strains	2.5–150	[333]
E. coli	5 isolates from domestic surfaces	3.4–6.8	[65]
E. coli	NCTC 10418	4–16[b]	[31]
E. coli	140 human isolates, 34 isolates from healthy chicken (all ESBL)	4–32	[84]
E. coli	153 blood culture isolates	4–64	[41]
E. coli	179 isolates from retail meats	4–64	[157]
E. coli	13 bovine and 7 equine strains	10	[348]
E. coli	ATCC 25922	11	[417]
E. coli	Strain HEC30	16	[77]
E. coli	ATCC 25922 and 9 avian and porcine strains	16–32	[360]
E. coli	202 isolates from livestock	16–128	[1]
E. coli	74 isolates from food contact surfaces	19.5–39.1	[154]
E. coli	IFO 14237	30	[421]
E. coli	27 isolates from hen egg shells	50–75	[124]
E. coli	ATCC 25922	156	[85]
Eubacterium spp.	Domestic drain biofilm isolate MBRG 4.14	31.2	[266]
F. oryzihabitans	Blood culture isolate from an oncology patient	75	[145]
G. haemolysans	1 cariogenic strain isolated from children	8	[188]
H. gallinarum	Domestic drain biofilm isolate MBRG 4.27	31.2	[266]
K. oxytoca	2 isolates from domestic surfaces	6.8	[65]
K. pneumoniae	37 isolates predominately from a variety of human infections pre-1949 ("Murray isolates") and 39 "modern strains" (2007–2012)	1–16 (old isolates)	[401]
		8–32 (modern isolates)	
K. pneumoniae	60 worldwide strains from hospital- and community-acquired infections	4–64	[271]
K. pneumoniae	27 carbapenem-resistant clinical isolates	4–64	[131]
K. pneumoniae	Strain 39.11	16	[77]
Klebsiella spp.[a]	30 isolates from food contact surfaces (16 *K. pneumoniae*, 14 *K. oxytoca*)	19.5–78.1	[154]
L. acidophilus	4 strains from different origins	1–2	[15]
L. amylovorus	7 strains from different origins	0.5–1	[15]

(continued)

Table 10.2 (continued)

Species	Strains/isolates	MIC value (mg/l)	References
L. brevis	13 strains from different origins	0.5–2	[15]
L. bulgaricus	6 strains from different origins	1–4	[15]
L. coryniformis	3 strains from different origins	1–2	[15]
L. fermentum	4 strains from different origins	0.25–2	[15]
L. garvieae	42 isolates from different origins	1–4	[15]
L. helveticus	39 strains from different origins	0.25–2	[15]
L. paracasei	75 strains from different origins	0.5–4	[15]
L. pentosus	60 strains from naturally fermented Aloreña green table olives	0.01–5.0	[50]
L. plantarum	43 strains from different origins	0.5–4	[15]
L. reuteri	42 strains from different origins	0.06–2	[15]
L. rhamnosus	9 strains from different origins	1–4	[15]
L. lactis	3 strains	1	[386]
L. pseudomesenteroides	13 strains from naturally fermented Aloreña green table olives	0.01–5.0	[50]
Leuconostoc spp.[a]	3 strains	0.5–2	[386]
L. monocytogenes	96 strains from frozen food	1.25–5	[315]
L. monocytogenes	14 strains (3 from blood, 6 from food, 5 from water)	1.25–10	[313]
L. monocytogenes	254 isolates from seafood products	1.9–15	[361]
L. monocytogenes	15 strains from pork processing plant	2.5–20	[304]
L. monocytogenes	LMG 16779	4	[297]
L. monocytogenes	114 isolates from food products	4–32	[2]
L. monocytogenes	142 strains from food processing	5–30	[91]
L. monocytogenes	71 isolates from commonly consumed food items	6–20	[414]
L. monocytogenes	9 strains from a cheese processing facility	7.8–31.3	[331]
L. monocytogenes	ATCC 7644	625	[85]
Listeria spp.[a]	127 isolates from food and food processing environments (75 *L. innocua*, 49 *L. welshimeri*, 2 *L. seeligeri* and 1 *L. grayi*)	2.5–40	[184]
M. phyllosphaerae	Domestic drain biofilm isolate MBRG 4.30	15.6	[266]
M. luteus	Human skin isolate MBRG 9.25	0.45	[266]
O. anthropi	Blood culture isolate from an oncology patient	10	[145]
P. aeruginosa	8 isolates from domestic surfaces	3.4–27.3	[65]
P. aeruginosa	ATCC 15442	12–60	[129]
P. aeruginosa	28 multidrug-resistant isolates from burn patients	30–1,000	[119]

(continued)

Table 10.2 (continued)

Species	Strains/isolates	MIC value (mg/l)	References
P. aeruginosa	NCTC 13359	32–128[b]	[31]
P. aeruginosa	175 isolates from veterinary sources	32–256	[22]
P. aeruginosa	91 clinical isolates, 37 hospital environmental isolates	$\leq$38.4	[301]
P. aeruginosa	ATCC 15442	60	[201]
P. aeruginosa	55 strains from various sources (20 clinical isolates, 19 industrial environmental isolates, 16 culture collection strains)	78–625	[198]
P. aeruginosa	ATCC 27853	128	[417]
P. aeruginosa	178 clinical isolates	312–5,000	[197]
P. alkylphenolia	2 isolates from meat chain production	0.25–2.5	[203]
P. fluorescens	3 isolates from meat chain production	0.025–2.5	[203]
P. fluorescens	5 isolates from chicken carcasses	70–200	[201]
P. fluorescens	9 of 14 strains from chill stored poultry carcasses	$\geq$200	[368]
P. fragi	4 isolates from meat chain production	0.0025–0.25	[203]
P. fragi	3 isolates from chicken carcasses	50–200	[201]
P. fragi	2 of 14 strains from chill stored poultry carcasses	$\geq$200	[368]
P. lundensis	34 isolates from meat chain production	0.0025–2.5	[203]
P. lundensis	4 isolates from chicken carcasses	50–200	[201]
P. lundensis	2 of 14 strains from chill stored poultry carcasses	$\geq$200	[368]
P. nitroreductans	Domestic drain biofilm isolate MBRG 4.6	31.2	[266]
Pseudomonas spp.	Domestic drain biofilm isolate MBRG 9.14	31.2	[266]
Pseudoxanthomonas spp.	Domestic drain biofilm isolate MBRG 9.20	31.2	[266]
P. mirabilis	52 isolates from cooked meat products	4–32	[159]
P. mirabilis	11 clinical strains	50–400	[364]
P. putida	9 isolates from meat chain production	0.0025–2.5	[203]
Ralstonia spp.	Domestic drain biofilm isolate MBRG 4.13	7.8	[266]
S. enterica	122 poultry isolates, 135 swine isolates	4–256	[63]
S. enterica	122 isolates from poultry and swine	8–256	[62]
S. Typhimurium	1 poultry isolate	8	[48]
Salmonella spp.[a]	112 isolates from meat	2.5–50	[115]
Salmonella spp.[a]	12 strains from various sources	4–32	[259]
Salmonella spp.[a]	375 avian isolates	8–128	[321]

(continued)

Table 10.2 (continued)

Species	Strains/isolates	MIC value (mg/l)	References
Salmonella spp.[a]	901 worldwide strains from hospital- and community-acquired infections	16–128	[271]
Salmonella spp.[a]	195 isolates from chicken and egg production	32–256	[223]
Salmonella spp.[a]	156 isolates from livestock	64–256	[1]
S. marcescens	18 clinical strains	87–139	[116]
S. multivorum	Domestic drain biofilm isolate MBRG 9.19	31.2	[266]
S. spiritivorum	Blood culture isolate from an oncology patient	10	[145]
S. aureus	7 isolates from domestic surfaces, 3 isolates from household individuals	<0.004–0.128	[65]
S. aureus	169 clinical MRSA isolates from community-acquired infections	0.1–5	[228]
S. aureus	ATCC 6538	0.25–2[b]	[31]
S. aureus	436 isolates without increased expression	0.3–2.5	[186]
	253 isolates with increased expression of ≥ 1 multidrug resistance efflux pump gene	1.25–5	
S. aureus	1,635 worldwide strains from hospital- and community-acquired infections	0.5–16	[271]
S. aureus	1,602 isolates from hospital- and community-acquired infections	0.5–16	[113]
S. aureus	54 clinical isolates	0.5–64	[370]
S. aureus	114 effluxing bloodstream isolates	0.6–5.0	[80]
S. aureus	65 clinical MRSA isolates	0.8–12.5	[343]
S. aureus	43 isolates from livestock	1–8	[1]
S. aureus	11 isolates from community environmental samples	1–32	[139]
S. aureus	60 clinical MRSA isolates	1–64	[370]
S. aureus	4 strains (CECT976, RN4220, SA1199b, XU212)	1.5–4	[123]
S. aureus	NCTC 6571, NCTC 83254 and 17 clinical MRSA isolates	1.5–4	[366]
S. aureus	11 strains with various resistance genes	<2–6	[217]
S. aureus	ATCC 6538	2	[417]
S. aureus	40 MRSA isolates from nursery pigs	2–5	[356]
S. aureus	28 isolates from automated teller machines	2–8	[425]
S. aureus	46 isolates from auricular infections	2–32	[428]
S. aureus	9 isolates from the oral cavity of children	2–128	[427]
S. aureus	22 isolates from food contact surfaces	2.44–9.77	[154]

(continued)

Table 10.2 (continued)

Species	Strains/isolates	MIC value (mg/l)	References
S. aureus	1 QAC-resistant isolate	2.5	[28]
S. aureus	ATCC 25923	4	[297]
S. aureus	100 ST9 MRSA strains from porcine carcasses	4–5	[412]
S. aureus	24 strains with various resistance genes (qacA, qacB, qacC, qacG or norA)	4–16	[242]
S. aureus	25 MSSA and 16 MRSA isolates from faecal samples	4–32	[8]
S. aureus	8 cariogenic strains isolated from children	8–128	[188]
S. aureus	ATCC 25923	16	[427]
S. aureus	ATCC 700698 (MRSA)	16	[417]
S. aureus	ATCC 25923	1,250	[85]
S. capitis	Human skin isolate MBRG 9.34	0.45	[266]
S. capitis	4 isolates from auricular infections	8–32	[428]
S. caprae	Human skin isolate MBRG 9.30	0.45	[266]
S. caprae	1 QAC-resistant isolate	3.5	[28]
S. caprae	1 isolate from community environmental samples	8	[139]
S. chromogene	3 isolates from auricular infections	2–16	[428]
S. cohnii	Human skin isolate MBRG 9.31	0.45	[266]
S. cohnii subsp. cohnii	1 methicillin-resistant isolate from a horse	2	[74]
S. cohnii	1 QAC-resistant isolate	5	[28]
S. delphini	1 QAC-resistant isolate	3.5	[28]
S. epidermidis	Human skin isolate M 9.33	0.45	[266]
S. epidermidis	65 isolates from food contact surfaces	1.22–9.77	[154]
S. epidermidis	32 isolates from auricular infections	2–32	[428]
S. epidermidis	2 QAC-resistant isolates	2.5–3.5	[28]
S. epidermidis	ATCC 12228	4	[417]
S. epidermidis	28 isolates from community environmental samples	4–32	[139]
S. equorum	2 isolates from auricular infections	8–16	[428]
S. fleurettii	1 methicillin-resistant isolate from a horse	0.5	[74]
S. haemolyticus	Human skin isolate MBRG 9.35	0.45	[266]
S. haemolyticus	14 QAC-resistant isolates	1–3.5	[28]
S. haemolyticus	1 methicillin-resistant isolate from a horse	2	[74]
S. haemolyticus	21 clinical isolates	4–8	[69]
S. haemolyticus	9 isolates from auricular infections	4–16	[428]
S. haemolyticus	9 isolates from community environmental samples	4–32	[139]
S. hominis	Human skin isolate MBRG 9.37	0.45	[266]

(continued)

Table 10.2 (continued)

Species	Strains/isolates	MIC value (mg/l)	References
S. hominis	2 QAC-resistant isolates	2–3	[28]
S. hominis	10 isolates from auricular infections	2–16	[428]
S. hominis	6 isolates from community environmental samples	4–32	[139]
S. hyicus	38 isolates from livestock	0.5–2	[1]
S. kloosii	Human skin isolate MBRG 9.28	1.3	[266]
S. lentus	1 methicillin-resistant isolate from a horse	0.5	[74]
S. lugdunensis	Human skin isolate MBRG 9.36	1.3	[266]
S. lugdunensis	11 clinical strains	7.8–62.5	[107]
S. pasteuri	1 QAC-resistant isolate	3	[28]
S. pseudointermedius	20 MRSP and 20 MSSP from dogs (35) and cats (5)	0.5–4	[72]
S. pseudointermedius	43 MSSP and 57 MRSP isolates from canine pyoderma	1–4	[278]
S. pseudointermedius	25 MSSP and 25 MRSP from dogs with skin and soft tissue infections	2–16	[390]
S. saprophyticus	Human skin isolate MBRG 9.29	0.45	[266]
S. saprophyticus	2 QAC-resistant isolates	2.5–4.5	[28]
S. saprophyticus	2 isolates from community environmental samples	4–16	[139]
S. schleiferi	12 clinical strains	15.6–31.2	[107]
S. sciuri	8 methicillin-resistant isolates from horses	0.5–2	[74]
S. simulans	6 isolates from auricular infections	4–32	[428]
S. simulans	1 isolate from community environmental samples	32	[139]
S. warneri	Human skin isolate MBRG 9.27	0.45	[266]
S. warneri	5 isolates from auricular infections	2–8	[428]
S. warneri	13 QAC-resistant isolates	2.5–3.5	[28]
S. warneri	5 isolates from community environmental samples	4–32	[139]
Staphyloccocus spp.[a]	78 coagulase-negative isolates from automated teller machines	0.25–16	[425]
Staphyloccocus spp.[a]	51 coagulase-negative clinical isolates	0.5–64	[370]
Staphyloccocus spp.[a]	181 isolates from meat and poultry plants	1–9	[367]
Staphyloccocus spp.[a]	69 clinical isolates (27 MRCNS, 19 MSSA, 13 MSCNS, 10 MRSA)	1–16	[151]
Staphyloccocus spp.[a]	4 QAC-resistant isolates	1.5–2.5	[28]
S. maltophilia	Domestic drain biofilm isolate MBRG 9.13	31.2	[266]
S. maltophilia	2 blood culture isolates from oncology patients	100	[145]

(continued)

Table 10.2 (continued)

Species	Strains/isolates	MIC value (mg/l)	References
S. agalactiae	52 strains from vaginal swabs of pregnant women	0.4–6.3	[272]
S. anginosus	1 cariogenic strain isolated from children	16	[188]
S. constellatus	1 cariogenic strain isolated from children	16	[188]
S. mitis	1 cariogenic strain isolated from children	8	[188]
S. mutans	1 cariogenic strain isolated from children	8	[188]
S. oralis	1 cariogenic strain isolated from children	>256	[188]
S. salivarus	44 strains from different origins	1–4	[15]
S. salivarius	1 cariogenic strain isolated from children	16	[188]
S. thermophilus	135 strains from different origins	0.125–4	[15]
Various species[a]	80 Gram-positive isolates from cow milk and goat cheese production	2.5–50	[104]
Various species[a]	378 isolates from organic food	10–100	[102]
Various species[a]	40 Gram-negative isolates from cow milk and goat cheese production	25–250	[104]

[a]No number of isolates per species; [b]Depending on the media composition and plate material

Table 10.3 MIC values obtained with biofilm grown cells of various bacterial species to benzalkonium chloride

Species	Number of strains/isolates	MIC value (mg/l)	References
E. coli	74 isolates from food contact surfaces	39.1–156.3	[154]
Klebsiella spp.	30 isolates from food contact surfaces	39.1–156.3	[154]
S. aureus	22 isolates from food contact surfaces	39.1–78.1	[154]
S. epidermidis	65 isolates from food contact surfaces	19.5–156.3	[154]

10.3.1.2 Bactericidal Activity (Suspension Tests)

The spectrum of the bactericidal activity of BAC is summarized in Table 10.4. Against Gram-positive bacterial species BAC is usually effective within 5 min at 0.02%, whereas some Gram-negative species may be more resistant to BAC, e.g. *S. marcescens* from an antiseptic footbath. BAC at 1% was found to be bactericidal in 5 min against various species. A product based on 0.00995% BAC with additional 0.00249% an N-(3-aminopropyl)-N-dodecylpropane-1,3-diamine was also effective against various multiresistant Gram-negative isolates within 1 h. A low concentration of 0.0014% BAC was effective against 9 biofilm-forming toilet bowl isolates within 5 h but it not sufficiently effective against 16 other species including *Pseudomonas* spp. and *S. maltophilia*.

Table 10.4 Bactericidal activity of benzalkonium chloride in suspension tests

Species	Strain/isolate	Exposure time	Concentration	log_{10} reduction	References
A. delafieldii	2 toilet bowl biofilm isolates	1 h	0.0014% (S)	0.9–1.5	[270]
		5 h		5.2–6.3	
A. baumannii	ATCC 19606 and 2 clinical isolates: antibiotic-susceptible and 4MRGN OXA-23	1 h	0.00995%[a] (P)	≥5.1	[320]
A. xylosoxidans	ATCC 27061	1 h	0.00995%[a] (P)	≥5.1	[320]
Blastomonas spp.	Toilet bowl biofilm isolate	1 h	0.0014% (S)	2.1	[270]
		5 h		5.2	
B. sanguinis	Toilet bowl biofilm isolate	1 h	0.0014% (S)	0.2	[270]
		5 h		1.2	
C. jejuni	ATCC BAA-1062, ATCC 33560 and 2 field strains	1 min	0.02% (S)	4.9–>6.0	[133]
Campylobacter spp.	46 strains from broilers and pigs	5 min	1% (S)	≥5.0	[16]
E. avium	Toilet bowl biofilm isolate	1 h	0.0014% (S)	1.6	[270]
		5 h		2.0	
E. coli	NCTC 86	30 s	1% (S)	≥5.0	[237]
E. coli	ATCC 11229	30 min	0.01% (S)	≥3.0	[276]
H. parasuis	2 strains (serovars 1 and 5)	1 min	0.02% (S)	>6.0 4.5–4.9[b]	[326]
H. pylori	NCTC 11637, NCTC 11916 and 7 clinical isolates	30 s	0.1% (S)	>5.0	[7]
			0.025% (S)		
H. flavidus	Toilet bowl biofilm isolate	1 h	0.0014% (S)	0.3	[270]
		5 h		5.0	
K. pneumoniae	ATCC 10031 and 4 clinical isolates: antibiotic-susceptible, 3MRGN, 4MRGN OXA-48 and 4MRGN KPC-2	1 h	0.00995%[a] (P)	≥5.1	[320]
K. oxytoca	ATCC 700324 and 4 clinical isolates: antibiotic-susceptible, 3MRGN, MRGN OXA-48 and 4MRGN KPC-2	1 h	0.00995%[a] (P)	≥5.1	[320]
L. monocytogenes	20 environmental and food isolates	5 min	0.0035–0.013% (P)	≥5.0	[76]
L. monocytogenes	10 clinical and 10 food strains	5 min	0.001% (S)	0.0–6.5[a]	[374]
L. brunescens	Toilet bowl biofilm isolate	1 h	0.0014% (S)	1.7	[270]
		5 h		6.8	

(continued)

Table 10.4 (continued)

Species	Strain/isolate	Exposure time	Concentration	log_{10} reduction	References
Luteimonas spp.	Toilet bowl biofilm isolate	1 h	0.0014% (S)	0.0	[270]
		5 h		0.2	
M. adhaesivum	Toilet bowl biofilm isolate	1 h	0.0014% (S)	0.3	[270]
		5 h		1.8	
M. aquaticum	Toilet bowl biofilm isolate	1 h	0.0014% (S)	0.0	[270]
		5 h		0.2	
Methylobacterium spp.	Toilet bowl biofilm isolate	1 h	0.0014% (S)	0.2	[270]
		5 h		0.9	
Microbacterium spp.	Toilet bowl biofilm isolate	1 h	0.0014% (S)	0.0	[270]
		5 h		2.6	
Paracoccus spp.	Toilet bowl biofilm isolate	1 h	0.0014% (S)	0.3	[270]
		5 h		1.3	
P. aeruginosa	NCTC 9027	30 s	1% (S)	≥5.0	[237]
P. aeruginosa	ATCC 15442	5 min	1% (S)	≥5.0	[16]
P. aeruginosa	ATCC 15442	5 min	0.02% (S)	>4.0	[397]
			0.01% (S)	3.6	
P. aeruginosa	NCIMB 10421 and 6 adapted strains	5 min	0.006% (S)	4.0–5.0	[377]
P. aeruginosa	ATCC 15442 and 3 clinical isolates: antibiotic-susceptible, 3MRGN and 4MRGN VIM-1	1 h	0.00995%[a] (P)	≥5.1	[320]
P. nitroreducens	Toilet bowl biofilm isolate	1 h	0.0014% (S)	0.0	[270]
		5 h		0.0	
Pseudomonas spp.	Toilet bowl biofilm isolate	1 h	0.0014% (S)	0.1	[270]
		5 h		0.3	
Pseudonocardia spp.	Toilet bowl biofilm isolate	1 h	0.0014% (S)	1.8	[270]
		5 h		5.2	
P. mexicana	Toilet bowl biofilm isolate	1 h	0.0014% (S)	1.5	[270]
		5 h		4.6	
S. marcescens	ATCC 13880	5 min	0.02% (S)	>5.0	[200]
S. marcescens	Isolates from contaminated alkylamine disinfectant foot baths (dairy)	5 min	0.02% (S)	0.5–3.0	[200]
S. marcescens	ATCC 14756 and 4 clinical isolates: antibiotic-susceptible, 3MRGN, 4MRGN OXA-48 and 4MRGN KPC-2	1 h	0.00995%[a] (P)	≥5.1	[320]

(continued)

Table 10.4 (continued)

Species	Strain/isolate	Exposure time	Concentration	$\log_{10}$ reduction	References
S. yanoikuyae	Toilet bowl biofilm isolate	1 h	0.0014% (S)	1.2	[270]
		5 h		4.7	
Sphingobium spp.	Toilet bowl biofilm isolate	1 h	0.0014% (S)	1.0	[270]
		5 h		2.4	
S. soli	Toilet bowl biofilm isolate	1 h	0.0014% (S)	3.2	[270]
		5 h		6.6	
S. wittichii	Toilet bowl biofilm isolate	1 h	0.0014% (S)	2.2	[270]
		5 h		5.6	
Sphingomonas spp.	3 toilet bowl biofilm isolates	1 h	0.0014% (S)	1.5–2.5	[270]
		5 h		4.4–6.6	
Sphingopyxis spp.	Toilet bowl biofilm isolate	1 h	0.0014% (S)	0.7	[270]
		5 h		2.6	
S. aureus	Strain RF3	30 s	1% (S)	4.8	[237]
		1 min			
		10 min			
S. aureus	ATCC 6538	5 min	1% (S)	≥ 5.0	[16]
S. aureus	Newman laboratory strain	5 min	1% (S)	≥ 5.0	[211]
S. aureus	IFO 13276	30 s	0.2% (S)	≥ 5.0	[418]
S. aureus	ATCC 6538	30 min	0.008% (S)	≥ 3.0	[276]
S. aureus	ATCC 6538	5 min	0.004% (S)	4.0	[397]
S. epidermidis	Toilet bowl biofilm isolate	1 h	0.0014% (S)	2.8	[270]
		5 h		5.3	
S. maltophilia	Toilet bowl biofilm isolate	1 h	0.0014% (S)	0.0	[270]
		5 h		0.1	
X. aerolatus	Toilet bowl biofilm isolate	1 h	0.0014% (S)	0.1	[270]
		5 h		1.4	

P Commercial product; *S* Solution; [a]In combination with 0.00249% an N-(3-aminopropyl)-N-dodecylpropane-1,3-diamine; [b]with organic load

The bactericidal activity of 0.2% BAC is largely neutralized in the presence of egg compounds, milk, beef gravy or tuna gravy [193, 194, 213], whereas the presence of serum albumin, starch or salad oil did not substantially reduce the bactericidal efficacy of BAC at 0.2%, only at 0.1% or 0.05% [213]. The bactericidal efficacy of BAC at 0.009% and 0.035% may be significantly lower when the bacterial cells of *S. aureus* or *P. aeruginosa* used for the suspension test are grown on agar instead of broth, a difference that cannot be found at higher BAC concentrations [40].

The MBC values obtained with different bacterial species are summarized in Table 10.5. The minimum bactericidal concentration depends on the species and begins at 0.0005% BAC (*S. aureus* and *S. epidermidis*) but can also be as high as

Table 10.5 MBC values of various bacterial species to benzalkonium chloride (5 min exposure time)

Species	Strains/isolates	MBC value	References
B. subtilis	ATCC 6633	0.0009%	[38]
E. faecalis	ATCC 19433	0.0052%	[38]
E. coli	74 isolates from food contact surfaces	0.00195–0.0156%	[154]
E. coli	Strain PHL 628	0.0024%	[38]
E. coli	ATCC 25922	0.0025%[a]	[417]
Klebsiella spp.	30 isolates from food contact surfaces	0.00195–0.0156%	[154]
L. monocytogenes	Strain EGDe	0.0028%	[38]
P. aeruginosa	ATCC 15442	0.0016%	[38]
P. aeruginosa	ATCC 15442	0.002–0.0075%	[129]
P. aeruginosa	ATCC 27853	0.0025%[a]	[417]
P. aeruginosa	ATCC 15442	0.008%	[201]
P. fluorescens	5 isolates from chicken carcasses	0.003–0.014%	[201]
P. fragi	3 isolates from chicken carcasses	0.002–0.006%	[201]
P. lundensis	4 isolates from chicken carcasses	0.004–0.009%	[201]
S. enterica	Strain S24	0.0042%	[38]
S. aureus	22 isolates from food contact surfaces	0.0005–0.00195%	[154]
S. aureus	56 isolates (QAC tolerant)	0.0008%–0.0064%[b]	[220]
S. aureus	ATCC 6538	0.0013%[a]	[417]
S. aureus	42 clinical MRSA isolates	0.0016–0.0128%	[286]
S. aureus	54 MRSA strains isolated in Canary black pigs	0.0039–0.0156%	[97]
S. aureus	ATCC 6538	0.007%	[38]
S. aureus	ATCC 6538 and 12 isolates from fishery products	0.1–0.4%[b]	[398]
S. epidermidis	65 isolates from food contact surfaces	0.00049–0.00195%	[154]

[a]10 min exposure time; [b]30 min exposure time

0.4% (*S. aureus*). It is noteworthy that the bactericidal concentration is for some species at the same level as the bacteriostatic concentration of benzalkonium chloride (see also Table 10.2).

10.3.1.3 Activity Against Bacteria in Biofilms

The activity of BAC against bacteria in biofilms has been investigated in numerous studies. The results are summarized in Table 10.6. At 1%, BAC was in some studies bactericidal within 30 min or 1 h (*S. aureus* and mixed biofilm) but much

Table 10.6 Bactericidal activity of benzalkonium chloride against bacterial cells in biofilms

Species	Strain/isolate	Type of biofilm	Exposure time	Concentration	log_{10} reduction	References
A. xylosoxidans	ATCC 27061	24-h incubation in lens cases	5 min	0.01% (S)	3.6	[55]
				0.005% (S)	2.7	
			4 h	0.01% (S)	4.1	
				0.005% (S)	3.9	
B. cepacia	6 isolates from disinfectant and aerosol solution	5-d incubation on silicone discs	1 h	0.5% (S)	≥ 5.0	[261]
				0.1% (S)	3.0	
E. coli	Strain O157, isolate from food poisoning outbreak	8-d incubation on stainless steel	5 min	0.1% (S)	≥ 5.2	[387]
				0.05% (S)		
E. coli	3 avian pathogenic strains	24-h incubation on polystyrene	30 min	0.01% (S)	≥ 4.2	[299]
				0.005% (S)	1.3–2.9	
E. coli	3 avian pathogenic strains	24-h incubation on PVC	30 min	0.01% (S)	≥ 4.3	[299]
				0.005% (S)	3.5–4.3	
E. coli O157:H7	ATCC 35150, ATCC 43889, ATCC 43890	24-h incubation on stainless steel	30 s	0.01% (P)	0.9	[19]
				0.005% (P)	0.7	
				0.002% (P)	0.6	
E. coli	MG 1655	24-h incubation in microtiter plates followed by 6-d incubation with and without 0.9 mM[a] BAC	30 min	1 mM[a] (S)	0.9 without adaptation	[235]
					4.1 log with adaptation	
L. monocytogenes	20 environmental and food isolates	48-h incubation in microtiter plates	5 min	0.19–0.23% (P)	≥ 5.0	[76]
L. monocytogenes	LO28 wild type and 8 acid resistant variants, each in mixture with *L. plantarum* WCFS1 wild type	24-h incubation at 20 °C in a 12-well plates	15 min	0.02% (S)	2.1–3.6	[258]
		12-h incubation at 30 °C in a 12-well plates			2.4–3.2	

(continued)

Table 10.6 (continued)

Species	Strain/isolate	Type of biofilm	Exposure time	Concentration	$\log_{10}$ reduction	References
L. monocytogenes	6 strains from various sources	24-h incubation in polystyrene microtiter plates	60 min	0.0125% (S)	≥ 6.2	[45]
				0.00125% (S)	0.1	
L. monocytogenes	ATCC 15315, ATCC 19114, ATCC 19115	24-h incubation on stainless steel	30 s	0.01% (P)	0.7	[19]
				0.005% (P)	0.6	
				0.002% (P)	0.4	
L. monocytogenes	11 strains from different origins	48-h incubation in polystyrene microtiter plates and on stainless steel	6 min	0.005% (S)	2.0	[187]
L. monocytogenes	3 strains from different origins (FMCC_B-125, MCC_B-129, MCC_B-169)	1- to 10-d incubation on stainless steel	6 min	0.005% (S)	1.0–2.7	[120]
P. aeruginosa	8 clinical isolates	24-h incubation on stainless steel, Teflon and polyethylene	24 h	1% (S)	0.5	[358]
P. aeruginosa	ATCC 700928	24-h incubation in microplates	1 min	0.1% (S)	0.6	[383]
			5 min		0.5	
			60 min		0.9	
P. aeruginosa	ATCC 10145 and a GI endoscope biofilm isolate	4-d incubation on polystyrene	30 min	0.036% (S)	2.2–3.0	[233]
P. aeruginosa	ATCC 9027	24-h incubation in lens cases	5 min	0.01% (S)	5.0	[55]
				0.005% (S)		
			4 h	0.01% (S)	5.0	
				0.005% (S)	4.9	
P. aeruginosa	ATCC 10154	24-h incubation in microtiter plates followed by 6-d incubation with and without 0.9 mM[a] BAC	30 min	1 mM[a] (S)	3.0 without adaptation	[235]
					3.7 with adaptation	

(continued)

Table 10.6 (continued)

Species	Strain/isolate	Type of biofilm	Exposure time	Concentration	log_{10} reduction	References
P. putida	3 strains from different origins (CK119, CK120, CK148)	1- to 10-d incubation on stainless steel	6 min	0.005% (S)	1.8–3.3	[120]
S. enterica	2 strains	2-d incubation in biofilm reactor	10 min	0.02% (S)	0.2	[68]
			45 min		0.3–0.4	
			90 min		0.8–1.0	
S. enterica	4 strains	7-d incubation in biofilm reactor	10 min	0.02% (S)	0.0–0.1	[68]
			45 min		0.0–0.2	
			90 min		0.1–0.3	
S. enterica	8 strains from different origins	48-h incubation in polystyrene microtiter plates and on stainless steel	6 min	0.005% (S)	3.0–3.8	[187]
S. Enteritidis	Isolate from food poisoning outbreak	8-d incubation on stainless steel	5 min	0.1% (S)	≥ 5.2	[387]
				0.05% (S)		
S. Typhimurium	ATCC 14028	3-d incubation on a 96-peg lid	1 min	1.5% (S)	3.0	[411]
				0.75% (S)	1.8	
			5 min	1.5% (S)	≥ 6.0	
				0.75% (S)	≥ 6.0	
				0.07% (S)	2.2	
S. Typhimurium	ATCC 19585, ATCC 43971, DT 104	24-h incubation on stainless steel	30 s	0.01% (P)	1.4	[19]
				0.005% (P)	1.0	
				0.002% (P)	0.5	
S. Typhimurium	3 strains (FMCC B-137, FMCC B-193, FMCC B-415)	6-d incubation on stainless steel	6 min	0.005% (S)	3.0–3.3	[121]
S. liquefaciens	Isolate from a raw chicken processing plant	3-d incubation on stainless steel	6 min	0.01% (S)	3.4	[218]

(continued)

Table 10.6 (continued)

Species	Strain/isolate	Type of biofilm	Exposure time	Concentration	log_{10} reduction	References
S. putrefaciens	Isolate from a raw chicken processing plant	3-d incubation on stainless steel	6 min	0.01% (S)	3.1	[218]
S. aureus	ATCC 6538 and 12 isolates from fishery products	48-h incubation on stainless steel coupons	30 min	1–2.6% (S)	≥ 5.0	[398]
S. aureus	8 clinical MRSA isolates	24-h incubation on stainless steel, Teflon and polyethylene	24 h	1% (S)	2.0	[358]
S. aureus	ATCC 6538	24-h incubation on glass coupons	5 min	0.5% (S)	≥ 4.0	[230]
				0.05% (S)	3.8–4.0	
				0.025% (S)	1.8–3.2	
S. aureus	Isolate from food poisoning outbreak	8-d incubation on stainless steel	5 min	0.1% (S)	≥ 5.2	[387]
				0.05% (S)	4.3	
S. aureus	3 strains (FMCC B-134, FMCC B-135, FMCC B-410)	6-d incubation on stainless steel	6 min	0.005% (S)	2.1–2.4	[121]
Mixed species	Mixed biofilm with isolates from lettuce, endives and cucumbers, mainly composed by *Pseudomonas* and *Stenotrophomonas* spp.	2-d incubation on stainless steel	1 h	1% (S)	4.2	[126]
				0.1% (S)	1.1	
				0.01% (S)	0.1	

S Solution; *P* Commercial product; [a]Molecular weight not described

Table 10.7 MBC values (5 min) obtained with biofilm grown cells of various bacterial species to benzalkonium chloride

Species	Strains/isolates	MBC value	References
E. coli	74 isolates from food contact surfaces	0.00781–0.0625%	[154]
Klebsiella spp.	30 isolates from food contact surfaces	0.01563–0.0625%	[154]
S. aureus	22 isolates from food contact surfaces	0.01563–0.0625%	[154]
S. epidermidis	65 isolates from food contact surfaces	0.00781–0.0625%	[154]

less effective with 0.5–2.0 log within 24 h against *S. aureus* and *P. aeruginosa*. At 0.1%, BAC was able to reduce bacterial counts of *E. coli*, *S. Enteritidis* and *S. aureus* within 5 min by ≥5.0 log but not *P. aeruginosa* (0.9 log in 60 min) or species in mixed biofilm (1.1 log in 60 min). At 0.01%, it usually required an exposure time of 4 h (*A. xylosoxidans*, *E. coli*, *P. aeruginosa*) whereas shorter times such as 30 s or 6 min did not achieve a sufficient bactericidal effect.

Some data are available to describe the MBC values for bacterial species that were obtained from biofilms. They are summarized in Table 10.7. Overall, the MBC values were higher compared those obtained with planktonic cells (Table 10.5).

The lower susceptibility of bacterial cells in biofilms or obtained from biofilms has been described in a few other studies. Tests with *E. coli* CIP 54127 obtained from culture on tryptic soy agar or in the form of biofilms showed a strong impairment of the bactericidal activity of BAC for biofilm cells. The reduction in sensitivity was attributed to a reduced accessibility of the bacterial cells to the disinfectants, due to the fact that the former adhered to a support [293]. The absence of oxygen increased the antimicrobial effect of BAC towards *E. coli* with both planktonic and sessile cells [26]. Eradication of biofilm cells of *P. aeruginosa* by BAC required much longer time than that of planktonic cells in suspensions [373]. A 24 h biofilm on polystyrene microtiter plates grown by 8 strains of *P. aeruginosa* was quite resistant to BAC requiring concentrations between 0.045 and 0.07% to achieve a bactericidal effect in 60 min [306]. Biofilm-grown *P. aeruginosa* cells (24 h in microtiterplates) were 100 times less susceptible to BAC (5 min exposure time) compared to planktonic cells [39]. These findings are supported by other authors. An alkyldimethyl BAC at 0.005% was 2,160 times less effective against *P. aeruginosa* in biofilm compared to planktonic cells; the resistance factor was only marginally lower at 0.01% (2,000 times less effective) and substantially lower at 0.025% (1,500 times less effective) [127]. For *P. aeruginosa* CIP A 22, the level of resistance of the bacteria in the biofilm relative to that of planktonic bacteria increased with the BAC C-chain length. For cells within the biofilm, the exopolysaccharide induced a characteristic increase in surface hydrophilicity. Three-dimensional structures (water channels) were also involved [47].

Similar results were found with *L. monocytogenes*. The susceptibility of biofilm grown cells (4 d on stainless steel) was 3.7 times lower to BAC (10 min exposure time) compared to planktonic cells. When the cells were grown for 11 d on stainless steel, the susceptibility was 6 times lower. And when the cells were grown for 11 d on polypropylene, the susceptibility was 36 times lower [335]. In single-species biofilms, *L. monocytogenes* developed higher tolerance to cleaning and disinfection over time for the quaternary ammonium compound disinfectant, indicating that a broad-spectrum mechanism was involved [100]. On a mature 6 d *L. monocytogenes* biofilm, BAC of at least 80 mg/l was necessary to reduce $\geq$ 80% of the metabolic activity [313]. Increased BAC tolerance was observed in *L. monocytogenes* biofilm (static or continous flow) after exposure to 20 mg/l peracetic acid in a wild-type strain only in static biofilm [394]. HrcA and DnaK play an important role in the resistance of *L. monocytogenes* planktonic and biofilm cells against disinfectants [394].

Studies with *S. aureus* show that BAC at 0.1% was ineffective for eradication of MRSA cells in biofilm even after 1 h but was effective for eradication of planktonic cells within 20 s [296]. Another study with 11 food-associated *Staphylococcus* spp. strains demonstrated that a similar bactericidal efficacy (0.3–3.5 log in 5 min) was achieved against planktonic cells with 0.001% BAC but against biofilm cells with 0.02% BAC [99]. In addition, it was observed that *L. monocytogenes* strain C719 in biofilms is at least 1,000 times more resistant to BAC than in planktonic form [327]. In contrast to these findings, it was reported that *S. aureus* cells taken from a 14 h biofilm are more susceptible to BAC at 10 and 20 mg/l compared to planktonic cells [46].

The efficacy of BAC against bacteria in biofilms depends on various parameters, e.g. the maturity of the biofilm. The resistance of 4 *L. monocytogenes* strains to 0.005% and 0.015% BAC was dependent on biofilm maturity (72 h vs. 24 h and 48 h) [289]. BAC is less effective against mixed biofilms. Mixed biofilm (*L. monocytogenes* and *L. plantarum*) was found to be less susceptible to the bactericidal activity of 0.01% BAC in 15 min (<2.0 log) compared to single-species biofilm of *L. monocytogenes* (4.5 log) and *L. plantarum* (3.3 log) [396]. Mixed biofilm (*S. liquefaciens* and *S. putrefaciens*) was also more difficult to inactivate by BAC at 0.01% compared to single-species biofilms [218]. The presence of *L. monocytogenes* in a *P. putida* biofilm strongly increases the resistance of the biofilm cells to BAC [120]. Biofilm formation with *P. putida* and *L. monocytogenes* depends on the bacterial species involved and the interactions between them and the environmental conditions. The resistance of mixed-species biofilms of *L. monocytogenes* and *P. putida* to BAC seems to be related to their microscopic structure and to the association between the involved species [337]. The formulation itself also seems to matter. Bacteria with lower antimicrobial susceptibility whose populations were enriched after low-level biocide exposure were more effectively suppressed by BAC at in-use concentrations (1%) in a formulation than in a simple aqueous solution [108].

BAC seems to encounter obstacles to penetration within the cluster. BAC treatment caused a non-uniform loss of fluorescence in *P. aeruginosa* ATCC 15442 biofilms. Cells in peripheral layers were inactivated first, and then the action of the biocide spread steadily into the cluster structure. This gradual inactivation of the structure together with the fact that disruption of the three-dimensional biofilm structure and elimination of the matrix led to a recovery of biocide efficiency, suggests that BAC encountered obstacles to penetration within the cluster, probably caused by interactions with biofilm components [39].

The reduced efficacy of an alkyldimethyl BAC against bacteria in biofilms (shown with *E. aerogenes*) is partly explained by a transport limitation of the biocide into the biofilm [363]. The ability of BAC to penetrate a biofilm, however, is not correlated with its killing or removal efficiency as shown with biofilms of *B. cereus* and *P. fluorescens* [11]. The addition of at least 0.5 mM copper to BAC was able to synergistically be effective against *P. aeruginosa* cells in biofilms [136].

Biofilm also enables microbial persistence. In a cold-smoked fish processing plant, a few clones of *L. monocytgenes* persisted. One persistent strain produced more biofilm than the transient strain in 48 h. In addition, the resistance to BAC was about 150-fold higher in the persistent strain than in the transient strain. The total amount of extracellular polymeric substances in the persistent strain biofilm was higher than that in the transient strain biofilm. These findings suggest that the persistent strain produces greater amounts of biofilm and extracellular polymeric substances than the transient strain, which results in greater resistance of the persistent strain to disinfectants [283].

Restricted penetration of BAC into biofilms might be one of the key processes explaining the resistance of *P. aeruginosa* biofilms to this biocide [39]. Reprocessed dispensers for surface disinfectants were refilled with use solutions of products based on BAC. Bacterial regrowth in the surface disinfectant solution was observed for some manual reprocessing protocols after 3 w at room temperature suggesting a niche for *Achromobacter* spp., e.g. in residual biofilm [165, 166]. Another explanation for the resistance of a *P. aeruginosa* ATCC 15442 biofilm to BAC is that few cells remained alive at different areas in the cluster, despite the apparent penetration of the biocide after 25 min of treatment. These cells may have been located in areas difficult for the biocides to attain; for example, the cells may have been located in areas protected by a large quantity of matrix and other cells. In addition, it cannot be excluded that these few cells expressed highly resistant phenotypes throughout physiological adaptations, e.g. persisters (20), or throughout genetic mutations [39].

In *E. coli* MG 1655 and a *L. innocua* field strain, it was shown that resistance to BAC is associated significantly increased stickiness in biofilm formation [372]. *L. monocytogenes* cells present in biofilms were shown to recover and grow after a 30 min BAC treatment with concentrations up to 10 mg/l thus providing a source of recontamination [327]. BAC-adapted *S. Enteritidis* biofilms acquired the ability to survive a normally lethal exposure to BAC (0.05%) and then regrew [240].

10.3.1.4 Carrier Test

Only few data were found on the efficacy of BAC on contaminated inanimate surfaces. *P. aeruginosa* NCIMB 10421, and 6 BAC-adapted strains were reduced by 0.006% BAC exposed for 5 min by 1.8–3.3 log [377]. Insufficient bactericidal activity with mostly between 1.0 and 2.0 log was described against *S. aureus* and *P. aeruginosa* in concentrations up to 0.3% BAC with an exposure time of 5 min [397]. Against 3 bacterial species (*S. aureus* strain RF3, *E. coli* NCTC 86 and

P. aeruginosa NCTC 9027), 1% BAC revealed a good bactericidal activity within 30 s on glass carriers with at least 4.2 log [237].

According to ASTM E 2967, the efficacy of disinfectant wipes soaked with 0.45% BAC plus 0.4% DDAC and 0.1% PHMB was determined on stainless steel carriers contaminated with *S. aureus* (ATCC 6538) or *A. baumannii* (ATCC 19568). With a 10 s wipe, the bacterial load was reduced by at least 5.0 log, the control wipe without the disinfectant yielded 3.0 log. A transfer of the test organism to another sterile surface was observed from all surfaces originally contaminated with *S. aureus* and from none of the surfaces originally contaminated with *A. baumannii* [339].

BAC is also strongly adsorbed to various types of tissue used in surface disinfection such as white pulp (up to 61% adsorption), viscose rayon (up to 70% adsorption) and mixed tissues (up to 54% adsorption) [30]. Cotton towels have also been described to bind between 82% and 85% of BAC after only 30 s of exposure [96]. Use of these tissues with surface disinfectants based on BAC will result in an insufficient bactericidal effect and in low-level exposure to the target micro-organisms [30, 96].

10.3.2 Fungicidal Activity

10.3.2.1 Fungistatic Activity (MIC Values)

An overview on published MIC values obtained with different fungal species can be found in Table 10.8. Most yeast isolates had MIC values below the proposed epidemiological cut-off value for *C. albicans* of 16 mg/l [271]. *Fusarium* spp. may have MIC values up to 64 mg/l, *Aspergillus* and *Alternaria* spp. up to 32 mg/l.

10.3.2.2 Fungicidal Activity (Suspension Tests)

Data from suspension tests to describe the fungicidal activity of products based on benzalkonium chloride are summarized in Table 10.9. The yeasticidal activity begins at 0.2% BAC within 5 min. *Aspergillus* spp. are overall sufficiently reduced by 0.5% BAC within 60 min. Food-borne fungi seem to be more resistant to benzalkonium chloride compared to the typical healthcare-associated fungi such as *Candida* spp. or *Aspergillus* spp. Some species are killed with 1.5% within 10 min by $\geq$4.0 log except *A. flavus*, *A. versicolor* and many *Penicillium* spp. Nevertheless, some authors describe that the fungicidal activity of 1% BAC is strong in 1–3 min against eight different fungal species including five strains of *A. niger* and five strains of *P. roquefortii* [185].

The yeasticidal activity of 0.2% BAC is largely neutralized in the presence of egg compounds and milk as shown in carrier tests [351]. In *S. cerevisiae,* it was shown that the mechanism of action of BAC is primarily metabolic inhibition rather than membrane damage [182].

Table 10.8 MIC values of various fungal species to benzalkonium chloride

Species	Strains/isolates	MIC value (mg/l)	References
A. alternata	10 fungal keratitis isolates	2–16	[416]
Alternaria spp.[a]	11 clean room isolates	8–32	[338]
A. flavipes	Isolate from the biodeteriorated mural paintings of an old church	0.6	[389]
A. flavus	3 clinical, 3 airborne and 2 food isolates	0.25–8	[163]
A. flavus	Isolate from the biodeteriorated mural paintings of an old church	>0.6	[389]
A. flavus	7 isolates from surfaces in a veterinary hospital	1.25–20	[244]
A. flavus	14 clean room isolates	2–8	[338]
A. flavus	61 fungal keratitis isolates	4–32	[416]
A. flavus	10 isolates from fungal keratitis cases	16	[78]
A. fumigatus	6 clinical and 14 airborne isolates	0.25–2	[163]
A. fumigatus	Isolate from the biodeteriorated mural paintings of an old church	0.6	[389]
A. fumigatus	9 isolates from surfaces in a veterinary hospital	1.25–5	[244]
A. fumigatus	11 clean room isolates	4–8	[338]
A. fumigatus	11 fungal keratitis isolates	4–16	[416]
A. nidulans	Isolate from the biodeteriorated mural paintings of an old church	>0.6	[389]
A. niger	2 airborne and 2 food isolates	0.5–2	[163]
A. niger	Isolate from the biodeteriorated mural paintings of an old church	>0.6	[389]
A. niger	Strain from cultural heritage objects in Serbia	1.25	[365]
A. niger	10 fungal keratitis isolates	2–8	[416]
A. niger	2 isolates from surfaces in a veterinary hospital	2.5	[244]
A. niger	11 clean room isolates	4–8	[338]
A. ochraceus	Isolate from the biodeteriorated mural paintings of an old church	>0.6	[389]
A. ochraceus	Strain from cultural heritage objects in Serbia	1	[365]
A. ochraceus	2 food isolates	4–8	[163]
A. parasiticus	Isolate from the biodeteriorated mural paintings of an old church	>0.6	[389]
A. terreus	Isolate from the biodeteriorated mural paintings of an old church	0.6	[389]
A. terreus	4 clean room isolates	4–8	[338]
A. versicolor	12 fungal keratitis isolates	4–32	[416]
B. spicifera	Strain from cultural heritage objects in Serbia	0.25	[365]
C. albicans	200 worldwide strains from hospital- and community-acquired infections	0.5–32	[271]
C. albicans	15 clinical isolates from patients with septicaemia	3.12	[152]

(continued)

Table 10.8 (continued)

Species	Strains/isolates	MIC value (mg/l)	References
C. albicans	ATCC 10231	27	[417]
Cladosporium spp.[a]	16 clean room isolates	4–16	[338]
Curvularia spp.[a]	16 clean room isolates	4–8	[338]
E. nigrum	Strain from cultural heritage objects in Serbia	0.2	[365]
Exserohilum spp.[a]	4 clean room isolates	8–16	[338]
F. oxysporum	10 fungal keratitis isolates	8–16	[416]
F. solani	82 fungal keratitis isolates	8–32	[416]
F. verticillioides	20 fungal keratitis isolates	8–32	[416]
Fusarium spp.[a]	10 clean room isolates	4–8	[338]
Fusarium spp.[a]	10 isolates from fungal keratitis cases	32–64	[78]
Mucor spp.[a]	2 clinical and 1 food isolates	1–8	[163]
P. aurantiogriseum	Food isolate	1	[163]
P. citrinum	15 airborne isolates	0.25–0.5	[163]
P. crysogenum	14 airborne isolates	0.25–1	[163]
P. paneum	2 food isolates	2–4	[163]
P. roquefortii	4 food isolates	2	[163]
Penicillium spp.[a]	Strain from cultural heritage objects in Serbia	0.25	[365]
Penicillium spp.[a]	15 clean room isolates	2–4	[338]
Rhizopus spp.[a]	2 clinical and 1 food isolates	0.5–16	[163]
T. viride	Strain from cultural heritage objects in Serbia	1.25	[365]
Trichoderma spp.[a]	Food isolate	4	[163]
Various species[a]	8 cleanroom fungal isolates incl. *Aspergillus* spp., *Penicillum* spp., *Curvularia* spp., *Cladosporium* spp. and *Alternaria* spp.	4–16	[399]

[a]No number of isolates per species

10.3.3 Mycobactericidal Activity

Since 1961, BAC is known to have no tuberculocidal activity [231, 413]. More recent data show that BAC at 0.1% has no activity in 120 min against *M. tuberculosis*, *M. kansasii* and *M. avium* [324]. Combinations of various QACs also revealed an insufficient activity against *M. tuberculosis* and *M. bovis* in 20 min [334]. At a low concentration of 0.0014%, the mycobactericidal activity of BAC was also poor within 5 h against toilet bowl biofilm isolates of *M. frederiksbergense* and another *Mycobacterium* spp. (1.3–1.9 log) [270].

Table 10.9 Fungicidal activity of benzalkonium chloride in suspension tests

Species	Strain/isolate	Exposure time	Concentration	$\log_{10}$ reduction	References
A. flavus	Bread isolate	10 min	1.5% (P)	2.0	[43]
A. fumigatus	15 clinical isolates	5 min	0.25% (P)	≥4.0	[382]
A. niger	Bread isolate	10 min	1.5% (P)	>5.2	[43]
A. niger	1 clinical isolate	1 h	0.2% (S)	≥4.0	[295]
A. ochraceus	2 clinical isolates	30 min	0.5% (S)	<4.0	[132]
		60 min		≥4.0	
A. terreus	2 clinical isolates	1 h	0.2% (S)	≥4.0	[295]
A. versicolor	2 cheese isolates	10 min	1.5% (P)	1.2–4.0	[43]
C. albicans	1 clinical isolate	15 min	0.5% (S)	≥4.0	[132]
C. albicans	IFO 1594	5 min	0.2% (S)	≥5.0	[418]
C. albicans	3 clinical isolates	15 min	0.2% (S)	≥4.0	[295]
C. krusei	1 clinical isolate	15 min	0.5% (S)	≥4.0	[132]
C. tropicalis	2 clinical isolates	15 min	0.2% (S)	≥4.0	[295]
C. parapsilosis	1 clinical isolate	15 min	0.5% (S)	≥4.0	[132]
C. parapsilosis	1 clinical isolate	15 min	0.2% (S)	≥4.0	[295]
Cladosporium spp.	Bread isolate	10 min	1.5% (P)	>4.1	[43]
D. hansenii	Cheese isolate	10 min	1.5% (P)	>4.5	[43]
E. repens	Bread factory isolate	10 min	1.5% (P)	>4.5	[43]
H. burtonii	Bread isolate	10 min	1.5% (P)	>5.2	[43]
M. ruber	Bread isolate	10 min	1.5% (P)	>4.1	[43]
M. suaveolens	Bread isolate	10 min	1.5% (P)	>4.5	[43]
N. pseudofischeri	Cherry filling isolate	10 min	1.5% (P)	>4.5	[43]
P. anomala	Bread isolate	10 min	1.5% (P)	>5.9	[43]
P. caseifulvum	Cheese isolate	10 min	1.5% (P)	2.7	[43]
P. chrysogenum	Cheese isolate	10 min	1.5% (P)	>5.2	[43]
P. commune	2 cheese and 1 bread isolates	10 min	1.5% (P)	2.7–4.0	[43]
P. corylophilum	Bread isolate	10 min	1.5% (P)	>4.8	[43]
P. crustosum	Cheese isolate	10 min	1.5% (P)	4.3	[43]
P. discolor	Cheese isolate	10 min	1.5% (P)	3.0	[43]
P. nalgiovense	2 cheese isolates	10 min	1.5% (P)	2.3–4.2	[43]
P. norvegensis	Cheese isolate	10 min	1.5% (P)	3.0	[43]
P. roqueforti	2 bread isolates	10 min	1.5% (P)	1.0–1.2	[43]
P. solitum	Cheese isolate	10 min	1.5% (P)	>4.8	[43]
P. verrucosum	Cheese isolate	10 min	1.5% (P)	3.1	[43]
S. brevicaulis	Cheese isolate	10 min	1.5% (P)	>4.2	[43]
T. delbrueckii	Cheese isolate	10 min	1.5% (P)	>4.8	[43]
T. rubrum	1 clinical isolate	30 min	0.4% (S)	≥4.0	[295]

S Solution; *P* Commercial product

10.4 Effect of Low-Level Exposure

Numerous studies show that low-level exposure to BAC has different effects on bacteria (Table 10.10). No adaptive response was found in isolates or strains from 12 Gram-negative species (*A. xylosoxidans*, *C. jejuni*, *C. indologenes*, *Chrysobacterium* spp., *C. sakazakii*, *H. gallinarum*, *M. osloensis*, *P. nitroreductans*, *S. enteritidis*, *Salmonella* spp., *S. multivorum*, *S. maltophilia*) and 7 Gram-positive species (*B. cereus*, *C. pseudogenitalum*, *E. saccharolyticus*, *S. cohnii*, *S. epidermidis*, *S. kloosii* and *S. lugdenensis*).

Some isolates or strains of 12 Gram-negative species were able to express a weak adaptive response (MIC increase $\leq$ 4-fold) such as *A. hydrophila*, *A. jandaei*, *C. coli*, *Citrobacter* spp., *E. coli*, *K. oxytoca*, *P. aeruginosa*, *P. putida*, *Pseudomonas* spp., *Pseudoxanthomonas* spp., *S.* Typhimurium and *Salmonella* spp. The same type of change was found in isolates or strains of 13 Gram-positive species such as *E. durans*, *E. faecalis*, *Eubacterium* spp., *L. monocytogenes*, *M. phyllosphaerae*, *M. luteus*, *S. aureus*, *S. capitis*, *S. caprae*, *S. hominis*, *S. saprophyticus*, *S. warneii* and *Staphylococcus* spp.

A strong but unstable MIC change (>4-fold) was found in isolates or strains of seven Gram-negative species (*E. cloacae*, *Enterobacter* spp., *Klebsiella* spp., *P. agglomerans*, *P. ananatis*, *Pantoea* spp. and *Salmonella* spp.) and 10 Gram-positive species (*B. cereus*, *B. licheniformis*, *Bacillus* spp., *E. casseliflavus*, *E. faecalis*, *E. faecium*, *Enterococcus* spp., *S. haemolyticus*, *S. saprophyticus* and *Staphylococcus* spp.).

A strong and stable MIC change (>4-fold) was described for isolates or strains of 12 Gram-negative species (*A. baumannii*, *Chryseobacterium* spp., *E. ludwigii*, *Enterobacter* spp., *E. coli*, *Pantoea* spp., *P. aeruginosa*, *S. enterica serovar Typhimurium*, *S. Enteritidis*, *S. Typhimurium*, *S. Virchow* and *Salmonella* spp.) and 2 Gram-positive species (*L. monocytogenes* and *S. aureus*).

In isolates or strains of 2 Gram-negative species (*A. proteolyticus* and *Ralstonia* spp.) and 1 Gram-positive species (*C. renale* group) the adaptive response was strong but its stability was not described.

Selected strains or isolates revealed substantial MIC changes: *Pantoea* spp. (up to 500-fold), *Enterobacter* spp. (up to 300-fold), *Salmonella* spp., *S. saprophyticus* and *B. cereus* (all up to 200-fold), *Staphylococcus* spp. (up to 150-fold), or *E. coli* (up to 100-fold). Other species still showed a strong but somewhat lower adaptive MIC increase such *Corynebacterium renale* group (up to 62.5-fold), *B. cereus*, *E. faecalis* and *E. faecium* (all up to 50-fold), *Klebsiella* spp. (up to 36-fold), *P. aeruginosa* (up to 33-fold) and *A. baumannii* (up to 31-fold).

In Gram-negative species, the highest MIC values after adaptation were 3,000 mg/l (*S. Typhimurium*), 2,500 mg/l (*P. aeruginosa* and *Pantoea* spp.), 1,500 mg/l (*Enterobacter* spp.), 1,000 mg/l (*E. coli*) and 500 mg/l (*B. cepacia complex*). Epidemiological cut-off values to determine resistance to BAC was proposed in 2014 for *Salmonella* spp. (128 mg/l), *E. coli* (64 mg/l), *K. pneumoniae* (32 mg/l) and *Enterobacter* spp. (32 mg/l) [271]. Based on this proposal, the

Table 10.10 Change of bacterial susceptibility to biocides and antimicrobials after low-level exposure to BAC

Species	Strain/isolate	Exposure time	Increase in MIC	MIC_{max} (mg/l)	Stability of MIC change	Associated changes	References
A. xylosoxidans	Domestic drain biofilm isolate MBRG 4.31	14 d at various concentrations	None	3.9	Not applicable	None reported	[266]
A. baumannii	Strain MBRG15.1 from a domestic kitchen drain biofilm	14 passages at various concentrations	31-fold	62.5	Stable for 14 d	None reported	[75]
A. hydrophila	Domestic drain biofilm isolate MBRG 4.3	14 d at various concentrations	4-fold	125	No data	None reported	[266]
A. jandaei	Domestic drain biofilm isolate MBRG 9.11	14 d at various concentrations	2-fold	62.5	No data	None reported	[266]
A. proteolyticus	Domestic drain biofilm isolate MBRG 9.12	14 d at various concentrations	32-fold	125	No data	None reported	[266]
B. cereus	Domestic drain biofilm isolate MBRG 4.21	14 d at various concentrations	None	7.8	Not applicable	None reported	[266]
B. cereus	5 biocide-sensitive strains from organic foods	Several passages with gradually higher concentrations	10-fold–200-fold	400	Unstable for 20 subcultures	Cross-adaptation[a] to chlorhexidine (10-fold–100-fold), triclosan (up to 100-fold), hexachlorophene (10-fold–100-fold) and DDAB[b] (10-fold–40-fold); cross-resistance[a] to ampicillin (2 strains), sulphamethoxazol (2 strains) and cefotaxime (1 strain)	[114]

(continued)

Table 10.10 (continued)

Species	Strain/isolate	Exposure time	Increase in MIC	MIC_{max} (mg/l)	Stability of MIC change	Associated changes	References
B. licheniformis	2 biocide-sensitive strains from organic foods	Several passages with gradually higher concentrations	25-fold–50-fold	25	Unstable for 20 subcultures	Cross-adaptation[a] to chlorhexidine (>100-fold), triclosan (5-fold–100-fold), hexachlorophene (>100-fold) and DDAB[b] (3-fold–7-fold); cross-resistance[a] to ceftazidime (1 strain) and cefotaxime (1 strain)	[114]
Bacillus spp.	4 biocide-sensitive strains from organic foods	Several passages with gradually higher concentrations	4-fold–25-fold	5	Unstable for 20 subcultures	Cross-adaptation[a] to chlorhexidine (>100-fold), triclosan (2-fold–100-fold), hexachlorophene ($\geq$ 100-fold) and DDAB[b] (2-fold–10-fold); cross-resistance[a] to sulphamethoxazol (2 strains), ampicillin (1 strain), and cefotaxime (1 strain)	[114]
B. cenocepacia	6 strains from clinical and environmental habitats	Up to 28 d at 50 mg/l	No data	200	No data	Survival; no degradation of BAC	[6]
B. cepacia complex	*B. lata* strain 383	5 min at 50 mg/l	No data	500	No data	Upregulation of transporter and efflux pump genes; resistance[a] to imipenem (3 of 4 experiments), meropenem and ciprofloxacin (2 of 4 experiments) and ceftazidime (1 of 4 experiments)	[181]

(continued)

Table 10.10 (continued)

Species	Strain/isolate	Exposure time	Increase in MIC	MIC_{max} (mg/l)	Stability of MIC change	Associated changes	References
C. coli	ATCC 33559 and a poultry isolate	Up to 15 passages with gradually higher concentrations	2-fold (only the ATCC strain)	4	Stable for 5 d, reverted after 10 d	None described	[246]
C. jejuni	NCTC 11168, ATCC 33560 and a poultry isolate	Up to 15 passages with gradually higher concentrations	None	1	Not applicable	None described	[246]
C. indologenes	Domestic drain biofilm isolate MBRG 9.15	14 d at various concentrations	None	31.2	Not applicable	None reported	[266]
Chryseobacterium spp.	1 biocide-sensitive strain from organic foods	Several passages with gradually higher concentrations	20-fold	200	Stable for 20 subcultures	Cross-adaptation[a] to chlorhexidine (40-fold), triclosan (100-fold), hexachlorophene (>100-fold) and DDAB[b] (>100-fold); cross-resistance[a] to ampicillin	[114]
C. pseudogenitalum	Human skin isolate MBRG 9.24	14 d at various concentrations	None	15.6	Not applicable	None reported	[266]
C. renale group	Human skin isolate MBRG 9.13	14 d at various concentrations	8-fold	62.5	No data	None reported	[266]

(continued)

Table 10.10 (continued)

Species	Strain/isolate	Exposure time	Increase in MIC	MIC_{max} (mg/l)	Stability of MIC change	Associated changes	References
C. sakazakii	Strain MBRG15.5 from a domestic kitchen drain biofilm	14 passages at various concentrations	None	51.2	Not applicable	None reported	[75]
E. cloacae	2 biocide-sensitive strains from organic foods	Several passages with gradually higher concentrations	12-fold–30-fold	150	Unstable for 20 subcultures	Cross-adaptation[a] to chlorhexidine ($\geq$ 100-fold), triclosan (5-fold), hexachlorophene (>100-fold) and DDAB[b] (>100-fold); cross-resistance[a] to cefotaxime (1 strain) and ampicillin (1 strain)	[114]
E. ludwigii	1 biocide-sensitive strain from organic foods	Several passages with gradually higher concentrations	30-fold	150	Stable for 20 subcultures	Cross-adaptation[a] to chlorhexidine (100-fold), triclosan (100-fold), hexachlorophene (>100-fold) and DDAB[b] (>100-fold); cross-resistance[a] to cefotaxime	[114]
Enterobacter spp.	6 biocide-sensitive strains from organic foods	Several passages with gradually higher concentrations	5-fold–300-fold	1,500	Unstable for 20 subcultures (3 strains), stable for 20 subcultures (3 strains)	Cross-adaptation[a] to chlorhexidine ($\geq$ 100-fold), triclosan (5-fold–100-fold), hexachlorophene (>100-fold) and DDAB[b] (3-fold–100-fold); cross-resistance[a] to ampicillin (4 strains), sulphamethoxazol (2 strains), ceftazidime (1 strain), cefotaxime (1 strain) and trimethoprim-sulphamethoxazol (1 strain)	[114]

(continued)

Table 10.10 (continued)

Species	Strain/isolate	Exposure time	Increase in MIC	MIC_{max} (mg/l)	Stability of MIC change	Associated changes	References
E. casseliflavus	2 biocide-sensitive strains from organic foods	Several passages with gradually higher concentrations	10-fold–20-fold	2	Unstable for 20 subcultures	Cross-adaptation[a] to chlorhexidine ($\geq$ 100-fold), triclosan ($\geq$ 100-fold), hexachlorophene (>100-fold) and DDAB[b] (2-fold–10-fold); cross-resistance[a] to ampicillin (1 strain)	[114]
E. durans	1 biocide-sensitive strain from organic foods	Several passages with gradually higher concentrations	4-fold	2	Unstable for 20 subcultures	Cross-adaptation[a] to chlorhexidine (100-fold), triclosan (>100-fold), hexachlorophene (>100-fold) and DDAB[b] (10-fold); cross-resistance[a] to ampicillin	[114]
E. faecalis	1 strain of unknown origin	14 passages at various concentrations	4-fold	7.8	Unstable for 14 d	None reported	[75]
E. faecalis	2 biocide-sensitive strains from organic foods	Several passages with gradually higher concentrations	5-fold–50-fold	2.5	Unstable for 20 subcultures	Cross-adaptation[a] to chlorhexidine (10-fold–100-fold), triclosan (20-fold–100-fold), hexachlorophene (20-fold–100-fold) and DDAB[b] (20-fold–40-fold); cross-resistance[a] to ceftazidime (2 strains), cefotaxime (1 strain) and sulphamethoxazol (1 strain)	[114]

(continued)

Table 10.10 (continued)

Species	Strain/isolate	Exposure time	Increase in MIC	MIC_{max} (mg/l)	Stability of MIC change	Associated changes	References
E. faecium	13 biocide-sensitive strains from organic foods	Several passages with gradually higher concentrations	4-fold–50-fold	5	Unstable for 20 subcultures	Cross-adaptation[a] to chlorhexidine (10-fold–200-fold), triclosan (40-fold–100-fold), hexachlorophene (10-fold–100-fold) and DDAB[b] (2-fold–20-fold); cross-resistance[a] to ampicillin (7 strains), cefotaxime (3 strains), ciprofloxacin (2 strains) and tetracycline (1 strain)	[114]
E. saccharolyticus	Domestic drain biofilm isolate MBRG 9.16	14 d at various concentrations	None	31.2	Not applicable	None reported	[266]
Enterococcus spp.	6 biocide-sensitive strains from organic foods	Several passages with gradually higher concentrations	4-fold–35-fold	7	Unstable for 20 subcultures	Cross-adaptation[a] to chlorhexidine (up to 100-fold), triclosan (up to 100-fold), hexachlorophene (10-fold–100-fold) and DDAB[b] (up to 10-fold); cross-resistance[a] to ampicillin (3 strains), cefotaxime (2 strains), ceftazidime (2 strains), and sulphamethoxazol (1 strain)	[114]
E. coli	4 BAC-susceptible and 4 BAC-resistant isolates from dairy	Several passages with gradually higher concentrations	1.3-fold–2.6-fold	340	Strain-dependent stability between 2–22 d	Some adaptive strains also exhibited enhanced biofilm formation potential, efflux pump activity and virulence potential (haemolysin activity).	[308]
E. coli	Mutant of strain O103	Not described	2-fold	20	Stable for 7 d	None reported	[348]

(continued)

Table 10.10 (continued)

Species	Strain/isolate	Exposure time	Increase in MIC	MIC_{max} (mg/l)	Stability of MIC change	Associated changes	References
E. coli	ATCC 25922 and 9 avian and porcine *E. coli* strains	7 d at various concentrations	2.6-fold	96–192	No data	2.9-fold increase in MIC to DDAC[f]; increased tolerance[a] to florfenicol (7-fold), cefotaxime (6.3-fold), chloramphenicol (6.1-fold), ceftazidime (4.8-fold), nalidixic acid (4.4-fold), ampicillin (4.3-fold), tetracycline (4.2-fold), ciprofloxacin (3.8-fold), sulphamethoxazole (3.7-fold) and trimethoprim (3.3-fold)	[360]
E. coli	ATCC 25922 and strain MBEG15 4 from a domestic kitchen drain biofilm	14 passages at various concentrations	3-fold–7-fold	31.3	Stable for 14 d	None reported	[75]
E. coli	ATCC 47076	30–40 d at variable concentrations	6-fold–7-fold	90	Stable for 4 passages	Increased tolerance[a] to chloramphenicol (up to 128 mg/l), florfenicol (up to 64 mg/l), ciprofloxacin (up to 0.25 mg/l), nalidixic acid (up to 64 mg/l), ampicillin (up to 8 mg/l) and cefotaxime (up to 0.5 mg/l); increased susceptibility was shown for gentamicin, streptomycin and kanamycin.	[32]

(continued)

Table 10.10 (continued)

Species	Strain/isolate	Exposure time	Increase in MIC	MIC_{max} (mg/l)	Stability of MIC change	Associated changes	References
E. coli	NCTC 12900 strain O157	6 passages at variable concentrations	Approximately. 100-fold	Approximately 1,000	Stable for 30 d	Increased tolerance[c] (>2 mm increase in zone of inhibition) to amoxicillin-clavulanic acid, amoxicillin, chloramphenicol, imipenem, tetracycline, trimethoprim, chlorhexidine and triclosan	[36]
E. coli	6 pan-susceptible strains	12 d at various concentrations	24% (mean)[e]	60	No data	Increased tolerance[d] to tetracycline (+776% to 23.3 mg/l), ciprofloxacin (+316% to 0.11 mg/l), chloramphenicol (+106% to 13.7 mg/l), trimethoprim/sulphamethoxazole (+58% to 0.14 mg/l), ampicillin (+35% to 12 mg/l) and gentamicin (+18% to 1.3 mg/l)	[288]
E. coli	Strain MG1655	1 d at 9 mg/l (25% of MIC value)	Survival of a small subpopulation (1–5%)	No data	Stable for 10 d	None	[263]
Eubacterium spp.	Domestic drain biofilm isolate MBRG 4.14	14 d at various concentrations	2-fold	31.2	No data	None reported	[266]
H. gallinarum	Domestic drain biofilm isolate MBRG 4.27	14 d at various concentrations	None	31.2	No data	None reported	[266]

(continued)

Table 10.10 (continued)

Species	Strain/isolate	Exposure time	Increase in MIC	MIC_{max} (mg/l)	Stability of MIC change	Associated changes	References
K. oxytoca	1 biocide-sensitive strain from organic foods	Several passages with gradually higher concentrations	3-fold	21	Unstable for 20 subcultures	Cross-adaptation[a] to chlorhexidine (>100-fold), triclosan (6-fold), hexachlorophene (>100-fold) and DDAB[b] (10-fold); no antibiotic cross-resistance[a]	[114]
Klebsiella spp.	1 biocide-sensitive strain from organic foods	Several passages with gradually higher concentrations	36-fold	90	Unstable for 20 subcultures	Cross-adaptation[a] to chlorhexidine (>100-fold), triclosan (40-fold), hexachlorophene (>100-fold) and DDAB[b] (>100-fold); cross-resistance[a] to ampicillin	[114]
L. pentosus	7 strains from naturally fermented Aloreña green table olives	48 h at 1 mg/l	No data	No data	Not applicable	Increased tolerance[a] to ampicillin (1-fold to 100-fold), chloramphenicol (2-fold–500-fold), ciprofloxacin (2-fold–14-fold), teicoplanin (1-fold–340-fold), tetracycline (2-fold–80-fold) and trimethoprim (1-fold–15-fold); no increase of MIC with clindamycin, erythromycin and streptomycin.	[50]
L. pentosus	Strain MP-10	48 h at 1 mg/l	No data	No data	Not applicable	Increase in growth rate, improved survival at pH 1.5 and in the presence of 2–3% bile	[49]

(continued)

Table 10.10 (continued)

Species	Strain/isolate	Exposure time	Increase in MIC	MIC_{max} (mg/l)	Stability of MIC change	Associated changes	References
L. pseudomesenteroides	1 strain from naturally fermented Aloreña green table olives	48 h at 1 mg/l	No data	No data	Not applicable	Increased tolerance[a] to chloramphenicol (2-fold), ciprofloxacin (3-fold) and tetracycline (2-fold); no increase of MIC with ampicillin, clindamycin, erythromycin, streptomycin, teicoplanin and trimethoprim.	[50]
L. monocytogenes	CECT 5873 and 5 strains from fish products or a fish processing plat	12–37 h at 0.88–8.33 mg/l	1.4-fold–3.7-fold	No data	No data	None reported	[336]
L. monocytogenes	2 food isolates (ice cream, poultry)	2 h at sublethal concentration	Up to 3-fold	5	Stable for 28 d	None reported	[229]
		2 h, followed by 24 h at sublethal concentration	Up to 4-fold	5		Cross-adaptation[a] to other QAC (4-fold to 8-fold), alkylamine (2-fold to 4-fold) and sodium hypochlorite (up to 2-fold)	
L. monocytogenes	4 isolates sensitive to BAC	2–3 w at variable concentrations	4-fold–6-fold	5	Stable for >1 y	Increased tolerance[a] to gentamicin (up to 5.5 mg/l) and kanamycin (up to 25 mg/l)	[328]
L. monocytogenes	4 BAC-sensitive strains	Several passages with gradually higher concentrations	5-fold–6-fold	6	Stable for 10 m	Increase of efflux pump activity in 3 of 4 strains	[380]

(continued)

Table 10.10 (continued)

Species	Strain/isolate	Exposure time	Increase in MIC	MIC_{max} (mg/l)	Stability of MIC change	Associated changes	References
L. monocytogenes	Strain EGD	48 h	No data	No data	Not applicable	Induction of virulence gene expression	[169]
L. monocytogenes	ATCC BAA-679 and 3 strains from food products	30 min at 1.25 mg/l	No data	5	Not applicable	Reduction of invasiveness, increase of intracellular proliferation; better survival	[317]
L. monocytogenes	Wild type outbreak strain	1 h at 10 mg/l	No data	30	Not applicable	49.6-fold upregulation of emrE (efflux function); upregulation of regulatory function genes (e.g. lmo1851 or lmo1861)	[189]
M. phyllosphaerae	Domestic drain biofilm isolate MBRG 4.30	14 d at various concentrations	2-fold	31.2	No data	None reported	[266]
M. luteus	Human skin isolate MBRG 9.25	14 d at various concentrations	2-fold	0.97	No data	None reported	[266]
M. osloensis	Strain MBRG15.3 from a domestic kitchen drain biofilm	14 passages at various concentrations	None	2	Not applicable	None reported	[75]
P. agglomerans	4 biocide-sensitive strains from organic foods	Several passages with gradually higher concentrations	20-fold–70-fold	35	Unstable for 20 subcultures	Cross-adaptation[a] to chlorhexidine (10-fold–100-fold), triclosan (>100-fold), hexachlorophene ($\geq$ 100-fold) and DDAB[b] (5-fold–20-fold); cross-resistance[a] to ampicillin (4 strains), ceftazidime (2 strains) and cefotaxime (2 strains)	[114]

(continued)

Table 10.10 (continued)

Species	Strain/isolate	Exposure time	Increase in MIC	MIC_{max} (mg/l)	Stability of MIC change	Associated changes	References
P. ananatis	1 biocide-sensitive strain from organic foods	Several passages with gradually higher concentrations	25-fold	25	Unstable for 20 subcultures	Cross-adaptation[a] to chlorhexidine (>100-fold), triclosan (50-fold), hexachlorophene (>100-fold) and DDAB[b] (>100-fold); cross-resistance[a] to ampicillin, cefotaxime and sulphamethoxazol	[114]
Pantoea spp.	3 biocide-sensitive strains from organic foods	Several passages with gradually higher concentrations	100-fold–500-fold	2,500	Unstable for 20 subcultures (2 strains), stable for 20 subcultures (1 strain)	Cross-adaptation[a] to chlorhexidine (>100-fold), triclosan (20-fold–100-fold), hexachlorophene (>100-fold) and DDAB[b] (20-fold–100-fold); cross-resistance[a] to ampicillin (1 strain), cefotaxime (1 strain) and sulphamethoxazol (1 strain)	[114]
P. aeruginosa	ATCC 15442, ATCC 15692 and 14 strains from hospitals	5 d exposure at 7.8 mg/l	2-fold–33-fold in 15 of 16 strains	500	Stable for 5 w	Increased tolerance[a] to other membrane-active agents (cetylpyridinium chloride and cetrimide); no change of susceptibility to chlorhexidine or triclosan	[225]
P. aeruginosa	22 isolates from biofilm samples in dairy	Several passages with variable concentrations	≤ 2.2-fold (baseline MICs were high with 100–350 mg/l)	430	Strain-dependent stability between 3–16 d	No conclusive cross-resistance to ciprofloxacin	[307]

(continued)

Table 10.10 (continued)

Species	Strain/isolate	Exposure time	Increase in MIC	MIC_{max} (mg/l)	Stability of MIC change	Associated changes	References
P. aeruginosa	ATCC 9027	14 passages at various concentrations	4-fold	62.5	Stable for 14 d	None reported	[75]
P. aeruginosa	Strain NCIMB 10421	27 passages with gradually higher concentrations	12-fold	580	Stable for 4 d, reverted after 8 d	Unchanged tolerance[d] to amikacin, ceftazidime, ciprofloxacin, gentamycin, imipenem, ticarcillin	[161]
P. aeruginosa	Strain NCIMB 10421	Several passages with gradually higher concentrations	>12-fold	350	Stable for 20 d	256-fold increase in tolerance[d] to ciprofloxacin (up to 32 mg/l)	[248]
P. aeruginosa	150 BAC-sensitive strains	Exposure to BAC	In 6 strains (4%): increase of MIC to 1,250–2,500 mg/l	2,500	No data	None reported	[197]
P. nitroreductans	Domestic drain biofilm isolate MBRG 4.6	14 d at various concentrations	None	31.2	Not applicable	None reported	[266]
P. putida	Strain MBRG15.2 from a domestic kitchen drain biofilm	14 passages at various concentrations	4-fold	62.5	Unstable for 14 d	None reported	[75]
Pseudomonas spp.	Domestic drain biofilm isolate MBRG 9.14	14 d at various concentrations	2-fold	62.5	No data	None reported	[266]

(continued)

Table 10.10 (continued)

Species	Strain/isolate	Exposure time	Increase in MIC	MIC_{max} (mg/l)	Stability of MIC change	Associated changes	References
Pseudoxanthomonas spp.	Domestic drain biofilm isolate MBRG 9.20	14 d at various concentrations	4-fold	31.2	No data	None reported	[266]
Ralstonia spp.	Domestic drain biofilm isolate MBRG 4.13	14 d at various concentrations	21-fold	167	No data	None reported	[266]
S. enterica serovar Typhimurium	Strain 14028S	5 min at 4 and 15 mg/l	20-fold–50-fold	2,000	Stable for 5 subcultures, reverted after 10 subcultures	13-fold–27-fold MIC increase[a] to chlorhexidine (up to 800 mg/l)	[180]
S. enterica serovar Typhimurium	Strain SL1344	5 min at 0.1, 1 and 4 mg/l	27-fold–100-fold	3,000	Stable for 5 subcultures, reverted after 10 subcultures	13-fold–27-fold MIC increase[a] to chlorhexidine (up to 800 mg/l)	[180]
S. Enteritidis	ATCC 13076	7 d of sublethal exposure	1.25	12.5	Unstable for 10 d	None reported	[322]
S. Enteritidis	ATCC 4931	6 d exposure at 0.0001%	3.2-fold and 18.3-fold more survivors of lethal challenge with 0.003% BAC (planktonic cells and biofilm cells, respectively)	35	No data	Various cellular changes in adapted biofilm cells (up-regulation of 17 unique proteins, increased expression of CspA, TrxA, Tsf, YjgF, a probable peroxidase, phenotype-specific alterations in cell surface roughness, and a shift in fatty acid composition)	[241]

(continued)

Table 10.10 (continued)

Species	Strain/isolate	Exposure time	Increase in MIC	MIC_{max} (mg/l)	Stability of MIC change	Associated changes	References
S. Enteritidis	Clinical isolate	Several passages with gradually higher concentrations	8-fold	256	Stable for 30 d	None described	[37]
S. Enteritidis	Clinical isolate	6 passages at variable concentrations	Approximately 200-fold	Approximately 250	Stable for 30 d	None reported	[36]
S. Typhimurium	NCTC 74	Several passages with gradually higher concentrations	2-fold	64	Stable for 30 d	None described	[37]
S. Typhimurium	1 poultry isolate	Not described	3.8-fold	>30.4	"stable"	None reported	[48]
S. Typhimurium	NCTC 74	6 passages at variable concentrations	Approximately 20-fold	Approximately 100	Stable for 30 d	Increased tolerance[c] to chlorhexidine (5 mm)	[36]
S. Typhimurium	Wild type strain 14028s	Gradually increasing levels of BAC	No data	No data	No data	Detection of five resistant mutants; 2-fold–64-fold higher MICs[a] to chloramphenicol, ciprofloxacin, nalidixic acid, and tetracycline	[130]
S. Virchow	Food isolate	Several passages with gradually higher concentrations	64-fold	256	Stable for 30 d	None described	[37]

(continued)

Table 10.10 (continued)

Species	Strain/isolate	Exposure time	Increase in MIC	MIC_{max} (mg/l)	Stability of MIC change	Associated changes	References
S. Virchow	Food isolate	6 passages at variable concentrations	Approximately 200-fold	Approximately 250	Stable for 30 d	Increased tolerance[c] to amoxicillin-clavulanic acid (0 mm), amoxicillin (1 mm), chloramphenicol (2 mm), imipenem (12 mm), trimethoprim (0 mm), chlorhexidine (4 mm) and triclosan (0 mm)	[36]
Salmonella spp.	6 strains with higher MICs to biocidal products	8 days at increasing concentrations	3.3-fold in 1 strain	50	No data	Increased tolerance[g] to ampicillin (16 mg/l), amoxicillin-clavulanic acid (4 mg/l), piperacillin (64 mg/l), cephalexin (16 mg/l), cefpodoxime (2 mg/l), ceftiofur (>8 mg/l), ceftriaxone (2 mg/l), tetracycline (8 mg/l), ciprofloxacin (0.5 mg/l), chloramphenicol (16 mg/l), cefoxitin (>32 mg/l) and nalidixic acid (32 mg/l); no change in 12 other antibiotics.	[66]
Salmonella spp.	3 biocide-sensitive strains from organic foods	Several passages with gradually higher concentrations	5-fold–70-fold	150	Unstable for 20 subcultures (2 strains), stable for 20 subcultures (1 strain)	Cross-adaptation[a] to chlorhexidine (>100-fold), triclosan ($\geq$ 100-fold), hexachlorophene (40-fold–100-fold) and DDAB[b] (10-fold–100-fold); cross-resistance[a] to ampicillin (2 strains), cefotaxime (2 strains), trimethoprim-sulphamethoxazol (2 strains), sulphamethoxazol (1 strain), tetracycline (1 strain) and nalidixic acid (1 strain)	[114]

(continued)

Table 10.10 (continued)

Species	Strain/isolate	Exposure time	Increase in MIC	MIC_{max} (mg/l)	Stability of MIC change	Associated changes	References
S. multivorum	Domestic drain biofilm isolate MBRG 9.19	14 d at various concentrations	None	31.2	Not applicable	None reported	[266]
S. aureus	ATCC 6538	7 d of sublethal exposure	2.5-fold	5	Unstable for 10 d	None reported	[322]
S. aureus	MRSA strain 48a isolated from a poultry hamburger	Several passages with gradually higher concentrations	2.5-fold	5.1	Unstable for 10 d	No enhancement of biofilm formation	[44]
S. aureus	ATCC 6538	14 passages at various concentrations	39-fold	3.9	Stable for 14 d	None reported	[75]
S. capitis	Human skin isolate MBRG 9.34	14 d at various concentrations	2-fold	0.97	No data	None reported	[266]
S. caprae	Human skin isolate MBRG 9.30	14 d at various concentrations	2-fold	0.97	No data	None reported	[266]
S. cohnii	Human skin isolate MBRG 9.31	14 d at various concentrations	None	0.45	Not applicable	None reported	[266]
S. epidermidis	Human skin isolate M 9.33	14 d at various concentrations	None	0.45	Not applicable	None reported	[266]

(continued)

Table 10.10 (continued)

Species	Strain/isolate	Exposure time	Increase in MIC	MIC_{max} (mg/l)	Stability of MIC change	Associated changes	References
S. epidermidis	CIP53124	1 d at various concentrations	No data	No data	Not applicable	Significant increase of biofilm formation at various sublethal concentrations	[149]
S. haemolyticus	Human skin isolate MBRG 9.35	14 d at various concentrations	35-fold	15.6	Unstable	MIC reverted in week 2 to 0.97 mg/l	[266]
S. hominis	Human skin isolate MBRG 9.37	14 d at various concentrations	2-fold	0.97	No data	None reported	[266]
S. kloosii	Human skin isolate MBRG 9.28	14 d at various concentrations	None	0.97	Not applicable	None reported	[266]
S. lugdunensis	Human skin isolate MBRG 9.36	14 d at various concentrations	None	0.97	Not applicable	None reported	[266]
S. saprophyticus	5 biocide-sensitive strains from organic foods	Several passages with gradually higher concentrations	10-fold–200-fold	1,000	Unstable for 20 subcultures	Cross-adaptation[a] to chlorhexidine ($\geq$ 100-fold), triclosan (20-fold–100-fold), hexachlorophene ($\geq$ 100-fold) and DDAB[b] (5-fold–20-fold); cross-resistance[a] to sulphamethoxazol (3 strains), ceftazidime (3 strains), ampicillin (2 strains) and tetracycline (1 strain)	[114]

(continued)

Table 10.10 (continued)

Species	Strain/isolate	Exposure time	Increase in MIC	MIC_{max} (mg/l)	Stability of MIC change	Associated changes	References
S. saprophyticus	Human skin isolate MBRG 9.29	14 d at various concentrations	2-fold	0.81	No data	None reported	[266]
S. warneri	Human skin isolate MBRG 9.27	14 d at various concentrations	2-fold	0.97	No data	None reported	[266]
Staphylococcus spp.	4 strains from meat and poultry plants	10 d with gradually higher concentrations	2-fold–4-fold	10	Stable for 10 subcultures	None reported	[367]
Staphylococcus spp.	4 biocide-sensitive strains from organic foods	Several passages with gradually higher concentrations	2-fold–150-fold	100	Unstable for 20 subcultures	Cross-adaptation[a] to chlorhexidine (up to 10-fold), triclosan (10-fold–100-fold), hexachlorophene (15-fold–100-fold) and DDAB[b] (up to 50-fold); cross-resistance[a] to sulphamethoxazol (3 strains), ampicillin (3 strains), ceftazidime (1 strain) and tetracycline (1 strain)	[114]
S. maltophilia	Domestic drain biofilm isolate MBRG 9.13	14 d at various concentrations	None	31.2	Not applicable	None reported	[266]

[a]Broth microdilution method; [b]Didodecyldimethylammonium bromide; [c]Disc diffusion method; [d]Etest; [e]Change more likely explained by BAC and not glutaraldehyde (product contained 15% BAC and 15% glutaraldehyde); [f]Didecyldimethylammonium chloride; [g]Agar dilution method (NARMS plates)

majority of *Salmonella* spp., *E. coli* and *Enterobacter* spp. isolates would be classified as resistant to BAC after low-level exposure. In Gram-positive species, the highest MIC *values* after adaptation were 1,000 mg/l in *S. saprophyticus*, 400 mg/l in *B. cereus*, 100 mg/l in *Staphylococcus* spp. and 15.6 mg/l in *S. haemolyticus*. Epidemiological cut-off values to determine resistance to BAC was proposed in 2014 for *S. aureus* (16 mg/l), *E. faecalis* and *E. faecium* (both 8 mg/l) [271]. Based on this proposal, the majority of *S. aureus* and *Enterococcus* spp. isolates would still have to be classified as susceptible to BAC after low-level exposure.

Cross-resistance to various antibiotics such as ampicillin, cefotaxime or ceftazidime was found in isolates of *B. cepacia complex*, *Chryseobacterium* spp., *Enterobacter* spp., *E. coli*, *Klebsiella* spp., *Pantoea* spp. and *Salmonella* spp. Cross-resistance to selected antibiotics was also detected in *B. cereus*, *B. licheniformis*, *Bacillus* spp., *E. casseliflavus*, *E. durans*, *E. faecalis*, *E. faecium*, *Enterococcus* spp., *S. saprophyticus* and *Staphylococcus* spp.

In addition, a lower susceptibility to other biocidal agents was described for some species to didecyldimethylammonium chloride or didecyldimethylammonium bromide, chlorhexidine, triclosan, other QACs, alkylamine and sodium hypochlorite.

Other adaptive changes include a significant up-regulation of transporter and efflux pump genes in *B. cepacia complex*, *E. coli* and *L. monocytogenes*. Enhanced biofilm formation was described for *E. coli* and *S. epidermidis*. In *S. epidermidis*, the effect depends on the BAC concentration. At 0.0001% BAC was also able to increase biofilm formation in three *S. epidermidis* strains but at 0.0002, 0.0003, 0.0004 and 0.0005% BAC biofilm formation was reduced [54].

A general adaptation to BAC by bacteria cannot be seen. Exposure of 7 species (*A. baumannii*, *C. sakazakii*, *E. faecalis*, *E. coli*, *P. aeruginosa*, *P. putida*, *S. aureus*) over 14 passages of 4 d each to increasing BAC concentrations on agar was associated with both increases and decreases in antibiotic susceptibility, but its effect was typically small relative to the differences observed among microbicides. Susceptibility changes resulting in resistance were not observed in this study [109].

Nevertheless, the data in Table 10.10 are in line with findings showing that BAC has a significant ermetic effect with *P. aeruginosa* and a less significant effect with *S. aureus* resulting in greater bacterial growth [267]. *P. fluorescens* in biofilm also exhibited adaptation to benzalkonium chloride at 0.001% [89]. When benzalkonium chloride deposits remain on polystyrene such in surface disinfection, *P. aeruginosa* readily acquired the ability to grow in BAC and also exhibited phyhysical–chemical surface changes. The existence of residues on polystyrene surfaces altered their hydrophobicity and favoured adhesion. Adapted bacteria revealed a higher ability to adhere to surfaces and to develop biofilms, especially on BAC-conditioned surfaces, which thereby could enhance resistance to sanitation attempts [234]. Adaptive resistance to BAC promoted some changes in *P. aeruginosa* in proteins previously described as involved in antibiotic resistance. These results contribute to the assumption that there are common resistance mechanisms, between adaptive and acquired resistance of *P. aeruginosa* [232].

Adaptation to BAC in *S. Enteritidis* ATCC 4931 occurred concurrently with the up-regulation of key proteins involved in the cold shock response, stress response, and detoxification and an overall increase in protein biosynthesis. Thus, the up-regulation of these important proteins explains the mechanisms responsible for adaptive resistance to BAC in *S. Enteritidis* biofilms [240].

Low-level exposure to BAC has also been described in food processing to enhance persistence of specific *L. monocytogenes* strains associated with a low-level resistance to BAC [303]. In five pork meat processing plants, the use of cleaning and disinfectant agents containing various agents such as 0.1–1% sodium hypochlorite (2 plants), 0.5–2% peracetic acid with hydrogen peroxide (2 plants) or 4% DDAC (1 plant) has been proposed to select for *L. monocytogenes* persisters by activating non-specific efflux pumps [67].

10.5 Resistance to BAC

BAC resistance among 1,325 food-associated Gram-negative bacteria and 500 *Enterococcus* spp. is not frequent, only 16 strains, mainly from meat retail shops, showed low-level resistance to BAC. No systematic cross-resistance between BAC and any of the other antimicrobial agents tested was detected. But resistance may develop to user concentrations after exposure to sublethal concentrations of BAC [353].

A study on 390 pigs from 26 farms revealed that frequent disinfection of nursery pens is significantly associated with MRSA shedding in nursery pigs. All MRSA isolates carried at least 1 QAC resistance gene. QAC-based disinfectants were described as important drivers in the selection and persistence of MRSA in commercial swine herds, and these agents may be co-selecting for other antimicrobial resistance genes [356].

Resistance to BAC was in one study determined by visible grow on agar with 3 mg/l BAC. In 653 *Staphylococcus* spp. strains from community environmental samples, a total of 63 (9.6%) were classified as BAC resistant based on this method [139]. Resistance was also determined by visible grow on agar with 10 mg/l BAC in three studies. Among 116 *L. monocytogenes* strains, a total of 71 (61.2%) were regarded as BAC resistant based on this method [88]. Among 123 *L. monocytogenes* isolates from turkey processing plants, a total of 57 (46.3%) were regarded as BAC resistant [273]. And in 138 *L. monocytogenes* isolates from processed foods and processing plants environments a total of 19 (13.8%) were regarded as BAC resistant [318]. Visible growth at 20 mg/l BAC was also used to determine BAC resistance. In 392 *L. monocytogenes* strains from various sources in Finland and Switzerland, a total of 45 (11.5%) were classified as BAC resistant mostly explained by the qacH efflux pump [255].

In order to facilitate the determination of MIC values, a disc diffusion test for BAC has been tested and validated to determine resistance to BAC using *S. aureus* and *E. coli* [140]. New QACs have been described to have a lower risk to trigger bacterial resistance as shown with MRSA [260].

10.5.1 High MIC Values

BAC is quite specific in its antimicrobial mechanism. Even very low concentrations cause damage to the cytoplasmic membrane due to perturbation of the bilayers by the molecules' alkyl chains [410]. Development of microbial resistance to BAC is therefore possible or even likely.

Mainly isolates of Gram-negative species have been described to be resistant to BAC as shown by high MIC values (see also Table 10.2). The highest MIC values were described with *A. hydrophila* (up to 31,300 mg/l), *B. cereus* and *E. meningoseptica* (up to 7,800 mg/l) *P. aeruginosa* (up to 5,000 mg/l), *L. monocytogenes* (up to 625 mg/l; proposed breakpoint: >7.5 mg/l) [361], *E. cloacae* (up to 512 mg/l but mostly below the recommended epidemiological cut-off value of 32 mg/l), *A. xylosoxidans* and *B. cepacia* (up to 500 mg/l) and *P. mirabilis* (up to 400 mg/l). *K. pneumoniae* was among the most susceptible Gram-negative species with MIC values up to 64 mg/l which is just above the recommended epidemiological cut-off value (32 mg/l). *E. coli* (up to 156 mg/l), *C. freundii* (up to 190 mg/l), *Acinetobacter* spp. (up to 200 mg/l), *Salmonella* spp. (up to 256 mg/l), *A. xylosoxidans* and *B. cepacia* (up to 500 mg/l), *Pseudomonas* spp. (up to 5,000 mg/l), *B. cereus* and *E. meningoseptica* (up to 7,800 mg/l) and *A. hydrophila* (up to 31,300 mg/l) were often susceptible although some isolates were above the recommended epidemiological cut-off value (64 mg/l for *E. coli*, 128 mg/l for *Salmonella* spp.).

10.5.2 Reduced Efficacy in Suspension Tests

Some data are available indicating BAC resistance by insufficient killing in suspension tests (≤ 5.0 log within the bactericidal exposure time), e.g. in *Achromobacter* spp. 3, *Methylobacterium* spp. and *S. marcescens* (Table 10.11). It is not surprising that an insufficient bactericidal activity of BAC was so far only described with Gram-negative bacterial species. In France, a clinical isolates of *P. cepacia* was identified with a MBC of >20% BAC whereas most other *P. cepacia* isolates from hospitals or veterinary care had MBC values between 0.05 and 0.1%. Five other *Pseudomonas* spp. were more susceptible than *P. cepacia* (MBC between 0.001 and 0.1%) [56].

Various *Methylobacterium* spp. strains isolated from pink biofilm in bathrooms were not reduced at all when exposed to 5% BAC for 5 min. Exposure for 24 h resulted in a 5.0 log reduction with BAC at 1%. Eleven other bacterial species isolated from the same biofilm were mostly killed by 0.1% BAC in 5 min, only a *Rhodococcus* spp. required either 1% BAC for 5 min or 0.1% BAC for 2 h [419].

Table 10.11 Bactericidal activity of BAC solutions (S) or commercial products (P)

Species	Strains/isolates	BAC concentration	Exposure time	Log_{10} reduction	References
Achromobacter spp. 3	1 isolate from contaminated surface disinfectant solution based on BAC	99.5 mg/l[a] (P) 49.8 mg/l[a] (P)	1 h 4 h	4.6/ $\geq$6.8[b] 2.4/4.4[b]	[165]
M. rhodesianum	1 isolate from a dairy production facility	200 mg/l (S)	5 min	0.6	[33]
S. marcescens	1 isolate from contaminated surface disinfectant solution based on BAC	99.5 mg/l[a] (P) 49.8 mg/l[a] (P)	1 h 4 h	0.1/2.4[b] 0.0/ <1.7[b]	[165]

[a]Plus an additional amine; [b]After 5 passages without selection pressure by BAC

10.5.3 Resistance Mechanisms

Various pathways and mechanisms of resistance to QACs have been described (Fig. 10.1). They are described below in more detail.

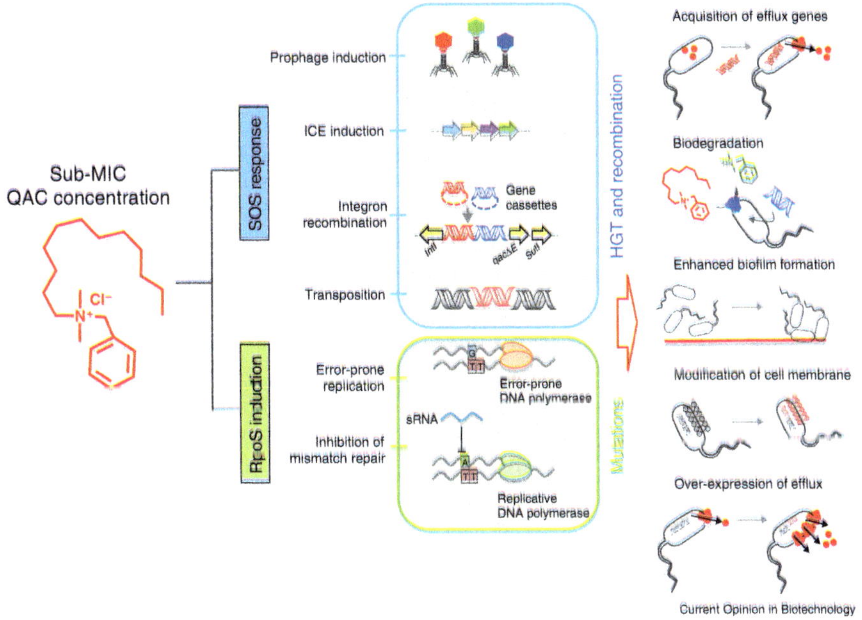

Fig. 10.1 Pathways and mechanisms of QAC resistance [376]. Reprinted from Current Opinion in Biotechnology, Volume number 33, Authors Tezel U and Pavlostathis SG, Quaternary ammonium disinfectants: microbial adaptation, degradation and ecology, pp. 296–304, Copyright 2015, with permission from Elsevier

10.5.4 Resistance Genes

Multidrug efflux pumps can be divided into five protein families depending on their energy requirements and structure [406]. Two of them are the "Major Facilitator Superfamily" (MFS) and the "Small Multidrug Resistance" (SMR) family [406]. MFS represents the largest known family of secondary transporter systems with at least 74 protein families including qacA and qacB [406]. Qac genes ("quaternary ammonium compound"), also described as biocide or antiseptic resistance genes [357, 424], are often found in isolates suspected to be BAC or CHG resistant.

10.5.4.1 qacA/B

The qacA/B gene is found in *S. aureus* and coagulase-negative staphylococci [71, 300, 375] in the chromosome and on plasmids including the pSK1-plasmid family [300]. Thirty three historical isolates of *S. epidermidis* from blood cultures did not carry qacA/B indicating that long-term use of specific biocides may select for the presence of qacA/B genes [355]. qacB transfers in *S. aureus* a resistance to monovalent organic cations and in addition on a low level to some bivalent substances [282, 300]. It can be found on various plasmids such as β-lactamase and on heavy metal resistance plasmids (pSK23) [300]. qacA and qacB are very similar and difficult to distinguish by PCR [300]. Many isolates are present with highly polymorphic forms of these genes. The functional differences between qacA and qacB originally reported by Paulsen et al. [310] are now less clear. qacA/B is considered to be the most frequent resistance gene for biocidal agents in disinfectants [220]. It has been claimed that the chronological emergence of qacA and qacB determinants in clinical isolates of *S. aureus* mirrors the introduction and usage of cationic biocides [332]. In *S. aureus,* it was shown that overexpression of qacA is only found when the culture exposed in the exponential phase of growth resulting in a 8-fold MIC increase [52]. In 2 *S. epidermidis* isolates from unpasteurized milk, the presence of qacA/B was associated with resistance to BAC allowing bacterial growth at 0.0002% BAC [250]. The presence of disinfectant resistance genes such as qacA/B significantly adversely affected disinfecting capacity of rigid gas permeable solutions for contact lenses against *Staphylococcus* spp. [350]. Twelve CNS isolates from invasive infections in very preterm infants were found with a reduced susceptibility to BAC, all of them carried qacA/B [212].

The qacA/B gene can be found in up to 100% of clinical MRSA isolates (Table 10.12). The trend in MRSA is increasing. In a pediatric oncology unit in the USA, the qacA/B rate among MRSA increased year by year [251].

The qacA/B gene can, however, also be found in MSSA and other bacterial species (Table 10.13). In carbapenem-resistant *K. pneumoniae,* it was 40.7%, in *S. aureus* between 0 and 65.4%, and in CNS between 8.6 and 92.3%. Detection of the qac gene correlated with a reduced susceptibility to biocidal agents such as BAC [131]. In addition, resistance to ampicillin, penicillin G and dyes was prevalent in staphylococcus strains from the food industry harbouring the qacA or qacB genes [143].

Table 10.12 Detection rates of qacA/B in MRSA isolates

Country	Number of isolates and source	qacA/B detection rate (%)	References
China	85 isolates from burn patients	100	[58]
Malaysia	60 clinical isolates	83.3	[346]
China	53 clinical isolates	83.0	[221]
Brazil	74 clinical isolates	81.1	[262]
Australia	151 clinical isolates from nosocomial infections	78.6	[148]
Mexico	21 strains from patients with catheter-related infections	76.2	[309]
Malaysia	95 clinical isolates	70.5	[118]
South Korea	62 clinical isolates with low level mupirocin resistance	64.5	[208]
Turkey	28 isolates from surgical site infections	64.3	[87]
Various European countries	297 isolates from blood cultures, skin or soft tissue infections	62.6	[247]
Iran	60 clinical isolates	61.7	[370]
China	131 clinical isolates	61.1	[402]
Japan	65 clinical isolates	52.3[a]	[343]
		1.5[b]	
Republic of Korea	174 isolates from surveillance and clinical cultures	46.6	[179]
Taiwan	96 isolates from chlorhexidine-impregnated catheter-related bloodstream infections	43.8	[147]
Various Asian countries	894 clinical isolates	41.6	[292]
Republic of Korea	169 isolates from patients on surgical intensive care units	37.7	[59]
Taiwan	206 clinical isolates	35.4	[347]
Japan	283 isolates from patients with impetigo and staphylococcal scalded skin syndrome	33.9	[291]
Republic of Korea	119 isolates from surveillance cultures and clinical samples on surgical intensive care units	32.8	[60]
Japan	334 clinical isolates	32.6	[9]
Serbia	50 clinical isolates	32.0	[300]
USA	98 isolates from pediatric patients with a nosocomial infection	26.5	[253]
Scotland	38 hospital-acquired and community-acquired isolates	26.3	[357]
Taiwan	240 clinical isolates	23.8	[403]
China	80 clinical isolates	23.8	[426]
Spain	182 clinical isolates	19.8[a]	[256]
USA	66 clinical isolates from pediatric oncology patients	18.0	[251]

(continued)

Table 10.12 (continued)

Country	Number of isolates and source	qacA/B detection rate (%)	References
China	414 hospital-acquired clinical isolates	15.7	[214]
Various European countries	223 isolates from patients on intensive care units	13.5	[83]
Turkey	69 clinical isolates	11.6	[17]
Kuwait	121 clinical isolates	10.7[a]	[392]
		0.8[b]	
Japan	98 clinical isolates	10.2[a]	[290]
Iran	100 clinical isolates	9.0	[137]
Scotland	120 clinical isolates	8.3	[393]
China	321 isolates from patients and the hospital environment	7.8	[227]
France	39 isolates from blood cultures and nasal swabs	7.7	[329]
Saudi Arabia	117 isolates from nosocomial infections	7.7[a]	[345]
USA	504 isolates from anterior nares among patients on a surgical intensive care unit	7.1	[404]
USA	281 clinical pediatric isolates	4.6	[160]
USA	28 clinical isolates, 17 of them from colonized patients	3.6	[226]
USA	250 isolates from surveillance cultures on a NICU	3.5	[319]
Iran	60 clinical isolates	3.3	[138]
Scotland	40 isolates from blood stream infections	2.5	[146]
Canada	40 nasal isolates in commercial swine herds	2.5	[356]
Spain	134 isolates from blood and nasal samples	2.2	[277]
Canada	334 clinical isolates from ICUs	2.1	[224]
USA	341 isolates from skin and soft tissue infections among infantry trainees	1.5	[341]
Various European countries	456 livestock-associated isolates	1.3	[12]
USA	458 clinical isolates	0.7	[269]
USA	1,968 clinical isolates	0.9	[249]
Egypt	40 isolates from milk and meat products	0	[10]
Turkey	10 clinical isolates	0	[151]

[a]qacA; [b]qacB

Table 10.13 Detection rates of qacA/B in other bacterial species

Species	Country	Number of isolates and source	qacA/B detection rate (%)	References
A. baumannii	Malaysia	122 multidrug-resistant clinical isolates	0[a]	[18]
E. faecalis	Germany	585 isolates from various sources	0.7	[25]
K. pneumoniae	China	27 carbapenem-resistant clinical isolates	40.7[a]	[131]
L. monocytogenes	China	71 isolates from retail food	1.4[a]	[414]
S. aureus	Norway	26 clinical isolates	65.4	[352]
S. aureus	China	152 isolates from male patients with urogenital tract infection	47.4	[420]
S. aureus	USA	11 community environmental isolates	45.5	[139]
S. aureus	USA	149 MSSA isolates from pediatric patients with a nosocomial infection	38.3	[253]
S. aureus	Iran	54 MSSA isolates from swabs	33.3	[370]
S. aureus	Hong Kong	28 isolates from automated teller machines	25.0	[425]
S. aureus	USA	183 isolates from infections of children with congenital heart disease	16.9	[254]
S. aureus	USA	506 clinical isolates, 377 of them community-acquired	15.6	[252]
S. aureus	Tunisia	46 clinical isolates	13.0[a] 8.7[b]	[428]
S. aureus	Various European countries	297 MSSA isolates from blood cultures, skin or soft tissue infections	12.0	[247]
S. aureus	Hong Kong	116 isolates from orthokeratology lens wearers	7.8	[128]
S. aureus	Japan	188 clinical MSSA isolates	7.5	[9]
S. aureus	Spain	111 isolates from young healthy carriers	2.7	[13]
S. aureus	Serbia	50 MSSA isolates from swabs	2.0	[300]
S. aureus	Iran	100 clinical MSSA isolates	0	[137]
S. aureus	Turkey	19 clinical MSSA isolates	0	[151]
S. epidermidis	Slovenia	57 clean room isolates	98.2	[323]
S. epidermidis	Sweden	143 clinical isolates mainly from prosthetic joint infections (61) and post-operative infections after cardiac surgery (31)	43.3	[316]
S. epidermidis	Denmark	75 isolates from the hands of nurses using chlorhexidine for surgical scrubbing (23) and patients (52)	22.7	[355]

(continued)

Table 10.13 (continued)

Species	Country	Number of isolates and source	qacA/B detection rate (%)	References
S. haemolyticus	Argentina	21 clinical isolates	23.8	[69]
S. pseudintermedius	Japan	100 isolates from cases of canine superficial pyoderma	0	[278]
S. pseudintermedius	USA	115 isolates from human and animal origin	0	[150]
Staphylococcus spp.	Norway	52 CNS clinical isolates	92.3	[352]
Staphylococcus spp.	USA	52 community environmental CNS isolates	69.2	[139]
Staphylococcus spp.	Thailand	41 MRCNS isolates from hospital environmental samples	63.4	[344]
Staphylococcus spp.	France	51 CNS isolates from invasive infections in very preterm neonates on a NICU	62.7	[212]
Staphylococcus spp.	Turkey	27 clinical MRCNS isolates	59.3	[151]
Staphylococcus spp.	Iran	51 CNS isolates from swabs	54.9	[370]
Staphylococcus spp.	Hong Kong	78 CNS isolates from automated teller machines	48.7	[425]
Staphylococcus spp.	Turkey	13 MSCNS clinical isolates	46.2	[151]
Staphylococcus spp.	Tunisia	71 clinical CNS isolates	32.4[a] 18.3[b]	[428]
Staphylococcus spp.	Norway	42 QAC-resistant isolates (35 bovine and 7 caprine)	28.6	[28]
Staphylococcus spp.	Hong Kong	67 CNS isolates from orthokeratology lens wearers	17.9	[128]
Staphylococcus spp.	Belgium	58 MRCNS isolates from veal calves	8.6[a]	[14]

[a]qacA; [b]qacB

In Australia, the qacA/B gene was detected in 52 of 78 skin site samples (66.7%) among 43 patients at catheter insertion sites in the arm that were covered with CHG dressings. A statistically greater proportion of specimens with greater than 72 h exposure to CHG dressings was qac-positive, suggesting that the patients were contaminated with bacteria or DNA containing qacA/B during their hospital stay. The presence of qac genes was not positively associated with the presence of DNA specific for *S. epidermidis* and *S. aureus* in these specimens suggesting that qacA/B genes are highly prevalent on hospital patients' skin, even in the absence of viable bacteria [61].

qacA/B-positive *S. aureus* isolates are common in children and are independently associated with nosocomial acquisition and underlying medical conditions as shown in an analysis of 506 *S. aureus* isolates obtained from paediatric patients with community- or hospital-acquired infections. These findings imply a role for the health care environment in acquisition of these organisms. However, genotypic antiseptic tolerance was seen in >25% of healthy children with an *S. aureus* infection, indicating that these organisms are prevalent in the community as well [252].

The presence of QAC resistance genes (mainly qacA/B) among clinical *S. epidermidis* isolates was found to be associated with deep surgical site infections [316]. In children with congenital heart disease and infections caused by *S. aureus,* the qacA/B gene was associated with bacteremia and prolonged hospitalization indicating adverse clinical outcomes [254].

10.5.4.2 smr (qacC)

The smr gene was first detected on a *S. aureus* plasmid and describes "staphylococcal multidrug resistance" [406]. It turned out to be identical with the qacC gene [406]. Irrespective of its description, it belongs to the smr protein family [406] and is also regarded as a biocide or antiseptic resistance gene [221, 349]. Today, both descriptions (smr gene and qacC gene) are used synonymously [406]. Table 10.14 summarizes the frequency of detection in various bacterial species. In *S. aureus,* it can be detected in up to 64.7%, in MRSA and in CNS in up to 100% (Table 10.14).

Table 10.14 Detection rates of smr or qacC in isolates from various bacterial species

Species	Country	Number of isolates and source	smr detection rate (%)	References
S. aureus	Mexico	21 MRSA strains from patients with catheter-related infections	100	[309]
S. aureus	China	53 clinical MRSA isolates	77.4	[221]
S. aureus	South Korea	62 clinical MRSA isolates with low level mupirocin resistance	71.0	[208]
S. aureus	Switzerland	34 isolates from chicken carcasses (neck samples)	64.7	[90]
S. aureus	USA	11 community environmental isolates	63.6	[139]
S. aureus	Canada	40 nasal MRSA isolates in commercial swine herds	62.5	[356]
S. aureus	USA	98 MRSA isolates from pediatric patients with a nosocomial infection	44.9	[253]
S. aureus	Scotland	120 clinical MRSA isolates	44.2	[393]
S. aureus	USA	149 MSSA isolates from pediatric patients with a nosocomial infection	43.6	[253]
S. aureus	Japan	98 clinical MRSA isolates	20.4	[290]
S. aureus	USA	506 clinical isolates, 377 of them community-acquired	19.8	[252]
S. aureus	Sweden	98 clinical isolates	19.4	[216]

(continued)

Table 10.14 (continued)

Species	Country	Number of isolates and source	smr detection rate (%)	References
S. aureus	Germany	68 MRSA isolates from broiler production chain	19.1	[190]
S. aureus	Portugal	74 isolates from pets, livestock, the environment and humans in contact with animals	18.9	[73]
S. aureus	Germany	88 MRSA isolates from turkey production chain	17.0	[190]
S. aureus	Turkey	19 clinical MSSA isolates	15.8	[151]
S. aureus	USA	281 clinical pediatric MRSA isolates	14.2	[160]
S. aureus	Germany	109 MSSA isolates from diseased turkeys and chicken	13.8	[265]
S. aureus	Tunisia	46 clinical isolates	10.9	[428]
S. aureus	Japan	65 clinical MRSA isolates	10.8	[343]
S. aureus	Iran	60 clinical MRSA isolates	10.0	[370]
S. aureus	Turkey	10 clinical MRSA isolates	10.0	[151]
S. aureus	Netherlands	37 MRSA isolates from 4 broiler farms	8.1	[409]
S. aureus	Hong Kong	100 MRSA isolates from pig carcasses	8.0	[412]
S. aureus	Canada	334 clinical MRSA isolates from ICUs	6.9	[224]
S. aureus	Japan	188 clinical MSSA isolates	5.9	[9]
S. aureus	Iran	54 clinical MSSA isolates	3.7	[370]
S. aureus	Hong Kong	28 isolates from automated teller machines	3.6	[425]
S. aureus	Hong Kong	116 isolates from orthokeratology lens wearers	3.4	[128]
S. aureus	Japan	334 clinical MRSA isolates	3.3	[9]
S. aureus	Various Asian countries	894 clinical MRSA isolates	3.1	[292]
S. aureus	Denmark	45 MRSA isolates from pig farms	2.2	[342]
S. aureus	Spain	182 clinical MRSA isolates	1.6	[256]
S. aureus	Japan	283 MRSA isolates from patients with impetigo and staphylococcal scalded skin syndrome	1.4	[291]
S. aureus	Australia	76 clinical MRSA isolates	1.3	[264]
S. aureus	Kuwait	121 clinical MRSA isolates	0.8	[392]
S. aureus	Norway	26 clinical isolates	0	[352]
S. aureus	Iran	100 clinical MSSA isolates	0	[137]
S. aureus	Iran	100 clinical MRSA isolates	0	[137]
S. aureus	Egypt	40 MRSA isolates from milk and meat products	0	[10]

(continued)

Table 10.14 (continued)

Species	Country	Number of isolates and source	smr detection rate (%)	References
S. aureus	Saudi Arabia	117 MRSA isolates from nosocomial infections	0	[345]
S. epidermidis	Slovenia	57 clean room isolates	98.2	[323]
S. epidermidis	Sweden	143 clinical isolates mainly from prosthetic joint infections (61) and post-operative infections after cardiac surgery (31)	5.6	[316]
S. haemolyticus	Argentina	21 clinical isolates	100	[69]
S. pseudointermedius	Japan	100 isolates from canine pyoderma	0	[278]
S. sciuri	Belgium	87 methicillin-resistant isolates from healthy chicken	12.6	[287]
Staphylococcus spp.	Norway	42 QAC-resistant isolates (35 bovine and 7 caprine)	64.3	[28]
Staphylococcus spp.	Tunisia	71 clinical CNS isolates	50.7	[428]
Staphylococcus spp.	Hong Kong	78 CNS isolates from automated teller machines	48.7	[425]
Staphylococcus spp.	USA	52 community environmental CNS isolates	44.2	[139]
Staphylococcus spp.	Turkey	13 clinical MSCNS isolates	23.1	[151]
Staphylococcus spp.	Turkey	27 clinical MRCNS isolates	14.8	[151]
Staphylococcus spp.	Hong Kong	67 CNS isolates from orthokeratology lens wearers	11.9	[128]
Staphylococcus spp.	Turkey	61 CNS isolates from surgical site infections	9.8	[87]
Staphylococcus spp.	Iran	51 clinical CNS isolates	5.9	[370]
Staphylococcus spp.	Belgium	58 MRCNS isolates from veal calves	5.2	[14]
Staphylococcus spp.	Norway	52 clinical CNS isolates	3.8	[352]

Evidence exists indicative of recent mobilization so that the genes have spread between different plasmid backgrounds. The lack of mutations in qacC suggests that the spread occurred relatively recently [405]. qacC is mobilized and transferred to acceptor RC plasmids without assistance of other genes, by means of its location in between the double-strand replication Origin (DSO) and the single-strand replication Origin (SSO) [407].

One of four QAC-resistant *Staphylococcus* strain harbouring the smr gene showed resistance to ampicillin, penicillin, tetracycline, erythromycin and trimethoprim [143]. In *S. epidermidis*, *E. coli* and *S. Typhimurium*, the qacC gene conferred to resistance to β-lactam antibiotics [112].

10.5.4.3 qacE and qacEΔ

The qacE gene was first detected in *E. coli* [406]. As shown in Table 10.15, it can be found quite commonly in Gram-negative species such as *A. baumannii* (31.4–93.8%), *E. coli* (up to 28.5%), *K. pneumoniae* (15.0–53.1%), *P. mirabilis* (53.8%) and *P. aeruginosa* (2.7–67.2%).

Table 10.15 Detection rates of qacE in isolates from various bacterial species

Species	Country	Number of isolates and source	qacE detection rate (%)	References
A. baumannii	USA	97 clinical multidrug-resistant isolates	93.8	[371]
A. baumannii	Malaysia	122 multidrug-resistant clinical isolates	73.0	[18]
A. baumannii	China	47 clinical isolates	70.2	[215]
A. baumannii	Egypt	22 metallo-β-lactamase positive clinical isolates	45.5	[122]
A. baumannii	Iran	5 clinical isolates from burn patients	40.0	[236]
A. baumannii	China	51 carbapenem-resistant clinical isolates	31.4	[222]
Chryseobacterium spp.	Spain	2 isolates from organic foods	0	[103]
C. freundii	Germany	32 clinical isolates	0	[192]
E. cloacae	Spain	2 isolates from organic foods	0	[103]
E. cloacae	Germany	21 clinical isolates	0	[192]
E. ludwigii	Spain	2 isolates from organic foods	0	[103]
Enterobacter spp.	Spain	7 isolates from organic foods	14.3	[103]
E. faecalis	Japan	45 clinical isolates	0	[173]
E. coli	China	179 isolates from retail meats	28.5	[157]
E. coli	USA	570 strains from retail meats	0	[429]
K. pneumoniae	Scotland	64 isolates from different infection sites	53.1	[4]
K. pneumoniae	China	27 carbapenem-resistant clinical isolates	15.0	[131]
K. oxytoca	Spain	2 isolates from organic foods	0	[103]
K. terrigena	Spain	1 isolate from organic foods	0	[103]
P. agglomerans	Spain	7 isolates from organic foods	0	[103]
P. ananatis	Spain	2 isolates from organic foods	0	[103]
P. mirabilis	China	52 isolates from cooked meat products	53.8	[159]
P. aeruginosa	Spain	61 carbapenem-resistant clinical isolates	67.2	[98]
P. aeruginosa	Egypt	36 multidrug-resistant clinical isolates	61.1	[218]

(continued)

Table 10.15 (continued)

Species	Country	Number of isolates and source	qacE detection rate (%)	References
P. aeruginosa	Iran	83 clinical isolates from burn patients	59.0	[236]
P. aeruginosa	Japan	63 clinical and 5 environmental isolates	22.1	[174]
P. aeruginosa	Germany	37 clinical isolates	2.7	[192]
P. putida	Japan	4 environmental isolates	0	[174]
Salmonella spp.	Spain	3 isolates from organic foods	0	[103]
S. aureus	Japan	91 clinical isolates	0	[173]
S. maltophilia	Germany	13 clinical isolates	0	[192]
V. alginolyticus	Japan	3 environmental isolates	0	[174]
V. cholerae	Japan	7 clinical and 1 environmental isolates	0	[174]
V. parahaemolyticus	Japan	10 environmental and 5 clinical isolates	0	[174]
Various Gram-negative species	Japan	5 environmental isolates (*P. vesicularis*, *P. diminuta*, *B. cepacia*, *F. indologenes*, *E. coli*)	0	[174]

A functional deletion variant exists ("qacEΔ") which can also mainly be found in different Gram-negative species such as *A. baumannii* (45.4–96.1%), *E. coli* (2.8–40.2%), *K. pneumoniae* (1.6–59.0%), *P. aeruginosa* (13.5–91.6%) and 66.7–100% in *Vibrio* spp. In addition, it has been detected in 5.9–39.6% of *S. aureus* isolates and 20% of *E. faecalis* isolates (Table 10.16).

Table 10.16 Detection rates of qacEΔ in isolates from various bacterial species

Species	Country	Number of isolates and source	qacEΔ detection rate (%)	References
A. baumannii	China	51 carbapenem-resistant clinical isolates	96.1	[222]
A. baumannii	Iran	5 clinical isolates from burn patients	80	[236]
A. baumannii	Egypt	22 metallo-β-lactamase positive clinical isolates	68.2	[122]
A. baumannii	China	47 clinical isolates	68.1	[215]
A. baumannii	USA	97 clinical multidrug-resistant isolates	45.4	[371]
C. freundii	Germany	32 clinical isolates	9.4	[192]
E. cloacae	Germany	21 clinical isolates	4.8	[192]
E. faecalis	Japan	45 clinical isolates	20.0	[173]
E. coli	China	179 isolates from retail meats	40.2	[157]
E. coli	Tunisia	13 ESBL-positive strains from food	38.5	[23]

(continued)

Table 10.16 (continued)

Species	Country	Number of isolates and source	qacEΔ detection rate (%)	References
E. coli	Nigeria	11 isolates from animal and human origin	36.4	[53]
E. coli	Portugal	144 faecal isolates from pets	2.8	[70]
K. pneumoniae	China	27 carbapenem-resistant clinical isolates	59.0	[131]
K. pneumoniae	Scotland	64 isolates from different infection sites	1.6	[4]
P. mirabilis	China	52 isolates from cooked meat products	53.8	[159]
P. aeruginosa	Iran	83 clinical isolates from burn patients	91.6	[236]
P. aeruginosa	Thailand	50 multidrug-resistant clinical isolates	82.0	[178]
P. aeruginosa	Costa Rica	198 clinical isolates, 125 of them carbapenem-resistant	68.7	[384]
P. aeruginosa	Japan	63 clinical and 5 environmental isolates	63.2	[174]
P. aeruginosa	Germany	37 clinical isolates	13.5	[192]
P. putida	Japan	4 environmental isolates	50.0	[174]
Salmonella spp.	China	152 isolates from retail foods of animal origins	8.6	[81]
S. aureus	Japan	91 clinical MRSA isolates	39.6	[173]
S. aureus	China	152 isolates from male patients with urogenital tract infection	5.9	[420]
S. maltophilia	Germany	13 clinical isolates	0	[192]
V. alginolyticus	Japan	3 environmental isolates	100	[174]
V. cholerae	Japan	7 clinical and 1 environmental isolates	87.5	[174]
V. parahaemolyticus	Japan	10 environmental and 5 clinical isolates	66.7	[174]
Various Gram-negative species	Japan	5 environmental isolates (*P. vesicularis*, *P. diminuta*, *B. cepacia*, *F. indologenes*, *E. coli*)	0	[174]

The presence of the qacE genes has an impact on the susceptibility to biocidal agents. In carbapenem-resistant *K. pneumoniae*, for example, detection of qacE or qacEΔ correlated with a reduced susceptibility to biocidal agents [131]. Another study describes that in 64 *K. pneumoniae* isolates a close link exists between carriage of efflux pump genes, cepA, qacΔE and qacE genes and a reduced benzalkonium chloride susceptibility [4]. In other species, no correlation was found. For example, in 122 *Salmonella* spp. from poultry and swine, an increased MIC value to BAC was independent of the presence of qacEΔ1 [62]. And in *E. coli* the presence of qacEΔ did not change the susceptibility to BAC significantly (range: 0.8–3.1 mg/l) or to chlorhexidine (range: 0.2–0.8 mg/l) [175].

Environmental presence of the qacE genes is an increasing concern. qacEΔ1 genes were detected in quite high levels in manure-treated and untreated soils, lettuce and potato rhizosphere, digestates and on-farm biopurification systems. The observed high prevalence of qacEΔ1 genes in the environment and their potential localization on broad host range plasmids may represent a constant reservoir for the spread of these genes into hospitals, food industry or other man-made environments where QACs are used for biocidal purposes, which may lead to a co-selection of class 1 integrons and associated antibiotic resistance genes [155]. An environmental *Aeromonas* spp. containing the plasmid pP2GI encoding resistance also to QAC via qacEΔI may potentially act as reservoirs of antibiotic resistance genes [243].

Preexposure of environmental bacteria to QAC has also an impact. Samples of effluent and soil were collected from a reed bed system used to remediate liquid waste from a wool finishing mill with a high use of quaternary ammonium compounds (QACs) and were compared with samples of agricultural soils. QAC resistance was higher in isolates from reed bed samples, and class 1 integron incidence was significantly higher for populations that were preexposed to QACs. This is the first study to demonstrate that QAC selection in the natural environment has the potential to co-select for antibiotic resistance, as class 1 integrons are well-established vectors for cassette genes encoding antibiotic resistance [117].

10.5.4.4 qacF

qacF has been detected in 1.8% in *E. coli* and in 18.4% among *Salmonella* spp. (Table 10.17).

10.5.4.5 qacG

The qacG gene has been detected in *S. aureus* in 0 to 90% of the isolates and in CNS in 7.4 to 52.4% of the isolates. Detection rate in Gram-negative species was lower with up to 0.4% in *E. coli*, 17.1% in carbapenemase-positive enterobacteriaceae and 23.5% in carbapenem-resistant *A. baumannii* (Table 10.18).

10.5.4.6 qacH

The qacH gene was detected in *S. aureus* with the highest rate in commercial swine heards (57.5%); it is also detected quite frequently in *L. monocytogenes* with 14.1 to 48.9% with a higher rate of 80.0% among BAC-tolerant isolates. In CNS isolates, the detection rate varies between 0% and 25.0% (Table 10.19). The presence of

Table 10.17 Detection rates of qacF in isolates from various bacterial species

Species	Country	Number of isolates	qacF detection rate (%)	References
E. coli	USA	570 strains from retail meats	1.8	[429]
Salmonella spp.	China	152 isolates from retail foods of animal origins	18.4	[81]

Table 10.18 Detection rates of qacG in isolates from various bacterial species

Species	Country	Number of isolates	qacG detection rate (%)	References
A. baumannii	China	51 carbapenem-resistant clinical isolates	23.5	[222]
Enterobacteriaceae	9 countries	35 carbapenemase-positive clinical strains	17.1	[311]
E. coli	USA	570 strains from retail meats	0.4	[429]
E. coli	China	179 isolates from retail meats	0	[157]
P. mirabilis	China	52 isolates from cooked meat products	0	[159]
S. aureus	Canada	40 nasal MRSA isolates in commercial swine herds	90.0	[356]
S. aureus	Turkey	10 clinical MRSA isolates	70.0	[151]
S. aureus	USA	11 community environmental isolates	45.5	[139]
S. aureus	Hong Kong	100 MRSA isolates from pig carcasses	45.0	[412]
S. aureus	Turkey	19 clinical MSSA isolates	21.1	[151]
S. aureus	Portugal	74 isolates from pets, livestock, the environment and humans in contact with animals	18.9	[73]
S. aureus	Denmark	45 MRSA isolates from pig farms	6.7	[342]
S. aureus	China	152 isolates from male patients with urogenital tract infection	0	[420]
S. haemolyticus	Argentina	21 clinical isolates	52.4	[69]
Staphylococcus spp.	USA	52 community environmental CNS isolates	38.5	[139]
Staphylococcus spp.	Turkey	13 clinical MSCNS isolates	23.1	[151]
Staphylococcus spp.	Turkey	27 clinical MRCNS isolates	7.4	[151]
Staphylococcus spp.	Norway	42 QAC-resistant isolates (35 bovine and 7 caprine)	4.8	[28]

qacH was shown to reduce to susceptibility of *S. saprophyticus* to BAC (MIC values: 10 vs. 4 mg/l) and *S. aureus* (MIC values: 10 vs. 2 mg/l) [142].

qacH and bcrABC confer resistance to BAC in *L. monocytogenes*. Six hundred and eighty isolates from nine Norwegian meat and salmon processing plants were investigated. QacH and bcrABC were detected in 101 isolates. Isolates with qacH and bcrABC showed increased tolerance to BAC with minimal inhibitory concentrations of 5–12, 10–13 and <5 mg/l for strains with qacH, bcrABC, and neither gene, respectively. Residuals of BAC may be present in concentrations after sanitation in the industry that results in a growth advantage for bacteria with such resistance genes [268].

Table 10.19 Detection rates of qacH in isolates from various bacterial species

Species	Country	Number of isolates and source	qacH detection rate (%)	References
E. coli	China	179 isolates from retail meats	2.2	[157]
L. monoctogenes	Switzerland and Finland	45 isolates from food, food production environment and humans	48.9	[255]
L. monocytogenes	Switzerland	142 isolates from food processing	14.1	[91]
		25 isolates with BAC MIC ≥ 10 mg/l	80.0	
P. mirabilis	China	52 isolates from cooked meat products	3.8	[159]
S. aureus	Canada	40 nasal MRSA isolates in commercial swine herds	57.5	[356]
S. aureus	USA	11 community environmental isolates	27.3	[139]
S. aureus	China	152 isolates from male patients with urogenital tract infection	1.3	[420]
S. aureus	Turkey	19 clinical MSSA isolates	0	[151]
S. aureus	Turkey	10 clinical MRSA isolates	0	[151]
S. epidermidis	Sweden	143 clinical isolates mainly from prosthetic joint infections (61) and post-operative infections after cardiac surgery (31)	0.7	[316]
Staphylococcus spp.	USA	52 community environmental CNS isolates	25.0	[139]
Staphylococcus spp.	Hong Kong	67 CNS isolates from orthokeratology lens wearers	7.5	[128]
Staphylococcus spp.	Turkey	13 clinical MSCNS isolates	0	[151]
Staphylococcus spp.	Turkey	27 clinical MRCNS isolates	0	[151]

10.5.4.7 qacJ

QacJ has been mainly found in staphylococci, both in *S. aureus* (0 to 40.0%) and CNS (9.6 to 46.2%). Among Gram-negative species from organic food, qacJ was only detected in one *E. cloacae* isolate (Table 10.20).

10.5.4.8 emrC

The BAC tolerance gene emrC can be found on the plasmid pLMST6 which is associated with unfavourable outcome in patients with meningitis caused by *L. monocytogenes*. Isolates harbouring emrC were growth inhibited at higher levels of benzalkonium chloride (median 60 mg/L vs. 15 mg/L; $p < 0.001$) and had higher MICs for amoxicillin and gentamicin compared with isolates without emrC (both $p < 0.001$). These findings warrant consideration of disinfectants used in the food

Table 10.20 Detection rates of qacJ in isolates from various bacterial species

Species	Country	Number of isolates and source	qacJ detection rate (%)	References
Chryseobacterium spp.	Spain	2 isolates from organic foods	0	[103]
E. cloacae	Spain	2 isolates from organic foods	50.0	[103]
E. ludwigii	Spain	2 isolates from organic foods	0	[103]
Enterobacter spp.	Spain	7 isolates from organic foods	0	[103]
K. oxytoca	Spain	2 isolates from organic foods	0	[103]
K. terrigena	Spain	1 isolate from organic foods	0	[103]
P. agglomerans	Spain	7 isolates from organic foods	0	[103]
P. ananatis	Spain	2 isolates from organic foods	0	[103]
Salmonella spp.	Spain	3 isolates from organic foods	0	[103]
S. aureus	Turkey	10 clinical MRSA isolates	40.0	[151]
S. aureus	USA	11 community environmental isolates	27.3	[139]
S. aureus	Turkey	19 clinical MSSA isolates	21.1	[151]
S. aureus	China	152 isolates from male patients with urogenital tract infection	0	[420]
Staphylococcus spp.	Turkey	13 clinical MSCNS isolates	46.2	[151]
Staphylococcus spp.	Turkey	27 clinical MRCNS isolates	14.8	[151]
Staphylococcus spp.	USA	52 community environmental CNS isolates	9.6	[139]
Staphylococcus spp.	Norway	42 QAC-resistant isolates (35 bovine and 7 caprine)	4.8	[28]

processing industry that selects for resistance mechanisms and may, inadvertently, lead to increased risk of poor disease outcome [191].

10.5.4.9 emrE

emrE has mainly been detected in *E. coli* (77.2%) and *P. mirabilis* (44.2%) (Table 10.21).

10.5.4.10 SigB

SigB encodes a major transcriptional regulator of stress response genes and is activated in static and continuous-flow biofilms. Disinfection treatments of planktonically grown cells and cells dispersed from static and continuous-flow *L. monocytogenes* biofilms showed that SigB is involved in the resistance of both planktonic cells and biofilms to the disinfectants benzalkonium chloride and peracetic acid [395].

Table 10.21 Detection rates of emrE in isolates from various bacterial species

Species	Country	Number of isolates	emrE detection rate (%)	References
E. coli	USA	570 strains from retail meats	77.2	[429]
L. monoctogenes	Switzerland and Finland	45 isolates from food, food production environment and humans	2.2	[255]
P. mirabilis	China	52 isolates from cooked meat products	44.2	[159]

10.5.5 Cell Membrane Changes

The outer membrane of *P. aeruginosa* can act as a barrier to prevent C16 benzalkonium but not C12 or C14 benzalkonium from entering the cell [129]. In *L. monocytogenes,* the adaptive response to BAC includes an increase in the cell membrane lipids in saturated-chain fatty acids (mainly C16:0 and C18:0) and unsaturated fatty acids (mainly C16:1 and C18:1) at the expense of branched-chain fatty acids (mainly Ca-15:0 and Ca-17:0) mainly because of neutral fatty acids, a decrease in lipid phosphorus, an obvious increase in the anionic phospholipids and a decrease in the amphiphilic phosphoaminolipid. These lipid changes could lead to decreased membrane fluidity and also to modifications of physicochemical properties of cell surface and thus changes in bacterial adhesion to abiotic surfaces [24].

Adaptive resistance to benzalkonium chloride was observed in *P. aeruginosa.* Analysis of the outer membrane protein of the resistant strain showed a significant increase in the level of expression of a protein (named OprR) which was expected to be an outer membrane-associated protein with homology to lipoproteins of other bacterial species. A correlation between the level of expression of OprR and the level of resistance of *P. aeruginosa* to BAC was observed [369].

High-level resistance to BAC in *P. fluorescens* was partly explained by reduced adsorption of BAC to the cell surface due to the decreased negative cell surface charge of the strain [280].

10.5.6 Efflux Pumps

Efflux pumps are one mechanism of resistance to BAC in *E. coli* [202]. They can induce changes in a small fraction of *E. coli* [288]. A multidrug efflux protein MdtM was detected in *E. coli* that belongs to the large and ubiquitous major facilitator superfamily (MFS). BAC, DDAC and some antibiotics are among the substrates transported by MdtM [205]. It was also shown with *E. coli* that many redundant multidrug resistance transporters enhance biofilm formation and drug tolerance including to BAC [21].

In *K. pneumoniae* the kpnGH efflux pump was described with a wide substrate specificity of the transporter including 14 antibiotics and BAC. kpnGH mediates antimicrobial resistance by active extrusion in *K. pneumoniae* [362].

Efflux pumps are a common mechanism for BAC resistance in *L. monocytogenes* [3]. Plasmid- and chromosome-encoded efflux pumps can mediate the BAC resistance of *L. monocytogenes* isolated from food [158]. The efflux pump Mdrl is at least partly responsible for the adaptation to BAC in *L. monocytogenes* [328]. bcrABC was detected in 17.8% of 45 strains of *L. monocytogenes* from Switzerland and Finland [255]. It was also detected in 70 of 71 BAC-resistant *L. monocytogenes* strains but in none of the BAC-susceptible ones [88]. The bcrABC cassette can lead to acquired BAC tolerance in *E. coli* by horizontal transfer [415]. BAC-resistant *L. welshimeri* and *L. innocua* harbouring bcrABC, along with the cadmium resistance determinant cadA2, were able to transfer resistance to other nonpathogenic listeriae as well as to *L. monocytogenes* of diverse serotypes [170]. It was also shown that efflux pumps play a role in plasmid-mediated tolerance to BAC in bcrABC positive *L. monocytogenes* [415]. bcrABC was detected in 3 of 142 *L. monocytogenes* isolates from food processing [91].

Efflux pumps explain resistance to BAC also in *P. aeruginosa* isolated from biofilm in dairy facilities [307]. High-level resistance to BAC in *P. fluorescens* was partly explained by putative efflux system which seemed to be unique in that it excretes only a certain range of cationic membrane-acting disinfectants belonging to QACs [280].

Overexpression of efflux pumps AcrAB or AcrEF was detected in BAC-resistant mutants of *S. Typhimurium* [130].

10.5.7 Plasmids for Resistance Transfer

Transposable elements can be thought of as keys that provide access to an accumulated stockpile of potentially advantageous functions, such as, but not restricted to, antimicrobial-resistance determinants. This stockpile is dynamic, responding not only to selective forces but also undergoing continuous competition-driven refinement by trial and error. The rapid emergence of multiple drug-resistant staphylococci represents a striking illustration of the evolutionary potential availed by the exploitation of such a resource [106]. Plasmid curing resulted in a reduction of MIC in all four BAC naturally resistant strains of *L. monocytogenes* [328]. *A. xylosoxidans* has been described as an emerging pathogen carrying different elements involved in horizontal gene transfer [385].

10.5.7.1 pSK1

Transcriptional profiling has revealed that plasmid carriage most likely has a minimal impact on the host, a factor that may contribute to the ability of pSK1 family plasmids to carry multiple resistance determinants [156]. Specific plasmids in *S. aureus* and *E. coli* (pSK1 and R471-1) influenced the bacteriostatic efficacy of BAC at 10 mg/l to some extent so that a concentration of 30 mg/l is necessary to comply with the criteria of the British Pharmacopeia for ophthalmic and parenteral products [79].

10.5.7.2 pSK41

pSK41 is a large, low-copy number, conjugative plasmid from *S. aureus* that is representative of a family of staphylococcal plasmids that confer multiple resistances to a wide range of antimicrobial agents. It carries the smr resistance determinant [219].

10.5.7.3 pSK108

The *S. epidermidis* plasmid pSK108 encodes a qacC multidrug resistance determinant. Sequence analysis suggests that the DNA segment containing qacC represents a resistance gene cassette that has undergone horizontal genetic exchange [210].

10.5.7.4 pLM80

pLM80 was associated with BAC resistance in a *L. monocytogenes* strain H7550 involved in a multistate outbreak involving contaminated hot dogs [94]. A putative BAC resistance cassette, known as bcrABC, was previously identified on pLM80 of *L. monocytogenes* H7550 strain involved in the 1998–1999 USA listeriosis outbreak, as well as in other *Listeria* sequenced genomes. BAC-associated resistance cassette is composed of TetR family transcriptional regulator (bcrA) and two SMR (small multidrug resistance) genes (bcrB and bcrC), all essential for imparting BAC resistance [374].

10.5.7.5 pSP187

pSP187 carrying smr encoding resistance to QACs was detected in *S. pasteuri* recovered from bulk milk in a dairy cattle herd [27].

10.5.7.6 pNVH01

pNVH01 harbouring qacJ was first described in three staphylococcal species (*S. aureus*, *S. intermedius*, *S. simulans*) associated with chronic infections in horses. Clonal spread of a qacJ-harbouring *S. aureus* strain and the horizontal transfer of pNVH01 were suggested. A recent gene transfer has probably occurred. In three of the horses, a skin preparation containing cetyltrimethylammonium bromide had been used extensively for several years. This might explain the selection of staphylococci harbouring the novel QAC resistance gene within and between different equine staphylococcal species [29].

10.5.7.7 pST94

The resistance plasmid pST94 was detected in staphylococci carrying qacG encoding resistance to BAC [144].

10.5.7.8 pST827

Plasmid pST827 is involved in resistance to benzalkonium chloride in meat-associated staphylococci as shown in 191 isolates from food processing plants [141].

10.5.7.9 pSx1

An extensively drug-resistant *S. xiamenensis* T17 isolated from hospital effluents in Algeria revealed the presence of a novel 268.4 kb plasmid designated pSx1 carrying also a class 1 integron with the qacG gene cassette [422].

10.5.8 Transposons for Resistance Transfer

Tn6188, an integrated chromosomally transposon, has been described to provide an increased tolerance of *L. monocytogenes* strains to BAC [274, 275, 374] thereby improving persistence despite use of disinfectants [302]. The tolerance is mediated via qacH, a small multidrug resistance protein family (SMR) transporter. Tn6188 confers tolerance of *L. monocytogenes* to ionic liquids based on imidazolium and ammonium cations [257].

Tn1546 was detected in a clinical isolate of *S. aureus* with high-level resistance to vancomycin (minimal inhibitory concentration = 1024 mg/l) in June 2002. This isolate harboured a 57.9-kilobase multiresistance conjugative plasmid within which Tn1546 (vanA) was integrated. Additional elements on the plasmid encoded resistance to trimethoprim (dfrA), beta-lactams (blaZ), aminoglycosides (aacA-aphD) and disinfectants (qacC). Genetic analyses suggest that the long anticipated transfer of vancomycin resistance to a methicillin-resistant *S. aureus* occurred in vivo by interspecies transfer of Tn1546 from a co-isolate of *E. faecalis* [408].

10.5.9 Class I Integrons

A class I integron was detected in 22 of 36 MDR *P. aeruginosa* isolates. Integron I-positive isolates showed reduced susceptibility to tested biocides including benzalkonium chloride. Class I integron may be responsible for generating MDR *P. aeruginosa* isolates with reduced susceptibility to biocides [162].

10.5.10 Infections Associated with Contaminated BAC Solutions or Products

Contaminated aqueous disinfectants or solutions based on BAC have resulted occasionally in outbreaks of healthcare-associated infections (Table 10.22). Almost all of them were caused by Gram-negative bacterial species. The most common types of infection were blood stream infections followed by septic arthritis or joint infections.

Table 10.22 Infections associated with contaminated BAC solutions or products adapted from [164]

Bacterial species	Type and number of infections	Patient population	Source of infection and role of BAC resistance	BAC concentration	References
Achromobacter spp.	1 case of infection after brain cyst surgery	Patient with *Achromobacter* isolates from clinical specimen	Identical strain was isolated from contaminated surface disinfectant based on BAC	No data	[196]
B. cepacia	34 cases, 21 with an infection, 13 with a colonization	Various hospital wards including haemato-oncology and endocrinology	Contaminated aqueous BAC solution used for treatment of skin, soft tissue and catheters	1:1000 of stock solution with unknown concentration	[207]
B. cepacia complex	46 cases, half of them colonization, other half blood stream infections or other infections	Patients from 9 institutions	Ready to use washing gloves preserved with BAC	0.1%	[359]
E. aerogenes	11 cases of infection, mainly blood stream infections	Patients from gastroenterology and haematology	Contaminated aqueous BAC solution with gauze was used for skin antisepsis	1:750 of stock solution with unknown concentration	[239]
M. abscessus	12 joint infections	Outpatients for intraarticular or periarticular injections	BAC-soaked cotton ball samples; relative resistance to commercial BAC solutions	0.13%	[378]
Pseudomonas-Achromobacter spp.	4 cases of blood stream infections, 2 of them fatal	General medical department	Container of cotton pledgets soaked in aqueous BAC	1:1000 of stock solution with unknown concentration	[209]
P. aeruginosa	11 cases of bacteraemia	Dialysis unit	Contaminated coils treated before reuse with BAC in tap water	1:750 of stock solution with unknown concentration	[400]

(continued)

Table 10.22 (continued)

Bacterial species	Type and number of infections	Patient population	Source of infection and role of BAC resistance	BAC concentration	References
P. aeruginosa	28 cases of abscess	Patients receiving intramuscular corticosteroid injections	Contaminated BAC solution used for wiping vial septa before puncturing with a needle	5.7%	[298]
P. cepacia	9 patients with septicaemia	General medicine	Contaminated aqueous BAC solution was used for skin antisepsis	1:750 of stock solution with unknown concentration	[111]
P. cepacia and *Enterobacter* spp.	Pseudooutbreak with 79 patients	Community hospital	Contaminated aqueous BAC solution was used for skin antisepsis	No data	[168]
P. kingii	Urinary tract infections	Various hospitals	Contaminated BAC solution as part of a catheter tray	No data	[51]
Pseudomonas EO-1	12 cases of urinary tract infection	Hospital	Contaminated BAC solution	No data	[135]
Pseudomonas spp.	40 cases of bacteraemia	County hospital	Contaminated BAC solution (0.1%) containing cotton swabs used for storage of needles and catheters	0.1%	[314]
S. marcescens	81 cases of infection in dogs and cats, mainly septicaemia and respiratory tract infection	Animal hospital	Contaminated BAC sponge pots located in the ICU, surgery rooms, and outpatient clinic areas	0.025%	[110]
S. marcescens	11 cases of septic arthritis	Office practice	Canister of cotton balls soaked in aqueous BAC; the strain was able to survive in the 1:100 BAC solution	1:750 of stock solution with unknown concentration	[284, 285]

(continued)

Table 10.22 (continued)

Bacterial species	Type and number of infections	Patient population	Source of infection and role of BAC resistance	BAC concentration	References
S. marcescens	1 case of nosocomial meningitis	Outpatient treatment	Contaminated skin antiseptic based on BAC was used prior to intrathecal infections for back pain	No data	[340]
S. marcescens	3 cases of sternal wound infection, 2 cases of bacteraemia, 1 case of endocarditis	Cardiac surgery	A spray disinfectant based on BAC was used before surgery in the cardiac operating room; no cleaning of spray bottles before refilling	0.045%	[93]

Another interesting observation was made in the Netherlands regarding the epidemiology of *L. monocytogenes* meningitis over 25 years. Peak incidence rates were observed in neonates (0.61 per 100,000 live births) and older adults (peak at 87 year; 0.53 cases per 100,000 population of the same age). Most clonal complexes (CC) decreased over time. Only CC 6 increased significantly from 2 to 26%. The emrC efflux transporter has been shown to be associated with the emergence of CC 6 in the Netherlands. The emrC gene encodes an efflux protein that pumps quaternary ammonium compounds out of the cell and increases the capacity to form a biofilm, resulting in benzalkonium chloride tolerance. Benzalkonium chloride is extensively used in the food processing industry as a disinfectant agent. Reduced susceptibility to benzalkonium chloride may explain the increasing incidence of CC 6 isolates in the Netherlands between 1985 and 2014 [183].

10.5.11 Contaminated BAC Solutions Without Evidence for Infections

P. fluorescens was isolated from a 10% aqueous BAC solution stored in a loosely capped bottle in the department of pharmacy of a university hospital. It was possible to show growth of the BAC-resistant *P. fluorescens* in 5% BAC, but other strains were not able to grow in 0.1% BAC. The strain was unable to decompose BAC [281]. A strain of *P. aeruginosa* and a waterborne strain of *Pseudomonas* spp. were also able to resist BAC at 0.36% or 0.4% [5]. The investigation of 20 samples of 0.02% BAC solutions used for intermittent self-catheterization revealed that 60% of them were contaminated, sometimes with 3×10^6 cells per ml, mainly with *B. cepacia* (9), *P. fluorescens* (4) and *Aeromonas* spp. (1) [134]. *B. cepacia* is known to be more resistant to BAC compared to other bacterial species [199].

10.6 Cross-Tolerance to Other Biocidal Agents

The spectrum of cross-resistance to other biocidal compounds depends mostly on the mode of resistance. Most efflux pumps are not substrate specific (see also Sect. 10.5.6) so that biocidal agents with a similar structure (e.g. cationic surface-active ingredients and other quaternary ammonium compounds) are also likely to be less effective. This effect has been shown in a study with 76 biocide-sensitive bacterial strains previously isolated from organic foods. They were exposed to BAC during several passages with gradually higher concentrations. Tolerance to BAC increased in 67 strains, and 97% of the adapted strains were more tolerant to DDAB, 95.5% increased their tolerance to triclosan and 94% to chlorhexidine [114]. A lower susceptibility to other biocidal agents was also described for some species to didecyldimethylammonium chloride other QACs, alkylamine and sodium hypochlorite (see also Table 10.10).

Cross-resistance has also been described with cadmium in *L. monocytogenes*. In 21 *L. monocytogenes* isolates, a correlation between resistance to BAC and cadmium was described [423]. Among 123 *L. monocytogenes* isolates from turkey processing plants, a total of 57 (46.3%) were regarded as BAC resistant, all of them were also resistant to cadmium [273].

10.7 Cross-Tolerance to Antibiotics

Some studies indicate a cross-tolerance between BAC and selected antibiotics. In 153 *E. coli* blood culture isolates, for example, a higher MIC of BAC was associated with a decreased susceptibility to cotrimoxazole [41]. In 52 *Pseudomonas* spp. from meat chain production, a correlation between resistance to BAC and ampicillin, amoxicillin, erythromycin and trimethoprim was found [203]. Repeated in vitro exposure of *S. Typhimurium* cells to QAC selects for reduced susceptibility to several antibiotics (chloramphenicol, tetracycline, ampicillin, acriflavine). This is associated with overexpression of AcrAB efflux pump [167]. In 1,632 clinical *S. aureus* isolates, a correlation of susceptibility profiles of at least 0.4 was found to BAC and the quinolones, beta-lactams and macrolides [294]. And a MIC value >2 mg/l for BAC was associated with multidrug antibiotic resistance in *S. aureus* as demonstrated in 1,632 human clinical *S. aureus* isolates from different geographical regions [64].

Other studies showed no cross-resistance to antibiotics. In 200 *L. monocytogenes* isolates, no association between resistance to BAC and antibiotics was found [3]. In 122 isolates of *Salmonella* spp. from poultry and swine, multiple antibiotic-resistant bacteria were no more tolerant to BAC than the non-multidrug-resistant strains [62]. In 103 Gram-negative clinical isolates, no association between resistance to multiple antibiotic and quaternary ammonium compounds was found [192].

Cross-resistance to various antibiotics such as ampicillin, cefotaxime or ceftazidime was found after low-level exposure in isolates of *B. cepacia complex*, *Chryseobacterium* spp., *Enterobacter* spp., *E. coli*, *Klebsiella* spp., *Pantoea* spp. and *Salmonella spp.* Cross-resistance to selected antibiotics was also detected in *B. cereus*, *B. licheniformis*, *Bacillus* spp., *E. casseliflavus*, *E. durans*, *E. faecalis*, *E. faecium*, *Enterococcus* spp., *S. saprophyticus* and *Staphylococcus* spp. (see also Table 10.10).

The unmet needs for adequate detection of reduced susceptibility to QACs and antibiotics include a consensus definition for resistance, epidemiological cut-off values and clinical resistance breakpoints [42].

10.8 Role of Biofilm

10.8.1 Effect on Biofilm Development

Biofilm development is inhibited by BAC in a few species. *L. monocytogenes* biofilm formation was inhibited by BAC at 1.25–10 mg/l when exposed for 48 h, and by

Table 10.23 Biofilm removal rates after exposure to BAC

Type of biofilm	Concentration	Exposure time	Biofilm removal rate	References
A. acidoterrestris biofilm on stainless steel, nylon and PVC surfaces	0.001562% (S)	10 min	Partial removal	[86]
E. coli MG 1655, 24-h incubation in microtiter plates followed by 6 d with and without 0.9 mM BAC	1 mM[a] (S)	30 min	No significant reduction	[235]
L. monocytogenes food processing plant isolate, 6-d incubation in polystyrene containers	0.04% (S)	1 min	0%	[381]
		5 min	16%	
		15 min	26%	
	0.01% (S)	1 min	1%	
		5 min	11%	
		15 min	20%	
P. aeruginosa ATCC 700928, 24-h incubation in microplates	0.1% (S)	60 min	0%	[383]
P. aeruginosa (8 dairy isolates exhibiting high biofilm formation, 24-h incubation in microtiter plates)	>0.09% (S)	5 min	"eradication"	[306]
	0.07–0.1% (S)	15 min		
	0.055–0.09% (S)	30 min		
	0.04–0.07% (S)	60 min		
P. aeruginosa ATCC 10145 and a GI endoscope biofilm isolate, 4-d incubation on polystyrene	0.036% (S)	30 min	22–38%	[233]
P. aeruginosa ATCC 10154, 24-h incubation in microtiter plates followed by 6-d incubation with and without 0.9 mM BAC	1 mM[a] (S)	30 min	No significant reduction	[235]
P. fluorescens ATCC 13525, grown for 7 d on stainless steel	0.9[a] mM	30 min	25%	[354]
	0.5[a] mM		14%	
	0.25[a] mM		9%	
	0.125[a] mM		0%	
S. Enteritidis ATCC 4931, 6-d incubation on polycarbonate	0.001% (S)	2 d	50%	[240]
S. liquefaciens raw-chicken plant isolate, 3-d incubation on stainless steel	0.01% (S)	6 min	21%	[218]
S. putrefaciens raw-chicken plant isolate, 3-d incubation on stainless steel	0.01% (S)	6 min	7%	[218]
S. aureus ATCC 6538, 72-h incubation in microplates	0.1% (S)	60 min	0%	[383]

[a]Molecular weight not described

BAC at 5–10 mg/l when exposed for 6 d [313]. *L. monocytogenes* biofilm formation measured with 4 strains was consistently lower during exposure to 5 mg/l BAC for 7 d. But with 2.5 mg/l BAC the biofilm formation was lower in 2 strains, and with 1.25 mg/l it was higher in 1 strain and lower in 2 strains [304]. BAC at 125 mg/l inhibited biofilm formation in two outbreak *S. Enteritidis* strains [125]. Biofilm formation by BAC was also inhibited at concentrations of the MIC or higher for *E. coli*, *S. epidermidis* and *P. aeruginosa* [149]. BAC was able to inhibit biofilm formation in 9 *S. aureus* isolates by 90% during 24-h incubation at concentrations between 16 and 200 mg/l, the ATCC strain 25923 required 1,015 mg/l [427].

Enhanced biofilm formation after low-level exposure, however, was described for *E. coli* and *S. epidermidis* (see also Table 10.10). In *S. epidermidis,* the effect depends on the BAC concentration. At 0.0001% was also able to increase biofilm formation in three *S. epidermidis* strains, but at 0.0002, 0.0003, 0.0004 and 0.0005%, biofilm formation was reduced [54]. Another study shows that biofilm formation was significantly enhanced by BAC at concentration below the MIC value of *S. aureus*, *S. agalactiae* (both isolates from mastitis cow milk) and *E. coli* (dead poultry isolate). There was a clear association higher biofilm formation and lower BAC concentrations [92].

10.8.2 Effect on Biofilm Removal

Biofilm removal by BAC is mostly poor with removal rates between 0 and 20% as shown with various species such as *E. coli*, *L. monocytogenes*, *P. aeruginosa*, *P. fluorescens*, *S. liquefaciens*, *S. putrefaciens* and *S. aureus*. Only with a *Salmonella* spp., the biofilm removal rate was higher with 50% but required a 2 d exposure. Biofilm eradication was described in one study for a *P. aeruginosa* biofilm (Table 10.23).

10.8.3 Effect on Biofilm Fixation

The biofilm mechanical stability (*P. fluorescens* ATCC 13525, grown for 7 d on stainless steel) is increased by BAC at 0.25 mM (+32%), 0.5 mM (+57%) and at 0.9 mM (+93%) [354].

10.9 Summary

The principal antimicrobial activity of BAC is summarized in Table 10.24.

The key findings on acquired resistance and cross-resistance including the role of biofilm for selecting resistant isolates are summarized in Table 10.25.

Table 10.24 Overview on the typical exposure times required for BAC to achieve sufficient biocidal activity against the different target micro-organisms

Target micro-organisms	Species	Concentration (%)	Exposure time (min)
Bacteria	Most species	1	5[a]
Fungi	Yeasts	0.2	5
	Aspergillus spp.	0.5	60
	Some food-associated fungi	>1.5	>10
Mycobacteria	Insufficient mycobactericidal activity (0.1% for 2 h)		

[a]In biofilm the efficacy will be lower

Table 10.25 Key findings on acquired BAC resistance, the effect of low level exposure, cross-tolerance to other biocides and antibiotics, and its effect on biofilm

Parameter	Species	Findings
Elevated MIC values	*A. hydrophila*	≤ 31,300 mg/l
	B. cereus, *E. meningoseptica*	≤ 7,800 mg/l
	P. aeruginosa	≤ 5,000 mg/l
	L. monocytogenes	≤ 625 mg/l
	E. cloacae	≤ 512 mg/l
	A. xylosoxidans, *B. cepacia*	≤ 500 mg/l
	P. mirabilis	≤ 400 mg/l
Proposed MIC values to determine resistance	*C. albicans*	16 mg/l
	Enterobacter spp.	32 mg/l
	E. faecium	8 mg/l
	E. faecalis	8 mg/l
	E. coli	64 mg/l
	K. pneumoniae	32 mg/l
	Salmonella spp.	128 mg/l
	S. aureus	16 mg/l
Cross-tolerance to biocides	*B. licheniformis*, *Bacillus* spp., *Chryseobacterium* spp., *E. cloacae*, *E. ludwigii*, *Enterobacter* spp., *E. casseliflavus*, *E. durans*, *E. faecalis*, *E. faecium*, *Enterococcus* spp., *E. coli*, *K. oxytoca*, *Klebsiella* spp., *P. agglomerans*, *P. ananatis*, *Pantoea* spp., *S.* Typhimurium, *S.* Virchow, *Salmonella* spp., *S. saprophyticus*, *Staphylococcus* spp.	Cross-tolerance to chlorhexidine
	B. cereus, *B. licheniformis*, *Bacillus* spp., *Chryseobacterium* spp., *E. ludwigii*, *Enterobacter* spp., *E. casseliflavus*, *E. durans*, *E. faecalis*, *E. faecium*, *Enterococcus* spp., *K. oxytoca*, *Klebsiella* spp., *P. agglomerans*, *P. ananatis*, *Pantoea* spp., *Salmonella* spp., *S. saprophyticus*, *Staphylococcus* spp.	Cross-tolerance to didecyldimethylammonium bromide
	E. coli	Cross-tolerance to didecyldimethylammonium chloride
	B. cereus, *B. licheniformis*, *Bacillus* spp., *Chryseobacterium* spp., *E. cloacae*, *E. ludwigii*, *Enterobacter* spp., *E. casseliflavus*, *E.*	Cross-tolerance to triclosan

(continued)

Table 10.25 (continued)

Parameter	Species	Findings
	durans, *E. faecalis*, *E. faecium*, *Enterococcus* spp., *E. coli*, *K. oxytoca*, *Klebsiella* spp., *P. agglomerans*, *P. ananatis*, *Pantoea* spp., *S.* Virchow, *Salmonella* spp., *S. saprophyticus*, *Staphylococcus* spp.	
	B. cereus, *B. licheniformis*, *Bacillus* spp., *Chryseobacterium* spp., *E. cloacae*, *E. ludwigii*, *Enterobacter* spp., *E. casseliflavus*, *E. durans*, *E. faecalis*, *E. faecium*, *Enterococcus* spp., *K. oxytoca*, *Klebsiella* spp., *P. agglomerans*, *P. ananatis*, *Pantoea* spp., *Salmonella* spp., *S. saprophyticus*, *Staphylococcus* spp.	Cross-tolerance to hexachlorophene
	L. monocytogenes	Cross-tolerance to other QAC, alkylamine and sodium hypochlorite
Cross-tolerance to antibiotics	*B. cepacia complex*, *Chryseobacterium* spp., *Enterobacter* spp., *E. coli*, *Klebsiella* spp., *Pantoea* spp., *Salmonella* spp., *B. cereus*, *B. licheniformis*, *Bacillus* spp., *E. casseliflavus*, *E. durans*, *E. faecalis*, *E. faecium*, *Enterococcus* spp., *S. saprophyticus*, *Staphylococcus* spp.	Cross-tolerance after low level exposure to various antibiotics such as ampicillin, cefotaxime or ceftazidime
Resistance mechanisms	Mainly *S. aureus*, CNS, *K. pneumoniae*	qacA/B resistance gene
	Mainly *S. aureus*, CNS	smr (qacC) resistance gene
	Mainly *A. baumannii*, *P. aeruginosa*, *P. mirabilis*, *K. pneumoniae*, *E. coli*, *Vibrio* spp.	qacE and qacEΔ resistance genes
	Mainly *Salmonella* spp.	qacF resistance gene
	Mainly *S. aureus*, CNS, *A. baumannii*	qacG resistance gene
	Mainly *S. aureus*, *L. monocytogenes*	qacH resistance gene
	Mainly *S. aureus*, CNS	qacJ resistance gene
	Mainly *E. coli*, *P. mirabilis*	emrE resistance gene
	P. aeruginosa, *P. fluorescens*	Cell membrane changes

(continued)

Table 10.25 (continued)

Parameter	Species	Findings
	E. coli, *K. pneumoniae*, *Listeria* spp., *P. aeruginosa*, *P. fluorescens*, *S.* Typhimurium	Efflux pumps
	S. aureus, *S. epidermidis*, *E. coli*, *L. monocytogenes*, *S. pasteuri*, *S. intermedius*, *S. simulans*, *S. xiamenensis*	Plasmids
	Achromobacter spp., *B. cepacia*, *B. cepacia complex*, *E. aerogenes*, *M. abscessus*, *Pseudomonas-Achromobacter* spp., *P. aeruginosa*, *P. cepacia*, *Enterobacter* spp. *P. kingii*, *Pseudomonas* spp., *S. marcescens*	Contaminated BAC solutions of products (up to 5.7% BAC) leading to various types of nosocomial infections, mainly blood stream infections, septic arthritis or joint infections
Effect of low-level exposure	*A. xylosoxidans*, *C. jejuni*, *C. indologenes*, *Chrysobacterium* spp., *C. sakazakii*, *H. gallinarum*, *M. osloensis*, *P. nitroreductans*, *S. enteritidis*, *Salmonella* spp., *S. multivorum*, *S. maltophilia*, *B. cereus*, *C. pseudogenitalum*, *E. saccharolyticus*, *S. cohnii*, *S. epidermidis*, *S. kloosii* and *S. lugdenensis*	No MIC increase
	A. hydrophila, *A. jandaei*, *C. coli*, *Citrobacter* spp., *E. coli*, *K. oxytoca*, *P. aeruginosa*, *P. putida*, *Pseudomonas* spp., *Pseudoxanthomonas* spp., *S.* Typhimurium, *Salmonella* spp., *E. durans*, *E. faecalis*, *Eubacterium* spp., *L. monocytogenes*, *M. phyllosphaerae*, *M. luteus*, *S. aureus*, *S. capitis*, *S. caprae*, *S. hominis*, *S. saprophyticus*, *S. warneii*, *Staphylococcus* spp.	Weak MIC increase ($\leq$4-fold)
	E. cloacae, *Enterobacter* spp., *Klebsiella* spp., *P. agglomerans*, *P. ananatis*, *Pantoea* spp., *Salmonella* spp., *B. cereus*, *B. licheniformis*, *Bacillus* spp., *E. casseliflavus*, *E. faecalis*, *E. faecium*, *Enterococcus* spp., *S. haemolyticus*, *S. saprophyticus*, *Staphylococcus* spp.	Strong (>4-fold) but unstable MIC increase
	A. baumannii, *Chryseobacterium* spp., *E. ludwigii*, *Enterobacter* spp., *E. coli*, *Pantoea* spp., *P. aeruginosa*, *S. enterica serovar* Typhimurium, *S.* Enteritidis, *S.* Typhimurium, *S.* Virchow, *Salmonella* spp., *L. monocytogenes*, *S. aureus*	Strong and stable MIC increase
	A. proteolyticus, *Ralstonia* spp., *C. renale* group	Strong MIC increase (unknown stability)

(continued)

Table 10.25 (continued)

Parameter	Species	Findings
	Pantoea spp. (up to 500-fold) *Enterobacter* spp. (up to 300-fold) *Salmonella* spp., *S. saprophyticus* and *B. cereus* (up to 200-fold) *Staphylococcus* spp. (up to 150-fold) *E. coli* (up to 100-fold) *C. renale* group (up to 62.5-fold) *B. cereus*, *E. faecalis* and *E. faecium* (up to 50-fold) *Klebsiella* spp. (up to 36-fold) *P. aeruginosa* (up to 33-fold) *A. baumannii* (up to 31-fold)	Strongest MIC change after low-level exposure
	S. Typhimurium (3,000 mg/l) *P. aeruginosa* and *Pantoea* spp. (2,500 mg/l) *Enterobacter* spp. (1,500 mg/l) *E. coli* and *S. saprophyticus* (1,000 mg/l) *B. cepacia complex* (500 mg/l) *B. cereus* (400 mg/l) *Staphylococcus* spp. (100 mg/l) *S. haemolyticus* (15.6 mg/l)	Highest MIC values after low-level exposure
	B. cepacia complex, *E. coli* and *L. monocytogenes*	Up-regulation of transporter and efflux pump genes
	E. coli and *S. epidermidis*	Enhanced biofilm formation
	L. monocytogenes	Activation of non-specific efflux pumps
	L. monocytogenes	Induction of virulence gene expression
	S. Enteritidis	Up-regulation of protective key proteins
Biofilm	Development	Impaired in *L. monocytogenes*, *S. Enteritidis*, *E. coli*, *S. epidermidis*, *S. aureus* and *P. aeruginosa*
		No change in *S. aureus*
		Enhanced in *E. coli*, *S. aureus*, *S. agalactiae* and *S. epidermidis*
	Removal	Mostly poor (*E. coli*, *L. monocytogenes*, *P. aeruginosa*, *P. fluorescens*, *S. liquefaciens*, *S. putrefaciens* and *S. aureus*)
	Fixation	Increase of biofilm mechanical stability in *P. fluorescens*

References

1. Aarestrup FM, Hasman H (2004) Susceptibility of different bacterial species isolated from food animals to copper sulphate, zinc chloride and antimicrobial substances used for disinfection. Vet Microbiol 100(1–2):83–89. https://doi.org/10.1016/j.vetmic.2004.01.013
2. Aarestrup FM, Knochel S, Hasman H (2007) Antimicrobial susceptibility of Listeria monocytogenes from food products. Foodborne Pathog Dis 4(2):216–221. https://doi.org/10.1089/fpd.2006.0078
3. Aase B, Sundheim G, Langsrud S, Rorvik LM (2000) Occurrence of and a possible mechanism for resistance to a quaternary ammonium compound in Listeria monocytogenes. Int J Food Microbiol 62(1–2):57–63
4. Abuzaid A, Hamouda A, Amyes SG (2012) Klebsiella pneumoniae susceptibility to biocides and its association with cepA, qacDeltaE and qacE efflux pump genes and antibiotic resistance. J Hosp Infect 81(2):87–91. https://doi.org/10.1016/j.jhin.2012.03.003
5. Adair FW, Geftic SG, Gelzer J (1969) Resistance of Pseudomonas to quaternary ammonium compounds. I. Growth in benzalkonium chloride solution. Appl Microbiol 18(3):299–302
6. Ahn Y, Kim JM, Lee YJ, LiPuma J, Hussong D, Marasa B, Cerniglia C (2017) Effects of extended storage of chlorhexidine gluconate and benzalkonium chloride solutions on the viability of Burkholderia cenocepacia. J Microbiol Biotechnol 27(12):2211–2220. https://doi.org/10.4014/jmb.1706.06034
7. Akamatsu T, Tabata K, Hironga M, Kawakami H, Uyeda M (1996) Transmission of Helicobacter pylori infection via flexible fiberoptic endoscopy. Am J Infect Control 24 (5):396–401
8. Akinkunmi EO, Lamikanra A (2012) Susceptibility of community associated methicillin resistant Staphylococcus aureus isolated from faeces to antiseptics. J Infect Developing Countries 6(4):317–323
9. Alam MM, Kobayashi N, Uehara N, Watanabe N (2003) Analysis on distribution and genomic diversity of high-level antiseptic resistance genes qacA and qacB in human clinical isolates of Staphylococcus aureus. Microb Drug Resist (Larchmont, NY) 9(2):109–121. https://doi.org/10.1089/107662903765826697
10. Ammar AM, Attia AM, Abd El-Hamid MI, El-Shorbagy IM, Abd El-Kader SA (2016) Genetic basis of resistance waves among methicillin resistant Staphylococcus aureus isolates recovered from milk and meat products in Egypt. Cell Mol Biol (Noisy-le-Grand, France) 62 (10):7–15
11. Araujo PA, Mergulhao F, Melo L, Simoes M (2014) The ability of an antimicrobial agent to penetrate a biofilm is not correlated with its killing or removal efficiency. Biofouling 30 (6):675–683. https://doi.org/10.1080/08927014.2014.904294
12. Argudin MA, Lauzat B, Kraushaar B, Alba P, Agerso Y, Cavaco L, Butaye P, Porrero MC, Battisti A, Tenhagen BA, Fetsch A, Guerra B (2016) Heavy metal and disinfectant resistance genes among livestock-associated methicillin-resistant Staphylococcus aureus isolates. Vet Microbiol 191:88–95. https://doi.org/10.1016/j.vetmic.2016.06.004
13. Argudin MA, Mendoza MC, Martin MC, Rodicio MR (2014) Molecular basis of antimicrobial drug resistance in Staphylococcus aureus isolates recovered from young healthy carriers in Spain. Microb Pathog 74:8–14. https://doi.org/10.1016/j.micpath.2014.06.005
14. Argudin MA, Vanderhaeghen W, Butaye P (2015) Diversity of antimicrobial resistance and virulence genes in methicillin-resistant non-Staphylococcus aureus staphylococci from veal calves. Res Vet Sci 99:10–16. https://doi.org/10.1016/j.rvsc.2015.01.004
15. Arioli S, Elli M, Ricci G, Mora D (2013) Assessment of the susceptibility of lactic acid bacteria to biocides. Int J Food Microbiol 163(1):1–5. https://doi.org/10.1016/j.ijfoodmicro.2013.02.002
16. Avrain L, Allain L, Vernozy-Rozand C, Kempf I (2003) Disinfectant susceptibility testing of avian and swine Campylobacter isolates by a filtration method. Vet Microbiol 96(1):35–40

17. Aykan SB, Caglar K, Engin ED, Sipahi AB, Sultan N, Yalinay Cirak M (2013) Investigation of the presence of disinfectant resistance genes qacA/B in nosocomial methicillin-resistant staphylococcus aureus isolates and evaluation of their in vitro disinfectant susceptibilities. Mikrobiyoloji Bul 47(1):1–10
18. Babaei M, Sulong A, Hamat R, Nordin S, Neela V (2015) Extremely high prevalence of antiseptic resistant Quaternary Ammonium Compound E gene among clinical isolates of multiple drug resistant Acinetobacter baumannii in Malaysia. Ann Clin Microbiol Antimicrob 14:11. https://doi.org/10.1186/s12941-015-0071-7
19. Ban GH, Kang DH (2016) Effect of sanitizer combined with steam heating on the inactivation of foodborne pathogens in a biofilm on stainless steel. Food Microbiol 55:47–54. https://doi.org/10.1016/j.fm.2015.11.003
20. Barroso JM (2013) COMMISSION DIRECTIVE 2013/7/EU of 21 February 2013 amending Directive 98/8/EC of the European Parliament and of the Council to include Alkyl (C12–16) dimethylbenzyl ammonium chloride as an active substance in Annex I thereto. Off J Eur Union 56(L 49):66-69
21. Bay DC, Stremick CA, Slipski CJ, Turner RJ (2017) Secondary multidrug efflux pump mutants alter Escherichia coli biofilm growth in the presence of cationic antimicrobial compounds. Res Microbiol 168(3):208–221. https://doi.org/10.1016/j.resmic.2016.11.003
22. Beier RC, Foley SL, Davidson MK, White DG, McDermott PF, Bodeis-Jones S, Zhao S, Andrews K, Crippen TL, Sheffield CL, Poole TL, Anderson RC, Nisbet DJ (2015) Characterization of antibiotic and disinfectant susceptibility profiles among Pseudomonas aeruginosa veterinary isolates recovered during 1994–2003. J Appl Microbiol 118(2):326–342. https://doi.org/10.1111/jam.12707
23. Ben Slama K, Jouini A, Ben Sallem R, Somalo S, Saenz Y, Estepa V, Boudabous A, Torres C (2010) Prevalence of broad-spectrum cephalosporin-resistant Escherichia coli isolates in food samples in Tunisia, and characterization of integrons and antimicrobial resistance mechanisms implicated. Int J Food Microbiol 137(2–3):281–286. https://doi.org/10.1016/j.ijfoodmicro.2009.12.003
24. Bisbiroulas P, Psylou M, Iliopoulou I, Diakogiannis I, Berberi A, Mastronicolis SK (2011) Adaptational changes in cellular phospholipids and fatty acid composition of the food pathogen Listeria monocytogenes as a stress response to disinfectant sanitizer benzalkonium chloride. Lett Appl Microbiol 52(3):275–280. https://doi.org/10.1111/j.1472-765X.2010.02995.x
25. Bischoff M, Bauer J, Preikschat P, Schwaiger K, Molle G, Holzel C (2012) First detection of the antiseptic resistance gene qacA/B in Enterococcus faecalis. Microb Drug Res (Larchmont, NY) 18(1):7–12. https://doi.org/10.1089/mdr.2011.0092
26. Bjergbaek LA, Haagensen JA, Molin S, Roslev P (2008) Effect of oxygen limitation and starvation on the benzalkonium chloride susceptibility of Escherichia coli. J Appl Microbiol 105(5):1310–1317. https://doi.org/10.1111/j.1365-2672.2008.03901.x
27. Bjorland J, Bratlie MS, Steinum T (2007) The smr gene resides on a novel plasmid pSP187 identified in a Staphylococcus pasteuri isolate recovered from unpasteurized milk. Plasmid 57(2):145–155. https://doi.org/10.1016/j.plasmid.2006.08.004
28. Bjorland J, Steinum T, Kvitle B, Waage S, Sunde M, Heir E (2005) Widespread distribution of disinfectant resistance genes among staphylococci of bovine and caprine origin in Norway. J Clin Microbiol 43(9):4363–4368. https://doi.org/10.1128/jcm.43.9.4363-4368.2005
29. Bjorland J, Steinum T, Sunde M, Waage S, Heir E (2003) Novel plasmid-borne gene qacJ mediates resistance to quaternary ammonium compounds in equine Staphylococcus aureus, Staphylococcus simulans, and Staphylococcus intermedius. Antimicrob Agents Chemother 47(10):3046–3052
30. Bloß R, Meyer S, Kampf G (2010) Adsorption of active ingredients from surface disinfectants to different types of fabrics. J Hosp Infect 75:56–61

31. Bock LJ, Hind CK, Sutton JM, Wand ME (2018) Growth media and assay plate material can impact on the effectiveness of cationic biocides and antibiotics against different bacterial species. Lett Appl Microbiol 66(5):368–377. https://doi.org/10.1111/lam.12863
32. Bore E, Hebraud M, Chafsey I, Chambon C, Skjaeret C, Moen B, Moretro T, Langsrud O, Rudi K, Langsrud S (2007) Adapted tolerance to benzalkonium chloride in Escherichia coli K-12 studied by transcriptome and proteome analyses. Microbiology (Reading, England) 153(Pt 4):935–946. https://doi.org/10.1099/mic.0.29288-0
33. Bore E, Langsrud S (2005) Characterization of micro-organisms isolated from dairy industry after cleaning and fogging disinfection with alkyl amine and peracetic acid. J Appl Microbiol 98(1):96–105. https://doi.org/10.1111/j.1365-2672.2004.02436.x
34. Braga TM, Marujo PE, Pomba C, Lopes MF (2011) Involvement, and dissemination, of the Enterococcal small multidrug resistance transporter QacZ in resistance to quaternary ammonium compounds. J Antimicrob Chemother 66(2):283–286. https://doi.org/10.1093/jac/dkq460
35. Braga TM, Pomba C, Lopes MF (2013) High-level vancomycin resistant Enterococcus faecium related to humans and pigs found in dust from pig breeding facilities. Vet Microbiol 161(3–4):344–349. https://doi.org/10.1016/j.vetmic.2012.07.034
36. Braoudaki M, Hilton AC (2004) Adaptive resistance to biocides in Salmonella enterica and Escherichia coli O157 and cross-resistance to antimicrobial agents. J Clin Microbiol 42 (1):73–78
37. Braoudaki M, Hilton AC (2005) Mechanisms of resistance in Salmonella enterica adapted to erythromycin, benzalkonium chloride and triclosan. Int J Antimicrob Agents 25(1):31–37. https://doi.org/10.1016/j.ijantimicag.2004.07.016
38. Bridier A, Briandet R, Thomas V, Dubois-Brissonnet F (2011) Comparative biocidal activity of peracetic acid, benzalkonium chloride and ortho-phthalaldehyde on 77 bacterial strains. J Hosp Infect 78(3):208–213. https://doi.org/10.1016/j.jhin.2011.03.014
39. Bridier A, Dubois-Brissonnet F, Greub G, Thomas V, Briandet R (2011) Dynamics of the action of biocides in Pseudomonas aeruginosa biofilms. Antimicrob Agents Chemother 55 (6):2648–2654. https://doi.org/10.1128/aac.01760-10
40. Brill F, Goroncy-Bermes P, Sand W (2006) Influence of growth media on the sensitivity of Staphylococcus aureus and Pseudomonas aeruginosa to cationic biocides. Int J Hyg Environ Health 209(1):89–95
41. Buffet-Bataillon S, Branger B, Cormier M, Bonnaure-Mallet M, Jolivet-Gougeon A (2011) Effect of higher minimum inhibitory concentrations of quaternary ammonium compounds in clinical E. coli isolates on antibiotic susceptibilities and clinical outcomes. J Hosp Infect 79 (2):141–146. https://doi.org/10.1016/j.jhin.2011.06.008
42. Buffet-Bataillon S, Tattevin P, Bonnaure-Mallet M, Jolivet-Gougeon A (2012) Emergence of resistance to antibacterial agents: the role of quaternary ammonium compounds–a critical review. Int J Antimicrob Agents 39(5):381–389. https://doi.org/10.1016/j.ijantimicag.2012.01.011
43. Bundgaard-Nielsen K, Nielsen PV (1996) Fungicidal effect of 15 disinfectants against 25 fungal contaminants commonly found in bread and cheese manufacturing. J Food Prot 59 (3):268–275
44. Buzón-Durán L, Alonso-Calleja C, Riesco-Peláez F, Capita R (2017) Effect of sub-inhibitory concentrations of biocides on the architecture and viability of MRSA biofilms. Food Microbiol 65(Supplement C):294–301. https://doi.org/10.1016/j.fm.2017.01.003
45. Caballero Gómez N, Abriouel H, Grande MJ, Pérez Pulido R, Gálvez A (2012) Effect of enterocin AS-48 in combination with biocides on planktonic and sessile Listeria monocytogenes. Food Microbiol 30(1):51–58. https://doi.org/10.1016/j.fm.2011.12.013
46. Cabo ML, Herrera JJ, Crespo MD, Pastoriza L (2009) Comparison among the effectiveness of ozone, nisin and benzalkonium chloride for the elimination of planktonic cells and

biofilms of Staphylococcus aureus CECT4459 on polypropylene. Food Control 20(5):521–525. https://doi.org/10.1016/j.foodcont.2008.08.002
47. Campanac C, Pineau L, Payard A, Baziard-Mouysset G, Roques C (2002) Interactions between biocide cationic agents and bacterial biofilms. Antimicrob Agents Chemother 46 (5):1469–1474
48. Capita R, Buzon-Duran L, Riesco-Pelaez F, Alonso-Calleja C (2017) Effect of sub-lethal concentrations of biocides on the structural parameters and viability of the biofilms formed by Salmonella Typhimurium. Foodborne Pathog Dis 14(6):350–356. https://doi.org/10.1089/fpd.2016.2241
49. Casado Munoz Mdel C, Benomar N, Ennahar S, Horvatovich P, Lavilla Lerma L, Knapp CW, Galvez A, Abriouel H (2016) Comparative proteomic analysis of a potentially probiotic Lactobacillus pentosus MP-10 for the identification of key proteins involved in antibiotic resistance and biocide tolerance. Int J Food Microbiol 222:8–15. https://doi.org/10.1016/j.ijfoodmicro.2016.01.012
50. Casado Munoz Mdel C, Benomar N, Lavilla Lerma L, Knapp CW, Galvez A, Abriouel H (2016) Biocide tolerance, phenotypic and molecular response of lactic acid bacteria isolated from naturally-fermented Alorena table to different physico-chemical stresses. Food Microbiol 60:1–12. https://doi.org/10.1016/j.fm.2016.06.013
51. CDC (1969) Food and drug administration warning: contaminated detergent solution. MMWR—Morb Mortal Wkly Rep 18(42):366
52. Cervinkova D, Babak V, Marosevic D, Kubikova I, Jaglic Z (2013) The role of the qacA gene in mediating resistance to quaternary ammonium compounds. Microb Drug Res (Larchmont, NY) 19(3):160–167. https://doi.org/10.1089/mdr.2012.0154
53. Chah KF, Agbo IC, Eze DC, Somalo S, Estepa V, Torres C (2010) Antimicrobial resistance, integrons and plasmid replicon typing in multiresistant clinical Escherichia coli strains from Enugu State, Nigeria. J Basic Microbiol 50(Suppl 1):S18–24. https://doi.org/10.1002/jobm.200900325
54. Chaieb K, Zmantar T, Souiden Y, Mahdouani K, Bakhrouf A (2011) XTT assay for evaluating the effect of alcohols, hydrogen peroxide and benzalkonium chloride on biofilm formation of Staphylococcus epidermidis. Microb Pathog 50(1):1–5. https://doi.org/10.1016/j.micpath.2010.11.004
55. Chang JM, McCanna DJ, Subbaraman LN, Jones LW (2015) Efficacy of antimicrobials against biofilms of Achromobacter and Pseudomonas. Optom Vis Sci: Official Publ Am Acad Optom 92(4):506–513. https://doi.org/10.1097/opx.0000000000000549
56. Chantefort A, Druilles J, Huet M (1990) Resistance de certains Pseudomonas aux antiseptiques et desinfectants. Med Mal Infectieuses 20(5):234–240. https://doi.org/10.1016/S0399-077X(05)81134-5
57. Chapuis A, Amoureux L, Bador J, Gavalas A, Siebor E, Chretien ML, Caillot D, Janin M, de Curraize C, Neuwirth C (2016) Outbreak of Extended-Spectrum Beta-Lactamase Producing Enterobacter cloacae with High MICs of Quaternary Ammonium Compounds in a Hematology Ward Associated with Contaminated Sinks. Frontiers Microbiol 7:1070. https://doi.org/10.3389/fmicb.2016.01070
58. Chen X, Wu Z, Zhou Y, Zhu J, Li K, Shao H, Wei L (2017) Molecular and virulence characteristics of methicillin-resistant Staphylococcus aureus in burn patients. Front Lab Med 1(1):43–47. https://doi.org/10.1016/j.flm.2017.02.010
59. Cho OH, Baek EH, Bak MH, Suh YS, Park KH, Kim S, Bae IG, Lee SH (2016) The effect of targeted decolonization on methicillin-resistant Staphylococcus aureus colonization or infection in a surgical intensive care unit. Am J Infect Control 44(5):533–538. https://doi.org/10.1016/j.ajic.2015.12.007
60. Cho OH, Park KH, Song JY, Hong JM, Kim T, Hong SI, Kim S, Bae IG (2017) Prevalence and microbiological characteristics of qacA/B-Positive Methicillin-Resistant Staphylococcus aureus Isolates in a Surgical Intensive Care Unit. Microb Drug Res (Larchmont, NY). https://doi.org/10.1089/mdr.2017.0072

61. Choudhury MA, Sidjabat HE, Rathnayake IU, Gavin N, Chan RJ, Marsh N, Banu S, Huygens F, Paterson DL, Rickard CM, McMillan DJ (2017) Culture-independent detection of chlorhexidine resistance genes qacA/B and smr in bacterial DNA recovered from body sites treated with chlorhexidine-containing dressings. J Med Microbiol 66(4):447–453. https://doi.org/10.1099/jmm.0.000463
62. Chuanchuen R, Khemtong S, Padungtod P (2007) Occurrence of qacE/qacEDelta1 genes and their correlation with class 1 integrons in salmonella enterica isolates from poultry and swine. Southeast Asian J Trop Med Public Health 38(5):855–862
63. Chuanchuen R, Pathanasophon P, Khemtong S, Wannaprasat W, Padungtod P (2008) Susceptibilities to antimicrobials and disinfectants in Salmonella isolates obtained from poultry and swine in Thailand. J Vet Med Sci 70(6):595–601
64. Coelho JR, Carrico JA, Knight D, Martinez JL, Morrissey I, Oggioni MR, Freitas AT (2013) The use of machine learning methodologies to analyse antibiotic and biocide susceptibility in Staphylococcus aureus. PLoS ONE 8(2):e55582. https://doi.org/10.1371/journal.pone.0055582
65. Cole EC, Addison RM, Rubino JR, Leese KE, Dulaney PD, Newell MS, Wilkins J, Gaber DJ, Wineinger T, Criger DA (2003) Investigation of antibiotic and antibacterial agent cross-resistance in target bacteria from homes of antibacterial product users and nonusers. J Appl Microbiol 95(4):664–676
66. Condell O, Iversen C, Cooney S, Power KA, Walsh C, Burgess C, Fanning S (2012) Efficacy of biocides used in the modern food industry to control salmonella enterica, and links between biocide tolerance and resistance to clinically relevant antimicrobial compounds. Appl Environ Microbiol 78(9):3087–3097. https://doi.org/10.1128/aem.07534-11
67. Conficoni D, Losasso C, Cortini E, Di Cesare A, Cibin V, Giaccone V, Corno G, Ricci A (2016) Resistance to Biocides in Listeria monocytogenes Collected in Meat-Processing Environments. Front Microbiol 7:1627. https://doi.org/10.3389/fmicb.2016.01627
68. Corcoran M, Morris D, De Lappe N, O'Connor J, Lalor P, Dockery P, Cormican M (2014) Commonly used disinfectants fail to eradicate Salmonella enterica biofilms from food contact surface materials. Appl Environ Microbiol 80(4):1507–1514. https://doi.org/10.1128/aem.03109-13
69. Correa JE, De Paulis A, Predari S, Sordelli DO, Jeric PE (2008) First report of qacG, qacH and qacJ genes in Staphylococcus haemolyticus human clinical isolates. J Antimicrob Chemother 62(5):956–960. https://doi.org/10.1093/jac/dkn327
70. Costa D, Poeta P, Saenz Y, Coelho AC, Matos M, Vinue L, Rodrigues J, Torres C (2008) Prevalence of antimicrobial resistance and resistance genes in faecal Escherichia coli isolates recovered from healthy pets. Vet Microbiol 127(1–2):97–105. https://doi.org/10.1016/j.vetmic.2007.08.004
71. Costa DM, Lopes LKO, Hu H, Tipple AFV, Vickery K (2017) Alcohol fixation of bacteria to surgical instruments increases cleaning difficulty and may contribute to sterilization inefficacy. Am J Infect Control 45(8):e81–e86. https://doi.org/10.1016/j.ajic.2017.04.286
72. Couto N, Belas A, Couto I, Perreten V, Pomba C (2014) Genetic relatedness, antimicrobial and biocide susceptibility comparative analysis of methicillin-resistant and -susceptible Staphylococcus pseudintermedius from Portugal. Microb Drug Res (Larchmont, NY) 20 (4):364–371. https://doi.org/10.1089/mdr.2013.0043
73. Couto N, Belas A, Kadlec K, Schwarz S, Pomba C (2015) Clonal diversity, virulence patterns and antimicrobial and biocide susceptibility among human, animal and environmental MRSA in Portugal. J Antimicrob Chemother 70(9):2483–2487. https://doi.org/10.1093/jac/dkv141
74. Couto N, Belas A, Tilley P, Couto I, Gama LT, Kadlec K, Schwarz S, Pomba C (2013) Biocide and antimicrobial susceptibility of methicillin-resistant staphylococcal isolates from horses. Vet Microbiol 166(1–2):299–303. https://doi.org/10.1016/j.vetmic.2013.05.011

75. Cowley NL, Forbes S, Amezquita A, McClure P, Humphreys GJ, McBain AJ (2015) Effects of formulation on microbicide potency and mitigation of the development of bacterial insusceptibility. Appl Environ Microbiol 81(20):7330–7338. https://doi.org/10.1128/aem.01985-15
76. Cruz CD, Fletcher GC (2012) Assessing manufacturers' recommended concentrations of commercial sanitizers on inactivation of Listeria monocytogenes. Food Control 26(1):194–199. https://doi.org/10.1016/j.foodcont.2012.01.041
77. Curiao T, Marchi E, Viti C, Oggioni MR, Baquero F, Martinez JL, Coque TM (2015) Polymorphic variation in susceptibility and metabolism of triclosan-resistant mutants of Escherichia coli and Klebsiella pneumoniae clinical strains obtained after exposure to biocides and antibiotics. Antimicrob Agents Chemother 59(6):3413–3423. https://doi.org/10.1128/aac.00187-15
78. Day S, Lalitha P, Haug S, Fothergill AW, Cevallos V, Vijayakumar R, Prajna NV, Acharya NR, McLeod SD, Lietman TM (2009) Activity of antibiotics against Fusarium and Aspergillus. Br J Ophthalmol 93(1):116–119. https://doi.org/10.1136/bjo.2008.142364
79. de Solis NMG, Davison AL, Pinney RJ (1994) Effect of plasmids conferring preservative resistance on performance of bacterial strains in compendial preservative efficacy tests. Eur J Pharm Sci 2(3):221–228. https://doi.org/10.1016/0928-0987(94)90026-4
80. DeMarco CE, Cushing LA, Frempong-Manso E, Seo SM, Jaravaza TA, Kaatz GW (2007) Efflux-related resistance to norfloxacin, dyes, and biocides in bloodstream isolates of Staphylococcus aureus. Antimicrob Agents Chemother 51(9):3235–3239. https://doi.org/10.1128/aac.00430-07
81. Deng W, Quan Y, Yang S, Guo L, Zhang X, Liu S, Chen S, Zhou K, He L, Li B, Gu Y, Zhao S, Zou L (2017) Antibiotic resistance in salmonella from retail foods of animal origin and its association with disinfectant and heavy metal resistance. Microb Drug Res (Larchmont, NY). https://doi.org/10.1089/mdr.2017.0127
82. Department of Health and Human Services; Food and Drug Administration (2015) Safety and effectiveness of healthcare antiseptics. Topical antimicrobial drug products for over-the-counter human use; proposed amendment of the tentative final monograph; reopening of administrative record; proposed rule. Fed Reg 80(84):25166–25205
83. Derde LP, Cooper BS, Goossens H, Malhotra-Kumar S, Willems RJ, Gniadkowski M, Hryniewicz W, Empel J, Dautzenberg MJ, Annane D, Aragao I, Chalfine A, Dumpis U, Esteves F, Giamarellou H, Muzlovic I, Nardi G, Petrikkos GL, Tomic V, Marti AT, Stammet P, Brun-Buisson C, Bonten MJ (2014) Interventions to reduce colonisation and transmission of antimicrobial-resistant bacteria in intensive care units: an interrupted time series study and cluster randomised trial. Lancet Infect Dis 14(1):31–39. https://doi.org/10.1016/s1473-3099(13)70295-0
84. Deus D, Krischek C, Pfeifer Y, Sharifi AR, Fiegen U, Reich F, Klein G, Kehrenberg C (2017) Comparative analysis of the susceptibility to biocides and heavy metals of extended-spectrum beta-lactamase-producing Escherichia coli isolates of human and avian origin, Germany. Diagn Microbiol Infect Dis 88(1):88–92. https://doi.org/10.1016/j.diagmicrobio.2017.01.023
85. Dominciano LCC, Oliveira CAF, Lee SH, Corassin CH (2016) Individual and combined antimicrobial activity of oleuropein and chemical sanitizers. J Food Chem Nanotechnol 2(3):124–127
86. dos Anjos MM, Ruiz SP, Nakamura CV, de Abreu Filho BA (2013) Resistance of Alicyclobacillus acidoterrestris spores and biofilm to industrial sanitizers. J Food Prot 76(8):1408–1413. https://doi.org/10.4315/0362-028x.jfp-13-020
87. Duran N, Temiz M, Duran GG, Eryilmaz N, Jenedi K (2014) Relationship between the resistance genes to quaternary ammonium compounds and antibiotic resistance in staphylococci isolated from surgical site infections. Med Sci Monit: Int Med J Exp Clin Res 20:544–550. https://doi.org/10.12659/msm.890177

88. Dutta V, Elhanafi D, Kathariou S (2013) Conservation and distribution of the benzalkonium chloride resistance cassette bcrABC in Listeria monocytogenes. Appl Environ Microbiol 79 (19):6067–6074. https://doi.org/10.1128/aem.01751-13
89. Dynes JJ, Lawrence JR, Korber DR, Swerhone GD, Leppard GG, Hitchcock AP (2009) Morphological and biochemical changes in Pseudomonas fluorescens biofilms induced by sub-inhibitory exposure to antimicrobial agents. Can J Microbiol 55(2):163–178. https://doi.org/10.1139/w08-109
90. Ebner R, Johler S, Sihto HM, Stephan R, Zweifel C (2013) Microarray-based characterization of Staphylococcus aureus isolates obtained from chicken carcasses. J Food Prot 76 (8):1471–1474. https://doi.org/10.4315/0362-028x.jfp-13-009
91. Ebner R, Stephan R, Althaus D, Brisse S, Maury M, Tasara T (2015) Phenotypic and genotypic characteristics of Listeria monocytogenes strains isolated during 2011–2014 from different food matrices in Switzerland. Food Control 57(Supplement C):321–326. https://doi.org/10.1016/j.foodcont.2015.04.030
92. Ebrahimi A, Hemati M, Shabanpour Z, Habibian Dehkordi S, Bahadoran S, Lotfalian S, Khubani S (2015) Effects of benzalkonium chloride on planktonic growth and biofilm formation by animal bacterial pathogens. Jundishapur J Microbiol 8(2):e16058. https://doi.org/10.5812/jjm.16058
93. Ehrenkranz NJ, Bolyard EA, Wiener M, Cleary TJ (1980) Antibiotic-sensitive Serratia marcescens infections complicating cardiopulmonary operations: contaminated disinfectant as a reservoir. Lancet 2(8207):1289–1292
94. Elhanafi D, Dutta V, Kathariou S (2010) Genetic characterization of plasmid-associated benzalkonium chloride resistance determinants in a Listeria monocytogenes strain from the 1998–1999 outbreak. Appl Environ Microbiol 76(24):8231–8238. https://doi.org/10.1128/aem.02056-10
95. Elli M, Arioli S, Guglielmetti S, Mora D (2013) Biocide susceptibility in Bifidobacteria of human origin. J Glob Antimicrob Resist 1(2):97–101. https://doi.org/10.1016/j.jgar.2013.03.007
96. Engelbrecht K, Ambrose D, Sifuentes L, Gerba C, Weart I, Koenig D (2013) Decreased activity of commercially available disinfectants containing quaternary ammonium compounds when exposed to cotton towels. Am J Infect Control 41(10):908–911. https://doi.org/10.1016/j.ajic.2013.01.017
97. Espigares E, Moreno Roldan E, Espigares M, Abreu R, Castro B, Dib AL, Arias A (2017) Phenotypic Resistance to Disinfectants and Antibiotics in Methicillin-Resistant Staphylococcus aureus Strains Isolated from Pigs. Zoonoses Public Health 64(4):272–280. https://doi.org/10.1111/zph.12308
98. Estepa V, Rojo-Bezares B, Azcona-Gutierrez JM, Olarte I, Torres C, Saenz Y (2017) Characterisation of carbapenem-resistance mechanisms in clinical Pseudomonas aeruginosa isolates recovered in a Spanish hospital. Enferm Infecc Microbiol Clin 35(3):141–147. https://doi.org/10.1016/j.eimc.2015.12.014
99. Fagerlund A, Langsrud S, Heir E, Mikkelsen MI, Moretro T (2016) Biofilm matrix composition affects the susceptibility of food associated staphylococci to cleaning and disinfection agents. Front Microbiol 7:856. https://doi.org/10.3389/fmicb.2016.00856
100. Fagerlund A, Moretro T, Heir E, Briandet R, Langsrud S (2017) Cleaning and disinfection of biofilms composed of Listeria monocytogenes and background microbiota from meat processing surfaces. Appl Environ Microbiol. https://doi.org/10.1128/aem.01046-17
101. Feder-Kubis J, Tomczuk K (2013) The effect of the cationic structures of chiral ionic liquids on their antimicrobial activities. Tetrahedron 69(21):4190–4198. https://doi.org/10.1016/j.tet.2013.03.107
102. Fernández-Fuentes MA, Ortega Morente E, Abriouel H, Pérez Pulido R, Gálvez A (2012) Isolation and identification of bacteria from organic foods: Sensitivity to biocides and antibiotics. Food Control 26(1):73–78. https://doi.org/10.1016/j.foodcont.2012.01.017

103. Fernández Fuentes MA, Ortega Morente E, Abriouel H, Pérez Pulido R, Gálvez A (2014) Antimicrobial resistance determinants in antibiotic and biocide-resistant gram-negative bacteria from organic foods. Food Control 37(Supplement C):9–14. https://doi.org/10.1016/j.foodcont.2013.08.041
104. Fernandez Marquez ML, Grande Burgos MJ, Lopez Aguayo MC, Perez Pulido R, Galvez A, Lucas R (2017) Characterization of biocide-tolerant bacteria isolated from cheese and dairy small-medium enterprises. Food Microbiol 62:77–81. https://doi.org/10.1016/j.fm.2016.10.008
105. Fidalgo SG, Longbottom CJ, Rjley TV (2002) Susceptibility of Erysipelothrix rhusiopathiae to antimicrobial agents and home disinfectants. Pathology 34(5):462–465
106. Firth N, Skurray RA (1998) Mobile elements in the evolution and spread of multiple-drug resistance in staphylococci. Drug Resist Updates: Rev Commentaries Antimicrob Anticancer Chemother 1(1):49–58
107. Fleurette J, Bes M, Brun Y, Freney J, Forey F, Coulet M, Reverdy ME, Etienne J (1989) Clinical isolates of Staphylococcus lugdunensis and S. schleiferi: bacteriological characteristics and susceptibility to antimicrobial agents. Res Microbiol 140(2):107–118
108. Forbes S, Cowley N, Humphreys G, Mistry H, Amezquita A, McBain AJ (2017) Formulation of Biocides Increases Antimicrobial Potency and Mitigates the Enrichment of Nonsusceptible Bacteria in Multispecies Biofilms. Appl Environ Microbiol 83(7). https://doi.org/10.1128/aem.03054-16
109. Forbes S, Knight CG, Cowley NL, Amezquita A, McClure P, Humphreys G, McBain AJ (2016) Variable effects of exposure to formulated microbicides on antibiotic susceptibility in firmicutes and proteobacteria. Appl Environ Microbiol 82(12):3591–3598. https://doi.org/10.1128/aem.00701-16
110. Fox JG, Beaucage CM, Folta CA, Thornton GW (1981) Nosocomial transmission of Serratia marcescens in a veterinary hospital due to contamination by benzalkonium chloride. J Clin Microbiol 14(2):157–160
111. Frank MJ, Schaffner W (1976) Contaminated aqueous benzalkonium chloride. An unnecessary hospital infection hazard. JAMA 236(21):2418–2419
112. Fuentes DE, Navarro CA, Tantalean JC, Araya MA, Saavedra CP, Perez JM, Calderon IL, Youderian PA, Mora GC, Vasquez CC (2005) The product of the qacC gene of Staphylococcus epidermidis CH mediates resistance to beta-lactam antibiotics in gram-positive and gram-negative bacteria. Res Microbiol 156(4):472–477. https://doi.org/10.1016/j.resmic.2005.01.002
113. Furi L, Ciusa ML, Knight D, Di Lorenzo V, Tocci N, Cirasola D, Aragones L, Coelho JR, Freitas AT, Marchi E, Moce L, Visa P, Northwood JB, Viti C, Borghi E, Orefici G, Morrissey I, Oggioni MR (2013) Evaluation of reduced susceptibility to quaternary ammonium compounds and bisbiguanides in clinical isolates and laboratory-generated mutants of Staphylococcus aureus. Antimicrob Agents Chemother 57(8):3488–3497. https://doi.org/10.1128/aac.00498-13
114. Gadea R, Fernandez Fuentes MA, Perez Pulido R, Galvez A, Ortega E (2017) Effects of exposure to quaternary-ammonium-based biocides on antimicrobial susceptibility and tolerance to physical stresses in bacteria from organic foods. Food Microbiol 63:58–71. https://doi.org/10.1016/j.fm.2016.10.037
115. Garrido AM, Burgos MJ, Marquez ML, Aguayo MC, Pulido RP, del Arbol JT, Galvez A, Lopez RL (2015) Biocide tolerance in Salmonella from meats in Southern Spain. Braz J Microbiol: [Publ Braz Soc Microbiol] 46(4):1177–1181. https://doi.org/10.1590/s1517-838246420140396
116. Gaston MA, Hoffman PN, Pitt TL (1986) A comparison of strains of Serratia marcescens isolated from neonates with strains isolated from sporadic and epidemic infections in adults. J Hosp Infect 8(1):86–95

117. Gaze WH, Abdouslam N, Hawkey PM, Wellington EM (2005) Incidence of class 1 integrons in a quaternary ammonium compound-polluted environment. Antimicrob Agents Chemother 49(5):1802–1807. https://doi.org/10.1128/aac.49.5.1802-1807.2005
118. Ghasemzadeh-Moghaddam H, van Belkum A, Hamat RA, van Wamel W, Neela V (2014) Methicillin-susceptible and -resistant Staphylococcus aureus with high-level antiseptic and low-level mupirocin resistance in Malaysia. Microb Drug Res (Larchmont, NY) 20 (5):472–477. https://doi.org/10.1089/mdr.2013.0222
119. Gholamrezazadeh M, Shakibaie MR, Monirzadeh F, Masoumi S, Hashemizadeh Z (2017) Effect of nano-silver, nano-copper, deconex and benzalkonium chloride on biofilm formation and expression of transcription regulatory quorum sensing gene (rh1R) in drug-resistance Pseudomonas aeruginosa burn isolates. Burns: J Int Soc Burn Injuries. https://doi.org/10.1016/j.burns.2017.10.021
120. Giaouris E, Chorianopoulos N, Doulgeraki A, Nychas GJ (2013) Co-culture with Listeria monocytogenes within a dual-species biofilm community strongly increases resistance of Pseudomonas putida to benzalkonium chloride. PLoS ONE 8(10):e77276. https://doi.org/10.1371/journal.pone.0077276
121. Gkana EN, Giaouris ED, Doulgeraki AI, Kathariou S, Nychas GJE (2017) Biofilm formation by Salmonella Typhimurium and Staphylococcus aureus on stainless steel under either mono- or dual-species multi-strain conditions and resistance of sessile communities to sub-lethal chemical disinfection. Food Control 73(Part B):838–846. https://doi.org/10.1016/j.foodcont.2016.09.038
122. Gomaa FAM, Helal ZH, Khan MI (2017) High Prevalence of blaNDM-1, blaVIM, qacE, and qacEDelta1 genes and their association with decreased susceptibility to antibiotics and common hospital biocides in clinical isolates of acinetobacter baumannii. Microorganisms 5 (2). https://doi.org/10.3390/microorganisms5020018
123. Gomes IB, Malheiro J, Mergulhao F, Maillard JY, Simoes M (2016) Comparison of the efficacy of natural-based and synthetic biocides to disinfect silicone and stainless steel surfaces. Pathog Dis 74(4):ftw014. https://doi.org/10.1093/femspd/ftw014
124. Grande Burgos MJ, Fernandez Marquez ML, Perez Pulido R, Galvez A, Lucas Lopez R (2016) Virulence factors and antimicrobial resistance in Escherichia coli strains isolated from hen egg shells. Int J Food Microbiol 238:89–95. https://doi.org/10.1016/j.ijfoodmicro.2016.08.037
125. Grande Burgos MJ, Lucas López R, López Aguayo M, Pérez Pulido R, Gálvez A (2013) Inhibition of planktonic and sessile Salmonella enterica cells by combinations of enterocin AS-48, polymyxin B and biocides. Food Control 30(1):214–221. https://doi.org/10.1016/j.foodcont.2012.07.011
126. Grande Burgos MJ, Pérez-Pulido R, Gálvez A, Lucas R (2017) Biofilms formed by microbiota recovered from fresh produce: Bacterial biodiversity, and inactivation by benzalkonium chloride and enterocin AS-48. LWT Food Sci Technol 77(Supplement C):80–84. https://doi.org/10.1016/j.lwt.2016.11.033
127. Grobe KJ, Zahller J, Stewart PS (2002) Role of dose concentration in biocide efficacy against Pseudomonas aeruginosa biofilms. J Ind Microbiol Biotechnol 29(1):10–15. https://doi.org/10.1038/sj.jim.7000256
128. Guang-Sen S, Boost M, Cho P (2016) Prevalence of antiseptic resistance genes increases in staphylococcal isolates from orthokeratology lens wearers over initial six-month period of use. Eur J Clin Microbiol Infect Dis 35(6):955–962. https://doi.org/10.1007/s10096-016-2622-z
129. Guerin-Mechin L, Leveau JY, Dubois-Brissonnet F (2004) Resistance of spheroplasts and whole cells of Pseudomonas aeruginosa to bactericidal activity of various biocides: evidence of the membrane implication. Microbiol Res 159(1):51–57. https://doi.org/10.1016/j.micres.2004.01.003

130. Guo W, Cui S, Xu X, Wang H (2014) Resistant mechanism study of benzalkonium chloride selected Salmonella Typhimurium mutants. Microb Drug Res (Larchmont, NY) 20(1):11–16. https://doi.org/10.1089/mdr.2012.0225
131. Guo W, Shan K, Xu B, Li J (2015) Determining the resistance of carbapenem-resistant Klebsiella pneumoniae to common disinfectants and elucidating the underlying resistance mechanisms. Pathog Glob Health 109(4):184–192. https://doi.org/10.1179/2047773215y.0000000022
132. Gupta AK, Ahmad I, Summerbell RC (2002) Fungicidal activities of commonly used disinfectants and antifungal pharmaceutical spray preparations against clinical strains of Aspergillus and Candida species. Med Mycol 40(2):201–208
133. Gutierrez-Martin CB, Yubero S, Martinez S, Frandoloso R, Rodriguez-Ferri EF (2011) Evaluation of efficacy of several disinfectants against Campylobacter jejuni strains by a suspension test. Res Vet Sci 91(3):e44–47. https://doi.org/10.1016/j.rvsc.2011.01.020
134. Hakuno H, Yamamoto M, Oie S, Kamiya A (2010) Microbial contamination of disinfectants used for intermittent self-catheterization. Jpn J Infect Dis 63(4):277–279
135. Hardy PC, Ederer GM, Matsen JM (1970) Contamination of commercially packaged urinary catheter kits with the pseudomonad EO-1. N Engl J Med 282(1):33–35. https://doi.org/10.1056/nejm197001012820108
136. Harrison JJ, Turner RJ, Joo DA, Stan MA, Chan CS, Allan ND, Vrionis HA, Olson ME, Ceri H (2008) Copper and quaternary ammonium cations exert synergistic bactericidal and antibiofilm activity against Pseudomonas aeruginosa. Antimicrob Agents Chemother 52 (8):2870–2881. https://doi.org/10.1128/aac.00203-08
137. Hasanvand A, Ghafourian S, Taherikalani M, Jalilian FA, Sadeghifard N, Pakzad I (2015) Antiseptic resistance in methicillin sensitive and methicillin resistant staphylococcus aureus isolates from some major hospitals, Iran. Recent Pat Anti-Infect Drug Dis 10(2):105–112
138. Hassanzadeh S, Mashhadi R, Yousefi M, Askari E, Saniei M, Pourmand MR (2017) Frequency of efflux pump genes mediating ciprofloxacin and antiseptic resistance in methicillin-resistant Staphylococcus aureus isolates. Microb Pathog 111:71–74. https://doi.org/10.1016/j.micpath.2017.08.026
139. He GX, Landry M, Chen H, Thorpe C, Walsh D, Varela MF, Pan H (2014) Detection of benzalkonium chloride resistance in community environmental isolates of staphylococci. J Med Microbiol 63(Pt 5):735–741. https://doi.org/10.1099/jmm.0.073072-0
140. He XF, Zhang HJ, Cao JG, Liu F, Wang JK, Ma WJ, Yin W (2017) A novel method to detect bacterial resistance to disinfectants. Genes Dis 4(3):163–169. https://doi.org/10.1016/j.gendis.2017.07.001
141. Heir E, Sundheim G, Holck AL (1995) Resistance to quaternary ammonium compounds in Staphylococcus spp. isolated from the food industry and nucleotide sequence of the resistance plasmid pST827. J Appl Bacteriol 79(2):149–156
142. Heir E, Sundheim G, Holck AL (1998) The Staphylococcus qacH gene product: a new member of the SMR family encoding multidrug resistance. FEMS Microbiol Lett 163 (1):49–56
143. Heir E, Sundheim G, Holck AL (1999) Identification and characterization of quaternary ammonium compound resistant staphylococci from the food industry. Int J Food Microbiol 48(3):211–219
144. Heir E, Sundheim G, Holck AL (1999) The qacG gene on plasmid pST94 confers resistance to quaternary ammonium compounds in staphylococci isolated from the food industry. J Appl Microbiol 86(3):378–388
145. Higgins CS, Murtough SM, Williamson E, Hiom SJ, Payne DJ, Russell AD, Walsh TR (2001) Resistance to antibiotics and biocides among non-fermenting Gram-negative bacteria. Clin Microbiol Infect 7(6):308–315
146. Hijazi K, Mukhopadhya I, Abbott F, Milne K, Al-Jabri ZJ, Oggioni MR, Gould IM (2016) Susceptibility to chlorhexidine amongst multidrug-resistant clinical isolates of

Staphylococcus epidermidis from bloodstream infections. Int J Antimicrob Agents 48(1):86–90. https://doi.org/10.1016/j.ijantimicag.2016.04.015

147. Ho CM, Li CY, Ho MW, Lin CY, Liu SH, Lu JJ (2012) High rate of qacA- and qacB-positive methicillin-resistant Staphylococcus aureus isolates from chlorhexidine-impregnated catheter-related bloodstream infections. Antimicrob Agents Chemother 56 (11):5693–5697. AAC.00761-12 [pii]. https://doi.org/10.1128/aac.00761-12 [doi]
148. Ho J, Branley J (2012) Prevalence of antiseptic resistance genes qacA/B and specific sequence types of methicillin-resistant Staphylococcus aureus in the era of hand hygiene. J Antimicrob Chemother 67(6):1549–1550. https://doi.org/10.1093/jac/dks035
149. Houari A, Di Martino P (2007) Effect of chlorhexidine and benzalkonium chloride on bacterial biofilm formation. Lett Appl Microbiol 45(6):652–656. https://doi.org/10.1111/j.1472-765X.2007.02249.x
150. Humphries RM, Wu MT, Westblade LF, Robertson AE, Burnham CA, Wallace MA, Burd EM, Lawhon S, Hindler JA (2016) In Vitro Antimicrobial Susceptibility of Staphylococcus pseudintermedius Isolates of Human and Animal Origin. J Clin Microbiol 54(5):1391–1394. https://doi.org/10.1128/jcm.00270-16
151. Ignak S, Nakipoglu Y, Gurler B (2017) Frequency of antiseptic resistance genes in clinical Staphycocci and Enterococci isolates in Turkey. Antimicrob Resist Infect Control 6:88. https://doi.org/10.1186/s13756-017-0244-6
152. Imbert C, Lassy E, Daniault G, Jacquemin JL, Rodier MH (2003) Treatment of plastic and extracellular matrix components with chlorhexidine or benzalkonium chloride: effect on Candida albicans adherence capacity in vitro. J Antimicrob Chemother 51(2):281–287
153. Italy (2015) Alkyl (C12–16) dimethylbenzyl ammonium chloride Product-type 8 (Wood preservative)
154. Jaglic Z, Červinková D, Vlková H, Michu E, Kunová G, Babák V (2012) Bacterial biofilms resist oxidising agents due to the presence of organic matter. Czech J Food Sci 30 (2):178–187
155. Jechalke S, Schreiter S, Wolters B, Dealtry S, Heuer H, Smalla K (2013) Widespread dissemination of class 1 integron components in soils and related ecosystems as revealed by cultivation-independent analysis. Front Microbiol 4:420. https://doi.org/10.3389/fmicb.2013.00420
156. Jensen SO, Apisiridej S, Kwong SM, Yang YH, Skurray RA, Firth N (2010) Analysis of the prototypical Staphylococcus aureus multiresistance plasmid pSK1. Plasmid 64(3):135–142. https://doi.org/10.1016/j.plasmid.2010.06.001
157. Jiang X, Xu Y, Li Y, Zhang K, Liu L, Wang H, Tian J, Ying H, Shi L, Yu T (2017) Characterization and horizontal transfer of qacH-associated class 1 integrons in Escherichia coli isolated from retail meats. Int J Food Microbiol 258:12–17. https://doi.org/10.1016/j.ijfoodmicro.2017.07.009
158. Jiang X, Yu T, Liang Y, Ji S, Guo X, Ma J, Zhou L (2016) Efflux pump-mediated benzalkonium chloride resistance in Listeria monocytogenes isolated from retail food. Int J Food Microbiol 217:141–145. https://doi.org/10.1016/j.ijfoodmicro.2015.10.022
159. Jiang X, Yu T, Liu L, Li Y, Zhang K, Wang H, Shi L (2017) Examination of quaternary ammonium compound resistance in proteus mirabilis isolated from cooked meat products in China. Frontiers in microbiology 8:2417. https://doi.org/10.3389/fmicb.2017.02417
160. Johnson JG, Saye EJ, Jimenez-Truque N, Soper N, Thomsen I, Talbot TR, Creech CB (2013) Frequency of disinfectant resistance genes in pediatric strains of methicillin-resistant Staphylococcus aureus. Infect Control Hosp Epidemiol 34(12):1326–1327. https://doi.org/10.1086/673983
161. Joynson JA, Forbes B, Lambert RJ (2002) Adaptive resistance to benzalkonium chloride, amikacin and tobramycin: the effect on susceptibility to other antimicrobials. J Appl Microbiol 93(1):96–107

162. Kadry AA, Serry FM, El-Ganiny AM, El-Baz AM (2017) Integron occurrence is linked to reduced biocide susceptibility in multidrug resistant Pseudomonas aeruginosa. Br J Biomed Sci 74(2):78–84. https://doi.org/10.1080/09674845.2017.1278884
163. Kalkanci A, Elli M, Adil Fouad A, Yesilyurt E, Jabban Khalil I (2015) Assessment of susceptibility of mould isolates towards biocides. J de mycologie medicale 25(4):280–286. https://doi.org/10.1016/j.mycmed.2015.08.001
164. Kampf G (2018) Adaptive microbial response to low level benzalkonium chloride exposure. J Hosp Infect. https://doi.org/10.1016/j.jhin.2018.05.019
165. Kampf G, Degenhardt S, Lackner S, Jesse K, von Baum H, Ostermeyer C (2014) Poorly processed reusable surface disinfection tissue dispensers may be a source of infection. BMC Infect Dis 14 (1):37. 1471–2334-14-37 [pii]. https://doi.org/10.1186/1471-2334-14-37 [doi]
166. Kampf G, Degenhardt S, Lackner S, Ostermeyer C (2014) Effective processing or reusable dispensers for surface disinfection tissues - the devil is in the details. GMS Hyg Infect Control 9(1):DOC09
167. Karatzas KA, Webber MA, Jorgensen F, Woodward MJ, Piddock LJ, Humphrey TJ (2007) Prolonged treatment of Salmonella enterica serovar Typhimurium with commercial disinfectants selects for multiple antibiotic resistance, increased efflux and reduced invasiveness. J Antimicrob Chemother 60(5):947–955. https://doi.org/10.1093/jac/dkm314
168. Kaslow RA, Mackel DC, Mallison GF (1976) Nosocomial pseudobacteremia. Positive blood cultures due to contaminated benzalkonium antiseptic. JAMA 236(21):2407–2409
169. Kastbjerg VG, Larsen MH, Gram L, Ingmer H (2010) Influence of sublethal concentrations of common disinfectants on expression of virulence genes in Listeria monocytogenes. Appl Environ Microbiol 76(1):303–309. https://doi.org/10.1128/aem.00925-09
170. Katharios-Lanwermeyer S, Rakic-Martinez M, Elhanafi D, Ratani S, Tiedje JM, Kathariou S (2012) Coselection of cadmium and benzalkonium chloride resistance in conjugative transfers from nonpathogenic Listeria spp. to other Listeriae. Appl Environ Microbiol 78 (21):7549–7556. https://doi.org/10.1128/aem.02245-12
171. Kawamura-Sato K, Wachino J, Kondo T, Ito H, Arakawa Y (2008) Reduction of disinfectant bactericidal activities in clinically isolated Acinetobacter species in the presence of organic material. J Antimicrob Chemother 61(3):568–576. https://doi.org/10.1093/jac/dkm498
172. Kawamura-Sato K, Wachino J, Kondo T, Ito H, Arakawa Y (2010) Correlation between reduced susceptibility to disinfectants and multidrug resistance among clinical isolates of Acinetobacter species. J Antimicrob Chemother 65(9):1975–1983. https://doi.org/10.1093/jac/dkq227
173. Kazama H, Hamashima H, Sasatsu M, Arai T (1998) Distribution of the antiseptic-resistance gene qacE delta 1 in gram-positive bacteria. FEMS Microbiol Lett 165(2):295–299
174. Kazama H, Hamashima H, Sasatsu M, Arai T (1998) Distribution of the antiseptic-resistance genes qacE and qacE delta 1 in gram-negative bacteria. FEMS Microbiol Lett 159 (2):173–178
175. Kazama H, Hamashima H, Sasatsu M, Arai T (1999) Characterization of the antiseptic-resistance gene qacE delta 1 isolated from clinical and environmental isolates of Vibrio parahaemolyticus and Vibrio cholerae non-O1. FEMS Microbiol Lett 174 (2):379–384
176. Khan AH, Macfie SM, Ray MB (2017) Sorption and leaching of benzalkonium chlorides in agricultural soils. J Environ Manage 196:26–35. https://doi.org/10.1016/j.jenvman.2017.02.065
177. Khan AH, Topp E, Scott A, Sumarah M, Macfie SM, Ray MB (2015) Biodegradation of benzalkonium chlorides singly and in mixtures by a Pseudomonas sp. isolated from returned activated sludge. J Hazard Mater 299:595–602. https://doi.org/10.1016/j.jhazmat.2015.07.073
178. Kiddee A, Henghiranyawong K, Yimsabai J, Tiloklurs M, Niumsup PR (2013) Nosocomial spread of class 1 integron-carrying extensively drug-resistant Pseudomonas aeruginosa

isolates in a Thai hospital. Int J Antimicrob Agents 42(4):301–306. https://doi.org/10.1016/j.ijantimicag.2013.05.009

179. Kim JS, Chung YK, Lee SS, Lee JA, Kim HS, Park EY, Shin KS, Kang BS, Lee HJ, Kang HJ (2016) Effect of daily chlorhexidine bathing on the acquisition of methicillin-resistant Staphylococcus aureus in a medical intensive care unit with methicillin-resistant S aureus endemicity. Am J Infect Control 44(12):1520–1525. https://doi.org/10.1016/j.ajic.2016.04.252
180. Knapp L, Amezquita A, McClure P, Stewart S, Maillard JY (2015) Development of a protocol for predicting bacterial resistance to microbicides. Appl Environ Microbiol 81(8):2652–2659. https://doi.org/10.1128/aem.03843-14
181. Knapp L, Rushton L, Stapleton H, Sass A, Stewart S, Amezquita A, McClure P, Mahenthiralingam E, Maillard JY (2013) The effect of cationic microbicide exposure against Burkholderia cepacia complex (Bcc); the use of Burkholderia lata strain 383 as a model bacterium. J Appl Microbiol 115(5):1117–1126. https://doi.org/10.1111/jam.12320
182. Kodedova M, Sigler K, Lemire BD, Gaskova D (2011) Fluorescence method for determining the mechanism and speed of action of surface-active drugs on yeast cells. Biotechniques 50(1):58–63. https://doi.org/10.2144/000113568
183. Koopmans MM, Bijlsma MW, Brouwer MC, van de Beek D, van der Ende A (2017) Listeria monocytogenes meningitis in the Netherlands, 1985–2014: a nationwide surveillance study. J Infection 75(1):12–19. https://doi.org/10.1016/j.jinf.2017.04.004
184. Korsak D, Szuplewska M (2016) Characterization of nonpathogenic Listeria species isolated from food and food processing environment. Int J Food Microbiol 238:274–280. https://doi.org/10.1016/j.ijfoodmicro.2016.08.032
185. Korukluoglu M, Sahan Y, Yigit A (2006) The fungicidal efficacy of various commercial disinfectants used in the food industry. Ann Microbiol 56(4):325–330
186. Kosmidis C, Schindler BD, Jacinto PL, Patel D, Bains K, Seo SM, Kaatz GW (2012) Expression of multidrug resistance efflux pump genes in clinical and environmental isolates of Staphylococcus aureus. Int J Antimicrob Agents 40(3):204–209. https://doi.org/10.1016/j.ijantimicag.2012.04.014
187. Kostaki M, Chorianopoulos N, Braxou E, Nychas GJ, Giaouris E (2012) Differential biofilm formation and chemical disinfection resistance of sessile cells of Listeria monocytogenes strains under monospecies and dual-species (with Salmonella enterica) conditions. Appl Environ Microbiol 78(8):2586–2595. https://doi.org/10.1128/aem.07099-11
188. Kouidhi B, Zmantar T, Jrah H, Souiden Y, Chaieb K, Mahdouani K, Bakhrouf A (2011) Antibacterial and resistance-modifying activities of thymoquinone against oral pathogens. Ann Clin Microbiol Antimicrob 10:29. https://doi.org/10.1186/1476-0711-10-29
189. Kovacevic J, Ziegler J, Walecka-Zacharska E, Reimer A, Kitts DD, Gilmour MW (2015) Tolerance of listeria monocytogenes to quaternary ammonium sanitizers is mediated by a Novel Efflux pump encoded by emrE. Appl Environ Microbiol 82(3):939–953. https://doi.org/10.1128/aem.03741-15
190. Kraushaar B, Ballhausen B, Leeser D, Tenhagen BA, Kasbohrer A, Fetsch A (2017) Antimicrobial resistances and virulence markers in Methicillin-resistant Staphylococcus aureus from broiler and turkey: a molecular view from farm to fork. Vet Microbiol 200:25–32. https://doi.org/10.1016/j.vetmic.2016.05.022
191. Kremer PH, Lees JA, Koopmans MM, Ferwerda B, Arends AW, Feller MM, Schipper K, Valls Seron M, van der Ende A, Brouwer MC, van de Beek D, Bentley SD (2017) Benzalkonium tolerance genes and outcome in Listeria monocytogenes meningitis. Clin Microbiol Infect 23(4):265.e261–265.e267. https://doi.org/10.1016/j.cmi.2016.12.008
192. Kucken D, Feucht H, Kaulfers P (2000) Association of qacE and qacEDelta1 with multiple resistance to antibiotics and antiseptics in clinical isolates of Gram-negative bacteria. FEMS Microbiol Lett 183 (1):95–98. doi:S0378109799006369 [pii]
193. Kuda T, Iwase T, Yuphakhun C, Takahashi H, Koyanagi T, Kimura B (2011) Surfactant-disinfectant resistance of Salmonella and Staphylococcus adhered and dried on

surfaces with egg compounds. Food Microbiol 28(5):920–925. https://doi.org/10.1016/j.fm.2010.12.006
194. Kuda T, Yano T, Kuda MT (2008) Resistances to benzalkonium chloride of bacteria dried with food elements on stainless steel surface. LWT Food Sci Technol 41(6):988–993. https://doi.org/10.1016/j.lwt.2007.06.016
195. Kummerer K, Eitel A, Braun U, Hubner P, Daschner F, Mascart G, Milandri M, Reinthaler F, Verhoef J (1997) Analysis of benzalkonium chloride in the effluent from European hospitals by solid-phase extraction and high-performance liquid chromatography with post-column ion-pairing and fluorescence detection. J Chromatogr A 774(1–2):281–286
196. Kupfahl C, Walther M, Wendt C, von Baum H (2015) Identical achromobacter strain in reusable surface disinfection tissue dispensers and a clinical Isolate. Infect Control Hosp Epidemiol 36(11):1362–1364. https://doi.org/10.1017/ice.2015.176
197. Kurihara T, Sugita M, Motai S, Kurashige S (1993) In vitro induction of chlorhexidine- and benzalkonium-resistance in clinically isolated Pseudomonas aeruginosa. Kansenshogaku zasshi Jpn Assoc Infect Dis 67(3):202–206
198. Lambert RJ, Joynson J, Forbes B (2001) The relationships and susceptibilities of some industrial, laboratory and clinical isolates of Pseudomonas aeruginosa to some antibiotics and biocides. J Appl Microbiol 91(6):972–984
199. Landes NJ, Livesay HN, Schaeffer F, Terry P, Trevino E, Weissfeld AS (2016) Burkholderia cepacia: a complex problem for more than Cystic Fibrosis Patients. Clin Microbiol Newsl 38 (18):147–150. https://doi.org/10.1016/j.clinmicnews.2016.08.003
200. Langsrud S, Moretro T, Sundheim G (2003) Characterization of Serratia marcescens surviving in disinfecting footbaths. J Appl Microbiol 95(1):186–195
201. Langsrud S, Sundheim G, Borgmann-Strahsen R (2003) Intrinsic and acquired resistance to quaternary ammonium compounds in food-related Pseudomonas spp. J Appl Microbiol 95 (4):874–882
202. Langsrud S, Sundheim G, Holck AL (2004) Cross-resistance to antibiotics of Escherichia coli adapted to benzalkonium chloride or exposed to stress-inducers. J Appl Microbiol 96 (1):201–208
203. Lavilla Lerma L, Benomar N, Casado Munoz Mdel C, Galvez A, Abriouel H (2015) Correlation between antibiotic and biocide resistance in mesophilic and Psychrotrophic Pseudomonas spp. isolated from slaughterhouse surfaces throughout meat chain production. Food Microbiol 51:33–44. https://doi.org/10.1016/j.fm.2015.04.010
204. Lavilla Lerma L, Benomar N, Valenzuela AS, Casado Munoz Mdel C, Galvez A, Abriouel H (2014) Role of EfrAB efflux pump in biocide tolerance and antibiotic resistance of Enterococcus faecalis and Enterococcus faecium isolated from traditional fermented foods and the effect of EDTA as EfrAB inhibitor. Food Microbiol 44:249–257. https://doi.org/10.1016/j.fm.2014.06.009
205. Law CJ, Alegre KO (2017) Clamping down on drugs: the Escherichia coli multidrug efflux protein MdtM. Res Microbiol. https://doi.org/10.1016/j.resmic.2017.09.006
206. Lear JC, Maillard JY, Dettmar PW, Goddard PA, Russell AD (2006) Chloroxylenol- and triclosan-tolerant bacteria from industrial sources—susceptibility to antibiotics and other biocides. Int Biodeter Biodegr 57(1):51–56. https://doi.org/10.1016/j.ibiod.2005.11.002
207. Lee CS, Lee HB, Cho YG, Park JH, Lee HS (2008) Hospital-acquired Burkholderia cepacia infection related to contaminated benzalkonium chloride. J Hosp Infect 68(3):280–282. https://doi.org/10.1016/j.jhin.2008.01.002
208. Lee H, Lim H, Bae IK, Yong D, Jeong SH, Lee K, Chong Y (2013) Coexistence of mupirocin and antiseptic resistance in methicillin-resistant Staphylococcus aureus isolates from Korea. Diagn Microbiol Infect Dis 75(3):308–312. https://doi.org/10.1016/j.diagmicrobio.2012.11.025
209. Lee JC, Fialkow PJ (1961) Benzalkonium chloride-source of hospital infection with gram-negative bacteria. JAMA 177:708–710

210. Leelaporn A, Firth N, Paulsen IT, Hettiaratchi A, Skurray RA (1995) Multidrug resistance plasmid pSK108 from coagulase-negative staphylococci; relationships to Staphylococcus aureus qacC plasmids. Plasmid 34(1):62–67. https://doi.org/10.1006/plas.1995.1034
211. Leitch CS, Leitch AE, Tidman MJ (2015) Quantitative evaluation of dermatological antiseptics. Clin Exp Dermatol 40(8):912–915. https://doi.org/10.1111/ced.12745
212. Lepainteur M, Royer G, Bourrel AS, Romain O, Duport C, Doucet-Populaire F, Decousser JW (2013) Prevalence of resistance to antiseptics and mupirocin among invasive coagulase-negative staphylococci from very preterm neonates in NICU: the creeping threat? J Hosp Infect 83(4):333–336. https://doi.org/10.1016/j.jhin.2012.11.025
213. Li R, Kuda T, Yano T (2014) Effect of food residues on efficiency of surfactant disinfectants against food related pathogens adhered on polystyrene and ceramic surfaces. LWT Food Sci Technol 57(1):200–206. https://doi.org/10.1016/j.lwt.2013.11.018
214. Li T, Song Y, Zhu Y, Du X, Li M (2013) Current status of Staphylococcus aureus infection in a central teaching hospital in Shanghai, China. BMC Microbiol 13:153. https://doi.org/10.1186/1471-2180-13-153
215. Lin F, Xu Y, Chang Y, Liu C, Jia X, Ling B (2017) Molecular characterization of reduced susceptibility to biocides in clinical isolates of acinetobacter baumannii. Frontiers Microbiol 8:1836. https://doi.org/10.3389/fmicb.2017.01836
216. Lindqvist M, Isaksson B, Swanberg J, Skov R, Larsen AR, Larsen J, Petersen A, Hallgren A (2015) Long-term persistence of a multi-resistant methicillin-susceptible Staphylococcus aureus (MR-MSSA) clone at a university hospital in southeast Sweden, without further transmission within the region. Eur J Clin Microbiol Infect Dis 34(7):1415–1422. https://doi.org/10.1007/s10096-015-2366-1
217. Littlejohn TG, Paulsen IT, Gillespie MT, Tennent JM, Midgley M, Jones IG, Purewal AS, Skurray RA (1992) Substrate specificity and energetics of antiseptic and disinfectant resistance in Staphylococcus aureus. FEMS Microbiol Lett 74(2–3):259–265
218. Liu J, Yu S, Han B, Chen J (2017) Effects of benzalkonium chloride and ethanol on dual-species biofilms of Serratia liquefaciens S1 and Shewanella putrefaciens S4. Food Control 78(Supplement C):196–202. https://doi.org/10.1016/j.foodcont.2017.02.063
219. Liu MA, Kwong SM, Jensen SO, Brzoska AJ, Firth N (2013) Biology of the staphylococcal conjugative multiresistance plasmid pSK41. Plasmid 70(1):42–51. https://doi.org/10.1016/j.plasmid.2013.02.001
220. Liu Q, Liu M, Wu Q, Li C, Zhou T, Ni Y (2009) Sensitivities to biocides and distribution of biocide resistance genes in quaternary ammonium compound tolerant Staphylococcus aureus isolated in a teaching hospital. Scand J Infect Dis 41(6–7):403–409. https://doi.org/10.1080/00365540902856545
221. Liu Q, Zhao H, Han L, Shu W, Wu Q, Ni Y (2015) Frequency of biocide-resistant genes and susceptibility to chlorhexidine in high-level mupirocin-resistant, methicillin-resistant Staphylococcus aureus (MuH MRSA). Diagn Microbiol Infect Dis 82(4):278–283. https://doi.org/10.1016/j.diagmicrobio.2015.03.023
222. Liu WJ, Fu L, Huang M, Zhang JP, Wu Y, Zhou YS, Zeng J, Wang GX (2017) Frequency of antiseptic resistance genes and reduced susceptibility to biocides in carbapenem-resistant Acinetobacter baumannii. J Med Microbiol 66(1):13–17. https://doi.org/10.1099/jmm.0.000403
223. Long M, Lai H, Deng W, Zhou K, Li B, Liu S, Fan L, Wang H, Zou L (2016) Disinfectant susceptibility of different Salmonella serotypes isolated from chicken and egg production chains. J Appl Microbiol 121(3):672–681. https://doi.org/10.1111/jam.13184
224. Longtin J, Seah C, Siebert K, McGeer A, Simor A, Longtin Y, Low DE, Melano RG (2011) Distribution of antiseptic resistance genes qacA, qacB, and smr in methicillin-resistant Staphylococcus aureus isolated in Toronto, Canada, from 2005 to 2009. Antimicrob Agents Chemother 55(6):2999–3001. https://doi.org/10.1128/aac.01707-10

225. Loughlin MF, Jones MV, Lambert PA (2002) Pseudomonas aeruginosa cells adapted to benzalkonium chloride show resistance to other membrane-active agents but not to clinically relevant antibiotics. J Antimicrob Chemother 49(4):631–639
226. Lowe CF, Lloyd-Smith E, Sidhu B, Ritchie G, Sharma A, Jang W, Wong A, Bilawka J, Richards D, Kind T, Puddicombe D, Champagne S, Leung V, Romney MG (2017) Reduction in hospital-associated methicillin-resistant Staphylococcus aureus and vancomycin-resistant Enterococcus with daily chlorhexidine gluconate bathing for medical inpatients. Am J Infect Control 45(3):255–259. https://doi.org/10.1016/j.ajic.2016.09.019
227. Lu Z, Chen Y, Chen W, Liu H, Song Q, Hu X, Zou Z, Liu Z, Duo L, Yang J, Gong Y, Wang Z, Wu X, Zhao J, Zhang C, Zhang M, Han L (2015) Characteristics of qacA/B-positive Staphylococcus aureus isolated from patients and a hospital environment in China. J Antimicrob Chemother 70(3):653–657. https://doi.org/10.1093/jac/dku456
228. Luna VA, Hall TJ, King DS, Cannons AC (2010) Susceptibility of 169 USA300 methicillin-resistant Staphylococcus aureus isolates to two copper-based biocides, CuAL42 and CuWB50. J Antimicrob Chemother 65(5):939–941. https://doi.org/10.1093/jac/dkq092
229. Lunden J, Autio T, Markkula A, Hellstrom S, Korkeala H (2003) Adaptive and cross-adaptive responses of persistent and non-persistent Listeria monocytogenes strains to disinfectants. Int J Food Microbiol 82(3):265–272
230. Luppens SB, Reij MW, van der Heijden RW, Rombouts FM, Abee T (2002) Development of a standard test to assess the resistance of Staphylococcus aureus biofilm cells to disinfectants. Appl Environ Microbiol 68(9):4194–4200
231. Lyon TC (1973) Quaternary ammonia compounds: Should they be used for disinfection in the dental office? Oral Surg Oral Med Oral Pathol 36(5):769–775. https://doi.org/10.1016/0030-4220(73)90154-0
232. Machado I, Coquet L, Jouenne T, Pereira MO (2013) Proteomic approach to Pseudomonas aeruginosa adaptive resistance to benzalkonium chloride. J Proteomics 89:273–279. https://doi.org/10.1016/j.jprot.2013.04.030
233. Machado I, Graca J, Lopes H, Lopes S, Pereira MO (2013) Antimicrobial pressure of ciprofloxacin and gentamicin on biofilm development by an endoscope-isolated pseudomonas Aeruginosa. ISRN Biotechnol 2013:178646. https://doi.org/10.5402/2013/178646
234. Machado I, Graca J, Sousa AM, Lopes SP, Pereira MO (2011) Effect of antimicrobial residues on early adhesion and biofilm formation by wild-type and benzalkonium chloride-adapted Pseudomonas aeruginosa. Biofouling 27(10):1151–1159. https://doi.org/10.1080/08927014.2011.636148
235. Machado I, Lopes SP, Sousa AM, Pereira MO (2012) Adaptive response of single and binary Pseudomonas aeruginosa and Escherichia coli biofilms to benzalkonium chloride. J Basic Microbiol 52(1):43–52. https://doi.org/10.1002/jobm.201100137
236. Mahzounieh M, Khoshnood S, Ebrahimi A, Habibian S, Yaghoubian M (2014) Detection of Antiseptic-resistance genes in pseudomonas and Acinetobacter spp. isolated from burn patients. Jundishapur J Nat Pharm Prod 9(2):e15402
237. Maillard JY, Messager S, Veillon R (1998) Antimicrobial efficacy of biocides tested on skin using an ex-vivo test. J Hosp Infect 40(4):313–323
238. Majtan V, Majtanova L (1999) The effect of new disinfectant substances on the metabolism of Enterobacter cloacae. Int J Antimicrob Agents 11(1):59–64
239. Malizia WF, Gangarosa EJ, Goley AF (1960) Benzalkonium chloride as a source of infection. N Engl J Med 263:800–802. https://doi.org/10.1056/nejm196010202631608 [doi]
240. Mangalappalli-Illathu AK, Korber DR (2006) Adaptive resistance and differential protein expression of Salmonella enterica serovar Enteritidis biofilms exposed to benzalkonium chloride. Antimicrob Agents Chemother 50(11):3588–3596. https://doi.org/10.1128/aac.00573-06
241. Mangalappalli-Illathu AK, Vidovic S, Korber DR (2008) Differential adaptive response and survival of Salmonella enterica serovar enteritidis planktonic and biofilm cells exposed to

benzalkonium chloride. Antimicrob Agents Chemother 52(10):3669–3680. https://doi.org/10.1128/aac.00073-08
242. Marchi E, Furi L, Arioli S, Morrissey I, Di Lorenzo V, Mora D, Giovannetti L, Oggioni MR, Viti C (2015) Novel insight into antimicrobial resistance and sensitivity phenotypes associated to qac and norA genotypes in Staphylococcus aureus. Microbiol Res 170:184–194. https://doi.org/10.1016/j.micres.2014.07.001
243. Marti E, Balcazar JL (2012) Multidrug resistance-encoding plasmid from Aeromonas sp. strain P2G1. Clin Microbiol Infect 18(9):E366–368. https://doi.org/10.1111/j.1469-0691.2012.03935.x
244. Mattei AS, Madrid IM, Santin R, Schuch LF, Meireles MC (2013) In vitro activity of disinfectants against Aspergillus spp. Braz J Microbiol: [Publ Braz Soc Microbiol] 44 (2):481–484. https://doi.org/10.1590/s1517-83822013000200024
245. Mavri A, Mozina SS (2012) Involvement of efflux mechanisms in biocide resistance of Campylobacter jejuni and Campylobacter coli. J Med Microbiol 61(Pt 6):800–808. https://doi.org/10.1099/jmm.0.041467-0
246. Mavri A, Smole Mozina S (2013) Development of antimicrobial resistance in Campylobacter jejuni and Campylobacter coli adapted to biocides. Int J Food Microbiol 160 (3):304–312. https://doi.org/10.1016/j.ijfoodmicro.2012.11.006
247. Mayer S, Boos M, Beyer A, Fluit AC, Schmitz FJ (2001) Distribution of the antiseptic resistance genes qacA, qacB and qacC in 497 methicillin-resistant and -susceptible European isolates of Staphylococcus aureus. J Antimicrob Chemother 47(6):896–897
248. Mc Cay PH, Ocampo-Sosa AA, Fleming GT (2010) Effect of subinhibitory concentrations of benzalkonium chloride on the competitiveness of Pseudomonas aeruginosa grown in continuous culture. Microbiology (Reading, England) 156(Pt 1):30–38. https://doi.org/10.1099/mic.0.029751-0
249. Mc Gann P, Milillo M, Kwak YI, Quintero R, Waterman PE, Lesho E (2013) Rapid and simultaneous detection of the chlorhexidine and mupirocin resistance genes qacA/B and mupA in clinical isolates of methicillin-resistant Staphylococcus aureus. Diagn Microbiol Infect Dis 77(3):270–272. https://doi.org/10.1016/j.diagmicrobio.2013.06.006
250. McKay AM (2008) Antimicrobial resistance and heat sensitivity of oxacillin-resistant, mecA-positive Staphylococcus spp. from unpasteurized milk. J Food Prot 71(1):186–190
251. McNeil JC, Hulten KG, Kaplan SL, Mahoney DH, Mason EO (2013) Staphylococcus aureus infections in pediatric oncology patients: high rates of antimicrobial resistance, antiseptic tolerance and complications. Pediatr Infect Dis J 32(2):124–128. https://doi.org/10.1097/INF.0b013e318271c4e0
252. McNeil JC, Hulten KG, Mason EO, Kaplan SL (2017) Impact of Health Care Exposure on Genotypic Antiseptic Tolerance in Staphylococcus aureus Infections in a Pediatric Population. Antimicrob Agents Chemother 61 (7). https://doi.org/10.1128/aac.00223-17
253. McNeil JC, Kok EY, Vallejo JG, Campbell JR, Hulten KG, Mason EO, Kaplan SL (2016) Clinical and molecular features of Decreased Chlorhexidine susceptibility among nosocomial staphylococcus aureus isolates at Texas Children's Hospital. Antimicrob Agents Chemother 60(2):1121–1128. https://doi.org/10.1128/aac.02011-15
254. McNeil JC, Ligon JA, Hulten KG, Dreyer WJ, Heinle JS, Mason EO, Kaplan SL (2013) Staphylococcus aureus Infections in Children With Congenital Heart Disease. J Pediatr Infect Dis Soc 2(4):337–344. https://doi.org/10.1093/jpids/pit037
255. Meier AB, Guldimann C, Markkula A, Pontinen A, Korkeala H, Tasara T (2017) Comparative phenotypic and genotypic analysis of swiss and finnish listeria monocytogenes isolates with respect to benzalkonium chloride resistance. Front Microbiol 8:397. https://doi.org/10.3389/fmicb.2017.00397
256. Menegotto F, Gonzalez-Cabrero S, Cubero A, Cuervo W, Munoz M, Gutierrez MP, Simarro M, Bratos MA, Orduna A (2012) Clonal nature and diversity of resistance, toxins and adhesins genes of meticillin-resistant Staphylococcus aureus collected in a Spanish

hospital. Infect, Genet Evol: J Mol Epidemiol Evol Genet Infect Dis 12(8):1751–1758. https://doi.org/10.1016/j.meegid.2012.07.020

257. Mester P, Gundolf T, Kalb R, Wagner M, Rossmanith P (2015) Molecular mechanisms mediating tolerance to ionic liquids in Listeria monocytogenes. Sep Purif Technol 155 (Supplement C):32–37. https://doi.org/10.1016/j.seppur.2015.01.017
258. Metselaar KI, Saa Ibusquiza P, Ortiz Camargo AR, Krieg M, Zwietering MH, den Besten HM, Abee T (2015) Performance of stress resistant variants of Listeria monocytogenes in mixed species biofilms with Lactobacillus plantarum. Int J Food Microbiol 213:24–30. https://doi.org/10.1016/j.ijfoodmicro.2015.04.021
259. Miladi H, Zmantar T, Chaabouni Y, Fedhila K, Bakhrouf A, Mahdouani K, Chaieb K (2016) Antibacterial and efflux pump inhibitors of thymol and carvacrol against food-borne pathogens. Microb Pathog 99:95–100. https://doi.org/10.1016/j.micpath.2016.08.008
260. Minbiole KPC, Jennings MC, Ator LE, Black JW, Grenier MC, LaDow JE, Caran KL, Seifert K, Wuest WM (2016) From antimicrobial activity to mechanism of resistance: the multifaceted role of simple quaternary ammonium compounds in bacterial eradication. Tetrahedron 72(25):3559–3566. https://doi.org/10.1016/j.tet.2016.01.014
261. Miyano N, Oie S, Kamiya A (2003) Efficacy of disinfectants and hot water against biofilm cells of Burkholderia cepacia. Biol Pharm Bull 26(5):671–674
262. Miyazaki NH, Abreu AO, Marin VA, Rezende CA, Moraes MT, Villas Boas MH (2007) The presence of qacA/B gene in Brazilian methicillin-resistant Staphylococcus aureus. Mem Inst Oswaldo Cruz 102(4):539–540
263. Moen B, Rudi K, Bore E, Langsrud S (2012) Subminimal inhibitory concentrations of the disinfectant benzalkonium chloride select for a tolerant subpopulation of Escherichia coli with inheritable characteristics. Int J Mol Sci 13(4):4101–4123. https://doi.org/10.3390/ijms13044101
264. Monecke S, Ehricht R, Slickers P, Tan HL, Coombs G (2009) The molecular epidemiology and evolution of the Panton-Valentine leukocidin-positive, methicillin-resistant Staphylococcus aureus strain USA300 in Western Australia. Clin Microbiol Infect 15(8):770–776. https://doi.org/10.1111/j.1469-0691.2009.02792.x
265. Monecke S, Ruppelt A, Wendlandt S, Schwarz S, Slickers P, Ehricht R, Jackel SC (2013) Genotyping of Staphylococcus aureus isolates from diseased poultry. Vet Microbiol 162(2–4):806–812. https://doi.org/10.1016/j.vetmic.2012.10.018
266. Moore LE, Ledder RG, Gilbert P, McBain AJ (2008) In vitro study of the effect of cationic biocides on bacterial population dynamics and susceptibility. Appl Environ Microbiol 74 (15):4825–4834. https://doi.org/10.1128/aem.00573-08
267. Morales-Fernandez L, Fernandez-Crehuet M, Espigares M, Moreno E, Espigares E (2014) Study of the hormetic effect of disinfectants chlorhexidine, povidone iodine and benzalkonium chloride. Eur J Clin Microbiol Infect Dis 33(1):103–109. https://doi.org/10.1007/s10096-013-1934-5
268. Moretro T, Schirmer BCT, Heir E, Fagerlund A, Hjemli P, Langsrud S (2017) Tolerance to quaternary ammonium compound disinfectants may enhance growth of Listeria monocytogenes in the food industry. Int J Food Microbiol 241:215–224. https://doi.org/10.1016/j.ijfoodmicro.2016.10.025
269. Morgan M, McGann P, Gierhart S, Chukwuma U, Richesson D, Hinkle M, Lesho E (2017) Consumption of chlorhexidine and mupirocin across the health system of the US Department of Defense (DOD) and the incidence of the qacA/B and mupA genes in the DOD Facilities of the national capital region. Clin Infect Dis: Official Publ Infect Dis Soc Am 64(12):1801–1802. https://doi.org/10.1093/cid/cix276
270. Mori M, Gomi M, Matsumune N, Niizeki K, Sakagami Y (2013) Biofilm-forming activity of bacteria isolated from toilet bowl biofilms and the bactericidal activity of disinfectants against the isolates. Biocontrol Sci 18(3):129–135
271. Morrissey I, Oggioni MR, Knight D, Curiao T, Coque T, Kalkanci A, Martinez JL (2014) Evaluation of epidemiological cut-off values indicates that biocide resistant subpopulations

are uncommon in natural isolates of clinically-relevant microorganisms. PLoS ONE 9(1): e86669. https://doi.org/10.1371/journal.pone.0086669
272. Mosca A, Russo F, Miragliotta G (2006) In vitro antimicrobial activity of benzalkonium chloride against clinical isolates of Streptococcus agalactiae. J Antimicrob Chemother 57 (3):566–568. https://doi.org/10.1093/jac/dki474
273. Mullapudi S, Siletzky RM, Kathariou S (2008) Heavy-metal and benzalkonium chloride resistance of Listeria monocytogenes isolates from the environment of turkey-processing plants. Appl Environ Microbiol 74(5):1464–1468. https://doi.org/10.1128/aem.02426-07
274. Muller A, Rychli K, Muhterem-Uyar M, Zaiser A, Stessl B, Guinane CM, Cotter PD, Wagner M, Schmitz-Esser S (2013) Tn6188 - a novel transposon in Listeria monocytogenes responsible for tolerance to benzalkonium chloride. PLoS ONE 8(10):e76835. https://doi.org/10.1371/journal.pone.0076835
275. Muller A, Rychli K, Zaiser A, Wieser C, Wagner M, Schmitz-Esser S (2014) The Listeria monocytogenes transposon Tn6188 provides increased tolerance to various quaternary ammonium compounds and ethidium bromide. FEMS Microbiol Lett 361(2):166–173. https://doi.org/10.1111/1574-6968.12626
276. Muller G, Kramer A (2008) Biocompatibility index of antiseptic agents by parallel assessment of antimicrobial activity and cellular cytotoxicity. J Antimicrob Chemother 61 (6):1281–1287. https://doi.org/10.1093/jac/dkn125
277. Munoz-Gallego I, Infiesta L, Viedma E, Perez-Montarelo D, Chaves F (2016) Chlorhexidine and mupirocin susceptibilities in methicillin-resistant Staphylococcus aureus isolates from bacteraemia and nasal colonisation. J Glob Antimicrob Resist 4:65–69. https://doi.org/10.1016/j.jgar.2015.11.005
278. Murayama N, Nagata M, Terada Y, Okuaki M, Takemura N, Nakaminami H, Noguchi N (2013) In vitro antiseptic susceptibilities for Staphylococcus pseudintermedius isolated from canine superficial pyoderma in Japan. Vet Dermatol 24(1):126–129.e129. https://doi.org/10.1111/j.1365-3164.2012.01103.x
279. Nagai I, Ogase H (1990) Absence of role for plasmids in resistance to multiple disinfectants in three strains of bacteria. J Hosp Infect 15(2):149–155
280. Nagai K, Murata T, Ohta S, Zenda H, Ohnishi M, Hayashi T (2003) Two different mechanisms are involved in the extremely high-level benzalkonium chloride resistance of a Pseudomonas fluorescens strain. Microbiol Immunol 47(10):709–715
281. Nagai K, Ohta S, Zenda H, Matsumoto H, Makino M (1996) Biochemical characterization of a Pseudomonas fluorescens strain isolated from a benzalkonium chloride solution. Biol Pharm Bull 19(6):873–875
282. Nakaminami H, Noguchi N, Nishijima S, Kurokawa I, Sasatsu M (2008) Characterization of the pTZ2162 encoding multidrug efflux gene qacB from Staphylococcus aureus. Plasmid 60 (2):108–117. https://doi.org/10.1016/j.plasmid.2008.04.003
283. Nakamura H, Takakura K, Sone Y, Itano Y, Nishikawa Y (2013) Biofilm formation and resistance to benzalkonium chloride in Listeria monocytogenes isolated from a fish processing plant. J Food Prot 76(7):1179–1186. https://doi.org/10.4315/0362-028x.jfp-12-225
284. Nakashima AK, Highsmith AK, Martone WJ (1987) Survival of Serratia marcescens in benzalkonium chloride and in multiple-dose medication vials: relationship to epidemic septic arthritis. J Clin Microbiol 25(6):1019–1021
285. Nakashima AK, McCarthy MA, Martone WJ, Anderson RL (1987) Epidemic septic arthritis caused by Serratia marcescens and associated with a benzalkonium chloride antiseptic. J Clin Microbiol 25(6):1014–1018
286. Narui K, Takano M, Noguchi N, Sasatsu M (2007) Susceptibilities of methicillin-resistant Staphylococcus aureus isolates to seven biocides. Biol Pharm Bull 30(3):585–587
287. Nemeghaire S, Argudin MA, Haesebrouck F, Butaye P (2014) Molecular epidemiology of methicillin-resistant Staphylococcus sciuri in healthy chickens. Vet Microbiol 171(3–4):357–363. https://doi.org/10.1016/j.vetmic.2014.01.041

288. Nhung NT, Thuy CT, Trung NV, Campbell J, Baker S, Thwaites G, Hoa NT, Carrique-Mas J (2015) Induction of antimicrobial resistance in Escherichia coli and non-typhoidal Salmonella strains after adaptation to disinfectant commonly used on farms in Vietnam. Antibiotics (Basel, Switzerland) 4(4):480–494. https://doi.org/10.3390/antibiotics4040480
289. Nilsson RE, Ross T, Bowman JP (2011) Variability in biofilm production by Listeria monocytogenes correlated to strain origin and growth conditions. Int J Food Microbiol 150 (1):14–24. https://doi.org/10.1016/j.ijfoodmicro.2011.07.012
290. Noguchi N, Hase M, Kitta M, Sasatsu M, Deguchi K, Kono M (1999) Antiseptic susceptibility and distribution of antiseptic-resistance genes in methicillin-resistant Staphylococcus aureus. FEMS Microbiol Lett 172(2):247–253
291. Noguchi N, Nakaminami H, Nishijima S, Kurokawa I, So H, Sasatsu M (2006) Antimicrobial agent of susceptibilities and antiseptic resistance gene distribution among methicillin-resistant Staphylococcus aureus isolates from patients with impetigo and staphylococcal scalded skin syndrome. J Clin Microbiol 44(6):2119–2125. https://doi.org/10.1128/jcm.02690-05
292. Noguchi N, Suwa J, Narui K, Sasatsu M, Ito T, Hiramatsu K, Song JH (2005) Susceptibilities to antiseptic agents and distribution of antiseptic-resistance genes qacA/B and smr of methicillin-resistant Staphylococcus aureus isolated in Asia during 1998 and 1999. J Med Microbiol 54(Pt 6):557–565. https://doi.org/10.1099/jmm.0.45902-0
293. Ntsama-Essomba C, Bouttier S, Ramaldes M, Dubois-Brissonnet F, Fourniat J (1997) Resistance of Escherichia coli growing as biofilms to disinfectants. Vet Res 28(4):353–363
294. Oggioni MR, Coelho JR, Furi L, Knight DR, Viti C, Orefici G, Martinez JL, Freitas AT, Coque TM, Morrissey I (2015) Significant differences characterise the correlation coefficients between biocide and antibiotic susceptibility profiles in Staphylococcus aureus. Curr Pharm Des 21(16):2054–2057
295. Ohta S, Makino M, Nagai K, Zenda H (1996) Comparative fungicidal activity of a new quaternary ammonium salt, N-alkyl-N-2-hydroxyethyl-N, N-dimethylammonium butyl phosphate and commonly used disinfectants. Biol Pharm Bull 19(2):308–310
296. Oie S, Huang Y, Kamiya A, Konishi H, Nakazawa T (1996) Efficacy of disinfectants against biofilm cells of methicillin-resistant *Staphylococcus aureus*. Microbios 85:223–230
297. Oliveira AR, Domingues FC, Ferreira S (2017) The influence of resveratrol adaptation on resistance to antibiotics, benzalkonium chloride, heat and acid stresses of Staphylococcus aureus and Listeria monocytogenes. Food Control 73:1420–1425. https://doi.org/10.1016/j.foodcont.2016.11.011
298. Olson RK, Voorhees RE, Eitzen HE, Rolka H, Sewell CM (1999) Cluster of postinjection abscesses related to corticosteroid injections and use of benzalkonium chloride. West J Med 170(3):143–147
299. Oosterik LH, Tuntufye HN, Butaye P, Goddeeris BM (2014) Effect of serogroup, surface material and disinfectant on biofilm formation by avian pathogenic Escherichia coli. Vet J (London, England: 1997) 202(3):561–565. https://doi.org/10.1016/j.tvjl.2014.10.001
300. Opacic D, Lepsanovic Z, Sbutega Milosevic G (2010) Distribution of disinfectant resistance genes qacA/B in clinical isolates of meticillin-resistant and -susceptible Staphylococcus aureus in one Belgrade hospital. J Hosp Infect 76(3):266–267. https://doi.org/10.1016/j.jhin.2010.04.019
301. Orsi GB, Tomao P, Visca P (1995) In vitro activity of commercially manufactured disinfectants against Pseudomonas aeruginosa. Eur J Epidemiol 11(4):453–457
302. Ortiz S, Lopez-Alonso V, Rodriguez P, Martinez-Suarez JV (2015) The connection between persistent, disinfectant-resistant listeria monocytogenes strains from two geographically separate Iberian pork processing plants: evidence from comparative genome analysis. Appl Environ Microbiol 82(1):308–317. https://doi.org/10.1128/aem.02824-15

303. Ortiz S, Lopez V, Martinez-Suarez JV (2014) Control of Listeria monocytogenes contamination in an Iberian pork processing plant and selection of benzalkonium chloride-resistant strains. Food Microbiol 39:81–88. https://doi.org/10.1016/j.fm.2013.11.007
304. Ortiz S, Lopez V, Martinez-Suarez JV (2014) The influence of subminimal inhibitory concentrations of benzalkonium chloride on biofilm formation by Listeria monocytogenes. Int J Food Microbiol 189:106–112. https://doi.org/10.1016/j.ijfoodmicro.2014.08.007
305. Ostman M, Lindberg RH, Fick J, Bjorn E, Tysklind M (2017) Screening of biocides, metals and antibiotics in Swedish sewage sludge and wastewater. Water Res 115:318–328. https://doi.org/10.1016/j.watres.2017.03.011
306. Pagedar A, Singh J (2015) Evaluation of antibiofilm effect of benzalkonium chloride, iodophore and sodium hypochlorite against biofilm of Pseudomonas aeruginosa of dairy origin. J Food Sci Technol 52(8):5317–5322. https://doi.org/10.1007/s13197-014-1575-4
307. Pagedar A, Singh J, Batish VK (2011) Efflux mediated adaptive and cross resistance to ciprofloxacin and benzalkonium chloride in Pseudomonas aeruginosa of dairy origin. J Basic Microbiol 51(3):289–295. https://doi.org/10.1002/jobm.201000292
308. Pagedar A, Singh J, Batish VK (2012) Adaptation to benzalkonium chloride and ciprofloxacin affects biofilm formation potential, efflux pump and haemolysin activity of Escherichia coli of dairy origin. J Dairy Res 79 (4):383–389. S0022029912000295 [pii]. https://doi.org/10.1017/s0022029912000295 [doi]
309. Paniagua-Contreras GL, Monroy-Perez E, Vaca-Paniagua F, Rodriguez-Moctezuma JR, Negrete-Abascal E, Vaca S (2014) Implementation of a novel in vitro model of infection of reconstituted human epithelium for expression of virulence genes in methicillin-resistant Staphylococcus aureus strains isolated from catheter-related infections in Mexico. Ann Clin Microbiol Antimicrob 13:6. https://doi.org/10.1186/1476-0711-13-6
310. Paulsen IT, Brown MH, Littlejohn TG, Mitchell BA, Skurray RA (1996) Multidrug resistance proteins QacA and QacB from Staphylococcus aureus: membrane topology and identification of residues involved in substrate specificity. Proc Natl Acad Sci USA 93 (8):3630–3635
311. Peirano G, Lascols C, Hackel M, Hoban DJ, Pitout JD (2014) Molecular epidemiology of Enterobacteriaceae that produce VIMs and IMPs from the SMART surveillance program. Diagn Microbiol Infect Dis 78(3):277–281. https://doi.org/10.1016/j.diagmicrobio.2013.11.024
312. Peyrat MB, Soumet C, Maris P, Sanders P (2008) Recovery of Campylobacter jejuni from surfaces of poultry slaughterhouses after cleaning and disinfection procedures: analysis of a potential source of carcass contamination. Int J Food Microbiol 124(2):188–194. https://doi.org/10.1016/j.ijfoodmicro.2008.03.030
313. Piercey MJ, Ells TC, Macintosh AJ, Truelstrup Hansen L (2017) Variations in biofilm formation, desiccation resistance and Benzalkonium chloride susceptibility among Listeria monocytogenes strains isolated in Canada. Int J Food Microbiol 257:254–261. https://doi.org/10.1016/j.ijfoodmicro.2017.06.025
314. Plotkin SA, Austrian R (1958) Bacteremia caused by Pseudomonas sp. following the use of materials stored in solutions of a cationic surface-active agent. Am J Med Sci 235 (6):621–627
315. Popowska M, Olszak M, Markiewicz Z (2006) Susceptibility of Listeria monocytogenes strains isolated from dairy products and frozen vegetables to antibiotics inhibiting murein synthesis and to disinfectants. Pol J Microbiol 55(4):279–288
316. Prag G, Falk-Brynhildsen K, Jacobsson S, Hellmark B, Unemo M, Soderquist B (2014) Decreased susceptibility to chlorhexidine and prevalence of disinfectant resistance genes among clinical isolates of Staphylococcus epidermidis. APMIS: acta pathologica, microbiologica, et immunologica Scandinavica 122(10):961–967. https://doi.org/10.1111/apm.12239

317. Pricope L, Nicolau A, Wagner M, Rychli K (2013) The effect of sublethal concentrations of benzalkonium chloride on invasiveness and intracellular proliferation of Listeria monocytogenes. Food Control 31(1):230–235. https://doi.org/10.1016/j.foodcont.2012.09.031
318. Ratani SS, Siletzky RM, Dutta V, Yildirim S, Osborne JA, Lin W, Hitchins AD, Ward TJ, Kathariou S (2012) Heavy metal and disinfectant resistance of Listeria monocytogenes from foods and food processing plants. Appl Environ Microbiol 78(19):6938–6945. https://doi.org/10.1128/aem.01553-12
319. Reich PJ, Boyle MG, Hogan PG, Johnson AJ, Wallace MA, Elward AM, Warner BB, Burnham CD, Fritz SA (2016) Emergence of community-associated methicillin-resistant Staphylococcus aureus strains in the neonatal intensive care unit: an infection prevention and patient safety challenge. Clin Microbiol Infect 22(7):645.e641–645.e648. https://doi.org/10.1016/j.cmi.2016.04.013
320. Reichel M, Schlicht A, Ostermeyer C, Kampf G (2014) Efficacy of surface disinfectant cleaners against emerging highly resistant gram-negative bacteria. BMC Infect Dis 14:292. https://doi.org/10.1186/1471-2334-14-292
321. Rensch U, Klein G, Schwarz S, Kaspar H, de Jong A, Kehrenberg C (2013) Comparative analysis of the susceptibility to triclosan and three other biocides of avian Salmonella enterica isolates collected 1979 through 1994 and 2004 through 2010. J Food Prot 76 (4):653–656. https://doi.org/10.4315/0362-028x.jfp-12-420
322. Riazi S, Matthews KR (2011) Failure of foodborne pathogens to develop resistance to sanitizers following repeated exposure to common sanitizers. Int Biodeter Biodegr 65 (2):374–378. https://doi.org/10.1016/j.ibiod.2010.12.001
323. Ribic U, Klancnik A, Jersek B (2017) Characterization of Staphylococcus epidermidis strains isolated from industrial cleanrooms under regular routine disinfection. J Appl Microbiol 122(5):1186–1196. https://doi.org/10.1111/jam.13424
324. Rikimaru T, Kondo M, Kondo S, Oizumi K (2000) Efficacy of common antiseptics against mycobacteria. Int J Tuberc Lung Dis: Official J Int Union Against Tuberc Lung Dis 4 (6):570–576
325. Rizzotti L, Rossi F, Torriani S (2016) Biocide and antibiotic resistance of Enterococcus faecalis and Enterococcus faecium isolated from the swine meat chain. Food Microbiol 60:160–164. https://doi.org/10.1016/j.fm.2016.07.009
326. Rodriguez Ferri EF, Martinez S, Frandoloso R, Yubero S, Gutierrez Martin CB (2010) Comparative efficacy of several disinfectants in suspension and carrier tests against Haemophilus parasuis serovars 1 and 5. Res Vet Sci 88(3):385–389. https://doi.org/10.1016/j.rvsc.2009.12.001
327. Romanova NA, Gawande PV, Brovko LY, Griffiths MW (2007) Rapid methods to assess sanitizing efficacy of benzalkonium chloride to Listeria monocytogenes biofilms. J Microbiol Methods 71(3):231–237. https://doi.org/10.1016/j.mimet.2007.09.002
328. Romanova NA, Wolffs PF, Brovko LY, Griffiths MW (2006) Role of efflux pumps in adaptation and resistance of Listeria monocytogenes to benzalkonium chloride. Appl Environ Microbiol 72(5):3498–3503. https://doi.org/10.1128/aem.72.5.3498-3503.2006
329. Rondeau C, Chevet G, Blanc DS, Gbaguidi-Haore H, Decalonne M, Dos Santos S, Quentin R, van der Mee-Marquet N (2016) Current molecular epidemiology of methicillin-resistant staphylococcus aureus in elderly French people: troublesome clones on the horizon. Frontiers Microbiol 7:31. https://doi.org/10.3389/fmicb.2016.00031
330. Rose H, Baldwin A, Dowson CG, Mahenthiralingam E (2009) Biocide susceptibility of the Burkholderia cepacia complex. J Antimicrob Chemother 63(3):502–510. https://doi.org/10.1093/jac/dkn540
331. Ruckerl I, Muhterem-Uyar M, Muri-Klinger S, Wagner KH, Wagner M, Stessl B (2014) L. monocytogenes in a cheese processing facility: learning from contamination scenarios over three years of sampling. Int J Food Microbiol 189:98–105. https://doi.org/10.1016/j.ijfoodmicro.2014.08.001

332. Russell AD (2002) Introduction of biocides into clinical practice and the impact on antibiotic-resistant bacteria. Symp Ser (Soc Appl Microbiol) 31:121s–135s
333. Russell AD, Furr JR (1986) Susceptibility of porin- and liposaccharide-deficient strains of *Escherichia coli* to some antiseptics and disinfectants. J Hosp Infect 8:47–56
334. Rutala WA, Cole EC, Wannamaker NS, Weber DJ (1991) Inactivation of Mycobacterium tuberculosis and Mycobacterium bovis by 14 hospital disinfectants. Am J Med 91(3b):267s–271s
335. Saa Ibusquiza P, Herrera JJ, Cabo ML (2011) Resistance to benzalkonium chloride, peracetic acid and nisin during formation of mature biofilms by Listeria monocytogenes. Food Microbiol 28(3):418–425. https://doi.org/10.1016/j.fm.2010.09.014
336. Saa Ibusquiza P, Herrera JJ, Vazquez-Sanchez D, Parada A, Cabo ML (2012) A new and efficient method to obtain benzalkonium chloride adapted cells of Listeria monocytogenes. J Microbiol Methods 91(1):57–61. https://doi.org/10.1016/j.mimet.2012.07.009
337. Saá Ibusquiza P, Herrera JJR, Vázquez-Sánchez D, Cabo ML (2012) Adherence kinetics, resistance to benzalkonium chloride and microscopic analysis of mixed biofilms formed by Listeria monocytogenes and Pseudomonas putida. Food Control 25(1):202–210. https://doi.org/10.1016/j.foodcont.2011.10.002
338. Sandle T, Vijayakumar R, Saleh Al Aboody M, Saravanakumar S (2014) In vitro fungicidal activity of biocides against pharmaceutical environmental fungal isolates. J Appl Microbiol 117(5):1267–1273. https://doi.org/10.1111/jam.12628
339. Sattar SA, Bradley C, Kibbee R, Wesgate R, Wilkinson MA, Sharpe T, Maillard JY (2015) Disinfectant wipes are appropriate to control microbial bioburden from surfaces: use of a new ASTM standard test protocol to demonstrate efficacy. J Hosp Infect 91(4):319–325. https://doi.org/10.1016/j.jhin.2015.08.026
340. Sautter RL, Mattman LH, Legaspi RC (1984) Serratia marcescens meningitis associated with a contaminated benzalkonium chloride solution. Infect Control: IC 5(5):223–225
341. Schlett CD, Millar EV, Crawford KB, Cui T, Lanier JB, Tribble DR, Ellis MW (2014) Prevalence of chlorhexidine-resistant methicillin-resistant Staphylococcus aureus following prolonged exposure. Antimicrob Agents Chemother 58(8):4404–4410. https://doi.org/10.1128/aac.02419-14
342. Seier-Petersen MA, Nielsen LN, Ingmer H, Aarestrup FM, Agerso Y (2015) Biocide Susceptibility of Staphylococcus aureus CC398 and CC30 Isolates from pigs and identification of the biocide resistance genes, qacG and qacC. Microb Drug Res (Larchmont, NY) 21(5):527–536. https://doi.org/10.1089/mdr.2014.0215
343. Sekiguchi J, Hama T, Fujino T, Araake M, Irie A, Saruta K, Konosaki H, Nishimura H, Kawano A, Kudo K, Kondo T, Sasazuki T, Kuratsuji T, Yoshikura H, Kirikae T (2004) Detection of the antiseptic- and disinfectant-resistance genes qacA, qacB, and qacC in methicillin-resistant Staphylococcus aureus isolated in a Tokyo hospital. Jpn J Infect Dis 57 (6):288–291
344. Seng R, Leungtongkam U, Thummeepak R, Chatdumrong W, Sitthisak S (2017) High prevalence of methicillin-resistant coagulase-negative staphylococci isolated from a university environment in Thailand. Int Microbiol: Official J Span Soc Microbiol 20 (2):65–73. https://doi.org/10.2436/20.1501.01.286
345. Senok A, Ehricht R, Monecke S, Al-Saedan R, Somily A (2016) Molecular characterization of methicillin-resistant Staphylococcus aureus in nosocomial infections in a tertiary-care facility: emergence of new clonal complexes in Saudi Arabia. New Microbes New Infect 14:13–18. https://doi.org/10.1016/j.nmni.2016.07.009
346. Shamsudin MN, Alreshidi MA, Hamat RA, Alshrari AS, Atshan SS, Neela V (2012) High prevalence of qacA/B carriage among clinical isolates of meticillin-resistant Staphylococcus aureus in Malaysia. J Hosp Infect 81(3):206–208. https://doi.org/10.1016/j.jhin.2012.04.015
347. Sheng WH, Wang JT, Lauderdale TL, Weng CM, Chen D, Chang SC (2009) Epidemiology and susceptibilities of methicillin-resistant Staphylococcus aureus in Taiwan: emphasis on

chlorhexidine susceptibility. Diagn Microbiol Infect Dis 63(3):309–313. https://doi.org/10.1016/j.diagmicrobio.2008.11.014
348. Sheridan A, Lenahan M, Duffy G, Fanning S, Burgess C (2012) The potential for biocide tolerance in Escherichia coli and its impact on the response to food processing stresses. Food Control 26(1):98–106
349. Shi GS, Boost M, Cho P (2015) Prevalence of antiseptic-resistance genes in staphylococci isolated from orthokeratology lens and spectacle wearers in Hong Kong. Invest Ophthalmol Vis Sci 56(5):3069–3074. https://doi.org/10.1167/iovs.15-16550
350. Shi GS, Boost MV, Cho P (2016) Does the presence of QAC genes in staphylococci affect the efficacy of disinfecting solutions used by orthokeratology lens wearers? Brit J Ophthalmol 100(5):708–712. https://doi.org/10.1136/bjophthalmol-2015-307811
351. Shikano A, Kuda T, Takahashi H, Kimura B (2017) Effect of quantity of food residues on resistance to desiccation, disinfectants, and UV-C irradiation of spoilage yeasts adhered to a stainless steel surface. LWT Food Sci Technol 80(Supplement C):169–177. https://doi.org/10.1016/j.lwt.2017.02.020
352. Sidhu MS, Heir E, Leegaard T, Wiger K, Holck A (2002) Frequency of disinfectant resistance genes and genetic linkage with beta-lactamase transposon Tn552 among clinical staphylococci. Antimicrob Agents Chemother 46(9):2797–2803
353. Sidhu MS, Sorum H, Holck A (2002) Resistance to quaternary ammonium compounds in food-related bacteria. Microb Drug Res (Larchmont, NY) 8(4):393–399. https://doi.org/10.1089/10766290260469679
354. Simoes M, Pereira MO, Vieira MJ (2005) Effect of mechanical stress on biofilms challenged by different chemicals. Water Res 39(20):5142–5152. https://doi.org/10.1016/j.watres.2005.09.028
355. Skovgaard S, Larsen MH, Nielsen LN, Skov RL, Wong C, Westh H, Ingmer H (2013) Recently introduced qacA/B genes in Staphylococcus epidermidis do not increase chlorhexidine MIC/MBC. J Antimicrob Chemother 68(10):2226–2233. https://doi.org/10.1093/jac/dkt182
356. Slifierz MJ, Friendship RM, Weese JS (2015) Methicillin-resistant Staphylococcus aureus in commercial swine herds is associated with disinfectant and zinc usage. Appl Environ Microbiol 81(8):2690–2695. https://doi.org/10.1128/aem.00036-15
357. Smith K, Gemmell CG, Hunter IS (2008) The association between biocide tolerance and the presence or absence of qac genes among hospital-acquired and community-acquired MRSA isolates. J Antimicrob Chemother 61(1):78–84. https://doi.org/10.1093/jac/dkm395
358. Smith K, Hunter IS (2008) Efficacy of common hospital biocides with biofilms of multi-drug resistant clinical isolates. J Med Microbiol 57(Pt 8):966–973. https://doi.org/10.1099/jmm.0.47668-0
359. Sommerstein R, Fuhrer U, Lo Priore E, Casanova C, Meinel DM, Seth-Smith HM, Kronenberg A, On Behalf Of A, Koch D, Senn L, Widmer AF, Egli A, Marschall J, On Behalf Of S (2017) Burkholderia stabilis outbreak associated with contaminated commercially-available washing gloves, Switzerland, May 2015 to August 2016. Euro Surveill 22(49). https://doi.org/10.2807/1560 7917.es.2017.22.49.17-00213
360. Soumet C, Fourreau E, Legrandois P, Maris P (2012) Resistance to phenicol compounds following adaptation to quaternary ammonium compounds in Escherichia coli. Vet Microbiol 158(1–2):147–152. https://doi.org/10.1016/j.vetmic.2012.01.030
361. Soumet C, Ragimbeau C, Maris P (2005) Screening of benzalkonium chloride resistance in Listeria monocytogenes strains isolated during cold smoked fish production. Lett Appl Microbiol 41(3):291–296. https://doi.org/10.1111/j.1472-765X.2005.01763.x
362. Srinivasan VB, Singh BB, Priyadarshi N, Chauhan NK, Rajamohan G (2014) Role of novel multidrug efflux pump involved in drug resistance in Klebsiella pneumoniae. PLoS ONE 9(5):e96288. https://doi.org/10.1371/journal.pone.0096288
363. Stewart PS, Grab L, Diemer JA (1998) Analysis of biocide transport limitation in an artificial biofilm system. J Appl Microbiol 85(3):495–500

364. Stickler DJ (1974) Chlorhexidine resistance in *Proteus mirabilis*. J Clin Pathol 27 (4):284–287
365. Stupar M, Grbić ML, Džamić A, Unković N, Ristić M, Jelikić A, Vukojević J (2014) Antifungal activity of selected essential oils and biocide benzalkonium chloride against the fungi isolated from cultural heritage objects. S Afr J Bot 93 (Supplement C):118–124. https://doi.org/10.1016/j.sajb.2014.03.016
366. Suller MTE, Russell AD (1999) Antibiotic and biocide resistance in methicillin-resistant *Staphylococcus aureus* and vancomycin-resistant enterococcus. J Hosp Infect 43:281–291
367. Sundheim G, Hagtvedt T, Dainty R (1992) Resistance of meat associated staphylococci to a quaternary ammonium compound. Food Microbiol 9(2):161–167. https://doi.org/10.1016/0740-0020(92)80023-W
368. Sundheim G, Langsrud S, Heir E, Holck AL (1998) Bacterial resistance to disinfectants containing quaternary ammonium compounds. Int Biodeter Biodegr 41(3):235–239. https://doi.org/10.1016/S0964-8305(98)00027-4
369. Tabata A, Nagamune H, Maeda T, Murakami K, Miyake Y, Kourai H (2003) Correlation between resistance of Pseudomonas aeruginosa to quaternary ammonium compounds and expression of outer membrane protein OprR. Antimicrob Agents Chemother 47(7):2093–2099
370. Taheri N, Ardebili A, Amouzandeh-Nobaveh A, Ghaznavi-Rad E (2016) Frequency of antiseptic resistance among Staphylococcus aureus and coagulase-negative Staphylococci isolated from a University Hospital in Central Iran. Oman Med J 31(6):426–432. https://doi.org/10.5001/omj.2016.86
371. Taitt CR, Leski TA, Stockelman MG, Craft DW, Zurawski DV, Kirkup BC, Vora GJ (2014) Antimicrobial resistance determinants in Acinetobacter baumannii isolates taken from military treatment facilities. Antimicrob Agents Chemother 58(2):767–781. https://doi.org/10.1128/aac.01897-13
372. Tajkarimi M, Harrison SH, Hung AM, Graves JL Jr (2016) Mechanobiology of antimicrobial resistant Escherichia coli and Listeria innocua. PLoS ONE 11(2):e0149769. https://doi.org/10.1371/journal.pone.0149769
373. Takeo Y, Oie S, Kamiya A, Konishi H, Nakazawa T (1994) Efficacy of disinfectants against biofilm cells of Pseudomonas aeruginosa. Microbios 79(318):19–26
374. Tamburro M, Ripabelli G, Vitullo M, Dallman TJ, Pontello M, Amar CF, Sammarco ML (2015) Gene expression in Listeria monocytogenes exposed to sublethal concentration of benzalkonium chloride. Comp Immunol Microbiol Infect Dis 40:31–39. https://doi.org/10.1016/j.cimid.2015.03.004
375. Teixeira CF, Pereira TB, Miyazaki NH, Villas Boas MH (2010) Widespread distribution of qacA/B gene among coagulase-negative Staphylococcus spp. in Rio de Janeiro, Brazil. J Hosp Infect 75(4):333–334. https://doi.org/10.1016/j.jhin.2010.01.011
376. Tezel U, Pavlostathis SG (2015) Quaternary ammonium disinfectants: microbial adaptation, degradation and ecology. Curr Opin Biotechnol 33:296–304. https://doi.org/10.1016/j.copbio.2015.03.018
377. Thomas L, Russell AD, Maillard JY (2005) Antimicrobial activity of chlorhexidine diacetate and benzalkonium chloride against Pseudomonas aeruginosa and its response to biocide residues. J Appl Microbiol 98(3):533–543. https://doi.org/10.1111/j.1365-2672.2004.02402.x
378. Tiwari TS, Ray B, Jost KC Jr, Rathod MK, Zhang Y, Brown-Elliott BA, Hendricks K, Wallace RJ Jr (2003) Forty years of disinfectant failure: outbreak of postinjection Mycobacterium abscessus infection caused by contamination of benzalkonium chloride. Clin infect Dis: Official Publ Infect Dis Soc Am 36(8):954–962. https://doi.org/10.1086/368192
379. Tluczkiewicz I, Bitsch A, Hahn S, Hahn T (2010) Emission of biocides from hospitals: comparing current survey results with European Union default values. Integr Environ Assess Manage 6(2):273–280. https://doi.org/10.1897/ieam_2009-046.1

380. To MS, Favrin S, Romanova N, Griffiths MW (2002) Postadaptational resistance to benzalkonium chloride and subsequent physicochemical modifications of Listeria monocytogenes. Appl Environ Microbiol 68(11):5258–5264
381. Torlak E, Sert D (2013) Combined effect of benzalkonium chloride and ultrasound against Listeria monocytogenes biofilm on plastic surface. Lett Appl Microbiol 57(3):220–226. https://doi.org/10.1111/lam.12100
382. Tortorano AM, Viviani MA, Biraghi E, Rigoni AL, Prigitano A, Grillot R (2005) In vitro testing of fungicidal activity of biocides against Aspergillus fumigatus. J Med Microbiol 54 (Pt 10):955–957. https://doi.org/10.1099/jmm.0.45997-0
383. Tote K, Horemans T, Vanden Berghe D, Maes L, Cos P (2010) Inhibitory effect of biocides on the viable masses and matrices of Staphylococcus aureus and Pseudomonas aeruginosa biofilms. Appl Environ Microbiol 76(10):3135–3142. https://doi.org/10.1128/aem.02095-09
384. Toval F, Guzman-Marte A, Madriz V, Somogyi T, Rodriguez C, Garcia F (2015) Predominance of carbapenem-resistant Pseudomonas aeruginosa isolates carrying blaIMP and blaVIM metallo-beta-lactamases in a major hospital in Costa Rica. J Med Microbiol 64 (Pt 1):37–43. https://doi.org/10.1099/jmm.0.081802-0
385. Traglia GM, Almuzara M, Merkier AK, Adams C, Galanternik L, Vay C, Centron D, Ramirez MS (2012) Achromobacter xylosoxidans: an emerging pathogen carrying different elements involved in horizontal genetic transfer. Curr Microbiol 65(6):673–678. https://doi.org/10.1007/s00284-012-0213-5 [doi]
386. Trauth E, Lemaître J-P, Rojas C, Diviès C, Cachon R (2001) Resistance of immobilized lactic acid bacteria to the inhibitory effect of quaternary ammonium sanitizers. LWT Food Sci Technol 34(4):239–243. https://doi.org/10.1006/fstl.2001.0759
387. Ueda S, Kuwabara Y (2007) Susceptibility of biofilm Escherichia coli, Salmonella enteritidis and Staphylococcus aureus to detergents and sanitizers. Biocontrol Sci 12(4):149–153
388. United States Environmental Protection Agency (2006) Reregistration eligibility decision for alkyl dimethyl benzyl ammonium chloride (ADBAC) https://archive.epa.gov/pesticides/reregistration/web/pdf/adbac_red.pdf
389. Unkovic N, Ljaljevic Grbic M, Stupar M, Vukojevic J, Jankovic V, Jovic D, Djordjevic A (2015) Aspergilli response to Benzalkonium Chloride and Novel-Synthesized Fullerenol/Benzalkonium Chloride Nanocomposite. Sci World J 2015:109262. https://doi.org/10.1155/2015/109262
390. Valentine BK, Dew W, Yu A, Weese JS (2012) In vitro evaluation of topical biocide and antimicrobial susceptibility of Staphylococcus pseudintermedius from dogs. Vet Dermatol 23(6):493–e495. https://doi.org/10.1111/j.1365-3164.2012.01095.x
391. Valenzuela AS, Benomar N, Abriouel H, Canamero MM, Lopez RL, Galvez A (2013) Biocide and copper tolerance in enterococci from different sources. J Food Prot 76 (10):1806–1809. https://doi.org/10.4315/0362-028x.jfp-13-124
392. Vali L, Dashti AA, Mathew F, Udo EE (2017) Characterization of Heterogeneous MRSA and MSSA with reduced Susceptibility to Chlorhexidine in Kuwaiti Hospitals. Front Microbiol 8:1359. https://doi.org/10.3389/fmicb.2017.01359
393. Vali L, Davies SE, Lai LL, Dave J, Amyes SG (2008) Frequency of biocide resistance genes, antibiotic resistance and the effect of chlorhexidine exposure on clinical methicillin-resistant Staphylococcus aureus isolates. J Antimicrob Chemother 61(3):524–532. https://doi.org/10.1093/jac/dkm520
394. van der Veen S, Abee T (2010) HrcA and DnaK are important for static and continuous-flow biofilm formation and disinfectant resistance in Listeria monocytogenes. Microbiology (Reading, England) 156(Pt 12):3782–3790. https://doi.org/10.1099/mic.0.043000-0
395. van der Veen S, Abee T (2010) Importance of SigB for Listeria monocytogenes static and continuous-flow biofilm formation and disinfectant resistance. Appl Environ Microbiol 76 (23):7854–7860. https://doi.org/10.1128/aem.01519-10

396. van der Veen S, Abee T (2011) Mixed species biofilms of Listeria monocytogenes and Lactobacillus plantarum show enhanced resistance to benzalkonium chloride and peracetic acid. Int J Food Microbiol 144(3):421–431. https://doi.org/10.1016/j.ijfoodmicro.2010.10.029
397. van Klingeren B (1995) Disinfectant testing on surfaces. J Hosp Infect 30(Suppl):397–408
398. Vázquez-Sánchez D, Cabo ML, Ibusquiza PS, Rodríguez-Herrera JJ (2014) Biofilm-forming ability and resistance to industrial disinfectants of Staphylococcus aureus isolated from fishery products. Food Control 39(Supplement C):8–16. https://doi.org/10.1016/j.foodcont.2013.09.029
399. Vijayakumar R, Kannan VV, Sandle T, Manoharan C (2012) In vitro antifungal efficacy of Biguanides and quaternary Ammonium Compounds against Cleanroom Fungal Isolates. PDA J Pharm Sci Technol 66(3):236–242. https://doi.org/10.5731/pdajpst.2012.00866
400. Wagnild JP, McDonald P, Craig WA, Johnson C, Hanley M, Uman SJ, Ramgopal V, Beirne GJ (1977) Pseudomonas aeruginosa bacteremia in a dialysis unit. 11. Relationship to reuse of coils. Am J Med 62(5):672–676
401. Wand ME, Baker KS, Benthall G, McGregor H, McCowen JW, Deheer-Graham A, Sutton JM (2015) Characterization of pre-antibiotic era Klebsiella pneumoniae isolates with respect to antibiotic/disinfectant susceptibility and virulence in Galleria mellonella. Antimicrob Agents Chemother 59(7):3966–3972. https://doi.org/10.1128/aac.05009-14
402. Wang C, Cai P, Zhan Q, Mi Z, Huang Z, Chen G (2008) Distribution of antiseptic-resistance genes qacA/B in clinical isolates of meticillin-resistant Staphylococcus aureus in China. J Hosp Infect 69(4):393–394. https://doi.org/10.1016/j.jhin.2008.05.009
403. Wang JT, Sheng WH, Wang JL, Chen D, Chen ML, Chen YC, Chang SC (2008) Longitudinal analysis of chlorhexidine susceptibilities of nosocomial methicillin-resistant Staphylococcus aureus isolates at a teaching hospital in Taiwan. J Antimicrob Chemother 62 (3):514–517. https://doi.org/10.1093/jac/dkn208
404. Warren DK, Prager M, Munigala S, Wallace MA, Kennedy CR, Bommarito KM, Mazuski JE, Burnham CA (2016) Prevalence of qacA/B Genes and Mupirocin Resistance Among Methicillin-Resistant Staphylococcus aureus (MRSA) Isolates in the Setting of Chlorhexidine Bathing Without Mupirocin. Infect Control Hosp Epidemiol 37(5):590–597. https://doi.org/10.1017/ice.2016.1
405. Wassenaar TM, Cabal A (2017) The mobile dso-gene-sso element in rolling-circle plasmids of staphylococci reflects the evolutionary history of its resistance gene. Lett Appl Microbiol 65(3):192–198. https://doi.org/10.1111/lam.12767
406. Wassenaar TM, Ussery D, Nielsen LN, Ingmer H (2015) Review and phylogenetic analysis of qac genes that reduce susceptibility to quaternary ammonium compounds in Staphylococcus species. Eur J Microbiol Immunol 5(1):44–61. https://doi.org/10.1556/eujmi-d-14-00038
407. Wassenaar TM, Ussery DW, Ingmer H (2016) The qacC Gene Has Recently Spread between Rolling Circle Plasmids of Staphylococcus, Indicative of a Novel Gene Transfer Mechanism. Front Microbiol 7:1528. https://doi.org/10.3389/fmicb.2016.01528
408. Weigel LM, Clewell DB, Gill SR, Clark NC, McDougal LK, Flannagan SE, Kolonay JF, Shetty J, Killgore GE, Tenover FC (2003) Genetic analysis of a high-level vancomycin-resistant isolate of Staphylococcus aureus. Science (New York, NY) 302 (5650):1569–1571. https://doi.org/10.1126/science.1090956
409. Wendlandt S, Kadlec K, Fessler AT, Mevius D, van Essen-Zandbergen A, Hengeveld PD, Bosch T, Schouls L, Schwarz S, van Duijkeren E (2013) Transmission of methicillin-resistant Staphylococcus aureus isolates on broiler farms. Vet Microbiol 167 (3–4):632–637. https://doi.org/10.1016/j.vetmic.2013.09.019
410. Wessels S, Ingmer H (2013) Modes of action of three disinfectant active substances: a review. Regul Toxicol Pharmacol: RTP 67(3):456–467. https://doi.org/10.1016/j.yrtph.2013.09.006

411. Wong HS, Townsend KM, Fenwick SG, Trengove RD, O'Handley RM (2010) Comparative susceptibility of planktonic and 3-day-old Salmonella Typhimurium biofilms to disinfectants. J Appl Microbiol 108(6):2222–2228. https://doi.org/10.1111/j.1365-2672.2009.04630.x
412. Wong TZ, Zhang M, O'Donoghue M, Boost M (2013) Presence of antiseptic resistance genes in porcine methicillin-resistant Staphylococcus aureus. Vet Microbiol 162(2–4):977–979. https://doi.org/10.1016/j.vetmic.2012.10.017
413. Wright ES, Mundy RA (1961) Studies on disinfection of clinical thermometers. II. Oral thermometers from a tuberculosis sanatorium. Appl Microbiol 9:508–510
414. Xu D, Li Y, Zahid MS, Yamasaki S, Shi L, Li JR, Yan H (2014) Benzalkonium chloride and heavy-metal tolerance in Listeria monocytogenes from retail foods. Int J Food Microbiol 190:24–30. https://doi.org/10.1016/j.ijfoodmicro.2014.08.017
415. Xu D, Nie Q, Wang W, Shi L, Yan H (2016) Characterization of a transferable bcrABC and cadAC genes-harboring plasmid in Listeria monocytogenes strain isolated from food products of animal origin. Int J Food Microbiol 217:117–122. https://doi.org/10.1016/j.ijfoodmicro.2015.10.021
416. Xu Y, He Y, Li X, Gao C, Zhou L, Sun S, Pang G (2013) Antifungal effect of ophthalmic preservatives phenylmercuric nitrate and benzalkonium chloride on ocular pathogenic filamentous fungi. Diagn Microbiol Infect Dis 75(1):64–67. https://doi.org/10.1016/j.diagmicrobio.2012.09.008
417. Yamamoto M, Takami T, Matsumura R, Dorofeev A, Hirata Y, Nagamune H (2016) In vitro evaluation of the biocompatibility of newly synthesized Bis-Quaternary Ammonium Compounds with spacer structures derived from Pentaerythritol or Hydroquinone. Biocontrol Cci 21(4):231–241. https://doi.org/10.4265/bio.21.231
418. Yanai R, Yamada N, Ueda K, Tajiri M, Matsumoto T, Kido K, Nakamura S, Saito F, Nishida T (2006) Evaluation of povidone-iodine as a disinfectant solution for contact lenses: antimicrobial activity and cytotoxicity for corneal epithelial cells. Cont Lens Anterior Eye 29 (2):85–91. https://doi.org/10.1016/j.clae.2006.02.006
419. Yano T, Kubota H, Hanai J, Hitomi J, Tokuda H (2013) Stress tolerance of Methylobacterium biofilms in bathrooms. Microbes Environ 28(1):87–95
420. Ye JZ, Yu X, Li XS, Sun Y, Li MM, Zhang W, Fan H, Cao JM, Zhou TL (2014) [Antimicrobial resistance characteristics of and disinfectant-resistant gene distribution in Staphylococcus aureus isolates from male urogenital tract infection]. Zhonghua nan ke xue = Nat J Androl 20(7):630–636
421. Yoshimatsu T, Hiyama K (2007) Mechanism of the action of didecyldimethylammonium chloride (DDAC) against Escherichia coil and morphological changes of the cells. Biocontrol Sci 12(3):93–99
422. Yousfi K, Touati A, Lefebvre B, Fournier E, Cote JC, Soualhine H, Walker M, Bougdour D, Tremblay C, Bekal S (2017) A novel plasmid, pSx1, harboring a new Tn1696 derivative from extensively drug-resistant Shewanella xiamenensis encoding OXA-416. Microb Drug Res (Larchmont, NY) 23(4):429–436. https://doi.org/10.1089/mdr.2016.0025
423. Zhang H, Zhou Y, Bao H, Zhang L, Wang R, Zhou X (2015) Plasmid-borne cadmium resistant determinants are associated with the susceptibility of Listeria monocytogenes to bacteriophage. Microbiol res 172(Supplement C):1–6. https://doi.org/10.1016/j.micres.2015.01.008
424. Zhang M, O'Donoghue MM, Ito T, Hiramatsu K, Boost MV (2011) Prevalence of antiseptic-resistance genes in Staphylococcus aureus and coagulase-negative staphylococci colonising nurses and the general population in Hong Kong. J Hosp Infect 78(2):113–117. https://doi.org/10.1016/j.jhin.2011.02.018
425. Zhang M, O'Dononghue M, Boost MV (2012) Characterization of staphylococci contaminating automated teller machines in Hong Kong. Epidemiol Infect 140(8):1366–1371. https://doi.org/10.1017/s095026881100207x

426. Zheng R, Wang M, He B, Li X, Cao H, Liang H, Qing Z, Tang A (2009) [Identification of active efflux system gene qacA/B in methicillin-resistant Staphylococcus aureus and its significance]. Zhong nan da xue xue bao Yi xue ban = J Central S Univ Med sci 34 (6):537–542
427. Zmantar T, Ben Slama R, Fdhila K, Kouidhi B, Bakhrouf A, Chaieb K (2017) Modulation of drug resistance and biofilm formation of Staphylococcus aureus isolated from the oral cavity of Tunisian children. Braz J Infect Dis: Official Publ Braz Soc Infect Dis 21(1):27–34. https://doi.org/10.1016/j.bjid.2016.10.009
428. Zmantar T, Kouidhi B, Miladi H, Bakhrouf A (2011) Detection of macrolide and disinfectant resistance genes in clinical Staphylococcus aureus and coagulase-negative staphylococci. BMC Res Notes 4:453. https://doi.org/10.1186/1756-0500-4-453
429. Zou L, Meng J, McDermott PF, Wang F, Yang Q, Cao G, Hoffmann M, Zhao S (2014) Presence of disinfectant resistance genes in Escherichia coli isolated from retail meats in the USA. J Antimicrob Chemother 69(10):2644–2649. https://doi.org/10.1093/jac/dku197

Didecyldimethylammonium Chloride 11

11.1 Chemical Characterization

Didecyldimethylammonium chloride (DDAC) belongs to the group of aliphatic alkyl quaternary chemicals that are structurally similar quaternary ammonium compounds characterized by having a positively charged nitrogen covalently bonded to two alkyl group substituents (at least one C8 or longer) and two methyl substituents. In finished form, these quats are salts with positively charged nitrogen (cation) balanced by a negatively charged molecule (anion) [43]. The basic chemical information on DDAC is summarized in Table 11.1.

11.2 Types of Application

DDAC is used as an antimicrobial in several types of applications, such as indoor and outdoor hard surfaces (e.g. walls, floors, tables, toilets and fixtures), eating utensils, laundry, carpets, agricultural tools and vehicles, egg shells, shoes, milking equipment and udders, humidifiers, medical instruments, human remains, ultrasonic tanks, reverse osmosis units and water storage tanks. There are also DDAC-containing products that are used in residential and commercial swimming pools, in aquatic areas such as decorative ponds and decorative fountains, and in industrial process and water systems such re-circulating cooling water systems, drilling muds and packer fluids, oil well injection and wastewater systems. Additionally, DDAC-containing products are used for wood preservation [43]. In healthcare products, it can be found in surface disinfectants, instrument disinfectants, antimicrobial soaps and alcohol-based hand rubs as a non-volatile active agent. As a wood preservative, it is usually applied at concentrations between 0.3 and 1.8% with the aim to act as a fungicidal or fungistatic agent [23].

G. Kampf, *Antiseptic Stewardship*, https://doi.org/10.1007/978-3-319-98785-9_11

Table 11.1 Basic chemical information on didecyldimethylammonium chloride [23, 32]

CAS number	7173-51-5
IUPAC name	N,N-Didecyl-N,N-dimethylammonium chloride
Synonyms	Bardac 22, deciquam 222
Molecular formula	$C_{22}H_{48}ClN$
Molecular weight (g/mol)	362.1

11.2.1 European Chemicals Agency (European Union)

DDAC has been approved for product type 8 (wood preservatives) [2]. It is still under review as active biocidal substances (June 2018) for product types 1 (human hygiene), 2 (disinfectants and algaecides not intended for direct application to humans or animals), 3 (veterinary hygiene), 4 (food and feed area), 6 (preservatives for products during storage), 10 (construction material preservatives), 11 (preservatives for liquid-cooling and processing systems) and 12 (slimicides).

11.2.2 Environmental Protection Agency (USA)

DDAC was the first active ingredient registered with the EPA in the group of aliphatic alkyl quaternary chemicals in 1962. The re-registration of DDAC was last approved in 2006 [43].

11.2.3 Overall Environmental Impact

DDAC is manufactured and/or imported in the European Economic Area in 100–1,000 t per year [12]. DDAC is hydrolytically and photolytically stable. Its half-life was determined to be 227 days with 7% degradation after 30 days. DDAC is stable and not subject to photodegradation on soil. It is well known that, because of their positive charge, the cationic surfactants adsorb strongly to the negatively charged surfaces of sludge, soil and sediments [22].

11.3 Spectrum of Antimicrobial Activity

DDAC is a membrane-active agent that interacted with the cytoplasmic membrane in *S. aureus*, inducing the immediate leakage of intracellular constituents [21, 24].

11.3.1 Bactericidal Activity

11.3.1.1 Bacteriostatic Activity (MIC Values)

The MIC values for DDAC obtained with different bacterial species are summarized in Table 11.2. *Enterococcus* spp. had MIC values between 0.1 and 3.5 mg/l similar to *Staphylococcus* spp. with 0.1–4.5 mg/l. Gram-negative species were less susceptible such as *E. cloacae* (0.01–512 mg/l), *P. aeruginosa* (4–128 mg/l) or *E. coli* (0.4–50 mg/l). For *L. monocytogenes,* it was proposed to classify isolates with an MIC >3 mg/l as resistant [41]. Based on this proposal most isolates detected in food (MIC range: 0.5–6.0) would have to be classified as susceptible to DDAC (Table 11.2).

Table 11.2 MIC values of various bacterial species to DDAC

Species	Strains/isolates	MIC value (mg/l)	References
A. xylosoxidans	Domestic drain biofilm isolate MBRG 4.31	3.9	[30]
A. hydrophila	Domestic drain biofilm isolate MBRG 4.3	15.6	[30]
A. jandaei	Domestic drain biofilm isolate MBRG 9.11	15.6	[30]
A. proteolyticus	Domestic drain biofilm isolate MBRG 9.12	3.9	[30]
B. cereus	Domestic drain biofilm isolate MBRG 4.21	3.9	[30]
C. coli	16 strains from pig faeces or pork meat	0.37–0.75	[40]
Citrobacter spp.	Domestic drain biofilm isolate MBRG 9.18	7.8	[30]
C. indologenes	Domestic drain biofilm isolate MBRG 9.15	15.6	[30]
Chrysobacterium spp.	Domestic drain biofilm isolate MBRG 9.17	3.9	[30]
C. pseudogenitalum	Human skin isolate MBRG 9.24	7.8	[30]
C. renale group	Human skin isolate MBRG 9.13	3.9	[30]
E. cloacae	Strain 17/97 (clinical isolate)	0.012–0.024	[29]
E. cloacae	43 ESBL patient isolates (haematology ward)	64–512	[8]
E. faecalis	68 isolates from different poultry sources	0.14–1.44	[45]
E. faecalis	824 isolates from various sources	≤3.5	[37]

(continued)

Table 11.2 (continued)

Species	Strains/isolates	MIC value (mg/l)	References
E. faecium	81 isolates from different poultry sources	0.14–1.44	[45]
E. faecium	130 isolates from various sources	≤3.5	[37]
E. coli	150 isolates from different poultry sources	<0.4–3.6	[45]
E. coli	54 strains from pig faeces or pork meat	1.5–4	[40]
E. coli	153 blood culture isolates	2–16	[4]
E. coli	IFO 14237	10	[46]
E. coli	ATCC 8739	10–50	[44]
	4 triclosan-resistant mutants	10–50	
E. saccharolyticus	Domestic drain biofilm isolate MBRG 9.16	31.2	[30]
Eubacterium spp.	Domestic drain biofilm isolate MBRG 4.14	15.6	[30]
H. gallinarum	Domestic drain biofilm isolate MBRG 4.27	15.6	[30]
L. pentosus	60 strains from naturally fermented Aloreña green table olives	0.001–5	[6]
L. lactis	3 strains	0.5–1	[42]
L. pseudomesenteroides	13 strains from naturally fermented Aloreña green table olives	0.01–5	[6]
Leuconostoc spp.	3 strains	1	[42]
L. monocytogenes	31 strains from pig faeces or pork meat	0.5–1.5	[40]
L. monocytogenes	254 isolates from seafood products	1–6	[41]
M. phyllosphaerae	Domestic drain biofilm isolate MBRG 4.30	3.9	[30]
M. luteus	Human skin isolate MBRG 9.25	0.45	[30]
P. aeruginosa	175 isolates from veterinary sources	4–128	[3]
P. aeruginosa	ATCC 15442	11	[17]
P. aeruginosa	ATCC 15442	20	[26]
P. aeruginosa	ATCC 15442, ATCC 47085	45–60	[9]
P. fluorescens	5 isolates from chicken carcasses	5–40	[26]
P. fragi	3 isolates from chicken carcasses	10–40	[26]
P. lundensis	4 isolates from chicken carcasses	5–40	[26]
P. nitroreductans	Domestic drain biofilm isolate MBRG 4.6	15.6	[30]

(continued)

Table 11.2 (continued)

Species	Strains/isolates	MIC value (mg/l)	References
Pseudomonas spp.	Domestic drain biofilm isolate MBRG 9.14	15.6	[30]
Pseudoxanthomonas spp.	Domestic drain biofilm isolate MBRG 9.20	7.8	[30]
Ralstonia spp.	Domestic drain biofilm isolate MBRG 4.13	7.8	[30]
S. enterica	35 strains from pig faeces or pork meat	4–8	[40]
S. multivorum	Domestic drain biofilm isolate MBRG 9.19	3.9	[30]
S. aureus	ATCC 6538	0.4–1.6	[21]
S. capitis	Human skin isolate MBRG 9.34	1.3	[30]
S. caprae	Human skin isolate MBRG 9.30	0.8	[30]
S. cohnii	Human skin isolate MBRG 9.31	0.45	[30]
S. epidermidis	57 clean room isolates	0.14–4.5	[36]
S. epidermidis	Human skin isolate M 9.33	0.48	[30]
S. haemolyticus	Human skin isolate MBRG 9.35	1.6	[30]
S. hominis	Human skin isolate MBRG 9.37	0.48	[30]
S. kloosii	Human skin isolate MBRG 9.28	0.65	[30]
S. lugdunensis	Human skin isolate MBRG 9.36	0.81	[30]
S. saprophyticus	Human skin isolate MBRG 9.29	0.45	[30]
S. warneri	Human skin isolate MBRG 9.27	1.3	[30]
S. maltophilia	Domestic drain biofilm isolate MBRG 9.13	7.8	[30]

11.3.1.2 Bactericidal Activity (Suspension Tests)

At exposure times of 1–5 h DDAC at 0.0014% may achieve a $\geq$ 5.0 log reduction in suspension tests but does not demonstrate this effect consistently against all selected bacterial species such as *E. avium*, *M. aquaticum*, *Methylobacterium* spp., *Microbacterium* spp. or *Pseudomonas* spp. At 1%, it was bactericidal within 1 min (*P. aeruginosa* and *S. aureus*; Table 11.3).

The MBC values obtained with different *Pseudomonas* spp. are summarized in Table 11.4. They were usually between 0.001 and 0.006% within a 5 min exposure time.

Cotton towels have been described to bind between 82.0 and 85.3% of DDAC in aqueous solution after only 30 s of exposure [11]. Use of these tissues soaked with a DDAC-based surface disinfectants will result in an insufficient bactericidal effect and in low-level exposure to the target micro-organisms [11]. This important finding should be taken into account when using DDAC for antiseptic purposes in combination with a towel or wipe.

Table 11.3 Bactericidal activity of DDAC in suspension tests

Species	Strain/isolate	Exposure time	Concentration	$\log_{10}$ reduction	References
A. delafieldii	2 toilet bowl biofilm isolates	1 h	0.0014% (S)	4.9–5.7	[31]
		5 h		6.3–6.7	
B. sanguinis	Toilet bowl biofilm isolate	1 h	0.0014% (S)	5.4	[31]
		5 h		6.7	
Blastomonas spp.	Toilet bowl biofilm isolate	1 h	0.0014% (S)	6.9	[31]
		5 h			
C. jejuni	ATCC BAA-1062, ATCC 33560 and 2 field strains	1 min	0.01125% (P)	4.9–6.0	[18]
E. avium	Toilet bowl biofilm isolate	1 h	0.0014% (S)	3.0	[31]
		5 h		3.3	
H. flavidus	Toilet bowl biofilm isolate	1 h	0.0014% (S)	4.8	[31]
		5 h		6.3	
Luteimonas spp.	Toilet bowl biofilm isolate	1 h	0.0014% (S)	4.1	[31]
		5 h		6.8	
L. brunescens	Toilet bowl biofilm isolate	1 h	0.0014% (S)	6.9	[31]
		5 h		6.8	
M. adhaesivum	Toilet bowl biofilm isolate	1 h	0.0014% (S)	0.9	[31]
		5 h		4.9	
M. aquaticum	Toilet bowl biofilm isolate	1 h	0.0014% (S)	0.0	[31]
		5 h		0.9	
Methylobacterium spp.	Toilet bowl biofilm isolate	1 h	0.0014% (S)	0.4	[31]
		5 h		3.0	
Microbacterium spp.	Toilet bowl biofilm isolate	1 h	0.0014% (S)	2.5	[31]
		5 h		4.2	
Paracoccus spp.	Toilet bowl biofilm isolate	1 h	0.0014% (S)	3.6	[31]
		5 h		5.3	
P. aeruginosa	179 clinical isolates	2 min	1% (S)	>5.0	[14]
		5 min	0.1% (S)		
P. nitroreducens	Toilet bowl biofilm isolate	1 h	0.0014% (S)	0.2	[31]
		5 h		4.8	
P. mexicana	Toilet bowl biofilm isolate	1 h	0.0014% (S)	7.0	[31]
		5 h		6.9	
Pseudomonas spp.	Toilet bowl biofilm isolate	1 h	0.0014% (S)	0.0	[31]
		5 h		0.7	
Pseudonocardia spp.	Toilet bowl biofilm isolate	1 h	0.0014% (S)	5.1	[31]
		5 h		5.2	
S. yanoikuyae	Toilet bowl biofilm isolate	1 h	0.0014% (S)	6.5	[31]
		5 h		6.9	

(continued)

Table 11.3 (continued)

Species	Strain/isolate	Exposure time	Concentration	log_{10} reduction	References
Sphingobium spp.	Toilet bowl biofilm isolate	1 h	0.0014% (S)	6.4	[31]
		5 h		6.6	
S. soli	Toilet bowl biofilm isolate	1 h	0.0014% (S)	6.6	[31]
		5 h			
S. wittichii	Toilet bowl biofilm isolate	1 h	0.0014% (S)	6.7	[31]
		5 h		6.6	
Sphingomonas spp.	2 toilet bowl biofilm isolates	1 h	0.0014% (S)	3.2–6.0	[31]
		5 h		6.4–6.6	
Sphingopyxis spp.	Toilet bowl biofilm isolate	1 h	0.0014% (S)	3.9	[31]
		5 h		5.3	
S. aureus	NBRC 12732	3 h	0.0003% (S)	5.6	[15]
			0.0001% (S)	3.4	
			0.00007% (S)	0.4	
			0.00005% (S)	0.0	
S. aureus	91 clinical MSSA isolates	1 min	1% (S)	>5.0	[14]
		2 min	0.1% (S)		
S. aureus	109 clinical MRSA isolates	1 min	1% (S)	>5.0	[14]
		2 min	0.1% (S)		
S. epidermidis	Toilet bowl biofilm isolate	1 h	0.0014% (S)	4.8	[31]
		5 h		5.3	
S. maltophilia	Toilet bowl biofilm isolate	1 h	0.0014% (S)	4.0	[31]
		5 h		6.3	
X. aerolatus	Toilet bowl biofilm isolate	1 h	0.0014% (S)	1.9	[31]
		5 h		6.1	

S Solution; *P* Commercial product

Table 11.4 MBC values of various bacterial species to DDAC (5 min exposure)

Species	Strains/isolates	MBC value	References
P. aeruginosa	ATCC 15442	0.006%	[26]
P. aeruginosa	ATCC 15442	0.0015%	[17]
P. fluorescens	5 isolates from chicken carcasses	0.003–0.006%	[26]
P. lundensis	4 isolates from chicken carcasses	0.003–0.006%	[26]
P. fragi	3 isolates from chicken carcasses	0.001–0.003%	[26]

11.3.1.3 Activity Against Bacteria in Biofilms

The efficacy of DDAC against bacteria biofilms has been described with *L. monocytogenes*. Even with 1 h exposure time, a concentration of 0.0025% was not sufficiently effective. Only 0.025% was able to reduce bacterial cells by at least 5.0 log within 1 h (Table 11.5).

Table 11.5 Efficacy of DDAC solutions (S) in against bacteria in biofilms

Species	Strains/isolates	Type of biofilm	Exposure time	Concentration	$\log_{10}$ reduction	References
L. monocytogenes	6 strains from various sources	24-h incubation in polystyrene microtiter plates	60 min	0.025% (S)	≥ 6.1	[5]
				0.0025% (S)	1.7	
				0.00025% (S)	0.4	
Mixed species	Various species from artificial wastewater and settled sewage	3-w incubation in a biological contactor unit	8 d	0.016% (S)	5.0	[27]
			10 d		6.0	
			8 d	0.012% (S)	2.6	
			10 d		3.0	

11.3.2 Fungicidal Activity

A fungistatic effect was described for some crop fungi. For *P. chlamydospora* and *P. aleophilum,* DDAC was described to have a fungistatic effect between 0.00006 and 0.00015% (mycelial growth) or at <0.00001% (conidial germination) [16]. For other crop fungi *C. liriodendri* and *C. macrodidymum,* DDAC was described to have a fungistatic effect between 0.00019 and 0.01% (mycelial growth) or between 0.00004 and 0.0001% (conidial germination) [1].

A yeasticidal activity of 0.0076% DDAC was described with *C. albicans* ATCC 10231 within 15 min [35]. At 0.000105%, DDAC reduced *C. albicans* on textiles by 4.5 log within 18 h [19]. DDAC at 0.000105% on textile was able to prevent growth of *T. rubrum* DSM 21146 but not of *T. mentagrophytes* ATCC 9533 [19]. Some authors showed that the effect of a product based on DDAC and diluted to 0.5% is rather weak against eight food-associated fungal species (*S. cerevisiae, S. uvarum, K. apiculata, C. oleophila, M. fructicola, S. pombe, A. niger* and *P. roqueforti*) in 8–60 min [25].

11.3.3 Mycobactericidal Activity

The mycobactericidal activity of 0.0014% DDAC is rather weak with log reductions <1.0 in 1 h and between 3.5 and 4.0 in 5 h as shown with two toilet bowl biofilm isolates (*M. frederiksbergense* and *Mycobacterium* spp.) [31].

11.4 Effect of Low-Level Exposure

Numerous studies show that low-level exposure to DACC has different effects on bacteria (Table 11.6). No adaptive response was found in isolates or strains from 23 species (*A. xylosoxidans, A. hydrophila, B. cereus, E. coli, L. monocytogenes, S. enterica, C. indologenes, Chrysobacterium* spp., *Citrobacter* spp., *E. saccharolyticus, H. gallinarum, M. phyllosphaerae, M. osloensis, P. nitroreductans, P. putida, Pseudoxanthomonas* spp., *S. cohnii, S. epidermidis, S. haemolyticus, S. kloosii, S. saprophyticus, S. warneii, S. maltophilia*).

Some isolates or strains of 18 species ware able to express a weak adaptive response (MIC increase $\leq$ 4-fold) such as *A. baumannii, C. coli, E. coli, L. monocytogenes, S. enterica, C. renale group, C. sakazakii, E. faecalis, Eubacterium* spp., *M. luteus, P. aeruginosa, Pseudomonas* spp., *S. multivorum, S. aureus, S. capitis, S. caprae, S. hominis* and *S. lugdenensis.*

Table 11.6 Change of susceptibility to DDAC and other antimicrobial agents after exposure to sublethal DDAC concentrations

Species	Strain/isolate	Exposure time	Increase in MIC	MIC_{max} (mg/l)	Stability of MIC change	Associated changes	References
A. xylosoxidans	Domestic drain biofilm isolate MBRG 4.31	14 d at various concentrations	None	3.9	Not applicable	None reported	[30]
A. baumannii	Strain MBRG15.1 from a domestic kitchen drain biofilm	14 passages at various concentrations	2-fold	31.3	Unstable for 14 d	None reported	[10]
A. hydrophila	Domestic drain biofilm isolate MBRG 4.3	14 d at various concentrations	None	15.6	Not applicable	None reported	[30]
A. proteolyticus	Domestic drain biofilm isolate MBRG 9.12	14 d at various concentrations	32-fold	125	No data	None reported	[30]
B. cereus	Domestic drain biofilm isolate MBRG 4.21	14 d at various concentrations	None	3.9	Not applicable	None reported	[30]
C. coli	16 strains from pig faeces or pork meat	7 d	2-fold (31% of strains)	2	No data	2 strains: new resistance[a] to tetracycline and streptomycin	[40]
C. pseudogenitalum	Human skin isolate MBRG 9.24	14 d at various concentrations	8-fold	62.5	No data	None reported	[30]
C. renale group	Human skin isolate MBRG 9.13	14 d at various concentrations	4-fold	15.6	No data	None reported	[30]
C. indologenes	Domestic drain biofilm isolate MBRG 9.15	14 d at various concentrations	None	15.6	Not applicable	None reported	[30]

(continued)

Table 11.6 (continued)

Species	Strain/isolate	Exposure time	Increase in MIC	MIC_{max} (mg/l)	Stability of MIC change	Associated changes	References
Chrysobacterium spp.	Domestic drain biofilm isolate MBRG 9.17	14 d at various concentrations	None	3.9	Not applicable	None reported	[30]
Citrobacter spp.	Domestic drain biofilm isolate MBRG 9.18	14 d at various concentrations	None	7.8	Not applicable	None reported	[30]
C. sakazakii	Strain MBRG15.5 from a domestic kitchen drain biofilm	14 passages at various concentrations	2-fold	15.6	Stable for 14 d	None reported	[10]
E. faecalis	1 strain of unknown origin	14 passages at various concentrations	2-fold	2.0	Stable for 14 d	None reported	[10]
E. faecalis	DSM 2570 and 3 field strains	Up to 70 d at various concentrations	6-fold	21.9	No data	None reported	[37]
E. saccharolyticus	Domestic drain biofilm isolate MBRG 9.16	14 d at various concentrations	None	7.8	Not applicable	None reported	[30]
Eubacterium spp.	Domestic drain biofilm isolate MBRG 4.14	14 d at various concentrations	2-fold	31.2	No data	None reported	[30]
E. coli	54 strains from pig faeces or pork meat	7 d at various concentrations	$\geq$3-fold (50% of strains)	24	No data	32 strains became multiresistant[a], most of them with a new resistance[a] to chloramphenicol, ampicillin, cefotaxime, ceftazidime and ciprofloxacin	[40]
E. coli	ATCC 25922 and 9 avian and porcine *E. coli* strains	7 d at various concentrations	3.5-fold	16.5	No data	2.7-fold increase of MIC[b] to dioctyl dimethyl ammonium chloride 2.6-fold increase of MIC[b] to benzalkonium chloride	[39]

(continued)

Table 11.6 (continued)

Species	Strain/isolate	Exposure time	Increase in MIC	MIC_{max} (mg/l)	Stability of MIC change	Associated changes	References
E. coli	ATCC 25922 and strain MBRG15.4 from a domestic kitchen drain biofilm	14 passages at various concentrations	1.5-fold–3-fold	15.6	Stable for 14 d (1 strain), unstable for 14 d (1 strain)	None reported	[10]
H. gallinarum	Domestic drain biofilm isolate MBRG 4.27	14 d at various concentrations	None	15.6	Not applicable	None reported	[30]
L. monocytogenes	31 strains from pig faeces or pork meat	7 d at various concentrations	≥ 3-fold (48% of strains)	3	No data	1 strain: new resistance[a] to tetracycline and streptomycin	[40]
L. monocytogenes	Strain LM 101	Biofilm exposed once per week for 60 min to in-use concentration (660 mg/l[c]) or 1:1 dilution	No data	No data	Not applicable	Bactericidal efficacy reduced by >50%	[34]
M. phyllosphaerae	Domestic drain biofilm isolate MBRG 4.30	14 d at various concentrations	None	3.9	No data	None reported	[30]
M. luteus	Human skin isolate MBRG 9.25	14 d at various concentrations	4-fold	1.6	No data	None reported	[30]
M. osloensis	Strain MBRG15.3 from a domestic kitchen drain biofilm	14 passages at various concentrations	None	1.0	Not applicable	None reported	[10]

(continued)

Table 11.6 (continued)

Species	Strain/isolate	Exposure time	Increase in MIC	MIC_{max} (mg/l)	Stability of MIC change	Associated changes	References
P. aeruginosa	ATCC 15442	1–2 d	≥ 11-fold	>375	Stable for 7 w	None described	[9]
		20 d	≥ 11-fold	>375			
		48–49 d	≥ 18-fold	800			
P. aeruginosa	ATCC 47085	1–2 d	≥ 11-fold	>375	Stable for 3 w, reverted after 7 w	None described	[9]
		20 d	≥ 13-fold	>375			
		48–49 d	≥ 5-fold	>1,000			
P. aeruginosa	ATCC 9027	14 passages at various concentrations	2-fold	31.3	Unstable for 14 d	None reported	[10]
P. fluorescens	Poultry isolate	2 passages at various concentrations	5-fold	>50	No data	Adapted cells were able to grow in the presence of 50 mg/l DDAC; associated cross-tolerance[b] to cocoamine acetate, BAC, amphoteric tenside and N,N-bis (3-aminopropyl) dodecylamin	[26]
P. nitroreductans	Domestic drain biofilm isolate MBRG 4.6	14 d at various concentrations	None	7.8	Not applicable	None reported	[30]
P. putida	Strain MBRG15.2 from a domestic kitchen drain biofilm	14 passages at various concentrations	None	31.3	Not applicable	None reported	[10]
Pseudomonas spp.	Domestic drain biofilm isolate MBRG 9.14	14 d at various concentrations	2-fold	31.2	No data	None reported	[30]

(continued)

Table 11.6 (continued)

Species	Strain/isolate	Exposure time	Increase in MIC	MIC_{max} (mg/l)	Stability of MIC change	Associated changes	References
Pseudoxanthomonas spp.	Domestic drain biofilm isolate MBRG 9.20	14 d at various concentrations	None	3.9	No data	None reported	[30]
Ralstonia spp.	Domestic drain biofilm isolate MBRG 4.13	14 d at various concentrations	16-fold	125	No data	None reported	[30]
S. enterica	35 strains from pig faeces or pork meat	7 d at various concentrations	≥3-fold (3% of strains)	24	No data	7 strains acquired a new resistance[a], mainly to chloramphenicol (3 strains)	[40]
S. multivorum	Domestic drain biofilm isolate MBRG 9.19	14 d at various concentrations	2-fold	7.8	No data	None reported	[30]
S. aureus	ATCC 6538	14 passages at various concentrations	2-fold	1.0	Stable for 14 d	None reported	[10]
S. capitis	Human skin isolate MBRG 9.34	14 d at various concentrations	2-fold	2.6	No data	None reported	[30]
S. caprae	Human skin isolate MBRG 9.30	14 d at various concentrations	1.6-fold	1.3	No data	None reported	[30]
S. cohnii	Human skin isolate MBRG 9.31	14 d at various concentrations	None	0.45	Not applicable	None reported	[30]
S. epidermidis	Human skin isolate M 9.33	14 d at various concentrations	None	0.45	Not applicable	None reported	[30]
S. haemolyticus	Human skin isolate MBRG 9.35	14 d at various concentrations	None	1.9	Not applicable	None reported	[30]
S. hominis	Human skin isolate MBRG 9.37	14 d at various concentrations	2-fold	0.81	No data	None reported	[30]

(continued)

Table 11.6 (continued)

Species	Strain/isolate	Exposure time	Increase in MIC	MIC_{max} (mg/l)	Stability of MIC change	Associated changes	References
S. kloosii	Human skin isolate MBRG 9.28	14 d at various concentrations	None	0.45	Not applicable	None reported	[30]
S. lugdunensis	Human skin isolate MBRG 9.36	14 d at various concentrations	2-fold	1.9	No data	None reported	[30]
S. saprophyticus	Human skin isolate MBRG 9.29	14 d at various concentrations	None	0.45	Not applicable	None reported	[30]
S. warneri	Human skin isolate MBRG 9.27	14 d at various concentrations	None	0.45	Not applicable	None reported	[30]
S. maltophilia	Domestic drain biofilm isolate MBRG 9.13	14 d at various concentrations	None	7.8	Not applicable	None reported	[30]

[a]Microtitre plates; [b]Macrodilution method; [c]10.14% DDAC and 6.76% BAC

A strong but unstable MIC change (>4-fold) was found in isolates or strains of *P. aeruginosa*. A strong and stable MIC change (>4-fold) was also described for isolates or strains of *P. aeruginosa*. In isolates or strains of *A. proteolyticus*, *C. pseudogenitalum*, *E. faecalis*, *P. fluorescens* and *Ralstonia* spp., the adaptive response was strong but its stability was not described.

Selected strains or isolates revealed strong MIC changes such as *A. proteolyticus* (32-fold), *P. aeruginosa* ($\geq$ 18-fold), *Ralstonia* spp. (16-fold), *C. pseudogenitalum* (8-fold) and *E. faecalis* (6-fold). The highest MIC values after adaptation were >1,000 mg/l (*P. aeruginosa*), 125 mg/l (*A. proteolyticus*), 62.5 mg/l (*C. pseudogenitalum*), >50 mg/l (*P. fluorescens*) and 31.3 mg/l (*A. baumannii*, *Eubacterium* spp., *P. putida*).

Increased tolerance to other biocidal agents was described for *E. coli* to dioctyl dimethyl ammonium chloride and benzalkonium chloride and for *P. fluorescens* to cocoamine acetate, BAC, amphoteric tenside and N,N-bis (3-aminopropyl) dodecylamin.

Resistance to antibiotics was also observed in few isolates. In *C. coli*, resistance to tetracycline and streptomycin was found in 2 of 16 strains after low-level exposure. In *E. coli*, multiresistance occurred in 32 of 54 strains. In *L. monocytogenes*, resistance to tetracycline and streptomycin was described in 1 of 31 strains, and in *S. enterica*, a new resistance to at least one antibiotic was detected in 7 of 35 strains, mainly to chloramphenicol.

11.5 Resistance to DDAC

11.5.1 Species with Resistance to DDAC

Four strains were isolated from activated sludge of a municipal sewage treatment plant with a resistance to DDAC defined as the ability to grow in the presence of 50 mg/l DDAC. Three strains were *P. fluorescens*, one was *A. xylosoxidans* subsp. *xylosoxidans*. One of the *P. fluorescens* strains was even able to multiply at 250 mg/l DDAC.

In France, a clinical isolate of *P. cepacia* was identified with a MBC of 20% DDAC whereas most other *P. cepacia* isolates from hospitals or veterinary care had MBC values between 0.05% and 0.5%. Five other *Pseudomonas* spp. were more susceptible (MBC between 0.001 and 0.05%) [7].

11.5.2 Resistance Mechanisms

A *P. fluorescens* strain was described to be able to metabolize DDAC and other quaternary ammonium compounds within 7 d [33].

Table 11.7 Outbreaks and pseudo-outbreaks caused by contaminated DDAC solutions or products

Bacterial species	Type and number of infections	Patient population	Source of infection and role of DDAC resistance	DDAC concentration	References
Achromobacter spp.	7 cases of bacteriaemia	Paediatric onco-haematology unit	Contaminated disinfectant atomizer; product based on DDAC	0.25%	[20]
A. xylosoxidans and *P. fluorescens*	Pseudo-outbreak involving 19 patients	Haematology unit	Contaminated disinfectant solution and associated liquid dispenser; the night staff were in the habit of immersing the blood culture bottles in the disinfectant solution before taking them into the protected area surrounding the neutropenic patients.	0.25%	[38]
B. cepacia complex	38 cases of bacteraemia in patients with central venous dialysis catheters	Dialysis unit	Contaminated napkins; catheter hubs were occasionally cleaned and wrapped with DDAC soaked napkins	No data	[28]
E. cloacae (ESBL)	43 patients (33 colonizations, 10 infections including urinary tract infection, thoracic wound infection and bloodstream infection)	Haematology unit	Contaminated sinks; disinfectant solution was used for cleaning of all surfaces surrounding the patient and poured daily into all sinks; presence of biofilm in sinks and exposure to subinhibitory DDAC concentrations; termination of outbreak after biofilm removal and use of sodium hypochlorite	0.25%	[8]

11.5.3 Resistance Genes

So far, no specific DDAC resistance genes have been detected. But many resistance genes have been described for quaternary ammonium compounds, especially for benzalkonium chloride. They are summarized in Sect. 10.5.4. As DDAC is also a cationic detergent, the BAC resistance genes may also be relevant for DDAC.

11.5.4 Infections and Pseudo-Outbreaks Associated with Tolerance to DDAC

A few outbreaks or pseudo-outbreaks have been described caused by contaminated DDAC solutions (Table 11.7). Only Gram-negative bacteria such as *A. xylosoxidans*, *P. fluorescens*, *B. cepacia complex* and ESBL *E. cloacae* have been isolated. It is of particular interest that one outbreak was presumably caused by low-level exposure of sink biofilm bacteria to DDAC finally resulting DDAC-adapted isolates causing infections in haematology patients.

11.6 Cross-Tolerance to Other Biocidal Agents

Cross-tolerance has been shown between DDAC and dioctyl dimethyl ammonium chloride and BAC (*E. coli*) and cocoamine acetate, BAC, amphoteric tenside and N, N-bis (3-aminopropyl) dodecylamin (*P. fluorescens*; see also Table 11.6).

11.7 Cross-Tolerance to Antibiotics

Some studies describe a cross-tolerance between DDAC and antibiotics. For example, in 153 *E. coli* blood culture isolates, a higher MIC of DDAC was associated with a decreased susceptibility to cotrimoxazole [4]. Another study showed that DDAC-MICs were positively correlated with several other antibiotic MICs (e.g. piperacillin and sulphamethoxazole/trimethoprim in *E. coli*, chloramphenicol in *E. faecalis*) and increased DDAC-MICs were statistically linked to high-level resistance to streptomycin in enterococci [45].

Low-level exposure resulted occasionally in cross-resistance to antibiotics (Table 11.6). In *C. coli,* resistance to tetracycline and streptomycin was found in 2 of 16 strains. In *E. coli,* multiresistance occurred in 32 of 54 strains. In *L. monocytogenes,* resistance to tetracycline and streptomycin was described in 1 of 31 strains, and in *S. enterica,* a new resistance to at least one antibiotic was detected in 7 of 35 strains, mainly to chloramphenicol.

Overall, however, exposure of 7 species (*A. baumannii*, *C. sakazakii*, *E. faecalis*, *E. coli*, *P. aeruginosa*, *P. putida*, *S. aureus*) over 14 passages of 4 d each to increasing DDAC concentrations on agar was associated with both increases and decreases in antibiotic susceptibility but its effect was typically small relative to the differences observed among microbicides. Susceptibility changes resulting in resistance were not observed [13].

11.8 Role of Biofilm

11.8.1 Effect on Biofilm Development

No studies were found on the effect of DDAC on biofilm development.

11.8.2 Effect on Biofilm Removal

No studies were found on the effect of DDAC on biofilm removal.

11.8.3 Effect on Biofilm Fixation

No studies were found on the effect of DDAC on biofilm fixation.

11.9 Summary

The principal antimicrobial activity of DDAC is summarized in Table 11.8.

The key findings on acquired resistance and cross-resistance including the role of biofilm for selecting resistant isolates are summarized in Table 11.9.

Table 11.8 Overview on the typical exposure times required for DDAC to achieve sufficient biocidal activity against the different target micro-organisms

Target micro-organisms	Species	Concentration (%)	Exposure time
Bacteria	*S. aureus*, *P. aeruginosa*	1	1 min[a]
	20 of 25 toilet bowel biofilm isolates	0.0014	5 h
Fungi	*C. albicans*	0.0076	15 min
Mycobacteria	*M. frederiksbergense*, *Mycobacterium* spp.	$\geq$0.0014	$\geq$5 h

[a]In biofilm the efficacy will be lower

Table 11.9 Key findings on acquired DDAC resistance, the effect of low-level exposure, cross-tolerance to other biocides and antibiotics, and its effect on biofilm

Parameter	Species	Findings
Elevated MIC values	*P. aeruginosa*	>1,000 mg/l
	P. fluorescens	>250 mg/l
	A. xylosoxidans	>50 mg/l
	P. cepacia	MBC of 20%
MIC value to determine resistance	Not proposed yet for bacteria, fungi or mycobacteria	
Cross-tolerance biocides	*E. coli*	Increased tolerance to dioctyl dimethyl ammonium chloride and benzalkonium chloride
	P. fluorescens	Increased tolerance to cocoamine acetate, BAC, amphoteric tenside and N,N-bis (3-aminopropyl) dodecylamin
Cross-resistance antibiotics	*C. coli* (2 of 16 strains)	Resistance to tetracycline and streptomycin
	E. coli (32 of 54 strains)	Multiresistance
	L. monocytogenes (1 of 31 strains)	Resistance to tetracycline and streptomycin
	S. enterica (7 of 35 strains)	New resistance to at least one antibiotic, mainly to chloramphenicol
Resistance mechanisms	*P. fluorescens*	Metabolization of DDAC
Effect of low-level exposure	*A. xylosoxidans, A. hydrophila, B. cereus, E. coli, L. monocytogenes, S. enterica, C. indologenes, Chrysobacterium* spp., *Citrobacter* spp., *E. saccharolyticus, H. gallinarum, M. phyllosphaerae, M. osloensis, P. nitroreductans, P. putida, Pseudoxanthomonas* spp., *S. cohnii, S. epidermidis, S. haemolyticus, S. kloosii, S. saprophyticus, S. warneii, S. maltophilia*	No MIC increase
	A. baumannii, C. coli, E. coli, L. monocytogenes, S. enterica, C. renale group, C. sakazakii, E. faecalis, Eubacterium spp., *M. luteus, P. aeruginosa, Pseudomonas* spp., *S. multivorum, S. aureus, S. capitis, S. caprae, S. hominis, S. lugdenensis*	Weak MIC increase ($\leq$4-fold)

(continued)

Table 11.9 (continued)

Parameter	Species	Findings
	P. aeruginosa	Strong (>4-fold) but unstable MIC increase
	P. aeruginosa	Strong and stable MIC increase
	A. proteolyticus, *C. pseudogenitalum*, *E. faecalis*, *P. fluorescens*, *Ralstonia* spp.	Strong MIC increase (unknown stability)
	A. proteolyticus (32-fold)	Strongest MIC change after low-level exposure
	P. aeruginosa ($\geq$ 18-fold)	
	Ralstonia spp. (16-fold)	
	C. pseudogenitalum (8-fold)	
	E. faecalis (6-fold)	
	P. aeruginosa (>1,000 mg/l)	Highest MIC values after low-level exposure
	A. proteolyticus (125 mg/l)	
	C. pseudogenitalum (62.5 mg/l)	
	P. fluorescens (>50 mg/l)	
	A. baumannii, *Eubacterium* spp., *P. putida* (31.3 mg/l)	
Biofilm	Development	Unknown
	Removal	Unknown
	Fixation	Unknown

References

1. Alaniz S, Abad-Campos P, García-Jiménez J, Armengol J (2011) Evaluation of fungicides to control Cylindrocarpon liriodendri and Cylindrocarpon macrodidymum in vitro, and their effect during the rooting phase in the grapevine propagation process. Crop Protect 30(4):489–494. https://doi.org/10.1016/j.cropro.2010.12.020
2. Barroso JM (2013) COMMISSION DIRECTIVE 2013/4/EU of 14 February 2013 amending Directive 98/8/EC of the European Parliament and of the Council to include Didecyldimethylammonium Chloride as an active substance in Annex I thereto. Off J Eur Union 56(L 44):10–11
3. Beier RC, Foley SL, Davidson MK, White DG, McDermott PF, Bodeis-Jones S, Zhao S, Andrews K, Crippen TL, Sheffield CL, Poole TL, Anderson RC, Nisbet DJ (2015) Characterization of antibiotic and disinfectant susceptibility profiles among Pseudomonas aeruginosa veterinary isolates recovered during 1994–2003. J Appl Microbiol 118(2):326–342. https://doi.org/10.1111/jam.12707
4. Buffet-Bataillon S, Branger B, Cormier M, Bonnaure-Mallet M, Jolivet-Gougeon A (2011) Effect of higher minimum inhibitory concentrations of quaternary ammonium compounds in clinical E. coli isolates on antibiotic susceptibilities and clinical outcomes. J Hosp Infect 79 (2):141–146. https://doi.org/10.1016/j.jhin.2011.06.008
5. Caballero Gómez N, Abriouel H, Grande MJ, Pérez Pulido R, Gálvez A (2012) Effect of enterocin AS-48 in combination with biocides on planktonic and sessile Listeria monocytogenes. Food Microbiol 30(1):51–58. https://doi.org/10.1016/j.fm.2011.12.013

6. Casado Munoz Mdel C, Benomar N, Lavilla Lerma L, Knapp CW, Galvez A, Abriouel H (2016) Biocide tolerance, phenotypic and molecular response of lactic acid bacteria isolated from naturally-fermented Alorena table to different physico-chemical stresses. Food Microbiol 60:1–12. https://doi.org/10.1016/j.fm.2016.06.013
7. Chantefort A, Druilles J, Huet M (1990) Resistance de certains Pseudomonas aux antiseptiques et desinfectants. Medecine et maladies infectieuses 20(5):234–240. https://doi.org/10.1016/S0399-077X(05)81134-5
8. Chapuis A, Amoureux L, Bador J, Gavalas A, Siebor E, Chretien ML, Caillot D, Janin M, de Curraize C, Neuwirth C (2016) Outbreak of extended-spectrum beta-lactamase producing enterobacter cloacae with High MICs of quaternary ammonium compounds in a hematology ward associated with contaminated sinks. Front Microbiol 7:1070. https://doi.org/10.3389/fmicb.2016.01070
9. Chojecka A, Wiercinska O, Rohm-Rodowald E, Kanclerski K, Jakimiak B (2014) Effect of adaptation process of Pseudomonas aeruginosa to didecyldimethylammonium chloride in 2-propanol on bactericidal efficiency of this active substance. Rocz Panstw Zakl Hig 65 (4):359–364
10. Cowley NL, Forbes S, Amezquita A, McClure P, Humphreys GJ, McBain AJ (2015) Effects of formulation on microbicide potency and mitigation of the development of bacterial insusceptibility. Appl Environ Microbiol 81(20):7330–7338. https://doi.org/10.1128/aem.01985-15
11. Engelbrecht K, Ambrose D, Sifuentes L, Gerba C, Weart I, Koenig D (2013) Decreased activity of commercially available disinfectants containing quaternary ammonium compounds when exposed to cotton towels. Am J Infect Control 41(10):908–911. https://doi.org/10.1016/j.ajic.2013.01.017
12. European Chemicals Agency (ECHA) Didecyldimethylammonium chloride. Substance information. https://echa.europa.eu/substance-information/-/substanceinfo/100.027.751. Accessed 25 Jan 2018
13. Forbes S, Knight CG, Cowley NL, Amezquita A, McClure P, Humphreys G, McBain AJ (2016) Variable effects of exposure to formulated microbicides on antibiotic susceptibility in Firmicutes and Proteobacteria. Appl Environ Microbiol 82(12):3591–3598. https://doi.org/10.1128/aem.00701-16
14. Giacometti A, Cirioni O, Greganti G, Fineo A, Ghiselli R, Del Prete MS, Mocchegiani F, Fileni B, Caselli F, Petrelli E, Saba V, Scalise G (2002) Antiseptic compounds still active against bacterial strains isolated from surgical wound infections despite increasing antibiotic resistance. Eur J Clin Microbiol Infect Dis 21(7):553–556. https://doi.org/10.1007/s10096-002-0765-6
15. Gomi M, Osaki Y, Mori M, Sakagami Y (2012) Synergistic bactericidal effects of a sublethal concentration of didecyldimethylammonium chloride (DDAC) and low concentrations of nonionic surfactants against Staphylococcus aureus. Biocontrol Sci 17(4):175–181
16. Gramaje D, Aroca Á, Raposo R, García-Jiménez J, Armengol J (2009) Evaluation of fungicides to control Petri disease pathogens in the grapevine propagation process. Crop Protect 28(12):1091–1097. https://doi.org/10.1016/j.cropro.2009.05.010
17. Guerin-Mechin L, Leveau JY, Dubois-Brissonnet F (2004) Resistance of spheroplasts and whole cells of Pseudomonas aeruginosa to bactericidal activity of various biocides: evidence of the membrane implication. Microbiol Res 159(1):51–57. https://doi.org/10.1016/j.micres.2004.01.003
18. Gutierrez-Martin CB, Yubero S, Martinez S, Frandoloso R, Rodriguez-Ferri EF (2011) Evaluation of efficacy of several disinfectants against Campylobacter jejuni strains by a suspension test. Res Vet Sci 91(3):e44–e47. https://doi.org/10.1016/j.rvsc.2011.01.020
19. Hammer TR, Mucha H, Hoefer D (2012) Dermatophyte susceptibility varies towards antimicrobial textiles. Mycoses 55(4):344–351. https://doi.org/10.1111/j.1439-0507.2011.02121.x

20. Hugon E, Marchandin H, Poiree M, Fosse T, Sirvent N (2015) Achromobacter bacteraemia outbreak in a paediatric onco-haematology department related to strain with high surviving ability in contaminated disinfectant atomizers. J Hosp Infect 89(2):116–122. https://doi.org/10.1016/j.jhin.2014.07.012
21. Ioannou CJ, Hanlon GW, Denyer SP (2007) Action of disinfectant quaternary ammonium compounds against Staphylococcus aureus. Antimicrob Agents Chemother 51(1):296–306. https://doi.org/10.1128/aac.00375-06
22. Italy (2012) Didecyldimethylammonium chloride. Product-type PT 8 (Wood preservative).92
23. Italy (2015) Didecyldimethylammonium chloride Product-type 8 (Wood preservative). 148
24. Jansen AC, Boucher CE, Coetsee E, Kock JLF, van Wyk PWJ, Swart HC, Bragg RR (2013) The influence of Didecyldimethylammonium Chloride on the morphology and elemental composition of Staphylococcus aureus as determined by NanoSAM. Sci Res Essays 8 (3):152–160
25. Korukluoglu M, Sahan Y, Yigit A (2006) The fungicidal efficacy of various commercial disinfectants used in the food industry. Ann Microbiol 56(4):325–330
26. Langsrud S, Sundheim G, Borgmann-Strahsen R (2003) Intrinsic and acquired resistance to quaternary ammonium compounds in food-related Pseudomonas spp. J Appl Microbiol 95 (4):874–882
27. Laopaiboon L, Hall SJ, Smith RN (2002) The effect of a quaternary ammonium biocide on the performance and characteristics of laboratory-scale rotating biological contactors. J Appl Microbiol 93(6):1051–1058
28. Lo Cascio G, Bonora MG, Zorzi A, Mortani E, Tessitore N, Loschiavo C, Lupo A, Solbiati M, Fontana R (2006) A napkin-associated outbreak of Burkholderia cenocepacia bacteraemia in haemodialysis patients. J Hosp Infect 64(1):56–62. https://doi.org/10.1016/j.jhin.2006.04.010
29. Majtan V, Majtanova L (1999) The effect of new disinfectant substances on the metabolism of Enterobacter cloacae. Int J Antimicrob Agents 11(1):59–64
30. Moore LE, Ledder RG, Gilbert P, McBain AJ (2008) In vitro study of the effect of cationic biocides on bacterial population dynamics and susceptibility. Appl Environ Microbiol 74 (15):4825–4834. https://doi.org/10.1128/aem.00573-08
31. Mori M, Gomi M, Matsumune N, Niizeki K, Sakagami Y (2013) Biofilm-forming activity of bacteria isolated from toilet bowl biofilms and the bactericidal activity of disinfectants against the isolates. Biocontrol Sci 18(3):129–135
32. National Center for Biotechnology Information Didecyl dimethyl ammonium chloride. PubChem Compound Database; CID=23558. https://pubchem.ncbi.nlm.nih.gov/compound/23558. Accessed 24 Jan 2018
33. Nishihara T, Okamoto T, Nishiyama N (2000) Biodegradation of didecyldimethylammonium chloride by Pseudomonas fluorescens TN4 isolated from activated sludge. J Appl Microbiol 88(4):641–647
34. Olszewska MA, Zhao T, Doyle MP (2016) Inactivation and induction of sublethal injury of Listeria monocytogenes in biofilm treated with various sanitizers. Food Control 70 (Supplement C):371–379. https://doi.org/10.1016/j.foodcont.2016.06.015
35. Rauwel G, Leclercq L, Criquelion J, Aubry JM, Nardello-Rataj V (2012) Aqueous mixtures of di-n-decyldimethylammonium chloride/polyoxyethylene alkyl ether: dramatic influence of tail/tail and head/head interactions on co-micellization and biocidal activity. J Colloid Interface Sci 374(1):176–186. https://doi.org/10.1016/j.jcis.2012.02.006
36. Ribic U, Klancnik A, Jersek B (2017) Characterization of Staphylococcus epidermidis strains isolated from industrial cleanrooms under regular routine disinfection. J Appl Microbiol 122 (5):1186–1196. https://doi.org/10.1111/jam.13424
37. Schwaiger K, Harms KS, Bischoff M, Preikschat P, Molle G, Bauer-Unkauf I, Lindorfer S, Thalhammer S, Bauer J, Holzel CS (2014) Insusceptibility to disinfectants in bacteria from animals, food and humans-is there a link to antimicrobial resistance? Front Microbiol 5:88. https://doi.org/10.3389/fmicb.2014.00088

38. Siebor E, Llanes C, Lafon I, Ogier-Desserrey A, Duez JM, Pechinot A, Caillot D, Grandjean M, Sixt N, Neuwirth C (2007) Presumed pseudobacteremia outbreak resulting from contamination of proportional disinfectant dispenser. Eur J Clin Microbiol Infect Dis 26 (3):195–198. https://doi.org/10.1007/s10096-007-0260-1
39. Soumet C, Fourreau E, Legrandois P, Maris P (2012) Resistance to phenicol compounds following adaptation to quaternary ammonium compounds in Escherichia coli. Vet Microbiol 158(1–2):147–152. https://doi.org/10.1016/j.vetmic.2012.01.030
40. Soumet C, Meheust D, Pissavin C, Le Grandois P, Fremaux B, Feurer C, Le Roux A, Denis M, Maris P (2016) Reduced susceptibilities to biocides and resistance to antibiotics in food-associated bacteria following exposure to quaternary ammonium compounds. J Appl Microbiol 121(5):1275–1281. https://doi.org/10.1111/jam.13247
41. Soumet C, Ragimbeau C, Maris P (2005) Screening of benzalkonium chloride resistance in Listeria monocytogenes strains isolated during cold smoked fish production. Lett Appl Microbiol 41(3):291–296. https://doi.org/10.1111/j.1472-765X.2005.01763.x
42. Trauth E, Lemaître J-P, Rojas C, Diviès C, Cachon R (2001) Resistance of immobilized lactic acid bacteria to the inhibitory effect of quaternary ammonium sanitizers. LWT Food Sci Technol 34(4):239–243. https://doi.org/10.1006/fstl.2001.0759
43. United States Environmental Protection Agency (2006) Reregistration eligibility decision for aliphatic alkyl quaternaries (DDAC) https://archive.epa.gov/pesticides/reregistration/web/pdf/ddac_red.pdf
44. Walsh SE, Maillard JY, Russell AD, Catrenich CE, Charbonneau DL, Bartolo RG (2003) Development of bacterial resistance to several biocides and effects on antibiotic susceptibility. J Hosp Infect 55(2):98–107
45. Wieland N, Boss J, Lettmann S, Fritz B, Schwaiger K, Bauer J, Holzel CS (2017) Susceptibility to disinfectants in antimicrobial-resistant and -susceptible isolates of Escherichia coli, Enterococcus faecalis and Enterococcus faecium from poultry-ESBL/AmpC-phenotype of E. coli is not associated with resistance to a quaternary ammonium compound, DDAC. J Appl Microbiol 122(6):1508–1517. https://doi.org/10.1111/jam.13440
46. Yoshimatsu T, Hiyama K (2007) Mechanism of the action of didecyldimethylammonium chloride (DDAC) against Escherichia coil and morphological changes of the cells. Biocontrol Sci 12(3):93–99

Polihexanide

12

12.1 Chemical Characterization

Polihexanide (PHMB) was firstly synthesized by Rose and Swain in 1954 [55] and introduced in the 1980s in Switzerland [75]. It is a cationic biguanide polymer. Preparations of PHMB are polydisperse mixtures of polymeric biguanides, with a weighted average number of 12 repeating hexamethylene biguanide units. The heterogeneity of the molecule is increased further by the presence of either amine, or cyanoguanidine or guanidine end-groups in any combination at the terminal positions of each chain [55]. It is freely water soluble. The basic chemical information on the most common PHMBs is summarized in Table 12.1.

12.2 Types of Application

As a preservative, PHMB is used in cosmetics, personal care products, fabric softeners, contact lens solutions, hand washes and more [73]. PHMB is also used to preserve wet wipes, to control odour in textiles; to prevent microbial contamination in wound irrigation (e.g. at 0.02–0.1%) and sterile dressings (e.g. at 0.2–0.5%) [15]; to disinfect medical or dental utensil and trays, farm equipment, animal drinking water, and hard surfaces for food handling institutions and hospitals; and to deodorize vacuums and toilets. PHMB is used in antimicrobial hand washes and rubs [28] and air filter treatments as an alternative to ozone. It is also used as an active ingredient for recreational water treatment, as a chlorine-free polymeric sanitizer. Further reported uses of PHMB are purification of swimming pool water, beer glass sanitisation, solid surface disinfection in breweries and short-term preservation of hides and skins [78].

PHMB is used for wound antisepsis and classified as the active agent of choice for critically colonized and infected chronic wounds as well as for burns [37, 51, 64]. In addition, PHMB may be used for coating of nitril examination gloves [44].

G. Kampf, *Antiseptic Stewardship*, https://doi.org/10.1007/978-3-319-98785-9_12

Table 12.1 Basic chemical information on polihexanide [63]

CAS number	27083-27-8	32289-58-0	28757-47-3	133029-32-0
IUPAC name	Homopolymer of N-(3-Aminopropyl)-Imidodicarbonimidic Diamide			
Synonyms	Lavasept, Baquacil, polyhexamethylen-biguanide hydrochloride			
Molecular formula	$C_{10}H_{19}ClN_8$	$C_{10}H_{23}N_5$	$C_8H_{19}N_5$	$(C_5H_{14}N_6)x$
Molecular weight (g/mol)	286.76446	213.32312	185.275	Variable

12.2.1 European Chemicals Agency (European Union)

PHMB with the CAS numbers 27083-27-8 and 32289-58-0 has been approved as an active biocidal substance for product types 2 (disinfectants and algaecides not intended for direct application to humans or animals), 3 (veterinary hygiene), 4 (food and feed area) and 11 (preservatives for liquid-cooling and processing systems) [41, 42]. It has not been approved for product types 1 (human hygiene), 5 (drinking water), 6 (preservatives for products during storage) and 9 (fibre, leather, rubber and polymerised materials preservatives) [40, 43].

In 2015, the Scientific Committee on Consumer Safety (SCCS) declared that PHMB up to 0.3% with the CAS numbers 27083-27-8 and 32289-58-0 were not considered safe in cosmetic spray formulations and all cosmetic products because of concerns regarding the acute toxicity by inhalation and insufficient data on dermal absorption [7].

12.2.2 Environmental Protection Agency (USA)

PHMB with the CAS number 32289-58-0 was first registered in the USA in 1982 and has last been approved in 2004. A risk was only seen for occupational handlers, especially pour liquid for drilling muds and workover fluids. The greatest risk for exposure was seen by inhalation and on the skin so that mitigation measures were enforced [82].

12.2.3 Overall Environmental Impact

The ECHA classified PHMB to be "very toxic to aquatic life" and "very toxic to aquatic life with long lasting effects" [20]. Nevertheless, the overall environmental impact of PHMB is considered to be low [82]. PHMB is stable in water. Soil with any humic matter binds approximately 80% of PHMB. The probability of PHMB leaching into ground water where any soil is present with any significant amount of humic matter is considered to be negligible [54]. Whilst amine and guanidine end-groups in PHMB are likely to be susceptible to biodegradation, cyanoguanidine end-groups are likely to be recalcitrant. In particular, a strain of *P. putida* was capable of extensive growth with 1,6-diguanidinohexane as a sole nitrogen source, with complete removal of guanidine groups from culture medium within 2 days, and with concomitant formation of unsubstituted urea, which in turn was also utilised by the organism [65].

12.3 Spectrum of Antimicrobial Activity

12.3.1 Bactericidal Activity

The mechanisms of antimicrobial activity have been described in various studies. PHMB disturbs the cell membrane's bilayer by interacting with it along the surface of the membrane [86]. PHMB molecules perturb *L. innocua* cytoplasmic membrane by interacting with the first layer of the membrane lipid bilayer [8]. Other authors reported that the electrostatic interaction with the cell membrane is a dominant factor in the antimicrobial activity of PHMB [92]. Hydrophobic interactions and dehydration have been described as relevant as electrostatic interactions to explain changes in membrane fluidity and permeability, believed to be responsible for the biocide action of PHMB [79].

12.3.1.1 Bacteriostatic Activity (MIC Values)

The MIC values for PHMB obtained with different bacterial species are summarized in Table 12.2. *Staphyloccocus* spp. (*S. aureus* and CNS) had MIC values between 0.25 and 8 mg/l, *Enterococcus* spp. between 1.8 and 31.2 mg/l. *B. cepacia* (58–256 mg/l), *K. pneumoniae* (1–25 mg/l), *H. influenzae* (2–32 mg/l), *P. aeruginosa* (2–32 mg/l) and *E. coli* (0.5–30 mg/l) were somewhat less susceptible to PHMB. The highest MIC values were found in *B. cepacia* (256 mg/l), *A. viscosus* (120 mg/l) and *N. asteroides* (100 mg/l). The bacteriostatic activity of PHMB at 1,000 mg/l is reduced in the presence of 0.25% mucin resulting in a bacteriostatic concentration of 4,000 mg/l PHMB [4].

Table 12.2 MIC values of various bacterial species to PHMB

Species	Strains/isolates	MIC value (mg/l)	References
A. baumannii	ICM 6841	43	[91]
A. hydrophila	Domestic drain biofilm isolate MBRG 4.3	31.2	[59]
A. jandaei	Domestic drain biofilm isolate MBRG 9.11	31.2	[59]
A. proteolyticus	Domestic drain biofilm isolate MBRG 9.12	7.8	[59]
A. viscosus	ATCC 15987	120	[83]
A. xylosoxidans	Domestic drain biofilm isolate MBRG 4.31	15.6	[59]
B. cereus	Domestic drain biofilm isolate MBRG 4.21	20.8	[59]
B. cereus	MRBG 4.21 (kitchen drain biofilm isolate)	58	[25]

(continued)

Table 12.2 (continued)

Species	Strains/isolates	MIC value (mg/l)	References
B. cepacia	ATCC BAA-245	58	[25]
B. cepacia	JCM 5964	256	[91]
Citrobacter spp.	Domestic drain biofilm isolate MBRG 9.18	31.2	[59]
C. pseudogenitalum	Human skin isolate MBRG 9.24	1.9	[59]
C. renale group	Human skin isolate MBRG 9.13	3.9	[59]
C. perfringens	ATCC 13124	2	[49]
C. indologenes	MRBG 4.29 (kitchen drain biofilm isolate)	0.9	[25]
C. indologenes	Domestic drain biofilm isolate MBRG 9.15	3.9	[59]
Chrysobacterium spp.	Domestic drain biofilm isolate MBRG 9.17	15.6	[59]
C. xerosis	WIBG 1.2 (wound isolate)	2.7	[25]
E. faecalis	WIBG 1.1 (wound isolate)	1.8	[25]
E. faecalis	ATCC 29212	2–16	[49]
E. faecalis	ATCC 29212	8	[91]
E. hirae	ATCC 10541	21	[91]
E. saccharolyticus	Domestic drain biofilm isolate MBRG 9.16	31.2	[59]
Enterococcus spp.	Clinical VRE isolate	4–8	[49]
E. coli	ATCC 35218	0.5–1	[49]
E. coli	50 clinical isolates	1–4	[24]
E. coli	ATCC 25922 and 4 clinical isolates	2	[5]
E. coli	ATCC 25922	3.3	[91]
E. coli	6 clinical ESBL isolates	4–8	[30]
E. coli	ATCC 25922	13.3	[25]
E. coli	IFO 14237	>30	[94]
Eubacterium spp.	Domestic drain biofilm isolate MBRG 4.14	7.8	[59]
H. gallinarum	Domestic drain biofilm isolate MBRG 4.27	7.8	[59]
H. influenzae	ATCC 49247	2	[49]
H. influenzae	50 clinical isolates	4–32	[24]
K. pneumoniae	50 clinical isolates	1–4	[24]
K. pneumoniae	DSM16609 and 3 clinical ESBL isolates	3.1–25	[30]
K. pneumoniae	ATCC 13883	7.3	[25]
L. acidophilus	ATCC 4356	30	[83]
L. rhamnosus	ATCC 7469	10	[83]
M. phyllosphaerae	Domestic drain biofilm isolate MBRG 4.30	7.8	[59]
M. luteus	Human skin isolate MBRG 9.25	1	[59]
M. luteus	MRBG 9.25 (skin isolate)	1.8	[25]
M. catarrhalis	50 clinical isolates	1–4	[24]
N. asteroides	Clinical isolates from a patient with keratitis	100	[53]
P. aeruginosa	ATCC 15442	2	[49]

(continued)

Table 12.2 (continued)

Species	Strains/isolates	MIC value (mg/l)	References
P. aeruginosa	50 clinical isolates	4–32	[24]
P. aeruginosa	ATCC 27853	21	[91]
P. aeruginosa	ATCC 9027	31.3	[25]
P. nitroreductans	Domestic drain biofilm isolate MBRG 4.6	15.6	[59]
Pseudomonas spp.	Domestic drain biofilm isolate MBRG 9.14	7.8	[59]
Pseudoxanthomonas spp.	Domestic drain biofilm isolate MBRG 9.20	15.6	[59]
Ralstonia spp.	Domestic drain biofilm isolate MBRG 4.13	7.8	[59]
S. marcescens	ATCC 13880	38.7	[25]
S. aureus	27 clinical MRSA isolates before decolonization with PHMB	0.25–1	[69]
	27 clinical isolates after decolonization with PHMB	0.25–1	
S. aureus	Clinical MRSA isolate	0.5	[49]
S. aureus	ATCC 6538	0.5–1	[49]
S. aureus	50 clinical isolates (MSSA)	0.5–2	[21]
	50 clinical isolates (MRSA)		
S. aureus	80 clinical strains (sporadic MSSA)	0.5–2	[22]
	80 clinical strains (sporadic MRSA)		
	6 clinical strains (epidemic MRSA)		
S. aureus	27 clinical MRSA isolates from patients with failed MRSA decolonization using PHMB	≤ 1	[52]
S. aureus	Strain RN4420, strain EMRSA 15, strain USA 300	1	[45]
S. aureus	ATCC 29213, 6 clinical MRSA strains and 6 clinical VISA strains	1–2	[5]
S. aureus	ATCC 6538	5.3	[91]
S. aureus	ATCC 6538	7.3	[25]
S. aureus	ATCC 700698 (MRSA)	8	[91]
S. capitis	MRBG 9.34 (skin isolate)	1.1	[25]
S. capitis	Human skin isolate MBRG 9.34	3.9	[59]
S. caprae	MRBG 9.3 (skin isolate)	6.7	[25]
S. caprae	Human skin isolate MBRG 9.30	7.8	[59]
S. cohnii	Human skin isolate MBRG 9.31	1.9	[59]
S. epidermidis	Human skin isolate M 9.33	1.9	[59]
S. epidermidis	MRBG 9.33 (skin isolate)	3	[25]
S. epidermidis	ATCC 12228	4	[91]
S. haemolyticus	MRBG 9.35 (skin isolate)	1.8	[25]
S. haemolyticus	Human skin isolate MBRG 9.35	7.8	[59]
S. hominis	Human skin isolate MBRG 9.37	7.8	[59]
S. kloosii	Human skin isolate MBRG 9.28	3.9	[59]

(continued)

Table 12.2 (continued)

Species	Strains/isolates	MIC value (mg/l)	References
S. lugdunensis	MRBG 9.36 (skin isolate)	3.6	[25]
S. lugdunensis	Human skin isolate MBRG 9.36	3.9	[59]
S. saprophyticus	Human skin isolate MBRG 9.29	3.9	[59]
S. warneri	MRBG 9.27 (skin isolate)	3.6	[25]
S. warneri	Human skin isolate MBRG 9.27	7.8	[59]
S. maltophilia	MRBG 4.17 (kitchen drain biofilm isolate)	3	[25]
S. maltophilia	Domestic drain biofilm isolate MBRG 9.13	3.9	[59]
S. mutans	ATCC 25175	60	[83]
S. pneumoniae	ATCC 49619	1–2	[49]
S. multivorum	Domestic drain biofilm isolate MBRG 9.19	20.8	[59]

12.3.1.2 Bactericidal Activity (Suspension Tests)

PHMB at 0.0014% had only limited bactericidal activity within 5 h against 20 of 25 toilet bowl biofilm isolates. At 0.016 or 0.02%, PHMB showed a mostly good bactericidal activity within 1 h, whereas at 5 min some studies indicate a lower efficacy (<5.0 log), e.g. against *E. faecium*, *E. coli*, *P. aeruginosa* or *S. aureus*. At 0.032 or 0.04%, PHMB was mostly bactericidal with 30 min against various bacterial species. At 5 min, the efficacy is less pronounced. When used at 0.1% for 7 d, a sufficient bactericidal efficacy was consistently observed (Table 12.3).

The MBC values for PHMB obtained with different bacterial species are summarized in Table 12.4. They were in the same range as the MIC values obtained with PHMB (see Table 12.2.). For *E. coli,* it was noteworthy that a shorter exposure time requires a higher PHMB concentration in order to reveal a bactericidal efficacy.

The bactericidal efficacy of PHMB is significantly reduced in the presence of albumin so that it is not possible to titrate PHMB to a concentration that fulfils effective requirements in the presence of albumin [46]. Thereby, the loss of antimicrobial effect in *S. aureus* was presented as a linear correlation to the rising concentration of albumin [47]. Wound fluid also reduced the efficacy, e.g. against *P. aeruginosa* (bactericidal concentration of 0.001% instead of 0.0001%) [85]. The bactericidal activity is abolished in the presence of chondroitin sulphate [61].

In contact lens solutions, the bactericidal efficacy of PHMB against *P. aeruginosa* was also largely impaired in the presence of organic soil allowing surviving bacteria even to multiply [10]. Wound dressings can also reduce the efficacy of PHMB against *S. aureus* as shown in 67.2% of 42 types of wound dressings [35]. The bactericidal activity of PHMB has been described to be higher at elevated pH values [87].

Table 12.3 Bactericidal activity of PHMB in suspension tests

Species	Strains/isolates	Exposure time	Concentration	$\log_{10}$ reduction	References
A. baumannii	ATCC 15149	7 d	0.1% (S)	5.5	[58]
A. delafieldii	2 toilet bowl biofilm isolates	1 h	0.0014% (S)	0.1–0.2	[60]
		5 h		0.6–0.9	
Blastomonas spp.	Toilet bowl biofilm isolate	1 h	0.0014% (S)	2.9	[60]
		5 h		5.7	
B. sanguinis	Toilet bowl biofilm isolate	1 h	0.0014% (S)	4.1	[60]
		5 h		5.2	
E. cloacae	ATCC 13047	7 d	0.1% (S)	5.5	[58]
E. avium	Toilet bowl biofilm isolate	1 h	0.0014% (S)	1.4	[60]
		5 h		3.3	
E. faecalis	ATCC 14508 and ATCC 51575 (VRE)	7 d	0.1% (S)	5.3–5.6	[58]
E. faecium	ATCC 6057	5 min	0.02% (P)	4.0–5.0	[61]
		60 min		≥6.0	
E. faecium	Not described	5 min	0.02% (P)	4.0–5.0[a]	[50]
		60 min		≥6.0	
E. hirae	Not described	1 min	0.05% (P)	3.4–3.6[a]	[50]
		5 min		≥6.0	
E. coli	ATCC 25922 and 1 clinical isolate	5 min	0.6% (P)	>5.0	[5]
			0.2% (P)	0.4–5.0	
			0.02% (P)	0.1–5.0	
E. coli	ATCC 25922	1 h	0.04% (S)	≥5.0	[24]
		30 min	0.02% (S)		
E. coli	Clinical ESBL isolate	1 min	0.035% (S)	1.1	[30]
		5 min		2.5	
		1 min	0.032% (S)	1.4	
		5 min		3.5	
		1 min	0.016% (S)	1.3	
		5 min		≥5.4	
E. coli	ATCC 11229	5 min	0.02% (P)	2.0–3.0	[61]
		60 min		≥6.0	
E. coli	ATCC 8739	7 d	0.1% (S)	5.4	[58]
E. coli	Not described	30 s	0.05% (P)	4.9–5.9[a]	[50]
		10 min		≥6.0	
		1 min	0.02% (P)	2.0–8.0[a]	
		5 min		≥6.0	
E. coli	ATCC 11229	30 min	0.009% (P)	≥3.0	[62]
H. influenzae	ATCC 49247	30 min	0.04% (S)	≥5.0	[24]
			0.02% (S)		

(continued)

Table 12.3 (continued)

Species	Strains/isolates	Exposure time	Concentration	log_{10} reduction	References
H. flavidus	Toilet bowl biofilm isolate	1 h	0.0014% (S)	3.2	[60]
		5 h		6.3	
K. pneumoniae	ATCC 4382	30 min	0.04% (S)	≥5.0	[24]
			0.02% (S)		
K. pneumoniae	DSM 16609	1 min	0.035% (S)	3.3	[30]
		5 min		5.3	
		1 min	0.032% (S)	3.8	
		5 min		6.1	
		1 min	0.016% (S)	3.8	
		5 min		6.2	
K. pneumoniae	Clinical ESBL isolate	1 min	0.035% (S)	2.0	[30]
		5 min		4.2	
		1 min	0.032% (S)	3.5	
		5 min		6.3	
		1 min	0.016% (S)	3.0	
		5 min		≥6.5	
Luteimonas spp.	Toilet bowl biofilm isolate	1 h	0.0014% (S)	2.5	[60]
		5 h		5.2	
L. brunescens	Toilet bowl biofilm isolate	1 h	0.0014% (S)	1.2	[60]
		5 h		1.9	
M. adhaesivum	Toilet bowl biofilm isolate	1 h	0.0014% (S)	0.0	[60]
		5 h		0.2	
M. aquaticum	Toilet bowl biofilm isolate	1 h	0.0014% (S)	0.0	[60]
		5 h			
Methylobacterium spp.	Toilet bowl biofilm isolate	1 h	0.0014% (S)	0.0	[60]
		5 h			
Microbacterium spp.	Toilet bowl biofilm isolate	1 h	0.0014% (S)	0.0	[60]
		5 h		0.8	
M. catarrhalis	ATCC 43617	15 min	0.04% (S)	≥5.0	[24]
			0.02% (S)		
Paracoccus spp.	Toilet bowl biofilm isolate	1 h	0.0014% (S)	1.7	[60]
		5 h		1.2	
P. mirabilis	ATCC 12453	7 d	0.1% (S)	5.6	[58]
P. aeruginosa	ATCC 9027	7 d	0.1% (S)	5.7	[58]
P. aeruginosa	ATCC 15442	1 min	0.05% (S)	5.3	[49]
		5 min	0.0125% (S)		
		10 min	0.005% (S)		
P. aeruginosa	ATCC 15442	1 h	0.04% (S)	≥5.0	[24]
		30 min	0.02% (S)		

(continued)

Table 12.3 (continued)

Species	Strains/isolates	Exposure time	Concentration	log_{10} reduction	References
P. aeruginosa	ATCC 15442	5 min	0.02% (P)	3.0–4.0	[61]
		60 min		≥ 5.0	
P. aeruginosa	Not described	5 min	0.02% (P)	≥ 6.0	[50]
		1 min	0.05% (P)		
P. aeruginosa	ATCC 9027	2–6 h	0.0002% (P)	3.8	[71]
			0.0001% (P)	4.2–4.3	
			0.00005% (P)	4.0	
P. nitroreducens	Toilet bowl biofilm isolate	1 h	0.0014% (S)	1.0	[60]
		5 h		2.0	
Pseudomonas spp.	Toilet bowl biofilm isolate	1 h	0.0014% (S)	0.0	[60]
		5 h		0.1	
Pseudonocardia spp.	Toilet bowl biofilm isolate	1 h	0.0014% (S)	1.0	[60]
		5 h		1.8	
P. mexicana	Toilet bowl biofilm isolate	1 h	0.0014% (S)	1.7	[60]
		5 h		4.1	
S. marcescens	ATCC 13880	7 d	0.1% (S)	5.6	[58]
S. marcescens	ATCC 13880	2–6 h	0.0002% (P)	3.6	[71]
			0.0001% (P)	3.3	
			0.00005% (P)	3.4	
S. soli	Toilet bowl biofilm isolate	1 h	0.0014% (S)	3.1	[60]
		5 h		4.7	
S. wittichii	Toilet bowl biofilm isolate	1 h	0.0014% (S)	2.9	[60]
		5 h		5.3	
Sphingomonas spp.	3 toilet bowl biofilm isolates	1 h	0.0014% (S)	0.1–0.6	[60]
		5 h		1.9–3.9	
S. yanoikuyae	Toilet bowl biofilm isolate	1 h	0.0014% (S)	0.2	[60]
		5 h		1.2	
Sphingobium spp.	Toilet bowl biofilm isolate	1 h	0.0014% (S)	1.3	[60]
		5 h		2.1	
Sphingopyxis spp.	Toilet bowl biofilm isolate	1 h	0.0014% (S)	0.7	[60]
		5 h		2.6	
S. aureus	IFO 13276	30 min	0.1% (S)	≥ 5.0	[93]
S. aureus	ATCC 6538 and ATCC 33591 (MRSA)	7 d	0.1% (S)	5.5	[58]
S. aureus	ATCC 29213 and 2 clinical MRSA strains	5 min	0.6% (P)	>5.0	[5]
			0.2% (P)	0.4–5.0	
			0.02% (P)	0.4–4.5	

(continued)

Table 12.3 (continued)

Species	Strains/isolates	Exposure time	Concentration	$\log_{10}$ reduction	References
S. aureus	Not described	30 s	0.05% (P)	3.2–6.0[a]	[50]
		5 min		4.5–6.0[a]	
		10 min		≥6.0	
		1 min	0.02% (P)	3.6–5.0[a]	
		5 min		4.1–6.0[a]	
		10 min		≥7.0	
S. aureus	ATCC 29213	5 min	0.04% (S)	4.0	[22]
		10 min		4.5	
		30 min		4.8	
		60 min		>5.0	
		5 min	0.02% (S)	3.0	
		10 min		3.5	
		30 min		4.0	
		60 min		4.5	
S. aureus	ATCC 6538	1 min	0.025% (S)	5.3	[49]
		10 min	0.005% (S)	5.3	
		60 min	0.0005% (S)	5.2	
S. aureus	ATCC 6538	5 min	0.02% (P)	≥6.0	[61]
S. aureus	ATCC 6538	30 min	0.01% (P)	≥3.0	[62]
S. aureus	ATCC 6538	2–6 h	0.0002% (P)	4.2	[71]
			0.0001% (P)	3.4–3.5	
			0.00005% (P)	3.4	
S. epidermidis	ATCC 12228	7 d	0.1% (S)	5.8	[58]
S. epidermidis	Toilet bowl biofilm isolate	1 h	0.0014% (S)	2.5	[60]
		5 h		1.3	
S. epidermidis	ATCC 17917	2–6 h	0.0002% (P)	2.8	[71]
			0.0001% (P)	3.0–4.7	
			0.00005% (P)	3.4	
S. maltophilia	Toilet bowl biofilm isolate	1 h	0.0014% (S)	2.5	[60]
		5 h		3.0	
X. aerolatus	Toilet bowl biofilm isolate	1 h	0.0014% (S)	0.0	[60]
		5 h		1.1	
Mixed anaerobic species	*A. actinomycetemcomitans* ATCC 43718, *A. viscosus* DSMZ 43798, *F. nucleatum* ATCC 10953, *P. gingivalis* ATCC 33277, *V. atypica* ATCC 17744 and *S. gordonii* ATCC 33399	30 s	0.1% (P)	6.3	[14]
			0.04% (P)	2.3	

S Solution; *P* Commercial product; [a]depending on the type of organic load

Table 12.4 MBC values of various bacterial species to PHMB at variable exposure times

Species	Strains/isolates	Exposure time	MBC value	References
E. coli	ATCC 25922	10 min	0.01%	[91]
E. coli	DSM 11250	6 h	<0.005%	[68]
E. coli	50 clinical isolates	24 h	0.0001–0.0008%	[24]
H. influenzae	50 clinical isolates	24 h	0.0004–0.0032%	[24]
K. pneumoniae	50 clinical isolates	24 h	0.0001–0.0004%	[24]
M. catarrhalis	50 clinical isolates	24 h	0.0001–0.0004%	[24]
P. aeruginosa	ATCC 27853	10 min	0.0042%	[91]
P. aeruginosa	DSM 939	6 h	<0.005%	[68]
P. aeruginosa	50 clinical isolates	24 h	0.0004–0.0032%	[24]
S. aureus	DSM 799	6 h	<0.005%	[68]
S. aureus	50 clinical isolates (MSSA)	24 h	0.00005–0.0002%	[21]
	50 clinical isolates (MRSA)		0.00005–0.0002%	

12.3.1.3 Activity Against Bacteria in Biofilms

Bacteria in biofilms required PHMB at 2% to achieve a strong bactericidal activity within 5 min as shown with *E. coli*, *S. Enteritidis* and *S. aureus*. The commonly used concentration of PHMB (0.04%) had only poor bactericidal activity against biofilm-grown bacterial species with log reductions $\leq$2.2. At 0.1%, the bactericidal activity against some single species biofilms was strong (*P. aeruginosa*) but not against other single species biofilms (*S. aureus* and *S. mutans*) and natural mixed biofilms (Table 12.5).

Similar results were described in other experimental settings. PHMB at 0.02 and 0.04% showed a low inhibition effect (65%) on the metabolic activity in a MRSA biofilm. The efficacy was strongly dependent on the applied exposure time [31]. In *S. epidermidis,* a 33× decrease of susceptibility of biofilm grown cells (48 h in microplates) was described compared to planktonic cells; in *E. coli,* it was only 2.4-fold [29]. In the presence of PHMB, the susceptibility of biofilm grown cells to the target microorganism was 3-fold lower (*E. coli*; exposed to 1.3 μmol/l) or 10-fold lower (*S. epidermidis*; exposed to 0.7 μmol/l) [12].

12.3.1.4 Bactericidal Activity on Skin

Some effect of PHMB at 0.02 and 0.04% was found within 30 min on the resident skin flora with 1.2 and 1.9 log [16]. For the decontamination of MRSA colonized skin, daily use of PHMB solution for at least 3 min over 10 days and treatment of the anterior nares with PHMB thrice daily did not result in a significant reduction of MRSA carriers compared to placebo [52].

Table 12.5 Efficacy of PHMB in against bacteria in biofilms

Species	Strains/isolates	Type of biofilm	Exposure time	Concentration	$\log_{10}$ reduction	References
E. coli	Strain O157, isolate from food poisoning outbreak	8-d incubation on stainless steel	5 min	2% (S)	≥ 5.1	[81]
				1% (S)		
				0.5% (S)		
P. aeruginosa	ATCC 25619	48-h incubation on polycarbonate coupons	15 min	0.1% (P)	7.0	[39]
P. aeruginosa	NCIMB 10434	48-h incubation in biofilm reactor	4 h	0.1% (P)	>6.0	[36]
			24 h			
			4 h	0.01% (P)	>6.0	
			24 h			
P. aeruginosa	Environmental strain SG81	44-h incubation on silicone swatches	30 min	0.04% (S)	2.2	[38]
				0.02% (S)	3.3	
P. aeruginosa	Environmental strain SG81	44-h incubation in polystyrene microtitre plates	30 min	0.04% (S)	1.4	[38]
				0.02% (S)	0.4	
S. Enteritidis	Isolate from food poisoning outbreak	8-d incubation on stainless steel	5 min	2% (S)	≥ 5.1	[81]
				1% (S)		
				0.5% (S)		
S. aureus	Isolate from food poisoning outbreak	8-d incubation on stainless steel	5 min	2% (S)	≥ 5.2	[81]
				1% (S)	4.3	
				0.5% (S)	1.8	
S. aureus	ATCC 25923	48-h incubation on polycarbonate coupons	15 min	0.1% (P)	0.8	[39]

(continued)

Table 12.5 (continued)

Species	Strains/isolates	Type of biofilm	Exposure time	Concentration	$\log_{10}$ reduction	References
S. aureus	ATCC 33593 (MRSA)	24-h incubation on partial thickness porcine wounds	2 irrigations per d for up to 6 d	0.1% (P)	1.7 (day 3)	[13]
					3.1 (day 6)	
S. mutans	DSM 20523	72-h incubation on titanium discs	30 min	0.1% (S)	2.5	[48]
Mixed species	Human saliva bacteria	72-h incubation on titanium discs	30 min	0.1% (S)	0.9	[48]
Mixed species	Subgingival plaque bacteria	Overnight incubation on titanium discs	30 min	0.1% (S)	1.0	[48]
Mixed species	*S. aureus* strain 308 (MRSA), *C. albicans* ATCC MYA 2876	48-h incubation in biofilm reactor	4 h	0.1% (P)	>5.0	[36]
			24 h			
			4 h	0.01% (P)	0.4	
			24 h		2.3	

P Commercial product; *S* Solution

12.3.1.5 Bactericidal Activity on Mucosa

At 0.2%, PHMB showed a similar efficacy in reducing bacterial counts on the oral mucosa as 0.12% chlorhexidine or 0.3% triclosan (approx. 1.5 log) [84]. The application of 3 drops of 0.2% PHMB showed also a good bactericidal efficacy when applied preoperatively in ophthalmic surgery [33]. A fair bactericidal efficacy was in addition described for 0.04% PHMB as a mouth rinse similar to 0.12% chlorhexidine when sampled on the mucosa [74]. In the oral cavity, an antiseptic mouth rinse based on PHMB at an unknown concentration was equally effective against three oral pathogens (*S. mutans*, *F. nucleatum*, *C. albicans*) compared to the positive control based on 0.2% chlorhexidine [70]. On porcine vaginal mucosa, 0.1% PHMB was able to reduce an artificial contamination of MRSA by 1.2 log (15 min) to 4.0 log (24 h) [3].

12.3.1.6 Bactericidal Activity on Wounds

In surgical wounds, application of 0.04% PHMB led to a significantly higher reduction of bacterial counts compared to the application of Ringer solution [23]. In acute traumatic wounds, the effect of 0.04% PHMB was only marginal [67]. In a proposed test to determine the efficacy of wound antiseptics (which is similar to a carrier test), 0.02% PHMB, however, did not show sufficient bactericidal activity within 24 h. At 0.04 and 0.1%, however, the efficacy was sufficient within 3 h with and without organic load [77]. When tie-over dressings were soaked with 0.1% PHMB in full thickness skin grafts, it had no effect on reducing bacterial loads in wounds and resulted in more surgical site infections compared to sterile water [76].

12.3.1.7 Bactericidal Activity of Impregnated Gloves

PHMB has some effect to reduce a contamination on gloves (*S. pyogenes*, carbapenem-resistant *E. coli*, MRSA, ESBL *K. pneumoniae*) within 10 min, but it is impaired in the presence of organic load. A lower bacterial transfer to other surfaces was found with a dry inoculums of all three species but not with a wet inoculum [1].

12.3.2 Fungicidal Activity

12.3.2.1 Fungistatic Activity (MIC Values)

The growth of various fungal species was inhibited by PHMB at up to 16 mg/l indicating an overall good susceptibility (Table 12.6). PHMB at 4,000 mg/l on textile, however, was not able to prevent growth of *T. rubrum* DSM 21146 or *T. mentagrophytes* ATCC 9533 [32].

12.3.2.2 Fungicidal Activity (Suspension Tests)

A yeasticidal activity was found for PHMB at 0.1% (5 min) and 0.02% (30 s). A general fungicidal activity of 0.1% PHMB does not exist, not even within 24 h. Some species can be reduced by $\geq$4.0 log in 24 h by 0.1% PHMB such as *A.*

Table 12.6 MIC values for different fungal species obtained with PHMB

Species	Strains/isolates	MIC value (mg/l)	References
A. flavus	16 clinical isolates	8–16	[89]
A. fumigatus	1 isolate from an ocular infection	3.1	[6]
A. niger	1 clinical isolate	6.1	[56]
C. albicans	ATCC 10231	1	[49]
C. albicans	ATCC 10231	8	[91]
Candida spp.[a]	25 *C. albicans* clinical isolates and ATCC 24433, *C. parapsilosis* ATCC 22019, *C. krusei* ATCC 6258	0.8–1.6	[56]
F. lichenicola	1 isolate from an ocular infection	1.6	[6]
F. oxysporum	2 isolates from ocular infections	1.6	[6]
F. proliferatum	1 isolate from an ocular infection	1.6	[6]
F. solani	5 isolates from ocular infections	1.6	[6]
F. solani	ATCC 44366	2.4	[56]
F. solani	24 clinical isolates	8–16	[89]
R. microsporus	1 isolate from an ocular infection	3.1	[6]
S. apiospermum	1 isolate from an ocular infection	1.6	[6]

[a]no data per species available

elegans, *A. fumigatus*, *Exophiala* spp., *F. oxysporum* or *M. circinelloides*, other species are resistant to 0.1% PHMB such as *Apophysomyces* spp., *A. brasiliensis*, *A. flavus*, *A. terreus* or *Lichtheimia* spp. (Table 12.7).

At 0.4%, PHMB reduced *C. albicans* on textiles by 3.1 log within 18 h [32]. The type of contact lens material may have an impact on the fungicidal activity. Some materials can bind between 30 and 60% of the PHMB within 6 h resulting in a lower efficacy against *F. solani* [72].

Table 12.7 Fungicidal activity of PHMB in suspension tests

Species	Strains/isolates	Exposure time	Concentration	$\log_{10}$ reduction	References
A. elegans	2 clinical isolates	12 h	0.1% (S)	>4.0	[90]
			0.04% (S)		
			0.01% (S)		
Apophysomyces spp.	1 clinical isolate	24 h	0.1% (S)	<2.0	[90]
			0.04% (S)		
			0.01% (S)		
A. brasiliensis	ATCC 16404	7 d	0.1% (S)	2.2	[58]
A. flavus	3 clinical isolates	24 h	0.1% (S)	<1.0	[90]
			0.04% (S)		
			0.01% (S)		

(continued)

Table 12.7 (continued)

Species	Strains/isolates	Exposure time	Concentration	$\log_{10}$ reduction	References
A. fumigatus	ATCC 10894	2–6 h	0.0002% (P)	0.1	[71]
			0.0001% (P)	0.1–0.3	
			0.00005% (P)	0.1	
A. fumigatus	1 clinical isolate	24 h	0.1% (S)	≥ 4.0	[90]
			0.04% (S)	<1.0	
			0.01% (S)	<1.0	
A. terreus	1 clinical isolate	24 h	0.1% (S)	<1.0	[90]
			0.04% (S)		
			0.01% (S)		
C. albicans	ATCC 10231	1 min	0.5% (S)	4.1	[49]
		5 min	0.05% (S)		
		60 min	0.01% (S)		
C. albicans	IFO 1594	1 h	0.1% (S)	≥ 5.0	[93]
C. albicans	ATCC 10231	7 d	0.1% (S)	5.7	[58]
C. albicans	Not described	30 s	0.02% (P)	4.0–5.0[a]	[50]
		10 min		≥ 6.0	
C. albicans	ATCC 10231	5 min	0.02% (P)	≥ 4.0	[61]
C. albicans	ATCC 10231	2–6 h	0.0002% (P)	0.3	[71]
			0.0001% (P)	0.3	
			0.00005% (P)	1.6	
Exophiala spp.	1 clinical isolate	5 min	0.1% (S)	≥ 5.0	[90]
			0.04% (S)		
			0.01% (S)		
F. oxysporum	1 clinical isolate	5 min	0.1% (S)	≥ 5.0	[90]
			0.04% (S)		
			0.01% (S)		
F. solani	ATCC 36031	2–6 h	0.0002% (P)	0.3	[71]
			0.0001% (P)	1.2–1.3	
			0.00005% (P)	0.3	
Lichtheimia spp.	1 clinical isolate	24 h	0.1% (S)	<1.0	[90]
			0.04% (S)		
			0.01% (S)		
M. circinelloides	2 clinical isolates	24 h	0.1% (S)	≥ 5.0	[90]
			0.04% (S)		
			0.01% (S)		

P Commercial product; *S* Solution

The PHMB treatment of the yeast cells *S. cerevisiae* activates the PKC1/Slt2 cell wall integrity pathway. In addition, it is suggested that HOG1 and YAP1 can play a role in the regulation of cell wall integrity genes [18].

12.3.3 Mycobactericidal Activity

PHMB was described with an MIC value of 5 mg/l for *M. smegmatis* [27]. The mycobactericidal activity of 0.0014% DDAC is rather weak with log reductions between 0.0 (*Mycobacterium* spp.) and 2.8 (*M. frederiksbergense*) in 1 h and between 2.8 (*Mycobacterium* spp.) and 5.3 (*M. frederiksbergense*) in 5 h as shown with two toilet bowl biofilm isolates [60].

12.4 Effect of Low-Level Exposure

Many studies show that low-level exposure to PHMB has different effects on bacteria (Table 12.8). No adaptive response was found in isolates or strains from 31 species (*A. xylosoxidans, A. hydrophila, A. jandaei, B. cereus, B. cepacia, C. indologenes, C. pseudogenitalum, Chrysobacterium* spp., *Citrobacter* spp., *E. saccharolyticus, Eubacterium* spp., *H. gallinarum, M. luteus, P. nitroreductans, P. putida, Pseudomonas* spp., *Pseudoxanthomonas* spp., *Ralstonia* spp., *S. marcescens, S. multivorum, S. aureus, S. capitis, S. caprae, S. epidermidis, S. haemolyticus, S. hominis, S. kloosii, S. lugdunensis, S. saprophyticus, S. warneri, S. maltophilia*).

Some isolates or strains of 18 species ware able to express a weak adaptive response (MIC increase $\leq$ 4-fold) such as *A. baumannii, C. indologenes, C. renale group, C. sakazakii, C. xerosis, E. faecalis, E. coli, K. pneumoniae, M. phyllosphaerae, M. luteus, M. osloensis, P. aeruginosa, S. cohnii, S. epidermidis, S. haemolyticus, S. lugdunensis, S. warneri* and *S. maltophilia.*

A strong but unstable MIC change (>4-fold) was found in isolates or strains of *S. capitis* and *S. epidermidis*. A strong and stable MIC change (>4-fold) was described for isolates or strains of *E. faecalis* and *S. aureus*. In isolates or strains of *A. proteolyticus* and *S. aureus,* the adaptive response was strong but its stability was not described.

Selected strains or isolates revealed strong MIC changes such as *A. proteolyticus* (16-fold), *E. faecalis* and *S. aureus* (8-fold), *S. capitis* (5.5-fold) and *S. epidermidis* (4.8-fold). The highest MIC values after adaptation were 125 mg/l (*A. proteolyticus*), 58 mg/l (*P. aeruginosa*), 31.3 mg/l (*A. baumannii, E. coli, E. faecalis, M. osloensis*), 29 mg/l (*K. pneumoniae*) and 23.5 mg/l (*S. aureus*).

No change of chlorhexidine susceptibility was described in MRSA after low-level PHMB exposure (Table 12.8). Exposure of 7 species (*A. baumannii, C. sakazakii, E. faecalis, E. coli, P. aeruginosa, P. putida, S. aureus*) over 14 passages

Table 12.8 Change of susceptibility to PHMB and other antimicrobial agents after exposure to sublethal PHMB concentrations

Species	Strains/isolates	Exposure time	Increase in MIC	MIC_{max} (mg/l)	Stability of MIC change	Associated changes	References
A. xylosoxidans	Domestic drain biofilm isolate MBRG 4.31	14 d at various concentrations	None	3.9	Not applicable	None reported	[59]
A. baumannii	Strain MBRG 15.1 from a domestic kitchen drain biofilm	14 passages at various concentrations	4-fold	31.3	Unstable for 14 d	None reported	[11]
A. hydrophila	Domestic drain biofilm isolate MBRG 4.3	14 d at various concentrations	None	31.4	Not applicable	None reported	[59]
A. jandaei	Domestic drain biofilm isolate MBRG 9.11	14 d at various concentrations	None	31.2	Not applicable	None reported	[59]
A. proteolyticus	Domestic drain biofilm isolate MBRG 9.12	14 d at various concentrations	16-fold	125	No data	None reported	[59]
B. cereus	MRBG 4.21 (kitchen drain biofilm isolate)	40 d at various concentrations	None	29	Not applicable	None described	[25]
B. cereus	Domestic drain biofilm isolate MBRG 4.21	14 d at various concentrations	None	20.8	Not applicable	None reported	[59]
B. cepacia	ATCC BAA-245	40 d at various concentrations	None	58	Not applicable	None described	[25]
C. indologenes	Domestic drain biofilm isolate MBRG 9.15	14 d at various concentrations	None	3.9	Not applicable	None reported	[59]
C. indologenes	MRBG 4.29 (kitchen drain biofilm isolate)	40 d at various concentrations	4-fold	3.6	Unstable for 14 d	None described	[25]
C. pseudogenitalum	Human skin isolate MBRG 9.24	14 d at various concentrations	None	1.9	Not applicable	None reported	[59]
C. renale group	Human skin isolate MBRG 9.13	14 d at various concentrations	2.5-fold	10	No data	None reported	[59]

(continued)

Table 12.8 (continued)

Species	Strains/isolates	Exposure time	Increase in MIC	MIC_{max} (mg/l)	Stability of MIC change	Associated changes	References
Chrysobacterium spp.	Domestic drain biofilm isolate MBRG 9.17	14 d at various concentrations	None	15.6	Not applicable	None reported	[59]
Citrobacter spp.	Domestic drain biofilm isolate MBRG 9.18	14 d at various concentrations	None	15.6	Not applicable	None reported	[59]
C. sakazakii	Strain MBRG 15.5 from a domestic kitchen drain biofilm	14 passages at various concentrations	2-fold	15.6	Stable for 14 d	None reported	[11]
C. xerosis	WIBG 1.2 (wound isolate)	40 d at various concentrations	2.7-fold	7.3	Unstable for 14 d	None described	[25]
E. faecalis	1 strain of unknown origin	14 passages at various concentrations	4-fold	31.3	Unstable for 14 d	None reported	[11]
E. faecalis	WIBG 1.1 (wound isolate)	40 d at various concentrations	8-fold	14.5	Stable for 14 d	None described	[25]
E. saccharolyticus	Domestic drain biofilm isolate MBRG 9.16	14 d at various concentrations	None	20.8	Not applicable	None reported	[59]
E. coli	ATCC 25922	40 d at various concentrations	1.8-fold	24.2	Unstable for 14 d	None described	[25]
E. coli	ATCC 25922 and strain MBRG 15.4 from a domestic kitchen drain biofilm	14 passages at various concentrations	$\leq$2-fold	31.3	Unstable for 14 d	None reported	[11]
Eubacterium spp.	Domestic drain biofilm isolate MBRG 4.14	14 d at various concentrations	None	7.8	No data	None reported	[59]
H. gallinarum	Domestic drain biofilm isolate MBRG 4.27	14 d at various concentrations	None	3.9	No data	None reported	[59]

(continued)

Table 12.8 (continued)

Species	Strains/isolates	Exposure time	Increase in MIC	MIC_{max} (mg/l)	Stability of MIC change	Associated changes	References
K. pneumoniae	ATCC 13883	40 d at various concentrations	4-fold	29	Unstable for 14 d	None described	[25]
M. phyllosphaerae	Domestic drain biofilm isolate MBRG 4.30	14 d at various concentrations	2-fold	15.6	No data	None reported	[59]
M. luteus	Human skin isolate MBRG 9.25	14 d at various concentrations	None	0.97	Not applicable	None reported	[59]
M. luteus	MRBG 9.25 (skin isolate)	40 d at various concentrations	4-fold	7.3	Unstable for 14 d	None described	[25]
M. osloensis	Strain MBRG 15.3 from a domestic kitchen drain biofilm	14 passages at various concentrations	4-fold	31.3	Unstable for 14 d	None reported	[11]
P. aeruginosa	ATCC 9027	14 passages at various concentrations	1.5-fold	31.3	Unstable for 14 d	None reported	[11]
P. aeruginosa	ATCC 9027	40 d at various concentrations	1.9-fold	58	Unstable for 14 d	None described	[25]
P. nitroreductans	Domestic drain biofilm isolate MBRG 4.6	14 d at various concentrations	None	15.6	Not applicable	None reported	[59]
P. putida	Strain MBRG 15.2 from a domestic kitchen drain biofilm	14 passages at various concentrations	None	31.3	Not applicable	None reported	[11]
Pseudomonas spp.	Domestic drain biofilm isolate MBRG 9.14	14 d at various concentrations	None	7.8	Not applicable	None reported	[59]
Pseudoxanthomonas spp.	Domestic drain biofilm isolate MBRG 9.20	14 d at various concentrations	None	7.8	Not applicable	None reported	[59]

(continued)

Table 12.8 (continued)

Species	Strains/isolates	Exposure time	Increase in MIC	MIC_{max} (mg/l)	Stability of MIC change	Associated changes	References
Ralstonia spp.	Domestic drain biofilm isolate MBRG 4.13	14 d at various concentrations	None	3.9	Not applicable	None reported	[59]
S. marcescens	ATCC 13880	40 d at various concentrations	None	29	Not applicable	None described	[25]
S. multivorum	Domestic drain biofilm isolate MBRG 9.19	14 d at various concentrations	None	7.8	Not applicable	None reported	[59]
S. aureus	ATCC 6538	100 d at various concentrations	None	0.9	Not applicable	None described	[88]
S. aureus	ATCC 6538	40 d at various concentrations	None	7.3	Not applicable	None described	[25]
S. aureus	ATCC 6538	14 passages at various concentrations	6-fold	23.5	Stable for 14 d	None reported	[11]
S. aureus	3 clinical MRSA strains	6 passages at various concentrations	8-fold (2 strains), none (1 strain)	8	No data	No change of chlorhexidine susceptibility	[69]
S. capitis	Human skin isolate MBRG 9.34	14 d at various concentrations	None	1.9	Not applicable	None reported	[59]
S. capitis	MRBG 9.34 (skin isolate)	40 d at various concentrations	5.5-fold	6.0	Unstable for 14 d	None described	[25]
S. caprae	MRBG 9.3 (skin isolate)	40 d at various concentrations	None	4.9	Not applicable	None described	[25]
S. caprae	Human skin isolate MBRG 9.30	14 d at various concentrations	None	3.9	Not applicable	None reported	[59]

(continued)

Table 12.8 (continued)

Species	Strains/isolates	Exposure time	Increase in MIC	MIC_{max} (mg/l)	Stability of MIC change	Associated changes	References
S. cohnii	Human skin isolate MBRG 9.31	14 d at various concentrations	3.4-fold	6.4	No data	None reported	[59]
S. epidermidis	ATCC 35983	20 passages at various concentrations	None	0.31	Not applicable	None reported	[80]
S. epidermidis	Human skin isolate M 9.33	14 d at various concentrations	2-fold	3.9	No data	None reported	[59]
S. epidermidis	MRBG 9.33 (skin isolate)	40 d at various concentrations	4.8-fold	14.5	Unstable for 14 d	None described	[25]
S. haemolyticus	Human skin isolate MBRG 9.35	14 d at various concentrations	None	7.8	Not applicable	None reported	[59]
S. haemolyticus	MRBG9.35 (skin isolate)	40 d at various concentrations	4-fold	7.3	Unstable for 14 d	None described	[25]
S. hominis	Human skin isolate MBRG 9.37	14 d at various concentrations	None	3.9	Not applicable	None reported	[59]
S. kloosii	Human skin isolate MBRG 9.28	14 d at various concentrations	None	3.9	Not applicable	None reported	[59]
S. lugdunensis	Human skin isolate MBRG 9.36	14 d at various concentrations	None	3.9	Not applicable	None reported	[59]
S. lugdunensis	MRBG 9.36 (skin isolate)	40 d at various concentrations	2-fold	7.3	Unstable for 14 d	None described	[25]
S. saprophyticus	Human skin isolate MBRG 9.29	14 d at various concentrations	None	3.9	No data	None reported	[59]

(continued)

Table 12.8 (continued)

Species	Strains/isolates	Exposure time	Increase in MIC	MIC_{max} (mg/l)	Stability of MIC change	Associated changes	References
S. warneri	Human skin isolate MBRG 9.27	14 d at various concentrations	None	7.8	No data	None reported	[59]
S. warneri	MRBG 9.27 (skin isolate)	40 d at various concentrations	1.7-fold	6.0	Unstable for 14 d	None described	[25]
S. maltophilia	MRBG 4.17 (kitchen drain biofilm isolate)	40 d at various concentrations	None	3.6	Not applicable	None described	[25]
S. maltophilia	Domestic drain biofilm isolate MBRG 9.13	14 d at various concentrations	2-fold	7.8	No data	None reported	[59]

of 4 d each to increasing PHMB concentrations on agar was associated with both increases and decreases in antibiotic susceptibility but its effect was typically small relative to the differences observed among microbiocides. Susceptibility changes resulting in resistance were not observed [26].

12.5 Resistance to PHMB

Resistance to PHMB appears not to develop despite many years of use in many fields [86]. Selective chromosome condensation provides an unanticipated paradigm for antimicrobial action that may not succumb to resistance [9].

12.5.1 Species with Resistance to PHMB

An isolate of *A. westerdijkiae* was detected in house dust and indoor air fallouts where the occupants suffered from building related ill health. The isolates showed the same growth in presence and absence of 100 or 1,000 mg/l PHMB [57]. In addition, *Shingomonas* spp. and *Azospirillum* spp. can be successfully enriched at the expense of PMHB (originally at 1,000 mg/l) leading finally to its biodegradation [66].

12.5.2 Resistance Mechanisms

Biodegradation of PHMB is one mechanism of resistance as shown with *A. westerdijkiae*, *Shingomonas* spp. and *Azospirillum* spp. [57, 66]. In addition, the rhs genes which are a widely distributed, enigmatic family of horizontally acquired genes [34] can be induced and enzymes can be involved in the repair or binding of nucleic acids in the generation of PHMB tolerance in *E. coli*, suggesting a novel dimension in the mechanism of action of PHMB based on its interaction with nucleic acids [2].

12.5.3 Resistance Genes

The NCW2 gene has been detected in *S. cerevisiae* enhancing tolerance to PHMB. It codes for a protein which participates in the cell wall biogenesis in yeasts [17].

12.5.4 Infections Associated with Resistance to PHMB

One outbreak of *Fusarium* keratitis with more than 250 cases worldwide was reported primarily associated with specific contact lens disinfecting solutions. The outbreak was explained by the contact lens material which was able to bind up to 60% of the PHMB within 6 h resulting in a diminished antimicrobial activity [19].

12.6 Cross-Tolerance to Other Biocidal Agents

So far no cross-resistance to other biocidal agents has been described. One study even shows in three clinical MRSA strains that low-level exposure to various PHMB concentrations for six passages increases the MIC value to PHMB up to 8-fold but does not change the susceptibility to chlorhexidine [69].

12.7 Cross-Tolerance to Antibiotics

So far, no cross-tolerance to antibiotics has been described.

12.8 Role of Biofilm

12.8.1 Effect on Biofilm Development

No studies were found on the effect of PHMB on biofilm development.

12.8.2 Effect on Biofilm Removal

One study was found on the effect of PHMB on biofilm removal. With commonly used PHMB concentrations (0.02 and 0.04%), no significant biofilm removal was found (Table 12.9).

Table 12.9 Biofilm removal rate (quantitative determination of biofilm matrix) by exposure to PHMB solutions (S)

Type of biofilm	Concentration	Exposure time (min)	Biofilm removal rate	References
P. aeruginosa environmental strain SG81, 44-h incubation in polystyrene microtitre plates	0.02% (S)	30	No significant reduction	[38]
	0.04% (S)			
P. aeruginosa environmental strain SG81, 44-h incubation on silicone swatches	0.02% (S)	30	No significant reduction	[38]
	0.04% (S)			

12.8.3 Effect on Biofilm Fixation

No studies were found on the effect of PHMB on biofilm fixation.

12.9 Summary

The principal antimicrobial activity of PHMB is summarized in Table 12.10.

The key findings on acquired resistance and cross-resistance including the role of biofilm for selecting resistant isolates are summarized in Table 12.11.

Table 12.10 Overview on the typical exposure times required for PHMB to achieve sufficient biocidal activity against the different target micro-organisms

Target micro-organisms	Species	Concentration	Exposure time
Bacteria	Most species	0.016%/0.02%	1 h[a]
		0.032%/0.04%	30 min
Fungi	*C. albicans*	0.1%	5 min
	Exophiala spp., *F. oxysporum*	0.1%	5 min
	A. fumigatus, *M. circinelloides*	0.1%	24 h
	Apophysomyces spp., *A. brasiliensis*, *A. flavus*, *A. terreus*, *Lichtheimia* spp.	0.1%	>24 h[b]
Mycobacteria	*M. frederiksbergense* but not *Mycobacterium* spp.	0.0014%	5 h

[a]in biofilm the bactericidal activity will be lower; [b]insufficient fungicidal activity within 24 h

Table 12.11 Key findings on acquired PHMB resistance, the effect of low-level exposure, cross-tolerance to other biocides and antibiotics, and its effect on biofilm

Parameter	Species	Findings
Elevated MIC values	*A. westerdijkiae*, *Shingomonas* spp. and *Azospirillum* spp.	>1,000 mg/l
MIC value to determine resistance	Not proposed yet for bacteria, fungi or mycobacteria	
Cross-tolerance biocides	None described	
Cross-tolerance antibiotics	None described	
Resistance mechanisms	*A. westerdijkiae*, *Shingomonas* spp. and *Azospirillum* spp.	Biodegradation of PHMB
	S. cerevisiae	NCW2 gene
	E. coli	rhs genes

(continued)

Table 12.11 (continued)

Parameter	Species	Findings
Effect of low-level exposure	*A. xylosoxidans*, *A. hydrophila*, *A. jandaei*, *B. cereus*, *B. cepacia*, *C. indologenes*, *C. pseudogenitalum*, *Chrysobacterium* spp., *Citrobacter* spp., *E. saccharolyticus*, *Eubacterium* spp., *H. gallinarum*, *M. luteus*, *P. nitroreductans*, *P. putida*, *Pseudomonas* spp., *Pseudoxanthomonas* spp., *Ralstonia* spp., *S. marcescens*, *S. multivorum*, *S. aureus*, *S. capitis*, *S. caprae*, *S. epidermidis*, *S. haemolyticus*, *S. hominis*, *S. kloosii*, *S. lugdunensis*, *S. saprophyticus*, *S. warneri*, *S. maltophilia*	No MIC increase
	A. baumannii, *C. indologenes*, *C. renale group*, *C. sakazakii*, *C. xerosis*, *E. faecalis*, *E. coli*, *K. pneumoniae*, *M. phyllosphaerae*, *M. luteus*, *M. osloensis*, *P. aeruginosa*, *S. cohnii*, *S. epidermidis*, *S. haemolyticus*, *S. lugdunensis*, *S. warneri*, *S. maltophilia*	Weak MIC increase ($\leq$ 4-fold)
	S. capitis, *S. epidermidis*	Strong (>4-fold) but unstable MIC increase
	E. faecalis, *S. aureus*	Strong and stable MIC increase
	A. proteolyticus, *S. aureus*	Strong MIC increase (unknown stability)
	A. proteolyticus (16-fold)	Strongest MIC change after low level exposure
	E. faecalis and *S. aureus* (8-fold)	
	S. capitis (5.5-fold)	
	S. epidermidis (4.8-fold)	
	A. proteolyticus (125 mg/l)	Highest MIC values after low level exposure
	P. aeruginosa (58 mg/l)	
	A. baumannii, *E. coli*, *E. faecalis*, *M. osloensis* (31.3 mg/l)	
	K. pneumoniae (29 mg/l)	
	S. aureus (23.5 mg/l)	
	MRSA	No change of chlorhexidine susceptibility
Biofilm	Development	Unknown
	Removal	No significant effect
	Fixation	Unknown

References

1. Ali S, Wilson APR (2017) Effect of poly-hexamethylene biguanide hydrochloride (PHMB) treated non-sterile medical gloves upon the transmission of Streptococcus pyogenes, carbapenem-resistant E. coli, MRSA and Klebsiella pneumoniae from contact surfaces. BMC Infect Dis 17(1):574. https://doi.org/10.1186/s12879-017-2661-9
2. Allen MJ, White GF, Morby AP (2006) The response of Escherichia coli to exposure to the biocide polyhexamethylene biguanide. Microbiology (Reading, England) 152(Pt 4):989–1000. https://doi.org/10.1099/mic.0.28643-0
3. Anderson MJ, Scholz MT, Parks PJ, Peterson ML (2013) Ex vivo porcine vaginal mucosal model of infection for determining effectiveness and toxicity of antiseptics. J Appl Microbiol 115(3):679–688. https://doi.org/10.1111/jam.12277
4. Ansorg R, Rath PM, Fabry W (2003) Inhibition of the anti-staphylococcal activity of the antiseptic polihexanide by mucin. Arzneimittelforschung 53(5):368–371. https://doi.org/10.1055/s-0031-1297121
5. Assadian O, Wehse K, Hubner NO, Koburger T, Bagel S, Jethon F, Kramer A (2011) Minimum inhibitory (MIC) and minimum microbicidal concentration (MMC) of polihexanide and triclosan against antibiotic sensitive and resistant Staphylococcus aureus and Escherichia coli strains. GMS Krankenhaushygiene interdisziplinar 6(1):Doc06. https://doi.org/10.3205/dgkh000163
6. Behrens-Baumann W, Seibold M, Hofmuller W, Walter S, Haeberle H, Wecke T, Tammer I, Tintelnot K (2012) Benefit of polyhexamethylene biguanide in Fusarium keratitis. Ophthalmic Res 48(4):171–176. https://doi.org/10.1159/000337140
7. Bernauer U (2015) Opinion of the scientific committee on consumer safety (SCCS)–2nd Revision of the safety of the use of poly(hexamethylene) biguanide hydrochloride or polyaminopropyl biguanide (PHMB) in cosmetic products. Regul Toxicol Pharmacol: RTP 73(3):885–886. https://doi.org/10.1016/j.yrtph.2015.09.035
8. Chadeau E, Dumas E, Adt I, Degraeve P, Noel C, Girodet C, Oulahal N (2012) Assessment of the mode of action of polyhexamethylene biguanide against Listeria innocua by Fourier transformed infrared spectroscopy and fluorescence anisotropy analysis. Can J Microbiol 58 (12):1353–1361. https://doi.org/10.1139/w2012-113
9. Chindera K, Mahato M, Sharma AK, Horsley H, Kloc-Muniak K, Kamaruzzaman NF, Kumar S, McFarlane A, Stach J, Bentin T, Good L (2016) The antimicrobial polymer PHMB enters cells and selectively condenses bacterial chromosomes. Scientific reports 6:23121. https://doi.org/10.1038/srep23121
10. Codling CE, Maillard JY, Russell AD (2003) Performance of contact lens disinfecting solutions against Pseudomonas aeruginosa in the presence of organic load. Eye & contact lens 29(2):100–102. https://doi.org/10.1097/01.icl.0000062347.66975.f1
11. Cowley NL, Forbes S, Amezquita A, McClure P, Humphreys GJ, McBain AJ (2015) Effects of formulation on microbicide potency and mitigation of the development of bacterial insusceptibility. Appl Environ Microbiol 81(20):7330–7338. https://doi.org/10.1128/aem.01985-15
12. Das JR, Bhakoo M, Jones MV, Gilbert P (1998) Changes in the biocide susceptibility of Staphylococcus epidermidis and Escherichia coli cells associated with rapid attachment to plastic surfaces. J Appl Microbiol 84(5):852–858
13. Davis SC, Harding A, Gil J, Parajon F, Valdes J, Solis M, Higa A (2017) Effectiveness of a polyhexanide irrigation solution on methicillin-resistant Staphylococcus aureus biofilms in a porcine wound model. Int Wound J 14(6):937–944. https://doi.org/10.1111/iwj.12734
14. Decker EM, Bartha V, Kopunic A, von Ohle C (2017) Antimicrobial efficiency of mouthrinses versus and in combination with different photodynamic therapies on periodontal pathogens in an experimental study. J Periodontal Res 52(2):162–175. https://doi.org/10.1111/jre.12379

15. Eberlein T, Assadian O (2010) Clinical use of polihexanide on acute and chronic wounds for antisepsis and decontamination. Skin Pharmacol Physiol 23(Suppl):45–51. https://doi.org/10.1159/000318267
16. Egli-Gany D, Brill FH, Hintzpeter M, Andree S, Pavel V (2012) Evaluation of the antiseptic efficacy and local tolerability of a polihexanide-based antiseptic on resident skin flora. Adv Skin Wound Care 25(9):404–408. https://doi.org/10.1097/01.ASW.0000419405.52570.3e
17. Elsztein C, de Lima Rde C, de Barros Pita W, de Morais MA, Jr (2016) NCW2, a Gene Involved in the Tolerance to Polyhexamethylene Biguanide (PHMB), May Help in the Organisation of beta-1,3-Glucan Structure of Saccharomyces cerevisiae Cell Wall. Curr Microbiol 73(3):341–345. https://doi.org/10.1007/s00284-016-1067-z
18. Elsztein C, de Lucena RM, de Morais MA, Jr. (2011) The resistance of the yeast Saccharomyces cerevisiae to the biocide polyhexamethylene biguanide: involvement of cell wall integrity pathway and emerging role for YAP1. BMC Mol Biol 12:38. https://doi.org/10.1186/1471-2199-12-38
19. Epstein AB (2007) In the aftermath of the Fusarium keratitis outbreak: what have we learned. Clin Ophtalm 1(4):355–366
20. European Chemicals Agency (ECHA) PHMB. Substance information. https://echa.europa.eu/substance-information/-/substanceinfo/100.115.789. Accessed 05 Feb 2018
21. Fabry W, Kock HJ (2014) In-vitro activity of polyhexanide alone and in combination with antibiotics against Staphylococcus aureus. J Hosp Infect 86(1):68–72. https://doi.org/10.1016/j.jhin.2013.10.002
22. Fabry W, Reimer C, Azem T, Aepinus C, Kock HJ, Vahlensieck W (2013) Activity of the antiseptic polyhexanide against meticillin-susceptible and meticillin-resistant Staphylococcus aureus. J Global Antimicrob Resist 1(4):195–199. https://doi.org/10.1016/j.jgar.2013.05.007
23. Fabry W, Trampenau C, Bettag C, Handschin AE, Lettgen B, Huber FX, Hillmeier J, Kock HJ (2006) Bacterial decontamination of surgical wounds treated with Lavasept. Int J Hyg Environ Health 209(6):567–573. https://doi.org/10.1016/j.ijheh.2006.03.008
24. Fabry WH, Kock HJ, Vahlensieck W (2014) Activity of the antiseptic polyhexanide against gram-negative bacteria. Microb Drug Resist (Larchmont, NY) 20(2):138–143. https://doi.org/10.1089/mdr.2013.0113
25. Forbes S, Dobson CB, Humphreys GJ, McBain AJ (2014) Transient and sustained bacterial adaptation following repeated sublethal exposure to microbicides and a novel human antimicrobial peptide. Antimicrob Agents Chemother 58(10):5809–5817. https://doi.org/10.1128/aac.03364-14
26. Forbes S, Knight CG, Cowley NL, Amezquita A, McClure P, Humphreys G, McBain AJ (2016) Variable effects of exposure to formulated microbicides on antibiotic susceptibility in firmicutes and proteobacteria. Appl Environ Microbiol 82(12):3591–3598. https://doi.org/10.1128/aem.00701-16
27. Frenzel E, Schmidt S, Niederweis M, Steinhauer K (2011) Importance of porins for biocide efficacy against Mycobacterium smegmatis. Appl Environ Microbiol 77(9):3068–3073. https://doi.org/10.1128/aem.02492-10
28. Geraldo IM, Gilman A, Shintre MS, Modak SM (2008) Rapid antibacterial activity of 2 novel hand soaps: evaluation of the risk of development of bacterial resistance to the antibacterial agents. Infect Control Hosp Epidemiol 29(8):736–741. https://doi.org/10.1086/589723
29. Gilbert P, Das JR, Jones MV, Allison DG (2001) Assessment of resistance towards biocides following the attachment of micro-organisms to, and growth on, surfaces. J Appl Microbiol 91 (2):248–254
30. Goroncy-Bermes P, Brill FHH, Brill H (2013) Antimicrobial activity of wound antiseptics against Extended-Spectrum Beta-Lactamase-producing bacteria. Wound Med 1(1):41–43
31. Gunther F, Blessing B, Tacconelli E, Mutters NT (2017) MRSA decolonization failure-are biofilms the missing link? Antimicrob Resist Infect Control 6:32. https://doi.org/10.1186/s13756-017-0192-1

32. Hammer TR, Mucha H, Hoefer D (2012) Dermatophyte susceptibility varies towards antimicrobial textiles. Mycoses 55(4):344–351. https://doi.org/10.1111/j.1439-0507.2011.02121.x
33. Hansmann F, Kramer A, Ohgke H, Strobel H, Muller M, Geerling G (2005) [Lavasept as an alternative to PVP-iodine as a preoperative antiseptic in ophthalmic surgery. Randomized, controlled, prospective double-blind trial]. Der Ophthalmologe: Zeitschrift der Deutschen Ophthalmologischen Gesellschaft 102(11):1043–1046, 1048–1050. https://doi.org/10.1007/s00347-004-1120-3
34. Hill CW, Sandt CH, Vlazny DA (1994) Rhs elements of Escherichia coli: a family of genetic composites each encoding a large mosaic protein. Mol Microbiol 12(6):865–871
35. Hirsch T, Limoochi-Deli S, Lahmer A, Jacobsen F, Goertz O, Steinau HU, Seipp HM, Steinstraesser L (2011) Antimicrobial activity of clinically used antiseptics and wound irrigating agents in combination with wound dressings. Plast Reconstr Surg 127(4):1539–1545. https://doi.org/10.1097/PRS.0b013e318208d00f
36. Hoekstra MJ, Westgate SJ, Mueller S (2017) Povidone-iodine ointment demonstrates in vitro efficacy against biofilm formation. Int Wound J 14(1):172–179. https://doi.org/10.1111/iwj.12578
37. Hubner NO, Kramer A (2010) Review on the efficacy, safety and clinical applications of polihexanide, a modern wound antiseptic. Skin Pharmacol Physiol 23(Suppl):17–27. https://doi.org/10.1159/000318264
38. Hubner NO, Matthes R, Koban I, Randler C, Muller G, Bender C, Kindel E, Kocher T, Kramer A (2010) Efficacy of chlorhexidine, polihexanide and tissue-tolerable plasma against Pseudomonas aeruginosa biofilms grown on polystyrene and silicone materials. Skin Pharmacol Physiol 23(Suppl):28–34. https://doi.org/10.1159/000318265
39. Johani K, Malone M, Jensen SO, Dickson HG, Gosbell IB, Hu H, Yang Q, Schultz G, Vickery K (2018) Evaluation of short exposure times of antimicrobial wound solutions against microbial biofilms: from in vitro to in vivo. J Antimicrob Chemother 73(2):494–502. https://doi.org/10.1093/jac/dkx391
40. Juncker JC (2016) COMMISSION IMPLEMENTING DECISION (EU) 2016/109 of 27 January 2016 not to approve PHMB (1600; 1.8) as an existing active substance for use in biocidal products for product-types 1, 6 and 9. Off J Eur Union 59 (L 21):84–85
41. Juncker JC (2016) COMMISSION IMPLEMENTING REGULATION (EU) 2016/124 of 29 January 2016 approving PHMB (1600; 1.8) as an existing active substance for use in biocidal products for product-type 4. Off J Eur Union 59 (L 24):1–5
42. Juncker JC (2016) COMMISSION IMPLEMENTING REGULATION (EU) 2016/125 of 29 January 2016 approving PHMB (1600; 1.8) as an existing active substance for use in biocidal products for product-types 2, 3, 11. Off J Eur Union 59 (L 24):6–11
43. Juncker JC (2017) COMMISSION IMPLEMENTING DECISION (EU) 2017/802 of 10 May 2017 not approving PHMB (1600; 1.8) as an existing active substance for use in biocidal products for product-type 5. Off J Eur Union 60 (L 120):29–30
44. Kahar Bador M, Rai V, Yusof MY, Kwong WK, Assadian O (2015) Evaluation of the efficacy of antibacterial medical gloves in the ICU setting. J Hosp Infect 90(3):248–252. https://doi.org/10.1016/j.jhin.2015.03.009
45. Kamaruzzaman NF, Firdessa R, Good L (2016) Bactericidal effects of polyhexamethylene biguanide against intracellular Staphylococcus aureus EMRSA-15 and USA 300. J Antimicrob Chemother 71(5):1252–1259. https://doi.org/10.1093/jac/dkv474
46. Kapalschinski N, Seipp HM, Kuckelhaus M, Harati KK, Kolbenschlag JJ, Daigeler A, Jacobsen F, Lehnhardt M, Hirsch T (2017) Albumin reduces the antibacterial efficacy of wound antiseptics against Staphylococcus aureus. J Wound Care 26(4):184–187. https://doi.org/10.12968/jowc.2017.26.4.184
47. Kapalschinski N, Seipp HM, Onderdonk AB, Goertz O, Daigeler A, Lahmer A, Lehnhardt M, Hirsch T (2013) Albumin reduces the antibacterial activity of polyhexanide-biguanide-based

antiseptics against Staphylococcus aureus and MRSA. Burns: J Int Soc Burn Injuries 39 (6):1221–1225. https://doi.org/10.1016/j.burns.2013.03.003
48. Koban I, Geisel MH, Holtfreter B, Jablonowski L, Hubner NO, Matthes R, Masur K, Weltmann KD, Kramer A, Kocher T (2013) Synergistic effects of nonthermal plasma and disinfecting agents against dental biofilms in vitro. ISRN Dent 2013:573262. https://doi.org/10.1155/2013/573262
49. Koburger T, Hübner N-O, Braun M, Siebert J, Kramer A (2010) Standardized comparison of antiseptic efficacy of triclosan, PVP-iodine, octenidine dihydrochloride, polyhexanide and chlorhexidine digluconate. J Antimicrob Chemother 65(8):1712–1719
50. Koburger T, Müller G, Eisenbeiß W, Assadian O, Kramer A (2007) Microbicidal activity of polihexanide. GMS Krankenhaushygiene interdisziplinar 2 (2):DOC44
51. Kramer A, Dissemond J, Kim S, Willy C, Mayer D, Papke R, Tuchmann F, Assadian O (2017) Consensus on wound antisepsis: update 2018. Skin Pharmacol Physiol 31(1):28–58. https://doi.org/10.1159/000481545
52. Landelle C, von Dach E, Haustein T, Agostinho A, Renzi G, Renzoni A, Pittet D, Schrenzel J, Francois P, Harbarth S (2016) Randomized, placebo-controlled, double-blind clinical trial to evaluate the efficacy of polyhexanide for topical decolonization of MRSA carriers. J Antimicrob Chemother 71(2):531–538. https://doi.org/10.1093/jac/dkv331
53. Lin JC, Ward TP, Belyea DA, McEvoy P, Kramer KK (1997) Treatment of Nocardia asteroides keratitis with polyhexamethylene biguanide. Ophthalmology 104(8):1306–1311
54. Lucas AD (2012) Environmental fate of polyhexamethylene biguanide. Bull Environ Contam Toxicol 88(3):322–325. https://doi.org/10.1007/s00128-011-0436-3
55. Mashat BH (2016) Polyhexamethylene biguanide hydrochloride: features and applications. Br J Environ Sci 4(1):49–55
56. Messick CR, Pendland SL, Moshirfar M, Fiscella RG, Losnedahl KJ, Schriever CA, Schreckenberger PC (1999) In-vitro activity of polyhexamethylene biguanide (PHMB) against fungal isolates associated with infective keratitis. J Antimicrob Chemother 44 (2):297–298
57. Mikkola R, Andersson MA, Hautaniemi M, Salkinoja-Salonen MS (2015) Toxic indole alkaloids avrainvillamide and stephacidin B produced by a biocide tolerant indoor mold Aspergillus westerdijkiae. Toxicon: Official J Int Soc Toxinol 99:58–67. https://doi.org/10.1016/j.toxicon.2015.03.011
58. Minnich KE, Stolarick R, Wilkins RG, Chilson G, Pritt SL, Unverdorben M (2012) The effect of a wound care solution containing polyhexanide and betaine on bacterial counts: results of an in vitro study. Ostomy/Wound Manage 58(10):32–36
59. Moore LE, Ledder RG, Gilbert P, McBain AJ (2008) In vitro study of the effect of cationic biocides on bacterial population dynamics and susceptibility. Appl Environ Microbiol 74 (15):4825–4834. https://doi.org/10.1128/aem.00573-08
60. Mori M, Gomi M, Matsumune N, Niizeki K, Sakagami Y (2013) Biofilm-forming activity of bacteria isolated from toilet bowl biofilms and the bactericidal activity of disinfectants against the isolates. Biocontrol Sci 18(3):129–135
61. Muller G, Kramer A (2000) In vitro action of a combination of selected antimicrobial agents and chondroitin sulfate. Chem Biol Interact 124(2):77–85
62. Muller G, Kramer A (2008) Biocompatibility index of antiseptic agents by parallel assessment of antimicrobial activity and cellular cytotoxicity. J Antimicrob Chemother 61(6):1281–1287. https://doi.org/10.1093/jac/dkn125
63. National Center for Biotechnology Information (2018) Polyhexanide. PubChem Compound Database; CID=20977. https://pubchem.ncbi.nlm.nih.gov/compound/20977. Accessed 25 Jan 2018
64. Norman G, Christie J, Liu Z, Westby MJ, Jefferies JM, Hudson T, Edwards J, Mohapatra DP, Hassan IA, Dumville JC (2017) Antiseptics for burns. Cochrane Database Syst Rev 7: Cd011821. https://doi.org/10.1002/14651858.cd011821.pub2

65. O'Malley LP, Collins AN, White GF (2006) Biodegradability of end-groups of the biocide polyhexamethylene biguanide (PHMB) assessed using model compounds. J Ind Microbiol Biotechnol 33(8):677–684. https://doi.org/10.1007/s10295-006-0103-6
66. O'Malley LP, Shaw CH, Collins AN (2007) Microbial degradation of the biocide polyhexamethylene biguanide: isolation and characterization of enrichment consortia and determination of degradation by measurement of stable isotope incorporation into DNA. J Appl Microbiol 103(4):1158–1169. https://doi.org/10.1111/j.1365-2672.2007.03354.x
67. Payne B, Simmen HP, Csuka E, Hintzpeter M, Pahl S, Brill FHH (2018) Randomised, double-blind, controlled clinical trial on the antiseptic efficacy by reduction of CFU with polihexanide 0.04% on acute traumatic wounds. J Hosp Infect 98(4):429–432. https://doi.org/10.1016/j.jhin.2017.12.020
68. Rembe JD, Fromm-Dornieden C, Schafer N, Bohm JK, Stuermer EK (2016) Comparing two polymeric biguanides: chemical distinction, antiseptic efficacy and cytotoxicity of polyaminopropyl biguanide and polyhexamethylene biguanide. J Med Microbiol 65(8):867–876. https://doi.org/10.1099/jmm.0.000294
69. Renzoni A, Von Dach E, Landelle C, Diene SM, Manzano C, Gonzales R, Abdelhady W, Randall CP, Bonetti EJ, Baud D, O'Neill AJ, Bayer A, Cherkaoui A, Schrenzel J, Harbarth S, Francois P (2017) Impact of exposure of methicillin-resistant staphylococcus aureus to polyhexanide in vitro and in vivo. Antimicrob Agents Chemother 61(10). https://doi.org/10.1128/aac.00272-17
70. Rohrer N, Widmer AF, Waltimo T, Kulik EM, Weiger R, Filipuzzi-Jenny E, Walter C (2010) Antimicrobial efficacy of 3 oral antiseptics containing octenidine, polyhexamethylene biguanide, or Citroxx: can chlorhexidine be replaced? Infect Control Hosp Epidemiol 31 (7):733–739. https://doi.org/10.1086/653822
71. Rosenthal RA, Bell WM, Abshire R (1999) Disinfecting action of a new multi-purpose disinfection solution for contact lenses. Cont Lens Anterior Eye 22(4):104–109
72. Rosenthal RA, Dassanayake NL, Schlitzer RL, Schlech BA, Meadows DL, Stone RP (2006) Biocide uptake in contact lenses and loss of fungicidal activity during storage of contact lenses. Eye contact Lens 32(6):262–266. https://doi.org/10.1097/ICL.0b013e31802b413f
73. Rosenthal RA, Stein JM, McAnally CL, Schlech BA (1995) A comparative study of the microbiologic effectiveness of chemical disinfectants and peroxide-neutralizer systems. CLAO J: Official Publ Contact Lens Assoc Ophthalmologists, Inc 21(2):99–110
74. Rosin M, Welk A, Bernhardt O, Ruhnau M, Pitten FA, Kocher T, Kramer A (2001) Effect of a polyhexamethylene biguanide mouthrinse on bacterial counts and plaque. J Clin Periodontol 28(12):1121–1126
75. Roth B, Brill FH (2010) Polihexanide for wound treatment–how it began. Skin Pharmacol Physiol 23(Suppl):4–6. https://doi.org/10.1159/000318236
76. Saleh K, Sonesson A, Persson K, Riesbeck K, Schmidtchen A (2016) Can dressings soaked with polyhexanide reduce bacterial loads in full-thickness skin grafting? A randomized controlled trial. J Am Acad Dermatol 75(6):1221–1228.e1224. https://doi.org/10.1016/j.jaad.2016.07.020
77. Schedler K, Assadian O, Brautferger U, Muller G, Koburger T, Classen S, Kramer A (2017) Proposed phase 2/ step 2 in-vitro test on basis of EN 14561 for standardised testing of the wound antiseptics PVP-iodine, chlorhexidine digluconate, polihexanide and octenidine dihydrochloride. BMC Infect Dis 17(1):143. https://doi.org/10.1186/s12879-017-2220-4
78. Scientific Committee On Consumer Safety S (2015) Opinion on the safety of poly (hexamethylene) biguanide hydrochloride (PHMB). https://eceuropaeu/health/scientific_committees/consumer_safety/docs/sccs_o_157pdf. Accessed 02 Feb 2018
79. Souza AL, Ceridorio LF, Paula GF, Mattoso LH, Oliveira ON Jr (2015) Understanding the biocide action of poly(hexamethylene biguanide) using Langmuir monolayers of dipalmitoyl phosphatidylglycerol. Colloids Surf B 132:117–121. https://doi.org/10.1016/j.colsurfb.2015.05.018

80. Tambe SM, Sampath L, Modak SM (2001) In vitro evaluation of the risk of developing bacterial resistance to antiseptics and antibiotics used in medical devices. J Antimicrob Chemother 47(5):589–598
81. Ueda S, Kuwabara Y (2007) Susceptibility of biofilm Escherichia coli, Salmonella Enteritidis and Staphylococcus aureus to detergents and sanitizers. Biocontrol Sci 12(4):149–153
82. United States Environmental Protection Agency (2004) Reregistration Eligibility Decision for PHMB. https://nepis.epa.gov/Exe/ZyPDF.cgi/P100142G.PDF?Dockey=P100142G.PDF
83. Uzer Celik E, Tunac AT, Ates M, Sen BH (2016) Antimicrobial activity of different disinfectants against cariogenic microorganisms. Braz Oral Res 30(1):e125. https://doi.org/10.1590/1807-3107BOR-2016.vol30.0125
84. Welk A, Splieth CH, Schmidt-Martens G, Schwahn C, Kocher T, Kramer A, Rosin M (2005) The effect of a polyhexamethylene biguanide mouthrinse compared with a triclosan rinse and a chlorhexidine rinse on bacterial counts and 4-day plaque re-growth. J Clin Periodontol 32 (5):499–505. https://doi.org/10.1111/j.1600-051X.2005.00702.x
85. Werthen M, Davoudi M, Sonesson A, Nitsche DP, Morgelin M, Blom K, Schmidtchen A (2004) Pseudomonas aeruginosa-induced infection and degradation of human wound fluid and skin proteins ex vivo are eradicated by a synthetic cationic polymer. J Antimicrob Chemother 54(4):772–779. https://doi.org/10.1093/jac/dkh407
86. Wessels S, Ingmer H (2013) Modes of action of three disinfectant active substances: a review. Regul Toxicol Pharmacol: RTP 67(3):456–467. https://doi.org/10.1016/j.yrtph.2013.09.006
87. Wiegand C, Abel M, Ruth P, Elsner P, Hipler UC (2015) pH influence on antibacterial efficacy of common antiseptic substances. Skin Pharmacol Physiol 28(3):147–158. https://doi.org/10.1159/000367632
88. Wiegand C, Abel M, Ruth P, Hipler UC (2012) Analysis of the adaptation capacity of Staphylococcus aureus to commonly used antiseptics by microplate laser nephelometry. Skin Pharmacol Physiol 25(6):288–297. https://doi.org/10.1159/000341222
89. Xu Y, He Y, Zhou L, Gao C, Sun S, Wang X, Pang G (2014) Effects of contact lens solution disinfectants against filamentous fungi. Optom Vis Sci: Official Publ Am Acad Optom 91 (12):1440–1445. https://doi.org/10.1097/opx.0000000000000407
90. Yabes JM, White BK, Murray CK, Sanchez CJ, Mende K, Beckius ML, Zera WC, Wenke JC, Akers KS (2017) In vitro activity of Manuka Honey and polyhexamethylene biguanide on filamentous fungi and toxicity to human cell lines. Med Mycol 55(3):334–343. https://doi.org/10.1093/mmy/myw070
91. Yamamoto M, Takami T, Matsumura R, Dorofeev A, Hirata Y, Nagamune H (2016) In vitro evaluation of the biocompatibility of newly synthesized bis-quaternary ammonium compounds with spacer structures derived from pentaerythritol or hydroquinone. Biocontrol science 21(4):231–241. https://doi.org/10.4265/bio.21.231
92. Yanai R, Ueda K, Nishida T, Toyohara M, Mori O (2011) Effects of ionic and surfactant agents on the antimicrobial activity of polyhexamethylene biguanide. Eye Contact Lens 37 (2):85–89. https://doi.org/10.1097/ICL.0b013e31820cebc3
93. Yanai R, Yamada N, Ueda K, Tajiri M, Matsumoto T, Kido K, Nakamura S, Saito F, Nishida T (2006) Evaluation of povidone-iodine as a disinfectant solution for contact lenses: antimicrobial activity and cytotoxicity for corneal epithelial cells. Cont Lens Anterior Eye 29 (2):85–91. https://doi.org/10.1016/j.clae.2006.02.006
94. Yoshimatsu T, Hiyama K (2007) Mechanism of the action of didecyldimethylammonium chloride (DDAC) against Escherichia coil and morphological changes of the cells. Biocontrol Sci 12(3):93–99

13 Chlorhexidine Digluconate

13.1 Chemical Characterization

Chlorhexidine is a cationic biguanide and was first described in 1954 as a promising new antibacterial agent, at that time as a diacetate and dihydrochloride [79]. It is still used primarily as its salts, today mostly as the digluconate (CHG) or the diacetate (CHA). The solubility of CHG in water is good with up to 50% (w/v), but high viscosity makes such concentrated solutions inconvenient to use. CHG is typically available at 20% (w/v). The water solubility of CHA is lower with up to 2% [422]. CHG at 2 and 5% but not at 0.2% has a detrimental effect on free available chlorine, e.g. from sodium hypochlorite. Their combined use should therefore be avoided [192]. The basic chemical information on CHG is summarized in Table 13.1.

13.2 Types of Application

CHG is used by consumers in washing and cleaning products, biocides (e.g. disinfectants, pest control products), perfumes and fragrances, cosmetics and personal care products, polishes, waxes and pharmaceuticals. It is in addition used in articles, e.g. in long-life materials with high release rate (e.g. release from fabrics, textiles during washing, removal of indoor paints). It can be found in products with material based on paper (e.g. tissues, feminine hygiene products, nappies, books, magazines, wallpaper). It is also used by professional workers in health care and in manufacturing food products. Examples for use in health care are use as hand scrub (0.1–4%), surgical site antiseptic (0.1–2%), mucosa and wound antiseptic (0.05%), surface disinfectant (0.05%) and instrument disinfectant (0.1–0.5%) [169, 198, 275]. It is also used for antiseptic treatment of burns [279] and as a non-volatile active ingredient in alcohol-based hand rubs (0.5–1%) [169]. Finally, it is used in

G. Kampf, *Antiseptic Stewardship*, https://doi.org/10.1007/978-3-319-98785-9_13

Table 13.1 Basic chemical information on CHG [276]

CAS number	18472-51-0
IUPAC name	(1E)-2-[6-[[amino-[(E)-[amino-(4-chloroanilino)methylidene]amino] methylidene]amino]hexyl]-1-[amino-(4-chloroanilino)methylidene] guanidine; (2R,3S,4R,5R)-2,3,4,5,6-pentahydroxyhexanoic acid
Synonyms	Chlorhexidine
Molecular formula	$C_{34}H_{54}Cl_2N_{10}O_{14}$
Molecular weight (g/mol)	897.762

formulation or repacking at industrial sites (e.g. for manufacturing pulp, paper and paper products) and in manufacturing [101].

13.2.1 European Chemicals Agency (European Union)

CHG is still under review as an active substance (June 2018) for product types 1 (human hygiene), 2 (disinfectants and algaecides not intended for direct application to humans or animals) and 3 (veterinary hygiene).

13.2.2 Food and Drug Administration (USA)

In 2015, CHG was described as ineligible for five types of application in health care: patient preoperative skin preparation, healthcare personnel hand wash, healthcare personnel hand rub, surgical hand scrub and surgical hand rub [86], in analogy to the classification as a new drug in the 1994 tentative final monograph for healthcare antiseptic products [85].

13.2.3 Overall Environmental Impact

CHG is manufactured and/or imported in the European Economic Area in 10–100 t per year [101]. It is described as "very toxic to aquatic life" and "very toxic to aquatic life with long-lasting effects" [101]. In Canada, it was found that CHG bioaccumulated extensively in lipid-rich regions of diatoms and bacteria of natural river biofilms [94]. First results indicate that a photocatalytic degradation process for 3 h is sufficient in reducing CHG concentration up to 90.7% from a reaction matrix. The products obtained after photomineralization of CHG upon releasing in environment are most unlikely to impose a relevant toxicity to the aquatic environment [78].

13.3 Spectrum of Antimicrobial Activity

13.3.1 Bactericidal Activity

13.3.1.1 Bacteriostatic Activity (MIC Values)

The highest MIC values were described for *E. faecalis* and *K. pneumoniae* (0.5–10,000 mg/l), *Proteus* spp. (2–10,000 mg/l) and *B. subtilis* (10,000 mg/l), followed by *P. aeruginosa* ($\leq$5,000 mg/l), *L. monocytogenes*, *E. faecium* and *S. aureus* ($\leq$2,500 mg/l), *Streptococcus* spp. ($\leq$2,000 mg/l), *S. marcescens* ($\leq$1,024 mg/l), *Acinetobacter* spp., *Citrobacter* spp. and *Enterobacter* spp. ($\leq$1,000 mg/l), *B. cepacia* ($\leq$700 mg/l), *Achromobacter* spp. ($\leq$500 mg/l), *E. coli* ($\leq$312 mg/l), *A. baumannii* ($\leq$256 mg/l), *E. cloacae* ($\leq$150 mg/l), *Salmonella* spp. ($\leq$100 mg/l) and coagulase-negative *Staphylococcus* spp. ($\leq$62.5 mg/l; Table 13.2). Taking into account the proposed epidemiological cut-off values such as 64 mg/l for *E. faecalis*,

Table 13.2 MIC values of various bacterial species to CHG

Species	Strains/isolates	MIC value (mg/l)	References
A. xylosoxidans	10 clinical isolates	4–128	[293]
A. xylosoxidans	Domestic drain biofilm isolate MBRG 4.31	15.6	[257]
A. xylosoxidans	2 clinical isolates	>125–500	[269]
A. baumannii	ATCC 19606 and 2 multidrug-resistant clinical isolates	4–16	[127]
A. baumannii	51 carbapenem-resistant clinical isolates	4–64	[218]
A. baumannii	98 carbapenem-resistant clinical isolates	8–64	[62]
A. baumannii	6 clinical isolates (colonization or infection) from hematopoietic stem cell transplantation patients during a study with CHG bathing	8–64	[246]
A. baumannii	149 clinical isolates during intervention with CHG whole-body washing in ICU patients	8–256	[247]
A. baumannii	236 non-repetitive clinical isolates	<10–150	[139]
A. baumannii	16 clinical isolates	16–128	[38]
A. baumannii	JCM 6841	53	[418]
A. baumannii	2 clinical strains	125–175	[144]
A. baumannii	2 blood culture isolates from oncology patients	125–175	[144]
A. calcoaceticus	10 clinical isolates	4–16	[293]
A. calcoaceticus	ATCC 19606	63	[290]
A. johnsonii	NCIMB 12460	2–2.2	[209]
	Triclosan-tolerant industrial strain	2.4–2.6	
A. viscosus	ATCC 15987	70	[392]

(continued)

Table 13.2 (continued)

Species	Strains/isolates	MIC value (mg/l)	References
Acinetobacter spp.[a]	21 clinical isolates	1–1,000	[248]
Acinetobacter spp.[a]	283 clinical isolates (273 *A. calcoaceticus-A. baumannii complex*, 7 *A. lwoffii*, 3 *A. junii*)	5–100	[177]
Acinetobacter spp.[a]	2 pan-susceptible clinical isolates	8	[229]
Acinetobacter spp.[a]	19 multidrug-resistant clinical isolates	8–256	[95]
Acinetobacter spp.[a]	283 clinical isolates (273 *A. calcoaceticus-A. baumannii complex*, 7 *A. lwoffii*, 3 *A. junii*)	10–400	[178]
Acinetobacter spp.[a]	69 non-repetitive, non-baumannii clinical isolates	>10–100	[139]
Acinetobacter spp.[a]	28 clinical MDR isolates	16–256	[229]
Acinetobacter spp.[a]	3 clinical XDR isolates	256	[229]
A. actinomycetemcomitans	71 clinical isolates, 9 control strains including ATCC 29524, NCTC 9710, Y4, SU-NyaB 75	2–32	[250]
A. actinomycetemcomitans	ATCC 29212	8	[9]
A. actinomycetemcomitans	ATCC 29523	16	[9]
A. actinomycetemcomitans	NCTC 10981, NCTC 10980, NCTC 10979, 1 dental school isolate	31–125	[356]
A. aphrophilus	NCTC 11098	8	[356]
A. israelii	ATCC 12102	8	[9]
A. odontolyticus	NCTC 9335	62	[356]
A. viscosus	NCTC 10951	125	[356]
A. hydrophila	10 clinical isolates	8–64	[293]
A. hydrophila	Domestic drain biofilm isolate MBRG 4.3	31.2	[257]
A. hydrophila	Water isolate	>200	[333]
A. hydrophilia	Blood culture isolate from an oncology patient	250	[144]
A. jandaei	Domestic drain biofilm isolate MBRG 9.11	7.8	[257]
Alcaligenes spp.	2 blood culture isolates from oncology patients	10–175	[144]
A. proteolyticus	Domestic drain biofilm isolate MBRG 9.12	7.8	[257]
B. cereus	VT 289	1	[375]
B. cereus	Domestic drain biofilm isolate MBRG 4.21	1.9	[257]
B. cereus	MRBG 4.21 (kitchen drain biofilm isolate)	14.5	[108]
B. subtilis var. globigii	ATCC 9372	10,000	[290]

(continued)

Table 13.2 (continued)

Species	Strains/isolates	MIC value (mg/l)	References
B. fragilis	NCTC 9343	250	[356]
B. gingivalis	11 dental school isolates	8–62	[356]
B. intermedius	11 clinical isolates and ATCC 33563	4–8	[250]
B. intermedius	NCTC 9336, 2 dental school isolates	15–62	[356]
B. melaninogenicus	NCTC 11321	62	[356]
B. adolescentis	4 isolates from faeces of healthy humans	16–64	[98]
B. animalis subsp. lactis	8 isolates from faeces of healthy humans	2–128	[98]
B. bifidum	31 isolates from faeces of healthy humans	2–32	[98]
B. breve	5 isolates from faeces of healthy humans	2–16	[98]
B. catenulatum	1 isolate from faeces of a healthy human	16	[98]
B. infantis	2 isolates from faeces of healthy humans	16	[98]
B. longum	25 isolates from faeces of healthy humans	2–128	[98]
B. pseudocatenulatum	15 isolates from faeces of healthy humans	2–128	[98]
B. pseudolongum	1 isolate from faeces of a healthy human	16	[98]
B. thermoacidophilum	6 isolates from faeces of healthy humans	16–128	[98]
B. suis	1 isolate from faeces of a healthy human	8	[98]
B. cepacia	1 clinical isolate	<1	[127]
B. cepacia	ATCC BAA-245	3.6	[108]
B. cepacia	10 clinical isolates	8–256	[293]
B. cepacia complex	38 clinical, non-clinical and environmental strains	10–100	[314]
B. cepacia	JCM 5964	149	[418]
B. cepacia	1 washbasin isolate	500	[269]
B. cepacia complex	*B. lata* strain 383	700	[183]
C. coli	8 strains from poultry	0.125–0.5	[237]
	6 strains from humans	0.125–0.5	
	4 strains from pigs	0.06–0.25	
	1 strain from water	0.5	
C. concisus	NCTC 11485	31	[356]
C. jajuni	5 strains from humans	0.125–0.5	[237]
	5 strains from water	0.25–0.5	
	3 strains from poultry	0.125–0.25	
C. ochracea	NCTC 11654, NCTC 11655	8–250	[356]
Capnocytophaga spp.	10 dental school isolates	250–500	[356]
C. trachomatis	ATCC VR-885	8	[286]
C. freundii	10 clinical isolates	1–32	[293]
C. freundii	NCIMB 11490	4–5	[209]
	Triclosan-tolerant industrial strain	8–11	

(continued)

Table 13.2 (continued)

Species	Strains/isolates	MIC value (mg/l)	References
C. koseri	10 clinical isolates	1–16	[293]
C. koseri	1 multidrug-resistant clinical isolate	<1	[127]
Citrobacter spp.	27 clinical isolates	1–1,000	[248]
C. acidivorans	Blood culture isolate from an oncology patient	50	[144]
C. perfringens	ATCC 13124	2.4	[313]
C. matruchotti	NCTC 10206, NCTC 10254	62–125	[356]
C. pseudogenitalum	Human skin isolate MBRG 9.24	0.9	[257]
C. renale group	Human skin isolate MBRG 9.13	7.8	[257]
C. indologenes	MRBG 4.29 (kitchen drain biofilm isolate)	7.3	[108]
C. indologenes	Domestic drain biofilm isolate MBRG 9.15	26	[257]
C. indologenes	Blood culture isolate from an oncology patient	175	[144]
C. meningosepticum	Blood culture isolate from an oncology patient	80	[144]
C. luteola	Blood culture isolate from an oncology patient	80	[144]
C. xerosis	WIBG 1.2 (wound isolate)	3.3	[108]
Chrysobacterium spp.	Domestic drain biofilm isolate MBRG 9.17	3.9	[257]
Citrobacter spp.	Domestic drain biofilm isolate MBRG 9.18	3.9	[257]
E. corrodens	NCTC 10596, NCTC 10647	62	[356]
E. aerogenes	10 clinical isolates	8–32	[293]
E. cloacae	3 multidrug-resistant clinical isolates	4–8	[127]
E. cloacae	10 clinical isolates	4–16	[293]
E. cloacae	Strain IAL 1976	71	[290]
E. cloacae	ATCC 13047	≤75	[50]
E. cloacae	4 isolates from the oral cavity of bone marrow transplant recipients	≤75–150	[50]
Enterobacter spp.[a]	54 worldwide strains from hospital- and community-acquired infections	1–64	[261]
Enterobacter spp.[a]	21 clinical isolates	1–1,000	[248]
Enterobacter spp.[a]	15 multidrug-resistant clinical isolates	8–128	[95]
Enterobacter spp.[a]	10 burn unit isolates	63–1,000	[142]
E. avium	1 vanB clinical isolate	1	[127]
E. casseliflavus	7 isolates from dust samples collected in breeding pig facilities	1–4	[45]
E. casseliflavus	1 vanC2/3 isolate	2	[127]

(continued)

Table 13.2 (continued)

Species	Strains/isolates	MIC value (mg/l)	References
E. casseliflavus	1 veterinary isolate	4	[44]
E. durans	5 dairy isolates	4–8	[44]
E. faecalis	52 isolates from livestock	0.5–8	[1]
E. faecalis	ATCC 29212 and 3 clinical isolates (2 of them vanA)	<1–2	[127]
E. faecalis	NCTC 775	2	[384]
E. faecalis	10 clinical isolates	2–8	[187]
E. faecalis	53 isolates from dust samples collected in breeding pig facilities	2–8	[45]
E. faecalis	18 clinical isolates, 11 veterinary isolates, 17 dairy isolates	2–8	[44]
E. faecalis	11 isolates from the urogenital tract of parturients	2–32	[60]
E. faecalis	ATCC 29212	2.5	[11]
E. faecalis	107 isolates from the environment and patients	2.5–2,500	[394]
E. faecalis	WIBG 1.1 (wound isolate)	3.6	[108]
E. faecalis	56 worldwide strains from hospital- and community-acquired infections	4–64	[261]
E. faecalis	9 isolates from swine meat production	8–12	[308]
E. faecalis	ATCC 29212	16	[186]
E. faecalis	ATCC 29212	27	[418]
E. faecalis	ATCC 29212	32	[16]
E. faecalis	ATCC 29212	156	[161]
E. faecalis	ATCC 29212	3,300	[417]
E. faecalis	ATCC 29212	≥ 10,000	[390]
E. faecium	22 isolates from dust samples collected in breeding pig facilities	0.5–4	[45]
E. faecium	78 isolates from livestock	0.5–8	[1]
E. faecium	6 vanA clinical isolates	<1–2	[127]
	2 vanB clinical isolates	8–16	
E. faecium	53 worldwide strains from hospital- and community-acquired infections	1–64	[261]
E. faecium	48 clinical VRE isolates (colonization or infection) from hematopoietic stem cell transplantation patients during a study with CHG bathing	1–32	[246]
E. faecium	10 clinical isolates, 4 veterinary isolates, 3 dairy isolates	2–8	[44]
E. faecium	165 isolates from the environment and patients	2.5–2,500	[394]
E. faecium	12 isolates from swine meat production	4–14	[308]

(continued)

Table 13.2 (continued)

Species	Strains/isolates	MIC value (mg/l)	References
E. gallinarum	2 isolates from dust samples collected in breeding pig facilities	1–2	[45]
E. gallinarum	1 vanC1 isolate	2	[127]
E. hirae	39 isolates from dust samples collected in breeding pig facilities	0.5–4	[45]
E. hirae	CIP 5855	2	[233]
E. hirae	2 clinical isolates, 1 dairy isolate	2–4	[44]
E. hirae	ATCC 10541	16	[418]
E. raffinosus	2 isolates from dust samples collected in breeding pig facilities	0.5	[45]
E. saccharolyticus	Domestic drain biofilm isolate MBRG 9.16	7.8	[257]
E. solitarius	1 veterinary isolate	4	[44]
Enterococcus spp.[a]	25 isolates from dust samples collected in breeding pig facilities	<0.25–4	[45]
Enterococcus spp.[a]	122 strains (*E. faecalis*, *E. faecium*) from different traditional fermented foods	0.25–2.5	[208]
Enterococcus spp.[a]	18 vancomycin-susceptible clinical isolates	1–8	[229]
Enterococcus spp.[a]	272 strains from various sources (165 *E. faecium*, 107 *E. faecalis*)	2.5–2,500	[394]
Enterococcus spp.[a]	5 clinical VRE isolates	4–6	[365]
Enterococcus spp.[a]	69 clinical isolates	6–12	[153]
Enterococcus spp.[a]	18 clinical VRE isolates	8–32	[229]
Enterococcus spp.	Clinical VRE isolate	16–32	[186]
E. coli	NCTC 10418	0.125–8[b]	[41]
E. coli	ATCC 25922	0.5–2	[229]
E. coli	140 human ESBL isolates, 34 isolates from healthy chicken	0.5–4	[87]
E. coli	39 clinical isolates	0.5–8	[293]
E. coli	306 worldwide strains from hospital- and community-acquired infections	0.5–32	[261]
E. coli	37 pan-sensitive clinical isolates	0.5–64	[229]
E. coli	13 bovine and 7 equine strains	0.78–12.5	[339]
E. coli	ATCC 25922 and 6 multidrug-resistant clinical isolates	<1–2	[127]
E. coli	202 isolates from livestock	1–2	[1]
E. coli	140 clinical isolates	1–100	[248]
E. coli	ATCC 25922	2	[233]
E. coli	10 clinical isolates	2–16	[187]
E. coli	10 burn unit isolates	2–16	[142]

(continued)

Table 13.2 (continued)

Species	Strains/isolates	MIC value (mg/l)	References
E. coli	ATCC 25922	2.5	[11]
E. coli	ATCC 25922	4	[418]
E. coli	25 NDM-positive clinical isolates	4–128	[229]
E. coli	ATCC 25922	≤4.7	[50]
E. coli	ATCC 25922	6.7	[108]
E. coli	Strain HEC30	8	[74]
E. coli	ATCC 35218	8–16	[186]
E. coli	369 clinical isolates from patients with urinary tract infection	<10–200	[360]
E. coli	6 clinical ESBL isolates	10–20	[124]
E. coli	ATCC 25922	71	[290]
E. coli	ATCC 25922	312	[90]
Eubacterium spp.	Domestic drain biofilm isolate MBRG 4.14	31.2	[257]
F. nucleatum	ATCC 25586	0.15	[313]
F. nucleatum	NCTC 10562	125	[356]
F. oryzihabitans	Blood culture isolate from an oncology patient	30	[144]
H. parainfluenzae	NCTC 7857	8	[356]
H. gallinarum	Domestic drain biofilm isolate MBRG 4.27	15.6	[257]
H. alvei	10 clinical isolates	2–8	[293]
K. oxytoca	ATCC 700324	4	[127]
K. oxytoca	9 clinical isolates	8–16	[293]
K. pneumoniae	37 isolates predominately from a variety of human infections pre-1949 ("Murray isolates") and 39 "modern strains" (2007–2012)	0.25–16 (old isolates) 8–32 (modern isolates)	[404]
K. pneumoniae	ATCC 13883	0.5–2	[229]
K. pneumoniae	26 pan-sensitive clinical isolates	0.5–64	[229]
K. pneumoniae	60 worldwide strains from hospital- and community-acquired infections	1–64	[261]
K. pneumoniae	102 clinical isolates	1–10,000	[248]
K. pneumoniae	3 clinical isolates (2 of them ESBL producer)	2–4	[127]
K. pneumoniae	25 clinical isolates	2–32	[293]
K. pneumoniae	32 clinical isolates	2–128	[38]
K. pneumoniae	ATCC 13883	2.1	[108]
K. pneumoniae	10 burn unit isolates	3–1,000	[142]

(continued)

Table 13.2 (continued)

Species	Strains/isolates	MIC value (mg/l)	References
K. pneumoniae	64 clinical isolates	4–128	[3]
K. pneumoniae	37 NDM1 positive clinical isolates	8–128	[229]
K. pneumoniae	126 XDR clinical isolates	8–256	[274]
K. pneumoniae	36 clinical isolates (ertapenem-resistant)	10–32	[262]
K. pneumoniae	5 clinical isolates	10–200	[363]
K. pneumoniae	6 isolates of strain ST395, 5 isolates of strain ST147, all non-susceptible to ertapenem	10.8–15.3	[262]
K. pneumoniae	10 clinical isolates	16–32	[187]
K. pneumoniae	NCTC 13368	16–32	[384]
K. pneumoniae	35 clinical ESBL isolates	16–64	[287]
K. pneumoniae	27 clinical isolates (colonization or infection) from hematopoietic stem cell transplantation patients during a study with CHG bathing	16–128	[246]
K. pneumoniae	DSM16609 and 3 clinical ESBL isolates	20–80	[124]
K. pneumoniae	Strain 39.11	32	[74]
K. pneumoniae	4 isolates from the oral cavity of bone marrow transplant recipients	≤ 75–300	[50]
K. pneumoniae	ATCC 10031	≤ 150	[50]
Klebsiella spp.[a]	14 multidrug-resistant clinical isolates	8–128	[95]
Klebsiella spp.[a]	167 clinical isolates from patients with urinary tract infection	<10–800	[360]
L. acidophilus	4 strains from different origins	0.5–2	[15]
L. acidophilus	ATCC 4356	4	[9]
L. acidophilus	ATCC 4356	70	[392]
L. amylovorus	7 strains from different origins	0.25–2	[15]
L. brevis	13 strains from different origins	0.5–2	[15]
L. bulgaricus	6 strains from different origins	1–2	[15]
L. coryniformis	3 strains from different origins	1	[15]
L. fermentum	4 strains from different origins	0.25–1	[15]
L. garvieae	42 isolates from different origins	2–4	[15]
L. helveticus	39 strains from different origins	0.5–8	[15]
L. odontolyticus	NCTC 1406	125	[356]
L. paracasei	75 strains from different origins	0.5–16	[15]
L. pentosus	60 strains from naturally fermented Aloreña green table olives	0.001–5	[55]
L. plantarum	43 strains from different origins	0.5–16	[15]
L. reuteri	42 strains from different origins	0.125–4	[15]
L. rhamnosus	9 strains from different origins	1–4	[15]
L. rhamnosus	ATCC 7469	150	[392]

(continued)

Table 13.2 (continued)

Species	Strains/isolates	MIC value (mg/l)	References
L. salivarius	ATCC 11741	2	[9]
L. lactis	1 strain	6	[89]
L. pseudomesenteroides	13 strains from naturally fermented Aloreña green table olives	0.01–5	[55]
L. monocytogenes	96 strains from frozen food	1.25–5	[295]
L. monocytogenes	ATCC 7644	2,500	[90]
M. phyllosphaerae	Domestic drain biofilm isolate MBRG 4.30	15.6	[257]
M. luteus	MRBG 9.25 (skin isolate)	3.6	[108]
M. luteus	Human skin isolate MBRG 9.25	3.9	[257]
M. morganii	ATCC 25830	2	[233]
M. morganii	2 clinical isolates (1 of them multidrug-resistant)	8–32	[127]
M. morganii	10 clinical isolates	16–64	[293]
N. subflava	VT 455	8	[375]
O. anthropi	Blood culture isolate from an oncology patient	30	[144]
P. anaerobius	NCTC 11460	8	[356]
P. micros	ATCC 33270	62	[356]
P. gingivalis	ATCC 33277	3.4	[88]
P. gingivalis	ATCC 33277	16	[9]
P. endodontalis	ATCC 35406	3.4	[88]
P. denticola	ATCC 35308	2.7	[88]
P. intermedia	ATCC 33563	3.4	[88]
P. intermedia	ATCC 25611	16	[9]
P. melaninogenica	ATCC 33563	3.4	[88]
P. nigrescens	ATCC 33563	0.15	[313]
P. acnes	NCTC 737	8	[356]
P. mirabilis	ATCC 43071 and 1 penicillinase-producing clinical isolate	2–16	[127]
P. mirabilis	31 clinical isolates	2–128	[293]
P. mirabilis	104 clinical isolates	10–800	[359]
P. mirabilis	11 clinical strains	20–800	[359]
P. mirabilis	Isolate from urine after repetitive bladder washouts with 100 ml of 0.02% chlorhexidine in patients with transurethral catheter	1,280	[361]
P. rettgeri	11 clinical isolates	8–32	[293]
P. vulgaris	1 clinical isolate	8	[127]
Proteus spp.[a]	181 clinical isolates from patients with urinary tract infection	<10–800	[360]

(continued)

Table 13.2 (continued)

Species	Strains/isolates	MIC value (mg/l)	References
Proteus spp.[a]	59 clinical isolates	10–10,000	[248]
Proteus spp.[a]	181 clinical isolates (139 *P. mirabilis*, 31 *P. morganii*, 10 *P. vulgaris*, 1 *P. rettgeri*)	10–1,600	[362]
Proteus spp.[a]	14 clinical isolates	10–1,600	[363]
Proteus spp.[a]	10 burn unit isolates	16–1,000	[142]
P. alcalifaciens	10 clinical isolates	4–64	[293]
P. stuartii	ATCC 33672 and 1 multidrug-resistant clinical isolate	<1–4	[127]
P. stuartii	10 clinical isolates	4–32	[293]
P. stuartii	24 clinical isolates from patients with urinary tract infection	<10–800	[360]
P. aeruginosa	ATCC 27853	0.2	[233]
P. aeruginosa	175 isolates from veterinary sources	1–16	[29]
P. aeruginosa	111 clinical isolates	1.6–25	[199]
P. aeruginosa	55 strains from various sources (20 clinical isolates, 19 industrial environmental isolates, 16 culture collection strains)	<2–39	[200]
P. aeruginosa	54 clinical isolates	2–32	[293]
P. aeruginosa	60 clinical isolates	2–128	[38]
P. aeruginosa	317 clinical isolates	3–400	[271]
P. aeruginosa	ATCC 27853 and 5 multidrug-resistant clinical isolates	4–16	[127]
P. aeruginosa	46 clinical isolates (colonization or infection) from hematopoietic stem cell transplantation patients during a study with chlorhexidine bathing	4–64	[246]
P. aeruginosa	NCTC 13359	4–64[b]	[41]
P. aeruginosa	ATCC 9027	7.3	[108]
P. aeruginosa	Strain PA01	8	[384]
P. aeruginosa	ATCC 27853	9	[418]
P. aeruginosa	35 clinical isolates from patients with urinary tract infection	<10–800	[360]
P. aeruginosa	35 clinical isolates	10–800	[362]
P. aeruginosa	21 multidrug-resistant clinical isolates	16–512	[95]
P. aeruginosa	ATCC 15442	32	[186]
P. aeruginosa	178 clinical strains	78–625	[194]
P. aeruginosa	91 clinical isolates, 37 hospital environmental isolates	125–156	[285]

(continued)

Table 13.2 (continued)

Species	Strains/isolates	MIC value (mg/l)	References
P. aeruginosa	20 burn unit isolates	125–1,000	[142]
P. aeruginosa	ATCC 14502	$\leq$ 150	[50]
P. aeruginosa	NCTC6749 and 3 extensively resistant clinical isolates	5,000	[412]
P. alkylphenolia	2 isolates from meat chain production	0.0025–0.025	[207]
P. fluorescens	3 isolates from meat chain production	0.0025–0.025	[207]
P. fragi	4 isolates from meat chain production	0.0025–0.025	[207]
P. lundensis	34 isolates from meat chain production	0.0025–0.25	[207]
P. nitroreductans	Domestic drain biofilm isolate MBRG 4.6	15.6	[257]
P. putida	9 isolates from meat chain production	0.0025–0.025	[207]
P. stutzeri	11 clinical strains	<10–50	[323]
Pseudomonas spp.	Domestic drain biofilm isolate MBRG 9.14	1.9	[257]
Pseudomonas spp.[a]	41 clinical isolates	10–1,000	[248]
Pseudoxanthomonas spp.	Domestic drain biofilm isolate MBRG 9.20	62.5	[257]
Ralstonia spp.	Domestic drain biofilm isolate MBRG 4.13	7.8	[257]
R. dentocariosa	NCTC 10918, NCTC 10917	8	[356]
S. enterica	122 poultry isolates, 135 swine isolates	2–64	[61]
Salmonella spp.[a]	375 avian isolates	0.5–32	[304]
Salmonella spp.[a]	901 worldwide strains from hospital- and community-acquired infections	1–64	[261]
Salmonella spp.[a]	156 isolates from livestock	2–64	[1]
Salmonella spp.[a]	195 isolates from chicken and egg production	<4–64	[219]
Salmonella spp.[a]	12 clinical isolates	10–100	[248]
S. marcescens	54 clinical isolates	2–128	[293]
S. marcescens	131 clinical strains	<3.1–400	[284]
S. marcescens	ATCC 13880	12.1	[108]
S. marcescens	1 clinical isolate	32	[127]
S. marcescens	18 clinical strains	32–64	[120]
S. marcescens	10 burn unit isolates	125–1,000	[142]
S. marcescens	Strain IAL 1478	141	[290]

(continued)

Table 13.2 (continued)

Species	Strains/isolates	MIC value (mg/l)	References
S. marcescens	25 isolates from 5 of 6 bottles of a 2% CHG stock solution	512–1,024	[234]
S. multivorum	Domestic drain biofilm isolate MBRG 9.19	20.8	[257]
S. spiritivorum	Blood culture isolate from an oncology patient	10	[144]
S. aureus	134 clinical MRSA isolates	<0.125–4	[267]
S. aureus	ATCC 6538	0.125–1	[41]
S. aureus	156 clinical isolates	0.125–4	[146]
S. aureus	206 clinical MRSA isolates	0.125–16	[338]
S. aureus	114 effluxing bloodstream isolates	0.16–1.25	[84]
S. aureus	54 clinical isolates	0.25–4	[367]
S. aureus	152 clinical isolates	0.25–16	[420]
S. aureus	256 clinical isolates (87 MRSA, 169 MSSA)	0.4–6.3	[199]
S. aureus	ATCC 9144	≤0.5	[384]
S. aureus	27 clinical MRSA isolates before decolonization with PHMB	<0.5–4	[305]
	27 clinical MRSA isolates after decolonization with PHMB	<0.5–4	
S. aureus	10 clinical isolates	0.5–1	[187]
S. aureus	121 clinical MRSA isolates	0.5–2	[396]
S. aureus	43 isolates from livestock	0.5–2	[1]
S. aureus	30 isolates from eyelids, eyelashes, and conjunctival sacs of lens wearers and spectacle wearers	0.5–2	[340]
S. aureus	829 clinical MRSA isolates	0.5–4	[241]
S. aureus	50 canine isolates (MSSA)	0.5–4	[63]
S. aureus	48 clinical isolates	0.5–4	[293]
S. aureus	1,635 worldwide strains from hospital- and community-acquired infections	0.5–8	[261]
S. aureus	1,602 isolates from hospital- and community-acquired infections	0.5–8	[114]
S. aureus	198 clinical isolates (161 MRSA, 37 MSSA)	0.5–8	[152]
S. aureus	60 clinical MRSA isolates	0.5–8	[367]
S. aureus	45 clinical isolates (MSSA)	0.5–8	[229]
S. aureus	24 clinical MRSA isolates	0.5–8	[38]
S. aureus	240 clinical MRSA isolates	0.5–16	[406]
S. aureus	56 clinical isolates (MSSA)	0.5–60	[396]
S. aureus	20 burn unit isolates	0.8–4	[142]

(continued)

Table 13.2 (continued)

Species	Strains/isolates	MIC value (mg/l)	References
S. aureus	ATCC 25923, ATCC 29213, 3 MSSA, 10 MRSA, 2 VISA	<1–2	[127]
S. aureus	10 isolates from the urogenital tract of parturients	1	[60]
S. aureus	21 multidrug-resistant clinical isolates	1–4	[95]
S. aureus	82 isolates from nurses	1–8	[423]
S. aureus	24 strains with various resistance genes (qacA, qacB, qacC, qacG or norA)	1–8	[231]
S. aureus	45 clinical MRSA isolates	1–16	[229]
S. aureus	36 clinical isolates (MSSA)	1–64	[38]
S. aureus	29 clinical strains	1.5–3	[140]
S. aureus	CIP 53154	2	[233]
S. aureus	Clinical MRSA isolate	2	[186]
S. aureus	174 MRSA isolates from surveillance and clinical cultures	2–4	[180]
S. aureus	100 ST9 MRSA strains from porcine carcasses	2–4	[414]
S. aureus	28 isolates from automated teller machines	2–4	[424]
S. aureus	ATCC 25923	2.5	[11]
S. aureus	27 clinical MRSA isolates from patients with failed MRSA decolonization using PHMB	≤ 4	[202]
S. aureus	ATCC 6538	4	[418]
S. aureus	50 canine MRSA isolates	4–8	[63]
S. aureus	41 isolates from faecal samples (25 MSSA, 16 MRSA)	4–32	[10]
S. aureus	ATCC 27212	<4.7	[50]
S. aureus	30 isolates during daily CHG bathing (28 MRSA, 2 MSSA)	≤ 8	[223]
S. aureus	ATCC 6538	8	[186]
S. aureus	ATCC 6538	8.5	[108]
S. aureus	259 clinical isolates (95 MRSA, 164 MSSA)	10.3–20.7	[121]
S. aureus	ATCC 700698 (MRSA)	12	[418]
S. aureus	ATCC 25923	16	[9]
S. aureus	ATCC 25923	32	[425]
S. aureus	9 isolates from the oral cavity of children	64–256	[425]
S. aureus	ATCC 25923	71	[290]
S. aureus	ATCC 25923	2,500	[90]
S. capitis	MRBG 9.34 (skin isolate)	3.6	[108]

(continued)

Table 13.2 (continued)

Species	Strains/isolates	MIC value (mg/l)	References
S. capitis	Human skin isolate MBRG 9.34	7.8	[257]
S. caprae	MRBG 9.3 (skin isolate)	3.6	[108]
S. caprae	Human skin isolate MBRG 9.30	7.8	[257]
S. cohnii	Human skin isolate MBRG 9.31	10	[257]
S. epidermidis	30 isolates; 10 antibiotic-resistant skin isolates, 10 repeatedly found skin isolates during CHG washing, 10 skin isolates from healthy volunteers	0.5–2	[141]
S. epidermidis	10 isolates from the urogenital tract of parturients	0.5–2	[60]
S. epidermidis	10 clinical isolates	0.5–2	[293]
S. epidermidis	85 isolates; 23 isolates from scrub nurses, 52 isolates from patients (26 before and 26 after hospitalization), 4 isolates from joint replacement infections, 6 isolates from endoscopes	0.9–3.8	[346]
S. epidermidis	ATCC 12228	<1	[127]
S. epidermidis	Strain RP62A and 1 clinical isolate	2	[176]
S. epidermidis	25 strains from blood cultures	2–4	[145]
S. epidermidis	10 burn unit isolates	2–8	[142]
S. epidermidis	Human skin isolate M 9.33	7.8	[257]
S. epidermidis	ATCC 12228	11	[418]
S. epidermidis	MRBG 9.33 (skin isolate)	13.3	[108]
S. haemolyticus	MRBG 9.35 (skin isolate)	1.4	[108]
S. haemolyticus	Human skin isolate MBRG 9.35	13	[257]
S. hominis	Human skin isolate MBRG 9.37	13	[257]
S. hyicus	38 isolates from livestock	0.5	[1]
S. kloosii	Human skin isolate MBRG 9.28	7.8	[257]
S. lugdunensis	MRBG 9.36 (skin isolate)	0.9	[108]
S. lugdunensis	Human skin isolate MBRG 9.36	13	[257]
S. lugdunensis	11 clinical strains	31.2–62.5	[107]
S. pseudointermedius	98 canine isolates (49 MRSP, 49 MSSP)	0.5–4	[63]
S. pseudointermedius	43 MSSP and 57 MRSP isolates from canine pyoderma	0.5–4	[268]
S. pseudointermedius	25 MSSP and 25 MRSP from dogs with skin and soft tissue infections	4–16	[393]
S. saprophyticus	Human skin isolate MBRG 9.29	13	[257]
S. schleiferi	12 clinical strains	15.6–62.5	[107]
S. warneri	Human skin isolate MBRG 9.27	7.8	[257]
S. warneri	MRBG 9.27 (skin isolate)	29	[108]

(continued)

Table 13.2 (continued)

Species	Strains/isolates	MIC value (mg/l)	References
Staphyloccocus spp.[a]	51 clinical CNS isolates	0.25–2	[367]
Staphyloccocus spp.[a]	78 CNS isolates from automated teller machines	0.25–4	[424]
Staphyloccocus spp.[a]	28 CNS isolates from eyelids, eyelashes, and conjunctival sacs of lens wearers and spectacle wearers	0.5–2	[340]
Staphyloccocus spp.[a]	51 clinical CNS isolates (25 *S. epidermidis*, 18 *S. capitis*, 2 *S. haemolyticus*, 6 other CNS species)	0.5–8	[213]
Staphyloccocus spp.[a]	146 CNS isolates from nurses and the general population	0.5–32	[423]
Staphyloccocus spp.[a]	48 clinical CNS isolates	0.5–32	[38]
Staphyloccocus spp.[a]	69 clinical isolates (10 MRSA, 19 MSSA, 27 MRCNS, 13 MSCNS)	0.75–12	[153]
Staphyloccocus spp.[a]	2 methicillin-susceptible and 10 methicillin-resistant CNS isolates	<1	[127]
S. maltophilia	2 multidrug-resistant clinical isolates	<1–32	[127]
S. maltophilia	12 clinical isolates	2–32	[293]
S. maltophilia	MRBG 4.17 (kitchen drain biofilm isolate)	4.8	[108]
S. maltophilia	Domestic drain biofilm isolate MBRG 9.13	15.6	[257]
S. maltophilia	13 multidrug-resistant clinical isolates	16–512	[95]
S. maltophilia	2 blood culture isolates from oncology patients	30–175	[144]
S. agalactiae	ATCC 13813	0.5	[118]
S. intermedius	NCTC 11324	125	[356]
S. mitis	VT 842	4	[375]
S. mutans	424 isolates from saliva samples	0.25–1	[159]
S. mutans	ATCC 25175	0.3	[368]
S. mutans	28 clinical isolates, ATCC 25175, NCTC 10449	0.5–8	[250]
S. mutans	863 clinical isolates from 58 subjects during short term oral CHG treatment	≤ 1	[158]
S. mutans	Strain UA159	1	[381]
S. mutans	Strain UA159	2.5	[91]
S. mutans	MTCC 890	5	[11]
S. mutans	NCTC 10449	8	[356]
S. mutans	ATCC 25175	16	[9]
S. mutans	ATCC 25175	70	[392]
S. mutans	ATCC 27351	500	[89]

(continued)

Table 13.2 (continued)

Species	Strains/isolates	MIC value (mg/l)	References
S. pneumoniae	ATCC 49619	<1	[127]
S. pneumoniae	ATCC 49619	4	[186]
S. salivarius	44 strains from different origins	0.5–16	[15]
S. salivarius	ATCC 25975	2,000	[89]
S. sanguis	ATCC 10556 and strain 804	16–32	[205]
S. sanguis	NCTC 10904	125	[356]
S. sobrinus	53 clinical isolates from 58 subjects during short term oral CHG treatment	$\leq$2.0	[158]
S. sobrinus	ATCC 33478	8	[9]
S. thermophilus	135 strains from different origins	0.125–2	[15]
Streptococcus spp.[a]	43 Group B isolates from the urogenital tract of parturients	0.5–1	[60]
V. parvula	NCTC 11463	62	[356]
Y. enterocolitica	ATCC 9610 and 1 clinical isolate	4–8	[127]
Various species[a]	378 isolates from organic food	10–5,000	[106]

[a]No data per species; [b]depending on the media composition and plate material

E. coli and *K. pneumoniae*, 32 mg/l for *E. faecium* and *Salmonella* spp., 16 mg/l for *Enterobacter* spp. and 8 mg/l for *S. aureus* to determine CHG resistance [261], it becomes obvious that among all these species resistant or even highly resistant isolates have been detected already. Some MIC values appear very high. Variations of MIC values may be explained by the inoculum size (a lower inoculum results in lower MIC values), the media composition and plate material showing the need to standardize biocide susceptibility testing [41, 403]. In *S. epidermidis*, the MIC of biofilm cells is 4-fold higher [176].

13.3.1.2 Bactericidal Activity (Suspension Tests)

Four per cent CHG has sufficient bactericidal activity ($\geq$5.0 log) against almost all bacterial species within 3–5 min except *Enterococcus* spp. with a $\leq$2.4 log (Table 13.3). Two per cent CHG is bactericidal within 5 min against most bacterial species apart *E. faecium*, MRSA and *S. epidermidis*. At lower concentrations (0.5% or 0.02%), the spectrum of bactericidal activity is not comprehensive in 5 min. Insufficient bactericidal activity can be found against *Enterococcus* spp., *P. aeruginosa* and *S. aureus* with 0.5% CHG and against *Acinetobacter* spp. and *P. aeruginosa* with 0.02% CHG.

Table 13.3 Bactericidal activity of CHG in suspension tests

Species	Strains/isolates	Exposure time	Concentration	log_{10} reduction	References
A. baumannii	20 clinical strains	15 s	4% (P)	>5.0	[411]
A. baumannii	13 clinical strains	5 min	4% (P)	6.2	[96]
A. baumannii	9 clinical strains	1 min	2.2% (P)	>3.1	[235]
A. baumannii	81 clinical and environmental isolates	24 h	0.5% (P)	>5.0	[203]
A. baumannii	1 MDR clinical isolate	2 h	0.006% (S)	≥ 5.0	[12]
A. lwoffii	2 clinical strains	5 min	4% (P)	5.6	[96]
Acinetobacter spp.[a]	43 non-repetitive clinical isolates	10 s	0.1% (S)	3.2–4.6	[139]
		30 s		4.6–5.1	
		60 s		≥ 5.0	
Acinetobacter spp.[a]	19 multidrug-resistant clinical isolates	5 min	0.5% (S)	6.6	[95]
			0.1% (S)	3.3	
B. cenocepacia	LMG 16656	15 min	0.05% (S)	1.9	[289]
B. cenocepacia	LMG 18828	15 min	0.05% (S)	≥ 5.0	[289]
			0.015% (S)		
C. jejuni	ATCC BAA-1062, ATCC 33560 and 2 field strains	1 min	2% (S)	>6.0	[133]
E. cloacae	3 clinical strains	5 min	4% (P)	6.2	[96]
E. cloacae	1 clinical strain	2 min	0.01% (S)	2.6	[115]
Enterobacter spp.[a]	15 multidrug-resistant clinical isolates	5 min	0.05% (S)	6.8	[95]
			0.02% (S)	4.2	
E. faecalis	Strain Q33	5 min	2% (S)	5.0	[249]
E. faecalis	ATCC 29212	2 min	0.01% (S)	1.6	[115]
E. faecium	VRE strain Z31901	5 min	2% (S)	4.7	[249]
E. faecium	ATCC 6057	5 min	0.5% (P)	1.0–2.0	[264]
		60 min		≥ 6.0	
E. faecium	ATCC 6057	30 s	0.2% (P)	≥ 5.9	[294]
E. hirae	ATCC 10541	3 min	4%[b] (P)	≥ 5.0	[230]
Enterococcus spp.	11 clinical and surveillance isolates; 8 *E. faecium*, 2 *E. faecalis*, 1 *E. gallinarum*; 4 vanA-positive, 3 vanB-positive, 4 vancomycin-susceptible	30 s	4% (S)	0.1–0.2	[170]
			0.5% (S)	0.0–0.2	
		60 s	4% (S)	0.2–0.3	
			0.5% (S)	0.0–0.2	
		5 min	4% (S)	1.7–2.4	
			0.5% (S)	0.7–1.3	
Enterococcus spp.[a]	17 multidrug-resistant clinical isolates	5 min	0.05% (S)	6.8	[95]
			0.02% (S)		

(continued)

Table 13.3 (continued)

Species	Strains/isolates	Exposure time	Concentration	$\log_{10}$ reduction	References
Enterococcus spp.[a]	ATCC 13820 plus a gentamicin- and a vancomycin-resistant strain	30 s	0.01% (S)	0.0–0.4	[365]
			0.001% (S)	0.0–0.3	
		20 min	0.01% (S)	2.5–2.9	
			0.001% (S)	1.0–1.2	
		60 min	0.01% (S)	>6.0	
			0.001% (S)	3.1–3.3	
E. coli	NCTC 10536	3 min	4%[b] (P)	≥ 5.0	[230]
E. coli	17 clinical strains	5 min	4% (P)	6.3	[96]
E. coli	NCTC 10538	5 min	2% (S)	5.0	[249]
E. coli	ATCC 11229	5 min	0.5% (P)	≥ 6.0	[264]
E. coli	Clinical ESBL isolate	1 min	0.45% (S)	3.9	[124]
		15 min		≥ 5.9	
		1 min	0.25% (S)	2.1	
		15 min		≥ 5.9	
		1 min	0.05% (S)	0.9	
		15 min		2.8	
E. coli	NCTC 10536	30 s	0.2% (P)	≥ 5.9	[294]
E. coli	3 clinical isolates from urine with MIC values < 10 mg/l	10 min	0.05% (S)	4.4–5.9	[377]
E. coli	ATCC 25922	2 min	0.01% (S)	2.5	[115]
E. coli	ATCC 11229	30 min	0.01% (S)	≥ 3.0	[265]
E. coli	1 cefotaxime-resistant clinical isolate	2 h	0.006% (S)	≥ 5.0	[12]
E. coli	ATCC 25922	24 h	0.005% (P)	>5.0	[203]
F. nucleatum	NCTC 10562	5 min	0.02% (P)	5.5	[179]
H. pylori	NCTC 11637, NCTC 11916 and 7 clinical isolates	30 s	0.1% (P)	>5.0	[8]
			0.05% (P)		
		0.5–5 min	0.1% (P)	>5.0[c]	
H. parasuis	2 strains (serovars 1 and 5)	1 min	2% (S)	>6.0	[311]
				5.6[c]	
K. eflonia	Clinical ESBL isolate	1 min	0.45% (S)	≥ 5.8	[124]
		15 min		≥ 6.5	
		1 min	0.25% (S)	3.8	
		15 min		≥ 6.5	
		1 min	0.05% (S)	3.2	
		15 min		≥ 6.5	
K. eflonia	NCIMB 13291	5 min	0.02% (P)	4.4	[179]
K. oxytoca	5 clinical strains	5 min	4% (P)	6.1	[96]

(continued)

Table 13.3 (continued)

Species	Strains/isolates	Exposure time	Concentration	$\log_{10}$ reduction	References
K. pneumoniae	15 clinical strains	5 min	4% (P)	6.1	[96]
K. pneumoniae	DSM 16609	1 min	0.45% (S)	≥ 6.0	[124]
		15 min		≥ 6.2	
		1 min	0.25% (S)	4.8	
		15 min		≥ 6.2	
		1 min	0.05% (S)	3.1	
		15 min		≥ 6.2	
K. pneumoniae	3 clinical isolates from urine with MIC values between 10 and 200 mg/l	10 min	0.05% (S)	4.2–5.8	[377]
K. pneumoniae	1 clinical strain	2 min	0.01% (S)	2.4	[115]
Klebsiella spp.[a]	14 multidrug-resistant clinical isolates	5 min	0.1% (S)	7.1	[95]
			0.05% (S)		
P. gingivalis	ATCC 53978	5 min	0.02% (P)	4.2	[179]
P. stuartii	5 clinical isolates from urine with MIC values between 800 and 1,600 mg/l	10 min	2% (S)	>6.0	[377]
	9 clinical isolates from urine with MIC values between 200 and 1,600 mg/l	60 min	0.05% (S)	0.1–0.9	
	1 clinical isolate from urine with a MIC value of 1,600 mg/l	10 min	0.05% (S)	0.2	
P. mirabilis	2 clinical isolates from urine with MIC values of 800 mg/l	10 min	0.05% (S)	2.1–2.5	[377]
	1 clinical isolate from urine with a MIC value of 20 mg/l	10 min	0.05% (S)	5.3	
	1 clinical isolate from urine with a MIC values of 800 mg/l	60 min	0.05% (S)	>6.0	
P. mirabilis	Isolate with a MIC value of 1,280 mg/l	30 min	0.02% (S)	7.8	[361]
P. mirabilis	1 clinical strain	2 min	0.01% (S)	1.6	[115]
P. aeruginosa	ATCC 15442	3 min	4%[b] (P)	≥ 5.0	[230]
P. aeruginosa	20 clinical strains	5 min	4% (P)	6.2	[96]
P. aeruginosa	21 multidrug-resistant clinical isolates	5 min	4% (S)	6.2	[95]
			0.5% (S)	4.3	
			0.1% (S)	1.1	

(continued)

Table 13.3 (continued)

Species	Strains/isolates	Exposure time	Concentration	$\log_{10}$ reduction	References
P. aeruginosa	NCIMB 10421	5 min	2% (S)	5.0	[249]
P. aeruginosa	ATCC 15442	1 min	0.5% (S)	4.9	[186]
		5 min	0.05% (S)	5.1	
		60 min	0.01% (S)	5.3	
P. aeruginosa	ATCC 15442	5 min	0.5% (P)	3.0–4.0	[264]
		60 min		4.0–5.0	
P. aeruginosa	ATCC 15442	30 s	0.2% (P)	1.2–7.1[d]	[294]
		1 min		3.0–7.1[d]	
		10 min		≥ 6.5	
P. aeruginosa	3 clinical isolates from urine with MIC values between 200 and 800 mg/l	10 min	0.05% (S)	2.7–4.9	[377]
	1 clinical isolate from urine with a MIC value of 800 mg/l	60 min	0.05% (S)	3.1	
P. aeruginosa	ATCC 15442	5 min	0.02% (S)	6.3	[95]
P. aeruginosa	ATCC 27853	2 min	0.01% (S)	2.5	[115]
P. aeruginosa	1 clinical isolate	2 h	0.006% (S)	≥ 5.0	[12]
S. marcescens	1 clinical strain	2 min	0.01% (S)	1.6	[115]
S. aureus	ATCC 6538	3 min	4%[b] (P)	≥ 5.0	[230]
S. aureus	15 clinical MRSA isolates	5 min	4% (S)	7.0	[95]
			2% (S)	4.1	
			0.5% (S)	3.4	
S. aureus	ATCC 6538 and 2 clonally distinct MSSA clinical isolates	30 s	2% (P)	4.0	[171]
			2% (S)	6.9	
			0.5% (S)	1.5	
		1 min	2% (P)	5.9	
			2% (S)	6.9	
			0.5% (S)	2.7	
		5 min	2% (S)	6.9	
			0.5% (S)	5.2	
S. aureus	ATCC 43300 and 2 clonally distinct clinical MRSA isolates	30 s	2% (P)	1.9	[171]
			2% (S)	4.5	
			0.5% (S)	0.4	
		1 min	2% (P)	2.7	
			2% (S)	6.2	
			0.5% (S)	1.0	
		5 min	2% (S)	7.2	
			0.5% (S)	4.2	
S. aureus	NCTC 6571	5 min	2% (S)	5.0	[249]

(continued)

Table 13.3 (continued)

Species	Strains/isolates	Exposure time	Concentration	$\log_{10}$ reduction	References
S. aureus	MRSA strain 9543	5 min	2% (S)	5.0	[249]
S. aureus	6 clinical MSSA isolates	5 min	2% (S)	6.8	[95]
			0.5% (S)	3.9	
S. aureus	ATCC 6538	5 min	0.5% (P)	1.0–2.0	[264]
		60 min		≥ 6.0	
S. aureus	ATCC 6538	30 s	0.2% (P)	3.3–6.4[d]	[294]
		1 min		4.1–6.3[d]	
		10 min		≥ 6.2	
S. aureus	ATCC 6538	1 min	0.2% (S)	5.3	[186]
		5 min	0.1% (S)	5.1	
		10 min	0.05% (S)	5.1	
		60 min	0.01% (S)	4.9	
S. aureus	ATCC 6538	5 min	0.02% (S)	6.5	[95]
S. aureus	2 clinical isolates (MSSA and MRSA)	2 h	0.012% (S)	≥ 5.0	[12]
S. aureus	ATCC 25913	2 min	0.01% (S)	2.7	[115]
S. aureus	NCTC 6571 plus 2 MRSA strains	30 s	0.01% (S)	1.3–6.0	[365]
			0.001% (S)	0.0–0.3	
		20 min	0.01% (S)	3.8–6.0	
			0.001% (S)	1.7–3.4	
		60 min	0.01% (S)	>6.0	
			0.001% (S)	2.8–6.0	
S. aureus	ATCC 6538	30 min	0.0085% (S)	≥ 3.0	[265]
S. aureus	Strain MN8 (clinical MRSA isolate)	2 h	0.0058% (S)	5.2	[13]
			0.0029% (S)	2.7	
S. chromogenes	4 bovine mastitis isolates	30 s	0.52% (P)	≥ 5.0	[387]
S. epidermidis	Strain P69	5 min	2% (S)	4.5	[249]
S. epidermidis	Bovine mastitis isolate	30 s	0.52% (P)	≥ 5.0	[387]
S. epidermidis	1 clinical strain	2 min	0.01% (S)	3.7	[115]
S. epidermidis	1 MRSE clinical isolate	2 h	0.004% (S)	≥ 5.0	[12]
S. haemolyticus	Bovine mastitis isolate	30 s	0.52% (P)	≥ 5.0	[387]
S. simulans	3 bovine mastitis isolates	30 s	0.52% (P)	≥ 5.0	[387]
S. xylosus	Bovine mastitis isolate	30 s	0.52% (P)	≥ 5.0	[387]
S. maltophilia	2 clinical strains	5 min	4% (P)	6.4	[96]
S. maltophilia	13 multidrug-resistant clinical isolates	5 min	0.05% (S)	7.9	[95]
			0.02% (S)	6.5	
S. mutans	NCTC 10449	5 min	0.02% (P)	5.4	[179]

(continued)

Table 13.3 (continued)

Species	Strains/isolates	Exposure time	Concentration	$\log_{10}$ reduction	References
Mixed anaerobic species	*A. actinomycetemcomitans* ATCC 43718, *A. viscosus* DSMZ 43798, *F. nucleatum* ATCC 10953, *P. gingivalis* ATCC 33277, *V. atypica* ATCC 17744 and *S. gordonii* ATCC 33399	30 s	0.2% (P)	>8.0	[83]
			0.06% (S)	3.8	

S solution; *P* commercial product; [a]no data per species; [b]diluted to 55%; [c]with organic load; [d]depending on the type of organic load

Table 13.4 MBC values of various bacterial species to CHG (5 min exposure)

Species	Strains/isolates	MBC value	References
Acinetobacter spp.	43 non-repetitive clinical isolates	0.05%–0.15%[a]	[139]
E. coli	ATCC 25922	0.0025%[b]	[418]
K. pneumoniae	7 « Murray isolates » from the pre-chlorhexidine era, 7 modern isolates/strains	0.0002%–0.0512%	[42]
P. aeruginosa	ATCC 15442	0.003%	[128]
S. aureus	54 MRSA strains isolated in Canary black pigs	0.0078%–0.125%	[99]
S. aureus	42 clinical MRSA isolates	0.0032%–0.0512%	[275]

[a]30 s exposure time; [b]10 min exposure time

The highest MBC values were described in clinical isolates of *Acinetobacter* spp. (up to 0.15%, 30 s), in pig isolates of *S. aureus* (up to 0.125%, 5 min) and in recent time isolates of *K. pneumoniae* (up to 0.05%, 5 min; Table 13.4).

The overall results are supported by data showing that between 7 and 20 of 20 clinical strains from 7 bacterial species are not sufficiently killed by 0.2% CHG within 10 min [273]. The bactericidal activity of 0.2% CHG has been described to be independent of the pH values between 5 and 9 [409]. CHG is less effective in the presence of saliva [353]. The bactericidal efficacy of CHG at 0.09 and 0.36% may be significantly lower when the bacterial cells of *S. aureus* or *P. aeruginosa* used for the suspension test are grown on agar instead of broth, a difference that cannot be found at higher CHG concentrations [47]. It is also impaired in the presence of chondroitin sulphate [264].

13.3.1.3 Activity Against Bacteria in Biofilm

The bactericidal activity of CHG against bacteria in biofilms is variable (Table 13.5). Four per cent CHG did not reach 4.0 log within 24 h against three

Table 13.5 Efficacy of CHG in against bacteria in biofilms

Species	Strains/isolates	Type of biofilm	Exposure time	Concentration	log_{10} reduction	References
A. naeslundii	Strain 631	3-d incubation on ceramic hydroxylapatite slabs	30 min	0.14% (S)	Complete kill	[347]
				0.07% (S)	Incomplete kill	
A. viscosus	Strain M-100B	3-d incubation on ceramic hydroxylapatite slabs	30 min	0.14% (S)	Complete kill	[347]
				0.07% (S)	Incomplete kill	
A. actinomycetemcomitans	DSMZ 11123	90-min incubation in 96-well plates	24 h	0.05% (S)	1.5	[19]
B. cepacia	6 isolates from disinfectants and aerosol solution	5-d incubation on silicone discs	1 h	0.5% (S)	3.0	[254]
				0.1% (S)	<1.0	
C. jejuni	30 strains from chicken carcasses	48-h incubation in 96-well plates	24 h	1% (S)	≥ 5.1	[245]
E. faecalis	ATCC 29212	8-w incubation in straight-rooted teeth root canals	4 w	5% (P)	≥ 4.4	[82]
E. faecalis	ATCC 29212	3-w incubation on pieces of cellulose nitrate membranes	1 s	2% (P)	"complete elimination"	[122]
E. faecalis	ATCC 29212	3-w incubation in single-rooted teeth canals	30 s	2% (P)	1.5	[380]
			1 min		1.6–1.7	
			5 min		2.2–2.3	
E. faecalis	ATCC 29212	24-h incubation min 48-well plates	1 min	2% (S)	1.5	[224]
				0.2% (S)	0.4	
E. faecalis	ATCC 29212	7-d incubation in root canals	2 min	2% (P)	0.5–0.6	[57]

(continued)

Table 13.5 (continued)

Species	Strains/isolates	Type of biofilm	Exposure time	Concentration	log_{10} reduction	References
E. faecalis	Strain A197A	3-w incubation in root samples	3 min	2% (P)	>7.0	[130]
E. faecalis	ATCC 29212	48-h incubation in canals of single-rooted teeth	5 min	2% (S)	3.0	[398]
E. faecalis	ATCC 29212	4-w incubation in roots of sterile teeth	10 min	2% (S)	0.3	[417]
E. faecalis	ATCC 29212	3-w incubation on dentin discs	10 min	2%[a] (S)	0.2	[51]
E. faecalis	Four species biofilm with *P. aeruginosa* PA01, *K. pneumoniae* NCTC 13368, *E. faecalis* NCTC 775, *S. aureus* ATCC 9144	24-h incubation on polypropylene	24 h	2% (S)	≥ 5.0	[384]
				1% (S)	3.9	
				0.1% (S)	2.2	
E. faecalis	Four species biofilm with *P. aeruginosa* PA01, *K. pneumoniae* NCTC 13368, *E. faecalis* NCTC 775, *S. aureus* ATCC 9144	24-h incubation on an artificial wound bed	24 h	0.5%[a] (S)	2.0	[384]
E. hirae	CIP 5855	48-h incubation on polypropylene, PVC and silicone	30 min	0.02% (S)	4.8–5.0	[233]
				0.002% (S)	3.2–5.0	
				0.0002% (S)	0.0–4.8	
E. coli	ATCC 25922	48-h incubation in microtiter plates	15 s	2% (S)	2.6	[375]
			30 s		2.8	
			60 s		3.0	
			15 s	0.05% (S)	2.4	
			30 s		2.6	
			60 s		2.8	

(continued)

Table 13.5 (continued)

Species	Strains/isolates	Type of biofilm	Exposure time	Concentration	$\log_{10}$ reduction	References
E. coli	Strain O157, isolate from food poisoning outbreak	8-d incubation on stainless steel	5 min	1% (S)	≥4.2	[389]
				0.1% (S)		
				0.05% (S)		
E. coli	Strain 416, urinary tract infection isolate	30-, 60- and 120-min incubation on silicone discs	30 min	0.02% (S)	0.5	[357]
			60 min		0.7	
			120 min		0.4	
E. coli	ATCC 25922	48-h incubation on polypropylene, PVC and silicone	30 min	0.02% (S)	4.9–5.0	[233]
				0.002% (S)	3.8–5.0	
				0.0002% (S)	0.0–5.0	
F. nucleatum	ATCC 25586	4-d incubation on glass slides	1 min	2% (P)	0.1	[18]
K. pneumoniae	Four species biofilm with *P. aeruginosa* PA01, *K. pneumoniae* NCTC 13368, *E. faecalis* NCTC 775, *S. aureus* ATCC 9144	24-h incubation on polypropylene	24 h	4% (S)	2.8	[384]
				2% (S)	1.6	
				1% (S)	1.7	
				0.1% (S)	1.1	
K. pneumoniae	Four species biofilm with *P. aeruginosa* PA01, *K. pneumoniae* NCTC 13368, *E. faecalis* NCTC 775, *S. aureus* ATCC 9144	24-h incubation on an artificial wound bed	24 h	0.5%[a] (S)	1.3	[384]
L. monocytogenes	6 strains from various sources	24-h incubation in polystyrene microtiter plates	60 min	5% (S)	≥6.1	[53]
				2% (S)	3.5	
				1% (S)	2.3	
M. morganii	ATCC 25830	48-h incubation on polypropylene, PVC and silicone	30 min	0.02% (S)	≥5.0	[233]
				0.002% (S)	3.0–5.0	
				0.0002% (S)	0.0–5.0	

(continued)

Table 13.5 (continued)

Species	Strains/isolates	Type of biofilm	Exposure time	Concentration	$\log_{10}$ reduction	References
P. gingivalis	Dual species biofilm with *P. gingivalis* W 83 and *S. gordonii* ATCC 10558	7-d incubation on ceramic hydroxyapatite discs	10 min	0.2% (S)	0.5–0.8	[206]
P. gingivalis	Dual species biofilm with *P. gingivalis* W 83 and *S. gordonii*	7-d incubation in a modified Robbins device	15 min	0.2% (S)	1.0	[31]
P. gingivalis	ATCC 33277	48-h incubation on titanium discs	24 h	0.1% (S)	0.1	[214]
P. aeruginosa	8 clinical isolates	24-h incubation on stainless steel, eflon and polyethylene	24 h	4% (P)	0.6	[349]
P. aeruginosa	Four species biofilm with *P. aeruginosa* PA01, *K. pneumoniae* NCTC 13368, *E. faecalis* NCTC 775, *S. aureus* ATCC 9144	24-h incubation on polypropylene	24 h	4% (S)	3.9	[384]
				2% (S)	2.4	
				1% (S)	2.0	
				0.1% (S)	0.8	
P. aeruginosa	ATCC 700928	24-h incubation in microplates	1 min	1% (S)	0.0	[383]
			5 min		0.0	
			60 min		0.4	
P. aeruginosa	Four species biofilm with *P. aeruginosa* PA01, *K. pneumoniae* NCTC 13368, *E. faecalis* NCTC 775, *S. aureus* ATCC 9144	24-h incubation on an artificial wound bed	24 h	0.5%[a] (S)	2.4	[384]
P. aeruginosa	ATCC 9027	16.5-, 40.5- and 64.5-h incubation on titan discs	1 min	0.2% (S)	1.1–2.9	[21]
				0.12% (S)	0.3–1.0	

(continued)

Table 13.5 (continued)

Species	Strains/isolates	Type of biofilm	Exposure time	Concentration	$\log_{10}$ reduction	References
P. aeruginosa	NCIME 10434	48-h incubation in biofilm reactor	4 h	0.1% (P)	>6.0	[148]
			24 h			
			4 h	0.01% (P)	>6.0	
			24 h			
P. aeruginosa	Environmental strain SG81	44-h incubation in polystyrene microtitre plates	30 min	0.1% (S)	1.3	[151]
P. aeruginosa	Environmental strain SG81	44-h incubation on silicone swatches	30 min	0.1% (S)	4.5	[151]
P. aeruginosa	ATCC 27853	48-h incubation on polypropylene, PVC and silicone	30 min	0.02% (S)	>5.0	[233]
				0.002% (S)	>5.0	
				0.0002% (S)	3.2–4.1	
P. aeruginosa	CIP 103 467	24- or 48-h incubation on glass slides	24 h	0.0125% (P)	2.2	[212]
S. Enteritidis	Isolate from food poisoning outbreak	8-d incubation on stainless steel	5 min	1% (S)	≥ 5.2	[389]
				0.1% (S)		
				0.05% (S)		
S. Typhimurium	ATCC 14028	3-d incubation on a 96-peg lid	1 min	0.5% (S)	3.3	[413]
				0.25% (S)	2.9	
				0.1% (S)	2.0	
			5 min	0.5% (S)	2.9	
				0.25% (S)	3.6	
				0.1% (S)	1.4	

(continued)

Table 13.5 (continued)

Species	Strains/isolates	Type of biofilm	Exposure time	Concentration	log_{10} reduction	References
S. aureus	AH 2547	Overnight incubation on porcine skin	4 lateral wipes with soaked pads	10% (S)	0.4	[407]
S. aureus	8 clinical MRSA isolates	24-h incubation on stainless steel, eflon and polyethylene	24 h	4% (P)	1.0	[349]
S. aureus	ATCC 25923	3-w incubation on pieces of cellulose nitrate membranes	1 s	2% (P)	"complete elimination"	[122]
S. aureus	ATCC 29213	48 h in microtiter plates	15 s	2% (S)	3.0	[375]
			30 s		3.1	
			60 s		3.1	
			15 s	0.05% (S)	2.8	
			30 s		3.0	
			60 s		3.2	
S. aureus	10 oral MRSA isolates, 18 bloodstream MRSA isolates	48-h incubation in microtitre plates	30 s–2 min	1% (P)	0.4	[350]
S. aureus	ATCC 6538	72-h incubation in microplates	1 min	1% (S)	0.8	[383]
			5 min		1.3	
			60 min		1.5	
S. aureus	Isolate from food poisoning outbreak	8-d incubation on stainless steel	5 min	1% (S)	≥4.2	[389]
				0.1% (S)		
				0.05% (S)		

(continued)

Table 13.5 (continued)

Species	Strains/isolates	Type of biofilm	Exposure time	Concentration	$\log_{10}$ reduction	References
S. aureus	Four species biofilm with *P. aeruginosa* PA01, *K. pneumoniae* NCTC 13368, *E. faecalis* NCTC 775, *S. aureus* ATCC 9144	24-h incubation on polypropylene	24 h	1% (S)	≥ 5.0	[384]
				0.1% (S)	2.9	
S. aureus	Four species biofilm with *P. aeruginosa* PA01, *K. pneumoniae* NCTC 13368, *E. faecalis* NCTC 775, *S. aureus* ATCC 9144	24-h incubation on an artificial wound bed	24 h	0.5%[a] (S)	≥ 8.0	[384]
S. aureus	ATCC 43387	16.5-, 40.5- and 64.5-h incubation on titan discs	1 min	0.2% (S)	≥ 3.5	[21]
				0.12% (S)	0.4–2.5	
S. aureus	Three veterinary wild type strains	24-h incubation on microscopy slides	3 s	0.1% (S)	0.2	[400]
S. aureus	CIP 53154	48-h incubation on polypropylene, PVC and silicone	30 min	0.02% (S)	>5.0	[233]
				0.002% (S)	≥ 5.0	
				0.0002% (S)	4.3– >5.0	
S. aureus	CIP 4.83	24- or 48-h incubation on glass slides	24 h	0.0125% (P)	2.1	[212]
S. capitis	CBS 517	24-h incubation in microtiter plates	30 s	2% (S)	0.2	[366]
S. chromogenes	4 bovine mastitis isolates	24-h incubation on pegs	0.5–2 min	0.52% (P)	≥ 5.0	[387]
S. epidermidis	Strain 9142	24-h incubation in microtiter plates	30 s	2% (S)	0.2	[366]
S. epidermidis	Bovine mastitis isolate	24-h incubation on pegs	≤ 0.5 min	0.52% (P)	≥ 5.0	[387]

(continued)

Table 13.5 (continued)

Species	Strains/isolates	Type of biofilm	Exposure time	Concentration	log_{10} reduction	References
S. haemolyticus	Bovine mastitis isolate	24-h incubation on pegs	≤ 0.5 min	0.52% (P)	≥ 5.0	[387]
S. simulans	3 bovine mastitis isolates	24-h incubation on pegs	≤ 0.5 min	0.52% (P)	≥ 5.0	[387]
S. xylosus	Bovine mastitis isolate	24-h incubation on pegs	≤ 0.5 min	0.52% (P)	≥ 5.0	[387]
S. mutans	NCTC 10449	3-d incubation on ceramic hydroxylapatite slabs	30 min	0.29% (S)	Complete kill	[347]
				0.14% (S)	Incomplete kill	
S. mutans	Strain JCM 5705	24-h incubation on a hydroxyapatite disc	4 min	0.2% (S)	0.4–1.7	[342]
S. mutans	Strain C180-2	24-h incubation on titanium discs	5 min	0.2% (P)	1.1	[280]
S. mutans	ATCC 25175	16.5-, 40.5- and 64.5-h incubation on titan discs	1 min	0.2% (S)	0.3–1.0	[21]
				0.12% (S)		
S. sanguis	Strain 804	2-d incubation on saliva-coated silicone discs	4 h	0.16% (S)	4.7	[205]
			24 h		≥ 5.0	
	ATCC 10556		4 h	0.016% (S)	2.8	
			24 h		>5.0	
S. sanguis	ATCC 10558	3-d incubation on ceramic hydroxylapatite slabs	30 min	0.14% (S)	Complete kill	[347]
				0.07% (S)	Incomplete kill	

(continued)

Table 13.5 (continued)

Species	Strains/isolates	Type of biofilm	Exposure time	Concentration	$\log_{10}$ reduction	References
S. mutans	ATCC 25175	24-h incubation on glass-based dishes	5 min	0.12% (P)	2.3	[402]
S. mutans	Strain UA 159	54-h incubation on glass microscope slides	5 times 1 min over 54 h	0.12% (S)	>4.0	[188]
S. mutans	DSM 20523	72-h incubation on titanium discs	30 min	0.1% (S)	1.8	[185]
S. mutans	DSMZ 20523	90-min incubation in 96-well plates	24 h	0.05% (S)	1.5	[19]
Mixed species	Polymicrobial samples from infected root canals	3-w incubation on teeth	3 min	2% (S)	0.0	[320]
Mixed species	Oral biofilm	12-h incubation in the oral cavity on titanium surfaces	1 min	0.2% (P)	"significant reduction"	[125]
Mixed species	*S. mutans* ATCC 25175, *S. aureus* ATCC 43387, *P. aeruginosa* ATCC 9027	16.5-, 40.5- and 64.5-h incubation on titan discs	1 min	0.2% (S)	0.0–1.0	[21]
				0.12% (S)	0.0–0.6	
Mixed species	Polymicrobial biofilm from saliva	48-h incubation on titanium discs	5 min	0.2% (P)	0.6	[280]
Mixed species	*P. gingivalis* ATCC 33277, *F. nucleatum* ATCC 10953, *A. actinomycetemcomitans* ATCC 43718 and *S. mitis* ATCC 12261	6-d incubation on cover slips	30 min	0.2% (S)	0.6	[251]
Mixed species	Natural biofilm from dental unit waterlines	Variable	2 d	0.2% (S)	1.0	[215]
Mixed species in oral biofilm	*S. oralis* ATCC 10557, *S. gordonii* ATCC 10558, and *A. naeslundii* ATCC 19039	20-h incubation in biofilm reactor	1 h	0.12% (S)	2.3	[72]

(continued)

Table 13.5 (continued)

Species	Strains/isolates	Type of biofilm	Exposure time	Concentration	$\log_{10}$ reduction	References
Mixed species	*S. gordonii* ATCC 10558, *P. gingivalis* ATCC 33277, *T. forsythia* ATCC 43037, *F. nucleatum* ATCC 25586, *A. naeslundii* ATCC 12104, and *P. micra* ATCC 33270	4-d incubation in 96-well plates	1 h	0.1% (S)	5.0	[162]
Mixed species	Subgingival plaque bacteria	Overnight incubation on titanium discs	30 min	0.1% (S)	0.7	[185]
Mixed species	Human saliva bacteria	72-h incubation on titanium discs	30 min	0.1% (S)	0.3	[185]
Mixed species	*S. aureus* strain 308 (MRSA), *C. albicans* ATCC MYA 2876	48-h incubation in biofilm reactor	4 h	0.1% (P)	>5.0	[148]
			24 h			
			4 h	0.01% (P)	1.5	
			24 h		2.2	
Mixed species	*C. diversus* R25.1, *P. aeruginosa* R1811, *E. faecalis* R812, all urinary catheter isolates	48-h incubation on silicone discs	30 min	0.02% (S)	<1.0	[358]
			60 min		0.8–1.4	
			120 min		0.6–1.8	

P commercial product; *S* solution; [a]gauze soaked solution bacterial species (*K. pneumoniae*, *P. aeruginosa*, *S. aureus*). Two per cent CHG was mostly partially effective (<4.0 log) between exposure times of 0.5 min to 24 h (*E. faecalis*, *E. coli*, *F. nucleatum*, *K. pneumoniae*, *L. monocytogenes*, *P. aeruginosa*, *S. aureus*, *S. capitis*, *S. epidermidis* and mixed species) although a $\geq$4.0 log reduction was found within 24 h against *E. faecalis*. One per cent CHG was mostly partially effective within 24 h against some strains of isolates of *E. faecalis*, *K. pneumoniae*, *L. monocytogenes*, *P. aeruginosa*, *S. aureus* although a better bactericidal activity was found against strains or isolates of *C. jejuni*, *E. coli*, *S. Enteritidis* and *S. aureus* ($\geq$4.0 log in 5 min–24 h). 0.5% or 0.52% CHG was mostly not effective enough although a $\geq$4.0 log was found against some strains or isolates of *S. aureus*, *S. chromogenes*, *S. epidermidis*, *S. haemolyticus*, *S. simulans* and *S. xylosus*.

Other studies support the findings of a reduced susceptibility of bacterial cells in biofilm towards CHG [43]. Bacteria from a polymicrobial mixed subgingival biofilm samples revealed MIC values between 0.06 and 0.3% [222]. The MIC values of ten bacterial species of oral pathogens were between 2-fold and 128-fold higher in biofilm-grown cells compared to planktonic cells [9]. In suspensions, *P. aeruginosa* and *S. sanguis* biofilm cells required much longer exposure times than planktonic cells [252, 370]. One per cent CHG showed only a moderate inhibition effect (91%) on the metabolic activity in a MRSA biofilm within 3 min [131]. When an *E. faecalis* biofilm attached to dentin (5-day incubation) was irrigated for 3 min with 2% CHG, the dead cells in biofilm increased from 13.8 to 26.4% [14]. The rather small effect was explained by a retarded penetration of CHG into the biofilm. In a *S. mutans* biofilm, it was shown that CHG from a product containing 0.12% CHG has an average penetration velocity of 6 µm per min [402]. In a non-typable *H. influenzae* biofilm, it was shown that resistance to CHG is mediated to a large part by the cohesive and protective properties of the biofilm matrix [157].

The maturity of the biofilm has also an impact on the bactericidal effect of CHG. In a multispecies biofilm grown from plaque bacteria on collagen-coated hydroxyapatite discs for time periods ranging from 2 days to several months, it was shown that bacteria in mature biofilms and nutrient-limited biofilms are more resistant to CHG killing than in young biofilms [337]. In a biofilm maturated for 6 h, the efficacy of 0.2% CHG (1 min exposure time) on plaque vitality was significant but in the 48 h biofilm only found in the outer layer [421].

Regrowth within 24 h has been described in a *B. cenocepacia* biofilm after treatments with 0.015–0.05% CHG for 15 min indicating that *B. cenocepacia* biofilms are highly resistant to CHG [289]. The expression of many genes is affected when biofilm-grown *B. cenocepacia* J2315 cells are treated with chlorhexidine. Several genes encoding membrane-related and regulatory proteins, as well as several genes coding for drug resistance determinants (including RND and MFS efflux systems), were up-regulated. The down-regulation of a gene encoding an adhesin and the up-regulation of many genes encoding chemotaxis and motility-related proteins indicate that sessile cells try to escape from the biofilm [66]. In a polymicrobial biofilm incubated for 3 w on teeth obtained from samples from infected root canals 2% CHG applied for 3 min left 62.9% viable cells in the biofilm indicating the potential for regrowth [320].

It has been suggested that the active subpopulation in *P. aeruginosa* biofilms is able to adapt to exposure to membrane-targeting agents through the use of different genetic determinants, dependent on the specific membrane-targeting compound. Development of CHG-tolerant subpopulations was found to depend on the mexCD-oprJ genes, but does not depend on the pmr, mexAB-oprM, mexPQ-opmE or muxABC-opmB genes [58].

13.3.1.4 Bactericidal Activity in Alcohol-Based Hand Rubs

CHG can be found at 0.5% in alcohol-based hand rubs based on 70% iso-propanol or at 1% in hand rubs based on 61% ethanol. When applied for 3–5 min for surgical hand disinfection, no superior bactericidal efficacy was found after 3 h under a surgical glove (EN 12791) when compared to the reference procedure [173, 318, 319]. Its overall contribution to the bactericidal efficacy of alcohol-based hand rubs is therefore very questionable [172].

13.3.1.5 Bactericidal Activity in Antiseptic Soaps

When applied for hygienic hand wash and tested in analogy to EN 1499, soaps based on 4% CHG and applied with 2 or 5 ml for 30 s reduced *S. marcescens* by 2.3–2.8 log, an effect which was not superior to non-medicated soap with 2.3 log [277]. Against *M. luteus* the effect was lower with 1.9–2.3 log but superior to simple soap with 1.5 log [277]. Against *E. coli* the efficacy was between 2.9 and 3.0 (2 × 3 ml for 1 min) which was not superior compared to plain soap with 3.0 log [26, 315, 317].

When applied for an antiseptic hand wash and tested according to ASTM E 1174 with *S. marcescens*, 3 ml soap based on 0.75–4% CHG revealed an effect between 1.9 and 2.0 log with a 10 s application. Larger volumes of 4% CHG soap such as 5 ml or 2 × 3 ml were slightly more effective in 30 or 60 s with 2.4–2.6 log [25, 167, 343].

When used for surgical scrubbing and tested according to EN 12791, soaps based on 4% CHG mostly showed a poor bactericidal immediate effect on the resident hand flora with 0.8–1.1 log (application times between 1 and 3 min). The 3 h efficacy was also low with 0.5–0.8 log [22, 174, 230, 316].

And when used for surgical scrubbing and tested according to ASTM E 1115, soaps based on 4% CHG applied for 6 min (2 × 5 ml à 3 min) or for 10 min (unknown volume for 2 × 5 min) showed an immediate efficacy against the resident hand flora between 1.4 and 1.9 log [147, 175, 226, 263, 352]. A higher reduction with 2.6 log was only found when neutralization was omitted during sampling resulting in false positive efficacy data [168, 175].

13.3.1.6 Bactericidal Activity in Carrier Tests

In carrier tests, CHG at 0.02% or 0.5% was not able to kill all ten bacterial species (*S. aureus*, *S. pyogenes*, *S. viridians*, *S. faecalis*, *E. coli*, *K. pneumoniae*, *P. vulgaris*, *P. pyocyanea*, *C. diphteriae* and *M. phlei*) in 15 min on dried films. *S. aureus* resisted most from all species [135]. When *S. aureus* is placed on a glass cup carrier and exposed to 0.0075% chlorhexidine, a 3.7 log is found after 1 min and >5.0 log after 10 min [35]. In a proposed test to determine the efficacy of wound antiseptics (which is similar to a carrier test), 0.05% CHG showed sufficient bactericidal activity within 10 h (no organic load) and within >24 h in the presence of organic load [329]. Against *L. innocua* and *L. monocytogenes*, 4% CHG was very effective within 1 min in a carrier test with >6.0 log [32]. Against seven strains from six

Table 13.6 Efficacy of CHG in against bacteria on skin

Species	Strains/isolates	Exposure time	Concentration	$\log_{10}$ reduction	References
E. coli	Strain Vogel	15 min	0.0036% (S)	6.1	[331]
			0.00045% (S)	1.3	
K. pneumoniae	Strain SWRI no. 87	15 min	0.0036% (S)	5.0	[331]
			0.00045% (S)	2.6	
			0.00009% (S)	0.4	
P. mirabilis	Strain MGH-1	15 min	0.0036% (S)	4.3	[331]
			0.00045% (S)	0.3	
P. aeruginosa	ATCC 15442	2 h	0.144% (S)	0.0	[266]
		4 h		0.8	
		24 h		1.6	
P. aeruginosa	ATCC 9027	15 min	0.0036% (S)	5.9	[331]
			0.00045% (S)	4.0	
			0.00009% (S)	1.6	
S. marcescens	ATCC 8195	15 min	0.0036% (S)	5.9	[331]
			0.00045% (S)	3.9	
			0.00009% (S)	0.2	
S. aureus	ATCC 6538	15 min	1% (S)	0.0–0.1[a]	[325]
		1 min		2.1[b]	
		5 min		3.0[b]	
		15 min		3.0[b]	
		15 min		1.9[c]	
S. aureus	ATCC 6538	2 h	0.144% (S)	1.2	[266]
		4 h		3.3	
		24 h		≥5.6	
S. aureus	ATCC 6538	15 min	0.0036% (S)	2.1	[331]
			0.00045% (S)	0.7	
			0.00009% (S)	0.0	
S. epidermidis	ATCC 17917	15 min	0.0036% (S)	5.2	[331]
			0.00045% (S)	5.2	
			0.00009% (S)	1.8	
S. pyogenes	ATCC 12384	15 min	0.0036% (S)	3.4	[331]
			0.00045% (S)	2.7	
			0.00009% (S)	0.7	

P commercial product; *S* solution; [a]dry contamination; [b]broth contamination; [c]dry contamination wetted with 5 µl water bacterial species (*E. faecalis*, *E. faecium* VRE, *E. coli*, *P. aeruginosa*, *S. aureus*, MRSA, *S. epidermidis*) the efficacy of 2% CHG on glass carriers was rather poor with 1.0–4.0 log in 1 min [249]. 0.5% CHG revealed a good bactericidal activity on stainless steel discs against ten strains of MSSA (3.8 log in 10 min), ten strains of MRSA (3.1 log in 10 min), ten strains of VSE (3.6 log in 7 min) and nine strains of VRE (3.4 log in 7 min) [39].

13.3.1.7 Bactericidal Activity on Skin

The efficacy against bacteria on skin is quite good even at low CHG concentrations, especially against selected Gram-negative species (Table 13.6).

Some studies suggest that CHG at 0.5 and 2% adds to the overall bactericidal activity of alcohols on the skin, but doubts have been raised especially since the identity of one formulation has been revised postpublication [4, 5, 166, 225]. Clinical data suggest that 2% CHG significantly contributes to the prevention of surgical site infections and catheter-associated bloodstream infection when used in alcohol-based skin antiseptics [77, 253, 388]. The overall evidence for surgical site infections, however, is questioned by some authors due to the limitations of study selection criteria and concerns regarding the applied controls in the studies [227, 228].

13.3.1.8 Activity Against Bacteria on Mucosa

Some studies addressed the efficacy of CHG in the oral cavity. An antiseptic mouth rinse based on 0.2% CHG was effective against two oral pathogens (*S. mutans* and *F. nucleatum*) [312]. 0.2% CHG was also quite effective within 1 min for disinfection of titanium implants contaminated with *S. sanguinis* but not against *S. epidermidis* [52]. Infected root canals from teeth with apical periodontitis were irrigated with 0.12% CHG. The mean bacterial cells' count was reduced by 3.6 log, with 8 of 16 canals yielding negative cultures [345].

On porcine vaginal mucosa, 3% CHG was able to reduce an artificial contamination of MRSA by approximately 2.0 log (15 min) to 5.0 log (24 h) [13].

13.3.2 Fungicidal Activity

13.3.2.1 Fungistatic Activity (MIC Values)

The CHG MIC values for *C. albicans* vary between 0.003 and 4,140 mg/l with the majority of them being below the proposed epidemiological cut-off value of 16 mg/l [261]. Similar results were found with the variety of other *Candida* spp. with most MIC values $\leq$ 8 mg/l with the exception of selected isolates of *C. krusei* ($\leq$400 mg/l), *C. tropicalis* ($\leq$75 mg/l) and *C. dubliniensis* ($\leq$ 15.6 mg/l). All other fungal species were described with MIC values between 0.1 and 32 mg/l (Table 13.7).

Table 13.7 MIC values of various fungal species to CHG

Species	Strains/isolates	MIC value (mg/l)	References
Alternaria spp.	11 clean room isolates	2–8	[327]
A. flavus	3 clinical, 3 airborne and 2 food isolates	0.25–16	[165]
A. flavus	14 clean room isolates	2–4	[327]
A. flavus	16 clinical isolates	2–32	[416]
A. fumigatus	6 clinical and 14 airborne isolates	0.25–4	[165]
A. fumigatus	11 clean room isolates	4–16	[327]
A. niger	2 airborne and 2 food isolates	1–8	[165]
A. niger	11 clean room isolates	4–16	[327]
A. ochraceus	2 food isolates	4–16	[165]
A. terreus	4 clean room isolates	16	[327]
C. albicans	11 isolates from periodontal pockets of patients with chronic periodontitis	0.003–1.9	[344]
C. albicans	ATCC 18804	0.5–16	[330]
C. albicans	83 oral cavity isolates from HIV patients	0.5–16	[385]
C. albicans	200 worldwide strains from hospital- and community-acquired infections	0.5–64	[261]
C. albicans	Not described	<0.63	[104]
C. albicans	ATCC 90028 and 10 clinical isolates	0.8–1.6	[112]
C. albicans	4 isolates from local culture collection	1–4	[123]
C. albicans	NYCY 1363, strain 135BM2/94	2–5	[335]
C. albicans	20 clinical isolates from oropharyngeal candidiasis cases	2–7.8	[300]
C. albicans	28 isolates from the oral cavity of bone marrow transplant recipients	≤2.5–20	[379]
C. albicans	ATCC 90028 and 31 clinical isolates	3.1–6.3	[326]
C. albicans	ATCC 10231	4	[186]
C. albicans	1 clinical isolate	4	[88]
C. albicans	4 clinical isolates	4.4[b]	[244]
C. albicans	ATCC 90028	5	[11]
C. albicans	15 clinical isolates from patients with septicemia	6.25–12.5	[154]
C. albicans	2 clinical isolates	8	[193]
C. albicans	ATCC 10231	16	[9]
C. albicans	35 clinical isolates	16–128	[38]
C. albicans	ATCC 10231	107	[418]
C. albicans	2 clinical strains from oral candidiasis	250	[23]
C. albicans	1 clinical isolate	400	[201]
C. albicans	ATCC 29212	512	[16]
C. albicans	Not described	4,140	[417]

(continued)

Table 13.7 (continued)

Species	Strains/isolates	MIC value (mg/l)	References
C. dubliniensis	20 fluconazole-susceptible clinical isolates from oropharyngeal candidiasis cases	1–7.8	[300]
	20 fluconazole-resistant clinical isolates from oropharyngeal candidiasis cases	2–15.6	
C. dubliniensis	1 oral cavity isolate from a HIV patient	2	[385]
C. dubliniensis	4 clinical isolates	2.4[b]	[244]
C. dubliniensis	10 clinical isolates	3.1	[326]
C. glabrata	3 isolates from local culture collection	1–2	[123]
C. glabrata	1 isolate from periodontal pockets of patients with chronic periodontitis	1.9	[344]
C. glabrata	2 oral cavity isolates from HIV patients	2–4	[385]
C. glabrata	13 clinical isolates	3.1–6.3	[326]
C. glabrata	4 clinical isolates	3.7[b]	[244]
C. guilliermondii	6 clinical isolates	0.8–1.6	[326]
C. kefyr	Clinical isolate	0.8	[326]
C. krusei	ATCC 6258	0.8	[112]
C. krusei	4 isolates from local culture collection	1–2	[123]
C. krusei	4 clinical isolates	2.9[b]	[244]
C. krusei	ATCC 6258 and 4 clinical isolates	3.1	[326]
C. krusei	1 NCAC strain	150	[103]
C. krusei	ATCC 14243	400	[17]
C. lusitaniae	4 clinical isolates	2.0[b]	[244]
C. parapsilosis	4 oral cavity isolates from HIV patients	1–2	[385]
C. parapsilosis	3 isolates from local culture collection	1–2	[123]
C. parapsilosis	4 clinical isolates	2.2[b]	[244]
C. parapsilosis	6 clinical isolates	3.1	[326]
C. parapsilosis	2 clinical isolates	8	[193]
C. tropicalis	4 isolates from periodontal pockets of patients with chronic periodontitis	0.003–1.9	[344]
C. tropicalis	3 isolates from local culture collection	0.75–1.5	[123]
C. tropicalis	ATCC 750 and 5 clinical isolates	0.8–3.1	[326]
C. tropicalis	4 clinical isolates	1.7[b]	[244]
C. tropicalis	1 oral cavity isolate from a HIV patient	4	[385]
C. tropicalis	1 NCAC strain	75	[103]
Cladosporium spp.[a]	16 clean room isolates	2–8	[327]
Curvularia spp.[a]	16 clean room isolates	2–4	[327]
Exserohilum spp.[a]	4 clean room isolates	8–16	[327]
F. solani	24 clinical isolates	8–32	[416]

(continued)

Table 13.7 (continued)

Species	Strains/isolates	MIC value (mg/l)	References
Fusarium spp.[a]	10 clean room isolates	2–8	[327]
M. canis	10 isolates from dogs, cats and humans	12.5–25	[292]
Mucor spp.[a]	2 clinical and 1 food isolates	1–16	[165]
P. aurantiogriseum	Food isolate	2	[165]
P. chrysogenum	14 airborne isolates	0.12–1	[165]
P. citrinum	15 airborne isolates	0.12–1	[165]
P. paneum	2 food isolates	4–8	[165]
P. roquefortii	4 food isolates	8	[165]
Penicillium spp.[a]	15 clean room isolates	4–8	[327]
Rhizopus spp.[a]	2 clinical and 1 food isolate	0.5–4	[165]
S. cerevisiae	Clinical isolate	2	[244]
T. harzianum	1 clinical isolate	8	[190]
T. longibrachiatum	ATCC 201044, ATCC 208859, 12 clinical and 3 environmental isolates	1–8	[190]
Trichoderma spp.[a]	Food isolate	8	[165]
Various species[a]	8 clean room fungal isolates incl. *Aspergillus* spp., *Penicillum* spp., *Curvularia* spp., *Cladosporium* spp. and *Alternaria* spp.	4–16	[401]

[a]No number of isolates per species; [b]mean

13.3.2.2 Fungicidal Activity (Suspension Tests)

Two per cent CHG is effective against *C. albicans* and *C. auris* but requires >2 min exposure time. Against other fungi, 2% CHG is mostly effective in 30 min except for *A. fumigatus*. 0.5% CHG was found to be effective in 5 min against various fungi including *Candida* spp., *Crytococcus* spp. and *R. rubra* (Table 13.8).

Overall, the activity of CHG has been described to be somewhat lower against *C. albicans* (effective at 12.7 mg/l) compared to other *Candida* species such as *C. parapsilosis* (effective at 7.2 mg/l), *C. glabrata* (effective at 5.2 mg/l), *C. krusei* (effective at 3.4 mg/l) or *C. tropicalis* (effective at 1.8 mg/l) [123]. 0.5% CHG is still fungicidal in 5 min (log $\geq$ 6.0) in mixed suspensions of environmental isolates (*R. rubra*, *C. albicans*, *C. uniguttulatus*) and clinical isolates (*R. rubra*, *C. albicans*, *C. neoformans*) although the effect was smaller against the clinical mix [376]. The activity of CHG against *C. albicans* is explained by a loss of cytoplasmic components and a coagulation of nucleoproteins [40].

Table 13.8 Fungicidal activity of CHG in suspension test

Species	Strains/isolates	Exposure time	Concentration	$\log_{10}$ reduction	References
A. brasiliensis	ATCC 16404	30 min	2% (P)	≥4.0	[221]
A. flavus	Animal unit isolate	30 min	2% (P)	≥4.0	[221]
A. fumigatus	Animal unit isolate	30 min	5% (P)	5.3	[221]
			2% (P)	3.6	
A. fumigatus	15 clinical isolates	30–60 min	0.06% (P)	≥4.0	[382]
A. niger	Clinical isolate	30 min	0.4% (S)	≥4.0	[283]
A. terreus	2 clinical isolates	30 min	0.4% (S)	≥4.0	[283]
C. albicans	ATCC 10231	1 min	2% (P)	2.8	[256]
		1.5 min		3.3–3.4	
		2 min		3.4–3.6	
C. albicans	ATCC 10231	30 min	2% (P)	≥4.0	[221]
C. albicans	ATCC 10231	5 min	0.5% (P)	5.0	[264]
C. albicans	1 human and 1 environmental isolate	5 min	0.5% (S)	>7.0	[376]
C. albicans	ATCC 10231	1 min	0.5% (S)	3.8	[186]
		5 min	0.05% (S)	4.1	
		60 min	0.01% (S)	4.1	
C. albicans	ATCC 10231	30 min	0.275% (S)	2.9–6.4[a]	[189]
		24 h		5.9–6.4[a]	
		10 min	0.275% (P)	2.7–6.3[a]	
		30 min		3.3–6.3[a]	
		60 min		4.4–6.3[a]	
C. albicans	3 clinical strains	15 s	0.2%	2.0–2.5	[244]
		30 s		1.9–2.4	
		1 min		2.3–2.4	
		2 min		2.7–3.1	
C. albicans	ATCC 10231	30 s	0.2% (P)	1.1–5.5[a]	[294]
		1 min		2.0–5.6[a]	
		10 min		≥5.3	
C. albicans	NCPF 3179	5 min	0.02% (P)	4.0	[179]
C. auris	3 clinical strains	1 min	2% (P)	1.1–2.3	[256]
		1.5 min		1.2–2.5	
		2 min		1.2–2.6	
C. auris	12 clinical isolates	3 min	0.125%–1.25% (P)	>5.0	[2]
C. dubliniensis	3 clinical strains	15 s	0.2%	2.7–3.1	[244]
		30 s		2.7–3.4	
		1 min		3.0–3.6	
		2 min		3.4–4.4	

(continued)

Table 13.8 (continued)

Species	Strains/isolates	Exposure time	Concentration	$\log_{10}$ reduction	References
Candida spp.	6 clinical isolates (3 *C. albicans*, 2 *C. tropicalis*, 1 *C. parapsilosis*)	1 h	0.1% (S)	≥4.0	[283]
C. neoformans	1 clinical isolate	5 min	0.5% (S)	>7.0	[376]
C. uniguttulatus	1 clinical isolate	5 min	0.5% (S)	>7.0	[376]
M. pachydermatis	1 veterinary clinical isolate	3 min	0.00146% (P)	>4.0	[391]
			0.01172% (P)		
		10 min	0.00073% (P)	>4.0	
			0.00586% (P)		
R. rubra	1 clinical isolate	5 min	0.5% (S)	>7.0	[376]
T. rubrum	1 clinical isolate	30 min	0.2% (S)	≥4.0	[283]

S solution; *P* commercial product; [a]depending on the type of organic load

13.3.2.3 Activity Against Fungi in Biofilms

Some studies indicate that the yeasticidal activity of 0.019–4% CHG is poor within 15 min against *C. albicans* in biofilms except on cellulose nitrate membranes (Table 13.9). The resistance of *C. albicans* cells in a biofilm can be explained by subpopulations that exhibit relative levels of phenotypic resistance to CHG [364].

0.5% CHG also reduced in a variable but sufficient degree (≥4.0 log) *R. rubra*, *C. albicans*, *C. uniguttulatus* or *C. neoformans* in 5 min in 24 h biofilms [376]. In a 48 h or 72 h biofilm, the susceptibility of *C. albicans* and *C. parapsilosis* may increase up to 8-fold depending on the strain [193, 201].

It has been shown that *C. albicans* biofilms may harbour subpopulations with phenotypic resistance to CHG suggesting that biofilms incorporate protective niches [364]. In a mature *C. albicans* biofilm, surviving persisters form a multidrug-tolerant subpopulation. Interestingly, surviving *C. albicans* persisters were detected only in biofilms and not in exponentially growing or stationary-phase planktonic populations. Attachment rather than formation of a complex biofilm architecture initiates persister formation [196]. An analysis of 150 *Candida* isolates from cancer patients suggests that antimicrobial therapy (e.g. with amphotericin B) selects for high-persister strains in vivo and that biofilms of the majority of high-persister strains showed an increased tolerance to chlorhexidine [197].

13.3.2.4 Fungicidal Activity in Carrier Tests

When spores of *T. mentagrophytes* are placed on a glass cup carrier and exposed to 0.0075% chlorhexidine, a < 1.0 log is found after 1 and 10 min indicating a limited fungicidal activity [35].

Table 13.9 Efficacy of CHG against fungi in biofilms

Species	Strains/isolates	Type of biofilm	Exposure time	Concentration	$\log_{10}$ reduction	References
C. albicans	ATCC 26790	72-h incubation on silicone specimen	10 min	4% (S)	0.3	[129]
C. albicans	ATCC 10231D-5	3-w incubation on pieces of cellulose nitrate membranes	1 s	2% (P)	"complete elimination"	[122]
C. albicans	ATCC 10231	3-w incubation in single-rooted teeth canals	30 s	2% (P)	0.9	[380]
			1 min		1.1–1.4	
			5 min		1.4–1.5	
C. albicans	ATCC 90028	14-d incubation in canals of single-rooted human teeth	3 min	2% (P)	4.0	[97]
C. albicans	Not described	4-w incubation in roots of sterile teeth	10 min	2% (S)	0.3	[417]
C. albicans	1 clinical isolate	2-, 6-, 24- or 72-h incubation on polymethylmethacrylate acrylic denture discs	5 min	0.019% (S)	0.7–1.7	[201]
			15 min		0.4–≥2.0	

S solution; *P* commercial product

13.3.2.5 Fungicidal Activity for Other Applications

Two per cent CHG was found to have little efficacy to reduce an artificial *C. albicans* contamination on fingertips within 20 s with a mean of 2.0 log; simple non-medicated soap reached the same reduction [386]. Two per cent CHG has some effect (1.4 log) within 1 min for disinfection of titanium implants contaminated with *C. albicans* [52]. On skin CHG at 0.00045% or 0.0036% showed only poor efficacy against *C. albicans* in 15 min with 1.2 and 2.7 log, respectively [331].

13.3.3 Mycobactericidal Activity

13.3.3.1 Mycobactericidal Activity (Suspension Tests)

In suspension tests, the mycobactericidal activity of 0.5–4% CHG is overall poor within 2 h with the exception of *M. smegmatis* (Table 13.10). The poor mycobactericidal activity may be explained by an intracellular sealing by CHG at concentrations from 25 to 500 mg/l [111].

13.3.3.2 Mycobactericidal Activity in Carrier Tests

The mycobactericidal activity of 0.5–4% CHG in carrier tests is rather poor, with the exception of *M. smegmatis* (Table 13.11).

Table 13.10 Mycobactericidal activity of CHG in suspension tests

Species	Strains/isolates	Exposure time	Concentration	$\log_{10}$ reduction	References
M. avium	ATCC 15769	10, 60 and 120 min	0.5% (P)	No effect	[307]
M. kansasii	ATCC 12478	10, 60 and 120 min	0.5% (P)	No effect	[307]
M. smegmatis	Strain TMC 1515	1 min	4% (S)	>6.0	[33]
M. tuberculosis	Strain H37Rv	1 min	4% (S)	2.8–2.9	[34]
M. tuberculosis	Strain H37Rv	10, 60 and 120 min	0.5% (P)	No effect	[307]

P commercial product; *S* solution

Table 13.11 Mycobactericidal activity of CHG in carrier tests

Species	Strains/isolates	Exposure time	Concentration	$\log_{10}$ reduction	References
M. bovis	ATCC 35743	1 and 10 min	0.0075% (S)	<1.0	[35]
M. smegmatis	Strain TMC 1515	1 min	4% (S)	>6.0	[33]
M. tuberculosis	Strain H37Rv	1 min	4% (S)	2.0	[34]
Mycobacterium spp.	6 strains	20 min	0.5% (S)	<3.0	[143]

S solution

13.4 Effect of Low-Level Exposure

Many studies show that low-level exposure to CHG has different effects on bacteria (Table 13.12). No adaptive response was found in isolates or strains from 33 species (*A. baumannii*, *A. hydrophila*, *B. cereus*, *C. coli*, *C. jejuni*, *C. indologenes*, *Citrobacter* spp., *C. xerosis*, *C. sakazakii*, *E. saccharolyticus*, *E. coli*, *Eubacterium* spp., *K. pneumoniae*, *M. phyllosphaerae*, *M. luteus*, *M. osloensis*, *P. aeruginosa*, *P. nitroreductans*, *P. putida*, *Pseudoxanthomonas* spp., *S. multivorum*, *S. aureus*, *S. capitis*, *S. caprae*, *S. cohnii*, *S. epidermidis*, *S. haemolyticus*, *S. hominis*, *S. kloosii*, *S. lugdenensis*, *S. saprophyticus*, *S. warneri* and *S. mutans*).

Some isolates or strains of 25 species were able to express a weak adaptive response (MIC increase $\leq$ 4-fold) such as *A. xylosoxidans*, *A. jandaei*, *B. cereus*, *C. albicans*, *Chrysobacterium* spp., *C. pseudogenitalum*, *C. renale* group, *E. cloacae*, *Enterobacter* spp., *E. casseliflavus*, *E. faecalis*, *E. faecium*, *E. coli*, *H. gallinarum*, *K. pneumoniae*, *M. luteus*, *P. aeruginosa*, *S.* Typhimurium, *Serratia* spp., *S. aureus*, *S. capitis*, *S. haemolyticus*, *S. lugdenensis*, *S. warneri* and *S. maltophilia*.

A strong but unstable MIC change (>4-fold) was found in isolates or strains of five species (*B. cepacia*, *E. faecalis*, *E. coli*, *S. enteritidis* and *S.* Typhimurium). A strong and stable MIC change (>4-fold) was described for isolates or strains of eight species (*E. coli*, *K. pneumoniae*, *P. aeruginosa*, *S.* Virchow, *Salmonella* spp., *S. marcescens*, *S. aureus* and *S. maltophilia*). In isolates or strains of seven species (*A. baylyi*, *A. proteolyticus*, *E. coli*, *Pseudomonas* spp., *Ralstonia* spp., *S. marcescens* and *S. aureus*), the adaptive response was strong, but its stability was not described (Table 13.12).

Selected strains or isolates revealed substantial MIC changes: *E. coli* (up to 500-fold), *Salmonella* spp. (up to 200-fold), *S. marcescens* (up to 128-fold), *P. aeruginosa* (up to 32-fold), or *A. proteolyticus*, *K. pneumoniae*, *Pseudomonas* spp. and *S. aureus* (all up to 16-fold).

The highest MIC values after adaptation were all found in Gram-negative species such as 2,048 mg/l (*S. marcescens*), 1,024 mg/l (*P. aeruginosa*), >1,000 mg/l (*Salmonella* spp.), 700 mg/l (*B. cepacia complex*), >512 mg/l (*K. pneumoniae*) and 500 mg/l (*E. coli*). This is in line with findings showing that CHG has a significant hormetic effect with *P. aeruginosa* and a less significant effect with *S. aureus* resulting in greater bacterial growth [258]. Epidemiological cut-off values to determine resistance to CHG were proposed in 2014 for some Gram-negative species such as *E. coli* and *K. pneumoniae* (64 mg/l), *Salmonella* spp. (32 mg/l) and *Enterobacter* spp. (16 mg/l) [261]. Based on this proposal, the majority of *Salmonella* spp., *E. coli* and *K. pneumoniae* isolates would be classified as resistant to CHG after low-level exposure.

Cross-resistance to various antibiotics such as tetracycline, gentamicin or meropenem was found in some isolates of *B. fragilis*, *B. cepacia complex*, *Salmonella* spp. and *S. aureus*. In addition, a lower susceptibility to other biocidal agents was

Table 13.12 Change of bacterial susceptibility to biocides and antimicrobials after low-level exposure to CHG

Species	Strains/isolates	Type of exposure	Increase in MIC	MIC_{max} (mg/l)	Stability of MIC change	Associated changes	References
A. xylosoxidans	Domestic drain biofilm isolate MBRG 4.31	14 d at various concentrations	2-fold	31.2	No data	None reported	[257]
A. baumannii	Strain MBRG 15.1 from a domestic kitchen drain biofilm	14 passages at various concentrations	None	7.8	Not applicable	None reported	[73]
A. baylyi	Strain ADP1	30 min at 0.000001%	Protection from lethal CHG concentration (0.00007%)	No data	No data	More resistance to a lethal hydrogen peroxide concentration (1%)	[113]
A. hydrophila	Domestic drain biofilm isolate MBRG 4.3	14 d at various concentrations	None	15.6	Not applicable	None reported	[257]
A. jandaei	Domestic drain biofilm isolate MBRG 9.11	14 d at various concentrations	2-fold	15.6	No data	None reported	[257]
A. proteolyticus	Domestic drain biofilm isolate MBRG 9.12	14 d at various concentrations	16-fold	125	No data	None reported	[257]
B. cereus	MRBG 4.21 (kitchen drain biofilm isolate)	40 d at various concentrations	None	14.5	Not applicable	None described	[108]

(continued)

Table 13.12 (continued)

Species	Strains/isolates	Type of exposure	Increase in MIC	MIC_{max} (mg/l)	Stability of MIC change	Associated changes	References
B. cereus	Domestic drain biofilm isolate MBRG 4.21	14 d at various concentrations	None	1.9	Not applicable	None reported	[257]
B. cereus	Organic food isolate	Several passages with gradually higher concentrations	"significant increase"	No data	No data	Decreased tolerance to sodium propionate	[116]
B. subtilis	2 strains and 3 derivates	2 h at 0.00005%	No data	No data	Not applicable	No increase of transfer of the mobile genetic element Tn916, a conjugative transposon	[332]
B. fragilis	ATCC 25285	12 h at 0.06%	No data	No data	Not applicable	Induction of multiple antibiotic resistance[a]; 2.7-fold–6-fold increase of 6 efflux pumps	[299]
B. cepacia	ATCC BAA-245	40 d at various concentrations	8-fold	29	Unstable for 14 d	Decrease biofilm formation	[108]
B. cenocepacia	6 strains from clinical and environmental habitats	Up to 28 d at 15 mg/l	Survival	100	No data	No degradation of CHG	[7]
B. cepacia complex	*B. lata* strain 383	5 min at 50 mg/l	No data	700	Not applicable	Reduced susceptibility[b] to ceftazidime (30–33 mm), ciprofloxacin (11–20 mm) and imipenem (15–21 mm; 2 of 4 experiments) and to meropenem (33 mm; 1 of 4 experiments); up-regulation of transporter and efflux pump genes	[183]

(continued)

Table 13.12 (continued)

Species	Strains/isolates	Type of exposure	Increase in MIC	MIC_{max} (mg/l)	Stability of MIC change	Associated changes	References
C. coli	ATCC 33559 and a poultry isolate	Up to 15 passages with gradually higher concentrations	None	0.031	Not applicable	None described	[238]
C. jejuni	NCTC 11168, ATCC 33560 and a poultry isolate	Up to 15 passages with gradually higher concentrations	None	1	Not applicable	None described	[238]
C. albicans	Laboratory strain	12 w at various concentrations	4-fold	No data	No data	None described	[278]
C. indologenes	MRBG 4.29 (kitchen drain biofilm isolate)	40 d at various concentrations	None	7.3	Not applicable	None described	[108]
C. indologenes	Domestic drain biofilm isolate MBRG 9.15	14 d at various concentrations	None	31.2	Not applicable	None reported	[257]
Chrysobacterium spp.	Domestic drain biofilm isolate MBRG 9.17	14 d at various concentrations	2-fold	7.8	No data	None reported	[257]
Citrobacter spp.	Domestic drain biofilm isolate MBRG 9.18	14 d at various concentrations	None	1.9	Not applicable	None reported	[257]
C. pseudogenitalum	Human skin isolate MBRG 9.24	14 d at various concentrations	4-fold	3.9	No data	None reported	[257]

(continued)

Table 13.12 (continued)

Species	Strains/isolates	Type of exposure	Increase in MIC	MIC_{max} (mg/l)	Stability of MIC change	Associated changes	References
C. renale group	Human skin isolate MBRG 9.13	14 d at various concentrations	4-fold	31.2	No data	None reported	[257]
C. xerosis	WIBG 1.2 (wound isolate)	40 d at various concentrations	None	3.6	Not applicable	None described	[108]
C. sakazakii	Strain MBRG 15.5 from a domestic kitchen drain biofilm	14 passages at various concentrations	None	7.8	Not applicable	None reported	[73]
E. cloacae	Organic food isolate	Several passages with gradually higher concentrations	"significant increase"	No data	No data	Decreased tolerance to sodium nitrite	[116]
Enterobacter spp.	Organic food isolate	Several passages with gradually higher concentrations	"significant increase"	No data	No data	Decreased tolerance to sodium nitrite and sodium propionate	[116]
E. casseliflavus	Organic food isolate	Several passages with gradually higher concentrations	"significant increase"	No data	No data	Decreased tolerance to sodium nitrite and sodium propionate	[116]
E. faecalis	1 strain of unknown origin	14 passages at various concentrations	2-fold	7.8	Stable for 14 d	None reported	[73]

(continued)

Table 13.12 (continued)

Species	Strains/isolates	Type of exposure	Increase in MIC	MIC_{max} (mg/l)	Stability of MIC change	Associated changes	References
E. faecalis	Strain SS497	10 passages at various concentrations	3.7-fold	11	No data	Significant increase of surface hydrophobicity	[181]
E. faecalis	WIBG 1.1 (wound isolate)	40 d at various concentrations	6.7-fold	24.2	Unstable for 14 d	None described	[108]
E. faecium	VRE strain 410 (skin and soft tissue infection isolate)	21 d at various concentrations	4-fold	19.6	No data	Subpopulation with reduced susceptibility[c] to daptomycin including significant alterations in membrane phospholipids	[36]
E. faecium	Organic food isolate	Several passages with gradually higher concentrations	"significant increase"	No data	No data	None reported	[116]
E. faecium	3 vanA VRE strains	15 min at MIC	No data	No data	Not applicable	$\geq$ 10-fold increase of vanHAX encoding VanA-type vancomycin resistance and of liaXYZ associated with reduced daptomycin susceptibility; vanA up-regulation was not strain or species specific; VRE was more susceptible to vancomycin in the presence of subinhibitory chlorhexidine	[37]
E. saccharolyticus	Domestic drain biofilm isolate MBRG 9.16	14 d at various concentrations	None	1.9	Not applicable	None reported	[257]

(continued)

Table 13.12 (continued)

Species	Strains/isolates	Type of exposure	Increase in MIC	MIC_{max} (mg/l)	Stability of MIC change	Associated changes	References
E. coli	ATCC 25922	40 d at various concentrations	None	7.3	Not applicable	None described	[108]
E. coli	NCIMB 8879	6 × 48 h at variable concentrations	None	0.7	Not applicable	None reported	[378]
E. coli	ATCC 25922 and strain MBRG 15.4 from a domestic kitchen drain biofilm	14 passages at various concentrations	1.5-fold–5-fold	11.7	Stable for 14 d	None reported	[73]
E. coli	NCIMB 8545	0.00005% for 30 s, 5 min and 24 h	≤ 6-fold	39	Unstable for 10 d	No increase of MBC; unstable resistance[b] to tobramycin	[408]
E. coli	NCTC 8196	12 w at various concentrations	32-fold	No data	No data	None described	[278]
E. coli	NCTC 12900 strain O157	6 passages at variable concentrations	Approximately 500-fold	Approximately 500	Stable for 30 d	Increased tolerance[b] to triclosan (15 mm)	[46]
E. coli	CV601	24.4 µg/l for 3 h	No data	4.9	Not applicable	Induction of horizontal gene transfer (sulfonamide resistance by conjugation)	[163]
Eubacterium spp.	Domestic drain biofilm isolate MBRG 4.14	14 d at various concentrations	None	31.2	Not applicable	None reported	[257]

(continued)

Table 13.12 (continued)

Species	Strains/isolates	Type of exposure	Increase in MIC	MIC_{max} (mg/l)	Stability of MIC change	Associated changes	References
H. gallinarum	Domestic drain biofilm isolate MBRG 4.27	14 d at various concentrations	2-fold	31.2	No data	None reported	[257]
K. pneumoniae	7 "Murray isolates" from the pre-CHG era	Up to 5 w at various concentrations	None (6 isolates)	256	Stable for 10 d	None reported	[42]
			4-fold (1 isolate)				
K. pneumoniae	7 modern isolates/strains	Up to 5 w at various concentrations	4-fold–16-fold (6 isolates)	>512	Stable for 10 d	None reported	[42]
K. pneumoniae	ATCC 13883	40 d at various concentrations	6.9-fold	14.5	Stable for 14 d	Increase of biofilm formation	[108]
M. phyllosphaerae	Domestic drain biofilm isolate MBRG 4.30	14 d at various concentrations	None	15.6	Not applicable	None reported	[257]
M. luteus	MRBG 9.25 (skin isolate)	40 d at various concentrations	None	3.6	Not applicable	None described	[108]
M. luteus	Human skin isolate MBRG 9.25	14 d at various concentrations	2-fold	7.8	No data	None reported	[257]
M. osloensis	Strain MBRG 15.3 from a domestic kitchen drain biofilm	14 passages at various concentrations	None	2.0	Not applicable	None reported	[73]

(continued)

Table 13.12 (continued)

Species	Strains/isolates	Type of exposure	Increase in MIC	MIC_{max} (mg/l)	Stability of MIC change	Associated changes	References
P. aeruginosa	178 CHG sensitive strains	Exposure to CHG	None	625	Not applicable	None reported	[194]
P. aeruginosa	ATCC 9027	40 d at various concentrations	2-fold	14.5	Unstable for 14 d	None described	[108]
P. aeruginosa	ATCC 9027	14 passages at various concentrations	4-fold	31.3	Stable for 14 d	None reported	[73]
P. aeruginosa	NCIMB 10421	6 × 48 h at variable concentrations	7-fold	70	Stable for 15 d	High MICs to BAC did not change in a relevant extent	[378]
P. aeruginosa	NCTC 6749	12 w at various concentrations	8-fold–32-fold	1,024	Stable for 7 w	None described	[278]
P. nitroreductans	Domestic drain biofilm isolate MBRG 4.6	14 d at various concentrations	None	3.9	Not applicable	None reported	[257]
P. putida	Strain MBRG 15.2 from a domestic kitchen drain biofilm	14 passages at various concentrations	None	7.8	Not applicable	None reported	[73]
Pseudomonas spp.	Domestic drain biofilm isolate MBRG 9.14	14 d at various concentrations	16-fold	15.6	No data	None reported	[257]

(continued)

Table 13.12 (continued)

Species	Strains/isolates	Type of exposure	Increase in MIC	MIC_{max} (mg/l)	Stability of MIC change	Associated changes	References
Pseudoxanthomonas spp.	Domestic drain biofilm isolate MBRG 9.20	14 d at various concentrations	None	0.97	Not applicable	None reported	[257]
Ralstonia spp.	Domestic drain biofilm isolate MBRG 4.13	14 d at various concentrations	21-fold	167	No data	None reported	[257]
S. enteritidis	ATCC 13076	7 d of sublethal exposure	≥ 10-fold	>50	Unstable for 10 d	None reported	[306]
S. Typhimurium	Strain 14028S	5 min at 1 and 5 mg/l	3-fold–33-fold	1,000	Unstable for 1 d	2.5-fold–20-fold increase of tolerance[c] to BAC	[182]
S. Typhimurium	Strain SL1344	5 min at 0.1, 0.5, 1 and 4 mg/l	13-fold–27-fold	800	Unstable for 1 d	3-fold–67-fold increase of tolerance[c] to BAC	[182]
S. Virchow	Food isolate	6 passages at variable concentrations	Approximately 120-fold	Approximately 120	Stable for 30 d	Increased tolerance[b] to triclosan (0 mm)	[46]
Salmonella spp.	6 strains with higher MICs to biocidal products	8 days at increasing concentrations	50-fold–200-fold (2 strains)	>1,000	"stable"	One strain with increased tolerance[c] to tetracycline (>16 mg/l), chloramphenicol (8 mg/l) and nalidixic acid (16 mg/l)	[68]
S. marcescens	Strain GSU 86-823	7 d exposure to CHG-containing contact lens solutions	8-fold	50	No data	Increased adherence to polyethylene	[119]

(continued)

Table 13.12 (continued)

Species	Strains/isolates	Type of exposure	Increase in MIC	MIC_{max} (mg/l)	Stability of MIC change	Associated changes	References
S. marcescens	ATCC 13880	40 d at various concentrations	9.6-fold	116	Stable for 14 d	Increase of biofilm formation	[108]
S. marcescens	Clinical isolate	12 w at various concentrations	32-fold–128-fold	2,048	Stable for 7 w	None described	[278]
Serratia spp.	Not described	5–8 transfers	"resistance to CHG"	No data	"stable"	None described	[298]
S. multivorum	Domestic drain biofilm isolate MBRG 9.19	14 d at various concentrations	None	15.6	Not applicable	None reported	[257]
S. aureus	ATCC 6538	40 d at various concentrations	None	3.6	Not applicable	None described	[108]
S. aureus	ATCC 6538	100 d at various concentrations	None	0.6	Not applicable	None described	[410]
S. aureus	NCTC 6571 plus 2 MRSA strains	Several passages with gradually higher concentrations	1.3-fold–2-fold	1	"unstable"	None described	[365]
S. aureus	NCIMB 9518	0.00005% for 30 s, 5 min and 24 h	2-fold–5-fold	20	Stable for 10 d	No increase of MBC	[408]
S. aureus	ATCC 6538	7 d of sublethal exposure	2.5-fold	2.5	Unstable for 10 d	None reported	[306]
S. aureus	3 clinical MRSA strains	10 passages at various concentrations	$\leq$ 4-fold	8	No data	No change of PHMB susceptibility[d]	[305]

(continued)

Table 13.12 (continued)

Species	Strains/isolates	Type of exposure	Increase in MIC	MIC_{max} (mg/l)	Stability of MIC change	Associated changes	References
S. aureus	ATCC 6538	14 passages at various concentrations	4-fold	7.8	Unstable for 14 d	None reported	[73]
S. aureus	ATCC 25923 and 14 clinical isolates	14 d at various sublethal concentrations	4-fold–6-fold (6 isolates)	6.3	No data	Increased tolerance[c] to ciprofloxacin (4-fold–64-fold; 10 isolates), tetracycline (4-fold–512-fold; all isolates), gentamicin (4-fold–512-fold; 8 isolates), amikacin (16-fold–512-fold; 11 isolates), cefepime (8-fold–64-fold; 11 isolates) and meropeneme (8-fold–64-fold; 9 isolates)	[415]
S. aureus	NCTC 4163	12 w at various concentrations	16-fold	No data	No data	None described	[278]
S. aureus	Strain SAU3 carrying plasmid pWG613	10 min at 0.00005%	No data	No data	Not applicable	No significant reduction of plasmid transfer frequency	[288]
S. capitis	Human skin isolate MBRG 9.34	14 d at various concentrations	None	7.8	Not applicable	None reported	[257]
S. capitis	MRBG 9.34 (skin isolate)	40 d at various concentrations	1.7-fold	6	Stable for 14 d	None described	[108]
S. caprae	MRBG 9.3 (skin isolate)	40 d at various concentrations	None	3.6	Not applicable	None described	[108]

(continued)

Table 13.12 (continued)

Species	Strains/isolates	Type of exposure	Increase in MIC	MIC_{max} (mg/l)	Stability of MIC change	Associated changes	References
S. caprae	Human skin isolate MBRG 9.30	14 d at various concentrations	None	7.8	No data	None reported	[257]
S. cohnii	Human skin isolate MBRG 9.31	14 d at various concentrations	None	3.9	Not applicable	None reported	[257]
S. epidermidis	MRBG 9.33 (skin isolate)	40 d at various concentrations	None	9.7	Not applicable	None described	[108]
S. epidermidis	Human skin isolate M 9.33	14 d at various concentrations	None	7.8	Not applicable	None reported	[257]
S. epidermidis	CIP53124	1 d at various concentrations	No data	No data	Not applicable	Significant increase of biofilm formation at various sublethal concentrations	[150]
S. haemolyticus	Human skin isolate MBRG 9.35	14 d at various concentrations	None	15.6	Not applicable	None reported	[257]
S. haemolyticus	MRBG 9.35 (skin isolate)	40 d at various concentrations	2.1-fold	3	Unstable for 14 d	None described	[108]
S. hominis	Human skin isolate MBRG 9.37	14 d at various concentrations	None	7.8	Not applicable	None reported	[257]
S. kloosii	Human skin isolate MBRG 9.28	14 d at various concentrations	None	7.8	Not applicable	None reported	[257]

(continued)

Table 13.12 (continued)

Species	Strains/isolates	Type of exposure	Increase in MIC	MIC_{max} (mg/l)	Stability of MIC change	Associated changes	References
S. lugdunensis	Human skin isolate MBRG 9.36	14 d at various concentrations	None	15.6	Not applicable	None reported	[257]
S. lugdunensis	MRBG 9.36 (skin isolate)	40 d at various concentrations	4-fold	3.6	Stable for 14 d	None described	[108]
S. saprophyticus	Human skin isolate MBRG 9.29	14 d at various concentrations	None	3.9	Not applicable	None reported	[257]
S. warneri	MRBG 9.27 (skin isolate)	40 d at various concentrations	None	29	Not applicable	None described	[108]
S. warneri	Human skin isolate MBRG 9.27	14 d at various concentrations	2-fold	15.6	No data	None reported	[257]
S. maltophilia	Domestic drain biofilm isolate MBRG 9.13	14 d at various concentrations	4-fold	62.5	No data	None reported	[257]
S. maltophilia	MRBG 4.17 (kitchen drain biofilm isolate)	40 d at various concentrations	6-fold	29	Stable for 14 d	None described	[108]
S. mutans	Strain UA159	10 passages at various concentrations	None	3	Not applicable	None reported	[181]

[a]spiral gradient endpoint method; [b]disc diffusion method; [c]broth microdilution; [d] macrodilution method

described for *E. coli* and *S. Virchow* to triclosan, for *A. baylyi* to hydrogen peroxide and for *S. Typhimurium* to benzalkonium chloride.

Other adaptive changes include a significant up-regulation of efflux pump genes in *B. fragilis* and *B. cepacia complex*. Enhanced biofilm formation was described for *K. pneumoniae*, *S. marcescens*, *S. epidermidis*, and adherence to polyethylene was increased in *S. marcescens*. Biofilm formation was decreased in *B. cepacia*. VanA-type vancomycin resistance gene expression was increased vanA *E. faecium* ($\geq$ 10-fold increase of vanHAX encoding). Horizontal gene transfer (sulphonamide resistance by conjugation) was induced in *E. coli*. No significant reduction of plasmid transfer frequency was detected in *S. aureus*.

Exposure of seven species (*A. baumannii*, *C. sakazakii*, *E. faecalis*, *E. coli*, *P. aeruginosa*, *P. putida*, *S. aureus*) over 14 passages of 4 d each to increasing CHG concentrations on agar was associated with both increases and decreases in antibiotic susceptibility, but its effect was typically small relative to the differences observed among microbicides. Susceptibility changes resulting in resistance were not observed [109].

In the UK, the MIC values of 251 clinical isolates to CHG were opposed to the magnitude of CHG exposure from different types of antiseptics (CHG in water, soap or alcohol solutions). A clear correlation between the exposure and the mean MIC was found. In isolates obtained from patients with low exposure, the mean MIC was 10 mg/l; in moderate exposure, it was 15 mg/l; and in high exposure, it was 25 mg/l [38].

13.5 Resistance to Chlorhexidine

13.5.1 High MIC Values

As summarized in Table 13.2, the highest MIC values were described for *E. faecalis, K. pneumoniae, Proteus* spp. and *B. subtilis* ($\leq$ 10,000 mg/l), followed by *P. aeruginosa* ($\leq$ 5,000 mg/l), *L. monocytogenes*, *E. faecium* and *S. aureus* ($\leq$ 2,500 mg/l), *Streptococcus* spp. ($\leq$ 2,000 mg/l), *S. marcescens* ($\leq$ 1,024 mg/l), *Acinetobacter* spp., *Citrobacter* spp. and *Enterobacter* spp. ($\leq$ 1,000 mg/l), *B. cepacia* ($\leq$ 700 mg/l), *Achromobacter* spp. ($\leq$ 500 mg/l), *E. coli* ($\leq$ 312 mg/l), *A. baumannii* ($\leq$ 256 mg/l), *E. cloacae* ($\leq$ 150 mg/l), *Salmonella* spp. ($\leq$ 100 mg/l), and coagulase-negative *Staphylococcus* spp. ($\leq$ 62.5 mg/l). Taking into account the proposed epidemiological cut-off values such as 64 mg/l for *E. faecalis*, *E. coli* and *K. pneumoniae*, 32 mg/l for *E. faecium* and *Salmonella* spp., 16 mg/l for *Enterobacter* spp. and 8 mg/l for *S. aureus* to determine CHG resistance [261], it becomes obvious that among all these species resistant or even highly resistant isolates have been detected already.

Some studies provide evidence that MRSA is less susceptible to CHG compared to MSSA. In 1996, it was described that the mean MIC value for MRSA is 5-fold to 10-fold higher compared to MSSA [155]. This difference was confirmed in other studies. In Nigeria, 41 isolates were found with an MIC_{50} of 2 mg/l (25 MSSA isolates) and 32 mg/l (16 MRSA isolates) [10]. From four hospitals in Iran, it was reported that 30% of 100 MSSA isolates have an MIC between 8 and 16 mg, whereas the rate was 70% in 100 MRSA isolates [136]. A higher MIC value to CHG in *S. aureus*, however, does not necessarily mean an impaired efficacy in clinical applications against these isolates or strains [71, 301]. Nevertheless, after 20 years of using a 4% CHG liquid soap in Taiwan, it was observed that the proportion of MRSA isolates with an MIC value $\geq$ 4 mg/l increased from 1.7% in 1990 to 50% in 1995, 40% in 2000 and 46.7% in 2005 [406].

13.5.2 Reduced Efficacy in Suspension Tests

A reduced bactericidal activity (<5.0 log) for 4% CHG was described for some isolates of strains of *Enterococcus* spp. with $\leq$2.4 log in 5 min (Table 13.4). Two per cent CHG has also a reduced bactericidal activity in 5 min against some isolates or strains of MRSA. At 0.5%, a reduced efficacy in 5 min was partially observed with some isolates or strains of *Enterococcus* spp., *P. aeruginosa* and *S. aureus*. These data indicate that the minimum requirement for a bactericidal activity ($\geq$5.0 log) is not achieved with some isolates or strains indicating CHG tolerance or even resistance.

13.5.3 Resistance Mechanisms

In *S. Typhimurium*, it was shown that CHG elicits a broad range of effects on *Salmonella*, with an impact on central cellular processes including aerobic energy production and protein synthesis [69]. In a CHG-resistant *S. marcescens* strain, an additional protein was detected in the outer membrane with an unknown function [282]. Some specific mechanisms for different bacterial species are described below.

13.5.4 Resistance Genes

13.5.4.1 qacA/B

The qacA/B gene confers to CHG resistance [301]. The detection rates of qacA/B are described in Table 10.14 (Chap. 10 on benzalkonium chloride). MRSA strains carrying the qacA/B gene have been described to have a CHG MIC value of 256 mg/l in the presence of 3% bovine serum albumin [217]. In combination with low-level resistance to mupirocin, the presence of qacA/B in MRSA significantly increases the risk of persistent MRSA carriage after decolonization therapy [210]. qacA/B- and smr-positive *S. aureus* isolates were more often associated with invasive bloodstream infections [243]. The authors thought it may be reflective of the use of CHG in the cleansing and maintenance of central venous catheters and the subsequent selection of antiseptic-tolerant organisms [243]. Detection of the qac gene is associated with a significantly higher MBC for 4% CHG which was determined in 94 *S. aureus* isolates with 38 of them being healthcare-associated MRSA, 25 community-acquired MRSA, 6 VISA and 25 MSSA [348].

13.5.4.2 qacE

The detection rates of qacE are described in Table 10.16 (Chap. 10 on benzalkonium chloride). Detection of qacE or qacEΔ correlated with a reduced susceptibility to biocidal agents including chlorhexidine acetate [132]. In 122 multiresistant *A. baumannii* isolates in Malaysia, qacE was detected in 73%. The MIC values for CHG, however, were all between 0.2 and 0.6 mg/l indicating phenotypical susceptibility to CHG [20]. The presence of qacE in a *K. oxytoca* isolate from a diabetic foot ulcer correlated with a reduced susceptibility to CHG (MIC of 30 mg/l) [395]. In 64 *K. pneumoniae*, a close link between carriage of efflux pump genes, cepA, qacΔE and qacE genes and reduced chlorhexidine susceptibility was also described [3].

13.5.4.3 smr (QacC)

The detection rates of smr are summarized in Table 10.15 (Chap. 10 on benzalkonium chloride). In MRSA, presence of the smr gene is associated with a phenotypically reduced susceptibility to CHG. In one study, the MBC to CHG was determined in 88 MRSA isolates. Whenever the MBC was 5 mg/l smr was present in 15% of the isolates. In isolates with a MBC of 10 mg/l, the proportion was 28%, and in isolates with a MBC of 20 mg/l, the proportion was even 50% [220]. In another study with 400 isolates of *S. aureus,* a similar finding was reported. Whenever the smr gene was present, the mean MBC was 16 mg/l. In isolates without the smr gene, the mean MBC was 2 mg/l [242]. And in 69 *S. aureus* isolates collected in 14 months from wound swabs, the qacC gene was detected in 18.8% with eleven of them having a MIC value $\geq$ 1 mg/l [93]. Most of the smr-positive MRSA isolates detected in China were not susceptible to CHG when exposed under high organic load (3% bovine serum albumin), and a mean MBC of 256 mg/l was described [217].

13.5.4.4 Other Resistance Genes

In *E. faecium,* a putative two-component system was identified, composed of a putative sensor histidine kinase (ChtS) and a cognate DNA-binding response regulator (ChtR), which contributed to CHG tolerance in *E. faecium.* The tolerance to both chlorhexidine and bacitracin provided by ChtRS in *E. faecium* highlights the overlap between responses to disinfectants and antibiotics and the potential for the development of cross-tolerance for these classes of antimicrobials [134].

13.5.5 Cell Membrane Changes

The outer membrane can act as a barrier in *P. aeruginosa* to prevent chlorhexidine from entering the cell [128]. In *P. stutzeri*, alterations in the cell envelope were suspected to be responsible for resistance to chlorhexidine diacetate including changes in the outer membrane proteins and the expression of two additional protein bands [371–373]. Resistance to CHG in *P. stuartii* and *S. marcescens* is probably mediated by the inner membrane [156, 204]. In *E. coli* distribution studies for the absorbed CHG indicated that it must saturate a number of envelope targets before penetration to the cytosol is possible [56].

Biofilm-forming antimicrobial-resistant *D. acidovorans* strains have been isolated, including ones displaying resistance to CHG. Multiple mechanisms involving both the cell envelope (and likely TolQ, a disrupted gene and a component of the tolQRAB gene cluster known to be involved in outer membrane stability) and panmetabolic regulation play roles in chlorhexidine tolerance in *D. acidovorans* [303]. It has been suggested that a subpopulation of cells that do not accumulate CHG appears to be responsible for greater CHG resistance in *D. acidovorans* WT15 biofilm in conjunction with the possible involvement of bacterial membrane stability [302].

13.5.6 Efflux Pumps

In *K. pneumoniae*, the kpnGH efflux pump was described with a wide substrate specificity of the transporter including 14 antibiotics and CHG. kpnGH mediates antimicrobial resistance by active extrusion in *K. pneumoniae*. smvA is an efflux pump gene in *K. pneumoniae* which is up-regulated 10- to 27-fold in the presence of CHG. In five of six strains, adaptation to CHG also led to resistance to the last-resort antibiotic colistin. The potential risk of colistin resistance emerging in *K. pneumoniae* as a consequence of exposure to chlorhexidine has important clinical implications for infection prevention procedures [405]. cepA is associated with CHG resistance in *K. pneumoniae* and may act as a cation efflux pump [102]. The KpnEF efflux pump might help transport polysaccharides to the outer layer of bacterial cell to form the slimy layer and is possibly under additional regulation by other transcriptional factors involved in modulating capsular polysaccharide synthesis and biofilm formation in *K. pneumoniae* [354].

In *P. aeruginosa*, the activity of the MexCD-OprJ multidrug efflux pump is induced upon subinhibitory CHG exposure [260] and is considered a determinant for CHG resistance [110]. It is also induced by BAC but not by norfloxacin, tetracycline, chloramphenicol, streptomycin, erythromycin or carbenicillin, although they are substrates for the pump [259]. In *S. marcescens*, SdeAB was detected as an efflux pump [236].

In *Acinetobacter* spp., the AceI efflux pump is associated with CHG resistance [138]. AceR, a putative transcriptional regulator of the chlorhexidine efflux pump gene aceI in *A. baumannii*, is an activator of aceI gene expression when challenged with chlorhexidine [216]. CHG adversely modified the expression and function of the RND-type efflux pump AdeABC in biofilm-associated *A. baumannii* cells. Furthermore, CHG decreased the negative charges on *A. baumannii* cell membranes, causing dysregulation of the efflux pump and leading to cell death [191].

Multidrug efflux pumps, such as CmeABC and CmeDEF, are involved in the resistance of *Campylobacter* to a broad spectrum of antimicrobials including CHG [239]. A 19-fold up-regulation of the gene CIN01S_RS05745 encoding the HlyD-like periplasmic adaptor protein of a tripartite efflux pump of *C. indologenes* was observed upon exposure to 16 mg/l CHG [105]. In *E. faecalis* and *E. faecium*, the EfrAB efflux pump was detected conferring to resistance to CHG and triclosan [208].

The proteobacterial antimicrobial compound efflux (PACE) family of transport proteins was only recently described. PACE family transport proteins can confer resistance to a range of biocides used as disinfectants and antiseptics and are encoded by many important Gram-negative human pathogens [137].

13.5.7 Plasmids

Studies of resistance to antimicrobials have revealed that resistance genes are probably moving to plasmids from chromosomes more rapidly than in the past and resistance genes are aggregating upon plasmids [24]. Some resistance genes may be transferred between bacterial species. The qacA gene, for example, is often carried on a plasmid from the pSK1 family of vector [195, 211], but other plasmids may also carry the resistance gene and can be transferred [160]. Another example is the plasmid pTZ2162qacB which was able to transfer the qacB gene horizontally to MRSA by transduction [272].

Plasmid pC3, a non-essential megaplasmid which confers virulence and both antifungal and proteolytic activity on several strains, increases the resistance of *B. cenocepacia* H111 to various stresses (oxidative, osmotic, high-temperature and chlorhexidine-induced stresses) [6]. Plasmid pSAJ1 from a methicillin- and gentamicin-resistant strain of *S. aureus* conferred resistance to CHG and in addition to kanamycin, gentamicin, tobramycin, amikacin, benzalkonium chloride, acriflavine and ethidium bromide [419].

13.5.8 Class I Integrons

A class I integron was detected in 22 of 36 MDR *P. aeruginosa* isolates. Integron I-positive isolates showed reduced susceptibility to tested biocides including chlorhexidine gluconate. Class I integron may be responsible for generating MDR *P. aeruginosa* isolates with reduced susceptibility to biocides [164].

Table 13.13 Infections associated with tolerance to CHG

Bacterial species	Type and number of infections	Patient population	Source of infection and role of CHG tolerance	CHG concentration	References
A. xylosoxidans	4 cases of long-term intravascular catheter-related bacteremia	Haemodialysis unit	Contaminated solution of 2.5% CHG in an atomizer used for skin disinfection	2.5%	[374]
A. xylosoxidans	8 cases of infection (5 blood stream infections, 3 cerebrospinal infections); 44 cases of colonization	Neonatal care unit	Contaminated CHG solutions in reusable containers probably contaminated by their handling by healthcare workers	0.5%	[255]
A. xylosoxidans	11 cases of bacterial ventriculitis	Patients on a neurosurgical unit after craniotomy or trepanation	11 strains of *A. xylosoxidans* were identified from cerebrospinal fluid. Off them, 7 strains were able to survive in a solution of 2% CHG, 3 strains in 1% CHG and 1 strain in 0.1% CHG. The species was isolated from a container for disposal of surgical instruments which contained a 0.1% CHG solution. CHG at 0.1% was in addition used for surgical scrubbing and for the preoperative treatment of skin	0.1%	[341]
B. cepacia complex	46 cases of pseudobacteremia	Patients on 15 different nursing units	Contaminated solution of 0.5% CHG used for skin antisepsis; 10 of 38 sealed bottles were culture positive	0.5%	[184]
B. cepacia	2 cases of fulminant sepsis	Severely ill patients	Contaminated mouthwash based on 0.2% CHG; the source was the rubber tubing in the pharmacy through which deionized water passed during the dilution of concentrated CHG (5%)	0.2%	[351]

(continued)

Table 13.13 (continued)

Bacterial species	Type and number of infections	Patient population	Source of infection and role of CHG tolerance	CHG concentration	References
B. cepacia	3 cases of ventilator-associated respiratory tract infection, 10 cases of colonization (respiratory secretions)	3 ICUs	Contaminated mouthwash based on 0.12% CHG and used for oral care in ventilated patients	0.12%	[426]
P. mirabilis	88 cases of urinary tract infections and two other types of infection.	Patients on general medical and geriatric wards; 75% of the urinary tract infections were catheter-associated	The isolate was multidrug- and chlorhexidine-resistant. Resistance to CHG was assumed when an isolates was able to survive in 200 mg/l CHG. The outbreak was suspected to be caused by the widespread use of CHG for various types of antiseptic treatment including hand hygiene	Not applicable	[76]
P. pickettii	4 cases, 2 of them died of septicaemia	General hospital	Contaminated aqueous solution of 0.02% CHG caused by contaminated ion-exchange resin in the deionization cartridges	0.02%	[296]
S. marcescens	5 cases of infection (3 blood stream infections, 1 pneumonia, 1 ventriculitis) and 11 cases of colonization	Patients in accident and emergency, patients in ICUs	Contaminated commercial product with 2% CHG in water used for skin antisepsis prior to blood sampling and catheter insertion	2%	[81]
S. marcescens	1 case of bacteremia	Cardiovascular ICU	Contaminated 2% CHG stock solution used for skin cleansing	2%	[234]
S. marcescens	Twelve cases of bacteremia in ten patients; three patients died.	Patients on a paediatric oncology unit with Hickman-lines	The source of the outbreak was a container close to the patients filled with 0.5% CHG in water. The container was used to store clamps which were used during disconnection to avoid air uptake. The solution was renewed daily. The container, however, was neither cleaned nor processed	0.5%	[240]

(continued)

Table 13.13 (continued)

Bacterial species	Type and number of infections	Patient population	Source of infection and role of CHG tolerance	CHG concentration	References
S. aureus (MRSA)	One case of recurrent cutaneous abscess	Patient with a first cutaneous infection on the left knee followed by a similar infection nine weeks later on the left foot	The patient was in a study group that received 4% chlorhexidine soap for weekly showering. He had used chlorhexidine once or twice before his first episode and four or five times prior to the second episode. The first clinical isolate (PFGE type USA 300) was negative for the chlorhexidine resistance genes (qacA/B), but the second one was positive (also PFGE type USA 300)	Not applicable	[160]
S. aureus (MRSA)	517 patients admitted with an MRSA infection, 347 patients acquired an MRSA infection	Two intensive care units	MRSA carrier was treated with an antiseptic protocol: 1% CHG applied to nostrils, around the mouth, and at tracheostomy sites 4 times daily; 1% CHA applied daily to groin, axillae and skinfolds; 4% CHG for daily washing. The outbreak strain carried qacA/B genes and demonstrated 3-fold increased chlorhexidine MBCs in vitro. The antiseptic protocol reduced acquisition of non-outbreak MRSA strains by 70% but significantly increased transmission of the outbreak MRSA strain	Not applicable	[27]

(continued)

Table 13.13 (continued)

Bacterial species	Type and number of infections	Patient population	Source of infection and role of CHG tolerance	CHG concentration	References
S. epidermidis	92 cases of joint or wound infections; 27 additional isolates came from the skin of the chest prior to cardiac surgery	Patients with prosthetic joint infection (61) or surgical site infections after cardiac surgery (31)	Depending on the type of wound infection the rate of resistance to CHG varied between 7 and 68%. Resistance to CHG was often associated with the presence of the qacA/B gene	Not applicable	[297]
S. haemolyticus (MRSH)	42 clinical isolates; 15 from blood cultures, 14 from vascular catheters, 11 from tracheal tubes and 2 from cerebrospinal fluid. Eight neonates died from the infection.	Patients on a neonatal intensive care unit	Two isolates were detected in open bottles of a multiple use disinfectant based on 1% CHG and 0.2% QAC. The commercial product was used for hand washing. Both isolates were qacA/B carrier and considered to be CHG-resistant. Even in four new unopened bottles of the same product, different gram-negative species and *S. hominis* were identified	1%	[30]

13.5.9 Infections Associated with Tolerance to Chlorhexidine

Various outbreaks of various types of infection have been described caused by contaminated CHG solutions or products or by frequent use of CHG products (Table 13.13). Most of them were caused by Gram-negative bacterial species.

A statistically significant negative association between the intensity of chlorhexidine use in clinical services in a large acute-care hospital and the chlorhexidine susceptibility of selected micro-organisms isolated from patients hospitalized in those areas has been demonstrated in 2002 [38].

13.5.10 Bacterial Contamination of CHG Products or Solutions

Some reports indicated that liquid soaps or aqueous solutions based on CHG at up to 2% may be contaminated indicating an adaptive response or even resistance but without any associated infections. This type of contamination was so far only found with various types of Gram-negative bacterial species (Table 13.14). When a 2% CHG stock solution in plastic bottles was contaminated with *S. marcescens* at 10^8 CFU per ml, the species was able to survive for up to 27 months probably due to biofilm formation inside the bottles [234].

Table 13.14 Bacterial contamination of CHG products or solutions

Bacterial species	Type of product	CHG concentration	Use of product	Frequency/bacterial load	References
A. baumannii (pan-resistant)	Dispensers containing liquid soap	2%	Hand washing	3 of 28 samples[a]	[49]
Flavimonas spp.	Dispensers containing liquid soap	2%	Hand washing	5 of 28 samples	[49]
P. aeruginosa	CHG solution	No data	Use in paediatrics, neonatology and surgery	11 of 120 samples	[117]
P. aeruginosa (multiresistant)	Dispensers containing liquid soap	2%	Hand washing	5 of 28 samples[a]	[49]
P. fluorescens	Dispensers containing liquid soap	2%	Hand washing	5 of 28 samples	[49]
S. marcescens (ESBL)	CHG disinfectant solutions	No data	Various applications	3 samples	[100]
S. marcescens	CHG stock solution	2%	Skin antisepsis	25 samples; 10^8 CFU per ml	[234]

[a]One isolate was able to multiply in 1% CHG

13.6 Cross-Tolerance to Other Biocidal Agents

The overview in Table 13.12 on the adaptive response of various bacterial species shows that a cross-tolerance has been described to triclosan (*E. coli* and *S. Virchow*), BAC (*S. Tyhimurium*) and to hydrogen peroxide (*A. baylyi*). No cross-tolerance was found to BAC (*P. aeruginosa*) and PHMB (*S. aureus*).

13.7 Cross-Tolerance to Antibiotics

A possible cross-resistance between CHG and antibiotics is discussed controversially [321, 324]. The widespread use of CHG has not yet resulted in a clinically relevant resistance to antibiotics [199, 281] even though the development of resistance to these agents is regarded as realistic [322]. Such a development is more likely to occur in clinical medicine but not in industry because the selection pressure by these substances is much higher in patient care [200].

Some studies have described that there is no cross-resistance between CHG and antibiotics. Among 101 genetically distinct isolates of the *B. cepacia complex*, no correlation was found between the susceptibility to CHG and 10 different antibiotics [314]. In 130 *Salmonella* spp. from two turkey farms no cross-resistance between CHG and five antibiotics was found [28]. In 52 *Pseudomonas* spp. from meat chain production, no correlation between resistance to chlorhexidine and 16 different antibiotics was found [207].

Other studies indicate that cross-resistance between CHG and antibiotics does occur. An analysis of 701 Gram-negative strains in 1991, representing 16 species or bacterial genera, showed that there is a positive correlation between resistance to antiseptics (cetrimide, chlorhexidine, hexachlorophene) and to antibiotics for *S. marcescens* and *Alcaligenes* spp. [232]. In 49 *A. baumannii* strains with a reduced susceptibility to CHG, a co-resistance to carbapenem, aminoglycoside, tetracycline and ciprofloxacin was found [105]. In *B. fragilis*, multiple antibiotic resistance was induced by a 2.7-fold–6-fold increase of six efflux pumps [299]. In an *E. coli* strain, an unstable resistance to tobramycin was detected after low-level exposure to CHG for up to 24 h [408]. In a food isolate of *S. Virchow*, an increased resistance to tetracycline was described after exposure to CHG for six passages at variable concentrations [46]. In Trinidad, 11 of 120 CHG solutions were found to be contaminated with *Pseudomonas* spp., with resistance rates to ciprofloxacin of 58.3%, to norfloxacin of 50.0%, to tobramycin of 45.8% and to gentamicin with 41.7% [117]. In a CHG-resistant *P. stutzeri* isolate, a cross-resistance to polymyxin and gentamicin was found [373]. And in 6 *P. stutzeri* strains, cross-resistance to ampicillin (5 strains), polymyxin (4 strains), erythromycin (3 strains), nalidixic acid and gentamicin (2 strains) was found after low-level exposure to CHG for 6 w but no transfer of resistance [372].

And it is also remarkable that the highest median MIC values for CHG were reported in XDR *K. pneumoniae*, especially since in 2014 a multidrug efflux pump was detected in *K. pneumoniae* which can eliminate a variety of antibiotics and biocidal agents out of the bacterial cell [355]. kpnEF is one SMR-type efflux pump in *K. pneumoniae* which is directly involved in capsule formation causing hypermucoviscosity [354]. In addition, it may cause resistance to some antibiotics such as cefepime, ceftriaxone, colistin, erythromycin, rifampin, tetracycline and streptomycin, and some biocidal agents such as benzalkonium chloride, chlorhexidine and triclosan [354]. A correlation was described in 27 carbapenem-resistant clinical *K. pneumoniae* isolates between the presence of drug resistance genes (qacA, qacΔE, qacE and acrA) and a higher tolerance to killing or growth inhibition by disinfectants including chlorhexidine acetate [132]. In Japan, an outbreak of seven cases of catheter-associated urinary tract infection caused by multiresistant *P. aeruginosa* was analysed. The outbreak strain was resistant to CHG and at the same time resistant to 25 of 27 tested antibiotics, whereas a CHG-resistant ATCC strain did not show a resistance to the antibiotics [334]. An analysis of 148 *E. coli* isolates from clinical lesions showed that 12.8% were classified as resistant to CHG (MIC $\geq$ 5 mg/l), and they were also multiple drug-resistant and multiple metal-resistant [270]. Exposure of *Burkholderia* spp. to 0.005% CHG for 5 min resulted in a significant reduction of susceptibility to ceftazidime, ciprofloxacin and imipenem in two of four experiments although a clinical interpretation was not possible for the authors [183].

Cross-resistance has also been described in Gram-positive species. When healthcare workers used a soap based on 2% CHG they had a relative risk of 1.9 to be colonized on their hand with a *S. epidermidis* resistant to oxacillin and 1.5 for resistance to gentamicin. In *S. warneri*, the relative risk for rifampicin resistance was even 7.2 [70]. An analysis of 301 *S. aureus* isolates from three African countries showed a significant association between specific resistance genes for biocidal agents (sepA, mepA, norA, lmrS, qacAB, smr) and resistance to antibiotics [67]. Recent data show that exposure of vancomycin-resistant *E. faecium* to CHG for only 15 min up-regulates the vanA-type vancomycin resistance gene (vanHAX) and genes associated with reduced daptomycin susceptibility (liaXYZ) [37]. In another VRE strain, a subpopulation with reduced daptomycin susceptibility including significant alterations in membrane phospholipids was detected after 21 d of CHG exposure at various concentrations [36]. In another study, 120 clinical MRSA isolates were exposed to various concentrations of CHG (range: 2.5–40 mg/ml) which was allowed to dry in a glass bottle. Possible changes in the susceptibility to eight antibiotics (ampicillin, tetracycline, vancomycin, gentamicin, oxacillin, cefotaxime, cefuroxime and ciprofloxacin) were determined. MICs of cefotaxime, vancomycin, gentamicin, cefuroxime and oxacillin increased in EMRSA-16 following 48 h of residue drying. There were also increases in the MICs of all tested antibiotics for the NCTC 6571, a *S. aureus* susceptible strain, following exposure to chlorhexidine residues that had been drying for 48 h (compared with the MICs for the strain before exposure). The increases in the MICs of all tested antibiotics for the susceptible control *S. aureus* strain following exposure to surface dried chlorhexidine residues are of interest as it suggests that the use of chlorhexidine in the hospital environment may be linked to increased resistance to antibiotics in previously susceptible strains [397]. An analysis of 247 nosocomial *S. aureus* isolates revealed that smr-positive *S. aureus* isolates (44.0%) were more often resistant to methicillin, ciprofloxacin and/or clindamycin [243]. The isolates positive for qacA/B (33.6%) had more often a vancomycin MIC of ≥ 2 mg/l [243]. An analysis of multiresistance plasmids found in 280 staphylococcal isolates from diverse geographical regions from the 1940s to the 2000 s suggested that enormous selective pressure has optimized the content of certain plasmids despite their large size and complex organization [404]. In 1,632 clinical *S. aureus* isolates, a correlation of susceptibility profiles of at least 0.4 was found to CHG and ciprofloxacin [281]. An analysis of 1,632 human clinical *S. aureus* isolates from different geographical regions shows that a MIC value > 2 for CHG is associated with multidrug antibiotic resistance in *S. aureus* [65]. Finally, various changes of antibiotic susceptibility were described in 14 clinical isolates of *S. aureus* after CHG exposure over 14 d at various sublethal concentrations: a 4-fold to 512-fold increase of tetracycline MIC in all isolates, a 16-fold to 512-fold increase of amikacin MIC in 11 isolates, a 8-fold to 64-fold increase of cefepime MIC in 11 isolates, a 4-fold to 64-fold increase of ciprofloxacin MIC in 10 isolates, a 8-fold to 64-fold increase of meropenem MIC in nine isolates and a 4-fold to 512-fold increase of gentamicin MIC in eight isolates [415].

In *B. subtilis*, CHG did not increase the transfer of the mobile genetic element Tn916, a conjugative transposon [332]. But in *E. coli*, horizontal gene transfer (sulphonamide resistance by conjugation) was induced by low-level exposure to 24.4 mg/l CHG for only 3 h [163].

13.8 Role of Biofilm

13.8.1 Effect on Biofilm Development

The majority of studies indicate that 0.0002–0.2% CHG can significantly inhibit biofilm formation of *C. albicans*, *E. faecalis*, *E. coli*, *S. aureus*, *S. mutans* and mixed-species biofilms. Few studies with *S. enteritidis* and *S. mutans*, however, suggest no significant biofilm formation inhibition by CHG (Table 13.15). One other study shows that chlorhexidine at 0.12% in a solution with or without 11.6% ethanol used as a mouth rinse for 4 days had also some preventive effect on subgingival biofilm formation [328]. Medically relevant concentrations of CHG were tested on single cells in an *E. coli* biofilm, and adhesion to the biofilm increased with exposure to 1% CHG, but not for the lower concentrations tested [310].

Low-level CHG exposure enhanced biofilm formation in *K. pneumoniae*, *S. marcescens* and *S. epidermidis*, and adherence to polyethylene was increased in *S. marcescens*. Biofilm formation was decreased in *B. cepacia* (Table 13.12).

Table 13.15 Effect of CHG on biofilm development

Species	Strains/isolates	Type of biofilm	Exposure time	Type of product	Inhibition of biofilm formation	References
C. albicans	ATCC 90028	24-h incubation in microtiter plates	4 h	0.2% (P)	66%	[11]
C. albicans	Clinical strain from intravascular line culture	48-h incubation at one-fourth of MIC on silicone elastomer discs	Overnight incubation	0.0002% (S)	"significantly lower"[a]	[193]
E. faecalis	ATCC 29212	24-h incubation in microtiter plates	4 h	0.2% (P)	82%	[11]
E. coli	ATCC 25922	24-h incubation in microtiter plates	4 h	0.2% (P)	87%	[11]
S. Enteritidis	Outbreak strain UJ3197	Up to 24-h incubation on polystyrene microtiter plates	5, 10 and 24 h	0.05% (S)	None	[126]
S. aureus	ATCC 25923	24-h incubation in microtiter plates	4 h	0.2% (P)	76%	[11]
S. aureus	ATCC 25923 and 9 oral cavity isolates from children	24-h incubation on glass cover slip	24 h	0.0064%–0.0556% (S)	90%	[425]
S. mutans	MTCC 890	24-h incubation in microtiter plates	4 h	0.2% (P)	84%	[11]
S. mutans	Strain UA159	54-h incubation on glass microscope slides	5 times 1 min over 54 h	0.12% (S)	"no further increase of biofilm mass"	[188]
S. mutans	Strain UA159	24-h incubation in polystyrene plates	24 h	0.03% (S)	None	[54]
S. mutans	ATCC 25175	24-h incubation on polystyrene cell culture plates	24 h	$\geq$0.0002% (S)	$\geq$95%	[368]
Mixed species	*S. mutans* strain UA159 and *C. albicans* strain SC5314	67-h incubation on saliva-coated hydroxyapatite discs	4 × 2 h on 2 days	0.12% (S)	"significant biofilm inhibition"	[309]
Mixed species	Mixed oral flora	7-d incubation of CHG-coated polyglactin sutures 3-0 in saliva collected from 10 chronic periodontitis patients	7 d	Unknown concentration	"substantial biofilm inhibition"	[336]

S solution; *P* commercial product; [a]measured as dry weight

13.8.2 Effect on Biofilm Removal

Overall, 0.015–4% CHG has mostly poor biofilm removal activity as shown with *B. cenocepacia*, *C. albicans*, *P. aeruginosa*, *S. aureus*, *S. epidermidis* and mixed-species biofilms. Only single-species *S. mutans* biofilm was removed by 80–97% by a 5-min treatment with a 0.12% CHG product (Table 13.16). In addition, when an *E. faecalis* biofilm attached to dentin (5-day incubation) was irrigated for 3 min with 2% CHG, the biovolume was only marginally removed from 63.5 to 61.6 mm^3 [14].

Table 13.16 Biofilm removal rate (quantitative determination of biofilm matrix) by exposure to products or solutions based on CHG

Type of biofilm	Concentration	Exposure time	Biofilm removal rate	References
B. cenocepacia LMG 18828, 4 h adhesion and 20-h incubation in polystyrene microtitre plates	0.05% (S)	15 min	25%	[289]
	0.015% (S)		5%	
C. albicans ATCC 90028, 24-h incubation on acrylic resin specimens	4% (P)	10 min	No significant reduction	[75]
P. aeruginosa ATCC 700928, 24-h incubation in microplates	1% (S)	60 min	0%	[383]
P. aeruginosa environmental strain SG81, 44-h incubation in polystyrene microtitre plates	0.1% (S)	30 min	No significant reduction	[151]
P. aeruginosa environmental strain SG81, 44-h incubation on silicone swatches	0.1% (S)	30 min	No significant reduction	[151]
S. aureus ATCC 6538, 72-h incubation in microplates	1% (S)	60 min	0%	[383]
S. epidermidis ATCC 35984, 24-h incubation in a glass capillary reactor	0.1% (S)	15 min	21%	[48]
S. mutans (ATCC 35688 and 7 oral cavity strains), 48-h incubation on sterile discs in microtitre plates	0.12% (P)	5 min	80%–97%	[291]
Mixed species: root canals from human mandibular premolars	2% (S)	1 min	No biofilm removal	[64]
Mixed species (*A. naeslundii*, *L. salivarius*, *S. mutans* and *E. faecalis*), 3-w incubation on sterile dentin blocks	2% (S)	7 d	"partial disruption"	[59]
Mixed species from dental plaques, 21-d incubation on large-grit, acid-etched (SLA) titanium implants	1% (P)	2 min	No superior effect to rinsing	[92]

(continued)

Table 13.16 (continued)

Type of biofilm	Concentration	Exposure time	Biofilm removal rate	References
Mixed-species biofilm (*A. naeslundii* and *S. oralis*), incubated for 2 h on titanium	0.2% (P)	10 min	40%	[399]
Mixed-species biofilm from fresh human saliva incubated for 16 h on titanium	0.2% (P)	10 min	40%	[399]
Mixed-species biofilm (*A. naeslundii* and *S. oralis*), incubated for 16 h on titanium	0.2% (P)	10 min	65%	[399]
Mixed species in a natural biofilm from dental unit waterlines	0.2% (S)	2 d	"no effective biofilm removal"	[215]
Mixed-species biofilm (*S. oralis* ATCC 10557, *S. gordonii* ATCC 10558, *A. naeslundii* ATCC 19039), 20-h incubation in a biofilm capillary reactor	0.12% (P)	20 min	No evidence for removal or detachment	[369]
Mixed-species biofilm: *S. oralis* (ATCC 10557), *S. gordonii* (ATCC 10558) and *A. naeslundii* (ATCC 19039), 20-h incubation in a biofilm capillary reactor	0.12% (S)	1 h	No removal	[72]
Mixed species in a natural biofilm on dentures worn for 5–10 y	0.12% (S)	20 min over 21 d in addition to brushing	Significantly less denture coverage with biofilm	[80]
Mixed species (*S. gordonii* ATCC 10558, *P. gingivalis* ATCC 33277, *T. forsythia* ATCC 43037, *F. nucleatum* ATCC 25586, *A. naeslundii* ATCC 12104, and *P. micra* ATCC 33270), 4-d incubation in 96-well plates	0.1%	1 h	No biofilm removal	[162]

P commercial product; *S* solution

13.8.3 Effect on Biofilm Fixation

No studies were found on the effect of CHG on biofilm fixation. But 0.2% CHG leads to a contraction of a mixed mature oral biofilm of 1.176 µm per min along the z axis and affects viability profiles through the biofilm after a delay of 3–5 min. 0.05% CHG exhibited barely detectable changes after 5 min in total fluorescence measurements indicating little change of viability [149]. Medically relevant concentrations of CHG were tested on single cells in an *E. coli* biofilm, and cells exposed to 1 and 0.1% CHG more than doubled in stiffness, while those exposed to 0.01% showed no change in elasticity [310].

13.9 Summary

The principal antimicrobial activity of CHG is summarized in Table 13.17.

The key findings on acquired resistance and cross-resistance including the role of biofilm for selecting resistant isolates are summarized in Table 13.18.

Table 13.17 Overview on the typical exposure times required for CHG to achieve sufficient biocidal activity against the different target micro-organisms

Target micro-organisms	Species	Concentration	Exposure time
Bacteria	Most species except *Enterococcus* spp.	4%	3–5 min[a]
	Most species except *E. faecium*, MRSA and *S. epidermidis*	2%	5 min[a]
Fungi	*C. albicans*	2%	30 min
	Most other fungi except *A. fumigates*	2%	30 min
Mycobacteria	Poor against most mycobacteria except *M. smegmatis* (4%, 1 min)	4%	>2 h

[a]in biofilm the bactericidal activity will be lower

Table 13.18 Key findings on acquired CHG resistance, the effect of low-level exposure, cross-tolerance to other biocides and antibiotics, and its effect on biofilm

Parameter	Species	Findings
Elevated MIC values	*B. subtilis*, *E. faecalis*, *K. pneumoniae*, *Proteus* spp.	≤ 10,000 mg/l
	P. aeruginosa	≤ 5,000 mg/l
	L. monocytogenes, *E. faecium*, *S. aureus*	≤ 2,500 mg/l
	Streptococcus spp.	≤ 2,000 mg/l
	S. marcescens	≤ 1,024 mg/l
	Acinetobacter spp., *Citrobacter* spp., *Enterobacter* spp.	≤ 1,000 mg/l
	B. cepacia	≤ 700 mg/l
	Achromobacter spp.	≤ 500 mg/l
	E. coli	≤ 312 mg/l
	A. baumannii	≤ 256 mg/l
	E. cloacae	≤ 150 mg/l
	Salmonella spp.	≤ 100 mg/l
	Coagulase-negative *Staphylococcus* spp.	≤ 62 mg/l
Proposed MIC value to determine resistance	*C. albicans*	16 mg/l
	Enterobacter spp.	16 mg/l
	E. faecium	32 mg/l
	E. faecalis	64 mg/l
	E. coli	64 mg/l
	K. pneumoniae	64 mg/l
	Salmonella spp.	32 mg/l
	S. aureus	8 mg/l

(continued)

Table 13.18 (continued)

Parameter	Species	Findings
Cross-tolerance biocides	*E. coli*, *S.* Virchow	Cross-tolerance to triclosan
	S. Tyhimurium	Cross-tolerance to BAC
	A. baylyi	Cross-tolerance to hydrogen peroxide
	P. aeruginosa	No cross-tolerance to BAC
	S. aureus	No cross-tolerance to PHMB
Cross-tolerance antibiotics	*B. cepacia*, *Salmonella* spp. and *Pseudomonas* spp.	No general correlation between CHG and antibiotic resistance
	Alcaligenes spp., *E. coli*, *S. marcescens*, *S. aureus*	General correlation between CHG and antibiotic resistance
	A. baumannii	Some strains with cross-tolerance to carbapenem, aminoglycoside, tetracycline and ciprofloxacin
	S. Virchow	Some strains with cross-tolerance to tetracycline
	P. stutzeri	Some strains with cross-tolerance to ampicillin, polymyxin, erythromycin, nalidixic acid and gentamicin.
	Burkholderia spp.	Some strains with cross-tolerance to ceftazidime, ciprofloxacin and imipenem
	K. pneumoniae	Some strains with cross-tolerance to carbapenem or pan-resistance
	S. aureus (MRSA)	Some strains with cross-tolerance to cefotaxime, vancomycin, gentamicin, cefuroxime and oxacillin.
	S. aureus (smr positive)	Some strains with cross-tolerance to methicillin, ciprofloxacin, and/or clindamycin
	S. aureus	Some strains with cross-tolerance to ciprofloxacin, tetracycline, gentamicin, amikacin, cefepime or meropenem after low-level exposure
Resistance mechanisms	*S. aureus*, MRSA	qacA/B and smr (qacC) resistance gene
	A. baumannii, *K. oxytoca*, *K. pneumoniae*	qacE resistance gene
	P. stutzeri, *D. acidovorans*	Cell membrane changes
	A. baumannii, *Campylobacter* spp., *C. indologenes*, *E. faecalis*, *E. faecium*, *K. pneumoniae*, *P. aeruginosa*, *S. marcescens*	Efflux pumps
	B. cenocepacia, *S. aureus*	Plasmids
	A. xylosoxidans, *B. cepacia*, *P. mirabilis*, *P. pickettii*, *S. marscescens*, *S. aureus* (MRSA), *S. epidermidis* and *S. haemolyticus* (MRSH)	Contaminated CHG solutions or products (up to 2.5% CHG) resulting in clinical infections such as ventriculitis, cerebrospinal infections, pseudobacteremia, blood stream infections, fulminant sepsis, ventilator-associated respiratory tract infection, urinary tract infections, recurrent cutaneous abscess and joint or wound infections

(continued)

Table 13.18 (continued)

Parameter	Species	Findings
Effect of low-level exposure	*A. baumannii, A. hydrophila, B. cereus, C. coli, C. jejuni, C. indologenes, Citrobacter* spp., *C. xerosis, C. sakazakii, E. saccharolyticus, E. coli, Eubacterium* spp., *K. pneumoniae, M. phyllosphaerae, M. luteus, M. osloensis, P. aeruginosa, P. nitroreductans, P. putida, Pseudoxanthomonas* spp., *S. multivorum, S. aureus, S. capitis, S. caprae, S. cohnii, S. epidermidis, S. haemolyticus, S. hominis, S. kloosii, S. lugdenensis, S. saprophyticus, S. warneri, S. mutans*	No MIC increase
	A. xylosoxidans, A. jandaei, B. cereus, C. albicans, Chrysobacterium spp., *C. pseudogenitalum, C. renale* group, *E. cloacae, Enterobacter* spp., *E. casseliflavus, E. faecalis, E. faecium, E. coli, H. gallinarum, K. pneumoniae, M. luteus, P. aeruginosa, S.* Typhimurium, *Serratia* spp., *S. aureus, S. capitis, S. haemolyticus, S. lugdenensis, S. warneri, S. maltophilia*	Weak MIC increase ($\leq$ 4-fold)
	B. cepacia, E. faecalis, E. coli, S. enteritidis, S. Typhimurium	Strong (>4-fold) but unstable MIC increase
	E. coli, K. pneumoniae, P. aeruginosa, S. Virchow, *Salmonella* spp., *S. marcescens, S. aureus, S. maltophilia*	Strong and stable MIC increase
	A. baylyi, A. proteolyticus, E. coli, Pseudomonas spp., *Ralstonia* spp., *S. marcescens, S. aureus*	Strong MIC increase (unknown stability)
	E. coli ($\leq$ 500-fold)	Strongest MIC change after low-level exposure
	Salmonella spp. ($\leq$ 200-fold)	
	S. marcescens ($\leq$ 128-fold)	
	P. aeruginosa ($\leq$ 32-fold)	
	A. proteolyticus, K. pneumoniae, Pseudomonas spp., *S. aureus* ($\leq$ 16-fold)	
	S. marcescens (2,048 mg/l)	Highest MIC values after low-level exposure
	P. aeruginosa (1,024 mg/l)	
	Salmonella spp. (>1,000 mg/l)	
	B. cepacia complex (700 mg/l)	
	K. pneumoniae (>512 mg/l)	
	E. coli (500 mg/l)	
	A. proteolyticus (125 mg/l)	
	S. maltophilia (62.5 mg/l)	
	A. xylosoxidans, C. indologenes, C. renale group, *Eubacterium* spp (31.2 mg/l)	

(continued)

Table 13.18 (continued)

Parameter	Species	Findings
	B. fragilis, *B. cepacia complex*	Up-regulation of efflux pump genes
	K. pneumoniae, *S. marcescens*, *S. epidermidis*	Enhanced biofilm formation
	B. cepacia	Decrease of biofilm formation
	E. faecium (vanA)	$\geq$ 10-fold increase of vanHAX encoding VanA-type vancomycin resistance
	E. coli	Induction of horizontal gene transfer (sulphonamide resistance by conjugation)
	B. subtilis	No increase of transfer of the mobile genetic element Tn916, a conjugative transposon
Biofilm	Development	Inhibition of biofilm formation of *C. albicans*, *E. faecalis*, *E. coli*, *S. aureus*, *S. mutans* and mixed-species biofilms
		No significant biofilm formation inhibition by CHG in *S. enteritidis* and *S. mutans*
	Removal	Mostly poor
	Fixation	Unknown; CHG can contract biofilm.

References

1. Aarestrup FM, Hasman H (2004) Susceptibility of different bacterial species isolated from food animals to copper sulphate, zinc chloride and antimicrobial substances used for disinfection. Vet Microbiol 100(1–2):83–89. https://doi.org/10.1016/j.vetmic.2004.01.013
2. Abdolrasouli A, Armstrong-James D, Ryan L, Schelenz S (2017) In vitro efficacy of disinfectants utilised for skin decolonisation and environmental decontamination during a hospital outbreak with Candida auris. Mycoses 60(11):758–763. https://doi.org/10.1111/myc.12699
3. Abuzaid A, Hamouda A, Amyes SG (2012) Klebsiella pneumoniae susceptibility to biocides and its association with cepA, qacDeltaE and qacE efflux pump genes and antibiotic resistance. J Hosp Infect 81(2):87–91. https://doi.org/10.1016/j.jhin.2012.03.003
4. Adams D, Quayum M, Worthington T, Lambert P, Elliott T (2005) Evaluation of a 2% chlorhexidine gluconate in 70% isopropyl alcohol skin disinfectant. J Hosp Infect 61 (4):287–290
5. Adams D, Quayum MH, Worthington T, Lambert PA, Elliott TS (2006) Are biofilms relevant for skin disinfection? Response to Dr Kampf. J Hosp Infect 63(4):480–481. https://doi.org/10.1016/j.jhin.2006.04.004
6. Agnoli K, Frauenknecht C, Freitag R, Schwager S, Jenul C, Vergunst A, Carlier A, Eberl L (2014) The third replicon of members of the Burkholderia cepacia Complex, plasmid pC3, plays a role in stress tolerance. Appl Environ Microbiol 80(4):1340–1348. https://doi.org/10.1128/aem.03330-13

7. Ahn Y, Kim JM, Lee YJ, LiPuma J, Hussong D, Marasa B, Cerniglia C (2017) Effects of extended storage of Chlorhexidine Gluconate and Benzalkonium chloride solutions on the viability of Burkholderia cenocepacia. J Microbiol Biotechnol 27(12):2211–2220. https://doi.org/10.4014/jmb.1706.06034
8. Akamatsu T, Tabata K, Hironga M, Kawakami H, Uyeda M (1996) Transmission of Helicobacter pylori infection via flexible fiberoptic endoscopy. Am J Infect Control 24 (5):396–401
9. Akca AE, Akca G, Topcu FT, Macit E, Pikdoken L, Ozgen IS (2016) The Comparative evaluation of the antimicrobial effect of Propolis with Chlorhexidine against oral pathogens: An in vitro study. Biomed Res Int 2016:3627463. https://doi.org/10.1155/2016/3627463
10. Akinkunmi EO, Lamikanra A (2012) Susceptibility of community associated methicillin resistant Staphylococcus aureus isolated from faeces to antiseptics. J Infect Develop Countries 6(4):317–323
11. Anand G, Ravinanthan M, Basaviah R, Shetty AV (2015) In vitro antimicrobial and cytotoxic effects of Anacardium occidentale and Mangifera indica in oral care. J Pharmacy Bioallied Sci 7(1):69–74. https://doi.org/10.4103/0975-7406.148780
12. Anderson MJ, Horn ME, Lin YC, Parks PJ, Peterson ML (2010) Efficacy of concurrent application of chlorhexidine gluconate and povidone iodine against six nosocomial pathogens. Am J Infect Control 38(10):826–831. https://doi.org/10.1016/j.ajic.2010.06.022
13. Anderson MJ, Scholz MT, Parks PJ, Peterson ML (2013) Ex vivo porcine vaginal mucosal model of infection for determining effectiveness and toxicity of antiseptics. J Appl Microbiol 115(3):679–688. https://doi.org/10.1111/jam.12277
14. Arias-Moliz MT, Ordinola-Zapata R, Baca P, Ruiz-Linares M, Garcia Garcia E, Hungaro Duarte MA, Monteiro Bramante C, Ferrer-Luque CM (2015) Antimicrobial activity of Chlorhexidine, Peracetic acid and Sodium hypochlorite/etidronate irrigant solutions against Enterococcus faecalis biofilms. Int Endod J 48(12):1188–1193. https://doi.org/10.1111/iej.12424
15. Arioli S, Elli M, Ricci G, Mora D (2013) Assessment of the susceptibility of lactic acid bacteria to biocides. Int J Food Microbiol 163(1):1–5. https://doi.org/10.1016/j.ijfoodmicro.2013.02.002
16. Arslan S, Ozbilge H, Kaya EG, Er O (2011) In vitro antimicrobial activity of propolis, BioPure MTAD, sodium hypochlorite, and chlorhexidine on Enterococcus faecalis and Candida albicans. Saudi Med J 32(5):479–483
17. Arzmi MH, Abdul Razak F, Yusoff Musa M, Wan Harun WH (2012) Effect of phenotypic switching on the biological properties and susceptibility to chlorhexidine in Candida krusei ATCC 14243. FEMS Yeast Res 12(3):351–358. https://doi.org/10.1111/j.1567-1364.2011.00786.x
18. Ashok R, Ganesh A, Deivanayagam K (2017) Bactericidal effect of different anti-microbial agents on Fusobacterium Nucleatum biofilm. Cureus 9(6):e1335. https://doi.org/10.7759/cureus.1335
19. Azzimonti B, Cochis A, Beyrouthy ME, Iriti M, Uberti F, Sorrentino R, Landini MM, Rimondini L, Varoni EM (2015) Essential oil from berries of Lebanese Juniperus excelsa M. Bieb displays Similar antibacterial activity to chlorhexidine but higher cytocompatibility with human oral primary cells. Molecules (Basel, Switzerland) 20(5):9344–9357. https://doi.org/10.3390/molecules20059344
20. Babaei M, Sulong A, Hamat R, Nordin S, Neela V (2015) Extremely high prevalence of antiseptic resistant quaternary ammonium compound E gene among clinical isolates of multiple drug resistant Acinetobacter baumannii in Malaysia. Ann Clin Microbiol Antimicrob 14:11. https://doi.org/10.1186/s12941-015-0071-7

21. Baffone W, Sorgente G, Campana R, Patrone V, Sisti D, Falcioni T (2011) Comparative effect of chlorhexidine and some mouthrinses on bacterial biofilm formation on titanium surface. Curr Microbiol 62(2):445–451. https://doi.org/10.1007/s00284-010-9727-x
22. Barbadoro P, Martini E, Savini S, Marigliano A, Ponzio E, Prospero E, D'Errico MM (2014) In vivo comparative efficacy of three surgical hand preparation agents in reducing bacterial count. J Hosp Infect 86(1):64–67. https://doi.org/10.1016/j.jhin.2013.09.013
23. Barkvoll P, Attramadal A (1989) Effect of nystatin and chlorhexidine digluconate on *Candida albicans*. Oral Surg Oral Med Oral Pathol 67(3):279–281
24. Barlow M (2009) What antimicrobial resistance has taught us about horizontal gene transfer. Meth Mol Biol (Clifton, NJ) 532:397–411. https://doi.org/10.1007/978-1-60327-853-9_23
25. Bartzokas CA, Corkill JE, Makin T (1987) Evaluation of the skin disinfecting activity and cumulative effect of chlorhexidine and triclosan handwash preparations on hands artificially contaminated with Serratia marcescens. Infection Control: IC 8(4):163–167
26. Bartzokas CA, Gibson MF, Graham R, Pinder DC (1983) A comparison of triclosan and chlorhexidine preparations with 60 per cent isopropyl alcohol for hygienic hand disinfection. J Hosp Infect 4:245–255
27. Batra R, Cooper BS, Whiteley C, Patel AK, Wyncoll D, Edgeworth JD (2010) Efficacy and limitation of a chlorhexidine-based decolonization strategy in preventing transmission of methicillin-resistant Staphylococcus aureus in an intensive care unit. Clinical Infect Dis: An Off Publ Infect Dis Soc Am 50(2):210–217. https://doi.org/10.1086/648717
28. Beier RC, Anderson PN, Hume ME, Poole TL, Duke SE, Crippen TL, Sheffield CL, Caldwell DJ, Byrd JA, Anderson RC, Nisbet DJ (2011) Characterization of Salmonella enterica isolates from turkeys in commercial processing plants for resistance to antibiotics, disinfectants, and a growth promoter. Foodborne Pathog Dis 8(5):593–600. https://doi.org/10.1089/fpd.2010.0702
29. Beier RC, Foley SL, Davidson MK, White DG, McDermott PF, Bodeis-Jones S, Zhao S, Andrews K, Crippen TL, Sheffield CL, Poole TL, Anderson RC, Nisbet DJ (2015) Characterization of antibiotic and disinfectant susceptibility profiles among Pseudomonas aeruginosa veterinary isolates recovered during 1994-2003. J Appl Microbiol 118(2):326–342. https://doi.org/10.1111/jam.12707
30. Ben Saida N, Marzouk M, Ferjeni A, Boukadida J (2009) A three-year surveillance of nosocomial infections by methicillin-resistant Staphylococcus haemolyticus in newborns reveals the disinfectant as a possible reservoir. Pathologie-biologie 57(3):e29–35. https://doi.org/10.1016/j.patbio.2008.02.019
31. Bercy P, Lasserre J (2007) Susceptibility to various oral antiseptics of Porphyromonas gingivalis W83 within a biofilm. Adv Therap 24(6):1181–1191
32. Best M, Kennedy ME, Coates F (1990) Efficacy of a variety of disinfectants against Listeria spp. Appl Environ Microbiol 56(2):377–380
33. Best M, Sattar SA, Springthorpe VS, Kennedy ME (1988) Comparative mycobactericidal efficacy of chemical disinfectants in suspension and carrier tests. Appl Environ Microbiol 54:2856–2858
34. Best M, Sattar SA, Springthorpe VS, Kennedy ME (1990) Efficacies of selected disinfectants against *Mycobacterium tuberculosis*. J Clin Microbiol 28(10):2234–2239
35. Best M, Springthorpe VS, Sattar SA (1994) Feasibility of a combined carrier test for disinfectants: studies with a mixture of five types of microorganisms. Am J Infect Control 22 (3):152–162
36. Bhardwaj P, Hans A, Ruikar K, Guan Z, Palmer KL (2018) Reduced chlorhexidine and daptomycin susceptibility in vancomycin-resistant Enterococcus faecium after serial chlorhexidine exposure. Antimicrob Agents Chemother 62(1). https://doi.org/10.1128/aac.01235-17

37. Bhardwaj P, Ziegler E, Palmer KL (2016) Chlorhexidine induces VanA-Type vancomycin resistance genes in Enterococci. Antimicrob Agents Chemother 60(4):2209–2221. https://doi.org/10.1128/aac.02595-15
38. Block C, Furman M (2002) Association between intensity of chlorhexidine use and microorganisms of reduced susceptibility in a hospital environment. J Hosp Infect 51:201–206
39. Block C, Robenshtok E, Simhon A, Shapiro M (2000) Evaluation of chlorhexidine and povidone iodine activity against methicillin-resistant Staphylococcus aureus and vancomycin-resistant Enterococcus faecalis using a surface test. J Hosp Infect 46(2):147–152. https://doi.org/10.1053/jhin.2000.0805
40. Bobichon H, Bouchet P (1987) Action of chlorhexidine on budding Candida albicans: scanning and transmission electron microscopic study. Mycopathologia 100(1):27–35
41. Bock LJ, Hind CK, Sutton JM, Wand ME (2018) Growth media and assay plate material can impact on the effectiveness of cationic biocides and antibiotics against different bacterial species. Lett Appl Microbiol 66(5):368–377. https://doi.org/10.1111/lam.12863
42. Bock LJ, Wand ME, Sutton JM (2016) Varying activity of chlorhexidine-based disinfectants against Klebsiella pneumoniae clinical isolates and adapted strains. J Hosp Infect 93(1):42–48. https://doi.org/10.1016/j.jhin.2015.12.019
43. Bonez PC, Dos Santos Alves CF, Dalmolin TV, Agertt VA, Mizdal CR, Flores Vda C, Marques JB, Santos RC, Anraku de Campos MM (2013) Chlorhexidine activity against bacterial biofilms. Am J Infect Control 41(12):e119–122. https://doi.org/10.1016/j.ajic.2013.05.002
44. Braga TM, Marujo PE, Pomba C, Lopes MF (2011) Involvement, and dissemination, of the enterococcal small multidrug resistance transporter QacZ in resistance to quaternary ammonium compounds. J Antimicrob Chemother 66(2):283–286. https://doi.org/10.1093/jac/dkq460
45. Braga TM, Pomba C, Lopes MF (2013) High-level vancomycin resistant Enterococcus faecium related to humans and pigs found in dust from pig breeding facilities. Vet Microbiol 161(3–4):344–349. https://doi.org/10.1016/j.vetmic.2012.07.034
46. Braoudaki M, Hilton AC (2004) Adaptive resistance to biocides in Salmonella enterica and Escherichia coli O157 and cross-resistance to antimicrobial agents. J Clin Microbiol 42(1):73–78
47. Brill F, Goroncy-Bermes P, Sand W (2006) Influence of growth media on the sensitivity of Staphylococcus aureus and Pseudomonas aeruginosa to cationic biocides. Int J Hyg Environ Health 209(1):89–95
48. Brindle ER, Miller DA, Stewart PS (2011) Hydrodynamic deformation and removal of Staphylococcus epidermidis biofilms treated with urea, chlorhexidine, iron chloride, or DispersinB. Biotechnol Bioeng 108(12):2968–2977. https://doi.org/10.1002/bit.23245
49. Brooks SE, Walczak MA, Hameed R, Coonan P (2002) Chlorhexidine resistance in antibiotic-resistant bacteria isolated from the surfaces of dispensers of soap containing chlorhexidine. Infect Control Hosp Epidemiol 23:692–695
50. Brown AT, Shupe JA, Sims RE, Matheny JL, Lillich TT, Douglass JB, Henslee PJ, Raybould TP, Ferretti GA (1990) In vitro effect of chlorhexidine and amikacin on oral gram-negative bacilli from bone marrow transplant recipients. Oral Surg Oral Med Oral Pathol 70(6):715–719
51. Bukhary S, Balto H (2017) Antibacterial efficacy of Octenisept, Alexidine, Chlorhexidine, and Sodium Hypochlorite against Enterococcus faecalis Biofilms. J Endod 43(4):643–647. https://doi.org/10.1016/j.joen.2016.09.013
52. Burgers R, Witecy C, Hahnel S, Gosau M (2012) The effect of various topical peri-implantitis antiseptics on Staphylococcus epidermidis, Candida albicans, and Streptococcus sanguinis. Arch Oral Biol 57(7):940–947. https://doi.org/10.1016/j.archoralbio.2012.01.015

53. Caballero Gómez N, Abriouel H, Grande MJ, Pérez Pulido R, Gálvez A (2012) Effect of enterocin AS-48 in combination with biocides on planktonic and sessile Listeria monocytogenes. Food Microbiol 30(1):51–58. https://doi.org/10.1016/j.fm.2011.12.013
54. Calixto GMF, Duque C, Aida KL, Dos Santos VR, Massunari L, Chorilli M (2018) Development and characterization of p1025-loaded bioadhesive liquid-crystalline system for the prevention of Streptococcus mutans biofilms. Int J Nanomed 13:31–41. https://doi.org/10.2147/ijn.s147553
55. Casado Munoz Mdel C, Benomar N, Lavilla Lerma L, Knapp CW, Galvez A, Abriouel H (2016) Biocide tolerance, phenotypic and molecular response of lactic acid bacteria isolated from naturally-fermented Alorena table to different physico-chemical stresses. Food Microbiol 60:1–12. https://doi.org/10.1016/j.fm.2016.06.013
56. Chawner JA, Gilbert P (1989) Adsorption of alexidine and chlorhexidine to Escherichia coli and membrane components. Int J Pharm 55(2):209–215. https://doi.org/10.1016/0378-5173(89)90043-4
57. Cherian B, Gehlot PM, Manjunath MK (2016) Comparison of the antimicrobial efficacy of octenidine dihydrochloride and chlorhexidine with and without passive ultrasonic irrigation—an invitro study. J Clin Diagn Res 10(6):Zc71–77. https://doi.org/10.7860/jcdr/2016/17911.8021
58. Chiang WC, Pamp SJ, Nilsson M, Givskov M, Tolker-Nielsen T (2012) The metabolically active subpopulation in Pseudomonas aeruginosa biofilms survives exposure to membrane-targeting antimicrobials via distinct molecular mechanisms. FEMS Immunol Med Microbiol 65(2):245–256. https://doi.org/10.1111/j.1574-695X.2012.00929.x
59. Choi YS, Kim C, Moon JH, Lee JY (2018) Removal and killing of multispecies endodontic biofilms by N-acetylcysteine. Brazilian J Microbiol: [Publication of the Brazilian Society for Microbiology] 49(1):184–188. https://doi.org/10.1016/j.bjm.2017.04.003
60. Christensen KK, Christensen P, Dykes AK, Kahlmeter G, Kurl DN, Linden V (1983) Chlorhexidine for prevention of neonatal colonization with group B streptococci. I. In vitro effect of chlorhexidine on group B streptococci. Eur J Obstet Gynecol Reprod Biol 16 (3):157–165
61. Chuanchuen R, Pathanasophon P, Khemtong S, Wannaprasat W, Padungtod P (2008) Susceptibilities to antimicrobials and disinfectants in Salmonella isolates obtained from poultry and swine in Thailand. Jveterinary Med Sci 70(6):595–601
62. Chung YK, Kim JS, Lee SS, Lee JA, Kim HS, Shin KS, Park EY, Kang BS, Lee HJ, Kang HJ (2015) Effect of daily chlorhexidine bathing on acquisition of carbapenem-resistant Acinetobacter baumannii (CRAB) in the medical intensive care unit with CRAB endemicity. Am J Infect Control 43(11):1171–1177. https://doi.org/10.1016/j.ajic.2015.07.001
63. Clark SM, Loeffler A, Bond R (2015) Susceptibility in vitro of canine methicillin-resistant and -susceptible staphylococcal isolates to fusidic acid, chlorhexidine and miconazole: opportunities for topical therapy of canine superficial pyoderma. J Antimicrob Chemother 70 (7):2048–2052. https://doi.org/10.1093/jac/dkv056
64. Coaguila-Llerena H, Stefanini da Silva V, Tanomaru-Filho M, Guerreiro Tanomaru JM, Faria G (2018) Cleaning capacity of octenidine as root canal irrigant: a scanning electron microscopy study. Microsc Res Tech. https://doi.org/10.1002/jemt.23007
65. Coelho JR, Carrico JA, Knight D, Martinez JL, Morrissey I, Oggioni MR, Freitas AT (2013) The use of machine learning methodologies to analyse antibiotic and biocide susceptibility in Staphylococcus aureus. PLoS ONE 8(2):e55582. https://doi.org/10.1371/journal.pone.0055582
66. Coenye T, Van Acker H, Peeters E, Sass A, Buroni S, Riccardi G, Mahenthiralingam E (2011) Molecular mechanisms of chlorhexidine tolerance in Burkholderia cenocepacia biofilms. Antimicrob Agents Chemother 55(5):1912–1919. https://doi.org/10.1128/aac.01571-10

67. Conceicao T, Coelho C, de Lencastre H, Aires-de-Sousa M (2015) High prevalence of biocide resistance determinants in Staphylococcus aureus isolates from three African countries. Antimicrob Agents Chemother. https://doi.org/10.1128/aac.02140-15
68. Condell O, Iversen C, Cooney S, Power KA, Walsh C, Burgess C, Fanning S (2012) Efficacy of biocides used in the modern food industry to control salmonella enterica, and links between biocide tolerance and resistance to clinically relevant antimicrobial compounds. Appl Environ Microbiol 78(9):3087–3097. https://doi.org/10.1128/aem.07534-11
69. Condell O, Power KA, Handler K, Finn S, Sheridan A, Sergeant K, Renaut J, Burgess CM, Hinton JC, Nally JE, Fanning S (2014) Comparative analysis of Salmonella susceptibility and tolerance to the biocide chlorhexidine identifies a complex cellular defense network. Front Microbiol 5:373. https://doi.org/10.3389/fmicb.2014.00373
70. Cook HA, Cimiotti JP, Della-Latta P, Saiman L, Larson EL (2007) Antimicrobial resistance patterns of colonizing flora on nurses' hands in the neonatal intensive care unit. Am J Infect Control 35(4):231–236. https://doi.org/10.1016/j.ajic.2006.05.291
71. Cookson BD, Bolton MC, Platt JH (1991) Chlorhexidine resistance in methicillin-resistant *Staphylococcus aureus* or just an elevated MIC? An in vitro and in vivo assessment. Antimicrob Agents Chemother 35(10):1997–2002
72. Corbin A, Pitts B, Parker A, Stewart PS (2011) Antimicrobial penetration and efficacy in an in vitro oral biofilm model. Antimicrob Agents Chemother 55(7):3338–3344. https://doi.org/10.1128/aac.00206-11
73. Cowley NL, Forbes S, Amezquita A, McClure P, Humphreys GJ, McBain AJ (2015) Effects of formulation on microbicide potency and mitigation of the development of bacterial insusceptibility. Appl Environ Microbiol 81(20):7330–7338. https://doi.org/10.1128/aem.01985-15
74. Curiao T, Marchi E, Viti C, Oggioni MR, Baquero F, Martinez JL, Coque TM (2015) Polymorphic variation in susceptibility and metabolism of triclosan-resistant mutants of Escherichia coli and Klebsiella pneumoniae clinical strains obtained after exposure to biocides and antibiotics. Antimicrob Agents Chemother 59(6):3413–3423. https://doi.org/10.1128/aac.00187-15
75. da Silva PM, Acosta EJ, Pinto Lde R, Graeff M, Spolidorio DM, Almeida RS, Porto VC (2011) Microscopical analysis of Candida albicans biofilms on heat-polymerised acrylic resin after chlorhexidine gluconate and sodium hypochlorite treatments. Mycoses 54(6): e712–717. https://doi.org/10.1111/j.1439-0507.2010.02005.x
76. Dance DAB, Pearson AD, Seal DV, Lowes JA (1987) A hospital outbreak caused by a chlorhexidine and antibiotic-resistant *Proteus mirabilis*. J Hosp Infect 10(1):10–16
77. Darouiche RO, Wall MJ, Itani KM, Otterson MF, Webb AL, Carrick MM, Miller HJ, Awad SS, Crosby CT, Mosier MC, Alsharif A, Berger DH (2010) Chlorhexidine-alcohol versus povidone-iodine for surgical-site antisepsis. N Engl J Med 362(1):18–26
78. Das R, Ghosh S, Bhattacharjee C (2015) A green practice for pharmaceutical drug chlorhexidine digluconate treatment and ecotoxicity assessment. J Water Process Eng 7:266–272. https://doi.org/10.1016/j.jwpe.2015.05.005
79. Davies GE, Francis J, Martin AR, Rose FL, Swain G (1954) 1:6-Di-4'-chlorophenyldiguanidohexane ("Hibitane"*). Laboratory investigation of a new antibacterial agent of high potency. Br J Pharmacol 9:192–196
80. de Andrade IM, Cruz PC, Silva-Lovato CH, de Souza RF, Souza-Gugelmin MC, Paranhos Hde F (2012) Effect of chlorhexidine on denture biofilm accumulation. J Prosthodontists: Off J Am College Prosthodontists 21(1):2–6. https://doi.org/10.1111/j.1532-849X.2011.00774.x

81. de Frutos M, Lopez-Urrutia L, Dominguez-Gil M, Arias M, Munoz-Bellido JL, Eiros JM, Ramos C (2017) Serratia marcescens outbreak due to contaminated 2% aqueous chlorhexidine. Enferm Infecc Microbiol Clin 35(10):624–629. https://doi.org/10.1016/j.eimc.2016.06.016
82. de Lucena JM, Decker EM, Walter C, Boeira LS, Lost C, Weiger R (2013) Antimicrobial effectiveness of intracanal medicaments on Enterococcus faecalis: chlorhexidine versus octenidine. Int Endod J 46(1):53–61. https://doi.org/10.1111/j.1365-2591.2012.02093.x
83. Decker EM, Bartha V, Kopunic A, von Ohle C (2017) Antimicrobial efficiency of mouthrinses versus and in combination with different photodynamic therapies on periodontal pathogens in an experimental study. J Periodontal Res 52(2):162–175. https://doi.org/10.1111/jre.12379
84. DeMarco CE, Cushing LA, Frempong-Manso E, Seo SM, Jaravaza TA, Kaatz GW (2007) Efflux-related resistance to norfloxacin, dyes, and biocides in bloodstream isolates of Staphylococcus aureus. Antimicrob Agents Chemother 51(9):3235–3239. https://doi.org/10.1128/aac.00430-07
85. Department of Health and Human Services; Food and Drug Administration (1994) Tentative final monograph for health care antiseptic products; proposed rule. Fed Reg 59(116):31401–31452
86. Department of Health and Human Services; Food and Drug Administration (2015) Safety and effectiveness of healthcare antiseptics. topical antimicrobial drug products for over-the-counter human use; proposed amendment of the tentative final monograph; reopening of administrative record; proposed rule. Fed Reg 80(84):25166–25205
87. Deus D, Krischek C, Pfeifer Y, Sharifi AR, Fiegen U, Reich F, Klein G, Kehrenberg C (2017) Comparative analysis of the susceptibility to biocides and heavy metals of extended-spectrum beta-lactamase-producing Escherichia coli isolates of human and avian origin, Germany. Diagnostic Microbiol Infect Dis 88(1):88–92. https://doi.org/10.1016/j.diagmicrobio.2017.01.023
88. do Amorim CV, Aun CE, Mayer MP (2004) Susceptibility of some oral microorganisms to chlorhexidine and paramonochlorophenol. Brazilian Oral Res 18(3):242–246
89. Dogan AA, Adiloglu AK, Onal S, Cetin ES, Polat E, Uskun E, Koksal F (2008) Short-term relative antibacterial effect of octenidine dihydrochloride on the oral microflora in orthodontically treated patients. Int J Infect Dis: IJID: Off Publ Int Soc Infect Dis 12(6): e19–25. https://doi.org/10.1016/j.ijid.2008.03.013
90. Dominciano LCC, Oliveira CAF, Lee SH, Corassin CH (2016) Individual and combined antimicrobial activity of Oleuropein and chemical sanitizers. J Food Chem Nanotechnol 2 (3):124–127
91. Dong L, Tong Z, Linghu D, Lin Y, Tao R, Liu J, Tian Y, Ni L (2012) Effects of sub-minimum inhibitory concentrations of antimicrobial agents on Streptococcus mutans biofilm formation. Int J Antimicrob Agents 39(5):390–395. https://doi.org/10.1016/j.ijantimicag.2012.01.009
92. Dostie S, Alkadi LT, Owen G, Bi J, Shen Y, Haapasalo M, Larjava HS (2017) Chemotherapeutic decontamination of dental implants colonized by mature multispecies oral biofilm. J Clin Periodontol 44(4):403–409. https://doi.org/10.1111/jcpe.12699
93. Duran N, Temiz M, Duran GG, Eryilmaz N, Jenedi K (2014) Relationship between the resistance genes to quaternary ammonium compounds and antibiotic resistance in Staphylococci isolated from surgical site infections. Medical Science Monitor: Int Med J Exp Clin Res 20:544–550. https://doi.org/10.12659/msm.890177
94. Dynes JJ, Lawrence JR, Korber DR, Swerhone GD, Leppard GG, Hitchcock AP (2006) Quantitative mapping of chlorhexidine in natural river biofilms. Sci Total Environ 369(1–3):369–383. https://doi.org/10.1016/j.scitotenv.2006.04.033

95. Ekizoglu M, Sagiroglu M, Kilic E, Hascelik AG (2016) An investigation of the bactericidal activity of chlorhexidine digluconate against multidrug-resistant hospital isolates. Turkish J Med Sci 46(3):903–909. https://doi.org/10.3906/sag-1503-140
96. Ekizoglu MT, Özalp M, Sultan N, Gür D (2003) An investigation of the bactericidal effect of certain antiseptics and disinfectants on some hospital isolates of gram-negative bacteria. Infect Control Hosp Epidemiol 24(3):225–227
97. Eldeniz AU, Guneser MB, Akbulut MB (2015) Comparative antifungal efficacy of light-activated disinfection and octenidine hydrochloride with contemporary endodontic irrigants. Lasers Med Sci 30(2):669–675. https://doi.org/10.1007/s10103-013-1387-1
98. Elli M, Arioli S, Guglielmetti S, Mora D (2013) Biocide susceptibility in bifidobacteria of human origin. J Global Antimicrob Resis 1(2):97–101. https://doi.org/10.1016/j.jgar.2013.03.007
99. Espigares E, Moreno Roldan E, Espigares M, Abreu R, Castro B, Dib AL, Arias A (2017) Phenotypic resistance to disinfectants and antibiotics in methicillin-resistant Staphylococcus aureus strains isolated from pigs. Zoonoses Public Health 64(4):272–280. https://doi.org/10.1111/zph.12308
100. Espinosa de los Monteros LE, Silva-Sanchez J, Jimenez LV, Rojas T, Garza-Ramos U, Valverde V (2008) Outbreak of infection by extended-spectrum beta-lactamase SHV-5-producing Serratia marcescens in a Mexican hospital. J Chemotherap (Florence, Italy) 20(5):586–592. https://doi.org/10.1179/joc.2008.20.5.586
101. European Chemicals Agency (ECHA) D-gluconic acid, compound with N,N″-bis (4-chlorophenyl)-3,12-diimino-2,4,11,13-tetraazatetradecanediamidine (2:1). Substance information. https://echa.europa.eu/de/substance-information/-/substanceinfo/100.038.489. Accessed 5 Feb 2018
102. Fang CT, Chen HC, Chuang YP, Chang SC, Wang JT (2002) Cloning of a cation efflux pump gene associated with chlorhexidine resistance in Klebsiella pneumoniae. Antimicrob Agents Chemother 46(6):2024–2028
103. Fathilah AR, Himratul-Aznita WH, Fatheen AR, Suriani KR (2012) The antifungal properties of chlorhexidine digluconate and cetylpyrinidinium chloride on oral Candida. J Dent 40(7):609–615. https://doi.org/10.1016/j.jdent.2012.04.003
104. Ferguson JW, Hatton JF, Gillespie MJ (2002) Effectiveness of intracanal irrigants and medications against the yeast Candida albicans. J Endod 28(2):68–71. https://doi.org/10.1097/00004770-200202000-00004
105. Fernandez-Cuenca F, Tomas M, Caballero-Moyano FJ, Bou G, Martinez-Martinez L, Vila J, Pachon J, Cisneros JM, Rodriguez-Bano J, Pascual A (2015) Reduced susceptibility to biocides in Acinetobacter baumannii: association with resistance to antimicrobials, epidemiological behaviour, biological cost and effect on the expression of genes encoding porins and efflux pumps. J Antimicrob Chemother 70(12):3222–3229. https://doi.org/10.1093/jac/dkv262
106. Fernández-Fuentes MA, Ortega Morente E, Abriouel H, Pérez Pulido R, Gálvez A (2012) Isolation and identification of bacteria from organic foods: sensitivity to biocides and antibiotics. Food Control 26(1):73–78. https://doi.org/10.1016/j.foodcont.2012.01.017
107. Fleurette J, Bes M, Brun Y, Freney J, Forey F, Coulet M, Reverdy ME, Etienne J (1989) Clinical isolates of Staphylococcus lugdunensis and S. schleiferi: bacteriological characteristics and susceptibility to antimicrobial agents. Res Microbiol 140(2):107–118
108. Forbes S, Dobson CB, Humphreys GJ, McBain AJ (2014) Transient and sustained bacterial adaptation following repeated sublethal exposure to microbicides and a novel human antimicrobial peptide. Antimicrob Agents Chemother 58(10):5809–5817. https://doi.org/10.1128/aac.03364-14

109. Forbes S, Knight CG, Cowley NL, Amezquita A, McClure P, Humphreys G, McBain AJ (2016) variable effects of exposure to formulated microbicides on antibiotic susceptibility in firmicutes and proteobacteria. Appl Environ Microbiol 82(12):3591–3598. https://doi.org/10.1128/aem.00701-16
110. Fraud S, Campigotto AJ, Chen Z, Poole K (2008) MexCD-OprJ multidrug efflux system of Pseudomonas aeruginosa: involvement in chlorhexidine resistance and induction by membrane-damaging agents dependent upon the AlgU stress response sigma factor. Antimicrob Agents Chemother 52(12):4478–4482. https://doi.org/10.1128/aac.01072-08
111. Fraud S, Hann AC, Maillard JY, Russell AD (2003) Effects of ortho-phthalaldehyde, glutaraldehyde and chlorhexidine diacetate on Mycobacterium chelonae and Mycobacterium abscessus strains with modified permeability. J Antimicrob Chemother 51(3):575–584
112. Fu J, Wei P, Zhao C, He C, Yan Z, Hua H (2014) In vitro antifungal effect and inhibitory activity on biofilm formation of seven commercial mouthwashes. Oral Dis 20(8):815–820. https://doi.org/10.1111/odi.12242
113. Fuangthong M, Julotok M, Chintana W, Kuhn K, Rittiroongrad S, Vattanaviboon P, Mongkolsuk S (2011) Exposure of Acinetobacter baylyi ADP1 to the biocide chlorhexidine leads to acquired resistance to the biocide itself and to oxidants. J Antimicrob Chemother 66 (2):319–322. https://doi.org/10.1093/jac/dkq435
114. Furi L, Ciusa ML, Knight D, Di Lorenzo V, Tocci N, Cirasola D, Aragones L, Coelho JR, Freitas AT, Marchi E, Moce L, Visa P, Northwood JB, Viti C, Borghi E, Orefici G, Morrissey I, Oggioni MR (2013) Evaluation of reduced susceptibility to quaternary ammonium compounds and bisbiguanides in clinical isolates and laboratory-generated mutants of Staphylococcus aureus. Antimicrob Agents Chemother 57(8):3488–3497. https://doi.org/10.1128/aac.00498-13
115. Fuursted K, Hjort A, Knudsen L (1997) Evaluation of bactericidal activity and lag of regrowth (postantibiotic effect) of five antiseptics on nine bacterial pathogens. J Antimicrob Chemother 40(2):221–226
116. Gadea R, Glibota N, Pérez Pulido R, Gálvez A, Ortega E (2017) Effects of exposure to biocides on susceptibility to essential oils and chemical preservatives in bacteria from organic foods. Food Control 80(Supplement C):176–182. https://doi.org/10.1016/j.foodcont.2017.05.002
117. Gajadhar T, Lara A, Sealy P, Adesiyun AA (2003) Microbial contamination of disinfectants and antiseptics in four major hospitals in Trinidad. Revista panamericana de salud publica. Pan American J Public Health 14(3):193–200
118. Galice DM, Bonacorsi C, Soares VC, Raddi MS, Fonseca LM (2006) Effect of subinhibitory concentration of chlorhexidine on Streptococcus agalactiae virulence factor expression. Int J Antimicrob Agents 28(2):143–146. https://doi.org/10.1016/j.ijantimicag.2006.03.024
119. Gandhi PA, Sawant AD, Wilson LA, Ahearn DG (1993) Adaption and growth of *Serratia marcescens* in contact lens disinfectant solutions containing chlorhexidine gluconate. Appl Environ Microbiol 59(1):183–188
120. Gaston MA, Hoffman PN, Pitt TL (1986) A comparison of strains of Serratia marcescens isolated from neonates with strains isolated from sporadic and epidemic infections in adults. J Hosp Infect 8(1):86–95
121. Ghasemzadeh-Moghaddam H, van Belkum A, Hamat RA, van Wamel W, Neela V (2014) Methicillin-susceptible and -resistant Staphylococcus aureus with high-level antiseptic and low-level mupirocin resistance in Malaysia. Microbial drug resistance (Larchmont, NY) 20 (5):472–477. https://doi.org/10.1089/mdr.2013.0222
122. Ghivari SB, Bhattacharya H, Bhat KG, Pujar MA (2017) Antimicrobial activity of root canal irrigants against biofilm forming pathogens-an in vitro study. J Conserv Dentistry: JCD 20 (3):147–151. https://doi.org/10.4103/jcd.jcd_38_16

123. Giuliana G, Pizzo G, Milici ME, Giangreco R (1999) In vitro activities of antimicrobial agents against Candida species. Oral Surg Oral Med Oral Pathol Oral Radiol Endod 87 (1):44–49. https://doi.org/10.1016/S1079-2104(99)70293-3
124. Goroncy-Bermes P, Brill FHH, Brill H (2013) Antimicrobial activity of wound antiseptics against extended-spectrum beta-lactamase-producing bacteria. Wound Med 1(1):41–43
125. Gosau M, Hahnel S, Schwarz F, Gerlach T, Reichert TE, Burgers R (2010) Effect of six different peri-implantitis disinfection methods on in vivo human oral biofilm. Clin Oral Implant Res 21(8):866–872. https://doi.org/10.1111/j.1600-0501.2009.01908.x
126. Grande Burgos MJ, Lucas López R, López Aguayo M, Pérez Pulido R, Gálvez A (2013) Inhibition of planktonic and sessile Salmonella enterica cells by combinations of enterocin AS-48, polymyxin B and biocides. Food Control 30(1):214–221. https://doi.org/10.1016/j.foodcont.2012.07.011
127. Grare M, Dibama HM, Lafosse S, Ribon A, Mourer M, Regnouf-de-Vains JB, Finance C, Duval RE (2010) Cationic compounds with activity against multidrug-resistant bacteria: interest of a new compound compared with two older antiseptics, hexamidine and chlorhexidine. Clin Microbiol Infect 16(5):432–438. https://doi.org/10.1111/j.1469-0691.2009.02837.x
128. Guerin-Mechin L, Leveau JY, Dubois-Brissonnet F (2004) Resistance of spheroplasts and whole cells of Pseudomonas aeruginosa to bactericidal activity of various biocides: evidence of the membrane implication. Microbiol Res 159(1):51–57. https://doi.org/10.1016/j.micres.2004.01.003
129. Guiotti AM, Cunha BG, Paulini MB, Goiato MC, Dos Santos DM, Duque C, Caiaffa KS, Brandini DA, Narciso de Oliveira DT, Brizzotti NS, Gottardo de Almeida MT (2016) Antimicrobial activity of conventional and plant-extract disinfectant solutions on microbial biofilms on a maxillofacial polymer surface. J Prosthet Dent 116(1):136–143. https://doi.org/10.1016/j.prosdent.2015.12.014
130. Guneser MB, Akbulut MB, Eldeniz AU (2016) Antibacterial effect of chlorhexidine-cetrimide combination, Salvia officinalis plant extract and octenidine in comparison with conventional endodontic irrigants. Dent Mater J 35(5):736–741. https://doi.org/10.4012/dmj.2015-159
131. Gunther F, Blessing B, Tacconelli E, Mutters NT (2017) MRSA decolonization failure-are biofilms the missing link? Antimicrob Resist Infect Control 6:32. https://doi.org/10.1186/s13756-017-0192-1
132. Guo W, Shan K, Xu B, Li J (2015) Determining the resistance of carbapenem-resistant Klebsiella pneumoniae to common disinfectants and elucidating the underlying resistance mechanisms. Pathogens Global Health 109(4):184–192. https://doi.org/10.1179/2047773215y.0000000022
133. Gutierrez-Martin CB, Yubero S, Martinez S, Frandoloso R, Rodriguez-Ferri EF (2011) Evaluation of efficacy of several disinfectants against Campylobacter jejuni strains by a suspension test. Res Vet Sci 91(3):e44–47. https://doi.org/10.1016/j.rvsc.2011.01.020
134. Guzman Prieto AM, Wijngaarden J, Braat JC, Rogers MRC, Majoor E, Brouwer EC, Zhang X, Bayjanov JR, Bonten MJM, Willems RJL, van Schaik W (2017) The two-component system ChtRS contributes to chlorhexidine tolerance in Enterococcus faecium. Antimicrob Agents Chemother 61(5). https://doi.org/10.1128/aac.02122-16
135. Hare R, Raik E, Gash S (1963) Efficiency of antiseptics when acting on dried organisms. BMJ 1(5329):496–500
136. Hasanvand A, Ghafourian S, Taherikalani M, Jalilian FA, Sadeghifard N, Pakzad I (2015) Antiseptic resistance in methicillin sensitive and methicillin resistant Staphylococcus aureus isolates from some major hospitals, Iran. Recent Patents Anti-infective Drug Discov 10 (2):105–112

137. Hassan KA, Liu Q, Elbourne LDH, Ahmad I, Sharples D, Naidu V, Chan CL, Li L, Harborne SPD, Pokhrel A, Postis VLG, Goldman A, Henderson PJF, Paulsen IT (2018) Pacing across the membrane: the novel PACE family of efflux pumps is widespread in gram-negative pathogens. Res Microbiol. https://doi.org/10.1016/j.resmic.2018.01.001
138. Hassan KA, Liu Q, Henderson PJ, Paulsen IT (2015) Homologs of the Acinetobacter baumannii AceI transporter represent a new family of bacterial multidrug efflux systems. mBio 6(1). https://doi.org/10.1128/mbio.01982-14
139. Hayashi M, Kawamura K, Matsui M, Suzuki M, Suzuki S, Shibayama K, Arakawa Y (2017) Reduction in chlorhexidine efficacy against multi-drug-resistant Acinetobacter baumannii international clone II. J Hosp Infect 95(3):318–323. https://doi.org/10.1016/j.jhin.2016.12.004
140. He XF, Zhang HJ, Cao JG, Liu F, Wang JK, Ma WJ, Yin W (2017) A novel method to detect bacterial resistance to disinfectants. Genes Dis 4(3):163–169. https://doi.org/10.1016/j.gendis.2017.07.001
141. Hedin G, Hambraeus A (1993) Daily scrub with chlorhexidine reduces skin colonization by antibiotic-resistant *Staphylococcus epidermidis*. J Hosp Infect 24(1):47–61
142. Herruzo-Cabrera R, Garcia-Torres V, Rey-Calero J, Vizcaino-Alcaide MJ (1992) Evaluation of the penetration strength, bactericidal efficacy and spectrum of action of several antimicrobial creams against isolated microorganisms in a burn centre. Burns: J Int Soc Burn Injuries 18(1):39–44
143. Herruzo-Cabrera R, Uriarte MC, Rey-Calero J (1999) Antimicrobial effectiveness of 2% glutaraldehyde versus other disinfectants for hospital equipment, in an in vitro test based on germ-carriers with a high microbial contamination. Rev Stomatol Chir Maxillofac 100 (6):299–305
144. Higgins CS, Murtough SM, Williamson E, Hiom SJ, Payne DJ, Russell AD, Walsh TR (2001) Resistance to antibiotics and biocides among non-fermenting Gram-negative bacteria. Clin Microbiol Infect 7(6):308–315
145. Hijazi K, Mukhopadhya I, Abbott F, Milne K, Al-Jabri ZJ, Oggioni MR, Gould IM (2016) Susceptibility to chlorhexidine amongst multidrug-resistant clinical isolates of Staphylococcus epidermidis from bloodstream infections. Int J Antimicrob Agents 48(1):86–90. https://doi.org/10.1016/j.ijantimicag.2016.04.015
146. Ho CM, Li CY, Ho MW, Lin CY, Liu SH, Lu JJ (2012) High rate of qacA- and qacB-positive methicillin-resistant Staphylococcus aureus isolates from chlorhexidine-impregnated catheter-related bloodstream infections. Antimicrob Agents Chemother 56(11):5693–5697. doi:AAC.00761-12 [pii]
147. Hobson DW, Woller W, Anderson L, Guthery E (1998) Development and evaluation of a new alcohol-based surgical hand scrub formulation with persistant antimicrobial characteristics and brushless application. Am J Infect Control 26(10):507–512
148. Hoekstra MJ, Westgate SJ, Mueller S (2017) Povidone-iodine ointment demonstrates in vitro efficacy against biofilm formation. Int Wound J 14(1):172–179. https://doi.org/10.1111/iwj.12578
149. Hope CK, Wilson M (2004) Analysis of the effects of chlorhexidine on oral biofilm vitality and structure based on viability profiling and an indicator of membrane integrity. Antimicrob Agents Chemother 48(5):1461–1468
150. Houari A, Di Martino P (2007) Effect of chlorhexidine and benzalkonium chloride on bacterial biofilm formation. Lett Appl Microbiol 45(6):652–656. https://doi.org/10.1111/j.1472-765X.2007.02249.x
151. Hubner NO, Matthes R, Koban I, Randler C, Muller G, Bender C, Kindel E, Kocher T, Kramer A (2010) Efficacy of chlorhexidine, polihexanide and tissue-tolerable plasma against Pseudomonas aeruginosa biofilms grown on polystyrene and silicone materials. Skin Pharmacol Physiol 23(Suppl):28–34. https://doi.org/10.1159/000318265

152. Hughes C, Ferguson J (2017) Phenotypic chlorhexidine and triclosan susceptibility in clinical Staphylococcus aureus isolates in Australia. Pathology 49(6):633–637. https://doi.org/10.1016/j.pathol.2017.05.008
153. Ignak S, Nakipoglu Y, Gurler B (2017) Frequency of antiseptic resistance genes in clinical staphycocci and enterococci isolates in Turkey. Antimicrob Resist Infect Control 6:88. https://doi.org/10.1186/s13756-017-0244-6
154. Imbert C, Lassy E, Daniault G, Jacquemin JL, Rodier MH (2003) Treatment of plastic and extracellular matrix components with chlorhexidine or benzalkonium chloride: effect on Candida albicans adherence capacity in vitro. J Antimicrob Chemother 51(2):281–287
155. Irizarry L, Merlin T, Rupp J, Griffith J (1996) Reduced susceptibility of methicillin-resistant Staphylococcus aureus to cetylpyridinium chloride and chlorhexidine. Chemotherapy 42 (4):248–252
156. Ismaeel N, Furr JR, Russell AD (1986) Sensitivity and resistance of some strains of *Providencia stuartii* to antiseptics, disinfectants and preservatives. Microbios Letter 33:59–64
157. Izano EA, Shah SM, Kaplan JB (2009) Intercellular adhesion and biocide resistance in nontypeable Haemophilus influenzae biofilms. Microb Pathog 46(4):207–213. https://doi.org/10.1016/j.micpath.2009.01.004
158. Järvinen H, Pienihäkkinen K, Huovinen P, Tenovuo J (1995) Susceptibility of *Streptococcus mutans* and *Streptococcus sobrinus* to antimicrobial agents after short-term oral chlorhexidine treatments. Eur J Oral Sci 103(1):32–35
159. Jarvinen H, Tenovuo J, Huovinen P (1993) In vitro susceptibility of Streptococcus mutans to chlorhexidine and six other antimicrobial agents. Antimicrob Agents Chemother 37 (5):1158–1159
160. Johnson RC, Schlett CD, Crawford K, Lanier JB, Merrell DS, Ellis MW (2015) Recurrent methicillin-resistant Staphylococcus aureus cutaneous abscesses and selection of reduced chlorhexidine susceptibility during chlorhexidine use. J Clin Microbiol 53(11):3677–3682. https://doi.org/10.1128/jcm.01771-15
161. Joy Sinha D, K DSN, Jaiswal N, Vasudeva A, Prabha Tyagi S, Pratap Singh U (2017) Antibacterial effect of Azadirachta indica (Neem) or curcuma longa (Turmeric) against Enterococcus faecalis compared with that of 5% sodium hypochlorite or 2% chlorhexidine in vitro. Bull Tokyo Dental College 58(2):103–109. https://doi.org/10.2209/tdcpublication.2015-0029
162. Jurczyk K, Nietzsche S, Ender C, Sculean A, Eick S (2016) In-vitro activity of sodium-hypochlorite gel on bacteria associated with periodontitis. Clin Oral Invest 20 (8):2165–2173. https://doi.org/10.1007/s00784-016-1711-9
163. Jutkina J, Marathe NP, Flach CF, Larsson DGJ (2017) Antibiotics and common antibacterial biocides stimulate horizontal transfer of resistance at low concentrations. Sci Total Environ 616–617:172–178. https://doi.org/10.1016/j.scitotenv.2017.10.312
164. Kadry AA, Serry FM, El-Ganiny AM, El-Baz AM (2017) Integron occurrence is linked to reduced biocide susceptibility in multidrug resistant Pseudomonas aeruginosa. Br J Biomed Sci 74(2):78–84. https://doi.org/10.1080/09674845.2017.1278884
165. Kalkanci A, Elli M, Adil Fouad A, Yesilyurt E, Jabban Khalil I (2015) Assessment of susceptibility of mould isolates towards biocides. Journal de mycologie medicale 25(4):280–286. https://doi.org/10.1016/j.mycmed.2015.08.001
166. Kampf G (2006) Are biofilms relevant for skin disinfection? J Hosp Infect 63(1):106–108. https://doi.org/10.1016/j.jhin.2005.11.007
167. Kampf G (2008) How effective are hand antiseptics for the post-contamination treatment of hands when used as recommended? Am J Infect Control 36(5):356–360
168. Kampf G (2008) What is left to justify the use of chlorhexidine in hand hygiene? J Hosp Infect 70(Suppl. 1):27–34

169. Kampf G (2016) Acquired resistance to chlorhexidine—is it time to establish an "antiseptic stewardship" initiative? J Hosp Infect 94(3):213–227
170. Kampf G, Höfer M, Wendt C (1999) Efficacy of hand disinfectants against vancomycin-resistant Enterococci in vitro. J Hosp Infect 42(2):143–150
171. Kampf G, Jarosch R, Rüden H (1998) Limited effectiveness of chlorhexidine based hand disinfectants against methicillin-resistant *Staphylococcus aureus* (MRSA). J Hosp Infect 38 (4):297–303
172. Kampf G, Kramer A, Suchomel M (2017) Lack of sustained efficacy for alcohol-based surgical hand rubs containing "residual active ingredients" according to EN 12791. J Hosp Infect 95(2):163–168
173. Kampf G, Ostermeyer C (2005) Efficacy of two distinct ethanol-based hand rubs for surgical hand disinfection—a controlled trial according to prEN 12791. BMC Infect Dis 5:17
174. Kampf G, Ostermeyer C, Heeg P (2005) Surgical hand disinfection with a propanol-based hand rub: equivalence of shorter application times. J Hosp Infect 59(4):304–310
175. Kampf G, Reichel M, Hollingsworth A, Bashir M (2013) Efficacy of surgical hand scrub products based on chlorhexidine is largely overestimated without neutralizing agents in the sampling fluid. Am J Infect Control 41(1):e1–5. doi:S0196-6553(12)01069-3 [pii]
176. Karpanen TJ, Worthington T, Hendry ER, Conway BR, Lambert PA (2008) Antimicrobial efficacy of chlorhexidine digluconate alone and in combination with eucalyptus oil, tea tree oil and thymol against planktonic and biofilm cultures of Staphylococcus epidermidis. J Antimicrob Chemother 62(5):1031–1036. https://doi.org/10.1093/jac/dkn325
177. Kawamura-Sato K, Wachino J, Kondo T, Ito H, Arakawa Y (2008) Reduction of disinfectant bactericidal activities in clinically isolated Acinetobacter species in the presence of organic material. J Antimicrob Chemother 61(3):568–576. https://doi.org/10.1093/jac/dkm498
178. Kawamura-Sato K, Wachino J, Kondo T, Ito H, Arakawa Y (2010) Correlation between reduced susceptibility to disinfectants and multidrug resistance among clinical isolates of Acinetobacter species. J Antimicrob Chemother 65(9):1975–1983. https://doi.org/10.1093/jac/dkq227
179. Kiesow A, Sarembe S, Pizzey RL, Axe AS, Bradshaw DJ (2016) Material compatibility and antimicrobial activity of consumer products commonly used to clean dentures. J Prosthet Dent 115(2):189–198.e188. https://doi.org/10.1016/j.prosdent.2015.08.010
180. Kim JS, Chung YK, Lee SS, Lee JA, Kim HS, Park EY, Shin KS, Kang BS, Lee HJ, Kang HJ (2016) Effect of daily chlorhexidine bathing on the acquisition of methicillin-resistant Staphylococcus aureus in a medical intensive care unit with methicillin-resistant S aureus endemicity. Am J Infect Control 44(12):1520–1525. https://doi.org/10.1016/j.ajic.2016.04.252
181. Kitagawa H, Izutani N, Kitagawa R, Maezono H, Yamaguchi M, Imazato S (2016) Evolution of resistance to cationic biocides in Streptococcus mutans and Enterococcus faecalis. J Dent 47:18–22. https://doi.org/10.1016/j.jdent.2016.02.008
182. Knapp L, Amezquita A, McClure P, Stewart S, Maillard JY (2015) Development of a protocol for predicting bacterial resistance to microbicides. Appl Environ Microbiol 81 (8):2652–2659. https://doi.org/10.1128/aem.03843-14
183. Knapp L, Rushton L, Stapleton H, Sass A, Stewart S, Amezquita A, McClure P, Mahenthiralingam E, Maillard JY (2013) The effect of cationic microbicide exposure against Burkholderia cepacia complex (Bcc); the use of Burkholderia lata strain 383 as a model bacterium. J Appl Microbiol 115(5):1117–1126. https://doi.org/10.1111/jam.12320
184. Ko S, An HS, Bang JH, Park SW (2015) An outbreak of Burkholderia cepacia complex pseudobacteremia associated with intrinsically contaminated commercial 0.5% chlorhexidine solution. Am J Infect Control 43(3):266–268. https://doi.org/10.1016/j.ajic.2014.11.010

185. Koban I, Geisel MH, Holtfreter B, Jablonowski L, Hubner NO, Matthes R, Masur K, Weltmann KD, Kramer A, Kocher T (2013) Synergistic effects of nonthermal plasma and disinfecting agents against dental biofilms in vitro. ISRN Dentistry 2013:573262. https://doi.org/10.1155/2013/573262
186. Koburger T, Hübner N-O, Braun M, Siebert J, Kramer A (2010) Standardized comparison of antiseptic efficacy of triclosan, PVP-iodine, octenidine dihydrochloride, polyhexanide and chlorhexidine digluconate. J Antimicrob Chemother 65(8):1712–1719
187. Koljalg S, Naaber P, Mikelsaar M (2002) Antibiotic resistance as an indicator of bacterial chlorhexidine susceptibility. J Hosp Infect 51(2):106–113
188. Koo H, Hayacibara MF, Schobel BD, Cury JA, Rosalen PL, Park YK, Vacca-Smith AM, Bowen WH (2003) Inhibition of Streptococcus mutans biofilm accumulation and polysaccharide production by apigenin and tt-farnesol. J Antimicrob Chemother 52 (5):782–789. https://doi.org/10.1093/jac/dkg449
189. Kramer A, Assadian O, Koburger-Janssen T (2016) Antimicrobial efficacy of the combination of chlorhexidine digluconate and dexpanthenol. GMS Hyg Infect Control 11: Doc24. https://doi.org/10.3205/dgkh000284
190. Kratzer C, Tobudic S, Schmoll M, Graninger W, Georgopoulos A (2006) In vitro activity and synergism of amphotericin B, azoles and cationic antimicrobials against the emerging pathogen Trichoderma spp. J Antimicrob Chemother 58(5):1058–1061. https://doi.org/10.1093/jac/dkl384
191. Krishnamoorthy S, Shah BP, Lee HH, Martinez LR (2015) Microbicides alter the expression and function of RND-Type efflux pump AdeABC in biofilm-associated cells of Acinetobacter baumannii clinical isolates. Antimicrob Agents Chemother 60(1):57–63. https://doi.org/10.1128/aac.01045-15
192. Krishnan U, Saji S, Clarkson R, Lalloo R, Moule AJ (2017) Free active chlorine in sodium hypochlorite solutions admixed with Octenidine, SmearOFF, Chlorhexidine, and EDTA. J Endod 43(8):1354–1359. https://doi.org/10.1016/j.joen.2017.03.034
193. Kuhn DM, George T, Chandra J, Mukherjee PK, Ghannoum MA (2002) Antifungal susceptibility of Candida biofilms: unique efficacy of amphotericin B lipid formulations and echinocandins. Antimicrob Agents Chemother 46(6):1773–1780
194. Kurihara T, Sugita M, Motai S, Kurashige S (1993) In vitro induction of chlorhexidine- and benzalkonium-resistance in clinically isolated Pseudomonas aeruginosa. Kansenshogaku zasshi J Jpn Assoc Infect Dis 67(3):202–206
195. Kwong SM, Lim R, Lebard RJ, Skurray RA, Firth N (2008) Analysis of the pSK1 replicon, a prototype from the staphylococcal multiresistance plasmid family. Microbiology (Reading, England) 154(Pt 10):3084–3094. https://doi.org/10.1099/mic.0.2008/017418-0
196. LaFleur MD, Kumamoto CA, Lewis K (2006) Candida albicans biofilms produce antifungal-tolerant persister cells. Antimicrob Agents Chemother 50(11):3839–3846. https://doi.org/10.1128/aac.00684-06
197. LaFleur MD, Qi Q, Lewis K (2010) Patients with long-term oral carriage harbor high-persister mutants of Candida albicans. Antimicrob Agents Chemother 54(1):39–44. https://doi.org/10.1128/aac.00860-09
198. Lam OL, McGrath C, Li LS, Samaranayake LP (2012) Effectiveness of oral hygiene interventions against oral and oropharyngeal reservoirs of aerobic and facultatively anaerobic gram-negative bacilli. Am J Infect Control 40(2):175–182. https://doi.org/10.1016/j.ajic.2011.03.004
199. Lambert RJ (2004) Comparative analysis of antibiotic and antimicrobial biocide susceptibility data in clinical isolates of methicillin-sensitive Staphylococcus aureus, methicillin-resistant Staphylococcus aureus and Pseudomonas aeruginosa between 1989 and 2000. J Appl Microbiol 97(4):699–711. https://doi.org/10.1111/j.1365-2672.2004.02345.x

200. Lambert RJ, Joynson J, Forbes B (2001) The relationships and susceptibilities of some industrial, laboratory and clinical isolates of Pseudomonas aeruginosa to some antibiotics and biocides. J Appl Microbiol 91(6):972–984
201. Lamfon H, Porter SR, McCullough M, Pratten J (2004) Susceptibility of Candida albicans biofilms grown in a constant depth film fermentor to chlorhexidine, fluconazole and miconazole: a longitudinal study. J Antimicrob Chemother 53(2):383–385. https://doi.org/10.1093/jac/dkh071
202. Landelle C, von Dach E, Haustein T, Agostinho A, Renzi G, Renzoni A, Pittet D, Schrenzel J, Francois P, Harbarth S (2016) Randomized, placebo-controlled, double-blind clinical trial to evaluate the efficacy of polyhexanide for topical decolonization of MRSA carriers. J Antimicrob Chemother 71(2):531–538. https://doi.org/10.1093/jac/dkv331
203. Lanjri S, Uwingabiye J, Frikh M, Abdellatifi L, Kasouati J, Maleb A, Bait A, Lemnouer A, Elouennass M (2017) In vitro evaluation of the susceptibility of Acinetobacter baumannii isolates to antiseptics and disinfectants: comparison between clinical and environmental isolates. Antimicrob Resist Infect Control 6:36. https://doi.org/10.1186/s13756-017-0195-y
204. Lannigan R, Bryan LE (1985) Decreased susceptibility of Serratia marcescens to chlorhexidine related to the inner membrane. J Antimicrob Chemother 15(5):559–565
205. Larsen T, Fiehn NE (1996) Resistance of Streptococcus sanguis biofilms to antimicrobial agents. APMIS: acta pathologica, microbiologica, et immunologica Scandinavica 104 (4):280–284
206. Lasserre JF, Leprince JG, Toma S, Brecx MC (2015) Electrical enhancement of chlorhexidine efficacy against the periodontal pathogen Porphyromonas gingivalis within a biofilm. The New Microbiologica 38(4):511–519
207. Lavilla Lerma L, Benomar N, Casado Munoz Mdel C, Galvez A, Abriouel H (2015) Correlation between antibiotic and biocide resistance in mesophilic and psychrotrophic Pseudomonas spp. isolated from slaughterhouse surfaces throughout meat chain production. Food Microbiol 51:33–44. https://doi.org/10.1016/j.fm.2015.04.010
208. Lavilla Lerma L, Benomar N, Valenzuela AS, Casado Munoz Mdel C, Galvez A, Abriouel H (2014) Role of EfrAB efflux pump in biocide tolerance and antibiotic resistance of Enterococcus faecalis and Enterococcus faecium isolated from traditional fermented foods and the effect of EDTA as EfrAB inhibitor. Food Microbiol 44:249–257. https://doi.org/10.1016/j.fm.2014.06.009
209. Lear JC, Maillard JY, Dettmar PW, Goddard PA, Russell AD (2006) Chloroxylenol- and triclosan-tolerant bacteria from industrial sources—susceptibility to antibiotics and other biocides. Int Biodeter Biodegr 57(1):51–56. https://doi.org/10.1016/j.ibiod.2005.11.002
210. Lee AS, Macedo-Vinas M, Francois P, Renzi G, Schrenzel J, Vernaz N, Pittet D, Harbarth S (2011) Impact of combined low-level mupirocin and genotypic chlorhexidine resistance on persistent methicillin-resistant Staphylococcus aureus carriage after decolonization therapy: a case-control study. Clinical Infect Dis: An Off Publ Infect Dis Soc Am 52(12):1422–1430. https://doi.org/10.1093/cid/cir233
211. Leelaporn A, Paulsen IT, Tennent JM, Littlejohn TG, Skurray RA (1994) Multidrug resistance to antiseptics and disinfectants in coagulase-negative staphylococci. J Med Microbiol 40(3):214–220. https://doi.org/10.1099/00222615-40-3-214
212. Lefebvre E, Vighetto C, Di Martino P, Larreta Garde V, Seyer D (2016) Synergistic antibiofilm efficacy of various commercial antiseptics, enzymes and EDTA: a study of Pseudomonas aeruginosa and Staphylococcus aureus biofilms. Int J Antimicrob Agents 48 (2):181–188. https://doi.org/10.1016/j.ijantimicag.2016.05.008
213. Lepainteur M, Royer G, Bourrel AS, Romain O, Duport C, Doucet-Populaire F, Decousser JW (2013) Prevalence of resistance to antiseptics and mupirocin among invasive coagulase-negative Staphylococci from very preterm neonates in NICU: the creeping threat? J Hosp Infect 83(4):333–336. https://doi.org/10.1016/j.jhin.2012.11.025

214. Li JY, Wang XJ, Wang LN, Ying XX, Ren X, Liu HY, Xu L, Ma GW (2015) High in vitro antibacterial activity of Pac-525 against Porphyromonas gingivalis biofilms cultured on titanium. Biomed Res Int 2015:909870. https://doi.org/10.1155/2015/909870
215. Liaqat I, Sabri AN (2008) Effect of biocides on biofilm bacteria from dental unit water lines. Curr Microbiol 56(6):619–624. https://doi.org/10.1007/s00284-008-9136-6
216. Liu Q, Hassan KA, Ashwood HE, Gamage H, Li L, Mabbutt BC, Paulsen IT (2018) Regulation of the aceI multidrug efflux pump gene in Acinetobacter baumannii. J Antimicrob Chemother. https://doi.org/10.1093/jac/dky034
217. Liu Q, Zhao H, Han L, Shu W, Wu Q, Ni Y (2015) Frequency of biocide-resistant genes and susceptibility to chlorhexidine in high-level mupirocin-resistant, methicillin-resistant Staphylococcus aureus (MuH MRSA). Diagn Microbiol Infect Dis 82(4):278–283. https://doi.org/10.1016/j.diagmicrobio.2015.03.023
218. Liu WJ, Fu L, Huang M, Zhang JP, Wu Y, Zhou YS, Zeng J, Wang GX (2017) Frequency of antiseptic resistance genes and reduced susceptibility to biocides in carbapenem-resistant Acinetobacter baumannii. J Med Microbiol 66(1):13–17. https://doi.org/10.1099/jmm.0.000403
219. Long M, Lai H, Deng W, Zhou K, Li B, Liu S, Fan L, Wang H, Zou L (2016) Disinfectant susceptibility of different Salmonella serotypes isolated from chicken and egg production chains. J Appl Microbiol 121(3):672–681. https://doi.org/10.1111/jam.13184
220. Longtin J, Seah C, Siebert K, McGeer A, Simor A, Longtin Y, Low DE, Melano RG (2011) Distribution of antiseptic resistance genes qacA, qacB, and smr in methicillin-resistant Staphylococcus aureus isolated in Toronto, Canada, from 2005 to 2009. Antimicrob Agents Chemother 55(6):2999–3001. https://doi.org/10.1128/aac.01707-10
221. Lorin D, Cristina RT, Teusdea V, Mitranescu E, Muselin F, Butnariu M, David G, Dumitrescu E (2017) Efficiency of four currently used decontamination conditionings in Romania against Aspergillus and Candida strains. Journal de mycologie medicale 27 (3):357–363. https://doi.org/10.1016/j.mycmed.2017.04.013
222. Lourenco TG, Heller D, do Souto RM, Silva-Senem MX, Varela VM, Torres MC, Feres-Filho EJ, Colombo AP (2015) Long-term evaluation of the antimicrobial susceptibility and microbial profile of subgingival biofilms in individuals with aggressive periodontitis. Brazilian J Microbiol [Publication of the Brazilian Society for Microbiology] 46(2):493–500. https://doi.org/10.1590/s1517-838246220131037
223. Lowe CF, Lloyd-Smith E, Sidhu B, Ritchie G, Sharma A, Jang W, Wong A, Bilawka J, Richards D, Kind T, Puddicombe D, Champagne S, Leung V, Romney MG (2017) Reduction in hospital-associated methicillin-resistant Staphylococcus aureus and vancomycin-resistant Enterococcus with daily chlorhexidine gluconate bathing for medical inpatients. Am J Infect Control 45(3):255–259. https://doi.org/10.1016/j.ajic.2016.09.019
224. Ma J, Tong Z, Ling J, Liu H, Wei X (2015) The effects of sodium hypochlorite and chlorhexidine irrigants on the antibacterial activities of alkaline media against Enterococcus faecalis. Arch Oral Biol 60(7):1075–1081. https://doi.org/10.1016/j.archoralbio.2015.04.008
225. Macias JH, Arreguin V, Munoz JM, Alvarez JA, Mosqueda JL, Macias AE (2013) Chlorhexidine is a better antiseptic than povidone iodine and sodium hypochlorite because of its substantive effect. Am J Infect Control 41(7):634–637. https://doi.org/10.1016/j.ajic.2012.10.002
226. Macinga DR, Edmonds SL, Campbell E, McCormack RR (2014) Comparative efficacy of alcohol-based surgical scrubs: the importance of formulation. AORN J 100(6):641–650. https://doi.org/10.1016/j.aorn.2014.03.013
227. Maiwald M (2017) Skin preparation for prevention of surgical site infection after Cesarean delivery: a randomized controlled trial. Obstet Gynecol 129(4):750–751. https://doi.org/10.1097/aog.0000000000001956
228. Maiwald M, Chan ES (2012) The forgotten role of alcohol: a systematic review and meta-analysis of the clinical efficacy and perceived role of chlorhexidine in skin antisepsis. PLoS ONE 7(9):e44277. https://doi.org/10.1371/journal.pone.0044277

229. Mal PB, Farooqi J, Irfan S, Hughes MA, Khan E (2016) Reduced susceptibility to chlorhexidine disinfectant among New Delhi metallo-beta-lactamase-1 positive Enterobacteriaceae and other multidrug-resistant organisms: Report from a tertiary care hospital in Karachi, Pakistan. Indian J Med Microbiol 34(3):346–349. https://doi.org/10.4103/0255-0857.188338
230. Marchetti MG, Kampf G, Finzi G, Salvatorelli G (2003) Evaluation of the bactericidal effect of five products for surgical hand disinfection according to prEN 12054 and prEN 12791. J Hosp Infect 54(1):63–67
231. Marchi E, Furi L, Arioli S, Morrissey I, Di Lorenzo V, Mora D, Giovannetti L, Oggioni MR, Viti C (2015) Novel insight into antimicrobial resistance and sensitivity phenotypes associated to qac and norA genotypes in Staphylococcus aureus. Microbiol Res 170:184–194. https://doi.org/10.1016/j.micres.2014.07.001
232. Maris P (1991) Resistance of 700 gram-negative bacterial strains to antiseptics and antibiotics. Annales de recherches veterinaires Ann Vet Res 22(1):11–23
233. Mariscal A, Lopez-Gigosos RM, Carnero-Varo M, Fernandez-Crehuet J (2009) Fluorescent assay based on resazurin for detection of activity of disinfectants against bacterial biofilm. Appl Microbiol Biotechnol 82(4):773–783. https://doi.org/10.1007/s00253-009-1879-x
234. Marrie TJ, Costerton JW (1981) Proplonged survival of *Serratia marcescens* in chlorhexidine. Appl Environ Microbiol 42(6):1093–1102
235. Martro E, Hernandez A, Ariza J, Dominguez MA, Matas L, Argerich MJ, Martin R, Ausina V (2003) Assessment of Acinetobacter baumannii susceptibility to antiseptics and disinfectants. J Hosp Infect 55(1):39–46
236. Maseda H, Hashida Y, Konaka R, Shirai A, Kourai H (2009) Mutational upregulation of a resistance-nodulation-cell division-type multidrug efflux pump, SdeAB, upon exposure to a biocide, cetylpyridinium chloride, and antibiotic resistance in Serratia marcescens. Antimicrob Agents Chemother 53(12):5230–5235. doi:AAC.00631-09 [pii]
237. Mavri A, Mozina SS (2012) Involvement of efflux mechanisms in biocide resistance of Campylobacter jejuni and Campylobacter coli. J Med Microbiol 61(Pt 6):800–808. https://doi.org/10.1099/jmm.0.041467-0
238. Mavri A, Smole Mozina S (2013) Development of antimicrobial resistance in Campylobacter jejuni and Campylobacter coli adapted to biocides. Int J Food Microbiol 160 (3):304–312. https://doi.org/10.1016/j.ijfoodmicro.2012.11.006
239. Mavri A, Smole Mozina S (2013) Effects of efflux-pump inducers and genetic variation of the multidrug transporter cmeB in biocide resistance of Campylobacter jejuni and Campylobacter coli. J Med Microbiol 62(Pt 3):400–411. https://doi.org/10.1099/jmm.0.052316-0
240. McAllister TA, Lucas CE, Mocan H, Liddell RHA, Gibson BES, Hann IM, Platt DJ (1989) *Serratia marcescens* outbreak in a paediatric oncology unit traced to contaminated chlorhexidine. Scott Med J 34:525–528
241. McDanel JS, Murphy CR, Diekema DJ, Quan V, Kim DS, Peterson EM, Evans KD, Tan GL, Hayden MK, Huang SS (2013) Chlorhexidine and mupirocin susceptibilities of methicillin-resistant staphylococcus aureus from colonized nursing home residents. Antimicrob Agents Chemother 57(1):552–558. https://doi.org/10.1128/aac.01623-12
242. McNeil JC, Hulten KG, Kaplan SL, Mason EO (2014) Decreased susceptibilities to Retapamulin, Mupirocin, and Chlorhexidine among Staphylococcus aureus isolates causing skin and soft tissue infections in otherwise healthy children. Antimicrob Agents Chemother 58(5):2878–2883. https://doi.org/10.1128/aac.02707-13
243. McNeil JC, Kok EY, Vallejo JG, Campbell JR, Hulten KG, Mason EO, Kaplan SL (2016) Clinical and Molecular features of decreased chlorhexidine susceptibility among nosocomial Staphylococcus aureus isolates at texas children's hospital. Antimicrob Agents Chemother 60(2):1121–1128. https://doi.org/10.1128/aac.02011-15

244. Meiller TF, Kelley JI, Jabra-Rizk MA, DePaola LG, Baqui AAMA, Falkler WA (2001) In vitro studies of the efficacy of antimicrobials against fungi. Oral Surg Oral Med Oral Pathol Oral Radiol Endod 91(6):663–670. https://doi.org/10.1067/moe.2001.113550
245. Melo RT, Mendonca EP, Monteiro GP, Siqueira MC, Pereira CB, Peres P, Fernandez H, Rossi DA (2017) Intrinsic and extrinsic aspects on Campylobacter jejuni biofilms. Front Microbiol 8:1332. https://doi.org/10.3389/fmicb.2017.01332
246. Mendes ET, Ranzani OT, Marchi AP, Silva MT, Filho JU, Alves T, Guimaraes T, Levin AS, Costa SF (2016) Chlorhexidine bathing for the prevention of colonization and infection with multidrug-resistant microorganisms in a hematopoietic stem cell transplantation unit over a 9-year period: Impact on chlorhexidine susceptibility. Medicine 95(46):e5271. https://doi.org/10.1097/md.0000000000005271
247. Mendoza-Olazaran S, Camacho-Ortiz A, Martinez-Resendez MF, Llaca-Diaz JM, Perez-Rodriguez E, Garza-Gonzalez E (2014) Influence of whole-body washing of critically ill patients with chlorhexidine on Acinetobacter baumannii isolates. Am J Infect Control 42 (8):874–878. https://doi.org/10.1016/j.ajic.2014.04.009
248. Mengistu Y, Erge W, Bellete B (1999) In vitro susceptibility of gram-negative bacterial isolates to chlorhexidine gluconate. East Afr Med J 76(5):243–246
249. Messager S, Goddard PA, Dettmar PW, Maillard JY (2001) Determination of the antibacterial efficacy of several antiseptics tested on skin by an 'ex-vivo' test. J Med Microbiol 50(3):284–292. https://doi.org/10.1099/0022-1317-50-3-284
250. Meurman JH, Jousimies-Somer H, Suomala P, Alaluusua S, Torkko H, Asikainen S (1989) Activity of amine-stannous fluoride combination and chlorhexidine against some aerobic and anaerobic oral bacteria. Oral Microbiol Immunol 4(2):117–119
251. Millhouse E, Jose A, Sherry L, Lappin DF, Patel N, Middleton AM, Pratten J, Culshaw S, Ramage G (2014) Development of an in vitro periodontal biofilm model for assessing antimicrobial and host modulatory effects of bioactive molecules. BMC oral health 14:80. https://doi.org/10.1186/1472-6831-14-80
252. Millward TA, Wilson M (1989) The effect of chlorhexidine on Streptococcus sanguis biofilms. Microbios 58(236–237):155–164
253. Mimoz O, Lucet JC, Kerforne T, Pascal J, Souweine B, Goudet V, Mercat A, Bouadma L, Lasocki S, Alfandari S, Friggeri A, Wallet F, Allou N, Ruckly S, Balayn D, Lepape A, Timsit JF (2015) Skin antisepsis with chlorhexidine-alcohol versus povidone iodine-alcohol, with and without skin scrubbing, for prevention of intravascular-catheter-related infection (CLEAN): an open-label, multicentre, randomised, controlled, two-by-two factorial trial. Lancet 386(10008):2069–2077. https://doi.org/10.1016/s0140-6736(15)00244-5
254. Miyano N, Oie S, Kamiya A (2003) Efficacy of disinfectants and hot water against biofilm cells of Burkholderia cepacia. Biol Pharm Bull 26(5):671–674
255. Molina-Cabrillana J, Santana-Reyes C, Gonzalez-Garcia A, Bordes-Benitez A, Horcajada I (2007) Outbreak of Achromobacter xylosoxidans pseudobacteremia in a neonatal care unit related to contaminated chlorhexidine solution. Eur J Clin Microbiol Infect Dis 26(6):435–437. https://doi.org/10.1007/s10096-007-0311-7
256. Moore G, Schelenz S, Borman AM, Johnson EM, Brown CS (2017) Yeasticidal activity of chemical disinfectants and antiseptics against Candida auris. J Hosp Infect 97(4):371–375. https://doi.org/10.1016/j.jhin.2017.08.019
257. Moore LE, Ledder RG, Gilbert P, McBain AJ (2008) In vitro study of the effect of cationic biocides on bacterial population dynamics and susceptibility. Appl Environ Microbiol 74 (15):4825–4834. https://doi.org/10.1128/aem.00573-08
258. Morales-Fernandez L, Fernandez-Crehuet M, Espigares M, Moreno E, Espigares E (2014) Study of the hormetic effect of disinfectants chlorhexidine, povidone iodine and benzalkonium chloride. Eur J Clin Microbiol Infect Dis 33(1):103–109. https://doi.org/10.1007/s10096-013-1934-5

259. Morita Y, Murata T, Mima T, Shiota S, Kuroda T, Mizushima T, Gotoh N, Nishino T, Tsuchiya T (2003) Induction of mexCD-oprJ operon for a multidrug efflux pump by disinfectants in wild-type Pseudomonas aeruginosa PAO1. J Antimicrob Chemother 51 (4):991–994. https://doi.org/10.1093/jac/dkg173
260. Morita Y, Tomida J, Kawamura Y (2014) Responses of Pseudomonas aeruginosa to antimicrobials. Front Microbiol 4:422. https://doi.org/10.3389/fmicb.2013.00422
261. Morrissey I, Oggioni MR, Knight D, Curiao T, Coque T, Kalkanci A, Martinez JL (2014) Evaluation of epidemiological cut-off values indicates that biocide resistant subpopulations are uncommon in natural isolates of clinically-relevant microorganisms. PLoS ONE 9(1): e86669. https://doi.org/10.1371/journal.pone.0086669
262. Muggeo A, Guillard T, Klein F, Reffuveille F, Francois C, Babosan A, Bajolet O, Bertrand X, de Champs C (2017) Spread of Klebsiella pneumoniae ST395 non-susceptible to carbapenems and resistant to fluoroquinolones in North-Eastern France. J Global Antimicrob Resis. https://doi.org/10.1016/j.jgar.2017.10.023
263. Mulberry G, Snyder AT, Heilman J, Pyrek J, Stahl J (2001) Evaluation of a waterless, scrubless chlorhexidine gluconate/ ethanol surgical scrub for antimicrobial efficacy. Am J Infect Control 29(12):377–382
264. Muller G, Kramer A (2000) In vitro action of a combination of selected antimicrobial agents and chondroitin sulfate. Chem Biol Interact 124(2):77–85
265. Muller G, Kramer A (2008) Biocompatibility index of antiseptic agents by parallel assessment of antimicrobial activity and cellular cytotoxicity. J Antimicrob Chemother 61 (6):1281–1287. https://doi.org/10.1093/jac/dkn125
266. Muller G, Langer J, Siebert J, Kramer A (2014) Residual antimicrobial effect of chlorhexidine digluconate and octenidine dihydrochloride on reconstructed human epidermis. Skin Pharmacol Physiol 27(1):1–8. https://doi.org/10.1159/000350172
267. Munoz-Gallego I, Infiesta L, Viedma E, Perez-Montarelo D, Chaves F (2016) Chlorhexidine and mupirocin susceptibilities in methicillin-resistant Staphylococcus aureus isolates from bacteraemia and nasal colonisation. J Global Antimicrob Resis 4:65–69. https://doi.org/10.1016/j.jgar.2015.11.005
268. Murayama N, Nagata M, Terada Y, Okuaki M, Takemura N, Nakaminami H, Noguchi N (2013) In vitro antiseptic susceptibilities for Staphylococcus pseudintermedius isolated from canine superficial pyoderma in Japan. Veterinary dermatology 24(1):126–129.e129. https://doi.org/10.1111/j.1365-3164.2012.01103.x
269. Nagai I, Ogase H (1990) Absence of role for plasmids in resistance to multiple disinfectants in three strains of bacteria. J Hosp Infect 15(2):149–155
270. Nakahara H, Kozukoe H (1981) Chlorhexidine resistance in *Escherichia coli* isolated from clinical lesions. Zentralblatt für Bakteriologie und Hygiene, I Abt Orig B 251(2):177–184
271. Nakahara H, Kozukue H (1982) Isolation of chlorhexidine-resistant *Pseudomonas aeruginosa* from clinical lesions. J Clin Microbiol 15(1):166–168
272. Nakaminami H, Noguchi N, Nishijima S, Kurokawa I, So H, Sasatsu M (2007) Transduction of the plasmid encoding antiseptic resistance gene qacB in Staphylococcus aureus. Biological & pharmaceutical bulletin 30(8).1412–1415
273. Namba Y, Suzuki A, Takeshima N, Kato N (1985) Comparative study of bactericidal activities of six different disinfectants. Nagoya J Med Sci 47(3–4):101–112
274. Naparstek L, Carmeli Y, Chmelnitsky I, Banin E, Navon-Venezia S (2012) Reduced susceptibility to chlorhexidine among extremely-drug-resistant strains of Klebsiella pneumoniae. J Hosp Infect 81(1):15–19. https://doi.org/10.1016/j.jhin.2012.02.007
275. Narui K, Takano M, Noguchi N, Sasatsu M (2007) Susceptibilities of methicillin-resistant Staphylococcus aureus isolates to seven biocides. Biol Pharm Bull 30(3):585–587
276. National Center for Biotechnology Information Chlorhexidine (digluconate). PubChem Compound Database; CID = 5360565. https://pubchem.ncbi.nlm.nih.gov/compound/5360565. Accessed 5 Feb 2018

277. Nicoletti G, Boghossian V, Borland R (1990) Hygienic hand disinfection: A comparative study with chlorhexidine detergents and soap. J Hosp Infect 15:323–337
278. Nicoletti G, Boghossian V, Gurevitch F, Borland R, Morgenroth P (1993) The antimicrobial activity *in vitro* of chlorhexidine, a mixture of isothiazolinones ('Kathon' CG) and cetyl trimethyl ammonium bromide (CTAB). J Hosp Infect 23:87–111
279. Norman G, Christie J, Liu Z, Westby MJ, Jefferies JM, Hudson T, Edwards J, Mohapatra DP, Hassan IA, Dumville JC (2017) Antiseptics for burns. Cochrane Database Syst Rev 7: Cd011821. https://doi.org/10.1002/14651858.cd011821.pub2
280. Ntrouka V, Hoogenkamp M, Zaura E, van der Weijden F (2011) The effect of chemotherapeutic agents on titanium-adherent biofilms. Clin Oral Implant Res 22 (11):1227–1234. https://doi.org/10.1111/j.1600-0501.2010.02085.x
281. Oggioni MR, Coelho JR, Furi L, Knight DR, Viti C, Orefici G, Martinez JL, Freitas AT, Coque TM, Morrissey I (2015) Significant differences characterise the correlation coefficients between biocide and antibiotic susceptibility profiles in Staphylococcus aureus. Curr Pharm Des 21(16):2054–2057
282. Ohta S (1990) Studies on resistant mechanisms in the resistant bacteria to chlorhexidine. II. Chemical components of the cell membrane and the electron microscopical observation of cell surface structure of chlorhexidine-resistant bacteria. Yakugaku Zasshi 110(6):414–425
283. Ohta S, Makino M, Nagai K, Zenda H (1996) Comparative fungicidal activity of a new quaternary ammonium salt, N-alkyl-N-2-hydroxyethyl-N, N-dimethylammonium butyl phosphate and commonly used disinfectants. Biol Pharm Bull 19(2):308–310
284. Okuda T, Endo N, Osada Y, Zen-Yoji H (1984) Outbreak of nosocomial urinary tract infections caused by *Serratia marcescens*. J Clin Microbiol 20(4):691–695
285. Orsi GB, Tomao P, Visca P (1995) In vitro activity of commercially manufactured disinfectants against Pseudomonas aeruginosa. Eur J Epidemiol 11(4):453–457
286. Parducz L, Eszik I, Wagner G, Burian K, Endresz V, Virok DP (2016) Impact of antiseptics on Chlamydia trachomatis growth. Lett Appl Microbiol 63(4):260–267. https://doi.org/10.1111/lam.12625
287. Pastrana-Carrasco J, Garza-Ramos JU, Barrios H, Morfin-Otero R, Rodriguez-Noriega E, Barajas JM, Suarez S, Diaz R, Miranda G, Solorzano F, Contreras J, Silva-Sanchez J (2012) [QacEdelta1 gene frequency and biocide resistance in extended-spectrum beta-lactamase producing enterobacteriaceae clinical isolates]. Revista de investigacion clinica; organo del Hospital de Enfermedades de la Nutricion 64(6 Pt 1):535–540
288. Pearce H, Messager S, Maillard JY (1999) Effect of biocides commonly used in the hospital environment on the transfer of antibiotic-resistance genes in Staphylococcus aureus. J Hosp Infect 43(2):101–107. https://doi.org/10.1053/jhin.1999.0250
289. Peeters E, Nelis HJ, Coenye T (2008) Evaluation of the efficacy of disinfection procedures against Burkholderia cenocepacia biofilms. J Hosp Infect 70(4):361–368. https://doi.org/10.1016/j.jhin.2008.08.015
290. Penna TC, Mazzola PG, Silva Martins AM (2001) The efficacy of chemical agents in cleaning and disinfection programs. BMC Infect Dis 1:16
291. Pereira CA, Costa AC, Liporoni PC, Rego MA, Jorge AO (2016) Antibacterial activity of Baccharis dracunculifolia in planktonic cultures and biofilms of Streptococcus mutans. J Infect Public Health 9(3):324–330. https://doi.org/10.1016/j.jiph.2015.10.012
292. Perrins N, Bond R (2003) Synergistic inhibition of the growth in vitro of Microsporum canis by miconazole and chlorhexidine. Vet Dermatol 14(2):99–102
293. Pitt TL, Gaston MA, Hoffman PN (1983) *In vitro* susceptibility of hospital isolates of various bacterial genera to chlorhexidine. J Hosp Infect 4(2):173–176
294. Pitten F-A, Werner H-P, Kramer A (2003) A standardized test to assess the impact of different organic challenges on the antimicrobial activity of antiseptics. J Hosp Infect 55 (2):108–115

295. Popowska M, Olszak M, Markiewicz Z (2006) Susceptibility of Listeria monocytogenes strains isolated from dairy products and frozen vegetables to antibiotics inhibiting murein synthesis and to disinfectants. Pol J Microbiol 55(4):279–288
296. Poty F, Denis C, Baufine-Ducrocq H (1987) [Nosocomial Pseudomonas pickettii infection. Danger of the use of ion-exchange resins]. Presse Medicale (Paris, France: 1983) 16 (24):1185–1187
297. Prag G, Falk-Brynhildsen K, Jacobsson S, Hellmark B, Unemo M, Soderquist B (2014) Decreased susceptibility to chlorhexidine and prevalence of disinfectant resistance genes among clinical isolates of Staphylococcus epidermidis. APMIS: acta pathologica, microbiologica, et immunologica Scandinavica 122(10):961–967. https://doi.org/10.1111/apm.12239
298. Prince HN, Nonemaker WS, Norgard RC, Prince DL (1978) Drug resistance studies with topical antiseptics. J Pharm Sci 67(11):1629–1631
299. Pumbwe L, Skilbeck CA, Wexler HM (2007) Induction of multiple antibiotic resistance in Bacteroides fragilis by benzene and benzene-derived active compounds of commonly used analgesics, antiseptics and cleaning agents. J Antimicrob Chemother 60(6):1288–1297. https://doi.org/10.1093/jac/dkm363
300. Reginato CF, Bandeira LA, Zanette RA, Santurio JM, Alves SH, Danesi CC (2017) Antifungal activity of synthetic antiseptics and natural compounds against Candida dubliniensis before and after in vitro fluconazole exposure. Rev Soc Bras Med Trop 50 (1):75–79. https://doi.org/10.1590/0037-8682-0461-2016
301. Reich PJ, Boyle MG, Hogan PG, Johnson AJ, Wallace MA, Elward AM, Warner BB, Burnham CD, Fritz SA (2016) Emergence of community-associated methicillin-resistant Staphylococcus aureus strains in the neonatal intensive care unit: an infection prevention and patient safety challenge. Clin Microbiol Infect 22(7):645.e641–645.e648. https://doi.org/10.1016/j.cmi.2016.04.013
302. Rema T, Lawrence JR, Dynes JJ, Hitchcock AP, Korber DR (2014) Microscopic and spectroscopic analyses of chlorhexidine tolerance in Delftia acidovorans biofilms. Antimicrob Agents Chemother 58(10):5673–5686. https://doi.org/10.1128/aac.02984-14
303. Rema T, Medihala P, Lawrence JR, Vidovic S, Leppard GG, Reid M, Korber DR (2016) Proteomic analyses of chlorhexidine tolerance mechanisms in delftia acidovorans Biofilms. mSphere 1(1). https://doi.org/10.1128/msphere.00017-15
304. Rensch U, Klein G, Schwarz S, Kaspar H, de Jong A, Kehrenberg C (2013) Comparative analysis of the susceptibility to triclosan and three other biocides of avian Salmonella enterica isolates collected 1979 through 1994 and 2004 through 2010. J Food Prot 76 (4):653–656. https://doi.org/10.4315/0362-028x.jfp-12-420
305. Renzoni A, Von Dach E, Landelle C, Diene SM, Manzano C, Gonzales R, Abdelhady W, Randall CP, Bonetti EJ, Baud D, O'Neill AJ, Bayer A, Cherkaoui A, Schrenzel J, Harbarth S, Francois P (2017) Impact of exposure of methicillin-resistant Staphylococcus aureus to polyhexanide in vitro and in vivo. Antimicrob Agents Chemother 61(10). https://doi.org/10.1128/aac.00272-17
306. Riazi S, Matthews KR (2011) Failure of foodborne pathogens to develop resistance to sanitizers following repeated exposure to common sanitizers. Int Biodeter Biodegr 65 (2):374–378. https://doi.org/10.1016/j.ibiod.2010.12.001
307. Rikimaru T, Kondo M, Kondo S, Oizumi K (2000) Efficacy of common antiseptics against mycobacteria. Int J Tuberculosis Lung Dis Off J Int Union Tuberculosis Lung Dis 4(6):570–576
308. Rizzotti L, Rossi F, Torriani S (2016) Biocide and antibiotic resistance of Enterococcus faecalis and Enterococcus faecium isolated from the swine meat chain. Food Microbiol 60:160–164. https://doi.org/10.1016/j.fm.2016.07.009
309. Rocha GR, Florez Salamanca EJ, de Barros AL, Lobo CIV, Klein MI (2018) Effect of tt-farnesol and myricetin on in vitro biofilm formed by Streptococcus mutans and Candida

albicans. BMC Complem Alternat Med 18(1):61. https://doi.org/10.1186/s12906-018-2132-x
310. Rodgers N, Murdaugh A (2016) Chlorhexidine-induced elastic and adhesive changes of Escherichia coli cells within a biofilm. Biointerphases 11(3):031011. https://doi.org/10.1116/1.4962265
311. Rodriguez Ferri EF, Martinez S, Frandoloso R, Yubero S, Gutierrez Martin CB (2010) Comparative efficacy of several disinfectants in suspension and carrier tests against Haemophilus parasuis serovars 1 and 5. Res Vet Sci 88(3):385–389. https://doi.org/10.1016/j.rvsc.2009.12.001
312. Rohrer N, Widmer AF, Waltimo T, Kulik EM, Weiger R, Filipuzzi-Jenny E, Walter C (2010) Antimicrobial efficacy of 3 oral antiseptics containing octenidine, polyhexamethylene biguanide, or Citroxx: can chlorhexidine be replaced? Infect Control Hosp Epidemiol 31 (7):733–739. https://doi.org/10.1086/653822
313. Rosa OP, Torres SA, Ferreira CM, Ferreira FB (2002) In vitro effect of intracanal medicaments on strict anaerobes by means of the broth dilution method. Pesquisa odontologica brasileira =. Brazilian Oral Res 16(1):31–36
314. Rose H, Baldwin A, Dowson CG, Mahenthiralingam E (2009) Biocide susceptibility of the Burkholderia cepacia complex. J Antimicrob Chemother 63(3):502–510. https://doi.org/10.1093/jac/dkn540
315. Rotter M, Koller W, Wewalka G (1980) Povidone-iodine and chlorhexidine gluconate detergents for disinfection of hands. J Hosp Infect 1:149–158
316. Rotter M, Kundi M, Suchomel M, Harke HP, Kramer A, Ostermeyer C, Rudolph P, Sonntag HG, Werner HP (2006) Reproducibility and workability of the European test standard EN 12791 regarding the effectiveness of surgical hand antiseptics: a randomized, multicenter trial. Infect Control Hosp Epidemiol 27(9):935–939
317. Rotter ML (1984) Hygienic hand disinfection. Infection Control: IC 5:18–22
318. Rotter ML, Kampf G, Suchomel M, Kundi M (2007) Population kinetics of the skin flora on gloved hands following surgical hand disinfection with 3 propanol-based hand rubs: a prospective, randomized, double-blind trial. Infect Control Hosp Epidemiol 28(3):346–350
319. Rotter ML, Koller W (1990) Surgical hand disinfection: effect of sequential use of two chlorhexidine preparations. J Hosp Infect 16:161–166
320. Ruiz-Linares M, Aguado-Perez B, Baca P, Arias-Moliz MT, Ferrer-Luque CM (2017) Efficacy of antimicrobial solutions against polymicrobial root canal biofilm. Int Endod J 50 (1):77–83. https://doi.org/10.1111/iej.12598
321. Russell AD (2000) Do biocides select for antibiotic resistance? J Pharm Pharmacol 52 (2):227–233
322. Russell AD (2002) Introduction of biocides into clinical practice and the impact on antibiotic-resistant bacteria. Symposium Series (Society for Applied Microbiology) 31:121s–135s
323. Russell AD, Mills AP (1974) Comparative sensitivity and resistance of some strains of Pseudomonas aeruginosa and Pseudomonas stutzeri to antibacterial agents. J Clin Pathol 27 (6):463–466
324. Russell AD, Tattawasart U, Maillard JY, Furr JR (1998) Possible link between bacterial resistance and use of antibiotics and biocides. Antimicrob Agents Chemother 42(8):2151
325. Rutter JD, Angiulo K, Macinga DR (2014) Measuring residual activity of topical antimicrobials: is the residual activity of chlorhexidine an artefact of laboratory methods? J Hosp Infect 88(2):113–115. https://doi.org/10.1016/j.jhin.2014.06.010
326. Salim N, Moore C, Silikas N, Satterthwaite J, Rautemaa R (2013) Chlorhexidine is a highly effective topical broad-spectrum agent against Candida spp. Int J Antimicrob Agents 41 (1):65–69. https://doi.org/10.1016/j.ijantimicag.2012.08.014
327. Sandle T, Vijayakumar R, Saleh Al Aboody M, Saravanakumar S (2014) In vitro fungicidal activity of biocides against pharmaceutical environmental fungal isolates. J Appl Microbiol 117(5):1267–1273. https://doi.org/10.1111/jam.12628

328. Santos GOD, Milanesi FC, Greggianin BF, Fernandes MI, Oppermann RV, Weidlich P (2017) Chlorhexidine with or without alcohol against biofilm formation: efficacy, adverse events and taste preference. Brazilian Oral Res 31:e32. https://doi.org/10.1590/1807-3107BOR-2017.vol31.0032
329. Schedler K, Assadian O, Brautferger U, Muller G, Koburger T, Classen S, Kramer A (2017) Proposed phase 2/step 2 in-vitro test on basis of EN 14561 for standardised testing of the wound antiseptics PVP-iodine, chlorhexidine digluconate, polihexanide and octenidine dihydrochloride. BMC Infect Dis 17(1):143. https://doi.org/10.1186/s12879-017-2220-4
330. Scheibler E, da Silva RM, Leite CE, Campos MM, Figueiredo MA, Salum FG, Cherubini K (2018) Stability and efficacy of combined nystatin and chlorhexidine against suspensions and biofilms of Candida albicans. Arch Oral Biol 89:70–76. https://doi.org/10.1016/j.archoralbio.2018.02.009
331. Sedlock DM, Bailey DM (1985) Microbicidal activity of octenidine hydrochloride, a new alkanediylbis[pyridine] germicidal agent. Antimicrob Agents Chemother 28(6):786–790
332. Seier-Petersen MA, Jasni A, Aarestrup FM, Vigre H, Mullany P, Roberts AP, Agerso Y (2014) Effect of subinhibitory concentrations of four commonly used biocides on the conjugative transfer of Tn916 in Bacillus subtilis. J Antimicrob Chemother 69(2):343–348. https://doi.org/10.1093/jac/dkt370
333. Sekavec JG, Moore WT, Gillock ET (2013) Chlorhexidine resistance in a Gram-negative bacterium isolated from an aquatic source. J Environ Sci Health Part A Toxic/Hazard Subst Environ Eng 48(14):1829–1834. https://doi.org/10.1080/10934529.2013.823338
334. Sekiguchi J, Asagi T, Miyoshi-Akiyama T, Fujino T, Kobayashi I, Morita K, Kikuchi Y, Kuratsuji T, Kirikae T (2005) Multidrug-resistant Pseudomonas aeruginosa strain that caused an outbreak in a neurosurgery ward and its aac(6')-Iae gene cassette encoding a novel aminoglycoside acetyltransferase. Antimicrob Agents Chemother 49(9):3734–3742. https://doi.org/10.1128/aac.49.9.3734-3742.2005
335. Serra E, Hidalgo-Bastida LA, Verran J, Williams D, Malic S (2018) Antifungal activity of commercial essential oils and biocides against Candida Albicans. Pathogens (Basel, Switzerland) 7(1). https://doi.org/10.3390/pathogens7010015
336. Sethi KS, Karde PA, Joshi CP (2016) Comparative evaluation of sutures coated with triclosan and chlorhexidine for oral biofilm inhibition potential and antimicrobial activity against periodontal pathogens: An in vitro study. Indian J Dental Res: Off Publ Indian Soc Dental Res 27(5):535–539. https://doi.org/10.4103/0970-9290.195644
337. Shen Y, Stojicic S, Haapasalo M (2011) Antimicrobial efficacy of chlorhexidine against bacteria in biofilms at different stages of development. J Endod 37(5):657–661. https://doi.org/10.1016/j.joen.2011.02.007
338. Sheng WH, Wang JT, Lauderdale TL, Weng CM, Chen D, Chang SC (2009) Epidemiology and susceptibilities of methicillin-resistant Staphylococcus aureus in Taiwan: emphasis on chlorhexidine susceptibility. Diagn Microbiol Infect Dis 63(3):309–313. https://doi.org/10.1016/j.diagmicrobio.2008.11.014
339. Sheridan A, Lenahan M, Duffy G, Fanning S, Burgess C (2012) The potential for biocide tolerance in Escherichia coli and its impact on the response to food processing stresses. Food Control 26(1):98–106
340. Shi GS, Boost M, Cho P (2015) Prevalence of antiseptic-resistance genes in staphylococci isolated from orthokeratology lens and spectacle wearers in Hong Kong. Invest Ophthalmol Vis Sci 56(5):3069–3074. https://doi.org/10.1167/iovs.15-16550
341. Shigeta S, Yasunaga Y, Honzumi K, Okamura H, Kumata R, Endo S (1978) Cerebral ventriculitis associated with Achromobacter xylosoxidans. J Clin Pathol 31(2):156–161
342. Shirato M, Nakamura K, Kanno T, Lingstrom P, Niwano Y, Ortengren U (2017) Time-kill kinetic analysis of antimicrobial chemotherapy based on hydrogen peroxide photolysis against Streptococcus mutans biofilm. J Photochem Photobiol, B 173:434–440. https://doi.org/10.1016/j.jphotobiol.2017.06.023

343. Sickbert-Bennett EE, Weber DJ, Gergen-Teague MF, Sobsey MD, Samsa GP, Rutala WA (2005) Comparative efficacy of hand hygiene agents in the reduction of bacteria and viruses. Am J Infect Control 33(2):67–77
344. Siqueira AB, Rodriguez LR, Santos RK, Marinho RR, Abreu S, Peixoto RF, Gurgel BC (2015) Antifungal activity of propolis against Candida species isolated from cases of chronic periodontitis. Brazilian Oral Res 29. https://doi.org/10.1590/1807-3107bor-2015.vol29.0083
345. Siqueira JF Jr, Rocas IN, Paiva SS, Guimaraes-Pinto T, Magalhaes KM, Lima KC (2007) Bacteriologic investigation of the effects of sodium hypochlorite and chlorhexidine during the endodontic treatment of teeth with apical periodontitis. Oral Surg Oral Med Oral Pathol Oral Radiol Endod 104(1):122–130. https://doi.org/10.1016/j.tripleo.2007.01.027
346. Skovgaard S, Larsen MH, Nielsen LN, Skov RL, Wong C, Westh H, Ingmer H (2013) Recently introduced qacA/B genes in Staphylococcus epidermidis do not increase chlorhexidine MIC/MBC. J Antimicrob Chemother 68(10):2226–2233. https://doi.org/10.1093/jac/dkt182
347. Slee AM, O'Connor JR (1983) In vitro antiplaque activity of octenidine dihydrochloride (WIN 41464-2) against preformed plaques of selected oral plaque-forming microorganisms. Antimicrob Agents Chemother 23(3):379–384
348. Smith K, Gemmell CG, Hunter IS (2008) The association between biocide tolerance and the presence or absence of qac genes among hospital-acquired and community-acquired MRSA isolates. J Antimicrob Chemother 61(1):78–84. https://doi.org/10.1093/jac/dkm395
349. Smith K, Hunter IS (2008) Efficacy of common hospital biocides with biofilms of multi-drug resistant clinical isolates. J Med Microbiol 57(Pt 8):966–973. https://doi.org/10.1099/jmm.0.47668-0
350. Smith K, Robertson DP, Lappin DF, Ramage G (2013) Commercial mouthwashes are ineffective against oral MRSA biofilms. Oral Surg Oral Med Oral Pathol Oral Radiol 115 (5):624–629. https://doi.org/10.1016/j.oooo.2012.12.014
351. Sobel JD, Hashman N, Reinherz G, Merzbach D (1982) Nosocomial Pseudomonas cepacia infection associated with chlorhexidine contamination. Am J Med 73(2):183–186
352. Soulsby ME, Barnett JB, Maddox S (1986) The antiseptic efficacy of chlorxylenol-containing vs. chlorhexidine gluconate-containing surgical scrub preparations. Infection Control: IC 7(4):223–226
353. Spijkervet FKL, van Saene JJM, van Saene HKF, Panders AK, Vermey A, Fidler V (1990) Chlorhexidine inactivation by saliva. Oral Surg Oral Med Oral Pathol 69(4):444–449. https://doi.org/10.1016/0030-4220(90)90377-5
354. Srinivasan VB, Rajamohan G (2013) KpnEF, a new member of the Klebsiella pneumoniae cell envelope stress response regulon, is an SMR-type efflux pump involved in broad-spectrum antimicrobial resistance. Antimicrob Agents Chemother 57(9):4449–4462. https://doi.org/10.1128/aac.02284-12
355. Srinivasan VB, Singh BB, Priyadarshi N, Chauhan NK, Rajamohan G (2014) Role of novel multidrug efflux pump involved in drug resistance in Klebsiella pneumoniae. PLoS ONE 9 (5):e96288. https://doi.org/10.1371/journal.pone.0096288
356. Stanley A, Wilson M, Newman HN (1989) The in vitro effects of chlorhexidine on subgingival plaque bacteria. J Clin Periodontol 16(4):259–264
357. Stickler D, Dolman J, Rolfe S, Chawla J (1989) Activity of antiseptics against Escherichia coli growing as biofilms on silicone surfaces. Eur J Clin Microbiol Infect Dis 8(11):974–978
358. Stickler D, Hewett P (1991) Activity of antiseptics against biofilms of mixed bacterial species growing on silicone surfaces. Eur J Clin Microbiol Infect Dis 10(5):416–421
359. Stickler DJ (1974) Chlorhexidine resistance in *Proteus mirabilis*. J Clin Pathol 27(4):284–287
360. Stickler DJ (2002) Susceptibility of antibiotic-resistant Gram-negative bacteria to biocides: a perspective from the study of catheter biofilms. J Appl Microbiol 92(Suppl):163s–170s
361. Stickler DJ, Clayton CL, Chawla JC (1987) The resistance of urinary tract pathogens to chlorhexidine bladder washouts. J Hosp Infect 10:28–39
362. Stickler DJ, Thomas B (1980) Antiseptic and antibiotic resistance in gram-negative bacteria causing urinary tract infections. J Clin Pathol 33(3):288–296

363. Stickler DJ, Thomas B, Chawla JC (1981) Antiseptic and antibiotic resistance in gram-negative bacteria causing urinary tract infection in spinal cord injured patients. Paraplegia 19:50–58
364. Suci PA, Tyler BJ (2003) A method for discrimination of subpopulations of Candida albicans biofilm cells that exhibit relative levels of phenotypic resistance to chlorhexidine. J Microbiol Methods 53(3):313–325
365. Suller MTE, Russell AD (1999) Antibiotic and biocide resistance in methicillin-resistant *Staphylococcus aureus* and vancomycin-resistant enterococcus. J Hosp Infect 43:281–291
366. Taha M, Kalab M, Yi QL, Landry C, Greco-Stewart V, Brassinga AK, Sifri CD, Ramirez-Arcos S (2014) Biofilm-forming skin microflora bacteria are resistant to the bactericidal action of disinfectants used during blood donation. Transfusion 54(11):2974–2982. https://doi.org/10.1111/trf.12728
367. Taheri N, Ardebili A, Amouzandeh-Nobaveh A, Ghaznavi-Rad E (2016) Frequency of antiseptic resistance among staphylococcus aureus and coagulase-negative Staphylococci isolated from a university hospital in Central Iran. Oman medical journal 31(6):426–432. https://doi.org/10.5001/omj.2016.86
368. Takahashi H, Nadres ET, Kuroda K (2017) Cationic amphiphilic polymers with antimicrobial activity for oral care applications: eradication of S. mutans biofilm. Biomacromol 18(1):257–265. https://doi.org/10.1021/acs.biomac.6b01598
369. Takenaka S, Trivedi HM, Corbin A, Pitts B, Stewart PS (2008) Direct visualization of spatial and temporal patterns of antimicrobial action within model oral biofilms. Appl Environ Microbiol 74(6):1869–1875. https://doi.org/10.1128/aem.02218-07
370. Takeo Y, Oie S, Kamiya A, Konishi H, Nakazawa T (1994) Efficacy of disinfectants against biofilm cells of Pseudomonas aeruginosa. Microbios 79(318):19–26
371. Tattawasart U, Hann AC, Maillard J-Y, Furr JR, Russell AD (2000) Cytological changes in chlorhexidine-resistant isolates of *Pseudomonas stutzeri*. J Antimicrob Chemother 45:145–152
372. Tattawasart U, Maillard J-Y, Furr JR, Russell AD (1999) Development of resistance to chlorhexidine diacetate and cetylpyridinium chloride in *Pseudomonas stutzeri* and changes in antibiotic susceptibility. J Hosp Infect 42(3):219–229
373. Tattawasart U, Maillard J-Y, Furr JR, Russell AD (2000) Outer membrane changes in *Pseudomonas stutzeri* resistant to chlorhexidine diacetate and cetylpyridinium chloride. Int J Antimicrob Agents 16:233–238
374. Tena D, Carranza R, Barbera JR, Valdezate S, Garrancho JM, Arranz M, Saez-Nieto JA (2005) Outbreak of long-term intravascular catheter-related bacteremia due to Achromobacter xylosoxidans subspecies xylosoxidans in a hemodialysis unit. Eur J Clin Microbiol Infect Dis 24(11):727–732. https://doi.org/10.1007/s10096-005-0028-4
375. Tetz G, Tetz V (2015) In vitro antimicrobial activity of a novel compound, Mul-1867, against clinically important bacteria. Antimicrob Resist Infect Control 4:45. https://doi.org/10.1186/s13756-015-0088-x
376. Theraud M, Bedouin Y, Guiguen C, Gangneux JP (2004) Efficacy of antiseptics and disinfectants on clinical and environmental yeast isolates in planktonic and biofilm conditions. J Med Microbiol 53(Pt 10):1013–1018. https://doi.org/10.1099/jmm.0.05474-0
377. Thomas B, Sykes L, Stickler DJ (1978) Sensitivity of urine-grown cells of *Providencia stuartii* to antiseptics. J Clin Pathol 31:929–932
378. Thomas L, Maillard JY, Lambert RJ, Russell AD (2000) Development of resistance to chlorhexidine diacetate in *Pseudomonas aeruginosa* and the effect of a "residual" concentration. J Hosp Infect 46:297–303
379. Thurmond JM, Brown AT, Sims RE, Ferretti GA, Raybould TP, Lillich TT, Henslee PJ (1991) Oral Candida albicans in bone marrow transplant patients given chlorhexidine rinses: occurrence and susceptibilities to the agent. Oral Surg Oral Med Oral Pathol 72(3):291–295
380. Tirali RE, Bodur H, Ece G (2012) In vitro antimicrobial activity of sodium hypochlorite, chlorhexidine gluconate and octenidine dihydrochloride in elimination of microorganisms

within dentinal tubules of primary and permanent teeth. Medicina oral, patologia oral y cirugia bucal 17(3):e517–522
381. Tong Z, Zhou L, Jiang W, Kuang R, Li J, Tao R, Ni L (2011) An in vitro synergetic evaluation of the use of nisin and sodium fluoride or chlorhexidine against Streptococcus mutans. Peptides 32(10):2021–2026. https://doi.org/10.1016/j.peptides.2011.09.002
382. Tortorano AM, Viviani MA, Biraghi E, Rigoni AL, Prigitano A, Grillot R (2005) In vitro testing of fungicidal activity of biocides against Aspergillus fumigatus. J Med Microbiol 54 (Pt 10):955–957. https://doi.org/10.1099/jmm.0.45997-0
383. Tote K, Horemans T, Vanden Berghe D, Maes L, Cos P (2010) Inhibitory effect of biocides on the viable masses and matrices of Staphylococcus aureus and Pseudomonas aeruginosa biofilms. Appl Environ Microbiol 76(10):3135–3142. https://doi.org/10.1128/aem.02095-09
384. Touzel RE, Sutton JM, Wand ME (2016) Establishment of a multi-species biofilm model to evaluate chlorhexidine efficacy. J Hosp Infect 92(2):154–160. https://doi.org/10.1016/j.jhin.2015.09.013
385. Traboulsi RS, Mukherjee PK, Ghannoum MA (2008) In vitro activity of inexpensive topical alternatives against Candida spp. isolated from the oral cavity of HIV-infected patients. Int J Antimicrob Agents 31(3):272–276. https://doi.org/10.1016/j.ijantimicag.2007.11.008
386. Traore O, Springthorpe VS, Sattar SA (2002) Testing chemical germicides against Candida species using quantitative carrier and fingerpad methods. J Hosp Infect 50(1):66–75. https://doi.org/10.1053/jhin.2001.1133
387. Tremblay YD, Caron V, Blondeau A, Messier S, Jacques M (2014) Biofilm formation by coagulase-negative staphylococci: impact on the efficacy of antimicrobials and disinfectants commonly used on dairy farms. Vet Microbiol 172(3–4):511–518. https://doi.org/10.1016/j.vetmic.2014.06.007
388. Tuuli MG, Liu J, Stout MJ, Martin S, Cahill AG, Odibo AO, Colditz GA, Macones GA (2016) A randomized trial comparing skin antiseptic agents at cesarean delivery. N Engl J Med 374(7):647–655. https://doi.org/10.1056/NEJMoa1511048
389. Ueda S, Kuwabara Y (2007) Susceptibility of biofilm Escherichia coli, Salmonella Enteritidis and Staphylococcus aureus to detergents and sanitizers. Biocontrol science 12(4):149–153
390. Ulusoy AT, Kalyoncuoglu E, Reis A, Cehreli ZC (2016) Antibacterial effect of N-acetylcysteine and taurolidine on planktonic and biofilm forms of Enterococcus faecalis. Dental Traumatol: Off Publ Int Assoc Dental Traumatol 32(3):212–218. https://doi.org/10.1111/edt.12237
391. Uri M, Buckley LM, Marriage L, McEwan N, Schmidt VM (2016) A pilot study comparing in vitro efficacy of topical preparations against veterinary pathogens. Vet Dermatol 27 (3):152–e139. https://doi.org/10.1111/vde.12306
392. Uzer Celik E, Tunac AT, Ates M, Sen BH (2016) Antimicrobial activity of different disinfectants against cariogenic microorganisms. Brazilian Oral Res 30(1):e125. https://doi.org/10.1590/1807-3107BOR-2016.vol30.0125
393. Valentine BK, Dew W, Yu A, Weese JS (2012) In vitro evaluation of topical biocide and antimicrobial susceptibility of Staphylococcus pseudintermedius from dogs. Vet Dermatol 23(6):493–e495. https://doi.org/10.1111/j.1365-3164.2012.01095.x
394. Valenzuela AS, Benomar N, Abriouel H, Canamero MM, Lopez RL, Galvez A (2013) Biocide and copper tolerance in enterococci from different sources. J Food Prot 76 (10):1806–1809. https://doi.org/10.4315/0362-028x.jfp-13-124
395. Vali L, Dashti AA, El-Shazly S, Jadaon MM (2015) Klebsiella oxytoca with reduced sensitivity to chlorhexidine isolated from a diabetic foot ulcer. Int J Infect Dis: IJID: Off Publ Int Soc Infect Dis 34:112–116. https://doi.org/10.1016/j.ijid.2015.03.021
396. Vali L, Dashti AA, Mathew F, Udo EE (2017) Characterization of heterogeneous MRSA and MSSA with reduced susceptibility to Chlorhexidine in Kuwaiti hospitals. Front Microbiol 8:1359. https://doi.org/10.3389/fmicb.2017.01359
397. Vali L, Davies SE, Lai LL, Dave J, Amyes SG (2008) Frequency of biocide resistance genes, antibiotic resistance and the effect of chlorhexidine exposure on clinical methicillin-resistant

Staphylococcus aureus isolates. J Antimicrob Chemother 61(3):524–532. https://doi.org/10.1093/jac/dkm520
398. Vaziri S, Kangarlou A, Shahbazi R, Nazari Nasab A, Naseri M (2012) Comparison of the bactericidal efficacy of photodynamic therapy, 2.5% sodium hypochlorite, and 2% chlorhexidine against Enterococcous faecalis in root canals; an in vitro study. Dental Res J 9(5):613–618
399. Verkaik MJ, Busscher HJ, Jager D, Slomp AM, Abbas F, van der Mei HC (2011) Efficacy of natural antimicrobials in toothpaste formulations against oral biofilms in vitro. J Dent 39 (3):218–224. https://doi.org/10.1016/j.jdent.2010.12.007
400. Vestby LK, Nesse LL (2015) Wound care antiseptics - performance differences against Staphylococcus aureus in biofilm. Acta Vet Scand 57:22. https://doi.org/10.1186/s13028-015-0111-5
401. Vijayakumar R, Kannan VV, Sandle T, Manoharan C (2012) In vitro antifungal efficacy of biguanides and quaternary ammonium compounds against cleanroom fungal isolates. PDA J Pharm Sci Technol 66(3):236–242. https://doi.org/10.5731/pdajpst.2012.00866
402. Wakamatsu R, Takenaka S, Ohsumi T, Terao Y, Ohshima H, Okiji T (2014) Penetration kinetics of four mouthrinses into Streptococcus mutans biofilms analyzed by direct time-lapse visualization. Clin Oral Invest 18(2):625–634. https://doi.org/10.1007/s00784-013-1002-7
403. Walker EM, Lowes JA (1985) An investigation into in vitro methods for the detection of chlorhexidine resistance. J Hosp Infect 6(4):389–397
404. Wand ME, Baker KS, Benthall G, McGregor H, McCowen JW, Deheer-Graham A, Sutton JM (2015) Characterization of pre-antibiotic era Klebsiella pneumoniae isolates with respect to antibiotic/disinfectant susceptibility and virulence in Galleria mellonella. Antimicrob Agents Chemother 59(7):3966–3972. https://doi.org/10.1128/aac.05009-14
405. Wand ME, Bock LJ, Bonney LC, Sutton JM (2017) Mechanisms of increased resistance to chlorhexidine and cross-resistance to colistin following exposure of klebsiella pneumoniae clinical isolates to chlorhexidine. Antimicrob Agents Chemother 61(1). https://doi.org/10.1128/aac.01162-16
406. Wang JT, Sheng WH, Wang JL, Chen D, Chen ML, Chen YC, Chang SC (2008) Longitudinal analysis of chlorhexidine susceptibilities of nosocomial methicillin-resistant Staphylococcus aureus isolates at a teaching hospital in Taiwan. J Antimicrob Chemother 62 (3):514–517. https://doi.org/10.1093/jac/dkn208
407. Wang Y, Leng V, Patel V, Phillips KS (2017) Injections through skin colonized with Staphylococcus aureus biofilm introduce contamination despite standard antimicrobial preparation procedures. Sci Rep 7:45070. https://doi.org/10.1038/srep45070
408. Wesgate R, Grasha P, Maillard JY (2016) Use of a predictive protocol to measure the antimicrobial resistance risks associated with biocidal product usage. Am J Infect Control 44 (4):458–464. https://doi.org/10.1016/j.ajic.2015.11.009
409. Wiegand C, Abel M, Ruth P, Elsner P, Hipler UC (2015) pH influence on antibacterial efficacy of common antiseptic substances. Skin Pharmacol Physiol 28(3):147–158. https://doi.org/10.1159/000367632
410. Wiegand C, Abel M, Ruth P, Hipler UC (2012) Analysis of the adaptation capacity of Staphylococcus aureus to commonly used antiseptics by microplate laser nephelometry. Skin Pharmacol Physiol 25(6):288–297. https://doi.org/10.1159/000341222
411. Wisplinghoff H, Schmitt R, Wohrmann A, Stefanik D, Seifert H (2007) Resistance to disinfectants in epidemiologically defined clinical isolates of Acinetobacter baumannii. J Hosp Infect 66(2):174–181. https://doi.org/10.1016/j.jhin.2007.02.016
412. Witney AA, Gould KA, Pope CF, Bolt F, Stoker NG, Cubbon MD, Bradley CR, Fraise A, Breathnach AS, Butcher PD, Planche TD, Hinds J (2014) Genome sequencing and characterization of an extensively drug-resistant sequence type 111 serotype O12 hospital outbreak strain of Pseudomonas aeruginosa. Clin Microbiol Infect 20(10):O609–618. https://doi.org/10.1111/1469-0691.12528

413. Wong HS, Townsend KM, Fenwick SG, Trengove RD, O'Handley RM (2010) Comparative susceptibility of planktonic and 3-day-old Salmonella Typhimurium biofilms to disinfectants. J Appl Microbiol 108(6):2222–2228. https://doi.org/10.1111/j.1365-2672.2009.04630.x
414. Wong TZ, Zhang M, O'Donoghue M, Boost M (2013) Presence of antiseptic resistance genes in porcine methicillin-resistant Staphylococcus aureus. Vet Microbiol 162(2–4):977–979. https://doi.org/10.1016/j.vetmic.2012.10.017
415. Wu D, Lu R, Chen Y, Qiu J, Deng C, Tan Q (2016) Study of cross-resistance mediated by antibiotics, chlorhexidine and Rhizoma coptidis in Staphylococcus aureus. J Global Antimicrob Resis 7:61–66. https://doi.org/10.1016/j.jgar.2016.07.011
416. Xu Y, He Y, Zhou L, Gao C, Sun S, Wang X, Pang G (2014) Effects of contact lens solution disinfectants against filamentous fungi. Optom Vision Sci: Off Publ Am Acad Optom 91 (12):1440–1445. https://doi.org/10.1097/opx.0000000000000407
417. Yadav P, Chaudhary S, Saxena RK, Talwar S, Yadav S (2017) Evaluation of antimicrobial and antifungal efficacy of chitosan as endodontic irrigant against Enterococcus Faecalis and Candida Albicans biofilm formed on tooth substrate. J Clin Exp Dent 9(3):e361–e367. https://doi.org/10.4317/jced.53210
418. Yamamoto M, Takami T, Matsumura R, Dorofeev A, Hirata Y, Nagamune H (2016) In Vitro evaluation of the biocompatibility of newly synthesized bis-quaternary ammonium compounds with spacer structures derived from pentaerythritol or hydroquinone. Biocontrol Sci 21(4):231–241. https://doi.org/10.4265/bio.21.231
419. Yamamoto T, Tamura Y, Yokota T (1988) Antiseptic and antibiotic resistance plasmid in *Staphylococcus aureus* that possesses ability to confer chlorhexidine and acrinol resistance. Antimicrob Agents Chemother 32(6):932–935
420. Ye JZ, Yu X, Li XS, Sun Y, Li MM, Zhang W, Fan H, Cao JM, Zhou TL (2014) [Antimicrobial resistance characteristics of and disinfectant-resistant gene distribution in Staphylococcus aureus isolates from male urogenital tract infection]. Zhonghua nan ke xue = Natl J Androl 20(7):630–636
421. Zaura-Arite E, van Marle J, ten Cate JM (2001) Conofocal microscopy study of undisturbed and chlorhexidine-treated dental biofilm. J Dent Res 80(5):1436–1440. https://doi.org/10.1177/00220345010800051001
422. Zeng P, Rao A, Wiedmann TS, Bowles W (2009) Solubility properties of chlorhexidine salts. Drug Dev Ind Pharm 35(2):172–176. https://doi.org/10.1080/03639040802220318
423. Zhang M, O'Donoghue MM, Ito T, Hiramatsu K, Boost MV (2011) Prevalence of antiseptic-resistance genes in Staphylococcus aureus and coagulase-negative staphylococci colonising nurses and the general population in Hong Kong. J Hosp Infect 78(2):113–117. https://doi.org/10.1016/j.jhin.2011.02.018
424. Zhang M, O'Dononghue M, Boost MV (2012) Characterization of staphylococci contaminating automated teller machines in Hong Kong. Epidemiol Infect 140(8):1366–1371. https://doi.org/10.1017/s095026881100207x
425. Zmantar T, Ben Slama R, Fdhila K, Kouidhi B, Bakhrouf A, Chaieb K (2017) Modulation of drug resistance and biofilm formation of Staphylococcus aureus isolated from the oral cavity of Tunisian children. Brazilian J Infect Dis: An Off Publ Brazilian Soc Infect Dis 21(1):27–34. https://doi.org/10.1016/j.bjid.2016.10.009
426. Zurita J, Mejia L, Zapata S, Trueba G, Vargas AC, Aguirre S, Falconi G (2014) Healthcare-associated respiratory tract infection and colonization in an intensive care unit caused by Burkholderia cepacia isolated in mouthwash. Int J Infect Dis: IJID: Off Publ Int Soc Infect Dis 29:96–99. https://doi.org/10.1016/j.ijid.2014.07.016

14 Octenidine Dihydrochloride

14.1 Chemical Characterization

Octenidine dihydrochloride (OCT) is a non-volatile, cationic surfactant which is able to lower the surface tension of the water. This is achieved by the fact that a hydrophilic end and a hydrophobic end are present in the molecule. In a pH range of 1.6–12.2, OCT is stable [42]. In the molecule, OCT has two cationic centres that do not interact with each other [42]. The basic chemical information on OCT is summarized in Table 14.1.

OCT has little effect on free available chlorine, e.g. from sodium hypochlorite, and can be used concurrently with sodium hypochlorite solutions for irrigation [51]. It can react with povidone iodine which releases iodine radicals resulting in a tissue irritation and a brown-to-violet discoloration [42, 83, 84]. OCT may also precipitate in the presence of sorbic acid, benzoic acid or parabens, all used as preservatives in cremes [55].

A related compound is octenidine (CAS number: 71251-02-0) with a molecular weight of 550.92 and the following molecular formula: $C_{36}H_{62}N_4$ [59].

14.2 Types of Application

OCT is found at 1% in antimicrobial washing lotions [76] and at 0.1% in alcohol-based hand rubs [20, 42, 66], mouth rinses [9] and skin disinfectants [14, 39, 54], and in an unknown concentration in a nasal ointment [65]. It is also used for the antisepsis of wounds (e.g. at 0.05%) or mucous membranes [6, 23, 37]. Specifically, the combination of 0.1% OCT with 2% phenoxyethanol has been found to be suitable for acute, contaminated, traumatic wounds, including MRSA-colonized wounds. For chronic wounds, preparations with 0.05% OCT are preferable [50]. For the decolonization of wounds colonized or infected with multidrug-resistant micro-organisms, the combination of 0.1% OCT with 2% phenoxyethanol is

G. Kampf, *Antiseptic Stewardship*, https://doi.org/10.1007/978-3-319-98785-9_14

Table 14.1 Basic chemical information on OCT [60]

CAS number	70775-75-6
IUPAC name	N-octyl-1-[10-(4-octyliminopyridin-1-yl)decyl]pyridin-4-imine dihydrochloride
Synonyms	Octenidine hydrochloride; N,N'-[Decane-1,10-diyldi-1(4H)-pyridyl-4-ylidene]bis(octylammonium) dichloride
Molecular formula	$C_{36}H_{64}Cl_2N_4$
Molecular weight (g/mol)	623.83

preferred [50]. The same combination has also been used for decolonization of MRSA from human skin [73]. OCT at 11 μg/cm has also been evaluated and proposed for antimicrobial coating of sutures [62]. Antimicrobial coating of tracheotomy tubes with OCT, however, is currently limited due to poor adhesive properties resulting in quick vanishing of OCT after reprocessing the tubes [89].

14.2.1 European Medicines Agency (European Union)

A total of 86 medicinal products with a national authorization containing OCT were listed in 2017 by the European Medicines Agency [28]. It was also included in 2011 as a pharmacologically active substance for skin and mucosal disinfection and short-term supportive antiseptic wound treatment in all mammalian food producing species [26]. In the paediatric population from 2 months to less than 18 years of age, the efficacy of OCT will be studied for skin antisepsis (cutaneous application) [27]. And in 2010, orphan designation was granted by the European Commission for OCT for the prevention of late-onset sepsis in premature infants of less than or equal to 32 weeks of gestational age [25].

14.2.2 Environmental Protection Agency (USA)

No public information was found on an evaluation of OCT by the EPA.

14.2.3 Food and Drug Administration (USA)

No public information was found on an evaluation of OCT by the FDA.

14.2.4 Overall Environmental Impact

No public information was found that may allow assessing the overall environmental of OCT.

14.3 Spectrum of Antimicrobial Activity

14.3.1 Bactericidal Activity

14.3.1.1 Bacteriostatic Activity (MIC Values)

The majority of bacterial species such as *E. faecalis* (4–16 mg/l), *E. coli* (0.25–8 mg/l), *P. aeruginosa* (1–8 mg/l), *S. aureus* (0.25–9.3 mg/l) and *S. pneumoniae* (8–32 mg/l) have low MIC value for OCT ($\leq$ 20 mg/l) indicating susceptible isolates or strains. Only few oral cavity species were less susceptible such as *S. mutans* ($\leq$ 120 mg/l) and *S. salivarius* ($\leq$ 800 mg/l; Table 14.2). Overall, it is important to know that with OCT the result of MIC testing depends to some extent on the media composition and plate material showing the need to standardize biocide susceptibility testing [10].

14.3.1.2 Bactericidal Activity (Suspension Tests)

OCT (0.1%), often in combination with 2% phenoxyethanol, has a broad bactericidal activity within 1 min. At 0.01%, an exposure time of 5 min still reveals sufficient bactericidal activity against *A. baumannii*, *B. afzelii*, *B. burgdorferi*, *B. garinii*, *E. cloacae*, *E. coli*, *K. pneumoniae*, *P. aeruginosa* and *S. aureus* (Table 14.3).

Few studies indicate a bactericidal activity of OCT (MBC values) within 10 min at concentrations between 5 (*E. coli*) and 27 mg/l (*P. aeruginosa*; Table 14.4).

The bactericidal efficacy of OCT is significantly reduced in the presence of 0.75% albumin [45] or by selected wound dressings as shown with *S. aureus* [40]. It has been described to be independent of the pH values between 5 and 9 [86]. The efficacy of OCT at 0.004% and 0.008% may be significantly affected when the bacterial cells of *S. aureus* or *P. aeruginosa* used for the suspension test are grown on agar instead of broth, a difference that cannot be found at higher OCT concentrations [11].

In a proposed test to determine the efficacy of wound antiseptics (which is similar to a carrier test), 0.05% and 0.1% OCT showed sufficient bactericidal activity within 10 h with and without organic load [69]. On cattle hides, OCT at 0.05, 0.15 and 0.25%, each in 95% ethanol, was able to reduce five isolates of *E. coli* O157:H7, *Salmonella* spp. and *L. monocytogenes* by at least 5.0 log within 2 min, whereas 95% ethanol alone revealed only 1.5 log [8].

14.3.1.3 Activity Against Bacteria in Biofilm

OCT (0.1%) has good bactericidal activity in 30 min ($\geq$ 4.0 log) against *A. viscosus*, *P. aeruginosa* and *S. aureus*, but other species are less susceptible in biofilms (*E. faecalis*, *S. mutans*), especially mixed-species biofilms. Higher concentrations reveal a stronger bactericidal effect against selected bacterial species such as *A. baumannii* (e.g. 0.3% in 60 min) or *S. aureus* (e.g. 3.1% in 5 min; Table 14.5).

The findings in Table 14.5 are supported by other studies. MIC values of *S. aureus*, *S. epidermidis* and *E. coli* were 8-fold to 16-fold higher in biofilm-grown

Table 14.2 MIC values of various bacterial species to OCT

Species	Strains/isolates	MIC value (mg/l)	References
A. baumannii	JCM 6841	13	[88]
A. viscosus	ATCC 15987	20	[82]
B. cepacia	JCM 5964	16	[88]
C. perfringens	ATCC 13124	1	[47]
Enterobacter spp.	1 strain from intraoperative metal orthopaedic components and a bone sequester	0.5	[7]
E. faecalis	ATCC 29212	4	[47]
E. faecalis	ATCC 29212	16	[88]
E. hirae	ATCC 10541	11	[88]
Enterococcus spp.	Clinical VRE isolate	4	[47]
E. coli	NCTC 10418	0.25–4[a]	[10]
E. coli	1 strain from intraoperative metal orthopaedic components and a bone sequester	1	[7]
E. coli	ATCC 35218	2	[47]
E. coli	ATCC 25922	4	[88]
E. coli	6 clinical ESBL isolates	4–8	[33]
H. influenzae	ATCC 49247	1	[47]
K. pneumoniae	DSM 16609 and 3 clinical ESBL isolates	4–8	[33]
L. acidophilus	ATCC 4356	10	[82]
L. lactis	1 strain	3	[21]
L. rhamnosus	ATCC 7469	10	[82]
P. aeruginosa	NCTC 13359	1–8[a]	[10]
P. aeruginosa	ATCC 15442	2–8	[47]
P. aeruginosa	ATCC 27853	8	[88]
S. aureus	ATCC 6538	0.25–2[a]	[10]
S. aureus	3 strains from intraoperative metal orthopaedic components and a bone sequester	0.5	[7]
S. aureus	Clinical MRSA isolate	1	[47]
S. aureus	ATCC 6538	2	[47]
S. aureus	ATCC 6538	2	[88]
S. aureus	100 clinical isolates (76 MRSA, 24 MSSA)	2–4	[1]
S. aureus	ATCC 700698	9.3	[88]
S. epidermidis	1 strain from intraoperative metal orthopaedic components and a bone sequester	1	[7]
S. epidermidis	ATCC 12228	8	[88]
S. mutans	ATCC 27351	100	[21]
S. mutans	ATCC 25175	120	[82]
S. pneumoniae	ATCC 49619	8–32	[47]
S. salivarius	ATCC 25975	800	[21]

[a]Depending on the media composition and plate material

Table 14.3 Bactericidal activity of OCT in suspension tests

Species	Strains/isolates	Exposure time	Concentration	$\log_{10}$ reduction	References
A. baumannii	5 clinical 3MRGN or 4MRGN strains	1 min	0.01% (S)	>5.0	[2]
B. afzelii	ATCC 51567	1 min	0.1% (S)	>7.0	[80]
		5 min	0.01% (S)		
		10 min	0.005% (S)		
B. burgdorferi	ATCC 35210	1 min	0.1% (S)	>7.0	[80]
		5 min	0.01% (S)		
		60 min	0.005% (S)		
B. garinii	ATCC 51383	1 min	0.1% (S)	>7.0	[80]
		5 min	0.01% (S)		
		5 min	0.005% (S)		
E. cloacae	5 clinical 3MRGN or 4MRGN strains	1 min	0.01% (S)	>5.0	[2]
E. faecalis	ATCC 29212	15 s	0.1%[a] (P)	≥5.0	[79]
E. faecium	ATCC 6057	30 s	0.1%[a] (P)	≥6.5	[64]
E. coli	NCTC 10536	30 s	0.1%[a] (P)	4.0–6.3[b]	[64]
		1 min		≥5.6	
E. coli	Clinical ESBL isolate	1 min	0.08% (S)	≥5.6	[33]
			0.05% (S)	≥5.6	
			0.025% (S)	5.0	
E. coli	5 clinical 3MRGN or 4MRGN strains	1 min	0.01% (S)	>5.0	[2]
E. coli	ATCC 11229	30 min	0.00225% (P)	≥3.0	[56]
K. pneumoniae	DSM 16609	1 min	0.08% (S)	≥6.3	[33]
			0.05% (S)	≥6.3	
			0.025% (S)	4.9	
K. pneumoniae	Clinical ESBL isolate	1 min	0.08% (S)	≥6.5	[33]
			0.05% (S)	≥6.5	
			0.025% (S)	3.5	
K. pneumoniae	5 clinical 3MRGN or 4MRGN strains	1 min	0.01% (S)	>5.0	[2]
L. monocytogenes	ATCC 19115	1 min	0.3125% (S)	6.8	[3]
		1 min	0.0625% (S)	4.3	
		2 min		5.8	
		5 min		6.8	
P. aeruginosa	ATCC 15442	30 s	0.1%[a] (P)	4.2–7.1[b]	[64]
		1 min			
P. aeruginosa	5 clinical 3MRGN or 4MRGN strains	1 min	0.01% (S)	>5.0	[2]

(continued)

Table 14.3 (continued)

Species	Strains/isolates	Exposure time	Concentration	$\log_{10}$ reduction	References
P. aeruginosa	ATCC 15442	1 min	0.005% (S)	5.0	[47]
		5 min	0.0025% (S)	5.2	
		10 min	0.001% (S)	5.2	
S. aureus	ATCC 29213	15 s	0.1%[a] (P)	≥5.0	[79]
S. aureus	ATCC 6538	30 s	0.1%[a] (P)	≥6.1	[64]
S. aureus	8 strains from clinical materials (6 MRSA, 2 MSSA)	30 s	0.1%[a] (P)	>6.0	[16]
			0.01% (S)		
S. aureus	ATCC 6538	30 min	0.00175% (P)	≥3.0	[56]
S. aureus	ATCC 6538	1 min	0.001% (S)	5.4	[47]
		10 min	0.0005% (S)	5.2	
		6 h	0.0001% (S)	5.8	
Mixed anaerobic species	*A. actinomycetemcomitans* ATCC 43718, *A. viscosus* DSMZ 43798, *F. nucleatum* ATCC 10953, *P. gingivalis* ATCC 33277, *V. atypica* ATCC 17744 and *S. gordonii* ATCC 33399	30 s	0.1%[a] (P)	>8.0	[19]

S solution; *P* commercial product; [a]plus 2% phenoxyethanol; [b]depending on the type of organic load

Table 14.4 MBC values of various bacterial species to OCT (10-min exposure)

Species	Strains/isolates	MBC value (mg/l)	References
E. coli	ATCC 25922	5	[88]
P. aeruginosa	ATCC 27853	27	[88]
S. aureus	ATCC 6538	12	[88]

cells compared to planktonic cells, but not cells of *Enterobacter* spp. [7]. In addition, 0.05–0.1% OCT showed a moderate inhibition effect (94%) on the metabolic activity in a MRSA biofilm. The efficacy began after 15 s and did not depend on the applied exposure time (15 s–20 min) or the concentration [35].

14.3.1.4 Bactericidal Activity of Mouth Rinse Solution

In the oral cavity, an antiseptic mouth rinse based on OCT was equally effective against three oral pathogens (*S. mutans*, *F. nucleatum*, *C. albicans*) compared to the positive control based on 0.2% CHG [67]. In another study, the salivary bacterial count was reduced by a 1-min mouth rinse with OCT at 0.1, 0.15 and 0.2% by 3.7, 3.7 and 4.2 log, respectively. A similar effect was seen on day 4 after the same type

Table 14.5 Efficacy of OCT in against bacteria in biofilms

Species	Strains/isolates	Type of biofilm	Exposure time	Concentration	$\log_{10}$ reduction	References
A. baumannii	ATCC 17978, ATCC 190451	24-h incubation on polystyrene tissue culture plates	5 min	0.9% (S)	≥5.0	[58]
			5 min	0.6% (S)	≥5.0	
			10 min	0.3% (S)	≥4.0	
A. baumannii	ATCC 17978, ATCC 190451	5-d incubation on Foley catheter pieces	15 min	0.9% (S)	≥5.0	[58]
			30 min	0.6% (S)	≥5.0	
			60 min	0.3% (S)	≥4.0	
A. baumannii	ATCC 17978, ATCC 190451	24-h incubation on stainless steel plates	1 min	0.9% (S)	≥4.0	[58]
			5 min	0.6% (S)	≥5.0	
			10 min	0.3% (S)	≥3.5	
A. naeslundii	Strain 631	3-d incubation on ceramic hydroxylapatite slabs	30 min	0.2% (S)	"Complete kill"	[72]
				0.1% (S)	"Incomplete kill"	
A. viscosus	Strain M-100B	3-d incubation on ceramic hydroxylapatite slabs	30 min	0.1% (S)	"Complete kill"	[72]
				0.05% (S)	"Incomplete kill"	
E. faecalis	ATCC 29212	8-w incubation in straight-rooted teeth root canals	4 w	5% (P)	≥4.4	[18]
E. faecalis	ATCC 29212	3-w incubation on pieces of cellulose nitrate membranes	30 s	0.1% (P)	1.2	[32]
E. faecalis	ATCC 29212	3-w incubation in single-rooted teeth canals	30 s	0.1%[a] (P)	1.8–1.9	[78]
			1 min		2.0	
			5 min		2.5–2.6	

(continued)

Table 14.5 (continued)

Species	Strains/isolates	Type of biofilm	Exposure time	Concentration	log_{10} reduction	References
E. faecalis	ATCC 29212	7-d incubation in root canals	2 min	0.1%[a] (P)	0.6–0.7	[13]
E. faecalis	Strain A197A	3-w incubation in root samples	3 min	0.1%[a] (P)	>7.0	[34]
E. faecalis	ATCC 29212	3-w incubation on dentin discs	10 min	0.1%[a] (P)	0.6	[12]
E. faecalis	ATCC 29212	4-w incubation in straight-rooted teeth root canals	1 min	0.05%[b] (P)	1.8	[75]
			10 min		3.4	
			7 d		3.0	
L. monocytogenes	ATCC 19115	24-h incubation in polystyrene tissue culture plates or on stainless steel	10 s	1.25% (S)	"complete inactivation"	[3]
				0.625% (S)		
P. aeruginosa	Leg ulcer isolate	24-h incubation in polystyrene 96-well plates	1 min	0.1%[a] (P)	4.4	[43]
			15 min		>8.0	
			30 min		>8.0	
P. aeruginosa	CIP 103.467	24- or 48-h incubation on glass slides	24 h	0.0125% (P)	5.7	[52]
P. aeruginosa	ATCC 15442	24-h incubation on agar disc	24 h	0.01% (S)	1.1	[44]
P. aeruginosa	NCIMB 10434	48-h incubation in biofilm reactor	4 h	Unknown (P)	>6.0	[41]
			24 h			
			4 h	10% of unknown (P)	>6.0	
			24 h			
S. aureus	ATCC 35556	24-h incubation on stainless steel plates	2 min	3.12% (S)	4.5	[4]
				1.56% (S)	1.0	
			5 min	3.12% (S)	>6.0	
				1.56% (S)	1.8	
			10 min	3.12% (S)	>6.0	
				1.56% (S)	2.7	

(continued)

Table 14.5 (continued)

Species	Strains/isolates	Type of biofilm	Exposure time	Concentration	$\log_{10}$ reduction	References
S. aureus	MRSA, strain NRS 123	24-h incubation on stainless steel plates	2 min	3.12% (S)	4.5	[4]
				1.56% (S)	1.0	
			5 min	3.12% (S)	>6.0	
				1.56% (S)	2.0	
			10 min	3.12% (S)	> 6.0	
				1.56% (S)	3.0	
S. aureus	VRSA, strain VRS 8	24-h incubation on stainless steel plates	2 min	3.12% (S)	4.0	[4]
				1.56% (S)	1.0	
			5 min	3.12% (S)	>6.0	
				1.56% (S)	2.5	
			10 min	3.12% (S)	>6.0	
				1.56% (S)	3.0	
S. aureus	ATCC 35556	5-d incubation on urinary catheter pieces	2 min	3.12% (S)	4.5	[4]
				1.56% (S)	1.0	
			5 min	3.12% (S)	>6.0	
				1.56% (S)	2.0	
			10 min	3.12% (S)	>6.0	
				1.56% (S)	3.0	
S. aureus	MRSA, strain NRS 123	5-d incubation on urinary catheter pieces	2 min	3.12% (S)	4.5	[4]
				1.56% (S)	0.5	
			5 min	3.12% (S)	>6.0	
				1.56% (S)	2.0	
			10 min	3.12% (S)	>6.0	
				1.56% (S)	2.5	

(continued)

Table 14.5 (continued)

Species	Strains/isolates	Type of biofilm	Exposure time	Concentration	log_{10} reduction	References
S. aureus	VRSA, strain VRS 8	5-d incubation on urinary catheter pieces	2 min	3.12% (S)	5.0	[4]
				1.56% (S)	1.5	
			5 min	3.12% (S)	>6.0	
				1.56% (S)	2.0	
			10 min	3.12% (S)	>6.0	
				1.56% (S)	2.5	
S. aureus	ATCC 25923	3-w incubation on pieces of cellulose nitrate membranes	30 s	0.1% (P)	1.1	[32]
S. aureus	Leg ulcer isolate	24-h incubation in polystyrene 96-well plates	1 min	0.1%[a] (P)	>6.0	[43]
			15 min			
			30 min			
S. aureus	ATCC 33593 (MRSA)	24-h incubation on partial thickness porcine wounds	2 irrigations per d for up to 6 d	0.1%[a] (P)	1.0 (day 3)	[17]
					2.7 (day 6)	
S. aureus	CIP 4.83	24- or 48-h incubation on glass slides	24 h	0.0125% (P)	>7.0	[52]
S. aureus	ATCC 6538	24-h incubation on agar disc	24 h	0.01% (S)	0.8–1.0	[44]
S. mutans	NCTC 10449	3-d incubation on ceramic hydroxylapatite slabs	30 min	0.2% (S)	"complete kill"	[72]
				0.1% (S)	"incomplete kill"	
S. mutans	DSM 20523	72-h incubation on titanium discs	30 min	0.1% (S)	1.8	[46]
S. sanguis	ATCC 10558	3-d incubation on ceramic hydroxylapatite slabs	30 min	0.2% (S)	"complete kill"	[72]
				0.1% (S)	"incomplete kill"	

(continued)

Table 14.5 (continued)

Species	Strains/isolates	Type of biofilm	Exposure time	Concentration	log_{10} reduction	References
Mixed species	Human saliva bacteria	72-h incubation on titanium discs	30 min	0.1% (S)	0.8	[46]
Mixed species	Subgingival plaque bacteria	Overnight incubation on titanium discs	30 min	0.1% (S)	1.5	[46]
Mixed species	*S. aureus* strain 308 (MRSA), *C. albicans* ATCC MYA 2876	48-h incubation in biofilm reactor	4 h	Unknown (P)	>5.0	[41]
			24 h			
			4 h	10% of unknown (P)	2.3	
			24 h		1.0	

P commercial product; *S* solution; [a]plus 2% phenoxyethanol; [b]plus 0.5% phenoxyethanol of treatment [53]. A 30-s application of a mouth rinse based on 0.1% OCT plus 2% phenoxyethanol revealed a reduction of bacterial load in saliva of 2.8 log (immediate effect). After 60 min, the mean CFU was still 1.8 log below baseline [63]. A strong bactericidal efficacy was also found in other studies with a 30-s or 2-min rinse using the same commercial solution [22, 48] or a solution based on 0.1% OCT alone [36, 85].

14.3.1.5 Bactericidal Activity on Skin

The data on the efficacy of OCT against bacteria on skin are variable. One study suggests a quite good bactericidal activity in 15 min even at 0.00012% OCT especially against various Gram-negative species. But all other studies show that 0.1% OCT has moderate bactericidal activity within 2 h (0.2–3.6 log; Table 14.6). For skin antisepsis during dressing changes around central venous catheter insertion sites among bone marrow transplant patients OCT (0.1%) in combination with 2% phenoxyethanol resulted in a continuous and substantial decline of bacterial density, most cultures were negative 2 weeks after insertion [77].

14.3.1.6 Bactericidal Activity in Other Applications

Against four strains of *L. monocytogenes*, *S. enterica* and *E. coli,* OCT (0.05% and 0.1%) washes on contaminated cantaloupe rinds reduced the bacterial load by >5.0 log within 3 min and below the level of detection within 5 min [81]. On porcine vaginal mucosa, OCT at 0.1% was able to reduce an artificial contamination of MRSA by approximately 1.8 log (15 min) to 5.2 log (24 h) [5].

14.3.2 Fungicidal Activity

14.3.2.1 Fungistatic Activity (MIC Values)

Table 14.7 shows that various fungal species were described with low OCT MIC values (0.4–6.7 mg/l) indicating susceptibility to the biocidal agent to the different yeasts.

14.3.2.2 Fungicidal Activity (Suspension Tests)

OCT (0.1%) in combination with 2% phenoxyethanol was mostly effective against *C. albicans* within 30 s although different types of organic load may substantially reduce the yeasticidal activity (Table 14.8). In *S. cerevisiae*, it was shown that OCT adheres fast and strongly to the cell surfaces [49]. At a concentration of 0.0002%, OCT permeabilizes the cells of *S. cerevisiae* in 3 min, longer exposure times resulted in full permeabilization [49].

14.3.2.3 Activity Against Yeasts in Biofilm

The yeasticidal activity of 0.1% OCT is variable depending on the type of biofilm. It may be effective and may completely eliminate *C. albicans* cells in 10 s on cellulose nitrate membranes, and it may also show only a 1.4 log reduction after 5 min in single-rooted teeth canals (Table 14.9).

Table 14.6 Efficacy of OCT against bacteria on skin

Species	Strains/isolates	Exposure time	Concentration	$\log_{10}$ reduction	References
E. coli	Strain Vogel	15 min	0.00012% (S)	6.1	[70]
			0.00009% (S)	2.1	
			0.00003% (S)	1.1	
K. pneumoniae	Strain SWRI no. 87	15 min	0.00012% (S)	5.0	[70]
			0.00009% (S)	4.8	
			0.00003% (S)	1.3	
P. mirabilis	Strain MGH-1	15 min	0.00012% (S)	6.2	[70]
			0.00009% (S)	2.9	
			0.00003% (S)	1.2	
P. aeruginosa	ATCC 15442	2 h	0.1%[a] (S)	0.3	[57]
		4 h		1.3	
		24 h		2.6	
P. aeruginosa	ATCC 9027	15 min	0.00012% (S)	5.8	[70]
			0.00009% (S)	3.4	
			0.00003% (S)	2.3	
S. marcescens	ATCC 8195	15 min	0.00012% (S)	5.8	[70]
			0.00009% (S)	4.6	
			0.00003% (S)	2.3	
S. aureus	ATCC 6538	2 h	0.1%[a] (S)	3.6	[57]
		4 h		4.2	
		24 h		≥ 5.6	
S. aureus	ATCC 6538	15 min	0.00012% (S)	5.0	[70]
			0.00009% (S)	3.8	
			0.00003% (S)	1.6	
S. epidermidis	ATCC 14990 (1,000 cells per cm^2; 1 min application)	1 min	0.1% (P)	0.1	[14]
		10 min		0.6	
		2 h		0.9	
	TCC 14990 (1,000,000 cells per cm^2; 1 min application)	1 min		0.1	
		10 min		0.3	
		2 h		0.2	
	ATCC 14990 (1,000,000 cells per cm^2; 10 min application)	1 min		0.7	
		10 min		1.5	
		2 h		1.1	
S. epidermidis	ATCC 17917	15 min	0.00012% (S)	5.0	[70]
			0.00009% (S)	5.2	
			0.00003% (S)	3.2	
S. pyogenes	ATCC 12384	15 min	0.00012% (S)	5.1	[70]
			0.00009% (S)	5.6	
			0.00003% (S)	2.2	
Mixed species	Resident skin flora	3 min	1.6% (S)	1.6	[70]
			0.8% (S)	1.5	
			0.4% (S)	1.3	
			0.2% (S)	1.0	

P commercial product; *S* solution; [a]exposed for 15 min to reconstructed human epidermis

Table 14.7 MIC values of various fungal species to OCT

Species	Strains/isolates	MIC value (mg/l)	References
C. albicans	ATCC 10231	0.8	[61]
C. albicans	ATCC 10231	1	[47]
C. albicans	KCCC 14172	1.5	[31]
C. albicans	ATCC 10231	3	[31]
C. albicans	ATCC 10231	6.7	[88]
C. pseudotropicalis	KCCC 13709	1.5	[31]
C. tropicalis	KCCC 13622	3	[31]
C. neoformans	ATCC 90112	0.4	[61]
S. cerevisae	NCYC 975	1.5–3	[30]

Table 14.8 Fungicidal activity of OCT in suspension tests

Species	Strains/isolates	Exposure time	Concentration	$\log_{10}$ reduction	References
C. albicans	ATCC 10231	15 s	0.1%[a] (P)	≥5.0	[79]
C. albicans	ATCC 10231	30 s	0.1%[a] (P)	1.8–6.2[b]	[64]
		1 min		2.8–6.3[b]	
		10 min		3.6–6.3[b]	
C. albicans	ATCC 10231	1 min	0.0025% (S)	4.0	[47]
		6 h	0.001% (S)	4.8	

S solution; *P* commercial product; [a]plus 2% phenoxyethanol; [b]depending on the type of organic load

Table 14.9 Efficacy of OCT in against yeasts in biofilms

Species	Strains/isolates	Type of biofilm	Exposure time	Concentration	$\log_{10}$ reduction	References
C. albicans	ATCC 10231D-5	3-w incubation on pieces of cellulose nitrate membranes	10 s	0.1% (P)	"complete elimination"	[32]
C. albicans	ATCC 10231	3-w incubation in single-rooted teeth canals	30 s	0.1%[a] (P)	1.0–1.2	[78]
			1 min		1.3	
			5 min		1.4	
C. albicans	ATCC 90028	14-d incubation in canals of single-rooted human teeth	3 min	0.1%[a] (P)	4.0	[24]

P commercial product; [a]plus 2% phenoxyethanol

14.3.2.4 Fungicidal Activity on Skin

On skin, OCT at 0.00012% was able to reduce *C. albicans* by >4.0 log within 15 min, whereas lower concentrations (0.00003% and 0.00009%) were less effective with 0.9 and 2.1 log, respectively [70].

14.3.3 Mycobactericidal Activity

A MIC value has only been described for *M. smegmatis* with 1 mg/l [29]. No further public information was found.

14.4 Effect of Low-Level Exposure

Low-level exposure experiments were so far only published with *S. aureus* and *P. aeruginosa* (Table 14.10). A weak adaptive response ($\leq$ 4-fold MIC increase) was observed in *S. aureus*, and a strong (>4-fold) and stable MIC increase was found in *P. aeruginosa*.

The strongest MIC change was found in isolates of *P. aeruginosa* ($\leq$32-fold) resulting in a MIC_{max} value of 128 mg/l. Cross-tolerance was described with CHG and some antibiotics (gentamicin, colistin, amikacin, tobramycin) but not benzalkonium chloride.

14.5 Resistance to OCT

14.5.1 High MIC Values

High MIC values have so far only been reported for *S. salivarius* ($\leq$800 mg/l), *P. aeruginosa* (128 mg/l after low-level exposure) and *S. mutans* ($\leq$120 mg/l). The frequent use of OCT for decolonization has significantly increased *S. aureus* MIC values from 0.49–0.56 to 0.86 (all mean) suggesting a correlation between its use and an increased tolerance [38]. No other bacterial or fungal isolates have been described with elevated MIC values suggesting tolerance to OCT.

14.5.2 Reduced Efficacy in Suspension Tests

So far no bacterial or fungal isolates have been described with reduced log reductions in suspension tests suggesting resistance to OCT.

Table 14.10 Change of bacterial susceptibility to biocides and antimicrobials after low-level exposure to OCT

Species	Strains/isolates	Exposure time	Increase in MIC	MIC_{max} (mg/l)	Stability of MIC change	Associated changes	References
P. aeruginosa	NCTC 13437 and 6 clinical isolates	12 d at various concentrations	4-fold–32-fold	128	Stable for 10 d	Increased tolerance[a] to chlorhexidine (8-fold–16-fold); no increased tolerance[a] to benzalkonium chloride; 1 strain with increased tolerance[a] to gentamicin and colistin (both 4-fold), amikacin and tobramycin (both 2-fold)	[71]
S. aureus	ATCC 6538	20–100 d at various concentrations	None	0.85	Not applicable	None described	[87]
S. aureus	5 international MRSA clones	3 m at sublethal concentrations	$\leq$ 2-fold	8	"Unstable"	None described	[1]

[a]Broth microdilution method

14.5.3 Resistance Mechanisms

No specific resistance mechanisms explaining a reduced susceptibility to OCT have been described so far.

14.5.4 Resistance Genes

No OCT resistance genes have been described so far.

14.6 Cross-Tolerance to Other Biocidal Agents

Cross-tolerance between OCT and chlorhexidine has been described in *P. aeruginosa* after low-level exposure.

14.7 Cross-Tolerance to Antibiotics

Cross-tolerance between OCT and gentamicin, colistin, amikacin and tobramycin has been described in a *P. aeruginosa* isolate. No other species with a cross-tolerance have so far been described

14.8 Role of Biofilm

14.8.1 Effect on Biofilm Development

Biofilm formation of *S. aureus* strains can be suppressed by OCT at 0.31–0.62% with an exposure time of 10 min resulting 0–25% biofilm formation compared to no treatment. On materials used in the oral cavity, it requires at least 3% OCT to partially inhibit biofilm formation over 7 d (Table 14.11). No data were found for OCT in concentrations typically used in clinical medicine (e.g. 0.1%).

14.8.2 Effect on Biofilm Removal

Biofilm removal seems to be effective in 1 min by 0.1% OCT in combination with 2% phenoxyethanol using a 24 h *S. aureus* biofilm, whereas it requires 30 min to be equally effective against a *P. aeruginosa* biofilm. A root canal biofilm was not removed by 0.1% OCT in combination with 2% phenoxyethanol within 1 min.

Table 14.11 Effect of OCT on biofilm development

Bacterial species	Strains/isolates	Type of biofilm	Exposure time	Type of product	Inhibition of biofilm formation	References
S. aureus	ATCC 35556	24-h incubation in polystyrene 96-well plates	2 min	1.25% (S)	100%	[4]
				0.62% (S)	85%	
				0.31% (S)	25%	
			5 min	1.25% (S)	100%	
				0.62% (S)	100%	
				0.31% (S)	50%	
			10 min	1.25% (S)	100%	
				0.62% (S)	100%	
				0.31% (S)	75%	
S. aureus	VRSA, strain VRS 8	24-h incubation in polystyrene 96-well plates	2 min	1.25% (S)	100%	[4]
				0.62% (S)	85%	
				0.31% (S)	25%	
			5 min	1.25% (S)	100%	
				0.62% (S)	100%	
				0.31% (S)	50%	
			10 min	1.25% (S)	100%	
				0.62% (S)	100%	
				0.31% (S)	80%	

(continued)

Table 14.11 (continued)

Bacterial species	Strains/isolates	Type of biofilm	Exposure time	Type of product	Inhibition of biofilm formation	References
S. aureus	MRSA, strain NRS 123	24-h incubation in polystyrene 96-well plates	2 min	1.25% (S)	100%	[4]
				0.62% (S)	85%	
				0.31% (S)	25%	
			5 min	1.25% (S)	100%	
				0.62% (S)	100%	
				0.31% (S)	50%	
			10 min	1.25% (S)	100%	
				0.62% (S)	100%	
				0.31% (S)	75%	
Mixed species	Mixed oral flora	Oral cavity exposure in healthy volunteers: one control resin without OCT, one with 3% OCT and one with 6% OCT	3 d	6%	Very few distinct pellicle layer	[68]
				3%	Few small microbial aggregations	
				0%	Established biofilm covering < 50% of the surface	
			7 d	6%	Distinct pellicle layer	
				3%	Few small microbial aggregations	
				0%	Established multilayer biofilm covering > 50% of the surface	

A vaginal biofilm was partially removed by a spray based on 0.1% OCT in combination with 2% phenoxyethanol (Table 14.12.).

A high non-response rate on biofilm removal by daily spray application for bacterial vaginosis and *Gardnerella* biofilm was accompanied by the persistence of the structured *Gardnerella* biofilm despite continuation of the antiseptic treatment [74].

Table 14.12 Biofilm removal rate (quantitative determination of biofilm matrix) by exposure to commercial products (P) based on OCT

Type of biofilm	Concentration	Exposure time	Biofilm removal rate	References
P. aeruginosa (14 clinical isolates and ATCC 15445), 24-h incubation in polystyrene microtitre plates	0.1%[a] (P)	1 min	0 of 15[b]	[43]
		15 min	7 of 15[b]	
		30 min	15 of 15[b]	
S. aureus (14 clinical isolates and ATCC 5638), 24-h incubation in polystyrene microtitre plates	0.1%[a] (P)	1 min	15 of 15[b]	[43]
		15 min		
		30 min		
Mixed species: root canals from human mandibular premolars	0.1%[a] (P)	1 min	None	[15]
Mixed-species biofilm: patients with symptomatic bacterial vaginosis and Gardnerella biofilm	0.1%[a] (P)	Daily spray application for 7 d	Biofilm undetectable in 21 of 24 patients (87.5%)	[74]
		Daily spray application for 28 d in 14 patients with relapse and 3 non-responsive patients	Biofilm undetectable in 11 of 17 patients (64.7%)	

[a]Plus 2% phenoxyethanol; [b]biofilm eradication rate

14.8.3 Effect on Biofilm Fixation

No data were found to evaluate the potential of OCT on biofilm fixation.

14.9 Summary

The principal antimicrobial activity of OCT, often in combination with 2% phenoxyethanol, is summarized in Table 14.13.

The key findings on acquired resistance and cross-resistance including the role of biofilm for selecting resistant isolates are summarized in Table 14.14.

Table 14.13 Overview on the typical exposure times required for OCT (often in combination with 2% phenoxyethanol) to achieve sufficient biocidal activity against the different target micro-organisms

Target micro-organisms	Species	Concentration	Exposure time
Bacteria	Most bacterial species	0.1%[a]	1 min
		0.01%[a]	5 min
Yeasts	Most yeasts	0.1%[a,b]	30 s
Mycobacteria	Unknown		

[a]In biofilm, the efficacy will be lower; [b]high organic load impairs the yeasticidal activity

Table 14.14 Key findings on acquired OCT resistance, the effect of low-level exposure, cross-tolerance to other biocides and antibiotics, and its effect on biofilm

Parameter	Species	Findings
Elevated MIC values	*S. salivarius*	$\leq$ 800 mg/l
	P. aeruginosa	$\leq$ 128 mg/l
	S. mutans	$\leq$ 120 mg/l
Proposed MIC value to determine resistance	Not proposed yet for bacteria, fungi or mycobacteria	
Cross-tolerance biocides	*P. aeruginosa*	Chlorhexidine
Cross-tolerance antibiotics	*P. aeruginosa*	Gentamicin, colistin, amikacin and tobramycin
Resistance mechanisms	Not described.	
Effect of low-level exposure	*S. aureus*	No MIC increase
	S. aureus	Weak MIC increase ($\leq$ 4-fold)
	P. aeruginosa	Strong and stable MIC increase (>4-fold)
	P. aeruginosa ($\leq$ 32-fold)	Strongest MIC change after low-level exposure
	P. aeruginosa (128 mg/l)	Highest MIC values after low-level exposure
Biofilm	Development	Inhibition of biofilm formation of *S. aureus* ($\geq$ 0.31% OCT) and mixed biofilm ($\geq$ 3% OCT)
	Removal	Strong removal (*S. aureus*, *P. aeruginosa*) in 30 min (0.1% OCT plus 2% phenoxyethanol)
		Poor removal in mixed-species biofilms in 1 min (0.1% OCT plus 2% phenoxyethanol)
	Fixation	Unknown

References

1. Al-Doori Z, Goroncy-Bermes P, Gemmell CG, Morrison D (2007) Low-level exposure of MRSA to octenidine dihydrochloride does not select for resistance. J Antimicrob Chemother 59(6):1280–1282
2. Alvarez-Marin R, Aires-de-Sousa M, Nordmann P, Kieffer N, Poirel L (2017) Antimicrobial activity of octenidine against multidrug-resistant Gram-negative pathogens. Eur J Clin Microbiol Infect Dis 36(12):2379–2383. https://doi.org/10.1007/s10096-017-3070-0
3. Amalaradjou MA, Norris CE, Venkitanarayanan K (2009) Effect of octenidine hydrochloride on planktonic cells and biofilms of Listeria monocytogenes. Appl Environ Microbiol 75 (12):4089–4092. https://doi.org/10.1128/aem.02807-08
4. Amalaradjou MA, Venkitanarayanan K (2014) Antibiofilm effect of octenidine hydrochloride on staphylococcus aureus, MRSA and VRSA. Pathogens (Basel, Switzerland) 3(2):404–416. https://doi.org/10.3390/pathogens3020404
5. Anderson MJ, Scholz MT, Parks PJ, Peterson ML (2013) Ex vivo porcine vaginal mucosal model of infection for determining effectiveness and toxicity of antiseptics. J Appl Microbiol 115(3):679–688. https://doi.org/10.1111/jam.12277
6. Assadian O (2016) Octenidine dihydrochloride: chemical characteristics and antimicrobial properties. J Wound Care 25(3 Suppl):S3–6. https://doi.org/10.12968/jowc.2016.25.Sup3.S3
7. Bartoszewicz M, Rygiel A, Krzeminski M, Przondo-Mordarska A (2007) Penetration of a selected antibiotic and antiseptic into a biofilm formed on orthopedic steel implants. Ortopedia, Traumatologia, Rehabilitacja 9(3):310–318
8. Baskaran SA, Upadhyay A, Upadhyaya I, Bhattaram V, Venkitanarayanan K (2012) Efficacy of octenidine hydrochloride for reducing Escherichia coli O157:H7, Salmonella spp., and Listeria monocytogenes on cattle hides. Appl Environ Microbiol 78(12):4538–4541. https://doi.org/10.1128/aem.00259-12
9. Beiswanger BB, Mallatt ME, Mau MS, Jackson RD, Hennon DK (1990) The clinical effects of a mouthrinse containing 0.1% octenidine. J Dent Res 69(2):454–457
10. Bock LJ, Hind CK, Sutton JM, Wand ME (2018) Growth media and assay plate material can impact on the effectiveness of cationic biocides and antibiotics against different bacterial species. Lett Appl Microbiol 66(5):368–377. https://doi.org/10.1111/lam.12863
11. Brill F, Goroncy-Bermes P, Sand W (2006) Influence of growth media on the sensitivity of Staphylococcus aureus and Pseudomonas aeruginosa to cationic biocides. Int J Hyg Environ Health 209(1):89–95
12. Bukhary S, Balto H (2017) Antibacterial efficacy of octenisept, alexidine, chlorhexidine, and Sodium hypochlorite against Enterococcus faecalis biofilms. J Endodontics 43(4):643–647. https://doi.org/10.1016/j.joen.2016.09.013
13. Cherian B, Gehlot PM, Manjunath MK (2016) Comparison of the antimicrobial efficacy of octenidine dihydrochloride and chlorhexidine with and without passive ultrasonic irrigation—an invitro study. J Clin Diagn Res 10(6):Zc71–77. https://doi.org/10.7860/jcdr/2016/17911.8021
14. Christiansen B (1988) The effectiveness of a skin disinfectant with a cation active additive. Zentralbl Bakteriol Hyg I Abt Orig B 186(4):368–374
15. Coaguila-Llerena H, Stefanini da Silva V, Tanomaru-Filho M, Guerreiro Tanomaru JM, Faria G (2018) Cleaning capacity of octenidine as root canal irrigant: a scanning electron microscopy study. Microsc Res Tech. https://doi.org/10.1002/jemt.23007
16. Conceicao T, de Lencastre H, Aires-de-Sousa M (2016) Efficacy of octenidine against antibiotic-resistant Staphylococcus aureus epidemic clones. J Antimicrob Chemother 71 (10):2991–2994. https://doi.org/10.1093/jac/dkw241
17. Davis SC, Harding A, Gil J, Parajon F, Valdes J, Solis M, Higa A (2017) Effectiveness of a polyhexanide irrigation solution on methicillin-resistant Staphylococcus aureus biofilms in a porcine wound model. Int Wound J 14(6):937–944. https://doi.org/10.1111/iwj.12734

18. de Lucena JM, Decker EM, Walter C, Boeira LS, Lost C, Weiger R (2013) Antimicrobial effectiveness of intracanal medicaments on Enterococcus faecalis: chlorhexidine versus octenidine. Int Endod J 46(1):53–61. https://doi.org/10.1111/j.1365-2591.2012.02093.x
19. Decker EM, Bartha V, Kopunic A, von Ohle C (2017) Antimicrobial efficiency of mouthrinses versus and in combination with different photodynamic therapies on periodontal pathogens in an experimental study. J Periodontal Res 52(2):162–175. https://doi.org/10.1111/jre.12379
20. Dettenkofer M, Wilson C, Gratwohl A, Schmoor C, Bertz H, Frei R, Heim D, Luft D, Schulz S, Widmer AF (2010) Skin disinfection with octenidine dihydrochloride for central venous catheter site care: a double-blind, randomized, controlled trial. Clin Microbiol Infect 16(6):600–606. https://doi.org/10.1111/j.1469-0691.2009.02917.x
21. Dogan AA, Adiloglu AK, Onal S, Cetin ES, Polat E, Uskun E, Koksal F (2008) Short-term relative antibacterial effect of octenidine dihydrochloride on the oral microflora in orthodontically treated patients. Int J Infect Dis: IJID: Official Publication of the International Society for Infectious Diseases 12(6):e19–25. https://doi.org/10.1016/j.ijid.2008.03.013
22. Dogan AA, Cetin ES, Hussein E, Adiloglu AK (2009) Microbiological evaluation of octenidine dihydrochloride mouth rinse after 5 days' use in orthodontic patients. The Angle orthodontist 79(4):766–772. https://doi.org/10.2319/062008-322.1
23. Eisenbeiß W, Siemers F, Amtsberg G, Hinz P, Hartmann B, Kohlmann T, Ekkernkamp A, Albrecht U, Assadian O, Kramer A (2012) Prospective, double-blinded, randomised controlled trial assessing the effect of an Octenidine-based hydrogel on bacterial colonisation and epithelialization of skin graft wounds in burn patients. Int J Burns Trauma 2(2):71–79
24. Eldeniz AU, Guneser MB, Akbulut MB (2015) Comparative antifungal efficacy of light-activated disinfection and octenidine hydrochloride with contemporary endodontic irrigants. Lasers Med Sci 30(2):669–675. https://doi.org/10.1007/s10103-013-1387-1
25. European Medicines Agency (2010) Public summary of opinion on orphan designation; octenidine dihydrochloride for the prevention of late-onset sepsis in premature infants of less than or equal to 32 weeks of gestational age. http://www.emaeuropaeu/docs/en_GB/document_library/Orphan_designation/2010/08/WC500095702pdf. Accessed 15 March 2018
26. European Medicines Agency (2011) Opinion of the Committee for Medicinal Products for Veterinary Use on the establishment of maximum residue limits; procedure no: EU/09/170/SCM; name of the substance: octenidine dihydrochloride (INN). http://www.emaeuropaeu/docs/en_GB/document_library/Maximum_Residue_Limits_-_Opinion/2012/03/WC500124460pdf. Accessed 15 March 2018
27. European Medicines Agency (2016) European Medicines Agency decision P/0360/2016. http://www.emaeuropaeu/docs/en_GB/document_library/PIP_decision/WC500220327pdf. Accessed 15 March 2018
28. European Medicines Agency (2017) List of nationally authorised medicinal products; active substance: octenidine dihydrochloride/phenoxyethanol; procedure no.: PSUSA/00002199/201701. http://www.emaeuropaeu/docs/en_GB/document_library/Periodic_safety_update_single_assessment/2017/10/WC500237398pdf. Accessed 15 March 2018
29. Frenzel E, Schmidt S, Niederweis M, Steinhauer K (2011) Importance of porins for biocide efficacy against Mycobacterium smegmatis. Appl Environ Microbiol 77(9):3068–3073. https://doi.org/10.1128/aem.02492-10
30. Ghannoum MA, Abu Elteen K, Ellabib M, Whittaker PA (1990) Antimycotic effects of octenidine and pirtenidine. J Antimicrob Chemother 25:237–245
31. Ghannoum MA, Abu Elteen K, Stretton RJ, Whittaker PA (1990) Effects of octendine and pirtenidine on adhesion of Candida species to human buccal epithelial cells *in vitro*. Arch Oral Biol 35(4):249–253
32. Ghivari SB, Bhattacharya H, Bhat KG, Pujar MA (2017) Antimicrobial activity of root canal irrigants against biofilm forming pathogens—an in vitro study. J Conserv Dentistry: JCD 20(3):147–151. https://doi.org/10.4103/jcd.jcd_38_16

33. Goroncy-Bermes P, Brill FHH, Brill H (2013) Antimicrobial activity of wound antiseptics against extended-spectrum beta-lactamase-producing bacteria. Wound Med 1(1):41–43
34. Guneser MB, Akbulut MB, Eldeniz AU (2016) Antibacterial effect of chlorhexidine-cetrimide combination, Salvia officinalis plant extract and octenidine in comparison with conventional endodontic irrigants. Dent Mater J 35(5):736–741. https://doi.org/10.4012/dmj.2015-159
35. Gunther F, Blessing B, Tacconelli E, Mutters NT (2017) MRSA decolonization failure-are biofilms the missing link? Antimicrob Resist Infect Control 6:32. https://doi.org/10.1186/s13756-017-0192-1
36. Gusic I, Medic D, Radovanovic Kanjuh M, Ethuric M, Brkic S, Turkulov V, Predin T, Mirnic J (2016) Treatment of periodontal disease with an octenidine-based antiseptic in HIV-positive patients. Int J Dental Hygiene 14(2):108–116. https://doi.org/10.1111/idh.12141
37. Hammerle G, Strohal R (2016) Efficacy and cost-effectiveness of octenidine wound gel in the treatment of chronic venous leg ulcers in comparison to modern wound dressings. Int Wound J 13(2):182–188. https://doi.org/10.1111/iwj.12250
38. Hardy K, Sunnucks K, Gil H, Shabir S, Trampari E, Hawkey P, Webber M, Wright GD (2018) Increased usage of antiseptics is associated with reduced susceptibility in clinicalisolates of Staphylococcus aureus. mBio 9(3):e00894-18. https://doi.org/10.1128/mBio.00894-18
39. Harke H-P (1989) Octenidindihydrochlorid, Eigenschaften eines neuen antimikrobiellen Wirkstoffes. Zentralbl Hyg Umweltmed 188(1–2):188–193
40. Hirsch T, Limoochi-Deli S, Lahmer A, Jacobsen F, Goertz O, Steinau HU, Seipp HM, Steinstraesser L (2011) Antimicrobial activity of clinically used antiseptics and wound irrigating agents in combination with wound dressings. Plast Reconstr Surg 127(4):1539–1545. https://doi.org/10.1097/PRS.0b013e318208d00f
41. Hoekstra MJ, Westgate SJ, Mueller S (2017) Povidone-iodine ointment demonstrates in vitro efficacy against biofilm formation. Int Wound J 14(1):172–179. https://doi.org/10.1111/iwj.12578
42. Hubner NO, Siebert J, Kramer A (2010) Octenidine dihydrochloride, a modern antiseptic for skin, mucous membranes and wounds. Skin Pharmacol Physiol 23(5):244–258. https://doi.org/10.1159/000314699
43. Junka A, Bartoszewicz M, Smutnicka D, Secewicz A, Szymczyk P (2014) Efficacy of antiseptics containing povidone-iodine, octenidine dihydrochloride and ethacridine lactate against biofilm formed by Pseudomonas aeruginosa and Staphylococcus aureus measured with the novel biofilm-oriented antiseptics test. Int Wound J 11(6):730–734. https://doi.org/10.1111/iwj.12057
44. Junka AF, Zywicka A, Szymczyk P, Dziadas M, Bartoszewicz M, Fijalkowski K (2017) A.D.A.M. test (Antibiofilm Dressing's Activity Measurement)—simple method for evaluating anti-biofilm activity of drug-saturated dressings against wound pathogens. J Microbiol Meth 143:6–12. https://doi.org/10.1016/j.mimet.2017.09.014
45. Kapalschinski N, Seipp HM, Kuckelhaus M, Harati KK, Kolbenschlag JJ, Daigeler A, Jacobsen F, Lehnhardt M, Hirsch T (2017) Albumin reduces the antibacterial efficacy of wound antiseptics against Staphylococcus aureus. J Wound Care 26(4):184–187. https://doi.org/10.12968/jowc.2017.26.4.184
46. Koban I, Geisel MH, Holtfreter B, Jablonowski L, Hubner NO, Matthes R, Masur K, Weltmann KD, Kramer A, Kocher T (2013) Synergistic effects of nonthermal plasma and disinfecting agents against dental biofilms in vitro. ISRN Dentistry 2013:573262. https://doi.org/10.1155/2013/573262
47. Koburger T, Hübner N-O, Braun M, Siebert J, Kramer A (2010) Standardized comparison of antiseptic efficacy of triclosan, PVP-iodine, octenidine dihydrochloride, polyhexanide and chlorhexidine digluconate. J Antimicrob Chemother 65(8):1712–1719
48. Kocak MM, Ozcan S, Kocak S, Topuz O, Erten H (2009) Comparison of the efficacy of three different mouthrinse solutions in decreasing the level of Streptococcus mutans in saliva. Eur J Dentistry 3(1):57–61

49. Kodedova M, Sigler K, Lemire BD, Gaskova D (2011) Fluorescence method for determining the mechanism and speed of action of surface-active drugs on yeast cells. Biotechniques 50 (1):58–63. https://doi.org/10.2144/000113568
50. Kramer A, Dissemond J, Kim S, Willy C, Mayer D, Papke R, Tuchmann F, Assadian O (2017) Consensus on wound antisepsis: update 2018. Skin Pharmacol Physiol 31(1):28–58. https://doi.org/10.1159/000481545
51. Krishnan U, Saji S, Clarkson R, Lalloo R, Moule AJ (2017) Free Active chlorine in sodium hypochlorite solutions admixed with octenidine, SmearOFF, chlorhexidine, and EDTA. J End 43(8):1354–1359. https://doi.org/10.1016/j.joen.2017.03.034
52. Lefebvre E, Vighetto C, Di Martino P, Larreta Garde V, Seyer D (2016) Synergistic antibiofilm efficacy of various commercial antiseptics, enzymes and EDTA: a study of Pseudomonas aeruginosa and Staphylococcus aureus biofilms. Int J Antimicrob Agents 48 (2):181–188. https://doi.org/10.1016/j.ijantimicag.2016.05.008
53. Lorenz K, Jockel-Schneider Y, Petersen N, Stolzel P, Petzold M, Vogel U, Hoffmann T, Schlagenhauf U, Noack B (2018) Impact of different concentrations of an octenidine dihydrochloride mouthwash on salivary bacterial counts: a randomized, placebo-controlled cross-over trial. Clin Oral Invest. https://doi.org/10.1007/s00784-018-2379-0
54. Lutz JT, Diener IV, Freiberg K, Zillmann R, Shah-Hosseini K, Seifert H, Berger-Schreck B, Wisplinghoff H (2016) Efficacy of two antiseptic regimens on skin colonization of insertion sites for two different catheter types: a randomized, clinical trial. Infection 44(6):707–712. https://doi.org/10.1007/s15010-016-0899-6
55. Melhorn S, Staubach P (2018) [Octenidine dihydrochloride: the antiseptic that does not like every base formulation]. Der Hautarzt; Zeitschrift fur Dermatologie, Venerologie, und verwandte Gebiete. https://doi.org/10.1007/s00105-018-4139-0
56. Muller G, Kramer A (2008) Biocompatibility index of antiseptic agents by parallel assessment of antimicrobial activity and cellular cytotoxicity. J Antimicrob Chemother 61(6):1281–1287. https://doi.org/10.1093/jac/dkn125
57. Muller G, Langer J, Siebert J, Kramer A (2014) Residual antimicrobial effect of chlorhexidine digluconate and octenidine dihydrochloride on reconstructed human epidermis. Skin Pharmacol Physiol 27(1):1–8. https://doi.org/10.1159/000350172
58. Narayanan A, Nair MS, Karumathil DP, Baskaran SA, Venkitanarayanan K, Amalaradjou MA (2016) Inactivation of acinetobacter baumannii biofilms on polystyrene, stainless steel, and urinary catheters by octenidine dihydrochloride. Front Microbiol 7:847. https://doi.org/10.3389/fmicb.2016.00847
59. National Center for Biotechnology Information (2018) Octenidine. PubChem Compound Database; CID = 51167. https://pubchem.ncbi.nlm.nih.gov/compound/51167. Accessed 15 March 2018
60. National Center for Biotechnology Information (2018) Octenidine hydrochloride PubChem Compound Database; CID = 51166. https://pubchem.ncbi.nlm.nih.gov/compound/51166. Accessed 15 March 2018
61. Ng CKL, Singhal V, Widmer F, Wright LC, Sorrell TC, Jolliffe KA (2007) Synthesis, antifungal and haemolytic activity of a series of bis(pyridinium)alkanes. Bioorg Med Chem 15(10):3422–3429
62. Obermeier A, Schneider J, Fohr P, Wehner S, Kuhn KD, Stemberger A, Schieker M, Burgkart R (2015) In vitro evaluation of novel antimicrobial coatings for surgical sutures using octenidine. BMC Microbiol 15:186. https://doi.org/10.1186/s12866-015-0523-4
63. Pitten F-A, Kramer A (1999) Antimicrobial efficacy of antiseptic mouthrinse solutions. Eur J Clin Pharmacol 55(2):95–100
64. Pitten F-A, Werner H-P, Kramer A (2003) A standardized test to assess the impact of different organic challenges on the antimicrobial activity of antiseptics. J Hosp Infect 55(2):108–115
65. Reiser M, Scherag A, Forstner C, Brunkhorst FM, Harbarth S, Doenst T, Pletz MW, Hagel S (2017) Effect of pre-operative octenidine nasal ointment and showering on surgical site

infections in patients undergoing cardiac surgery. J Hosp Infect 95(2):137–143. https://doi.org/10.1016/j.jhin.2016.11.004
66. Rochon-Edouard S, Pons JL, Veber B, Larkin M, Vassal S, Lemeland JF (2004) Comparative in vitro and in vivo study of nine alcohol-based handrubs. Am J Infect Control 32(4):200–204. https://doi.org/10.1016/j.ajic.2003.08.003
67. Rohrer N, Widmer AF, Waltimo T, Kulik EM, Weiger R, Filipuzzi-Jenny E, Walter C (2010) Antimicrobial efficacy of 3 oral antiseptics containing octenidine, polyhexamethylene biguanide, or Citroxx: can chlorhexidine be replaced? Infect Control Hosp Epidemiol 31 (7):733–739. https://doi.org/10.1086/653822
68. Rupf S, Balkenhol M, Sahrhage TO, Baum A, Chromik JN, Ruppert K, Wissenbach DK, Maurer HH, Hannig M (2012) Biofilm inhibition by an experimental dental resin composite containing octenidine dihydrochloride. Dental Materials: Official Publication of the Academy of Dental Materials 28(9):974–984. https://doi.org/10.1016/j.dental.2012.04.034
69. Schedler K, Assadian O, Brautferger U, Muller G, Koburger T, Classen S, Kramer A (2017) Proposed phase 2/step 2 in-vitro test on basis of EN 14561 for standardised testing of the wound antiseptics PVP-iodine, chlorhexidine digluconate, polihexanide and octenidine dihydrochloride. BMC Infect Dis 17(1):143. https://doi.org/10.1186/s12879-017-2220-4
70. Sedlock DM, Bailey DM (1985) Microbicidal activity of octenidine hydrochloride, a new alkanediylbis[pyridine] germicidal agent. Antimicrob Agents Chemother 28(6):786–790
71. Shepherd MJ, Moore G, Wand ME, Sutton JM, Bock LJ (2018) Pseudomonas aeruginosa adapts to octenidine in the laboratory and a simulated clinical setting, leading to increased tolerance to chlorhexidine and other biocides. J Hosp Infect. https://doi.org/10.1016/j.jhin.2018.03.037
72. Slee AM, O'Connor JR (1983) In vitro antiplaque activity of octenidine dihydrochloride (WIN 41464-2) against preformed plaques of selected oral plaque-forming microorganisms. Antimicrob Agents Chemother 23(3):379–384
73. Sloot N, Siebert J, Höffler U (1999) Eradication of MRSA from carriers by means of whole-body washing with an antiseptic in combination with mupirocin nasal ointment. Zentralbl Hyg Umweltmed 202(6):513–523
74. Swidsinski A, Loening-Baucke V, Swidsinski S, Verstraelen H (2015) Polymicrobial Gardnerella biofilm resists repeated intravaginal antiseptic treatment in a subset of women with bacterial vaginosis: a preliminary report. Arch Gynecol Obstet 291(3):605–609. https://doi.org/10.1007/s00404-014-3484-1
75. Tandjung L, Waltimo T, Hauser I, Heide P, Decker EM, Weiger R (2007) Octenidine in root canal and dentine disinfection ex vivo. Int Endod J 40(11):845–851. https://doi.org/10.1111/j.1365-2591.2007.01279.x
76. Tanner J, Gould D, Jenkins P, Hilliam R, Mistry N, Walsh S (2012) A fresh look at preoperative body washing. J Infect Prev 13(1):11–15. https://doi.org/10.1177/1757177411428095
77. Tietz A, Frei R, Dangel M, Bolliger D, Passweg JR, Gratwohl A, Widmer AF (2005) Octenidine hydrochloride for the care of central venous catheter insertion sites in severely immunocompromised patients. Infect Control Hosp Epidemiol 26(8):703–707
78. Tirali RE, Bodur H, Ece G (2012) In vitro antimicrobial activity of sodium hypochlorite, chlorhexidine gluconate and octenidine dihydrochloride in elimination of microorganisms within dentinal tubules of primary and permanent teeth. Medicina oral, patologia oral y cirugia bucal 17(3):e517–522
79. Tirali RE, Turan Y, Akal N, Karahan ZC (2009) In vitro antimicrobial activity of several concentrations of NaOCl and octenisept in elimination of endodontic pathogens. Oral Surg Oral Med Oral Pathol Oral Radiol Endod 108(5):e117–120. https://doi.org/10.1016/j.tripleo.2009.07.012
80. Tylewska-Wierzbanowska S, Rogulska U, Lewandowska G, Chmielewski T (2017) Bactericidal activity of octenidine to various genospecies of Borrelia burgdorferi, Sensu Lato Spirochetes in vitro and in vivo. Pol J Microbiol 66(2):259–263

81. Upadhyay A, Chen C, Yin H, Upadhyaya I, Fancher S, Liu Y, Nair MS, Jankelunas L, Patel JR, Venkitanarayanan K (2016) Inactivation of Listeria monocytogenes, Salmonella spp. and Escherichia coli O157:H7 on cantaloupes by octenidine dihydrochloride. Food Microbiol 58:121–127. https://doi.org/10.1016/j.fm.2016.04.007
82. Uzer Celik E, Tunac AT, Ates M, Sen BH (2016) Antimicrobial activity of different disinfectants against cariogenic microorganisms. Brazilian Oral Res 30(1):e125. https://doi.org/10.1590/1807-3107BOR-2016.vol30.0125
83. van Meurs SJ, Gawlitta D, Heemstra KA, Poolman RW, Vogely HC, Kruyt MC (2014) Selection of an optimal antiseptic solution for intraoperative irrigation: an in vitro study. J Bone Joint Surgery American 96(4):285–291. https://doi.org/10.2106/jbjs.m.00313
84. Varelmann D, Hostmann F, Stüber F, Schroeder S (2004) Livide Verfärbung der Hand als unerwünschtes Ereignis bei axillärer Plexusanästhesie. Der Anaesthesist 53:441–444
85. Welk A, Zahedani M, Beyer C, Kramer A, Muller G (2016) Antibacterial and antiplaque efficacy of a commercially available octenidine-containing mouthrinse. Clin Oral Invest 20 (7):1469–1476. https://doi.org/10.1007/s00784-015-1643-9
86. Wiegand C, Abel M, Ruth P, Elsner P, Hipler UC (2015) pH influence on antibacterial efficacy of common antiseptic substances. Skin Pharmacol Physiol 28(3):147–158. https://doi.org/10.1159/000367632
87. Wiegand C, Abel M, Ruth P, Hipler UC (2012) Analysis of the adaptation capacity of Staphylococcus aureus to commonly used antiseptics by microplate laser nephelometry. Skin Pharmacol Physiol 25(6):288–297. https://doi.org/10.1159/000341222
88. Yamamoto M, Takami T, Matsumura R, Dorofeev A, Hirata Y, Nagamune H (2016) In vitro evaluation of the biocompatibility of newly synthesized bis-quaternary ammonium compounds with spacer structures derived from pentaerythritol or hydroquinone. Biocontrol Sci 21(4):231–241. https://doi.org/10.4265/bio.21.231
89. Zumtobel M, Assadian O, Leonhard M, Stadler M, Schneider B (2009) The antimicrobial effect of Octenidine-dihydrochloride coated polymer tracheotomy tubes on Staphylococcus aureus and Pseudomonas aeruginosa colonisation. BMC Microbiol 9:150. https://doi.org/10.1186/1471-2180-9-150

Silver

15

15.1 Chemical Characterization

Silver is a naturally occurring element and can be found in four oxidative states: Ag^0, Ag^+, Ag^{++} and Ag^{+++}. The two latter states produce complexes that are insoluble or less antimicrobial than the former. The antimicrobial action of silver is dependent upon the bioavailability of the silver ion (Ag^+). Silver compounds ionize in the presence of water, bodily fluids and other exudates [36]. Silver oxynitrate (Ag_7NO_{11}) is another potential antimicrobial substance [74] but is not reviewed here in detail.

The basic chemical information on silver and silver nitrate is summarized in Table 15.1.

15.2 Types of Application

Silver is used as an antiseptic agent in various forms. Probably the earliest medical use of silver was for water disinfection and storage [7]. The Romans included silver in their official book of medicines and were known to have used silver nitrate [7]. In the 1880s, the German obstetrician Carl Credé found that dilute solutions of silver nitrate reduced the incidence of neonatal eye infections from 10.8% to less than 2% [7].

Silver is used today in a wide range of medical applications. Examples are the use of silver preparations as topical cream in the treatment of burn wounds, in dental amalgams, in preventative eye care and the use of silver-impregnated polymers to prevent bacterial (biofilm) growth on medical devices such as catheters and heart valves [139]. It is also used for antiseptic treatment of burns [109] and as a (co-)disinfectant of water systems such as swimming pool water, hospital hot water systems and potable water systems [93]. It is also considered for use in health

G. Kampf, *Antiseptic Stewardship*, https://doi.org/10.1007/978-3-319-98785-9_15

Table 15.1 Basic chemical information on silver and silver nitrate [104, 105]

CAS number	7440-22-4	7761-88-8
IUPAC name	Silver	Silver nitrate
Molecular formula	Ag	$AgNO_3$
Molecular weight (g/mol)	107.868	169.872

care for self-disinfecting surfaces although its impact on healthcare-associated infections is unknown [165]. This substance is also used in articles, by professional workers (widespread uses), in formulation or repacking, at industrial sites and in manufacturing [40].

15.2.1 European Chemicals Agency (European Union)

Silver is currently under review (June 2018) as a biocidal agent for product types 2 (disinfectants and algaecides not intended for direct application to humans or animals), 4 (food and feed area), 5 (drinking water) and 11 (preservatives for liquid-cooling and processing systems). Silver nitrate is under review (June 2018) as a biocidal agent for product types 1 (human hygiene), 2 (disinfectants and algaecides not intended for direct application to humans or animals), 3 (veterinary hygiene), 4 (food and feed area), 5 (drinking water), 7 (film preservatives), 9 (fibre, leather, rubber and polymerized materials preservatives) and 11 (preservatives for liquid-cooling and processing systems). Silver chloride was not approved in 2014 for product types 3 (veterinary hygiene), 4 (food and feed area), 5 (drinking water) and 13 (working or cutting fluid preservatives) [8] but is still under review (June 2018) for product types 6 (preservatives for products during storage), 7 (film preservatives) and 9 (fibre, leather, rubber and polymerized materials preservatives).

15.2.2 Environmental Protection Agency (USA)

Silver was first registered as a pesticide in the USA in 1954 for use in disinfectants, sanitizers and fungicides [161]. In 1993, silver was registered for use in water filters to inhibit the growth of bacteria within the filter unit of water filter systems designed to remove objectionable taste, odours and colour from municipally treated tap water accounting for over 90% of its pesticidal use. Only about 3% was used to control several types of algae in swimming pool water systems [161]. In 2009, a registration review was announced for silver [38].

15.2.3 Overall Environmental Impact

Silver is manufactured and/or imported in the European Economic Area in 100,000–1,000,000 t per year [40]. Samples from 11 Swedish sewage treatment plants revealed that silver is detected in 67% of the samples with levels between 10.9 and 560 μg per g [111].

Silver nanoparticles (Ag-NP) discharged to the wastewater stream will become sulphidized to various degrees in the sewer system and are efficiently transported to the wastewater treatment plants. The sulphidation of the Ag-NP will continue in the wastewater treatment plants but may not be complete, primarily depending on the size the Ag-NP. Very high removal efficiencies in the wastewater treatment plants will divert most of the Ag-NP mass flow to the digester, and only a small fraction of the silver will be released to surface waters [73]. Ag NPs caused the shifts in microbial community structures and changed the relative abundances of key functional bacteria, which finally resulted in a lower efficiency of biological nitrogen and phosphorus removal [19].

15.3 Spectrum of Antimicrobial Activity

The mode of action has been described for silver in various studies. Silver reacts with the cell membrane resulting in uncoupling of the respiratory electron transport system from oxidative phosphorylation, also interfering with membrane permeability and the proton motive force [34, 132], inhibiting respiratory chain enzymes [23, 134], inhibiting intracellular enzymes reacting with electron donor groups, especially sulphydral groups and interchelation with DNA (Fig. 15.1) [42, 95, 127]. In addition, a silver ion solution exerts its antibacterial effect as shown with *E. coli* and *S. aureus* by inducing bacteria into a state of VBNC, in which the mechanisms required for the uptake and utilization of substrates leading to cell division were disrupted at the initial stage and caused the cells to undergo morphological changes and die at the later stage [72].

15.3.1 Bactericidal Activity

15.3.1.1 Bacteriostatic Activity (MIC Values)

The MIC values depend mainly on the presence of absence of sil genes in the bacterial species. Various isolates of species without sil genes revealed MIC values of 1–52 mg/l (*Citrobacter* spp.), 1–170 mg/l (*E. cloacae*), 2–16 mg/l (*Enterococcus* spp.), 0.004–512 mg/l (*E. coli*), 1–64 mg/l (*Klebsiella* spp.), 1–39 mg/l (*Proteus* spp.) and 0.016–100 mg/l (*P. aeruginosa*). Isolates of the same species harbouring sil genes were less susceptible with MIC values $\leq$ 250 mg/l (*Citrobacter* spp.), $\leq$ 512,000 mg/l (*E. cloacae*), $\leq$ 300 mg/l (*Enterococcus* spp.), $\leq$ 512,000 mg/l (*E. coli*), $\leq$ 5,500 mg/l (*Klebsiella* spp.), 250 mg/l (*Proteus* spp.) and $\leq$ 128,000 mg/l (*P. aeruginosa*) (Table 15.2). A total of 77 *Halococcus* spp. isolates were all described to be susceptible against silver [108]. Both the type of broth and the light may have an impact on the MIC value obtained with silver

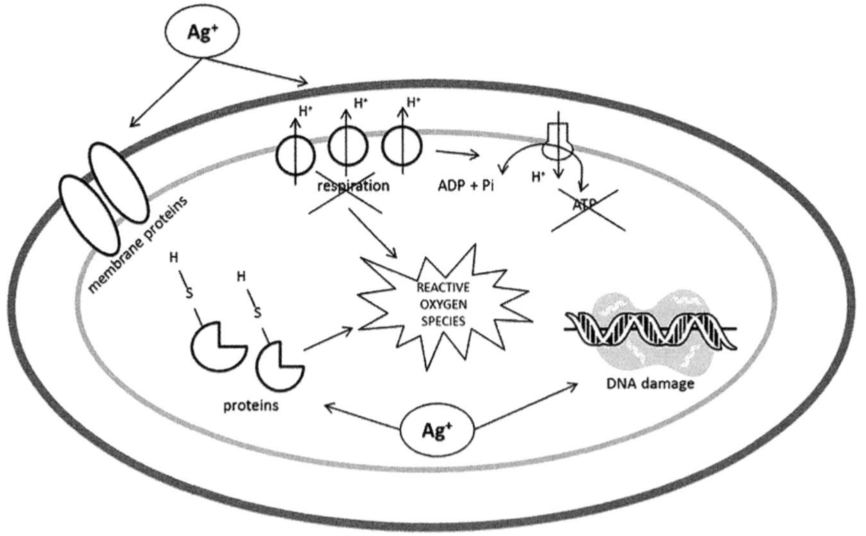

Fig. 15.1 Antimicrobial effects of Ag^+. Interaction with membrane proteins and blocking respiration and electron transfer; inside the cell, Ag^+ ions interact with DNA, proteins and induce reactive oxygen species production [93]. Reprinted by permission from Springer Nature, Biometals (Mijnendonckx K, Leys N, Mahillon J, Silver S, Van Houdt R. Antimicrobial silver: uses, toxicity and potential for resistance. Biometals. 2013; 26: 609–21)

nitrate so that a standardization of broth and light was suggested for the determination of MIC values [175].

Various silver nanoparticles showed mostly low MIC values in *A. baumannii* (0.4–15.6 mg/l for ATCC 19606 and 17 clinical isolates), *A. nosocomialis* (0.4–0.8 mg/l in ten clinical isolates), *E. coli* (3.8–140 mg/l in ATCC 10536, MTCC 443, MTCC 739, MTCC 1302, MTCC 1687 and one clinical isolate), *M. morganii* (10 mg/l in one clinical isolate), *P. aeruginosa* (1.0–15.6 mg/l in ATCC 27853 and two clinical isolates), *B. subtilis* (10 mg/l in one clinical isolate), *C. striatum* (10 mg/l in one clinical isolate), *E. faecalis* (5 mg/l in one clinical isolate), *S. aureus* (0.9–125 mg/l in ATCC 25923, ATCC 33591, NCIM 2079, NCIM 5021, NCIM 5022 and 33 isolates), *S. epidermidis* (62.5 mg/l in ATCC 14990), *S. salivarius* (12–25 mg/l in four clinical isolates), *S. sanguinis* (25 mg/l in four clinical isolates), *S. mitis* (50 mg/l in four clinical isolates), *S. agalactiae* (10 mg/l in one clinical isolate) and *S. mutans* (4–50 mg/l in PTCC 1683 and five clinical isolates) [1, 69, 87, 92, 96, 106, 118, 120, 128, 135, 154, 155].

15.3.1.2 Bactericidal Activity (Suspension Tests)

Silver nitrate at 0.032 mg/l revealed a ≥ 3.0 log reduction within 24 h against various bacterial species. At shorter application times such as 3 h (0.009 mg/l) or 30 min (10,000 mg/l), silver nitrate showed only a partial bactericidal activity (Table 15.3). The bactericidal activity of silver NPs seems to be size dependent

Table 15.2 MIC values of various bacterial species to silver nitrate or [b]silver

Species	Strains/isolates	MIC value (mg/l)	References
Acinetobacter spp.[a]	27 clinical isolates	1–8	[37]
B. pumulis	One isolate of unknown source	2.1	[88]
B. diminuta	Strain from a biofilm model	0.064	[160]
C. meningosepticum	Strain from a biofilm model	0.064	[160]
C. freundii	1 strain from a UTI patient with a silver-coated catheter	≤ 16	[129]
C. freundii	1 clinical isolate with various sil genes	250	[43]
C. intermedius	Environmental isolate	52[b]	[48]
Citrobacter spp.[a]	5 clinical isolates	1–8	[37]
Coliform bacteria	33 isolates from burn patients		[20]
	- 29 of the isolates	10–39	
	- 4 of the isolates	>5,000	
C. metallidurans	4 isolates from the space industry and the International Space Station (ISS)	0.05–0.4	[94]
E. aerogenes	1 clinical isolate with various sil genes	300	[43]
E. aerogenes	29 blood culture isolates; 2 of them with elevated MIC of ≥ 64	16–64	[149]
E. cloacae	2 resistant strains	>0.064	[160]
E. cloacae	4 sil-negative strains from human and equine wounds	1–2.5	[168]
	6 sil-positive strains from human and equine wounds	≥ 5	
E. cloacae	99 blood culture isolates; 15 of them with elevated MIC of ≥ 64	16–64	[149]
E. cloacae complex	3 strains without sil genes	≤ 100	[79]
	2 strains with silS, silR, silC and silP genes	800–1,000	
E. cloacae	2 strains from extracted teeth	170[b]	[29]
E. cloacae	7 clinical isolates with various sil genes	300–5,500	[43]
E. cloacae	Clinical isolate from burns unit	>1,000	[2]
E. cloacae	Silver-resistant control strain	512,000	[70]
Enterobacter spp.[a]	75 clinical isolates	1–8	[37]
E. faecalis	One isolate of unknown source	2.4	[88]
E. faecalis	13 isolates from wounds	6–16[b]	[70]
Enterococcus spp.[a]	8 strains from UTI patients with silver-coated catheters	≤ 16	[129]
Enterococcus spp.[a]	3 clinical isolates with various sil genes	250–300	[43]
E. amylovora	Strain Ea1189	6.2	[119]
	Strain Ea1189 with MdtABC efflux pump	25	
E. coli	ATCC 11775	0.004	[160]

(continued)

Table 15.2 (continued)

Species	Strains/isolates	MIC value (mg/l)	References
E. coli	Strain MG1655	0.1	[148]
E. coli	ATCC 25922 and 1 dental isolate	0.5–1[b]	[10]
E. coli	135 clinical isolates	1–8	[37]
E. coli	140 human ESBL isolates, 34 ESBL isolates from healthy chicken	2–4	[33]
E. coli	Multidrug-resistant clinical isolate MC-2	4	[28]
E. coli	ATCC 23848	5.4	[176]
E. coli	244 isolates (6 from wounds, 34 from bacteremia, 34 from healthy volunteers, 34 from broiler chicken meat, 34 from boiler chicken faecal, 34 from pork, 34 from pigs faecal)	6–16[b]	[70]
E. coli	3 clinical strains	8	[84]
E. coli	186 urine isolates	8–512	[152]
E. coli	ATCC 35218, ATCC 25922, 18 strains from UTI patients with silver-coated catheters	≤ 16	[129]
E. coli	ATCC 10536	18.4	[96]
E. coli	154 strains	26–204[b]	[91]
E. coli	1 isolate from burn patient	>170	[144]
E. coli	1 clinical isolate with various sil genes	300	[43]
E. coli	Silver-resistant control strain	512,000	[70]
K. oxytoca	59 blood culture isolates; 2 of them with elevated MIC of ≥ 64	16–64	[149]
K. oxytoca	2 clinical isolates with various sil genes	300	[43]
K. pneumoniae	95 blood culture isolates; 2 of them with elevated MIC of ≥ 64	16–64	[149]
K. pneumoniae	10 clinical isolates with various sil genes	250–5,500	[43]
K. pneumoniae	10 isolates from the alimentary canal and gills of shrimps	$\geq 1{,}080$	[26]
Klebsiella spp.[a]	105 clinical isolates	1–8	[37]
Klebsiella spp.[a]	8 strains from UTI patients with silver-coated catheters	≤ 16	[129]
L. pneumophila	Strain Corby	0.064	[160]
M. testaceum	Strain PCSB7	1	[148]
Morganella spp.	Insect gut isolate	>85	[114]
P. mirabilis	1 clinical isolate	0.1[b]	[10]
P. mirabilis	1 strain from a UTI patient with silver-coated catheters	≤ 16	[129]
P. mirabilis	1 clinical isolate with a sil gene	250	[43]
P. vulgaris	One isolate of unknown source	2.5	[88]
Proteus spp.[a]	46 isolates from burn patients	10–39	[20]
Proteus spp.[a]	6 clinical isolates	1–8	[37]

(continued)

Table 15.2 (continued)

Species	Strains/isolates	MIC value (mg/l)	References
P. rettgeri	1 clinical isolate with a sil gene	250	[43]
P. stuartii	Strain A 21471	0.1[b]	[10]
P. stuartii	2 strains from UTI patients with silver-coated catheters	≤16	[129]
P. aeruginosa	ATCC 27857	0.016	[160]
P. aeruginosa	Strain DS10-129	0.1	[148]
P. aeruginosa	ATCC 27853	0.3[b]	[10]
P. aeruginosa	91 clinical isolates	1–8	[37]
P. aeruginosa	100 clinical isolates	5–100	[164]
P. aeruginosa	24 isolates from wounds	6–16[b]	[70]
P. aeruginosa	Approximately 100 strains	8–70	[25]
P. aeruginosa	92 isolates from burn patients	10–39	[20]
P. aeruginosa	ATCC 27853	≤16	[129]
P. aeruginosa	Clinical isolate	18.4	[96]
P. aeruginosa	Strain PA14	20	[103]
P. aeruginosa	1 clinical isolate with various sil genes	250	[43]
P. aeruginosa	"several strains"	>5,000	[18]
P. aeruginosa	Silver-resistant control strain	128,000	[70]
P. fluorescens	Strains OS8 and KC1	1	[148]
P. stutzeri	Isolate from soil of silver mine	>4,250	[60]
Pseudomonas spp.[a]	7 strains from UTI patients with silver-coated catheters	≤16	[129]
R. pickettii	8 isolates from the space industry and the International Space Station (ISS)	0.1–0.2	[94]
S. Enteritidis	Strain 3546/6 2012	10	[86]
S. Hadar	Strain 2507/5 2009	20	[86]
S. Senftenberg	Strain 3014/3 2012	>20	[86]
S. Typhimurium	ATCC 14028	25	[138]
S. Typhimurium	10 from an outbreak in Tehran	102	[91]
	3 isolates from burn patients treated with topical silver	1,700	
S. paucimobilis	Strain from a biofilm model	0.064	[160]
S. aureus	ATCC 25923 and 1 dental isolate	0.03–0.3[b]	[10]
S. aureus	ATCC 29213	0.064	[160]
S. aureus	Strain RN4220	0.1–1	[148]
S. aureus	238 MSSA isolates (38 from wounds, 200 from unknown origin)	6–16[b]	[70]
S. aureus	Multidrug-resistant clinical isolate MMC-20	8	[28]
S. aureus	846 clinical isolates	8–16	[124]

(continued)

Table 15.2 (continued)

Species	Strains/isolates	MIC value (mg/l)	References
S. aureus	Approximately 100 strains	8–80	[25]
S. aureus	ATCC 25923	≤ 16	[129]
S. aureus	6 isolates from leg ulcers	16–32	[151]
S. aureus	52 isolates from burn patients	20–39	[20]
S. aureus	ATCC 25923	25	[138]
S. aureus	5 clinical isolates	54	[159]
S. aureus	1 clinical isolate with various sil genes	300	[43]
S. capitis	Clinical isolate	54	[159]
S. chromogenes	Clinical isolate	54	[159]
S. epidermidis	Clinical isolate	54	[159]
S. lentus	Clinical isolate	54	[159]
S. sciuri	Clinical isolate	54	[159]
S. xylosus	8 clinical isolates	54	[159]
Staphylococcus spp.[a] (coagulase-negative)	160 clinical isolates	4–16	[124]
Staphylococcus spp.[a] (coagulase-negative)	2 strains from UTI patients with silver-coated catheters	≤ 16	[129]
Staphylococcus spp.[a]	MSSE 1457, MRSE ATCC 35984, MSSA ATCC 13420, MRSA ATCC 43300, copper-resistant *S. aureus* ATCC 12600, and MRSA USA300 and its putative ΔsilE mutant	15.6–31.2	[80]
Staphylococcus spp.	4 non-identifiable clinical isolates	54	[159]
S. maltophilia	1 clinical isolate with various sil genes	300	[43]
S. mitis	1 dental isolate	0.3[b]	[10]
S. mutans	Strains GS-5 and GS-7	0.6[b]	[10]
S. pyogenes	ATCC 19615	0.2[b]	[10]
S. salivarius	1 dental isolate	1[b]	[10]
Streptococcus spp.	Group B strain 296	0.6[b]	[10]
Streptococcus spp.	Group B strain from a UTI patient with a silver-coated catheter	≤ 16	[129]
V. parahaemolyticus	One isolate of unknown source	3.7	[88]

[a]No MIC values per species

suggesting that NPs with a diameter of 1–10 nm can have a direct interaction with the bacteria [100].

15.3.1.3 Activity Against Bacteria in Biofilm

The efficacy of silver nitrate or silver NPs at common concentrations against bacteria in biofilms is overall poor (Table 15.4). In addition, the effect of the silver in silver-containing wound dressings against bacteria in biofilms depends on the

Table 15.3 Bactericidal activity of silver nitrate or [a]silver in suspension tests

Species	Strains/isolates	Exposure time	Concentration (mg/l)	$\log_{10}$ reduction	References
B. diminuta	Isolate from biofilm	24 h	0.016 (S)	≥ 3.0	[160]
C. meningosepticum	Isolate from biofilm	24 h	0.016 (S)	≥ 3.0	[160]
E. cloacae	2 resistant strains	24 h	0.032 (S)	≥ 3.0	[160]
E. coli	ATCC 11229	30 min	10,000 (S)	< 3.0	[102]
E. coli	ATCC 23848	35 min	482 (S)[a]	5.0	[176]
		50 min	120.5 (S)[a]	5.0	
E. coli	IFO 3301	8 h	0.1 (S)[a]	>7.2	[66]
E. coli	Strain WR1 and strain K12	3 h	0.009 (S)	1.6	[162]
			0.002 (S)	1.0	
E. coli	ATCC 11775	24 h	0.004 (S)	≥ 3.0	[160]
L. pneumophila	ATCC 33152	8 h	0.1 (S)[a]	>7.2	[66]
L. pneumophila	Strain Corby	24 h	0.064 (S)	≥ 3.0	[160]
			0.032 (S)		
P. aeruginosa	ATCC 10145	8 h	0.1 (S)[a]	>7.2	[66]
P. aeruginosa	ATCC 27857	24 h	0.016 (S)	≥ 3.0	[160]
S. paucimobilis	Isolate from biofilm	24 h	0.016 (S)	≥ 3.0	[160]
S. aureus	ATCC 6538	30 min	10,000 (S)	< 3.0	[102]
S. aureus	54 MRSA strains	5 min	6,250 (P)	≥ 5.0	[39]
S. aureus	ATCC 29213	24 h	0.016 (S)	≥ 3.0	[160]

S Solution; *P* commercial product

type of dressing material and structure [115]. The combination of ionic silver with a metal chelating agent and a surfactant can substantially improve the antimicrobial efficacy of ionic silver against biofilm pathogens (MRSA and *P. aeruginosa*) in a simulated wound biofilm model [13]. Similar favourable results were found with *S. aureus* and the combination of silver, EDTA and benzethonium chloride [131].

Data obtained with ≥ 1000 mg/l silver NPs suggest an effect (≥ 3.0 log) against bacterial biofilm cells within 24 h (Table 15.4). These findings are supported by data obtained with original wastewater biofilms. They were highly tolerant to silver NPs. However, accumulated silver NPs in wastewater biofilms may impact their microbial activity [136]. Susceptibility to silver NPs is different for each micro-organism in the biofilm microbial community. Thiotrichales, in one study, is more sensitive than other biofilm bacteria [136].

Some factors with an impact of the bactericidal effect in biofilm have been evaluated. The effect of silver NPs (total Ag concentration: 27.3 mg/l; released Ag^{+}: 1.5 mg/l) on *P. putida* biofilms was low (1.0 log) when the biofilms had high biomass amount, high thickness, high biomass volume, low surface-to-volume ratio and low roughness coefficient [156]. Mature biofilms have greatly reduced

Table 15.4 Efficacy of silver nitrate or [a]silver nanoparticles against bacteria in biofilms

Species	Strains/isolates	Type of biofilm	Exposure time	Concentration	$\log_{10}$ reduction	References
L. pneumophila	Strain Corby	2–3-w incubation of stagnant drinking water from a large building water conduit	24 h	0.016 mg/l (S)	0.0	[160]
P. aeruginosa	NCIMB 10434	48-h incubation in biofilm reactor	4 h	Unknown (P)[a]	0.5	[64]
			24 h		1.8	
S. Typhimurium	ATCC 14028	24-h incubation in polystyrene microtiter plates	15 min	25 mg/l (P)[a]	0.3	[138]
S. aureus	ATCC 25923	24-h incubation in polystyrene microtiter plates	15 min	25 mg/l (P)[a]	0.5	[138]
S. mutans	1 clinical isolate	24-h incubation on hydroxyapatite coupons in biofilm reactor	24 h	1,000 mg/l (S)[a]	7.0	[118]
				500 mg/l (S)[a]	3.8	
				250 mg/l (S)[a]	2.5	
				100 mg/l (S)[a]	2.3	
Mixed species	*P. aeruginosa* and *S. aureus* (MRSA), both isolates from chronic wounds	48-h incubation in biofilm reactor	24 h	1,000 mg/l (S)[a]	3.3–6.0	[117]
Mixed species	Biofilms from a wastewater treatment plant	Natural biofilm	24 h	200 mg/l (S)[a]	0.1	[137]
Mixed species	Activated sludge from a wastewater treatment plant	24-h incubation in polystyrene microtiter plates	30 min	0.01 mg/l (S)	0.0	[171]
			1 h		0.1	
			2 h		0.1	
Mixed species	*S. aureus* strain 308 (MRSA), *C. albicans* ATCC MYA 2876	48-h incubation in biofilm reactor	4 h	Unknown (P)[a]	>5.0	[64]
			24 h			

P Commercial product; *S* Solution

susceptibility to silver NPs compared to immature biofilms. Silver NPs were less toxic in steady-state systems with mature biofilms, but systems during start-up, when biofilms are becoming established, will be vulnerable to silver NPs [157]. Short-term studies also showed sequential dose-dependent toxic effects of silver NPs on *P. putida* biofilm morphology (with impacts characterized from 0.01 mg/l), then activity (from 1 to 10 mg/l) and viability (from 10 mg/l) via a single pulse of 24 h in artificial wastewater [89].

Biofilm can provide physical protections for bacteria under silver NP treatment, and extracellular polymeric substances may play an important role in this protection. Biofilm bacteria with loosely bound extracellular polymeric substances removed are more sensitive to silver NP [136].

15.3.1.4 Bactericidal Activity in Wound Dressings

A study with 130 wound isolates from 12 bacterial species revealed an overall good phenotypic susceptibility to a silver dressing [44]. Dressing containing silver has been described to have a significant effect on the cells' density of *A. baumannii*, *A. calcoaceticus*, *E. coli*, *K. pneumoniae*, *P. acnes*, *P. aeruginosa*, *S. aureus*, MRSA and *S. epidermidis* within 24 h although the concentrations of the applied silver remain usually unknown [14, 32, 76]. Only *E. faecalis* was resistant with basically 0 log reduction in 24 h [76]. There is, however, some variation of the bactericidal activity between different types of commercially available silver wound dressings [3, 21, 22, 67].

Preclinical and clinical study data suggest that silver dissociation is affected by the test medium used. The bactericidal activity differences may be a function of the bacterial strain used for testing. Higher rather than lower levels of silver may be needed because Ag^+ binds to proteins and nucleic acids, and rapid delivery of silver (i.e. rate of kill) may be a positive factor when considering prevention of silver resistance and biofilm formation [17].

15.3.1.5 Bactericidal Activity in Other Applications

Silver on door handles across a college campus resulted in lower bacterial populations compared to control handles after 3 years. However, bacteria were consistently isolated from silver-coated door handles suggesting that the silver zeolite was only effective against a portion of the bacterial populations [121]. Silver at 50 mg/l was able to reduce *P. aeruginosa* on ceramic tiles by 2.2 (30 min) to 4.4 log (4 h) and *S. aureus* by 3.3 (30 min) to 4.1 log (2 h) [16]. On collagen-coated polyester vascular grafts, silver coating was able to reduce MRSA ATCC 33591 by 4.2 log within 24 h [125].

15.3.2 Fungicidal Activity

15.3.2.1 Fungistatic Activity (MIC Values)

Most yeasts are quite susceptible to silver or silver nitrate with MIC values between 0.5 and 75 mg/l, and only some isolates of the yeasts *S. carlsbergensis* and

Table 15.5 MIC values of various fungal species to silver nitrate or [a]silver

Species	Strains/isolates	MIC value (mg/l)	References
A. fumigatus	1 wastewater isolate deriving from jewellery industry	648[a]	[130]
C. albicans	2 clinical isolates	0.5–3.5[a]	[9]
C. albicans	2 clinical urine strains	100	[62]
C. argentea	5 environmental isolates	≥ 42.5	[65]
C. glabrata	1 clinical isolate	1.6[a]	[9]
C. parapsilosis	2 clinical urine strains	50–75	[62]
C. parapsilosis	1 clinical isolate	4.7[a]	[9]
C. pseudotropicalis	1 clinical isolate	1.6[a]	[9]
C. tropicalis	1 clinical isolate	1[a]	[9]
C. tropicalis	2 clinical urine strains	50–75	[62]
Candida spp.[b]	4 strains from UTI patients with silver-coated catheters	≤ 16	[129]
S. carlsbergensis	2 isolates from oranges and pineapples	216–756	[110]
S. cerevisiae	Strain BY4741	1	[148]
S. cerevisiae	7 isolates from oranges, palm wine and pineapples	108–972	[110]

[b]No MIC values per species

S. cerevisiae had higher MIC values (≤ 972 mg/l) as well as *A. fumigatus* (648 mg/l; Table 15.5).

Silver NPs showed rather high MIC values against some fungi, e.g. against *M. canis* (200 mg/l in PTCC 5069), *M. gypseum* (170 mg/l in PTCC 5070) and *T. mentagrophytes* (180 mg/l in PTCC 5054) [5]. With other fungi, quite low MIC_{50} values were described, e.g. with *Fusarium* spp. (1 mg/l in 112 clinical isolates), *Aspergillus* spp. (0.5 mg/l in 94 clinical isolates) and *A. alternata* (0.5 in 10 clinical isolates) [172].

15.3.2.2 Fungicidal Activity (Suspension Tests)

No published data were found to evaluate the fungicidal activity of silver or silver nitrate in suspension tests

15.3.2.3 Activity Against Yeasts in Biofilm

No data were found to describe the fungicidal activity of silver or silver nitrate against biofilm-grown cells of fungi. The activity of silver NPs, however, was evaluated in one study. The yeasticidal activity as demonstrated with *C. albicans* and *C. glabrata* was overall poor (Table 15.6). The susceptibility to silver is already reduced to some degree within the first 2 h of attachment to silicone as shown with *C. albicans*, *C. glabrata* and *C. krusei* [174].

Table 15.6 Efficacy of silver NPs solutions (S) against fungi in biofilms

Species	Strains/isolates	Type of biofilm	Exposure time	Concentration	$\log_{10}$ reduction	References
C. albicans	ATCC 10231 and 1 oral clinical isolate	24-h incubation on acryl resin specimens	5 h	54 mg/l (S)	0.3–1.0	[99]
		48-h incubation on acryl resin specimens			0.6–1.4	
C. glabrata	ATCC 90030 and 1 oral clinical isolate	24-h incubation on acryl resin specimens	5 h	54 mg/l (S)	1.0–1.6	[99]
		48-h incubation on acryl resin specimens			1.3	

15.3.3 Mycobactericidal Activity

The mycobactericidal activity of silver NPs within 48 h has been evaluated with a strain of *M. smegmatis*, *M. avium* and *M. marinum*. A 1.9 log was found against *M. smegmatis* at 100 μM silver NPs. The effect was lower against *M. avium* with 1.3 log using silver NPs at 270 μM. The most resistant species was *M. marinum* with 0.8 log using 860 μM [68]. One study with *M. phlei* indicates that the susceptibility of biofilm-grown cells to silver nitrate is lower (MBEC: 313 mg/l in 30 min) compared to planktonic cells (MBC: 26 mg/l in 30 min) [6]. A silver-containing wound dressing was able to reduce the cell number of *M. fortuitum* within 7 d by 4.0 log [15].

15.4 Effect of Low-Level Exposure

The effect of exposure to sublethal silver concentrations depends mainly on the presence or absence of sil genes. In most bacterial isolates from nine species without sil genes no adaptive response was found (*Acinetobacter* spp., *Citrobacter* spp., *E. cloacae*, *E. coli*, *K. pneumoniae*, *K. oxytoca*, *Proteus* spp., *P. aeruginosa*, *S. aureus*). Some isolates or strains of three species were able to express a weak adaptive response (MIC increase $\leq$ 4-fold) such as *E. coli*, *M. smegmatis* and *S. aureus*.

A strong MIC change (>4-fold) was found in isolates or strains of 6 species. It was unstable in isolates or strains of *E. coli*, it was stable in isolates or strains of *E. cloacae*, *E. coli*, *K. pneumoniae* and *K. oxytoca*, and the stability was sometimes unknown, e.g. in isolates or strains of *A. ferrooxidans*, *Enterobacter* spp. and *E. coli* (Table 15.7).

Table 15.7 Change of bacterial susceptibility to biocides and antimicrobials after low-level exposure to silver nitrate, silver NPs or silver sulphadiazine

Species	Strains/isolates	Exposure time	Increase in MIC	MIC_{max} (mg/l)	Stability of MIC change	Associated changes	References
A. ferrooxidans	18 strains from acid mine drainage water samples	Multiple passages at various silver nitrate concentrations	12-fold–48-fold	240	No data	No detection of silC gene in two strains with MIC values of 60 and 240 mg/l	[170]
Acinetobacter spp.	27 clinical isolates	Plating saturated cultures onto MHA containing 128 mg/l silver nitrate	None	8	Not applicable	None described	[37]
Citrobacter spp.	5 clinical isolates	Plating saturated cultures onto MHA containing 128 mg/l silver nitrate	None	8	Not applicable	None described	[37]
E. cloacae	ATCC 23355 without known silver resistance	50 passages at various concentrations	None	31.2	Not applicable	None described	[80]
E. cloacae	5 blood culture isolates without silE, silS and silP	Up to 10 passages at various concentrations of silver nitrate	None	32	Not applicable	None described	[149]
	5 blood culture isolates with silE, silS and silP		≥ 16-fold	≥ 512	"mostly stable"		
E. cloacae	5 clinical wound isolates	Up to 10 passages at various concentrations of silver nitrate	32-fold (2 isolates)	≥ 512	Stable for 6 d	Increased tolerance[a] to imipenem (32-fold; 8 mg/l) and meropenem (16-fold; 2 mg/l) in 1 isolate	[151]
E. cloacae	ATCC 13047 harbouring the chromosomal silver	5 passages at various concentrations	>32-fold	>1,000	No data	None described	[80]

(continued)

Table 15.7 (continued)

Species	Strains/isolates	Exposure time	Increase in MIC	MIC_{max} (mg/l)	Stability of MIC change	Associated changes	References
	resistance cassette SilPABCRSE						
Enterobacter spp.	75 clinical isolates	Plating saturated cultures onto MHA containing 128 mg/l silver nitrate	Selection for silver resistance in 57 isolates (76%)	>128	No data	None described	[37]
E. coli	5 blood culture isolates without silE, silS and silP	Repeated exposure to various concentrations of silver nitrate	None	32	Not applicable	None described	[149]
	2 blood culture isolates with silE, silS and silP		≥ 16-fold	≥ 512	"mostly stable"		
E. coli	Strain MG1655	Repeated exposure to silver NPs at various concentrations	1.4-fold–4.7-fold	No data	No data	Three mutations had swept to high frequency in the silver nanoparticles resistance stocks	[50]
E. coli	13 human faecal isolates, all silE-positive	Up to 10 passages at various concentrations of silver nitrate	≥ 16-fold	≥ 512	Stable for 5 d	Cross-tolerance[b] to ceftibuten (3), piperacillin-tazobactam (3), cotrimoxazole (2), ciprofloxacin (2) and gentamicin (1)	[150]
E. coli	ATCC 23848	24- or 48-h exposure to various concentrations of silver nitrate	20-fold–60-fold	723	No data	None described	[176]

(continued)

Table 15.7 (continued)

Species	Strains/isolates	Exposure time	Increase in MIC	MIC_{max} (mg/l)	Stability of MIC change	Associated changes	References
E. coli	ATCC 25922 and 3 clinical wound isolates	Up to 10 passages at various concentrations of silver nitrate	32-fold (2 isolates)	≥ 512	Unstable for 2 and 5 d	None described	[151]
E. coli	Strain BW25113	6-d exposure to various concentrations of silver nitrate	64-fold	>256	No data	None described	[123, 124]
E. coli	3 clinical isolates	Repeated exposure to increasing concentrations of silver nitrate or silver sulphadiazine	64-fold–128-fold	>1,024	No data	Increase of active silver efflux	[84]
E. coli	135 clinical isolates	Plating saturated cultures onto MHA containing 128 mg/l silver nitrate	Selection for silver resistance in 1 isolate (0.7%)	>128	No data	None described	[37]
K. pneumoniae	5 blood culture isolates without silE, silS and silP	Repeated exposure to various concentrations of silver nitrate	None (4 isolates)	32	Not applicable	None described	[149]
	5 blood culture isolates with silE, silS and silP		$\geq$ 16-fold	≥ 512	"mostly stable"		

(continued)

Table 15.7 (continued)

Species	Strains/isolates	Exposure time	Increase in MIC	MIC_{max} (mg/l)	Stability of MIC change	Associated changes	References
K. pneumoniae	2 clinical wound isolates	Up to 10 passages at various concentrations of silver nitrate	32-fold (1 isolate)	≥512	Stable for 6 d	None described	[151]
K. oxytoca	5 blood culture isolates without silE, silS and silP	Repeated exposure to various concentrations of silver nitrate	None	32	Not applicable	None described	[149]
	5 blood culture isolates with silE, silS and silP		≥16-fold	≥512	"mostly stable"		
Klebsiella spp.	105 clinical isolates	Plating saturated cultures onto MHA containing 128 mg/l silver nitrate	Selection for silver resistance in 61 isolates (58%)	>128	No data	None described	[37]
M. smegmatis	4 isolates of strain mc[2] 155 preselected on agar containing 430 µM silver NP	48-h exposure to various concentration of silver NPs and silver nitrate	"significant increase of resistance"	>3.4 >100 µM[a]	No data	Increased tolerance[c] to isoniazid (4-fold)	[83]
Proteus spp.	6 clinical isolates	Plating saturated cultures onto MHA containing 128 mg/l silver nitrate	None	8	Not applicable	None described	[37]
P. aeruginosa	91 clinical isolates	Plating saturated cultures onto MHA	None	8	Not applicable	None described	[37]

(continued)

Table 15.7 (continued)

Species	Strains/isolates	Exposure time	Increase in MIC	MIC_{max} (mg/l)	Stability of MIC change	Associated changes	References
		containing 128 mg/l silver nitrate					
P. aeruginosa	3 clinical wound isolates	Up to 10 passages at various concentrations of silver nitrate	None	16	Not applicable	None described	[151]
P. aeruginosa	Strain PAO1	42 days at various concentrations of silver nitrate	None	4	Not applicable	None described	[124]
S. aureus	Strain MRSA 252 and strain SH 1000	42 days at various concentrations of silver nitrate	None	16	Not applicable	None described	[124]
S. aureus	ATCC 6538	100 d at various concentrations	"significant increase"	40	No data	None described	[166]
Staphylococcus spp.	MSSE 1457, MRSE ATCC 35984, MSSA ATCC 13420, MRSA ATCC 43300, copper-resistant *S. aureus* ATCC 12600, and MRSA USA300 and its putative ΔsilE mutant	50 passages at various concentrations	None	31.2	Not applicable	None described	[80]
Mixed species	Activated sludge	65 d at 0.1 mg/l silver supplied as silver NPs	No data	No data	Not applicable	silE gene copy number increased 50-fold within 41 d and decreased on d 65	[173]

[a]Etest; [b]Decreased inhibition zone (>5 mm) in disk diffusion test; [c]Macrodilution method

Selected strains or isolates revealed substantial MIC increases such as *E. coli* ($\leq$ 128-fold), *E. cloacae* and *K. pneumoniae* ($\geq$ 32-fold) and *K. oxytoca* ($\geq$ 16-fold). The highest MIC values after adaptation were all found in Gram-negative species such as 1,024 mg/l (*E. coli*), 1,000 mg/l (*E. cloacae*), 512 mg/l (*K. pneumoniae* and *K. oxytoca*) and 240 mg/l (*A. ferrooxidans*). A cut-off value to determine resistance to silver was proposed for Gram-negative species with >8 mg/l [37]. Based on this proposal, all adapted isolates would be classified as resistant to silver after low-level exposure.

Cross-tolerance to various antibiotics such as imipenem, meropenem, ceftibuten, piperacillin–tazobactam, cotrimoxazole, ciprofloxacin and gentamicin was found in some isolates of *E. cloacae* and *E. coli*. Increase of silver efflux after low-level exposure was detected in *E. coli* (Table 15.7).

One more study describes that a silver-resistant mutant of *K. pneumoniae* B-5 was produced by passaging in nutrient broth containing graded concentrations of silver nitrate up to 150 mg/l. The development of silver resistance in the strain resulted in rough colonies, decrease in cell size, carbohydrate content and a change in the klebocin pattern [58].

15.5 Resistance to Silver

The cut-off value to determine silver resistance is variable in the literature. In one hospital laboratory, 85 mg/l silver nitrate was included in the agar [63]. Other authors used MHA containing silver nitrate at 128 mg/l [37], isolates with visible growth were regarded as silver resistant. Another hospital laboratory used lysogeny broth agar supplemented with 27 mg/l Ag^+ [43].

15.5.1 High MIC Values

Isolates of various species harbouring sil genes were tolerant to silver with MIC values of 250 mg/l (*Citrobacter* spp.), 5–512,000 mg/l (*E. cloacae*), 250–300 mg/l (*Enterococcus* spp.), 300–512,000 mg/l (*E. coli*), 250–5,500 mg/l (*Klebsiella* spp.), 250 mg/l (*Proteus* spp.) and 250–128,000 mg/l (*P. aeruginosa*; see also Table 15.2). Even if some bacterial species with various sil genes are initially silver susceptible, exposure to silver increased the MIC value in *A. ferrooxidans* (12-fold–48-fold), *E. cloacae* (16-fold–32-fold), *K. pneumoniae* (16-fold–32-fold) and *E. coli* (16-fold–128-fold). The MIC value may be as high as >1,024 mg/l in these isolates after silver exposure (see also Table 15.8). The findings are not surprising. A study published in 1983 has suggested already that silver resistance may occur among Gram-negative bacterial species [75].

Table 15.8 Detection rates of silE in isolates of various bacterial species

Species	Country	Number of isolates	silE detection rate	References
E. aerogenes	Sweden	32	12.5%	[149]
E. cloacae	Sweden	131	54.2%	[149]
Enterobacter spp.	USA	44	4.5%	[43]
Enterococcus spp.	USA	64	1.6%	[43]
E. coli	Sweden	223	4.5%	[149]
E. coli	Sweden	216	6.0%[a]	[150]
Escherichia spp.	USA	256	0.4%	[43]
K. oxytoca	Sweden	79	49.4%	[149]
K. pneumoniae	Sweden	129	36.4%	[149]
Klebsiella spp.	USA	69	5.8%	[43]
Pseudomonas spp.	USA	54	0%	[43]
S. aureus (MRSA)	UK	33	6.1%	[85]
Staphylococcus spp. (coagulase-negative and methicillin-resistant)	UK	8	12.5%	[85]
Staphylococcus spp.	USA	148	0%	[43]

[a]All 13 were among the 105 human faecal isolates

15.5.2 Reduced Efficacy in Suspension Tests

No studies were found to describe a reduced efficacy of silver in suspension tests to indicate phenotypic resistance

15.5.3 Resistance Mechanisms

Silver resistance was studied in a silver-resistant *P. stutzeri* AG259 strain and compared to a silver-sensitive *P. stutzeri* JM303 strain. Silver resistance was not due to silver complexation to intracellular polyphosphate or the presence of low molecular weight metal-binding protein(s). Both the silver-resistant and silver-sensitive *P. stutzeri* strains produced hydrogen sulphide, with the silver-resistant AG259 strain producing lower amounts of hydrogen sulphide than the silver-sensitive JM303 strain. However, intracellular acid-labile sulphide levels were generally higher in the silver-resistant *P. stutzeri* AG259 strain. Silver resistance may be due to formation of silver–sulphide complexes in the silver-resistant *P. stutzeri* AG259 strain [141]. Pyocyanin confers resistance by *P. aeruginosa* to Ag^+. The conversion of toxic Ag^+ to insoluble non-toxic Ag^0 by pyocyanin effectively reduces the bioavailable concentration of Ag^+ [103]. In *E. coli,* a silver-binding peptide was identified. Cells secreting the peptide into the periplasm exhibited silver tolerance in a batch culture, while those expressing a cytoplasmic version of the fusion protein or maltose-binding protein alone did not [133].

15.5.4 Resistance Genes

Contrary to current dogma, the original *E. coli* strain NCTC 86 described by Theodor Escherich in 1885 includes a nine gene sil locus that encodes a silver-resistant efflux pump acquired before the current widespread use of silver nanoparticles as an antibacterial agent, possibly resulting from the widespread use of silver utensils and currency in Germany in the 1800s [35]. Silver resistance genes are part of a plasmid-associated gene cluster (Fig. 15.2) that encodes a silver-binding protein (silE), efflux pump (silA and silP) and a membrane sensor kinase (silS) [139].

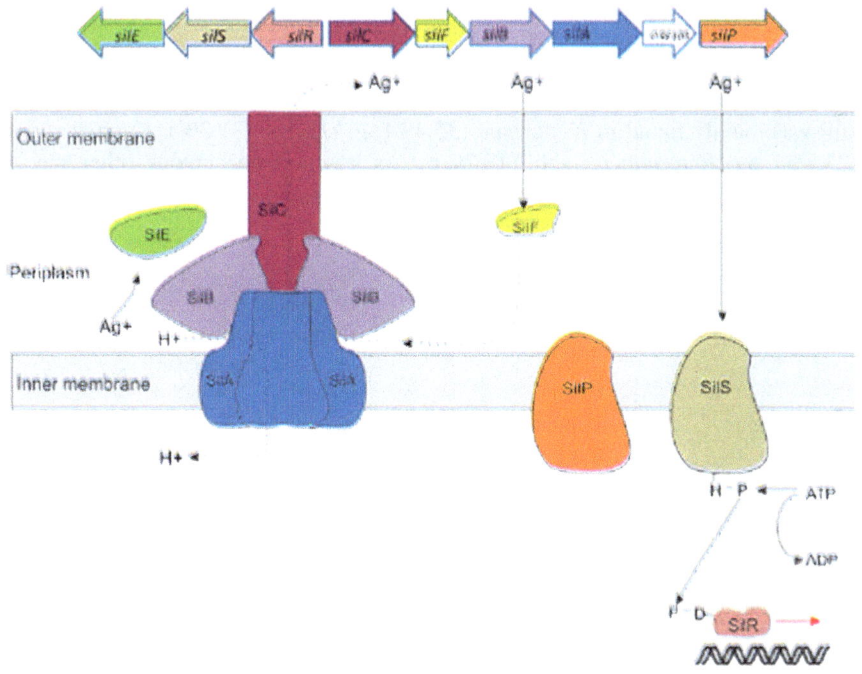

Fig. 15.2 Genetic architecture of the sil operon [123]; reproduced in parts without change from Randall CP, Gupta A, Jackson N, Busse D, O'Neill AJ. Silver resistance in Gram-negative bacteria: a dissection of endogenous and exogenous mechanisms. J Antimicrob Chemother. 2015; 70: 1037–46; the article is distributed under the terms of the Creative Commons CC BY licence

Table 15.9 Detection rates of silA in isolates of various bacterial species

Species	Country	Number of isolates	silA detection rate	References
Enterobacter spp.	USA	44	18.2%	[43]
Enterococcus spp.	USA	64	3.1%	[43]
Escherichia spp.	USA	256	0.4%	[43]
Klebsiella spp.	USA	69	15.9%	[43]
Pseudomonas spp.	USA	54	1.9%	[43]
Staphylococcus spp.	USA	148	0.7%	[43]

15.5.4.1 silE

The silE gene is mostly found in *E. cloacae* (54.2%), *K. oxytoca* (49.4%) and *K. pneumoniae* (36.4%). In other bacterial species, the silE gene is less common (Table 15.8). It was also detected in a clinical isolate of *C. tropicalis* [43].

15.5.4.2 silA

The silA gene is less common and was so far mainly found in *Enterobacter* spp. (18.2%) and *Klebsiella* spp. (15.9%; Table 15.9).

15.5.4.3 silP

silP was mainly found in *K. oxytoca* (35.4%), *E. cloacae* (31.3%), *K. pneumoniae* (23.7%) and *Enterobacter* spp. (18.2%). It is less common among other species (Table 15.10). silP was also detected in a clinical isolate of *C. tropicalis* [43].

Table 15.10 Detection rates of silP in isolates of various bacterial species

Species	Country	Number of isolates	silP detection rate	References
E. aerogenes	Sweden	32	3.1%	[149]
E. cloacae	Sweden	131	31.3%	[149]
Enterobacter spp.	USA	44	18.2%	[43]
Enterococcus spp.	USA	64	1.6%	[43]
E. coli	Sweden	223	0.9%	[149]
E. coli	Sweden	216	0%	[150]
Escherichia spp.	USA	256	0.4%	[43]
K. oxytoca	Sweden	79	35.4%	[149]
K. pneumoniae	Sweden	129	23.7%	[149]
Klebsiella spp.	USA	69	15.9%	[43]
Pseudomonas spp.	USA	54	0%	[43]
S. aureus (MRSA)	UK	33	0%	[85]
Staphylococcus spp. (coagulase-negative and methicillin-resistant)	UK	8	0%	[85]
Staphylococcus spp.	USA	148	0.7%	[43]

Table 15.11 Detection rates of silS in isolates of various bacterial species

Species	Country	Number of isolates	silS detection rate	References
E. aerogenes	Sweden	32	9.4%	[149]
E. cloacae	Sweden	131	47.3%	[149]
E. coli	Sweden	223	1.8%	[149]
E. coli	Sweden	216	0%	[150]
K. oxytoca	Sweden	79	44.3%	[149]
K. pneumoniae	Sweden	129	29.5%	[149]
S. aureus (MRSA)	UK	33	0%	[85]
Staphylococcus spp. (coagulase-negative and methicillin-resistant)	UK	8	0%	[85]

15.5.4.4 silS

The silS gene was most frequently found in *E. cloacae* (47.3%), *K. oxytoca* (44.3%) and *K. pneumoniae* (29.5%) whereas it is less common in other bacterial species (Table 15.11). In 119 Gram-negative clinical bacterial isolates with cryptic silver resistance (initially susceptible but upon silver exposure resistant), all of them were carriers of silS whereas all 30 isolates obtained from a cross section without silver-resistant mutants were silS negative [37].

15.5.4.5 Various Sil Genes

One hundred sixty-four clinical isolates of all genotypes of the *E. cloacae complex* were screened for silS, silR, silC and silP. Of these isolates, 63% were positive in all sil PCRs, suggesting that about two-thirds of clinical isolates of the *E. cloacae complex* harbour the complete silver-resistant determinant [79]. An analysis of 172 bacterial isolates from human and equine wounds revealed that six of them contained the silver resistance genes silE, silRS, silCBA, silF, silB, silA and silP, all of which were strains of *E. cloacae* [168]. In 131 isolates from various sources and European countries, the silA-silE genes were detected in 79.4% [101]. It was concluded that metal toxic concentrations in food–animal environments can contribute to persistence of genetic platforms carrying metal/antibiotic resistance genes in this foodborne zoonotic pathogen [101]. Among 112 bacterial isolates from diabetic foot ulcers silS, silE and silP genes were detected in 1.8%, both were *E. cloacae* [116].

Despite being ubiquitous in domestic wastewater treatment plants in the USA, sil silver resistance genes do not appear to correlate with total silver concentrations in activated sludge. This lack of association may be due to the low concentrations of the most toxic form of silver (Ag^{+}). The maintenance of silver resistance genes in the absence of a strong selective pressure may be a result of their known co-location with antibiotic resistance genes [59].

15.5.5 Efflux Pumps

Experimental results with *E. coli* showed that the genetic mechanism for silver resistance includes up-regulation of efflux pumps as well as up-regulation of metal oxidoreductases. The gene, copA, a P-type ATPase efflux flux, was up-regulated in response to silver exposure, and the gene of CusCFBA, a Cu(I) efflux pump, was also up-regulated. The gene of CueO, a robust cuprous oxidase, was also up-regulated and may have reduced silver toxicity through oxidation of silver ions [169]. In *E. coli* strain BW25113 exogenous resistance involved derepression of the SilCFBA efflux transporter as a consequence of mutation in silS, but was additionally contingent on expression of the periplasmic silver-sequestration protein SilE [123]. In *E. hirae,* CopB ATPase is a pump for the extrusion of monovalent copper and silver ions [143]. In *A. baumannii,* harbouring plasmid pUPI199 activation of an endogenous silver efflux system together with porin mutations provides the basis for silver resistance [25]. And in *C. albicans,* an eukaryotic copper pump was detected which provides the primary source of cellular copper resistance, and it was able to confer silver resistance [126].

15.5.6 Plasmids

Mijnendonckx et al. summarized in 2013 that the sil gene cluster is highly conserved in several other plasmids of the IncHI-2 incompatibility group such as plasmids MIP233, MIP235 and WR23 of various *Salmonella* serovars and plasmids pR47b and pR478 of *S. marcescens* [57, 93]. In *E. cloacae*, the major difference between virulent and avirulent genotypes appears to be the presence of a large plasmid that also belongs to the IncHI-2 incompatibility group, which contains, besides several antibiotic-resistant determinants, a functional sil gene cluster [79]. In *P. stutzeri* AG259, isolated from the soil of a silver mine, silver resistance was also mediated by one of its plasmids [60]. This strain was able to grow on rich medium with 8,750 mg/l silver nitrate by accumulation of Ag and Ag_2S crystals in its periplasm [77]. Isolation of plasmids from all six sil-positive and silver-resistant *E. cloacae* strains from human and equine wounds provided evidence that these genes were present extrachromosomally [168]. Transferable plasmids have also been described in *P. stutzeri* to harbour silver resistance [60].

In *D. acidovorans* and *B. petrii*, silCBA is located on an integrative conjugative element (ICE) belonging to the Tn4371 family. This family refers to a group of mobile genetic elements that carry functional modules involved in conjugative transfer, integration, maintenance/stability and accessory genes conferring a special phenotype to the host bacteria [93]. All together, in many strains, the silver-resistant determinants are located on mobile genetic elements, facilitating the spread of these traits to other members of the population [93].

15.5.6.1 pMG101

Ag^+ resistance was initially found on the *S. Typhimurium* multiresistance plasmid pMG101 isolated from patients with burns in 1975 [4]. The silver-resistant determinant from plasmid pMG101 contains nine genes, and the functions for eight named genes and their corresponding protein products were reviewed by Silver et al. [140]. It mediates silver resistance, e.g. in *E. coli* J 53 and *S. Typhimurium* [55, 56]. pMG101 belongs to the IncHI2 incompatibility group of plasmids which are large multi-antibiotic resistance plasmids found widely in the enterobacteriaceae and that are transferred by conjugation only at lower temperatures. The identification of new sil genes on five additional plasmids, all of which are IncHI2 or IncHI3, and homologous genes on the chromosomes of *E. coli* K-12 and O157:H7 and other bacteria raises important concerns about the development of Ag^+-resistant bacteria [57].

15.5.6.2 pJT1 and pJT2

Plasmids pJT1 (83 kb) and pJT2 (77 kb) were found in *E. coli* and are transferable yielding silver-resistant transconjugants [145]. *E. coli* C600 containing PJT1 and PJT2 displayed decreased accumulation of Ag^+ similar to *E. coli* R1. *E. coli* C600 could not tolerate 11 and 54 mg/l Ag^+, rapidly accumulated Ag^+ and became non-viable [145]. The plasmid pJT1 of *E. coli* R1, isolated from patients with burn wounds, conferred resistance up to 170 mg/l silver nitrate [144].

15.5.6.3 pSTM6-275

pSTM6-275 is a IncHI2 plasmid from *S. enterica*. The plasmid was thermosensitive for transfer to *E. coli* and conferred reduced susceptibility to antibiotics, copper sulphate and silver nitrate. Metal ion susceptibility was dependent on physiological conditions, giving an insight into the environments where this trait might confer a fitness advantage [12]. IncHI2 plasmids from *E. coli* isolates of food-producing animals carried pco and sil which contributed to increasing in the MICs of copper sulphate and silver nitrate. Co-existence of the pco and sil operons, and oqxAB/$bla_{CTX\text{-}M}$ as well as other antibiotic resistance genes on IncHI2 plasmids may promote the development of multidrug-resistant bacteria [41].

15.5.6.4 pUPI199

Deshpande and Chopade discovered a 54-kb plasmid (pUPI199) encoding resistance to silver nitrate in an environmental isolate of *A. baumannii* that was transferable to *E. coli* by conjugation [31]. The isolate tolerated up to 128 mg/l silver nitrate and contains, in addition, resistance determinants for 13 different metals and 10 antibiotics [31]. *A. baumannii* was found to accumulate and retain silver, whereas *E. coli* (pUPI199) effluxed 63% of the accumulated silver ions [31].

15.5.6.5 pKQPS142

A carbapenem-resistant virulent *K. quasipneumoniae subsp. similipneumoniae* isolate from Brazil harboured two plasmids (pKQPS142a and pKQPS142b) and an integrative conjugative element ICEPm1 which is a chromosomal mobile pathogenicity island common to *P. mirabilis*, *P. stuartii* and *M. morganii* [45, 46, 107]. It could be involved in the mobilization of pKQPS142b and determinants of resistance to other classes of antimicrobials, including aminoglycoside and silver [107].

15.5.6.6 pLVPK

In *K. pneumoniae* CG43, a large virulence plasmid pLVPK was described with several gene clusters homologous with copper, silver, lead and tellurite resistance genes of other bacteria [24]. The plasmid was recently detected during an outbreak caused by a hypervirulent carbapenem-resistant *K. pneumoniae* causing fatal pneumonia in five ventilated patients [51].

15.5.6.7 pUUH239.2

This is a 20-kbp multidrug resistance plasmid, first isolated in *K. pneumoniae* and *E. coli* in 2005 from a large nosocomial outbreak. Besides the genes that confer resistance to antibiotics (β-lactams, tetracyclines, aminoglycosides, macrolides, sulphonamides, trimethoprim and ciprofloxacin) and biocides, the plasmid also carries genes conferring resistance to silver, copper and arsenic [53].

15.5.6.8 Megaplasmids

Type strain *C. metallidurans* CH34 harbours resistance determinants for at least 20 different metal ions [71], mainly located on its two megaplasmids [97], although chromosomally encoded metal responsive clusters have also been identified [98]. *C. metallidurans* is specialized in metal resistance and is often associated with industrial sites linked to mining, metallurgical and chemical industries [49] but is also isolated from different spacecraft-related environments [81, 112], from patients with cystic fibrosis [27] or as the causative agent of an invasive human infection [82]. Recent analysis of *C. metallidurans* isolates from different potable water management systems of the International Space Station and from the air of the Kennedy Space Center Payload Hazardous Servicing Facility during assembly of the Mars Exploration Rover indicated that each isolate harbours at least one megaplasmid. Moreover, PCR analysis of the plasmid extracts showed that the silCBA operon is located on one of the megaplasmids [94]. Among others, the presence of the sil gene cluster in the potable water isolates gives them the ability to withstand the sanitation procedure in which silver is used [94].

15.5.7 Silver Uptake and Accumulation

In *C. intermedius* and *P. stutzeri*, but not in *E. coli*, it was found that a silver-resistant strain was capable to accumulate silver resulting in removal from the solution [47, 48, 142, 146]. A nucleation core initiates Ag^+-mediated folding of SilE which is a "molecular sponge" for absorbing metal ions [4]. Incubation of a silver-resistant *K. pneumoniae* on a silver-containing agar resulted in dark metallic colonies [43]. Silver uptake in a strain of *A. fumigatus* isolated from wastewater deriving from the jewellery industry and rich of various metal ions explains tolerance to a high silver concentration of 648 mg/l [130].

15.6 Cross-Tolerance to Other Biocidal Agents

Some efflux pumps have been described in *E. faecium*, *E. hirae*, *E. coli*, *P. putida* and *S. enteritidis* mediating resistance to silver and copper ions [30, 52, 143, 147, 158]. A cross-resistance between silver and copper was also described in five environmental isolates of *C. argentea* [65].

15.7 Cross-Tolerance to Antibiotics

Silver may also contribute to the promotion of antibiotic resistance through co-selection. This may occur when resistance genes to both antibiotics and silver are co-located together in the same cell (co-resistance), or a single resistance mechanism (e.g. an efflux pump) confers resistance to both antibiotics and silver (cross-resistance), leading to co-selection of bacterial strains, or mobile genetic elements that they carry [113].

15.7.1 Clinical Isolates

Cross-resistance has been described in various clinical isolates. Two silver-resistant strains of *E. cloacae* isolated from extracted teeth were also resistant to ampicillin, erythromycin and clindamycin [29]. Five clinical wound isolates were exposed to various concentrations of silver nitrate. Two of them became resistant to silver, and one of them to imipenem and meropenem [151]. Three *S. Typhimurium* strains from burn patients treated topically with 0.5% silver nitrate solution were silver resistant and had cross-resistance to ampicillin, chloramphenicol, tetracycline, streptomycin and sulphonamides [91]. And in 13 human faecal silE-positive *E. coli* isolates, low-level silver exposure resulted in phenotypic silver resistance including cross-resistance to ceftibuten (three isolates), piperacillin-tazobactam (three isolates), cotrimoxazole (two isolates), ciprofloxacin (two isolates) and gentamicin (one isolate) [150].

15.7.2 Environmental Isolates

Cross-resistance has also been described in environmental isolates. A surface water isolate of *R. planticola* was isolated having both multidrug- and multimetal-resistant ability. It displayed resistance to 15 antibiotics like ampicillin, amoxicillin/clavulanic acid, aztreonam, erythromycin, imipenem, oxacillin, pefloxacin, penicillin, piperacillin, piperacillin/tazobactam, rifampin, sulbactam/cefoperazone, ticarcillin, ticarcillin/clavulanic acid, vancomycin, and to 11 heavy metals like aluminium, barium, copper, iron, lead, lithium, manganese, nickel, silver, strontium and tin. The multidrug and multimetal-resistant *R. planticola* may remain present in the environment for a long time [78]. Ten strains of *K. pneumoniae* were isolated from the alimentary canal and gills of shrimps. They were resistant to erythromycin, ampicillin, furazolidone and penicillin and were able to grow in the presence of 1,080 mg/l silver (Ag^+) [26]. In the environmental *M. smegmatis* strain mc^{2}155, a 4-fold MIC increase to isoniazid was detected after exposure to silver NPs which has resulted in silver resistance [83].

15.7.3 Plasmids

Some plasmids have been described conferring resistance to silver and various antibiotics. pMG101 belongs to the IncHI2 incompatibility group of plasmids which are large multi-antibiotic resistance plasmids found widely in the enterobacteriaceae [57]. pSTM6-275 is a IncHI2 plasmid from *S. enterica* confers reduced susceptibility to antibiotics, copper sulphate and silver nitrate [12]. pUPI199 encodes resistance to silver nitrate in *A. baumannii* and contains resistance determinants for 13 different metals and 10 antibiotics [31]. pKQPS142a and pKQPS142b were described in a carbapenem-resistant virulent *K. quasipneumoniae subsp. similipneumoniae* isolate [31]. pLVPK was described in *K. pneumoniae* CG43 with copper, silver, lead and tellurite resistance genes of other bacteria [24]. The plasmid was recently detected in a hypervirulent carbapenem-resistant *K. pneumoniae* [51]. And another multidrug resistance plasmid in *K. pneumoniae* and *E. coli* confers resistance to antibiotics (β-lactams, tetracyclines, aminoglycosides, macrolides, sulphonamides, trimethoprim and ciprofloxacin), silver, copper and arsenic [53].

15.8 Role of Biofilm

15.8.1 Effect on Biofilm Development

Silver as NPs or on impregnated surfaces mostly inhibits single-species biofilm formation by 57–97%, although few studies indicate no such effect, e.g. with *P. aeruginosa* or *S. aureus* on fluoroplastic tympanostomy tubes. Silver alone requires a concentration of at least 0.1 mg/l to inhibit biofilm formation at >50% within 24 h (Table 15.12). A comparison of seven different types of silver-coated dressings showed that there is a large variation in their ability to prevent biofilm formation of *P. aeruginosa* and *A. baumannii* over 72 h, with a number of them not being able to prevent biofilm formation so that they are considered not to be better

Table 15.12 Effect of silver on biofilm development

Bacterial species	Strains/isolates	Type of biofilm	Exposure time	Type of product	Inhibition of biofilm formation	References
C. albicans	ATCC 24433 and 2 clinical urine strains	24-h incubation in microtiter plates	24 h	100 mg/l[a] (S)	83–97%	[62]
C. parapsilosis	2 clinical urine strains	24-h incubation in microtiter plates	24 h	100 mg/l[a] (S)	62–67%	[62]
C. tropicalis	2 clinical urine strains	24-h incubation in microtiter plates	24 h	100 mg/l[a] (S)	57%	[62]
E. coli	1 multidrug-resistant clinical isolate (strain MC-2)	22-h incubation in test tubes	22 h	4 mg/l[a] (S)	65.2%	[28]
P. aeruginosa	Not described	5-d incubation	5 d	Silver oxide–impregnated fluoroplastic tympanostomy tube	None	[11]
P. fluorescens	ATCC 13525	48-h incubation on glass cover slips	48 h	Glass cover slips coated with silver NP	Prevention of biofilm formation only when 100% of planktonic cells were killed by silver NP	[167]
S. aureus	Not described	5-d incubation	5 d	Silver oxide–impregnated fluoroplastic tympanostomy tube	None	[11]
S. aureus	Strain AMC201 (MRSA)	24- and 48-h incubation on titanium plates	24 and 48 h	Titanium implants with embedded silver nanoparticles (approximately 0.1 mg/l)	Partial effect	[163]

(continued)

Table 15.12 (continued)

Bacterial species	Strains/isolates	Type of biofilm	Exposure time	Type of product	Inhibition of biofilm formation	References
S. aureus	1 multidrug-resistant clinical isolate (strain MMC-20)	22 h in test tubes	22 h	8 mg/l[a] (S)	82.6%	[28]
S. epidermidis	ATCC 35984	6-, 12- or 24-h incubation on titanium plates	6, 12 or 24 h	Titanium plates with Ag NPs, fabricated and immobilized in situ by a cathodic arc silver plasma immersion ion implantation	60–80%	[122]
Mixed species	Activated sludge from a wastewater treatment plant	24-h incubation in polystyrene microtiter plates	24 h	1 mg/l (S)	69%	[171]
				0.1 mg/l (S)	70%	
				0.05 mg/l (S)	23%	
				0.01 mg/l (S)	0%	

[a]Nanoparticles

than non-antimicrobial dressings [61]. In *E. coli* MG 1655 and a *L. innocua* field strain, it was shown that resistance to silver nanoparticle is associated significantly increased stickiness in biofilm formation [153].

15.8.2 Effect on Biofilm Removal

Silver alone does not remove mixed-species biofilm when used at 0.01 mg/l for 24 h. Silver NPs at concentrations between 25 and 100 mg/l have some single-species biofilm removal activity (23–93%) beginning after 15-min exposure time (Table 15.13). The single-species biofilm removal effect of silver NPs can be enhanced by 17% EDTA as shown with *S. aureus* and *S. Typhimurium* [90]. In silver-containing wound dressings, there seems to be some biofilm removal effect of

Table 15.13 Biofilm removal rate (quantitative determination of biofilm matrix) by exposure to products or solutions based on silver

Type of biofilm	Concentration	Exposure time	Biofilm removal rate	References
C. albicans (ATCC 10231), 24-h incubation on acryl resin specimens	54 mg/l[a] (S)	5 h	23%	[99]
C. albicans (ATCC 10231), 48-h incubation on acryl resin specimens	54 mg/l[a] (S)	5 h	47%	[99]
C. albicans (1 oral clinical isolate), 24-h incubation on acryl resin specimens	54 mg/l[a] (S)	5 h	23%	[99]
C. albicans (1 oral clinical isolate), 48-h incubation on acryl resin specimens	54 mg/l[a] (S)	5 h	36%	[99]
C. glabrata (ATCC 90030), 24-h incubation on acryl resin specimens	54 mg/l[a] (S)	5 h	43%	[99]
C. glabrata (ATCC 90030), 48-h incubation on acryl resin specimens	54 mg/l[a] (S)	5 h	52%	[99]
C. glabrata (1 oral clinical isolate), 24-h incubation on acryl resin specimens	54 mg/l[a] (S)	5 h	28%	[99]
C. glabrata (1 oral clinical isolate), 48-h incubation on acryl resin specimens	54 mg/l[a] (S)	5 h	37%	[99]
S. Typhimurium (ATCC 14028), 24-h incubation in polystyrene microtiter plates	100 mg/l[a] (P)	15 min	58%	[138]
	50 mg/l[a] (P)		79%	
	25 mg/l[a] (P)		82%	
S. aureus (ATCC 25923), 24-h incubation in polystyrene microtiter plates	100 mg/l[a] (P)	15 min	71%	[138]
	50 mg/l[a] (P)		87%	
	25 mg/l[a] (P)		93%	
Mixed species (activated sludge from a wastewater treatment plant), 24-h incubation in polystyrene microtiter plates	0.01 mg/l (S)	24 h	0%	[171]

S Solution; *P* Commercial product; [a]Silver NPs

silver, but it depends on the type of dressing material and its structure [115]. A functionalized silver nanocomposite with a biocompatible carbohydrate polymer (PAGA) and a membrane-disrupting cationic polymer (PDMAEMA-C4) was described as a potent antibiofilm agent (*P. aeruginosa*, *E. coli*, *S. aureus* and *B. amyloliquefaciens*) [54].

15.8.3 Effect on Biofilm Fixation

No studies were found to assess a potential biofilm fixation by exposure to silver, silver nitrate or silver NPs.

15.9 Summary

The principal antimicrobial activity of silver is summarized in Table 15.14.

The key findings on acquired resistance and cross-resistance including the role of biofilm for selecting resistant isolates are summarized in Table 15.15.

Table 15.14 Overview on the typical exposure times required for silver to achieve sufficient biocidal activity against the different target micro-organisms

Target micro-organisms	Species	Concentration	Exposure time
Bacteria	Moderate bactericidal activity (3.0 log) against selected bacterial species	0.032 mg/l[a]	24 h
	Insufficient bactericidal activity	10,000 mg/l	30 min
Fungi	Insufficient data		
Mycobacteria	Insufficient data		

[a]in biofilm the bactericidal activity will be lower

Table 15.15 Key findings on acquired silver resistance, the effect of low-level exposure, cross-tolerance to other biocides and antibiotics, and its effect on biofilm

Parameter	Species	Findings
Elevated MIC values	*E. coli*, *E. cloacae*	$\leq$ 512,000 mg/l
	P. aeruginosa	$\leq$ 128,000 mg/l
	Klebsiella spp.	$\leq$ 5,500 mg/l
	Enterococcus spp.	$\leq$ 300 mg/l
	Citrobacter spp.	$\leq$ 250 mg/l
	Proteus spp.	$\leq$ 250 mg/l

(continued)

Table 15.15 (continued)

Parameter	Species	Findings
Proposed MIC value to determine resistance	Gram-negative species	>8 mg/l
		27 mg/l silver and 85 or 128 mg/l silver nitrate was also used for silver resistance screening
Cross-tolerance biocides	*E. faecium*, *E. hirae*, *E. coli*, *P. putida*, *S. enteritidis*, *C. argentea*	Cross-tolerance to copper via specific efflux pumps
Cross-tolerance antibiotics	*E. cloacae*	Some clinical strains with cross-resistance to ampicillin, erythromycin and clindamycin or imipenem and meropenem
	E. coli	Some clinical strains with cross-resistance to ceftibuten, piperacillin-tazobactam, cotrimoxazole, ciprofloxacin and gentamicin
	K. pneumoniae	Some shrimp isolates with cross-resistance to erythromycin, ampicillin, furazolidone, and penicillin
	R. planticola	Environmental isolate with multidrug- and multimetal-resistance
	S. Typhimurium	Some clinical strains with cross-resistance to cross-resistance to ampicillin, chloramphenicol, tetracycline, streptomycin and sulphonamides
Resistance mechanisms	*E. cloacae*, *K. oxytoca*, *K. pneumoniae*, *E. aerogenes*, *Enterobacter* spp., *Enterococcus* spp., *E. coli*, *Escherichia* spp., *Klebsiella* spp., *S. aureus*, *CNS*	SilE gene
	Klebsiella spp., *Enterobacter* spp., *Enterococcus* spp., *Escherichia* spp., *Pseudomonas* spp., *Staphylococcus* spp.	SilA gene
	E. aerogenes, *E. cloacae*, *Enterobacter* spp., *Enterococcus* spp., *E. coli*, *Escherichia* spp., *K. oxytoca*, *K. pneumoniae*, *Klebsiella* spp., *Staphylococcus* spp.	SilP gene
	E. aerogenes, *E. cloacae*, *E. coli*, *K. oxytoca*, *K. pneumoniae*	SilS
	A. baumannii, *C. metallidurans*, *E. cloacae*, *Klebsiella* spp., *P. stutzeri*, *Salmonella* spp., *S. marcescens*	Plasmids
	A. baumannii, *E. coli*, *E. hirae*, *C. albicans*	Efflux pumps
	A. fumigatus, *C. intermedius*, *K. pneumoniae*, *P. stutzeri*	Silver uptake and accumulation

(continued)

Table 15.15 (continued)

Parameter	Species	Findings
Effect of low-level exposure	*Acinetobacter* spp., *Citrobacter* spp., *E. cloacae, E. coli, K. pneumoniae, K. oxytoca, Proteus* spp., *P. aeruginosa, S. aureus* (mostly sil negative)	No MIC increase
	E. coli, M. smegmatis, S. aureus	Weak MIC increase ($\leq$ 4-fold)
	E. coli	Strong and unstable MIC increase (>4-fold)
	E. cloacae, E. coli, K. pneumoniae, K. oxytoca	Strong and stable MIC increase (>4-fold)
	A. ferrooxidans, Enterobacter spp., *E. coli*	Strong MIC increase (>4-fold; unknown stability)
	E. coli (128-fold)	Strongest MIC change after low-level exposure
	E. cloacae, K. pneumoniae ($\geq$ 32-fold)	
	K. oxytoca ($\geq$ 16-fold)	
	E. coli (1,024 mg/l)	Highest MIC values after low-level exposure
	E. cloacae ($\geq$ 1,000 mg/l)	
	K. pneumoniae, K. oxytoca ($\geq$ 512 mg/l)	
	A. ferrooxidans (240 mg/l)	
	E. coli	Increase of silver efflux
	E. coli, E. cloacae	Antibiotic tolerance in some isolates to selected agents, e.g. imipenem, meropenem, ceftibuten, piperacillin-tazobactam, cotrimoxazole, ciprofloxacin and gentamicin
Biofilm	Development	Mostly moderate inhibition
	Removal	Some biofilm removal activity for silver NPs at 25–100 mg/l
	Fixation	Unknown

References

1. Abdel Rahim KA, Ali Mohamed AM (2015) Bactericidal and antibiotic synergistic effect of nanosilver against methicillin-resistant Staphylococcus aureus. Jundishapur J Microbiol 8 (11):e25867. https://doi.org/10.5812/jjm.25867
2. Annear DI, Mee BJ, Bailey M (1976) Instability and linkage of silver resistance, lactose fermentation, and colony structure in Enterobacter cloacae from burn wounds. J Clin Pathol 29(5):441–443
3. Aramwit P, Muangman P, Namviriyachote N, Srichana T (2010) In vitro evaluation of the antimicrobial effectiveness and moisture binding properties of wound dressings. Int J Mol Sci 11(8):2864–2874. https://doi.org/10.3390/ijms11082864
4. Asiani KR, Williams H, Bird L, Jenner M, Searle MS, Hobman JL, Scott DJ, Soultanas P (2016) SilE is an intrinsically disordered periplasmic "molecular sponge" involved in

bacterial silver resistance. Mol Microbiol 101(5):731–742. https://doi.org/10.1111/mmi.13399

5. Ayatollahi Mousavi SA, Salari S, Hadizadeh S (2015) Evaluation of antifungal effect of silver nanoparticles against microsporum canis, trichophyton mentagrophytes and microsporum gypseum. Iranian J Biotechnol 13(4):38–42. https://doi.org/10.15171/ijb.1302
6. Bardouniotis E, Huddleston W, Ceri H, Olson ME (2001) Characterization of biofilm growth and biocide susceptibility testing of Mycobacterium phlei using the MBEC assay system. FEMS Microbiol Lett 203(2):263–267
7. Barillo DJ, Marx DE (2014) Silver in medicine: a brief history BC 335 to present. Burns: J Int Soc Burn Injuries 40(Suppl 1):S3–8. https://doi.org/10.1016/j.burns.2014.09.009
8. Barroso JM (2014) COMMISSION IMPLEMENTING DECISION of 24 April 2014 on the non-approval of certain biocidal active substances pursuant to Regulation (EU) No 528/2012 of the European Parliament and of the Council. Off J Eur Union 57(L 124):27–29
9. Berger TJ, Spadaro JA, Bierman R, Chapin SE, Becker RO (1976) Antifungal properties of electrically generated metallic ions. Antimicrob Agents Chemother 10(5):856–860
10. Berger TJ, Spadaro JA, Chapin SE, Becker RO (1976) Electrically generated silver ions: quantitative effects on bacterial and mammalian cells. Antimicrob Agents Chemother 9 (2):357–358
11. Berry JA, Biedlingmaier JF, Whelan PJ (2000) In vitro resistance to bacterial biofilm formation on coated fluoroplastic tympanostomy tubes. Otolaryngol–Head Neck Surgery: Official J Am Acad Otolaryngol-Head Neck Surgery 123(3):246–251. https://doi.org/10.1067/mhn.2000.107458
12. Billman-Jacobe H, Liu Y, Haites R, Weaver T, Robinson L, Marenda M, Dyall-Smith M (2018) pSTM6-275, a conjugative IncHI2 plasmid of Salmonella that confers antibiotic and heavy metal resistance under changing physiological conditions. Antimicrob Agents Chemother. https://doi.org/10.1128/aac.02357-17
13. Bowler PG, Parsons D (2016) Combatting wound biofilm and recalcitrance with a novel anti-biofilm Hydrofiber® wound dressing. Wound Med 14:6–11. https://doi.org/10.1016/j.wndm.2016.05.005
14. Bowler PG, Welsby S, Hogarth A, Towers V (2013) Topical antimicrobial protection of postoperative surgical sites at risk of infection with Propionibacterium acnes: an in-vitro study. J Hosp Infect 83(3):232–237. https://doi.org/10.1016/j.jhin.2012.11.018
15. Bowler PG, Welsby S, Towers V (2013) In vitro antimicrobial efficacy of a silver-containing wound dressing against mycobacteria associated with atypical skin ulcers. Wounds: A Compend Clin Res Pract 25(8):225–230
16. Brady MJ, Lisay CM, Yurkovetskiy AV, Sawan SP (2003) Persistent silver disinfectant for the environmental control of pathogenic bacteria. Am J Infect Control 31(4):208–214
17. Brett DW (2006) A discussion of silver as an antimicrobial agent: alleviating the confusion. Ostomy/wound Manag 52(1):34–41
18. Bridges K, Kidson A, Lowbury EJ, Wilkins MD (1979) Gentamicin- and silver-resistant pseudomonas in a burns unit. BMJ 1(6161):446–449
19. Cao C, Huang J, Yan C, Liu J, Hu Q, Guan W (2018) Shifts of system performance and microbial community structure in a constructed wetland after exposing silver nanoparticles. Chemosphere 199:661–669. https://doi.org/10.1016/j.chemosphere.2018.02.031
20. Cason JS, Jackson DM, Lowbury EJ, Ricketts CR (1966) Antiseptic and aseptic prophylaxis for burns: use of silver nitrate and of isolators. BMJ 2(5525):1288–1294
21. Castellano JJ, Shafii SM, Ko F, Donate G, Wright TE, Mannari RJ, Payne WG, Smith DJ, Robson MC (2007) Comparative evaluation of silver-containing antimicrobial dressings and drugs. Int Wound J 4(2):114–122. https://doi.org/10.1111/j.1742-481X.2007.00316.x
22. Cavanagh MH, Burrell RE, Nadworny PL (2010) Evaluating antimicrobial efficacy of new commercially available silver dressings. Int Wound J 7(5):394–405. https://doi.org/10.1111/j.1742-481X.2010.00705.x

23. Chappell JB, Greville GD (1954) Effect of silver ions on mitochondrial adenosine triphosphatase. Nature 174(4437):930–931
24. Chen YT, Chang HY, Lai YC, Pan CC, Tsai SF, Peng HL (2004) Sequencing and analysis of the large virulence plasmid pLVPK of Klebsiella pneumoniae CG43. Gene 337:189–198. https://doi.org/10.1016/j.gene.2004.05.008
25. Chopra I (2007) The increasing use of silver-based products as antimicrobial agents: a useful development or a cause for concern? J Antimicrob Chemother 59(4):587–590. https://doi.org/10.1093/jac/dkm006
26. Choudhury P, Kumar R (1998) Multidrug- and metal-resistant strains of Klebsiella pneumoniae isolated from Penaeus monodon of the coastal waters of deltaic Sundarban. Can J Microbiol 44(2):186–189
27. Coenye T, Spilker T, Reik R, Vandamme P, Lipuma JJ (2005) Use of PCR analyses to define the distribution of Ralstonia species recovered from patients with cystic fibrosis. J Clin Microbiol 43(7):3463–3466. https://doi.org/10.1128/jcm.43.7.3463-3466.2005
28. Das BC, Dash SK, Mandal D, Ghosh T, Chattopadhyay S, Tripathy S, Das S, Dey SK, Das D, Roy S (2017) Green synthesized silver nanoparticles destroy multidrug resistant bacteria via reactive oxygen species mediated membrane damage. Arab J Chem 10(6):862–876. https://doi.org/10.1016/j.arabjc.2015.08.008
29. Davis IJ, Richards H, Mullany P (2005) Isolation of silver- and antibiotic-resistant enterobacter cloacae from teeth. Oral Microbiol Immunol 20(3):191–194. https://doi.org/10.1111/j.1399-302X.2005.00218.x
30. Delmar JA, Su CC, Yu EW (2014) Bacterial multidrug efflux transporters. Ann Rev Biophys 43:93–117. https://doi.org/10.1146/annurev-biophys-051013-022855
31. Deshpande LM, Chopade BA (1994) Plasmid mediated silver resistance in Acinetobacter baumannii. Biometals: An Int J Role Metal Ions Biol Biochem Med 7(1):49–56
32. Desroche N, Dropet C, Janod P, Guzzo J (2016) Antibacterial properties and reduction of MRSA biofilm with a dressing combining polyabsorbent fibres and a silver matrix. J Wound Care 25(10):577–584. https://doi.org/10.12968/jowc.2016.25.10.577
33. Deus D, Krischek C, Pfeifer Y, Sharifi AR, Fiegen U, Reich F, Klein G, Kehrenberg C (2017) Comparative analysis of the susceptibility to biocides and heavy metals of extended-spectrum beta-lactamase-producing Escherichia coli isolates of human and avian origin, Germany. Diagnostic Microbiol Infect Dis 88(1):88–92. https://doi.org/10.1016/j.diagmicrobio.2017.01.023
34. Dibrov P, Dzioba J, Gosink KK, Hase CC (2002) Chemiosmotic mechanism of antimicrobial activity of Ag(+) in Vibrio cholerae. Antimicrob Agents Chemother 46 (8):2668–2670
35. Dunne KA, Chaudhuri RR, Rossiter AE, Beriotto I, Browning DF, Squire D, Cunningham AF, Cole JA, Loman N, Henderson IR (2017) Sequencing a piece of history: complete genome sequence of the original Escherichia coli strain. Microbial Genom 3(3): mgen000106. https://doi.org/10.1099/mgen.0.000106
36. Edwards-Jones V (2009) The benefits of silver in hygiene, personal care and healthcare. Lett Appl Microbiol 49(2):147–152. https://doi.org/10.1111/j.1472-765X.2009.02648.x
37. Elkrewi E, Randall CP, Ooi N, Cottell JL, O'Neill AJ (2017) Cryptic silver resistance is prevalent and readily activated in certain gram-negative pathogens. J Antimicrob Chemother 72(11):3043–3046. https://doi.org/10.1093/jac/dkx258
38. Environmental Protection Agency (2009) 2-(Decylthio)ethanamine hydrochloride; and silver and compounds registration review; antimicrobial pesticide dockets opened for review and comment. Fed Reg 74(120):30070–30073
39. Espigares E, Moreno Roldan E, Espigares M, Abreu R, Castro B, Dib AL, Arias A (2017) Phenotypic resistance to disinfectants and antibiotics in methicillin-resistant Staphylococcus aureus strains isolated from pigs. Zoonoses Public Health 64(4):272–280. https://doi.org/10.1111/zph.12308

40. European Chemicals Agency (ECHA) (2018) Silver. Substance information. https://echa.europa.eu/de/substance-information/-/substanceinfo/100.028.301. Accessed 27 March 2018
41. Fang L, Li X, Li L, Li S, Liao X, Sun J, Liu Y (2016) Co-spread of metal and antibiotic resistance within ST3-IncHI2 plasmids from E. coli isolates of food-producing animals. Scientific Reports 6:25312. https://doi.org/10.1038/srep25312
42. Feng QL, Wu J, Chen GQ, Cui FZ, Kim TN, Kim JO (2000) A mechanistic study of the antibacterial effect of silver ions on Escherichia coli and Staphylococcus aureus. J Biomed Mater Res 52(4):662–668
43. Finley PJ, Norton R, Austin C, Mitchell A, Zank S, Durham P (2015) Unprecedented silver resistance in clinically isolated enterobacteriaceae: major implications for burn and wound management. Antimicrob Agents Chemother 59(8):4734–4741. https://doi.org/10.1128/aac.00026-15
44. Finley PJ, Peterson A, Huckfeldt RE (2013) The prevalence of phenotypic silver resistance in clinical isolates. Wounds: A Compend Clin Res Pract 25(4):84–88
45. Flannery EL, Antczak SM, Mobley HL (2011) Self-transmissibility of the integrative and conjugative element ICEPm1 between clinical isolates requires a functional integrase, relaxase, and type IV secretion system. J Bacteriol 193(16):4104–4112. https://doi.org/10.1128/jb.05119-11
46. Flannery EL, Mody L, Mobley HL (2009) Identification of a modular pathogenicity island that is widespread among urease-producing uropathogens and shares features with a diverse group of mobile elements. Infect Immun 77(11):4887–4894. https://doi.org/10.1128/iai.00705-09
47. Gadd GM, Laurence OS, Briscoe PA, Trevors JT (1989) Silver accumulation in Pseudomonas stutzeri AG259. Biol Metals 2(3):168–173
48. Goddard PA, Bull AT (1989) The isolation and characterisation of bacteria capable of accumulating silver. Appl Microbiol Biotechnol 31(3):308–313
49. Goris J, De Vos P, Coenye T, Hoste B, Janssens D, Brim H, Diels L, Mergeay M, Kersters K, Vandamme P (2001) Classification of metal-resistant bacteria from industrial biotopes as Ralstonia campinensis sp. nov., Ralstonia metallidurans sp. nov. and Ralstonia basilensis Steinle et al. 1998 emend. Int J Syst Evol Microbiol 51(Pt 5):1773–1782. https://doi.org/10.1099/00207713-51-5-1773
50. Graves JL Jr, Tajkarimi M, Cunningham Q, Campbell A, Nonga H, Harrison SH, Barrick JE (2015) Rapid evolution of silver nanoparticle resistance in Escherichia coli. Front Genet 6:42. https://doi.org/10.3389/fgene.2015.00042
51. Gu D, Dong N, Zheng Z, Lin D, Huang M, Wang L, Chan EW, Shu L, Yu J, Zhang R, Chen S (2018) A fatal outbreak of ST11 carbapenem-resistant hypervirulent Klebsiella pneumoniae in a Chinese hospital: a molecular epidemiological study. Lancet Infect Dis 18(1):37–46. https://doi.org/10.1016/s1473-3099(17)30489-9
52. Gudipaty SA, Larsen AS, Rensing C, McEvoy MM (2012) Regulation of Cu(I)/Ag(I) efflux genes in escherichia coli by the sensor kinase CusS. FEMS Microbiol Lett 330(1):30–37. https://doi.org/10.1111/j.1574-6968.2012.02529.x
53. Gullberg E, Albrecht LM, Karlsson C, Sandegren L, Andersson DI (2014) Selection of a multidrug resistance plasmid by sublethal levels of antibiotics and heavy metals. mBio 5(5): e01918–01914. https://doi.org/10.1128/mbio.01918 14
54. Guo Q, Zhao Y, Dai X, Zhang T, Yu Y, Zhang X, Li C (2017) Functional silver nanocomposites as broad-spectrum antimicrobial and biofilm-disrupting agents. ACS Appl Mater Interf 9(20):16834–16847. https://doi.org/10.1021/acsami.7b02775
55. Gupta A, Matsui K, Lo JF, Silver S (1999) Molecular basis for resistance to silver cations in salmonella. Nat Med 5(2):183–188. https://doi.org/10.1038/5545
56. Gupta A, Maynes M, Silver S (1998) Effects of halides on plasmid-mediated silver resistance in escherichia coli. Appl Environ Microbiol 64(12):5042–5045

57. Gupta A, Phung LT, Taylor DE, Silver S (2001) Diversity of silver resistance genes in IncH incompatibility group plasmids. Microbiology (Reading, England) 147(Pt 12):3393–3402. https://doi.org/10.1099/00221287-147-12-3393
58. Gupta LK, Jindal R, Beri HK, Chhibber S (1992) Virulence of silver-resistant mutant of Klebsiella pneumoniae in burn wound model. Folia Microbiol 37(4):245–248
59. Gwin CA, Gunsch CK (2018) Examining relationships between total silver concentration and Sil silver resistance genes in domestic wastewater treatment plants. J Appl Microbiol. https://doi.org/10.1111/jam.13731
60. Haefeli C, Franklin C, Hardy K (1984) Plasmid-determined silver resistance in pseudomonas stutzeri isolated from a silver mine. J Bacteriol 158(1):389–392
61. Halstead FD, Rauf M, Bamford A, Wearn CM, Bishop JRB, Burt R, Fraise AP, Moiemen NS, Oppenheim BA, Webber MA (2015) Antimicrobial dressings: comparison of the ability of a panel of dressings to prevent biofilm formation by key burn wound pathogens. Burns: J Int Soc Burn Injuries 41(8):1683–1694. https://doi.org/10.1016/j.burns.2015.06.005
62. Hamid S, Zainab S, Faryal R, Ali N (2017) Deterrence in metabolic and biofilms forming activity of Candida species by mycogenic silver nanoparticles. J Appl Biomed 15(4):249–255. https://doi.org/10.1016/j.jab.2017.02.003
63. Hendry AT, Stewart IO (1979) Silver-resistant Enterobacteriaceae from hospital patients. Canadian J Microbiol 25(8):915–921
64. Hoekstra MJ, Westgate SJ, Mueller S (2017) Povidone-iodine ointment demonstrates in vitro efficacy against biofilm formation. Int Wound J 14(1):172–179. https://doi.org/10.1111/iwj.12578
65. Holland SL, Dyer PS, Bond CJ, James SA, Roberts IN, Avery SV (2011) Candida argentea sp. nov., a copper and silver resistant yeast species. Fungal Biol 115(9):909–918. https://doi.org/10.1016/j.funbio.2011.07.004
66. Hwang MG, Katayama H, Ohgaki S (2007) Inactivation of Legionella pneumophila and pseudomonas aeruginosa: evaluation of the bactericidal ability of silver cations. Water Res 41(18):4097–4104. https://doi.org/10.1016/j.watres.2007.05.052
67. Ip M, Lui SL, Poon VK, Lung I, Burd A (2006) Antimicrobial activities of silver dressings: an in vitro comparison. J Med Microbiol 55(Pt 1):59–63. https://doi.org/10.1099/jmm.0.46124-0
68. Islam MS, Larimer C, Ojha A, Nettleship I (2013) Antimycobacterial efficacy of silver nanoparticles as deposited on porous membrane filters. Mater Sci Eng C, Mater Biol Appl 33 (8):4575–4581. https://doi.org/10.1016/j.msec.2013.07.013
69. Jadhav K, Dhamecha D, Bhattacharya D, Patil M (2016) Green and ecofriendly synthesis of silver nanoparticles: characterization, biocompatibility studies and gel formulation for treatment of infections in burns. J Photochem Photobiol b, Biology 155:109–115. https://doi.org/10.1016/j.jphotobiol.2016.01.002
70. Jakobsen L, Andersen AS, Friis-Moller A, Jorgensen B, Krogfelt KA, Frimodt-Moller N (2011) Silver resistance: an alarming public health concern? Int J Antimicrob Agents 38 (5):454–455. https://doi.org/10.1016/j.ijantimicag.2011.07.005
71. Janssen PJ, Van Houdt R, Moors H, Monsieurs P, Morin N, Michaux A, Benotmane MA, Leys N, Vallaeys T, Lapidus A, Monchy S, Medigue C, Taghavi S, McCorkle S, Dunn J, van der Lelie D, Mergeay M (2010) The complete genome sequence of Cupriavidus metallidurans strain CH34, a master survivalist in harsh and anthropogenic environments. PLoS One 5(5):e10433. https://doi.org/10.1371/journal.pone.0010433
72. Jung WK, Koo HC, Kim KW, Shin S, Kim SH, Park YH (2008) Antibacterial activity and mechanism of action of the silver ion in Staphylococcus aureus and Escherichia coli. Appl Environ Microbiol 74(7):2171–2178. https://doi.org/10.1128/aem.02001-07
73. Kaegi R, Voegelin A, Ort C, Sinnet B, Thalmann B, Krismer J, Hagendorfer H, Elumelu M, Mueller E (2013) Fate and transformation of silver nanoparticles in urban wastewater systems. Water Res 47(12):3866–3877. https://doi.org/10.1016/j.watres.2012.11.060

74. Kalan LR, Pepin DM, Ul-Haq I, Miller SB, Hay ME, Precht RJ (2017) Targeting biofilms of multidrug-resistant bacteria with silver oxynitrate. Int J Antimicrob Agents 49(6):719–726. https://doi.org/10.1016/j.ijantimicag.2017.01.019
75. Khor SY, Jegathesan M (1983) Heavy metal and disinfectant resistance in clinical isolates of gram-negative rods. Southeast Asian J Trop Med Public Health 14(2):199–203
76. Kim H, Makin I, Skiba J, Ho A, Housler G, Stojadinovic A, Izadjoo M (2014) Antibacterial efficacy testing of a bioelectric wound dressing against clinical wound pathogens. Open Microbiol J 8:15–21. https://doi.org/10.2174/1874285801408010015
77. Klaus T, Joerger R, Olsson E, Granqvist CG (1999) Silver-based crystalline nanoparticles, microbially fabricated. Proc Natl Acad Sci USA 96(24):13611–13614
78. Koc S, Kabatas B, Icgen B (2013) Multidrug and heavy metal-resistant Raoultella planticola isolated from surface water. Bull Environ Contamin Toxicol 91(2):177–183. https://doi.org/10.1007/s00128-013-1031-6
79. Kremer AN, Hoffmann H (2012) Subtractive hybridization yields a silver resistance determinant unique to nosocomial pathogens in the Enterobacter cloacae complex. J Clin Microbiol 50(10):3249–3257. https://doi.org/10.1128/jcm.00885-12
80. Kuehl R, Brunetto PS, Woischnig AK, Varisco M, Rajacic Z, Vosbeck J, Terracciano L, Fromm KM, Khanna N (2016) Preventing implant-associated infections by silver coating. Antimicrob Agents Chemother 60(4):2467–2475. https://doi.org/10.1128/aac.02934-15
81. La Duc MT, Nicholson W, Kern R, Venkateswaran K (2003) Microbial characterization of the Mars Odyssey spacecraft and its encapsulation facility. Environ Microbiol 5(10):977–985
82. Langevin S, Vincelette J, Bekal S, Gaudreau C (2011) First case of invasive human infection caused by Cupriavidus metallidurans. J Clin Microbiol 49(2):744–745. https://doi.org/10.1128/jcm.01947-10
83. Larimer C, Islam MS, Ojha A, Nettleship I (2014) Mutation of environmental mycobacteria to resist silver nanoparticles also confers resistance to a common antibiotic. Biometals: An Int J Role Metal Ions Biol Biochem Med 27(4):695–702. https://doi.org/10.1007/s10534-014-9761-4
84. Li XZ, Nikaido H, Williams KE (1997) Silver-resistant mutants of escherichia coli display active efflux of Ag^+ and are deficient in porins. J Bacteriol 179(19):6127–6132
85. Loh JV, Percival SL, Woods EJ, Williams NJ, Cochrane CA (2009) Silver resistance in MRSA isolated from wound and nasal sources in humans and animals. Int Wound J 6(1):32–38. https://doi.org/10.1111/j.1742-481X.2008.00563.x
86. Losasso C, Belluco S, Cibin V, Zavagnin P, Micetic I, Gallocchio F, Zanella M, Bregoli L, Biancotto G, Ricci A (2014) Antibacterial activity of silver nanoparticles: sensitivity of different Salmonella serovars. Front Microbiol 5:227. https://doi.org/10.3389/fmicb.2014.00227
87. Lysakowska ME, Ciebiada-Adamiec A, Klimek L, Sienkiewicz M (2015) The activity of silver nanoparticles (Axonnite) on clinical and environmental strains of acinetobacter spp. Burns: J Int Soc Burn Injuries 41(2):364–371. https://doi.org/10.1016/j.burns.2014.07.014
88. Malaikozhundan B, Vijayakumar S, Vaseeharan B, Jenifer AA, Chitra P, Prabhu NM, Kannapiran E (2017) Two potential uses for silver nanoparticles coated with Solanum nigrum unripe fruit extract: biofilm inhibition and photodegradation of dye effluent. Microb pathogenesis 111:316–324. https://doi.org/10.1016/j.micpath.2017.08.039
89. Mallevre F, Fernandes TF, Aspray TJ (2016) Pseudomonas putida biofilm dynamics following a single pulse of silver nanoparticles. Chemosphere 153:356–364. https://doi.org/10.1016/j.chemosphere.2016.03.060
90. Martinez-Andrade JM, Avalos-Borja M, Vilchis-Nestor AR, Sanchez-Vargas LO, Castro-Longoria E (2018) Dual function of EDTA with silver nanoparticles for root canal treatment-A novel modification. PLoS One 13(1):e0190866. https://doi.org/10.1371/journal.pone.0190866

91. McHugh GL, Moellering RC, Hopkins CC, Swartz MN (1975) Salmonella typhimurium resistant to silver nitrate, chloramphenicol, and ampicillin. Lancet 1(7901):235–240
92. Mekkawy AI, El-Mokhtar MA, Nafady NA, Yousef N, Hamad MA, El-Shanawany SM, Ibrahim EH, Elsabahy M (2017) In vitro and in vivo evaluation of biologically synthesized silver nanoparticles for topical applications: effect of surface coating and loading into hydrogels. Int J Nanomed 12:759–777. https://doi.org/10.2147/ijn.s124294
93. Mijnendonckx K, Leys N, Mahillon J, Silver S, Van Houdt R (2013) Antimicrobial silver: uses, toxicity and potential for resistance. Biometals: An Int J Role Metal Ions Biol Biochem Med 26(4):609–621. https://doi.org/10.1007/s10534-013-9645-z
94. Mijnendonckx K, Provoost A, Ott CM, Venkateswaran K, Mahillon J, Leys N, Van Houdt R (2013) Characterization of the survival ability of Cupriavidus metallidurans and Ralstonia pickettii from space-related environments. Microbial Ecol 65(2):347–360. https://doi.org/10.1007/s00248-012-0139-2
95. Modak SM, Fox CL Jr (1973) Binding of silver sulfadiazine to the cellular components of Pseudomonas aeruginosa. Biochem Pharmacol 22(19):2391–2404
96. Mohan S, Oluwafemi OS, George SC, Jayachandran VP, Lewu FB, Songca SP, Kalarikkal N, Thomas S (2014) Completely green synthesis of dextrose reduced silver nanoparticles, its antimicrobial and sensing properties. Carbohydr Polym 106:469–474. https://doi.org/10.1016/j.carbpol.2014.01.008
97. Monchy S, Benotmane MA, Janssen P, Vallaeys T, Taghavi S, van der Lelie D, Mergeay M (2007) Plasmids pMOL28 and pMOL30 of Cupriavidus metallidurans are specialized in the maximal viable response to heavy metals. J Bacteriol 189(20):7417–7425. https://doi.org/10.1128/jb.00375-07
98. Monsieurs P, Moors H, Van Houdt R, Janssen PJ, Janssen A, Coninx I, Mergeay M, Leys N (2011) Heavy metal resistance in Cupriavidus metallidurans CH34 is governed by an intricate transcriptional network. Biometals: An Int J Role Metal Ions Biol Biochem Med 24(6):1133–1151. https://doi.org/10.1007/s10534-011-9473-y
99. Monteiro DR, Takamiya AS, Feresin LP, Gorup LF, de Camargo ER, Delbem AC, Henriques M, Barbosa DB (2015) Susceptibility of Candida albicans and Candida glabrata biofilms to silver nanoparticles in intermediate and mature development phases. J Prosthodontic Res 59(1):42–48. https://doi.org/10.1016/j.jpor.2014.07.004
100. Morones JR, Elechiguerra JL, Camacho A, Holt K, Kouri JB, Ramírez JT, Yacaman MJ (2005) The bactericidal effect of silver nanoparticles. Nanotechnol 16(10):2346
101. Mourao J, Novais C, Machado J, Peixe L, Antunes P (2015) Metal tolerance in emerging clinically relevant multidrug-resistant Salmonella enterica serotype 4,[5],12:i:- clones circulating in Europe. Int J Antimicrob Agents 45(6):610–616. https://doi.org/10.1016/j.ijantimicag.2015.01.013
102. Muller G, Kramer A (2008) Biocompatibility index of antiseptic agents by parallel assessment of antimicrobial activity and cellular cytotoxicity. J Antimicrob Chemother 61(6):1281–1287. https://doi.org/10.1093/jac/dkn125
103. Muller M, Merrett ND (2014) Pyocyanin production by Pseudomonas aeruginosa confers resistance to ionic silver. Antimicrob Agents Chemother 58(9):5492–5499. https://doi.org/10.1128/aac.03069-14
104. National Center for Biotechnology Information Silver. PubChem Compound Database; CID = 23954. https://pubchem.ncbi.nlm.nih.gov/compound/23954. Accessed 27 March 2018
105. National Center for Biotechnology Information Silver nitrate. PubChem Compound Database; CID = 24470. https://pubchem.ncbi.nlm.nih.gov/compound/24470. Accessed 27 March 2018
106. Neethu S, Midhun SJ, Radhakrishnan EK, Jyothis M (2018) Green synthesized silver nanoparticles by marine endophytic fungus Penicillium polonicum and its antibacterial efficacy against biofilm forming, multidrug-resistant Acinetobacter baumanii. Microb Pathogen 116:263–272. https://doi.org/10.1016/j.micpath.2018.01.033

107. Nicolas MF, Ramos PIP, Marques de Carvalho F, Camargo DRA, de Fatima Morais Alves C, Loss de Morais G, Almeida LGP, Souza RC, Ciapina LP, Vicente ACP, Coimbra RS, Ribeiro de Vasconcelos AT (2018) Comparative genomic analysis of a clinical isolate of klebsiella quasipneumoniae subsp. similipneumoniae, a KPC-2 and OKP-B-6 beta-lactamases producer harboring two drug-resistance plasmids from southeast Brazil. Front Microbiol 9:220. https://doi.org/10.3389/fmicb.2018.00220
108. Nieto JJ, Ventosa A, Montero CG, Ruiz-Berraquero F (1989) Toxicity of heavy metals to archaebacterial halococci. Syst Appl Microbiol 11(2):116–120. https://doi.org/10.1016/S0723-2020(89)80049-9
109. Norman G, Christie J, Liu Z, Westby MJ, Jefferies JM, Hudson T, Edwards J, Mohapatra DP, Hassan IA, Dumville JC (2017) Antiseptics for burns. The cochrane database of systematic reviews 7:Cd011821. https://doi.org/10.1002/14651858.cd011821.pub2
110. Olasupo NA, Scott-Emuakpor MB, Ogunshola RA (1993) Resistance to heavy metals by some Nigerian yeast strains. Folia Microbiol 38(4):285–287
111. Ostman M, Lindberg RH, Fick J, Bjorn E, Tysklind M (2017) Screening of biocides, metals and antibiotics in Swedish sewage sludge and wastewater. Water Res 115:318–328. https://doi.org/10.1016/j.watres.2017.03.011
112. Ott CM, Bruce RJ, Pierson DL (2004) Microbial characterization of free floating condensate aboard the Mir space station. Microb Ecol 47(2):133–136. https://doi.org/10.1007/s00248-003-1038-3
113. Pal C, Asiani K, Arya S, Rensing C, Stekel DJ, Larsson DGJ, Hobman JL (2017) Metal resistance and its association with antibiotic resistance. Adv Microb Physiol 70:261–313. https://doi.org/10.1016/bs.ampbs.2017.02.001
114. Parikh RY, Singh S, Prasad BL, Patole MS, Sastry M, Shouche YS (2008) Extracellular synthesis of crystalline silver nanoparticles and molecular evidence of silver resistance from Morganella sp.: towards understanding biochemical synthesis mechanism. Chembiochem: A Eur J Chem Biol 9(9):1415–1422. https://doi.org/10.1002/cbic.200700592
115. Parsons D, Meredith K, Rowlands VJ, Short D, Metcalf DG, Bowler PG (2016) Enhanced performance and mode of action of a novel antibiofilm hydrofiber(R) wound dressing. BioMed Res Int 2016:7616471. https://doi.org/10.1155/2016/7616471
116. Percival SL, Woods E, Nutekpor M, Bowler P, Radford A, Cochrane C (2008) Prevalence of silver resistance in bacteria isolated from diabetic foot ulcers and efficacy of silver-containing wound dressings. Ostomy/Wound Manag 54(3):30–40
117. Perez-Diaz M, Alvarado-Gomez E, Magana-Aquino M, Sanchez-Sanchez R, Velasquillo C, Gonzalez C, Ganem-Rondero A, Martinez-Castanon G, Zavala-Alonso N, Martinez-Gutierrez F (2016) Anti-biofilm activity of chitosan gels formulated with silver nanoparticles and their cytotoxic effect on human fibroblasts. Materials Sci Eng C, Mater Biol Appl 60:317–323. https://doi.org/10.1016/j.msec.2015.11.036
118. Perez-Diaz MA, Boegli L, James G, Velasquillo C, Sanchez-Sanchez R, Martinez-Martinez RE, Martinez-Castanon GA, Martinez-Gutierrez F (2015) Silver nanoparticles with antimicrobial activities against streptococcus mutans and their cytotoxic effect. Mater Sci Eng C, Mater Biol Appl 55:360–366. https://doi.org/10.1016/j.msec.2015.05.036
119. Pletzer D, Weingart H (2014) Characterization and regulation of the resistance-nodulation-cell division-type multidrug efflux pumps MdtABC and MdtUVW from the fire blight pathogen Erwinia amylovora. BMC Microbiol 14:185. https://doi.org/10.1186/1471-2180-14-185
120. Pokrowiecki R, Zareba T, Mielczarek A, Opalinska A, Wojnarowicz J, Majkowski M, Lojkowski W, Tyski S (2013) Evaluation of biocidal properties of silver nanoparticles against cariogenic bacteria. Medycyna doswiadczalna i mikrobiologia 65(3):197–206
121. Potter BA, Lob M, Mercaldo R, Hetzler A, Kaistha V, Khan H, Kingston N, Knoll M, Maloy-Franklin B, Melvin K, Ruiz-Pelet P, Ozsoy N, Schmitt E, Wheeler L, Potter M, Rutter MA, Yahn G, Parente DH (2015) A long-term study examining the antibacterial

effectiveness of Agion silver zeolite technology on door handles within a college campus. Lett Appl Microbiol 60(2):120–127. https://doi.org/10.1111/lam.12356
122. Qin H, Cao H, Zhao Y, Zhu C, Cheng T, Wang Q, Peng X, Cheng M, Wang J, Jin G, Jiang Y, Zhang X, Liu X, Chu PK (2014) In vitro and in vivo anti-biofilm effects of silver nanoparticles immobilized on titanium. Biomaterials 35(33):9114–9125. https://doi.org/10.1016/j.biomaterials.2014.07.040
123. Randall CP, Gupta A, Jackson N, Busse D, O'Neill AJ (2015) Silver resistance in Gram-negative bacteria: a dissection of endogenous and exogenous mechanisms. J Antimicrob Chemother 70(4):1037–1046. https://doi.org/10.1093/jac/dku523
124. Randall CP, Oyama LB, Bostock JM, Chopra I, O'Neill AJ (2013) The silver cation (Ag^+): antistaphylococcal activity, mode of action and resistance studies. J Antimicrob Chemother 68(1):131–138. https://doi.org/10.1093/jac/dks372
125. Ricco JB, Assadian A, Schneider F, Assadian O (2012) In vitro evaluation of the antimicrobial efficacy of a new silver-triclosan vs a silver collagen-coated polyester vascular graft against methicillin-resistant Staphylococcus aureus. J Vascular Surgery 55(3):823–829. https://doi.org/10.1016/j.jvs.2011.08.015
126. Riggle PJ, Kumamoto CA (2000) Role of a Candida albicans P1-type ATPase in resistance to copper and silver ion toxicity. J Bacteriol 182(17):4899–4905
127. Rosenkranz HS, Rosenkranz S (1972) Silver sulfadiazine: interaction with isolated deoxyribonucleic acid. Antimicrob Agents Chemother 2(5):373–383
128. Ruparelia JP, Chatterjee AK, Duttagupta SP, Mukherji S (2008) Strain specificity in antimicrobial activity of silver and copper nanoparticles. Acta biomaterialia 4(3):707–716. https://doi.org/10.1016/j.actbio.2007.11.006
129. Rupp ME, Fitzgerald T, Marion N, Helget V, Puumala S, Anderson JR, Fey PD (2004) Effect of silver-coated urinary catheters: efficacy, cost-effectiveness, and antimicrobial resistance. Am J Infect Control 32(8):445–450. https://doi.org/10.1016/s0196655304004742
130. Sabatini L, Battistelli M, Giorgi L, Iacobucci M, Gobbi L, Andreozzi E, Pianetti A, Franchi R, Bruscolini F (2016) Tolerance to silver of an Aspergillus fumigatus strain able to grow on cyanide containing wastes. J Hazard Mater 306:115–123. https://doi.org/10.1016/j.jhazmat.2015.12.014
131. Said J, Walker M, Parsons D, Stapleton P, Beezer AE, Gaisford S (2014) An in vitro test of the efficacy of an anti-biofilm wound dressing. Int J Pharm 474(1–2):177–181. https://doi.org/10.1016/j.ijpharm.2014.08.034
132. Schreurs WJ, Rosenberg H (1982) Effect of silver ions on transport and retention of phosphate by Escherichia coli. J Bacteriol 152(1):7–13
133. Sedlak RH, Hnilova M, Grosh C, Fong H, Baneyx F, Schwartz D, Sarikaya M, Tamerler C, Traxler B (2012) Engineered Escherichia coli silver-binding periplasmic protein that promotes silver tolerance. Appl Environ Microbiol 78(7):2289–2296. https://doi.org/10.1128/aem.06823-11
134. Semeykina AL, Skulachev VP (1990) Submicromolar Ag^+ increases passive Na^+ permeability and inhibits the respiration-supported formation of Na^+ gradient in Bacillus FTU vesicles. FEBS letters 269(1):69–72
135. Sheikholeslami S, Mousavi SE, Ahmadi Ashtiani HR, Hosseini Doust SR, Mahdi Rezayat S (2016) Antibacterial activity of silver nanoparticles and their combination with zataria multiflora essential oil and methanol extract. Jundishapur J Microbiol 9(10):e36070. https://doi.org/10.5812/jjm.36070
136. Sheng Z, Liu Y (2011) Effects of silver nanoparticles on wastewater biofilms. Water Res 45(18):6039–6050. https://doi.org/10.1016/j.watres.2011.08.065
137. Sheng Z, Van Nostrand JD, Zhou J, Liu Y (2015) The effects of silver nanoparticles on intact wastewater biofilms. Front Microbiol 6:680. https://doi.org/10.3389/fmicb.2015.00680
138. Shirdel M, Tajik H, Moradi M (2017) Combined activity of colloid nanosilver and zataria multi flora boiss essential oil-mechanism of action and biofilm removal activity. Adv Pharmac Bull 7(4):621–628. https://doi.org/10.15171/apb.2017.074

139. Silver S (2003) Bacterial silver resistance: molecular biology and uses and misuses of silver compounds. FEMS Microbiol Rev 27(2–3):341–353
140. Silver S, le Phung T, Silver G (2006) Silver as biocides in burn and wound dressings and bacterial resistance to silver compounds. J Indus Microbiol Biotechnol 33(7):627–634. https://doi.org/10.1007/s10295-006-0139-7
141. Slawson RM, Lohmeier-Vogel EM, Lee H, Trevors JT (1994) Silver resistance in Pseudomonas stutzeri. Biometals: An Int J Role Metal Ions Biol Biochem Med 7(1):30–40
142. Slawson RM, Van Dyke MI, Lee H, Trevors JT (1992) Germanium and silver resistance, accumulation, and toxicity in microorganisms. Plasmid 27(1):72–79
143. Solioz M, Odermatt A (1995) Copper and silver transport by CopB-ATPase in membrane vesicles of Enterococcus hirae. J Biol Chem 270(16):9217–9221
144. Starodub ME, Trevors JT (1989) Silver resistance in Escherichia coli R1. J Med Microbiol 29(2):101–110. https://doi.org/10.1099/00222615-29-2-101
145. Starodub ME, Trevors JT (1990) Mobilization of Escherichia coli R1 silver-resistance plasmid pJT1 by Tn5-Mob into Escherichia coli C600. Biol Metals 3(1):24–27
146. Starodub ME, Trevors JT (1990) Silver accumulation and resistance in Escherichia coli R1. J Inorganic Biochem 39(4):317–325
147. Su CC, Long F, Yu EW (2011) The Cus efflux system removes toxic ions via a methionine shuttle. Protein Sci: A Publication Protein Soc 20(1):6–18. https://doi.org/10.1002/pro.532
148. Suppi S, Kasemets K, Ivask A, Kunnis-Beres K, Sihtmae M, Kurvet I, Aruoja V, Kahru A (2015) A novel method for comparison of biocidal properties of nanomaterials to bacteria, yeasts and algae. J Hazard Mater 286:75–84. https://doi.org/10.1016/j.jhazmat.2014.12.027
149. Sutterlin S, Dahlo M, Tellgren-Roth C, Schaal W, Melhus A (2017) High frequency of silver resistance genes in invasive isolates of Enterobacter and Klebsiella species. J Hosp Infect 96 (3):256–261. https://doi.org/10.1016/j.jhin.2017.04.017
150. Sutterlin S, Edquist P, Sandegren L, Adler M, Tangden T, Drobni M, Olsen B, Melhus A (2014) Silver resistance genes are overrepresented among Escherichia coli isolates with CTX-M production. Appl Environ Microbiol 80(22):6863–6869. https://doi.org/10.1128/aem.01803-14
151. Sutterlin S, Tano E, Bergsten A, Tallberg AB, Melhus A (2012) Effects of silver-based wound dressings on the bacterial flora in chronic leg ulcers and its susceptibility in vitro to silver. Acta dermato-venereologica 92(1):34–39. https://doi.org/10.2340/00015555-1170
152. Sutterlin S, Tellez-Castillo CJ, Anselem L, Yin H, Bray JE, Maiden MCJ (2018) Heavy metal susceptibility on Escherichia coli from urine samples from Sweden, Germany and Spain. Antimicrob Agents Chemother. https://doi.org/10.1128/aac.00209-18
153. Tajkarimi M, Harrison SH, Hung AM, Graves JL Jr (2016) Mechanobiology of antimicrobial resistant Escherichia coli and Listeria innocua. PLoS One 11(2):e0149769. https://doi.org/10.1371/journal.pone.0149769
154. Tavaf Z, Tabatabaei M, Khalafi-Nezhad A, Panahi F (2017) Evaluation of antibacterial, antibofilm and antioxidant activities of synthesized silver nanoparticles (AgNPs) and casein peptide fragments against streptococcus mutans. Eur J Integr Med 12:163–171. https://doi.org/10.1016/j.eujim.2017.05.011
155. Thanganadar Appapalam S, Panchamoorthy R (2017) Aerva lanata mediated phytofabrication of silver nanoparticles and evaluation of their antibacterial activity against wound associated bacteria. J Taiwan Inst Chem Eng 78:539–551. https://doi.org/10.1016/j.jtice.2017.06.035
156. Thuptimdang P, Limpiyakorn T, Khan E (2017) Dependence of toxicity of silver nanoparticles on Pseudomonas putida biofilm structure. Chemosphere 188:199–207. https://doi.org/10.1016/j.chemosphere.2017.08.147

157. Thuptimdang P, Limpiyakorn T, McEvoy J, Pruss BM, Khan E (2015) Effect of silver nanoparticles on Pseudomonas putida biofilms at different stages of maturity. J Hazard Mater 290:127–133. https://doi.org/10.1016/j.jhazmat.2015.02.073
158. Torres-Urquidy O, Bright K (2012) Efficacy of multiple metals against copper-resistant bacterial strains. J Appl Microbiol 112(4):695–704. https://doi.org/10.1111/j.1365-2672.2012.05245.x
159. Ug A, Ceylan Ö (2003) Occurrence of resistance to antibiotics, metals, and plasmids in clinical strains of Staphylococcus spp. Arch Med Res 34(2):130–136. https://doi.org/10.1016/S0188-4409(03)00006-7
160. Unger C, Luck C (2012) Inhibitory effects of silver ions on Legionella pneumophila grown on agar, intracellular in Acanthamoeba castellanii and in artificial biofilms. J Appl Microbiol 112(6):1212–1219. https://doi.org/10.1111/j.1365-2672.2012.05285.x
161. United States Environmental Protection Agency (1993) EPA Reregistration Eligibility Document (RED) Silver. https://nepis.epa.gov/Exe/ZyPDF.cgi/9101UJL9103.PDF?Dockey=9101UJL9103.PDF
162. van der Laan H, van Halem D, Smeets PW, Soppe AI, Kroesbergen J, Wubbels G, Nederstigt J, Gensburger I, Heijman SG (2014) Bacteria and virus removal effectiveness of ceramic pot filters with different silver applications in a long term experiment. Water Res 51:47–54. https://doi.org/10.1016/j.watres.2013.11.010
163. van Hengel IAJ, Riool M, Fratila-Apachitei LE, Witte-Bouma J, Farrell E, Zadpoor AA, Zaat SAJ, Apachitei I (2017) Selective laser melting porous metallic implants with immobilized silver nanoparticles kill and prevent biofilm formation by methicillin-resistant Staphylococcus aureus. Biomaterials 140:1–15. https://doi.org/10.1016/j.biomaterials.2017.02.030
164. Vasishta R, Chhibber S, Saxena M (1989) Heavy metal resistance in clinical isolates of Pseudomonas aeruginosa. Folia Microbiol 34(5):448–452
165. Weber DJ, Rutala WA (2013) Self-disinfecting surfaces: review of current methodologies and future prospects. Am J Infect Control 41(5 Suppl):S31–35. https://doi.org/10.1016/j.ajic.2012.12.005
166. Wiegand C, Abel M, Ruth P, Hipler UC (2012) Analysis of the adaptation capacity of Staphylococcus aureus to commonly used antiseptics by microplate laser nephelometry. Skin Pharmacol Physiol 25(6):288–297. https://doi.org/10.1159/000341222
167. Wirth SM, Bertuccio AJ, Cao F, Lowry GV, Tilton RD (2016) Inhibition of bacterial surface colonization by immobilized silver nanoparticles depends critically on the planktonic bacterial concentration. J Colloid Interf Sci 467:17–27. https://doi.org/10.1016/j.jcis.2015.12.049
168. Woods EJ, Cochrane CA, Percival SL (2009) Prevalence of silver resistance genes in bacteria isolated from human and horse wounds. Vet Microbiol 138(3–4):325–329. https://doi.org/10.1016/j.vetmic.2009.03.023
169. Wu MY, Suryanarayanan K, van Ooij WJ, Oerther DB (2007) Using microbial genomics to evaluate the effectiveness of silver to prevent biofilm formation. Water Sci Technol 55(8–9):413–419
170. Wu XL, Qiu GZ, Gao J, Ding JN, Kang J, Liu XX (2007) Mutagenic breeding of silver-resistant Acidithiobacillus ferrooxidans and exploration of resistant mechanism. Trans Nonferrous Met Soc China 17(2):412–417
171. Wu Y, Quan X, Si X, Wang X (2016) A small molecule norspermidine in combination with silver ion enhances dispersal and disinfection of multi-species wastewater biofilms. Appl Microbiol Biotechnol 100(12):5619–5629. https://doi.org/10.1007/s00253-016-7394-y
172. Xu Y, Gao C, Li X, He Y, Zhou L, Pang G, Sun S (2013) In vitro antifungal activity of silver nanoparticles against ocular pathogenic filamentous fungi. J Ocular Pharmacol Therapeutics: Official J Assoc Ocular Pharmacol Therap 29(2):270–274. https://doi.org/10.1089/jop.2012.0155

173. Zhang C, Liang Z, Hu Z (2014) Bacterial response to a continuous long-term exposure of silver nanoparticles at sub-ppm silver concentrations in a membrane bioreactor activated sludge system. Water Res 50:350–358. https://doi.org/10.1016/j.watres.2013.10.047
174. Zhang S, Ahearn DG, Mateus C, Crow SA Jr (2006) In vitro effects of Ag^{+} on planktonic and adhered cells of fluconazole-resistant and susceptible strains of Candida albicans, C. glabrata and C. krusei. Biomaterials 27(13):2755–2760. https://doi.org/10.1016/j.biomaterials.2005.12.010
175. Zhang S, Liu L, Pareek V, Becker T, Liang J, Liu S (2014) Effects of broth composition and light condition on antimicrobial susceptibility testing of ionic silver. J Microbiol Meth 105:42–46. https://doi.org/10.1016/j.mimet.2014.07.009
176. Zhao G, Stevens SE Jr (1998) Multiple parameters for the comprehensive evaluation of the susceptibility of Escherichia coli to the silver ion. Biometals: An Int J Role Metal Ions Biol Biochem Med 11(1):27–32

Povidone Iodine

16

16.1 Chemical Characterization

Povidone iodine is a stable chemical complex of polyvinylpyrrolidone (povidone, PVP) and elemental iodine. It contains from 9.0 to 12.0% available iodine, calculated on a dry basis [82]. When iodine is complexed with surfactants, the complexed iodine is used for the manufacturing of the biocidal product (the premix may be either prepared on site or bought from suppliers) [98].

In principle, iodine should be regarded as the active substance as long as an iodophor is not considered as discrete active substances. Iodophors are substances which are capable of taking up iodine and transport it. The carrier does not react with the substance taken up via a stable chemical bond but rather takes it up due to its electrochemical configuration in its scaffold. The chemical properties of the individual substances are essentially maintained, the physical properties, i.e. solubility, can in contrast change. In addition, the iodophor affects the content of reactive iodine in the formulation, thereby preventing negative effects such as irritation, but keeping sufficient free iodine in the formulation to ensure its efficacy. Povidone and surfactants are used in the first place to bring iodine into the formulation in a soluble form [98].

The basic chemical information on iodine and povidone iodine is summarized in Table 16.1.

16.2 Types of Application

Iodine, typically as povidone iodine, is used in biocidal products for hand hygiene (e.g., surgical scrubbing) and in embalming fluids for the short-term preservation and hygienisation of cadavers until burial or cremation. Iodine, typically complexed with a surfactant, is also used in biocidal products for disinfection of milking equipment and bulk milk tanks. Iodine, typically complexed with surfactant or

G. Kampf, *Antiseptic Stewardship*, https://doi.org/10.1007/978-3-319-98785-9_16

Table 16.1 Basic chemical information on iodine and povidone iodine [82, 98]

	Iodine	Povidone iodine
CAS number	7553-56-2	25655-41-8
IUPAC name	Iodine	Polyvinylpyrrolidone iodine
Synonyms	None	None
Molecular formula	I_2	$C_6H_9I_2NO$
Molecular weight (g/mol)	253.81	364.953

povidone, is also used in biocidal products for the disinfection of animals' teats or udder and animal houses [98]. Other types of application are its use as an antiseptic agent for bite, stab, puncture or gunshot wounds where povidone iodine is the first choice [16, 17, 62]. It is also used for mucosal antisepsis, e.g., for oral hygiene, and drinking water disinfection [65, 104].

16.2.1 European Chemicals Agency (European Union)

In 2014, iodine including polyvinylpyrrolidone iodine was approved as an existing active substance for use in biocidal products for product types 1 (human hygiene), 3 (veterinary hygiene), 4 (food and feed area) and 22 (embalming and taxidermist fluids) [9].

16.2.2 Environmental Protection Agency (USA)

Products containing iodine as the active ingredient were initially registered in the USA by the US Department of Agriculture beginning in 1948. Iodine and iodophor complexes were last reregistered in 2006, e.g., for emergency drinking water purification, fresh food sanitization, food contact surface sanitization, hospital surface disinfection, materials preservation, and commercial and industrial water cooling tower systems [104].

16.2.3 Food and Drug Administration (USA)

In 2015, povidone iodine between 5 and 10% was eligible for three types of application in health care: patient preoperative skin preparation, healthcare personnel hand wash and surgical hand scrub [27]. It is classified in category IIISE indicating that available data are insufficient to classify povidone iodine as safe and effective, and further testing is required [27]. The main aspect on safety is human pharmacokinetics [27].

16.2.4 Overall Environmental Impact

The oceans are the most important source of natural iodine in the air, water and soil. Iodine in the oceans enters the air from sea spray or as iodine gases. Once in the air, iodine can combine with water or with particles in the air and can enter the soil and surface water, or land on vegetation when these particles fall to the ground or when it rains. Iodine can remain in soil for a long time because it combines with organic material in the soil. It can also be taken up by plants that grow in the soil. Cows or other animals that eat these plants will take up the iodine in the plants. Iodine that enters surface water can re-enter the air as iodine gases. Iodine can enter the air when coal or fuel oil is burned for energy; however, the amount of iodine that enters the air from these activities is very small compared to the amount that comes from the oceans [103]. The EPA summarized in 2006 that the use of iodine and iodophor complexes makes it unlikely that any appreciable exposure to terrestrial or aquatic organisms would occur [104].

16.3 Spectrum of Antimicrobial Activity

The mode of action of iodine is non-selective and is based on the following mechanisms. Iodine rapidly penetrates into micro-organisms showing a high affinity pattern of adsorption. It combines with protein substances in the bacterial cell; these could be peptidoglycans in the cell walls or enzymes in the cytoplasm. This results in irreversible coagulation of the protein and consequent loss of function. It is also known to act on thiol groups in the cell. If a thiol enzyme is part of a metabolic chain, then metabolic inhibition will result. Iodine reacts with key groups of proteins, in particular the free-sulphur amino acids cysteine and methionine, nucleotides and fatty acids. And it interferes at the level of the respiratory chain of the aerobic micro-organisms by blocking the transport of electrons through electrophilic reactions with the enzymes of the respiratory chain [98].

Especially *C. albicans* exhibited a rapid, dose-dependent "loosening" of the cell wall. Cells remained intact without lysis, rupture or wall breakage. Changes in beta-galactosidase and nucleotide concentrations were measured in *E. coli*. A rapid and dose dependent loss of cellular beta-galactosidase activity was found, with no increase in the supernatant. Loss of cellular nucleotides corresponded with an increase in the supernatant. Electron microscopy and biochemical observations support the conclusion that povidone iodine interacts with cell walls of micro-organisms causing pore formation or generating solid–liquid interfaces at the lipid membrane level which lead to loss of cytosol material, in addition to enzyme denaturation. The chemical mechanism of action is assumed to explain the fact that povidone iodine has so far not generated resistance in micro-organisms [91].

16.3.1 Bactericidal Activity

16.3.1.1 Bacteriostatic Activity (MIC Values)

Gram-positive species seem to be more susceptible to povidone iodine with MIC values of 80–2,344 mg/l in *Enterococcus* spp., 80–4,688 mg/l in *Streptococcus* spp., 400–5,000 mg/l in *S. epidermidis* and 8–10,000 mg/l in *S. aureus*. Among Gram-negative species, the range of MIC values begins at 8 mg/l in *K. pneumoniae*, 40 mg/l in *E. coli*, 250 mg/l in *S. marcescens*, 400 mg/l in *P. aeruginosa* and 2,344 mg/l in *Enterobacter* spp. and can be as high as 10,000 mg/l in all of them (Table 16.2).

16.3.1.2 Bactericidal Activity (Suspension Tests)

The bactericidal activity of povidone iodine is comprehensive at 7.5–10% within 30 s although strains of *E. faecium* and *S. epidermidis* have been described to require $\geq$30 s. At 2%, the bactericidal effect is largely achieved within 5 min although some isolates of *E. faecium*, *E. coli*, *S. aureus* and *S. epidermidis* have been described with <5.0 log in 5 min, indicating that a longer exposure time is necessary. Data obtained with povidone iodine at 0.6% indicate that an exposure time of 2 h should be adequate to achieve a sufficient bactericidal activity (Table 16.3).

Table 16.2 MIC values of various bacterial species to povidone iodine

Species	Strains/isolates	MIC value (mg/l)	References
A. xylosoxidans	2 clinical isolates	190–1,560	[80]
A. anitratus	ATCC 49137	2,344	[54]
B. fragilis	ATCC 25285	4,688	[54]
B. subtilis	ATCC 9372	100[a]	[21]
Bacillus spp.[b]	15 hospital strains	75–250[a]	[21]
C. trachomatis	ATCC VR-885	1,562	[86]
		>800[a]	
C. perfringens	ATCC 13124	1,024	[60]
E. cloacae	ATCC 13047	2,344	[54]
Enterobacter spp.[b]	10 burn unit isolates	10,000	[46]
E. faecalis	ATCC 29212	80	[3]
E. faecalis	ATCC 29212	1,024	[60]
E. faecalis	ATCC 29212, ATCC 51575	2,344	[54]
E. faecium	ATCC 49224	2,344	[54]
Enterococcus spp.	Clinical VRE isolate	1,024	[60]
E. coli	ATCC 25922	40	[3]
E. coli	ATCC 25922	75[a]	[21]
E. coli	ATCC 35218	1,024	[60]

(continued)

Table 16.2 (continued)

Species	Strains/isolates	MIC value (mg/l)	References
E. coli	ATCC 11229, ATCC 25922	2,344	[54]
E. coli	10 burn unit isolates	10,000	[46]
Escherichia spp.	Hospital strain	150[a]	[21]
H. influenzae	ATCC 49247	512	[60]
H. influenzae	ATCC 19418	2,344	[54]
K. oxytoca	ATCC 15764	2,344	[54]
K. pneumoniae	35 carbapenem-resistant clinical isolates	8–32[a]	[42]
K. pneumoniae	ATCC 27736	2,344	[54]
K. pneumoniae	10 burn unit isolates	10,000	[46]
L. lactis	1 strain	30,000	[28]
M. luteus	ATCC 7468	2,344	[54]
Micrococcus spp.[b]	6 hospital strains	75–200[a]	[21]
P. mirabilis	ATCC 4630	2,344	[54]
Proteus spp.[b]	10 burn unit isolates	5,000	[46]
P. aeruginosa	ATCC 27853	400–3,200	[95]
P. aeruginosa	ATCC 15442	1,024	[60]
P. aeruginosa	175 isolates from veterinary sources	2,048–8,192	[10]
P. aeruginosa	ATCC 15442	2,344	[54]
P. aeruginosa	ATCC 27853	4,688	[54]
P. aeruginosa	NCTC6749 and 3 extensively resistant clinical isolates	6,250	[112]
P. aeruginosa	20 burn unit isolates	10,000	[46]
P. cepacia	1 wash basin isolate	3,130	[80]
S. marcescens	18 clinical strains	250[a]	[38]
S. marcescens	ATCC 14756	2,344	[54]
S. marcescens	10 burn unit isolates	10,000	[46]
S. aureus	8 clinical MSSA isolates	7.8	[114]
	12 clinical MRSA isolates	31.3	
S. aureus	ATCC 6538	8–512	[60]
S. aureus	ATCC 25923	40	[3]
S. aureus	EMRSA 15	64[a]	[20]
S. aureus	ATCC 6538	100[a]	[21]
S. aureus	Clinical MRSA isolate	256	[60]
S. aureus	ATCC 25923	800–1,600	[95]
S. aureus	ATCC 6538, ATCC 29213, ATCC 33591, ATCC 33592, ATCC 33594, ATCC 43300	1,172–2,344	[54]
S. aureus	20 burn unit isolates	5,000–10,000	[46]
S. epidermidis	TISTR17	400–3,200	[95]
S. epidermidis	5 clinical isolates	781–1,562	[92]

(continued)

Table 16.2 (continued)

Species	Strains/isolates	MIC value (mg/l)	References
S. epidermidis	ATCC 12288, ATCC 51624, ATCC 51625	1,172–2,344	[54]
S. epidermidis	10 burn unit isolates	5,000	[46]
S. haemolyticus	ATCC 29970	1,172	[54]
S. hominis	ATCC 25615	2,344	[54]
S. lugdunensis	11 clinical strains	250–1,000	[34]
S. saprophyticus	ATCC 15305	2,344	[54]
S. schleiferi	12 clinical strains	500–1,000	[34]
Staphylococcus spp.[b]	3 hospital strains	75–100[a]	[21]
S. mitis	4 clinical isolates	3,124	[92]
S. mutans	MTCC 890	80	[3]
S. mutans	ATCC 27351	150	[28]
S. mutans	1 clinical isolate	3,124	[92]
S. pneumoniae	ATCC 35088	586	[54]
S. pneumoniae	ATCC 49619	>1,024	[60]
S. pyogenes	ATCC 12351	4,688	[54]
S. salivarius	ATCC 25975	150	[28]

[a]Available iodine; [b]no MIC values per species

Table 16.3 Bactericidal activity of povidone iodine in suspension tests

Species	Strains/isolates	Exposure time	Concentration	$\log_{10}$ reduction	References
A. anitratus	ATCC 49137	15 s	7.5% (P)	>5.4	[54]
A. baumannii	20 clinical strains	15 s	10%[a] (P)	>5.0	[111]
A. baumannii	1 multiresistant clinical strain	1 min	8% (P)	>6.2	[88]
A. baumannii	81 clinical and environmental isolates	24 h	4% (P)	>5.0	[66]
A. baumannii	9 ICU outbreak strains	5 min	2.2% (P)	>3.1	[72]
A. baumannii	1 MDR clinical isolate	2 h	0.6% (S)	≥5.0	[4]
			0.3% (S)	<3.0	
A. salmonicida	ATCC 14174	30 min	0.0078%[b] (P)	≥5.0	[105]
			0.0056%[b] (P)	<3.8	
B. fragilis	ATCC 25285	15 s	7.5% (P)	4.8	[54]
		30 s		>5.5	
B. cepacia	1 multiresistant clinical isolate	1 min	1% (P)	>5.0	[101]
C. jejuni	ATCC 33560, ATCC BAA-1062, 2 field strains from broiler flocks	1 min	1%[b] (S)	>6.0	[44]
			0.1%[b] (S)	1.2–4.4	

(continued)

Table 16.3 (continued)

Species	Strains/isolates	Exposure time	Concentration	$\log_{10}$ reduction	References
C. piscicola	ATCC 35586	30 min	0.011%[b] (P)	≥5.0	[105]
			0.0078%[b] (P)	<4.5	
C. diversus	Clinical multiresistant isolate	5 min	10% (P)	>5.0	[84]
			5% (P)		
			2.5% (P)		
E. cloacae	Clinical multiresistant isolate	5 min	10% (P)	>5.0	[84]
			5% (P)		
			2.5% (P)		
E. cloacae	ATCC 13047	15 s	7.5% (P)	>6.0	[54]
E. cloacae	1 clinical strain	2 min	0.5% (S)	1.8	[37]
Enterobacter spp.	Clinical strain	1 min	8% (P)	>6.4	[88]
E. faecalis	13 clinical isolates	3 min	10% (P)	>5.0	[31]
			7.5% (P)		
E. faecalis	1 VRE strain	1 min	8% (P)	1.0	[88]
		5 min		≥5.3	
E. faecalis	ATCC 29212	15 s	7.5% (P)	>6.0	[54]
E. faecalis	ATCC 51575	15 s	7.5% (P)	4.6	[54]
		30 s		>6.0	
E. faecalis	Strain Q33	5 min	2% (S)	2.8	[74]
E. faecalis	ATCC 29212	2 min	0.5% (S)	0.8	[37]
E. faecium	5 strains	30 s	10% (P)	≥5.0	[108]
E. faecium	ATCC 6057	30 s	10% (P)	1.3–5.8[c]	[89]
		1 min		1.4–5.8[c]	
		10 min		≥5.5	
E. faecium	7 clinical isolates	3 min	10% (P)	>5.0	[31]
			7.5% (P)		
E. faecium	ATCC 49224	15 s	7.5% (P)	3.8	[54]
		30 s		5.0	
E. faecium	VRE strain Z31901	5 min	2% (S)	1.0	[74]
E. faecium	ATCC 6057	5 min	0.5% (P)	≥6.0	[78]
E. hirae	ATCC 10541	3 min	4.1% (P)	>5.0	[71]
Enterococcus spp.	Non-typable clinical strain	3 min	10% (P)	>5.0	[31]
			7.5% (P)		
Enterococcus spp.[d]	6 multiresistant clinical isolates	1 min	1% (P)	>5.0	[101]
E. coli	NCTC 10536	30 s	10% (P)	≥6.4	[89]
E. coli	ATCC 25922 and clinical multiresistant isolate	5 min	10% (P)	>5.0	[84]
			5% (P)		
			2.5% (P)		

(continued)

Table 16.3 (continued)

Species	Strains/isolates	Exposure time	Concentration	log_{10} reduction	References
E. coli	1 multiresistant clinical strain	1 min	8% (P)	>6.3	[88]
E. coli	ATCC 11229	15 s	7.5% (P)	>7.7	[54]
E. coli	ATCC 25922	15 s	7.5% (P)	>5.8	[54]
E. coli	NCTC 86	30 s	2% (S)	≥5.0	[70]
E. coli	NCTC 10538	5 min	2% (S)	4.3	[74]
E. coli	ATCC 25922	24 h	1.3% (P)	>5.0	[66]
E. coli	ATCC 11229	30 min	0.7% (P)	≥3.0	[79]
E. coli	1 cefotaxime-resistant clinical isolate	2 h	0.6% (S)	≥5.0	[4]
			0.3% (S)	<3.0	
E. coli	ATCC 25922	2 min	0.5% (S)	2.3	[37]
E. coli	ATCC 11229	5 min	0.5% (P)	≥6.0	[78]
G. vaginalis	1 clinical strain	1 min	8% (P)	>3.9	[88]
H. influenzae	ATCC 33533	15 s	7.5% (P)	>5.7	[54]
H. parasuis	2 strains (serovars 1 and 5)	1 min	1%[b] (S)	4.4–6.0	[94]
			0.1%[b] (S)	0.1–0.7	
K. oxytoca	ATCC 15764	15 s	7.5% (P)	>5.7	[54]
K. pneumoniae	Clinical multiresistant isolate	5 min	10% (P)	>5.0	[84]
			5% (P)		
			2.5% (P)		
K. pneumoniae	1 multiresistant clinical strain	1 min	8% (P)	>6.3	[88]
K. pneumoniae	ATCC 27736	15 s	7.5% (P)	4.0	[54]
		30 s		>5.6	
K. pneumoniae	DSM 16609	15 s	0.7% (P)	>5.5	[30]
			0.23% (P)	>5.4	
			0.07% (P)	>2.8	
K. pneumoniae	1 clinical strain	2 min	0.5% (S)	3.4	[37]
L. garvieae	NCIMB 702927	30 min	0.011%[b] (P)	≥5.0	[105]
			0.0078%[b] (P)	< 4.1	
L. innocua	LCDC 86-417	1 min	0.008%[b] (P)	<1.0	[13]
L. monocytogenes	LCDC 88-702	1 min	0.008%[b] (P)	≤2.0	[13]
M. luteus	ATCC 7468	15 s	7.5% (P)	>4.6	[54]
P. acnes	1 clinical strain	1 min	8% (P)	>4.0	[88]
P. mirabilis	1 clinical strain	1 min	8% (P)	≥6.3	[88]
P. mirabilis	ATCC 4630	15 s	7.5% (P)	>6.1	[54]
P. mirabilis	1 clinical strain	2 min	0.5% (S)	3.4	[37]
P. aeruginosa	ATCC 15442	30 s	10% (P)	≥7.0	[89]

(continued)

Table 16.3 (continued)

Species	Strains/isolates	Exposure time	Concentration	log_{10} reduction	References
P. aeruginosa	ATCC 27853 and clinical multiresistant isolate	5 min	10% (P)	>5.0	[84]
			5% (P)		
			2.5% (P)		
P. aeruginosa	ATCC 15442 and 1 gentamicin-resistant strain	1 min	8% (P)	3.7–6.4	[88]
		5 min		4.0–6.4	
P. aeruginosa	ATCC 15442	15 s	7.5% (P)	>6.0	[54]
P. aeruginosa	ATCC 27853	15 s	7.5% (P)	>5.8	[54]
P. aeruginosa	ATCC 15442	3 min	4.1% (P)	>5.0	[71]
P. aeruginosa	NCTC 9027	30 s	2% (S)	≥ 5.0	[70]
P. aeruginosa	NCIMB 10421	5 min	2% (S)	4.3	[74]
P. aeruginosa	179 clinical isolates	10 min	1% (P)	>5.0	[39]
		20 min	0.1% (P)		
P. aeruginosa	1 clinical isolate	2 h	0.6% (S)	≥ 5.0	[4]
			0.3% (S)	<3.0	
P. aeruginosa	ATCC 27853	2 min	0.5% (S)	2.2	[37]
P. aeruginosa	ATCC 15442	5 min	0.5% (P)	7.0	[78]
P. aeruginosa	ATCC 15442	1 min	0.025% (S)	5.2	[60]
		6 h	0.0125% (S)		
P. putida	Surface water isolate	3 min	0.002% (P)	>7.0	[107]
Salmonella spp.	Clinical strain	1 min	8% (P)	>6.4	[88]
S. marcescens	Clinical multiresistant isolate	5 min	10% (P)	>5.0	[84]
			5% (P)		
			2.5% (P)		
S. marcescens	1 clinical strain	1 min	8% (P)	>6.3	[88]
S. marcescens	ATCC 14756	15 s	7.5% (P)	>6.0	[54]
S. marcescens	1 clinical strain	2 min	0.5% (S)	2.5	[37]
S. aureus	ATCC 6538	30 s	10% (P)	1.1–6.2[c]	[89]
		1 min		1.2–6.2[c]	
		10 min		≥ 5.9	
S. aureus	10 MRSA strains	30 s	10% (P)	≥ 5.0	[108]
S. aureus	30 clinical isolates (16 MRSA, 14 MSSA)	3 min	10% (P)	>5.0	[31]
			7.5% (P)		
S. aureus	ATCC 25923 and clinical MRSA isolate	5 min	10% (P)	>5.0	[84]
			5% (P)		
			2.5% (P)		
S. aureus	ATCC 6538 and EMRSA 15	1 min	8% (P)	1.5–2.1	[88]
		5 min		3.9–4.7	
S. aureus	ATCC 6538	15 s	7.5% (P)	7.3	[54]

(continued)

Table 16.3 (continued)

Species	Strains/isolates	Exposure time	Concentration	$\log_{10}$ reduction	References
S. aureus	ATCC 33594	15 s	7.5% (P)	>6.0	[54]
S. aureus	ATCC 29213	15 s	7.5% (P)	>5.9	[54]
S. aureus	ATCC 33593	15 s	7.5% (P)	>5.7	[54]
S. aureus	ATCC 33592	15 s	7.5% (P)	2.2	[54]
		30 s		>6.2	
S. aureus	ATCC 43300	15 s	7.5% (P)	2.9	[54]
		30 s		>5.8	
S. aureus	ATCC 33591	15 s	7.5% (P)	2.1	[54]
		30 s		4.3	
S. aureus	ATCC 6538	3 min	4.1% (P)	>5.0	[71]
S. aureus	Strain RF3	30 s	2% (S)	4.6	[70]
		1 min		4.3	
		10 min		4.7	
S. aureus	NCTC 6571	5 min	2% (S)	4.1	[74]
S. aureus	MRSA strain 9543	5 min	2% (S)	4.8	[74]
S. aureus	3 multiresistant clinical isolates	1 min	1% (P)	>5.0	[101]
S. aureus	91 clinical MSSA isolates	5 min	1% (P)	>5.0	[39]
		10 min	0.1% (P)		
S. aureus	109 clinical MRSA isolates	5 min	1% (P)	>5.0	[39]
		15 min	0.1% (P)		
S. aureus	ATCC 6538	30 min	0.7% (P)	≥ 3.0	[79]
S. aureus	54 MRSA strains	5 min	0.625% (P)	≥ 5.0	[32]
S. aureus	Strain MN8 (clinical MRSA isolate)	2 h	0.625% (S)	6.9	[5]
			0.31% (S)	0.9	
S. aureus	2 clinical isolates (MSSA and MRSA)	2 h	0.6% (S)	≥ 5.0	[4]
			0.3% (S)	<3.0	
S. aureus	ATCC 25913	2 min	0.5% (S)	2.8	[37]
S. aureus	ATCC 6538	5 min	0.5% (P)	≥ 6.0	[78]
S. aureus	33 clinical isolates	30 s	0.4% (P)	≥ 5.4	[73]
S. aureus	42 clinical MRSA isolates	5 min	0.1% (S)	≥ 5.0	[81]
S. aureus	ATCC 25923	5 min	0.0256% (S)	≥ 5.0	[69]
S. aureus	ATCC 6538	1 min	0.025% (S)	5.1	[60]
		10 min	0.0125% (S)		
S. aureus	IFO 13276	30 s	0.005% (S)	≥ 5.0	[113]
S. chromogenes	4 bovine mastitis isolates	30 s	1% (P)	≥ 5.0	[102]
			0.4% (P)		
S. epidermidis	Strain RP62A	30 s	10% (P)	6.3–6.5	[2]
S. epidermidis	ATCC 12228	15 s	7.5% (P)	>5.4	[54]
S. epidermidis	ATCC 51625	15 s	7.5% (P)	>5.3	[54]
S. epidermidis	ATCC 51624	15 s	7.5% (P)	2.9	[54]
		30 s			

(continued)

Table 16.3 (continued)

Species	Strains/isolates	Exposure time	Concentration	log_{10} reduction	References
S. epidermidis	Strain P69	5 min	2% (S)	2.7	[74]
S. epidermidis	Bovine mastitis isolate	30 s	1% (P)	≥ 5.0	[102]
			0.4% (P)		
S. epidermidis	1 MRSE clinical isolate	2 h	0.6% (S)	≥ 5.0	[4]
			0.3% (S)	<3.0	
S. epidermidis	1 clinical strain	2 min	0.5% (S)	3.9	[37]
S. haemolyticus	ATCC 29970	15 s	7.5% (P)	2.2	[54]
		30 s		>5.1	
S. haemolyticus	Bovine mastitis isolate	30 s	1% (P)	≥ 5.0	[102]
			0.4% (P)		
S. hominis	ATCC 25615	15 s	7.5% (P)	1.8	[54]
		30 s		>5.2	
S. saprophyticus	ATCC 15305	15 s	7.5% (P)	2.3	[54]
		30 s		>5.6	
S. simulans	3 bovine mastitis isolates	30 s	1% (P)	≥ 5.0	[102]
			0.4% (P)		
S. xylosus	Bovine mastitis isolate	30 s	1% (P)	≥ 5.0	[102]
			0.4% (P)		
S. agalactiae	5 isolates from fish aquaculture outbreaks	1 min	1% (P)	0.0–2.8[c]	[75]
		10 min		≥ 5.0	
S. pyogenes	1 group A clinical strain	1 min	8% (P)	>6.2	[88]
S. pyogenes	ATCC 12351	15 s	7.5% (P)	>4.5	[54]
S. pneumoniae	ATCC 35088	15 s	7.5% (P)	3.1	[54]
		30 s		>4.4	
S. pneumoniae	ATCC 49619	15 s	0.7% (P)	>5.2	[30]
			0.23% (P)	>5.2	
			0.07% (P)	4.9	
V. cholerae	NCTC 10225	1 min	8% (P)	>6.3	[88]
V. indigofera	Surface water isolate	3 min	0.002% (P)	>6.0	[107]
Y. ruckeri	ATCC 29473	30 min	0.0078%[b] (P)	≥ 5.0	[105]
			0.0056%[b] (P)	< 4.6	
Mixed anaerobic species	*A. actinomycetemcomitans* ATCC 43718, *A. viscosus* DSMZ 43798, *F. nucleatum* ATCC 10953, *P. gingivalis* ATCC 33277, *V. atypica* ATCC 17744 and *S. gordonii* ATCC 33399	30 s	10% (P)	>8.0	[26]

S solution; *P* commercial product; [a]1.1% iodine; [b]iodine; [c]depending on the type of organic load; [d]no MIC values per species

The bactericidal activity of povidone iodine has been described to be distinctly lower at elevated pH values [109]. It is significantly reduced in the presence of 0.019% albumin [59]. Wound dressings can also reduce the efficacy of povidone iodine against *S. aureus* as shown in 33.3% of 42 types of wound dressings [49]. It is not impaired in the presence of chondroitin sulphate [78].

16.3.1.3 Activity Against Bacteria in Biofilm

Povidone iodine at 7.5–10% has overall a good bactericidal activity against bacteria in biofilms, e.g., within 1 min (*S. aureus*), 15 min (*P. aeruginosa*) or 4 h (mixed species biofilm). Data on the efficacy of lower povidone iodine concentrations are rather incomplete so that general statements are not justified (Table 16.4). In a non-typable *H. influenzae* biofilm, it was shown that resistance to povidone iodine is mediated to a large part by the cohesive and protective properties of the biofilm matrix [53].

16.3.1.4 Bactericidal Activity in Surgical Scrubbing

Povidone iodine, e.g., at 7.5%, has traditionally been used in surgical hand scrubs [33]. Its bactericidal efficacy on the resident hand flora has often been described to be inferior to the use of alcohol-based hand rubs both in clinical practice and according to international test methods such as EN 12791 [7, 8, 19, 35, 43, 48, 58, 64, 71]. On hands artificially contaminated with MRSA use of a soap based on 10% povidone iodine resulted in a 3.8 log reduction, similar to the effect of 70% ethanol [41].

16.3.1.5 Bactericidal Activity in Carrier Tests

In a proposed test to determine the efficacy of wound antiseptics (which is similar to a carrier test), 10% povidone iodine showed sufficient bactericidal activity within 30 min with or without organic load [96]. Another study revealed that 10% povidone iodine has a good bactericidal activity within 1.5 min on stainless steel discs against 10 MSSA strains (3.8 log), 10 MRSA strains (3.5 log), 10 VSE strains (3.5 log) and 9 VRE strains (3.1 log) [18].

Povidone iodine with 1% available iodine reduced *L. innocua* and *L. monocytogenes* in 1 min by at least 6.0 log, whereas a formulation with 0.008% available iodine had only little effect (<1.0 log) [13]. Another povidone iodine solution with 1% available iodine was also very effective without organic load (>5.0 log). The presence of serum, however, impaired its efficacy (2.1–2.3 log) [94].

Against three bacterial species (*S. aureus* strain RF3, *E. coli* NCTC 86 and *P. aeruginosa* NCTC 9027), 2% povidone iodine revealed a good bactericidal activity within 30 s on glass carriers with log reductions of at least 5.0 [70]. Two per cent povidone iodine was only partially effective against seven strains from six bacterial species (*E. faecalis*, *E. faecium* VRE, *E. coli*, *P. aeruginosa*, *S. aureus*, MRSA, *S. epidermidis*) on glass carriers with 1.0 to 1.8 log in 1 min [74].

Against *H. parasuis* serovar 1 and 5 a 1% iodophor solution with 0.1% available iodine reduced the bacterial cell number by 2.7–2.9 log without organic load and by 1.0–6.9 with serum as organic load [94].

Table 16.4 Efficacy of povidone iodine against bacteria in biofilms

Species	Strains/isolates	Type of biofilm	Exposure time	Concentration	log_{10} reduction	References
P. gingivalis	Dual species biofilm with *P. gingivalis* W 83 and *S. gordonii*	7-d incubation in a modified Robbins device	15 min	1% (S)	1.0	[11]
P. aeruginosa	ATCC 25619	48-h incubation on polycarbonate coupons	15 min	10% (P)	7.0	[55]
P. aeruginosa	NCIMB 10434	48-h incubation in biofilm reactor	4 h	10% (P)	>6.0	[50]
			24 h			
			4 h	3% (P)	>6.0	
			24 h			
P. aeruginosa	Leg ulcer isolate	24-h incubation in polystyrene 96-well plates	1 min	7.5% (P)	4.3	[56]
			15 min		5.5	
			30 min		>8.0	
P. aeruginosa	CIP 103.467	24- or 48-h incubation on glass slides	24 h	2.5% (P)	$\geq$5.0	[68]
P. aeruginosa	ATCC 15442	24-h incubation on agar disc	24 h	0.75% (S)	0.7	[57]
P. putida	Surface water isolate	24-h incubation on silicone discs	3 min	0.004% (P)	4.0	[107]
				0.002% (P)	3.3–3.5	
S. aureus	ATCC 25923	48-h incubation on polycarbonate coupons	15 min	10% (P)	6.0	[55]
S. aureus	AH 2547	Overnight incubation on porcine skin	4 lateral wipes with soaked pads	10% (S)	0.4	[106]

(continued)

Table 16.4 (continued)

Species	Strains/isolates	Type of biofilm	Exposure time	Concentration	log_{10} reduction	References
S. aureus	Leg ulcer isolate	24-h incubation in polystyrene 96-well plates	1 min	7.5% (P)	>6.0	[56]
			15 min			
			30 min			
S. aureus	CIP 4.83	24- or 48-h incubation on glass slides	24 h	2.5% (P)	>5.0	[68]
S. aureus	ATCC 6538	24-h incubation on agar disc	24 h	0.75% (S)	0.7–0.8	[57]
S. chromogenes	4 bovine mastitis isolates	24-h incubation on pegs	0.5–5 min	1%[a] (P) 0.4%[a] (P)	≥ 5.0	[102]
S. epidermidis	Strain RP62A	24-h incubation in microtiter plates	30 s	10% (P)	4.4–5.9	[2]
S. epidermidis	Bovine mastitis isolate	24-h incubation on pegs	1–2 min	1%[a] (P) 0.4%[a] (P)	≥ 5.0	[102]
S. haemolyticus	Bovine mastitis isolate	24-h incubation on pegs	≤ 0.5–1 min	1%[a] (P) 0.4%[a] (P)	≥ 5.0	[102]
S. simulans	3 bovine mastitis isolates	24-h incubation on pegs	≤ 0.5–2 min	1%[a] (P) 0.4%[a] (P)	≥ 5.0	[102]
S. xylosus	Bovine mastitis isolate	24-h incubation on pegs	≤ 0.5 min	1%[a] (P) 0.4%[a] (P)	≥ 5.0	[102]
V. indigofera	Surface water isolate	24-h incubation on silicone discs	3 min	0.004% (P)	4.1	[107]
				0.002% (P)	1.8–2.6	
Mixed species	*S. aureus* strain 308 (MRSA), *C. albicans* ATCC MYA 2876	48-h incubation in biofilm reactor	4 h	10% (P)	>5.0	[50]
			24 h			
			4 h	3% (P)	>5.0	

(continued)

Table 16.4 (continued)

Species	Strains/isolates	Type of biofilm	Exposure time	Concentration	$\log_{10}$ reduction	References
			24 h			
Mixed species	*C. diversus* R25.1, *P. aeruginosa* R1811, *E. faecalis* R812 (all urinary catheter isolates)	48-h incubation on silicone discs	30 min	1% (S)	<1.0	[97]
			60 min		≤ 1.0	
			120 min		0.4–1.2	
Mixed species	*S. aureus* strain D76 (MSSA), *M. luteus* strain B81, *S. oralis* strain B52 and *P. aeruginosa* strain D40 (all wound isolates)	24-h incubation in a constant-depth film fermenter	Up to 8 d	1% (S)	0.0–1.4	[47]
Mixed species	*V. indigofera* and *P. putida* (both surface water isolates)	24-h incubation on silicone discs	3 min	0.004% (P)	2.8–3.0	[107]
				0.002% (P)	2.5–3.0	
Mixed species	20 different species (all surface water isolates)	24-h incubation on silicone discs	3 min	0.004% (P)	3.0–4-0	[107]
				0.002% (P)	2.5–3.5	

P commercial product; *S* solution; [a]iodine

16.3.1.6 Bactericidal Activity in Other Applications

Povidone iodine is also used for mucosal antisepsis, e.g., at 10% prior to vaginal surgery. Its bactericidal efficacy was described to be lower compared to 4% CHG [25]. On porcine vaginal mucosa, povidone iodine at 7.5% was able to reduce an artificial contamination of MRSA by approximately 4.0 log (15 min), but the effect diminished with time resulting in no MRSA reduction after 24 h [5]. The application of three drops of 1.25% povidone iodine showed a moderate bactericidal efficacy when applied preoperatively in ophthalmic surgery [45].

16.3.2 Fungicidal Activity

16.3.2.1 Fungistatic Activity (MIC-Values)

The susceptibility of most yeasts to povidone iodine is variable but in a similar range for different species, e.g., for *C. albicans* (12.5–5,000 mg/l), *C. glabrata* (10–5,000 mg/l), *C. parapsilosis* (300–5,000 mg/l) and *C. tropicalis* (312–5,000 mg/l). The MIC values of other fungal species are within the same range (Table 16.5).

16.3.2.2 Fungicidal Activity (Suspension Tests)

Povidone iodine at 7.5–10% has yeasticidal activity, mostly within 2 min. *Malassezia* spp. and *Rhodotorula* spp. are sufficiently killed by 0.5% povidone iodine in 1 min. *A. fumigatus* required 1% available iodine in 5 min to achieve sufficient fungicidal activity (Table 16.6).

16.3.3 Mycobactericidal Activity

16.3.3.1 Mycobactericidal Activity (Suspension Tests)

Povidone iodine at 1% has mostly shown sufficient mycobactericidal activity within 1 min (Table 16.7).

16.3.3.2 Mycobactericidal Activity (Carrier Tests)

On carriers povidone iodine with 1% available iodine reduced *M. tuberculosis* strain H37Rv in 1 min by at least 2.9 log, whereas a formulation with 0.008% available iodine had only little effect (0.5 log) [15]. Similar results were described for the same two formulations with *M. smegmatis* strain TMC 1515 [14]. Against *M. abscessus* ATCC 19977 and *M. bolletii* BCRC 16915, a similar and strong mycobactericidal activity was found with 10% povidone iodine within 2 min (2.0–5.5 log) but not against an outbreak strain of *M. massiliense* (0.9–1.2 log) [22].

Table 16.5 MIC values of various fungal species to povidone iodine

Species	Strains/isolates	MIC value (mg/l)	References
C. albicans	ATCC 90028 and 10 clinical isolates	12.5–25	[36]
C. albicans	ATCC 90028	40	[3]
C. albicans	ATCC 10231	256	[60]
C. albicans	ATCC A 18804	312	[29]
C. albicans	33 clinical isolates	600–5,000	[61]
C. albicans	83 oral cavity isolates from HIV patients	700–2,500	[100]
C. albicans	ATCC 24433	1,200–2,500	[61]
C. albicans	ATCC 10231	2,344	[54]
C. ciferrii	Clinical isolate	1,200–5,000	[61]
C. dubliniensis	1 oral cavity isolate from a HIV patient	1,200	[100]
C. famata	Clinical isolate	5,000	[61]
C. glabrata	2 oral cavity isolates from HIV patients	10–100	[100]
C. glabrata	ATCC G 2001	625	[29]
C. glabrata	11 clinical isolates	1,200–5,000	[61]
C. guilliermondii	2 clinical isolates	5,000	[61]
C. intermedia	Clinical isolate	5,000	[61]
C. krusei	ATCC 6258	25	[36]
C. krusei	ATCC 6258 and 1 clinical isolate	1,200–5,000	[61]
C. lusitaniae	3 clinical isolates	1,200–2,500	[61]
C. melibiosica	Clinical isolate	5,000	[61]
C. norvegensis	Clinical isolate	5,000	[61]
C. parapsilosis	4 oral cavity isolates from HIV patients	300–1,200	[100]
C. parapsilosis	ATCC 20019 and ATCC 90018	600–2,500	[61]
C. parapsilosis	30 clinical isolates	600–5,000	[61]
C. pelliculosa	Clinical isolate	2,500	[61]
C. sake	Clinical isolate	5,000	[61]
C. tropicalis	ATCC T 750	312	[29]
C. tropicalis	1 oral cavity isolate from a HIV patient	1,200	[100]
C. tropicalis	ATCC 750	2,344	[54]
C. tropicalis	7 clinical isolates	5,000	[61]
C. utilis	Clinical isolate	5,000	[61]
Candida spp.[a]	11 cattle otitis strains	156–625	[29]
Cryptococcus spp.	Clinical isolate	2,500	[61]
M. furfur	15 cattle otitis strains	78–1,250	[29]

(continued)

Table 16.5 (continued)

Species	Strains/isolates	MIC value (mg/l)	References
M. furfur	CBS 1878	312	[29]
M. pachydermatis	CBS 1879	625	[29]
M. slooffiae	12 cattle otitis strains	39–1,250	[29]
M. sympodialis	12 cattle otitis strains	39–625	[29]
P. ohmeri	Clinical isolate	5,000	[61]
R. mucilaginosa	12 cattle otitis strains	39	[29]
Rhodotorula spp.	Clinical isolate	2,500	[61]
S. cerevisiae	Clinical isolate	1,200	[61]
Trichosporon spp.	3 clinical isolates	5,000	[61]

[a]No MIC values per species

Table 16.6 Fungicidal activity of povidone iodine in suspension tests

Species	Strains/isolates	Exposure time	Concentration	$\log_{10}$ reduction	References
A. fumigatus	15 clinical isolates	5 min	1%[a] (P)	>4.0	[99]
C. albicans	ATCC 10231	30 s	10% (P)	3.8–6.2[b]	[89]
		1 min		4.2–6.2[b]	
		10 min		≥ 5.8	
C. albicans	ATCC 10231	2 min	10% (P)	2.6–3.2	[76]
C. albicans	Clinical isolate	2 min	10% (P)	>4.5	[76]
C. albicans	1 clinical strain	1 min	8% (P)	>6.1	[88]
C. albicans	ATCC 10231	15 s	7.5% (P)	2.0	[54]
		30 s		5.0	
C. albicans	ATCC A 18804	1 min	0.5% (S)	8.6	[29]
C. albicans	ATCC 10231	5 min	0.5% (P)	5.0	[78]
C. albicans	IFO 1594	30 s	0.05% (S)	≥ 5.0	[113]
C. albicans	ATCC 10231	1 min	0.05% (S)	4.2	[60]
		5 min	0.025% (S)		
C. auris	4 clinical strains	2 min	10% (P)	>4.7	[76]
C. auris	12 clinical isolates	3 min	0.07–1.25% (P)	>5.0	[1]
C. glabrata	ATCC G 2001	1 min	0.5% (S)	8.5	[29]
C. tropicalis	ATCC 750	15 s	7.5% (P)	1.4	[54]
		30 s		3.4	
C. tropicalis	ATCC T 750	1 min	0.5% (S)	8.2	[29]

(continued)

Table 16.6 (continued)

Species	Strains/isolates	Exposure time	Concentration	log_{10} reduction	References
Candida spp.	11 cattle otitis strains	1 min	0.5% (S)	7.8–8.7	[29]
M. furfur	CBS 1878 and 15 cattle otitis strains	1 min	0.5% (S)	6.9–7.7	[29]
M. pachydermatis	CBS 1879	1 min	0.5% (S)	7.1	[29]
M. slooffiae	12 cattle otitis strains	1 min	0.5% (S)	6.8–7.6	[29]
M. sympodialis	12 cattle otitis strains	1 min	0.5% (S)	6.9–7.5	[29]
R. mucilaginosa	12 cattle otitis strains	1 min	0.5% (S)	6.9–7.7	[29]

S solution; *P* commercial product; [a]available iodine; [b]depending on the type of organic load

Table 16.7 Mycobactericidal activity of povidone iodine in suspension tests

Species	Strains/isolates	Exposure time	Concentration	log_{10} reduction	References
M. abscessus	ATCC 19977, BCRC 16915, outbreak strain TPE 101	30 s	0.4% (P)	4.1–5.4	[22]
			0.2% (P)	3.6–5.4	
			0.1% (P)	3.4–5.4	
			0.05% (P)	3.6–5.4	
M. smegmatis	TMC 1515	1 min	1%[a] (P)	>6.0	[14]
			0.008%[a] (P)	≤2.0	
M. tuberculosis	17 drug-resistant clinical isolates	30 s	0.2% (S)	>3.0	[93]
		1 min		>4.0	
M. tuberculosis	Strain H37Rv	1 min	1%[a] (P)	>5.0	[15]
			0.008%[a] (P)	≤1.8	

S solution; *P* commercial product; [a]available iodine

16.4 Effect of Low-Level Exposure

Low-level povidone iodine exposure did not increase the MIC values in four species (*E. coli*, *K. aerogenes*, *S. marcescens* and *S. aureus*). Only in an isolate of *P. aeruginosa* a weak adaptive change was observed (≤4-fold MIC increase). One study suggests that biofilm formation can be reduced in *S. aureus* and *S. epidermidis* during low-level exposure. The growth rate of *P. aeruginosa* could be enhanced but not of *S. aureus*. No cross-tolerance to other biocidal agents or antibiotics has so far been described after low-level exposure (Table 16.8).

Already in 1978, it was shown that resistance to povidone iodine was not encountered in species of *Proteus*, *Serratia* and *Pseudomonas* after up to eight transfers [90]. This is in line with another finding. The catheter exit sites of patients with continuous ambulatory peritoneal dialysis were sampled over at least

Table 16.8 Change of bacterial susceptibility to biocides and antimicrobials after low-level exposure to povidone iodine

Species	Strains/isolates	Exposure time	Increase in MIC	MIC_{max} (mg/l)	Stability of MIC change	Associated changes	References
E. coli	2 clinical isolates (strains 0111/B4/H2 and 0141/K85/H4)	20 subcultures at various concentrations	None	4.9	Not applicable	None described	[51]
K. aerogenes	2 clinical isolates	20 subcultures at various concentrations	None	2.4	Not applicable	None described	[51]
P. aeruginosa	NCTC 5525 and 1 environmental strain	20 subcultures at various concentrations	2-fold (1 strain)	39	No data	None described	[51]
P. aeruginosa	CIP A22	9 h at subinhibitory concentrations	None	No data	Not applicable	Increase of growth rate	[77]
S. marcescens	1 clinical isolate	20 subcultures at various concentrations	None	1.2	Not applicable	None described	[51]
S. aureus	ATCC 6538	100 d at various concentrations	None	1,000	Not applicable	None described	[110]
S. aureus	RN 4220	Overnight incubation supplemented with sublethal povidone iodine concentrations	No data	14,000	Not applicable	Significant inhibition of biofilm formation; decreased icaA transcription with unchanged expression of icaR	[83]
S. aureus	ATCC 9144	9 h at subinhibitory concentrations	None	No data	Not applicable	No increase of growth rate	[77]
S. epidermidis	Strain 1457	Overnight incubation supplemented with sublethal povidone iodine concentrations	No data	14,000	Not applicable	Significant inhibition of biofilm formation; increased icaR expression and decreased transcription of the icaADBC biofilm locus	[83]

6 months. Twenty-three CNS isolates were sampled from patients using povidone iodine as a disinfectant. No development of resistance was found [67].

P. cepacia cells taken directly from contaminated povidone iodine, however, survived for significantly longer periods of time. Large numbers of *P. cepacia* were found embedded in extracellular material and among strands of glycocalyx between cells as shown by scanning electron microscopy [6].

16.5 Resistance to Povidone Iodine

One clinical report claims povidone iodine resistance. From a pediatric burn unit, a total of 34 wound infections caused by *P. aeruginosa* were described. Fifty-three isolates were assessed for susceptibility to povidone iodine with 49 of them described as resistant (92.5%) [23]. Without a cut-off value, however, it is impossible to assess if the susceptibility was indeed significantly lower compared to other *P. aeruginosa* strains. During another outbreak of infections caused by *P. aeruginosa,* resistance to povidone iodine was suspected. Fifteen episodes of infection due to *P. aeruginosa,* including peritonitis and catheter site infections, occurred in nine patients receiving continuous ambulatory peritoneal dialysis over a 27-month period. Eight episodes were associated with catheter loss. Occurrence of *P. aeruginosa* infection was significantly associated with use of povidone iodine solution to cleanse the catheter site. There was no association with use of povidone iodine solution to disinfect tubing connections, use of other skin care products or exposure to other environmental sources of *P. aeruginosa.* Cultures of available povidone iodine products were negative. Local irritation and alteration in skin flora caused by antiseptic solution or low-level contamination of povidone iodine solution were considered to be potential mechanisms of infection [40].

16.5.1 High MIC Values

The highest MIC values were found in *L. lactis* (30,000 mg/l), in *S. aureus* and *S. epidermidis* (14,000 mg/l), as well as in *Enterobacter* spp., *E. coli*, *K. pneumoniae*, *P. aeruginosa* and *S. marcescens* (all 10,000 mg/l; Tables 16.2 and 16.8). Without epidemiological cut-off values it is currently not possible to further classify the high MIC values.

16.5.2 Reduced Efficacy in Suspension Tests

Few studies indicate a tolerance to 2, 7.5 or 10% povidone iodine by an insufficient bactericidal activity in suspension tests, especially in *E. faecium*, *E. coli*, *S. aureus* or *S. epidermidis* (Table 16.3). It is more likely to be an intrinsic tolerance because the lower efficacy against *E. faecium* and *S. epidermidis* was found at different concentrations of povidone iodine.

16.5.3 Infections Associated with Contaminated Povidone Iodine Solutions or Products

Four cases of peritonitis have been reported in chronic peritoneal dialysis patients caused by *P. aeruginosa* which was detected in one open and two closed bottles of povidone iodine solution of unknown strength [87]. *P. cepacia* has also been described to be a possible contaminant of a 10% povidone iodine solution which has resulted in at least 52 cases of pseudobacteraemia when applied before taking blood cultures [12, 24].

16.5.4 Contaminated Povidone Iodine Solutions Without Evidence for Infections

At very low levels of povidone iodine *Pseudomonas* spp. may indeed persist. *P. cepacia* survived in a iodophor antiseptic up to 68 weeks from the date of manufacture. A uniform concentration of 1% available iodine was found in all lots of povidone iodine tested as specified on the product label, but free iodine values varied greatly. Low free iodine levels of 0.23–0.46 mg/l were associated with the contaminated lot of povidone iodine [6].

16.5.5 Resistance Mechanisms

Taking into account the mode of action of iodine which is non-selective, development of resistance against iodine is unlikely. Iodine and iodophors have been used for over 170 years as disinfectants for a variety of applications. Such applications include disinfection of skin in the human hygiene and medical area but also skin of animals using teat dips as well as surfaces such as milk tanks [98]. So far no resistance genes, efflux pumps or plasmids were described explaining a reduced susceptibility to povidone iodine.

16.6 Cross-Tolerance to Other Biocidal Agents

Cross-tolerance of povidone iodine to other biocidal agents such as chlorhexidine or alkyldiaminoethylglycine hydrochloride has so far not been described [63, 67].

16.7 Cross-Tolerance to Antibiotics

One study addressed the antibiotic susceptibility of bacterial isolates of conjunctival cultures. Ocular surface preparation for intravitreal injection using povidone iodine 5% alone in the absence of postinjection topical antibiotics did not appear to promote bacterial resistance [52]. No other studies on cross-tolerance to antibiotics have so far been published.

16.8 Role of Biofilm

16.8.1 Effect on Biofilm Development

Only few data are available suggesting that povidone iodine has mostly a poor inhibitory effect on biofilm formation (1–38%) which is dependent on the species (Table 16.9).

16.8.2 Effect on Biofilm Removal

Only few studies are available suggesting that biofilm removal by povidone iodine is overall good and depends on the exposure time and its concentrations (Table 16.10).

16.8.3 Effect on Biofilm Fixation

No studies were found to describe a possible biofilm fixation by povidone iodine.

Table 16.9 Effect of povidone iodine on biofilm development

Species	Strains/isolates	Type of biofilm	Exposure time	Type of product	Inhibition of biofilm formation	References
C. albicans	ATCC 90028	24-h incubation in microtiter plates	4 h	0.2% (P)	38%	[3]
E. faecalis	ATCC 29212	24-h incubation in microtiter plates	4 h	0.2% (P)	22%	[3]
E. coli	ATCC 25922	24-h incubation in microtiter plates	4 h	0.2% (P)	1%	[3]
S. aureus	ATCC 25923	24-h incubation in microtiter plates	4 h	0.2% (P)	29%	[3]
S. mutans	MTCC 890	24-h incubation in microtiter plates	4 h	0.2% (P)	6%	[3]

P commercial product

Table 16.10 Biofilm removal rate (quantitative determination of biofilm matrix) by exposure to products or solutions based on povidone iodine

Type of biofilm	Concentration	Exposure time	Biofilm removal rate	References
P. aeruginosa (14 clinical isolates and ATCC 15445), 24-h incubation in polystyrene microtitre plates	7.5% (P)	1 min	0 of 15[a]	[56]
		15 min	5 of 15[a]	
		30 min	10 of 15[a]	
P. aeruginosa (8 dairy isolates exhibiting high biofilm formation, 24-h incubation in microtiter plates)	0.015–0.0375% (S)	5 min	"eradication"	[85]
	0.01–0.0325% (S)	15 min		
	0.0075–0.0175% (S)	30 min		
	0.0025–0.01% (S)	60 min		
S. aureus (14 clinical isolates and ATCC 5638), 24-h incubation in polystyrene microtitre plates	7.5% (P)	1 min	15 of 15[a]	[56]
		15 min	15 of 15[a]	
		30 min	15 of 15[a]	

S solution; *P* commercial product; [a]biofilm eradication rate

16.9 Summary

The principal antimicrobial activity of povidone iodine is summarized in Table 16.11.

The key findings on acquired resistance and cross-resistance including the role of biofilm for selecting resistant isolates are summarized in Table 16.12.

Table 16.11 Overview on the typical exposure times required for povidone iodine to achieve sufficient biocidal activity against the different target micro-organisms

Target micro-organisms	Species	Concentration	Exposure time
Bacteria	Most bacterial species except selected isolates of *E. faecium* and *E. epidermidis* (7.5–10%) and *E. faecium, E. coli, S. aureus* and *S. epidermidis* (2%)	7.5–10%[a]	30 s
		2%[a]	5 min
		0.6%[a]	2 h
Fungi	*Malassezia* spp. and *Rhodotorula* spp.	0.5%	1 min
	Candida spp.	7.5–10%	2 min
	A. fumigatus	1%[b]	5 min
Mycobacteria	*M. tuberculosis, M. smegmatis*	1%[b]	1 min

[a]In biofilm, the efficacy will be lower; [b]available iodine

Table 16.12 Key findings on acquired povidone iodine resistance, the effect of low-level exposure, cross-tolerance to other biocides and antibiotics, and its effect on biofilm

Parameter	Species	Findings
Elevated MIC values	*L. lactis*	$\leq$ 30,000 mg/l
	S. aureus, S. epidermidis	$\leq$ 14,000 mg/l
	Enterobacter spp., *E. coli, K. pneumoniae, P. aeruginosa, S. marcescens*	$\leq$ 10,000 mg/l
Proposed MIC value to determine resistance	None proposed yet for bacteria, fungi or mycobacteria	
Cross-tolerance biocides	None	
Cross-tolerance antibiotics	None	
Resistance mechanisms	Unknown	
	P. aeruginosa, P. cepacia	Contaminated solutions or products based on povidone iodine partly associated with infections (e.g., peritonitis) or pseudo-outbreaks
Effect of low-level exposure	*E. coli, K. aerogenes, S. marcescens, S. aureus*	No MIC increase
	P. aeruginosa	Weak MIC increase ($\leq$ 4-fold)
	None	Strong MIC increase (>4-fold)
	P. aeruginosa (2-fold)	Strongest MIC change after low-level exposure
	S. aureus, S. epidermidis (14,000 mg/l)	Highest MIC values after low-level exposure
	S. aureus, S. epidermidis	Inhibition of biofilm formation
	P. aeruginosa	Increase of growth rate
	S. aureus	No increase of growth rate
Biofilm	Development	Inhibition of biofilm formation in *C. albicans, E. faecalis* and *S. aureus*
	Removal	Mostly good removal of *S. aureus* or *P. aeruginosa* biofilm
	Fixation	Unknown

References

1. Abdolrasouli A, Armstrong-James D, Ryan L, Schelenz S (2017) In vitro efficacy of disinfectants utilised for skin decolonisation and environmental decontamination during a hospital outbreak with Candida auris. Mycoses 60(11):758–763. https://doi.org/10.1111/myc.12699
2. Adams D, Quayum M, Worthington T, Lambert P, Elliott T (2005) Evaluation of a 2% chlorhexidine gluconate in 70% isopropyl alcohol skin disinfectant. J Hosp Infect 61 (4):287–290
3. Anand G, Ravinanthan M, Basaviah R, Shetty AV (2015) In vitro antimicrobial and cytotoxic effects of Anacardium occidentale and Mangifera indica in oral care. J Pharmacy Bioallied Sci 7(1):69–74. https://doi.org/10.4103/0975-7406.148780
4. Anderson MJ, Horn ME, Lin YC, Parks PJ, Peterson ML (2010) Efficacy of concurrent application of chlorhexidine gluconate and povidone iodine against six nosocomial pathogens. Am J Infect Control 38(10):826–831. https://doi.org/10.1016/j.ajic.2010.06.022
5. Anderson MJ, Scholz MT, Parks PJ, Peterson ML (2013) Ex vivo porcine vaginal mucosal model of infection for determining effectiveness and toxicity of antiseptics. J Appl Microbiol 115(3):679–688. https://doi.org/10.1111/jam.12277
6. Anderson RL, Vess RW, Carr JH, Bond WW, Panlilio AL, Favero MS (1991) Investigations of intrinsic Pseudomonas cepacia contamination in commercially manufactured povidone-iodine. Infect Control Hosp Epidemiol 12(5):297–302
7. Babb JR, Davies JG, Ayliffe GAJ (1991) A test procedure for evaluating surgical hand disinfection. J Hosp Infect 18(suppl. B):41–49
8. Barbadoro P, Martini E, Savini S, Marigliano A, Ponzio E, Prospero E, D'Errico MM (2014) In vivo comparative efficacy of three surgical hand preparation agents in reducing bacterial count. J Hosp Infect 86(1):64–67. https://doi.org/10.1016/j.jhin.2013.09.013
9. Barroso JM (2014) COMMISSION IMPLEMENTING REGULATION (EU) No 94/2014 of 31 January 2014 approving iodine, including polyvinylpyrrolidone iodine, as an existing active substance for use in biocidal products for product-types 1, 3, 4 and 22. Off J Eur Union 57(L 32):23–26
10. Beier RC, Foley SL, Davidson MK, White DG, McDermott PF, Bodeis-Jones S, Zhao S, Andrews K, Crippen TL, Sheffield CL, Poole TL, Anderson RC, Nisbet DJ (2015) Characterization of antibiotic and disinfectant susceptibility profiles among Pseudomonas aeruginosa veterinary isolates recovered during 1994-2003. J Appl Microbiol 118(2):326–342. https://doi.org/10.1111/jam.12707
11. Bercy P, Lasserre J (2007) Susceptibility to various oral antiseptics of Porphyromonas gingivalis W83 within a biofilm. Adv Ther 24(6):1181–1191
12. Berkelman RL, Lewin S, Allen JR, Anderson RL, Budnick LD, Shapiro S, Friedman SM, Nicholas P, Holzman RS, Haley RW (1981) Pseudobacteremia attributed to contamination of povidone-iodine with Pseudomonas cepacia. Ann Intern Med 95(1):32–36
13. Best M, Kennedy ME, Coates F (1990) Efficacy of a variety of disinfectants against Listeria spp. Appl Environ Microbiol 56(2):377–380
14. Best M, Sattar SA, Springthorpe VS, Kennedy ME (1988) Comparative mycobactericidal efficacy of chemical disinfectants in suspension and carrier tests. Appl Environ Microbiol 54:2856–2858
15. Best M, Sattar SA, Springthorpe VS, Kennedy ME (1990) Efficacies of selected disinfectants against *Mycobacterium tuberculosis*. J Clin Microbiol 28(10):2234–2239
16. Bigliardi P, Langer S, Cruz JJ, Kim SW, Nair H, Srisawasdi G (2017) An Asian perspective on povidone iodine in wound healing. Dermatol (Basel, Switzerland) 233(2–3):223-233. https://doi.org/10.1159/000479150

17. Bigliardi PL, Alsagoff SAL, El-Kafrawi HY, Pyon JK, Wa CTC, Villa MA (2017) Povidone iodine in wound healing: a review of current concepts and practices. Int J Surgery (London, England) 44:260–268. https://doi.org/10.1016/j.ijsu.2017.06.073
18. Block C, Robenshtok E, Simhon A, Shapiro M (2000) Evaluation of chlorhexidine and povidone iodine activity against methicillin-resistant Staphylococcus aureus and vancomycin-resistant Enterococcus faecalis using a surface test. J Hosp Infect 46(2):147–152. https://doi.org/10.1053/jhin.2000.0805
19. Carro C, Camilleri L, Traore O, Badrikian L, Legault B, Azarnoush K, Dualé C, De Riberolles C (2007) An in-use microbiological comparison of two surgical hand disinfection techniques in cardiothoracic surgery: hand rubbing versus hand scrubbing. J Hosp Infect 67 (1):62–66
20. Casey AL, Karpanen TJ, Nightingale P, Conway BR, Elliott TS (2015) Antimicrobial activity and skin permeation of iodine present in an iodine-impregnated surgical incise drape. J Antimicrob Chemother 70(8):2255–2260. https://doi.org/10.1093/jac/dkv100
21. Chen L, Lu X, Cao W, Zhang C, Xu R, Meng X, Chen K (2015) An investigation and evaluation on species and characteristics of pathogenic microorganisms in Chinese local hospital settings. Microb Pathog 89:154–160. https://doi.org/10.1016/j.micpath.2015.10.015
22. Cheng A, Sun HY, Tsai YT, Wu UI, Chuang YC, Wang JT, Sheng WH, Hsueh PR, Chen YC, Chang SC (2018) In vitro evaluation of povidone-iodine and chlorhexidine against outbreak and nonoutbreak strains of mycobacterium abscessus using standard quantitative suspension and carrier testing. Antimicrob Agents Chemother 62(1). https://doi.org/10.1128/aac.01364-17
23. Coetzee E, Rode H, Kahn D (2013) Pseudomonas aeruginosa burn wound infection in a dedicated paediatric burns unit. South African J Surgery Suid-Afrikaanse tydskrif vir chirurgie 51(2):50–53. https://doi.org/10.7196/sajs.1134
24. Craven DE, Moody B, Connolly MG, Kollisch NR, Stottmeier KD, McCabe WR (1981) Pseudobacteremia caused by povidone-iodine solution contaminated with Pseudomonas cepacia. N Engl J Med 305(11):621–623. https://doi.org/10.1056/nejm198109103051106
25. Culligan PJ, Kubik K, Murphy M, Blackwell L, Snyder J (2005) A randomized trial that compared povidone iodine and chlorhexidine as antiseptics for vaginal hysterectomy. Am J Obstet Gynecol 192(2):422–425. https://doi.org/10.1016/j.ajog.2004.08.010
26. Decker EM, Bartha V, Kopunic A, von Ohle C (2017) Antimicrobial efficiency of mouthrinses versus and in combination with different photodynamic therapies on periodontal pathogens in an experimental study. J Periodontal Res 52(2):162–175. https://doi.org/10.1111/jre.12379
27. Department of Health and Human Services; Food and Drug Administration (2015) Safety and effectiveness of healthcare antiseptics. Topical antimicrobial drug products for over-the-counter human use; proposed amendment of the tentative final monograph; reopening of administrative record; proposed rule. Fed Reg 80(84):25166–25205
28. Dogan AA, Adiloglu AK, Onal S, Cetin ES, Polat E, Uskun E, Koksal F (2008) Short-term relative antibacterial effect of octenidine dihydrochloride on the oral microflora in orthodontically treated patients. Int J Infect Dis: IJID: Official Publication of the Int Soc Infect Dis 12(6):e19–25. https://doi.org/10.1016/j.ijid.2008.03.013
29. Duarte ER, Hamdan JS (2006) Susceptibility of yeast isolates from cattle with otitis to aqueous solution of povidone iodine and to alcohol-ether solution. Med Mycol 44(4):369–373. https://doi.org/10.1080/13693780500064623
30. Eggers M, Koburger-Janssen T, Eickmann M, Zorn J (2018) In vitro bactericidal and virucidal efficacy of povidone-iodine gargle/mouthwash against respiratory and oral tract pathogens. Infect Dis Ther. https://doi.org/10.1007/s40121-018-0200-7
31. Eryilmaz M, Akin A, Arikan Akan O (2011) Investigation of the efficacy of some disinfectants against nosocomial Staphylococcus aureus and Enterococcus spp. isolates. Mikrobiyoloji Bulteni 45(3):454–460

32. Espigares E, Moreno Roldan E, Espigares M, Abreu R, Castro B, Dib AL, Arias A (2017) Phenotypic resistance to disinfectants and antibiotics in methicillin-resistant Staphylococcus aureus strains isolated from pigs. Zoonoses Public Health 64(4):272–280. https://doi.org/10.1111/zph.12308
33. Faoagali J, Fong J, George N, Mahoney P, O'Rourke V (1995) Comparison of the immediate, residual, and cumulative antibacterial effects of novaderm, novascrub, betadine surgical scrub, hibiclens, and liquid soap. Am J Infect Control 23(6):337–343
34. Fleurette J, Bes M, Brun Y, Freney J, Forey F, Coulet M, Reverdy ME, Etienne J (1989) Clinical isolates of staphylococcus lugdunensis and S. schleiferi: bacteriological characteristics and susceptibility to antimicrobial agents. Res Microbiol 140(2):107–118
35. Forer Y, Block C, Frenkel S (2017) Preoperative hand decontamination in ophthalmic surgery: a comparison of the removal of bacteria from surgeons' hands by routine antimicrobial scrub versus an alcoholic hand rub. Curr Eye Res 42(9):1333–1337. https://doi.org/10.1080/02713683.2017.1304559
36. Fu J, Wei P, Zhao C, He C, Yan Z, Hua H (2014) In vitro antifungal effect and inhibitory activity on biofilm formation of seven commercial mouthwashes. Oral Dis 20(8):815–820. https://doi.org/10.1111/odi.12242
37. Fuursted K, Hjort A, Knudsen L (1997) Evaluation of bactericidal activity and lag of regrowth (postantibiotic effect) of five antiseptics on nine bacterial pathogens. J Antimicrob Chemother 40(2):221–226
38. Gaston MA, Hoffman PN, Pitt TL (1986) A comparison of strains of Serratia marcescens isolated from neonates with strains isolated from sporadic and epidemic infections in adults. J Hosp Infect 8(1):86–95
39. Giacometti A, Cirioni O, Greganti G, Fineo A, Ghiselli R, Del Prete MS, Mocchegiani F, Fileni B, Caselli F, Petrelli E, Saba V, Scalise G (2002) Antiseptic compounds still active against bacterial strains isolated from surgical wound infections despite increasing antibiotic resistance. Eur J Clin Microbiol Infect Dis 21(7):553–556. https://doi.org/10.1007/s10096-002-0765-6
40. Goetz A, Muder RR (1989) Pseudomonas aeruginosa infections associated with use of povidone-iodine in patients receiving continuous ambulatory peritoneal dialysis. Infect Control Hosp Epidemiol 10(10):447–450
41. Guilhermetti M, Hernandes SE, Fukushigue Y, Garcia LB, Cardoso CL (2001) Effectiveness of hand-cleansing agents for removing methicillin-resistant *Staphylococcus aureus* from contaminated hands. Infect Control Hosp Epidemiol 22(2):105–108
42. Guo W, Shan K, Xu B, Li J (2015) Determining the resistance of carbapenem-resistant Klebsiella pneumoniae to common disinfectants and elucidating the underlying resistance mechanisms. Pathog Global Health 109(4):184–192. https://doi.org/10.1179/2047773215y.0000000022
43. Gupta C, Czubatyj AM, Briski LE, Malani AK (2007) Comparison of two alcohol-based surgical scrub solutions with an iodine-based scrub brush for presurgical antiseptic effectiveness in a community hospital. J Hosp Infect 65:65–71
44. Gutierrez-Martin CB, Yubero S, Martinez S, Frandoloso R, Rodriguez-Ferri EF (2011) Evaluation of efficacy of several disinfectants against Campylobacter jejuni strains by a suspension test. Res Vet Sci 91(3):e44–47. https://doi.org/10.1016/j.rvsc.2011.01.020
45. Hansmann F, Kramer A, Ohgke H, Strobel H, Muller M, Geerling G (2005) [Lavasept as an alternative to PVP-iodine as a preoperative antiseptic in ophthalmic surgery. Randomized, controlled, prospective double-blind trial]. Der Ophthalmologe: Zeitschrift der Deutschen Ophthalmologischen Gesellschaft 102(11):1043–1046, 1048–1050. https://doi.org/10.1007/s00347-004-1120-3

46. Herruzo-Cabrera R, Garcia-Torres V, Rey-Calero J, Vizcaino-Alcaide MJ (1992) Evaluation of the penetration strength, bactericidal efficacy and spectrum of action of several antimicrobial creams against isolated microorganisms in a burn centre. Burns: J Int Soc Burn Injuries 18(1):39–44
47. Hill KE, Malic S, McKee R, Rennison T, Harding KG, Williams DW, Thomas DW (2010) An in vitro model of chronic wound biofilms to test wound dressings and assess antimicrobial susceptibilities. J Antimicrob Chemother 65(6):1195–1206. https://doi.org/10.1093/jac/dkq105
48. Hingst V, Juditzki I, Heeg P, Sonntag H-G (1992) Evaluation of the efficacy of surgical hand disinfection following a reduced application time of 3 instead of 5 min. J Hosp Infect 20:79–86
49. Hirsch T, Limoochi-Deli S, Lahmer A, Jacobsen F, Goertz O, Steinau HU, Seipp HM, Steinstraesser L (2011) Antimicrobial activity of clinically used antiseptics and wound irrigating agents in combination with wound dressings. Plast Reconstr Surg 127(4):1539–1545. https://doi.org/10.1097/PRS.0b013e318208d00f
50. Hoekstra MJ, Westgate SJ, Mueller S (2017) Povidone-iodine ointment demonstrates in vitro efficacy against biofilm formation. Int Wound J 14(1):172–179. https://doi.org/10.1111/iwj.12578
51. Houang ET, Gilmore OJ, Reid C, Shaw EJ (1976) Absence of bacterial resistance to povidone iodine. J Clin Pathol 29(8):752–755
52. Hsu J, Gerstenblith AT, Garg SJ, Vander JF (2014) Conjunctival flora antibiotic resistance patterns after serial intravitreal injections without postinjection topical antibiotics. American J Ophthalmol 157(3):514–518.e511. https://doi.org/10.1016/j.ajo.2013.10.003
53. Izano EA, Shah SM, Kaplan JB (2009) Intercellular adhesion and biocide resistance in nontypeable Haemophilus influenzae biofilms. Microb Pathog 46(4):207–213. https://doi.org/10.1016/j.micpath.2009.01.004
54. Jeng DK, Severin JE (1998) Povidone iodine gel alcohol: a 30-second, onetime application preoperative skin preparation. Am J Infect Control 26(5):488–494
55. Johani K, Malone M, Jensen SO, Dickson HG, Gosbell IB, Hu H, Yang Q, Schultz G, Vickery K (2018) Evaluation of short exposure times of antimicrobial wound solutions against microbial biofilms: from in vitro to in vivo. J Antimicrob Chemother 73(2):494–502. https://doi.org/10.1093/jac/dkx391
56. Junka A, Bartoszewicz M, Smutnicka D, Secewicz A, Szymczyk P (2014) Efficacy of antiseptics containing povidone-iodine, octenidine dihydrochloride and ethacridine lactate against biofilm formed by Pseudomonas aeruginosa and Staphylococcus aureus measured with the novel biofilm-oriented antiseptics test. Int Wound J 11(6):730–734. https://doi.org/10.1111/iwj.12057
57. Junka AF, Zywicka A, Szymczyk P, Dziadas M, Bartoszewicz M, Fijalkowski K (2017) A.D.A.M. test (Antibiofilm Dressing's Activity Measurement)—simple method for evaluating anti-biofilm activity of drug-saturated dressings against wound pathogens. J Microbiol Meth 143:6–12. https://doi.org/10.1016/j.mimet.2017.09.014
58. Kac G, Masmejean E, Gueneret M, Rodi A, Peyrard S, Podglajen I (2009) Bactericidal efficacy of a 1.5 min surgical hand-rubbing protocol under in-use conditions. J Hosp Infect 72(2):135–139
59. Kapalschinski N, Seipp HM, Kuckelhaus M, Harati KK, Kolbenschlag JJ, Daigeler A, Jacobsen F, Lehnhardt M, Hirsch T (2017) Albumin reduces the antibacterial efficacy of wound antiseptics against Staphylococcus aureus. J Wound Care 26(4):184–187. https://doi.org/10.12968/jowc.2017.26.4.184
60. Koburger T, Hübner N-O, Braun M, Siebert J, Kramer A (2010) Standardized comparison of antiseptic efficacy of triclosan, PVP-iodine, octenidine dihydrochloride, polyhexanide and chlorhexidine digluconate. J Antimicrob Chemother 65(8):1712–1719

61. Kondo S, Tabe Y, Yamada T, Misawa S, Oguri T, Ohsaka A, Miida T (2012) Comparison of antifungal activities of gentian violet and povidone-iodine against clinical isolates of Candida species and other yeasts: a framework to establish topical disinfectant activities. Mycopathologia 173(1):21–25. https://doi.org/10.1007/s11046-011-9458-y
62. Kramer A, Dissemond J, Kim S, Willy C, Mayer D, Papke R, Tuchmann F, Assadian O (2017) Consensus on wound antisepsis: update 2018. Skin Pharmacol Physiol 31(1):28–58. https://doi.org/10.1159/000481545
63. Kunisada T, Yamada K, Oda S, Hara O (1997) Investigation on the efficacy of povidone-iodine against antiseptic-resistant species. Dermatology (Basel, Switzerland) 195 (suppl. 2):14–18
64. Lai KW, Foo TL, Low W, Naidu G (2012) Surgical hand antisepsis-a pilot study comparing povidone iodine hand scrub and alcohol-based chlorhexidine gluconate hand rub. Ann Acad Med Singapore 41(1):12–16
65. Lam OL, McGrath C, Li LS, Samaranayake LP (2012) Effectiveness of oral hygiene interventions against oral and oropharyngeal reservoirs of aerobic and facultatively anaerobic gram-negative bacilli. Am J Infect Control 40(2):175–182. https://doi.org/10.1016/j.ajic.2011.03.004
66. Lanjri S, Uwingabiye J, Frikh M, Abdellatifi L, Kasouati J, Maleb A, Bait A, Lemnouer A, Elouennass M (2017) In vitro evaluation of the susceptibility of Acinetobacter baumannii isolates to antiseptics and disinfectants: comparison between clinical and environmental isolates. Antimicrob Resist Infect Control 6:36. https://doi.org/10.1186/s13756-017-0195-y
67. Lanker Klossner B, Widmer HR, Frey F (1997) Nondevelopment of resistance by bacteria during hospital use of povidone-iodine. Dermatology (Basel, Switzerland) 195(Suppl 2):10–13. https://doi.org/10.1159/000246024
68. Lefebvre E, Vighetto C, Di Martino P, Larreta Garde V, Seyer D (2016) Synergistic antibiofilm efficacy of various commercial antiseptics, enzymes and EDTA: a study of pseudomonas aeruginosa and staphylococcus aureus biofilms. Int J Antimicrob Agents 48 (2):181–188. https://doi.org/10.1016/j.ijantimicag.2016.05.008
69. Liu Q, Liu M, Wu Q, Li C, Zhou T, Ni Y (2009) Sensitivities to biocides and distribution of biocide resistance genes in quaternary ammonium compound tolerant Staphylococcus aureus isolated in a teaching hospital. Scand J Infect Dis 41(6–7):403–409. https://doi.org/10.1080/00365540902856545
70. Maillard JY, Messager S, Veillon R (1998) Antimicrobial efficacy of biocides tested on skin using an ex-vivo test. J Hosp Infect 40(4):313–323
71. Marchetti MG, Kampf G, Finzi G, Salvatorelli G (2003) Evaluation of the bactericidal effect of five products for surgical hand disinfection according to prEN 12054 and prEN 12791. J Hosp Infect 54(1):63–67
72. Martro E, Hernandez A, Ariza J, Dominguez MA, Matas L, Argerich MJ, Martin R, Ausina V (2003) Assessment of acinetobacter baumannii susceptibility to antiseptics and disinfectants. J Hosp Infect 55(1):39–46
73. McLure AR, Gordon J (1992) In-vitro evaluation of povidone-iodine and chlorhexidine against methicillin-resistant *Staphylococcus aureus*. J Hosp Infect 21:291–299
74. Messager S, Goddard PA, Dettmar PW, Maillard JY (2001) Determination of the antibacterial efficacy of several antiseptics tested on skin by an 'ex-vivo' test. J Med Microbiol 50(3):284–292. https://doi.org/10.1099/0022-1317-50-3-284
75. Mon-On N, Surachetpong W, Mongkolsuk S, Sirikanchana K (2018) Roles of water quality and disinfectant application on inactivation of fish pathogenic Streptococcus agalactiae with povidone iodine, quaternary ammonium compounds and glutaraldehyde. J Fish Dis 41 (5):783–789. https://doi.org/10.1111/jfd.12776
76. Moore G, Schelenz S, Borman AM, Johnson EM, Brown CS (2017) Yeasticidal activity of chemical disinfectants and antiseptics against Candida auris. J Hosp Infect 97(4):371–375. https://doi.org/10.1016/j.jhin.2017.08.019

77. Morales-Fernandez L, Fernandez-Crehuet M, Espigares M, Moreno E, Espigares E (2014) Study of the hormetic effect of disinfectants chlorhexidine, povidone iodine and benzalkonium chloride. Eur J Clin Microbiol Infect Dis 33(1):103–109. https://doi.org/10.1007/s10096-013-1934-5
78. Muller G, Kramer A (2000) In vitro action of a combination of selected antimicrobial agents and chondroitin sulfate. Chem Biol Interact 124(2):77–85
79. Muller G, Kramer A (2008) Biocompatibility index of antiseptic agents by parallel assessment of antimicrobial activity and cellular cytotoxicity. J Antimicrob Chemother 61 (6):1281–1287. https://doi.org/10.1093/jac/dkn125
80. Nagai I, Ogase H (1990) Absence of role for plasmids in resistance to multiple disinfectants in three strains of bacteria. J Hosp Infect 15(2):149–155
81. Narui K, Takano M, Noguchi N, Sasatsu M (2007) Susceptibilities of methicillin-resistant Staphylococcus aureus isolates to seven biocides. Biol Pharmac Bull 30(3):585–587
82. National Center for Biotechnology Information Betadine. PubChem Compound Database; CID = 410087. https://pubchem.ncbi.nlm.nih.gov/compound/410087. Accessed 12 April 2018
83. Oduwole KO, Glynn AA, Molony DC, Murray D, Rowe S, Holland LM, McCormack DJ, O'Gara JP (2010) Anti-biofilm activity of sub-inhibitory povidone-iodine concentrations against Staphylococcus epidermidis and Staphylococcus aureus. J Orthop Res: Off Publication Orthop Res Soc 28(9):1252–1256. https://doi.org/10.1002/jor.21110
84. Ozkurt Z, Altoparlak U, Erol S, Celebi S (2003) Activity of frequently used disinfectants and antiseptics against nosocomial bacterial types. Mikrobiyoloji bulteni 37(2–3):157–162
85. Pagedar A, Singh J (2015) Evaluation of antibiofilm effect of benzalkonium chloride, iodophore and sodium hypochlorite against biofilm of Pseudomonas aeruginosa of dairy origin. J Food Sci Technol 52(8):5317–5322. https://doi.org/10.1007/s13197-014-1575-4
86. Parducz L, Eszik I, Wagner G, Burian K, Endresz V, Virok DP (2016) Impact of antiseptics on chlamydia trachomatis growth. Lett Appl Microbiol 63(4):260–267. https://doi.org/10.1111/lam.12625
87. Parrott PL, Terry PM, Whitworth EN, Frawley LW, Coble RS, Wachsmuth IK, McGowan JE Jr (1982) Pseudomonas aeruginosa peritonitis associated with contaminated poloxamer-iodine solution. Lancet 2(8300):683–685
88. Payne DN, Babb JR, Bradley CR (1999) An evaluation of the suitability of the European suspension test to reflect in vitro activity of antiseptics against clinically significant organisms. Lett Appl Microbiol 28(1):7–12
89. Pitten F-A, Werner H-P, Kramer A (2003) A standardized test to assess the impact of different organic challenges on the antimicrobial activity of antiseptics. J Hosp Infect 55 (2):108–115
90. Prince HN, Nonemaker WS, Norgard RC, Prince DL (1978) Drug resistance studies with topical antiseptics. J Pharm Sci 67(11):1629–1631
91. Reimer K, Schreier H, Erdos G, Konig B, Konig W, Fleischer W (1998) Molecular effects of a microbicidal substance on relevant microorganisms: electron microscopic and biochemical studies on povidone-iodine. Zentralbl Hyg Umweltmed 200(5–6):423–434
92. Reynolds MM, Greenwood-Quaintance KE, Patel R, Pulido JS (2016) Selected antimicrobial activity of topical ophthalmic anesthetics. Transl Vision Sci Technol 5(4):2. https://doi.org/10.1167/tvst.5.4.2
93. Rikimaru T, Kondo M, Kajimura K, Hashimoto K, Oyamada K, Sagawa K, Tanoue S, Oizumi K (2002) Bactericidal activities of commonly used antiseptics against multidrug-resistant mycobacterium tuberculosis. Dermatology (Basel, Switzerland) 204 (Suppl 1):15–20. https://doi.org/10.1159/000057719
94. Rodriguez Ferri EF, Martinez S, Frandoloso R, Yubero S, Gutierrez Martin CB (2010) Comparative efficacy of several disinfectants in suspension and carrier tests against haemophilus parasuis serovars 1 and 5. Res Vet Sci 88(3):385–389. https://doi.org/10.1016/j.rvsc.2009.12.001

95. Sa A, Sawatdee S, Phadoongsombut N, Buatong W, Nakpeng T, Sritharadol R, Srichana T (2017) Quantitative analysis of povidone-iodine thin films by X-ray photoelectron spectroscopy and their physicochemical properties. Acta Pharmaceutica (Zagreb, Croatia) 67(2):169–186. https://doi.org/10.1515/acph-2017-0011
96. Schedler K, Assadian O, Brautferger U, Muller G, Koburger T, Classen S, Kramer A (2017) Proposed phase 2/ step 2 in-vitro test on basis of EN 14561 for standardised testing of the wound antiseptics PVP-iodine, chlorhexidine digluconate, polihexanide and octenidine dihydrochloride. BMC Infect Dis 17(1):143. https://doi.org/10.1186/s12879-017-2220-4
97. Stickler D, Hewett P (1991) Activity of antiseptics against biofilms of mixed bacterial species growing on silicone surfaces. Eur J Clin Microbiol Infect Dis 10(5):416–421
98. Sweden (2013) Assessment report. Iodine (including PVP-iodine). Product-types 1, 3, 4 and 22
99. Tortorano AM, Viviani MA, Biraghi E, Rigoni AL, Prigitano A, Grillot R (2005) In vitro testing of fungicidal activity of biocides against aspergillus fumigatus. J Med Microbiol 54 (Pt 10):955–957. https://doi.org/10.1099/jmm.0.45997-0
100. Traboulsi RS, Mukherjee PK, Ghannoum MA (2008) In vitro activity of inexpensive topical alternatives against Candida spp. isolated from the oral cavity of HIV-infected patients. Int J Antimicrob Agents 31(3):272–276. https://doi.org/10.1016/j.ijantimicag.2007.11.008
101. Traoré O, Fayard SF, Laveran H (1996) An in-vitro evaluation of the activity of povidone-iodine against nosocomial bacterial strains. J Hosp Infect 34(3):217–222
102. Tremblay YD, Caron V, Blondeau A, Messier S, Jacques M (2014) Biofilm formation by coagulase-negative staphylococci: impact on the efficacy of antimicrobials and disinfectants commonly used on dairy farms. Vet Microbiol 172(3–4):511–518. https://doi.org/10.1016/j.vetmic.2014.06.007
103. United States Department of Health and Human Services (2004) Toxicological profile of iodine. https://www.atsdr.cdc.gov/toxprofiles/tp158.pdf
104. United States Environmental Protection Agency (2006) Reregistration eligibility decision for iodine and iodophor complexes. https://nepis.epa.gov/Exe/ZyPDF.cgi/P100L2D2.PDF?Dockey=P100L2D2.PDF
105. Verner–Jeffreys DW, Joiner CL, Bagwell NJ, Reese RA, Husby A, Dixon PF (2009) Development of bactericidal and virucidal testing standards for aquaculture disinfectants. Aquaculture 286(3):190–197. https://doi.org/10.1016/j.aquaculture.2008.10.001
106. Wang Y, Leng V, Patel V, Phillips KS (2017) Injections through skin colonized with Staphylococcus aureus biofilm introduce contamination despite standard antimicrobial preparation procedures. Scientific Reports 7:45070. https://doi.org/10.1038/srep45070
107. Whiteley M, Ott JR, Weaver EA, McLean RJ (2001) Effects of community composition and growth rate on aquifer biofilm bacteria and their susceptibility to betadine disinfection. Environ Microbiol 3(1):43–52
108. Wichelhaus TA, Schafer V, Hunfeld KP, Reimer K, Fleischer W, Brade V (1998) Antibacterial effectiveness of povidone-iodine (Betaisodona) against highly resistance gram positive organisms. Zentralbl Hyg Umweltmed 200(5–6):435–442
109. Wiegand C, Abel M, Ruth P, Elsner P, Hipler UC (2015) pH influence on antibacterial efficacy of common antiseptic substances. Skin Pharmacol Physiol 28(3):147–158. https://doi.org/10.1159/000367632
110. Wiegand C, Abel M, Ruth P, Hipler UC (2012) Analysis of the adaptation capacity of Staphylococcus aureus to commonly used antiseptics by microplate laser nephelometry. Skin Pharmacol Physiol 25(6):288–297. https://doi.org/10.1159/000341222
111. Wisplinghoff H, Schmitt R, Wohrmann A, Stefanik D, Seifert H (2007) Resistance to disinfectants in epidemiologically defined clinical isolates of Acinetobacter baumannii. J Hosp Infect 66(2):174–181. https://doi.org/10.1016/j.jhin.2007.02.016
112. Witney AA, Gould KA, Pope CF, Bolt F, Stoker NG, Cubbon MD, Bradley CR, Fraise A, Breathnach AS, Butcher PD, Planche TD, Hinds J (2014) Genome sequencing and characterization of an extensively drug-resistant sequence type 111 serotype O12 hospital

outbreak strain of pseudomonas aeruginosa. Clin Microbiol Infect 20(10):O609–618. https://doi.org/10.1111/1469-0691.12528

113. Yanai R, Yamada N, Ueda K, Tajiri M, Matsumoto T, Kido K, Nakamura S, Saito F, Nishida T (2006) Evaluation of povidone-iodine as a disinfectant solution for contact lenses: antimicrobial activity and cytotoxicity for corneal epithelial cells. Cont Lens Anterior Eye 29 (2):85–91. https://doi.org/10.1016/j.clae.2006.02.006
114. Zisi AP, Exindari MK, Siska EK, Koliakos GG (2018) Iodine-lithium-alpha-dextrin (ILalphaD) against Staphylococcus aureus skin infections: a comparative study of in-vitro bactericidal activity and cytotoxicity between ILalphaD and povidone-iodine. J Hosp Infect 98(2):134–140. https://doi.org/10.1016/j.jhin.2017.07.013

Antiseptic Stewardship for Alcohol-Based Hand Rubs

17

17.1 Composition and Intended Use

Alcohol-based hand rubs are usually based in ethanol, propan-2-ol, propan-1-ol or a combination of the three alcohols. Typical alcohol concentrations are 70–95%. Some commercially available hand rubs contain additional non-volatile biocidal agents, e.g. 0.1% benzalkonium chloride [13], 0.1–1% chlorhexidine digluconate [12, 13], 0.3–0.5% triclosan [12], 0.1% octenidine dihydrochloride [12], hydrogen peroxide, DDAC, polihexanide or peracetic acid. Most of them also contain emollients as auxiliary agents to reduce skin dryness especially under frequent use conditions [6, 8, 14]. Non-volatile antiseptic agents will remain for some time on the skin when applied with an alcohol-based hand rub although the duration of persistence and the concentrations on the skin are unknown and will largely depend on the frequency of use and other hand hygiene activities such as hand washing.

These products are used in health care, nursing homes, veterinary medicine, food processing and manufacturing and occasionally also in the domestic setting. There are two typical applications: hygienic hand disinfection according to the five indications for hand hygiene and surgical hand disinfection before surgical procedures [16]. The summary below is an extract of previous book chapters on the biocidal agents.

17.2 Selection Pressure Associated with Commonly Used Biocidal Agents

17.2.1 Change of Susceptibility by Low-Level Exposure

Any adaptive effects were classified as "no MIC increase", "weak MIC increase" with a $\leq$4-fold MIC increase, and "strong MIC increase" with a >4-fold MIC increase. The last category was divided into an unstable or stable MIC increase;

G. Kampf, *Antiseptic Stewardship*, https://doi.org/10.1007/978-3-319-98785-9_17

sometimes the stability was unknown. A species may be found in two or more categories indicating that the adaptive response depends on the type of isolate and not primarily on the species itself. Most data on different adaptive effects caused by low-level exposure were found for triclosan (90 species), followed by chlorhexidine digluconate and benzalkonium chloride (both 78 species), polihexanide (55 species) and DDAC (48 species). Only few data were found for hydrogen peroxide (8 species), peracetic acid, ethanol and octenidine dihydrochloride (3 species) and propan-2-ol (1 species). No data were found for propan-1-ol.

Figure 17.1 shows the distribution of adaptive response categories for the different biocidal agents. The majority of species did not show any MIC change or only a weak MIC increase ($\leq$ 4-fold). A strong adaptive response was most frequently seen in benzalkonium chloride (44% of the evaluated species), followed by triclosan (34%), chlorhexidine digluconate (26%), DDAC (15%) and polihexanide (11%). With octenidine dihydrochloride, one species showed a strong adaptive response. The strong MIC increase was stable in 42% (triclosan), 41% (benzalkonium chloride) and 40% (chlorhexidine digluconate) of the species. Hydrogen peroxide, ethanol and propan-2-ol have so far not shown a strong adaptive response.

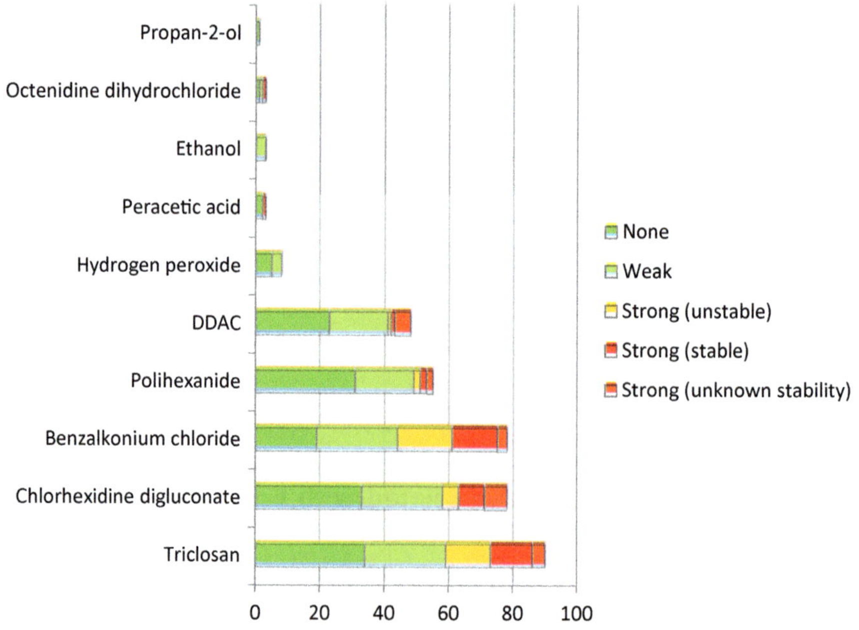

Fig. 17.1 Number of species with no, a weak or a strong adaptive MIC increase after low-level exposure to biocidal agents that may be found in alcohol-based hand rubs

Table 17.1 Examples for healthcare-associated bacterial species with a strong (>4-fold MIC increase) and stable adaptive response after low-level exposure to selected biocidal agents

Biocidal agent	Bacterial species with a strong and stable adaptive MIC increase
Benzalkonium chloride	*Enterobacter* spp. ($\leq$ 300-fold)
	E. coli ($\leq$ 100-fold)
	S. aureus ($\leq$ 39-fold)
	P. aeruginosa ($\leq$ 33-fold)
	A. baumannii ($\leq$ 31-fold)
Triclosan	*E. coli* ($\leq$ 8,192-fold)
	S. aureus ($\leq$ 313-fold)
	K. pneumoniae ($\leq$ 129-fold)
	A. baumannii ($\leq$ 16-fold)
	S. epidermidis ($\leq$ 8-fold)
Chlorhexidine digluconate	*E. coli* ($\leq$ 500-fold)
	S. marcescens ($\leq$ 128-fold)
	P. aeruginosa ($\leq$ 32-fold)
	K. pneumoniae ($\leq$ 16-fold)
	S. aureus ($\leq$ 16-fold)
Polihexanide	*E. faecalis* ($\leq$ 8-fold)
	S. aureus ($\leq$ 8-fold)
DDAC	*P. aeruginosa* ($\geq$ 18-fold)
Octenidine dihydrochloride	*P. aeruginosa* ($\leq$ 32-fold)

A strong and stable MIC increase after low-level exposure is probably the most critical adaptive response. Some species can be found in this group that have a high relevance for infection control (Table 17.1). Most of them belong to the group of Gram-negative species.

The effect on biofilm is not covered in this chapter because it was assumed that it has only minor relevance for alcohol-based hand rubs.

17.2.2 Cross-Tolerance to Other Biocidal Agents

Other risks may also be relevant when the agents are used in alcohol-based hand rubs. Cross-tolerance between alcohols and other biocidal agents is very uncommon. A primarily ethanol-tolerant *L. monocytogenes* has been described to be cross-tolerant to hydrogen peroxide. With propan-1-ol and propan-2-ol, no cross-tolerance to other biocidal agents has so far been reported.

Cross-tolerance to other biocidal agents is more common in non-volatile biocidal agents. Isolates of 22 primarily benzalkonium chloride-tolerant species were cross-tolerant to chlorhexidine digluconate and triclosan. An isolate of a benzalkonium chloride-tolerant *E. coli* was cross-tolerant to DDAC, and a benzalkonium chloride-tolerant *L. monocytogenes* was cross-tolerant to another QAC, alkylamine and sodium hypochlorite. Similar results were found with triclosan. Isolates of 17 primarily triclosan-tolerant species were cross-tolerant to chlorhexidine digluconate and 13 species to benzalkonium chloride. Primarily, chlorhexidine digluconate-tolerant isolates of *E. coli* and *S. Virchow* were cross-tolerant to triclosan, isolates of *S. Tyhimurium* were cross-tolerant to benzalkonium chloride, and isolates of *A. baylyi* were cross-tolerant to hydrogen peroxide. Isolates of primarily DDAC-tolerant *E. coli and P. fluorescens* can be cross-tolerant to benzalkonium chloride, isolates of primarily octenidine-tolerant *P. aeruginosa* can be cross-tolerant to chlorhexidine digluconate, isolates of primarily peracetic acid-tolerant *B. subtilis* can be cross-tolerant to other oxidizing agents, isolates of primarily hydrogen peroxide-tolerant *E. coli* can be cross-tolerant to aldehyde, and isolates of primarily hydrogen peroxide-tolerant *S. cerevisiae* can be cross-tolerant to ethanol. Especially, the rather frequently observed cross-tolerance between benzalkonium chloride, triclosan and chlorhexidine digluconate is a clear indication to carefully select biocidal agents in order to reduce this type of cross-tolerance to a minimum.

17.2.3 Cross-Tolerance to Antibiotics

Ethanol, propan-1-ol, propan-2-ol, peracetic acid, hydrogen peroxide and polihexanide have so far never been described with a cross-tolerance to antibiotics. A cross-tolerance between triclosan, chlorhexidine digluconate and benzalkonium chloride and selected antibiotics can occur in numerous species. Occasional cross-resistance between DDAC and selected antibiotics was found in *C. coli, E. coli, L. monocytogenes* and *S. enterica*. Cross-tolerance between octenidine dihydrochloride and selected antibiotics can occur in *P. aeruginosa*.

17.2.4 Efflux Pump Genes

Transporter and efflux pump genes were up-regulated after benzalkonium chloride exposure in *B. cepacia complex*, *E. coli* and *L. monocytogenes*, and after chlorhexidine digluconate exposure in *B. fragilis* and *B. cepacia complex*.

17.2.5 Horizontal Gene Transfer

Horizontal gene transfer can be successfully induced by chlorhexidine digluconate and triclosan in *E. coli* (sulphonamide resistance by conjugation). In *B. subtilis*, ethanol at 4% can cause a 5-fold increase in mobile genetic element transfer (resistance genes).

17.2.6 Antibiotic Resistance Gene Expression

In a vanA *E. faecium*, chlorhexidine digluconate was able to induce a $\geq$ 10-fold increase in vanHAX encoding VanA-type vancomycin resistance.

17.2.7 Viable but not Culturable

Peracetic acid is able in *S. Typhimurium* to induce the VBNC state.

17.2.8 Other Risks Associated with Additional Biocidal Agents

Other risks may also be relevant when the agents are used in alcohol-based hand rubs. They are not covered in detail. Sensitization to the agent may occur possibly resulting in local or systemic allergic reactions up to anaphylactic reactions. This has been described at least for chlorhexidine digluconate and polihexanide [7]. Cationic surfactants or peracetic acid may cause a higher degree of skin irritation [7]. This may also be an aspect to consider in alcohol-based hand rubs.

17.3 Health Benefit of Biocidal Agents in Alcohol-Based Hand Rubs

The main benefit of the alcohols is the strong and immediate bactericidal and yeasticidal activity [4]. Non-volatile biocidal agents in alcohol-based hand rubs are expected to exhibit an ongoing antimicrobial activity after evaporation of the alcohol. This may be particularly useful in surgical hand disinfection once the sterile glove has been donned. During surgery, the antimicrobial effect may slow down or even revert the recolonization of the skin so that the glove juice has a lower microbial count which is expected to be relevant for the prevention of surgical site infections in case of glove punctures [10].

When used for hygienic hand disinfection in defined clinical situations alcohol-based hand rubs can reduce the rate of healthcare-associated infections. In 2000, Pittet et al. were able to show that an increase in hand hygiene compliance from 48 to 66%, mainly by using the alcohol-based hand rub more frequently, was able to reduce the rate of healthcare-associated infections over 3 years significantly from 16.9 to 9.9% [11]. Alcohol-based hand rubs are therefore recommended for use in patient care according to the five moments for hand hygiene [15]. But a health benefit such as the prevention of any type of healthcare-associated infection has never been shown for any of the additional biocidal substances in alcohol-based hand rubs. In addition, a lack of efficacy has been described for benzalkonium chloride, chlorhexidine, triclosan and mecetronium etilsulphate in hygienic hand disinfection [9, 13]. 0.2% peracetic acid may improve the virucidal activity of an alcohol-based hand rub, especially against non-enveloped viruses although the dermal tolerance is likely to be worse compared to formulations without peracetic acid [18].

For surgical hand disinfection, the WHO recommended in 2009 using "a suitable hand rub preferably with a product ensuring sustained activity" suggesting that it should contain e.g. chlorhexidine digluconate [16]. Based on the lack of evidence for the prevention of surgical site infections, the WHO recommended in 2016 to use a "suitable hand rub" for surgical hand disinfection taking into account that there is no evidence for the prevention of surgical site infections by using the combination of alcohol and chlorhexidine digluconate [17].

The efficacy of additional biocidal agents on the hand flora is overall doubtful. An easy method to determine a possible persistent antimicrobial activity is to measure the long-term efficacy in surgical hand disinfection according to EN 12791. The efficacy of a hand rub is compared with the reference alcohol. In case of a "long-term efficacy" or "persistent efficacy", the hand rub would reveal an effect after 3 h under the surgical glove which is superior to the reference alcohol ($p < 0.01$). For some biocidal agents such as chlorhexidine digluconate (0.5 or 1%) or mecetronium etilsulphate (0.1%), there is convincing evidence that hand rubs containing these agents do not have a superior efficacy in surgical hand disinfection after 3 h when applied for up to 2 min [1, 2, 5]. One study addressed the efficacy of an alcohol-based hand rub containing 0.1% octenidine dihydrochloride. When applied for 3 min, the immediate efficacy was 0.5 log better; after 3 h, the efficacy was 1.3 log higher compared to the reference treatment. When applied for 5 min, the immediate efficacy was 0.8 log better; after 3 h, the efficacy was 0.5 log higher compared to the reference treatment. A comparative statistical evaluation was not done by the authors so that it remains unclear if the effect can be considered to be superior to the reference treatment indicative of a sustained effect [1].

17.4 Antiseptic Stewardship Implications

Overall, the probability for a clinically relevant selection pressure caused by low-level exposure to alcohols is very small. The main reason is the volatility of the alcohols. An appropriate aliquot for hygienic hand disinfection is typically 2–3 ml. After 30–45 s, the hands will be dry again [3]. For surgical hand disinfection the applied volume will be 6–12 ml, depending on the size of the hands and the recommended application time of the hand rub. The contact time between the alcohols at an adequate concentration (70–95%) and the micro-organisms is too short for any adaptive response caused by a low alcohol concentration during its evaporation possibly resulting in a lower susceptibility of micro-organisms to the alcohols.

Additional biocidal agents in alcohol-based hand rubs have mostly no relevant antimicrobial efficacy on hands. In addition, there is no evidence for these agents to show a health benefit (prevention of infection). In this situation the known risks of these agents come into the focus. Some of them (benzalkonium chloride, triclosan, chlorhexidine digluconate and DDAC) can cause a strong and stable MIC increase in numerous mainly Gram-negative bacterial species. Biocide cross-tolerance is frequently found between benzalkonium chloride, triclosan and chlorhexidine digluconate. Some biocidal agents can enhance antibiotic resistance development. Horizontal gene transfer can be successfully induced by chlorhexidine digluconate and triclosan in *E. coli*. Antibiotic resistance gene expression can be increased by chlorhexidine digluconate in a vanA *E. faecium*. And efflux pump genes can be up-regulated in some species by benzalkonium chloride and chlorhexidine digluconate. The overall balance provides evidence for a number of relevant risks but no evidence for a relevant benefit.

For professional users, alcohol-based hand rubs containing any of these additional biocidal agents such as chlorhexidine digluconate, triclosan, benzalkonium chloride, hydrogen peroxide, DDAC, polihexanide, peracetic acid and octenidin dihydrochloride without convincing evidence to a support a health benefit should be replaced by formulations based on alcohol(s) alone as active agent(s). These formulations should have at least an equivalent spectrum of antimicrobial efficacy, an equivalent in vivo efficacy and a comparable user acceptability. The WHO provides tools to determine the user acceptability of hand rubs [16].

For non-professional use, alcohol-based hand rubs containing any of these additional biocidal agents without convincing evidence to a support a health benefit should be banned.

References

1. Hingst V, Juditzki I, Heeg P, Sonntag H-G (1992) Evaluation of the efficacy of surgical hand disinfection following a reduced application time of 3 instead of 5 min. J Hosp Infect 20:79–86
2. Kampf G (2017) Lack of antimicrobial efficacy of mecetronium etilsulfate in propanol-based hand rubs for surgical hand disinfection. J Hosp Infect 96(2):189–191
3. Kampf G (2017) The puzzle of volume, coverage and application time in hand disinfection. Infect Control Hosp Epidemiol 38(7):880–881
4. Kampf G, Kramer A (2004) Epidemiologic background of hand hygiene and evaluation of the most important agents for scrubs and rubs. Clin Microbiol Rev 17(4):863–893
5. Kampf G, Kramer A, Suchomel M (2017) Lack of sustained efficacy for alcohol-based surgical hand rubs containing "residual active ingredients" according to EN 12791. J Hosp Infect 95(2):163–168
6. Kampf G, Wigger-Alberti W, Schoder V, Wilhelm KP (2005) Emollients in a propanol-based hand rub can significantly decrease irritant contact dermatitis. Contact Dermatitis 53:344–349
7. Lachapelle JM (2014) A comparison of the irritant and allergenic properties of antiseptics. Eur J Dermatol: EJD 24(1):3–9. https://doi.org/10.1684/ejd.2013.2198
8. Löffler H, Kampf G, Schmermund D, Maibach HI (2007) How irritant is alcohol? Br J Dermatol 157(1):74–81
9. Lopez-Gigosos RM, Mariscal-Lopez E, Gutierrez-Bedmar M, Garcia-Rodriguez A, Mariscal A (2017) Evaluation of antimicrobial persistent activity of alcohol-based hand antiseptics against bacterial contamination. Eur J Clin Microbiol Infect Dis 36(7):1197–1203. https://doi.org/10.1007/s10096-017-2908-9
10. Misteli H, Weber WP, Reck S, Rosenthal R, Zwahlen M, Füglistaler P, Bolli MK, Örtli D, Widmer AF, Marti WR (2009) Surgical glove perforation and the risk of surgical site infection. Arch Surg 144(6):553–558
11. Pittet D, Hugonnet S, Harbarth S, Monronga P, Sauvan V, Touveneau S, Perneger TV (2000) Effectiveness of a hospital-wide programme to improve compliance with hand hygiene. Lancet 356:1307–1312
12. Rochon-Edouard S, Pons JL, Veber B, Larkin M, Vassal S, Lemeland JF (2004) Comparative in vitro and in vivo study of nine alcohol-based handrubs. Am J Infect Control 32(4):200–204. https://doi.org/10.1016/j.ajic.2003.08.003

13. Rosas-Ledesma P, Mariscal A, Carnero M, Munoz-Bravo C, Gomez-Aracena J, Aguilar L, Granizo JJ, Lafuente A, Fernandez-Crehuet J (2009) Antimicrobial efficacy in vivo of a new formulation of 2-butanone peroxide in n-propanol: comparison with commercial products in a cross-over trial. J Hosp Infect 71(3):223–227. https://doi.org/10.1016/j.jhin.2008.11.007
14. Rotter ML, Koller W, Neumann R (1991) The influence of cosmetic additives on the acceptability of alcohol-based hand disinfectants. Journal of Hospital Infection 18 (suppl. B): 57–63
15. Sax H, Allegranzi B, Uçkay I, Larson E, Boyce J, Pittet D (2007) 'My five moments for hand hygiene': a user-centred design approach to understand, train, monitor and report hand hygiene. J Hosp Infect 67(1):9–21
16. WHO (2009) WHO guidelines on hand hygiene in health care. First Global Patient Safety Challenge Clean Care is Safer Care, WHO, Geneva
17. WHO (2016) Global guidelines for the prevention of surgical site infections. WHO, Geneva
18. Wutzler P, Sauerbrei A (2000) Virucidal efficacy of a combination of 0.2% peracetic acid and 80% (v/v) ethanol (PAA-ethanol) as a potential hand disinfectant. J Hosp Infect 46(4):304–308. https://doi.org/10.1053/jhin.2000.0850

Antiseptic Stewardship for Skin Antiseptics

18

18.1 Composition and Intended Use

Skin antiseptics based on povidone iodine have been used in some parts of the world for decades. They are now less common as alcohol-based formulations are mostly recommended [3, 14, 25] which are usually based on ethanol, propan-2-ol, propan-1-ol or a combination of them. Typical alcohol concentrations are 63–75%. Some skin antiseptics contain additional non-volatile biocidal agents such as 0.1% benzalkonium chloride [23], 0.1–1% chlorhexidine digluconate [22, 23], 0.1% octenidine dihydrochloride [8], 8.3% povidone iodine [5] or 0.125–0.45% hydrogen peroxide [11]. Some of the products contain dyes with the aim to ensure easy visibility of the treated skin area. Some skin antiseptics also even contain fragrances or other compounds with an unknown function [11]. Non-volatile antiseptic agents will remain for some time on the skin when applied with an alcohol-based skin antiseptic although the duration of persistence and the concentrations on the skin are largely unknown.

They are used in health care on intact skin prior to a surgical intervention and before the insertion of vascular catheters or other invasive procedures. They are also used for antisepsis of vascular catheter puncture sites [1, 13, 21]. The summary below is an extract of previous book chapters on the biocidal agents.

18.2 Selection Pressure Associated with Commonly Used Biocidal Agents

18.2.1 Change of Susceptibility by Low-Level Exposure

The adaptive effects were classified as "no MIC increase", "weak MIC increase" with a $\leq$4-fold MIC increase, and "strong MIC increase" with a >4-fold MIC increase. The last category was divided in an unstable or stable MIC increase,

G. Kampf, *Antiseptic Stewardship*, https://doi.org/10.1007/978-3-319-98785-9_18

sometime the stability was unknown. A species may be found in two or more categories indicating that the adaptive response depends on the type of isolate. Most data on different adaptive effects caused by low-level exposure were found for chlorhexidine digluconate and benzalkonium chloride (both 78 species). Only few data were found for hydrogen peroxide (8 species), povidone iodine (5 species), ethanol and octenidine dihydrochloride (both 3 species) and propan-2-ol (1 species). No data were found for propan-1-ol.

Figure 18.1 shows the distribution of adaptive response categories for the selected biocidal agents. The majority of species did not show any MIC increase or only a weak MIC increase ($\leq$4-fold). A strong adaptive response was most frequently seen in benzalkonium chloride (44% of the evaluated species) and chlorhexidine digluconate (26%). The strong MIC increase was stable in 41% (benzalkonium chloride) and 40% (chlorhexidine digluconate) of species. With octenidine dihydrochloride one species was found with a strong and stable adaptive response. Hydrogen peroxide, ethanol, propan-2-ol and povidone iodine have so far not shown a strong adaptive response.

A strong and stable MIC increase after low-level exposure is the most critical adaptive response. Some species can be found in this group that have certainly a high relevance for infection control (Table 18.1). Most of them are among the group of Gram-negative species.

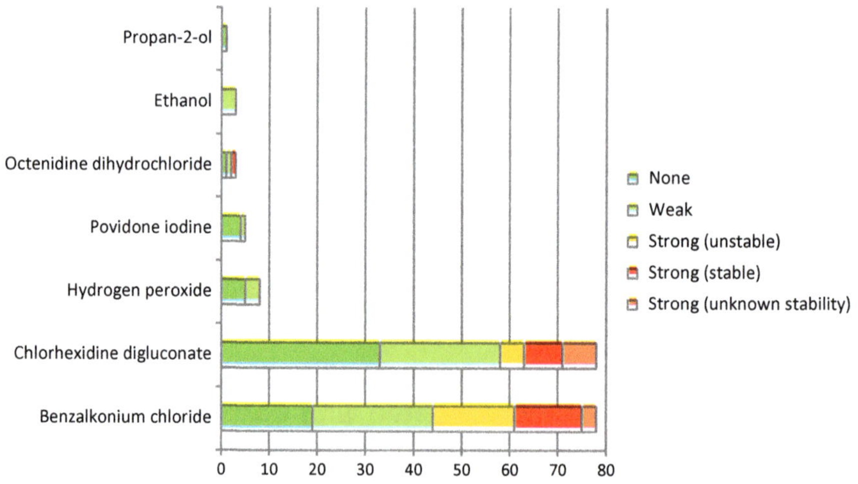

Fig. 18.1 Number of species with no, a weak or a strong adaptive MIC increase after low level exposure to biocidal agents that may be found in skin antiseptics

Table 18.1 Bacterial species with a strong (>4-fold MIC increase) and stable adaptive response after low level exposure to selected biocidal agents sometimes found in alcohol-based skin antiseptics

Biocidal agent	Bacterial species with a strong and stable adaptive MIC increase
Benzalkonium chloride	*Enterobacter* spp. ($\leq$ 300-fold)
	E. coli ($\leq$ 100-fold)
	S. aureus ($\leq$ 39-fold)
	P. aeruginosa ($\leq$ 33-fold)
	A. baumannii ($\leq$ 31-fold)
Chlorhexidine digluconate	*E. coli* ($\leq$ 500-fold)
	S. marcescens ($\leq$ 128-fold)
	P. aeruginosa ($\leq$ 32-fold)
	K. pneumoniae ($\leq$ 16-fold)
	S. aureus ($\leq$ 16-fold)
Octenidine dihydrochloride	*P. aeruginosa* ($\leq$ 32-fold)

18.2.2 Cross-Tolerance to Other Biocidal Agents

Other risks may also be relevant when the agents are used in alcohol-based skin antiseptics. The most common effect is cross-tolerance to other biocidal agents. Isolates of 22 primarily benzalkonium chloride-tolerant species were cross-tolerant to chlorhexidine digluconate and triclosan. An isolate of a benzalkonium chloride-tolerant *E. coli* was cross-tolerant to DDAC, and a benzalkonium chloride-tolerant *L. monocytogenes* was cross-tolerant to another QAC, alkylamine and sodium hypochlorite. Primarily chlorhexidine digluconate-tolerant isolates of *E. coli* and *S. Virchow* were cross-tolerant to triclosan, isolates of *S. Tyhimurium* were cross-tolerant to benzalkonium chloride, and isolates of *A. baylyi* were cross-tolerant to hydrogen peroxide. Isolates of primarily hydrogen peroxide-tolerant *E. coli* can be cross-tolerant to aldehyde, and isolates of primarily hydrogen peroxide-tolerant *S. cerevisiae* can be cross-tolerant to ethanol. A primarily octenidine dihydrochloride-tolerant *P. aeruginosa* was cross-tolerant to chlorhexidine digluconate after low-level exposure. And ethanol-adapted isolates of *L. monocytogenes* can be cross-tolerant to hydrogen peroxide. No cross-tolerance has so far been described between povidone iodine, propan-1-ol and propan-2-ol and other biocidal agents.

18.2.3 Cross-Tolerance to Antibiotics

Ethanol, propan-1-ol, propan-2-ol, povidone iodine and hydrogen peroxide have so far never been described with a cross-tolerance to antibiotics. A cross-tolerance between chlorhexidine digluconate and benzalkonium chloride and selected antibiotics can occur in numerous species. Cross-tolerance between octenidine dihydrochloride and selected antibiotics can occur in *P. aeruginosa*.

18.2.4 Efflux Pump Genes

Transporter and efflux pump genes were up-regulated after benzalkonium chloride exposure in *B. cepacia complex*, *E. coli* and *L. monocytogenes*, and after chlorhexidine digluconate exposure in *B. fragilis* and *B. cepacia complex*. No data were found for octenidine dihydrochloride, hydrogen peroxide and povidone iodine.

18.2.5 Horizontal Gene Transfer

Horizontal gene transfer can be successfully induced by chlorhexidine digluconate in *E. coli* (sulphonamide resistance by conjugation). In *B. subtilis,* ethanol at 4% can cause a 5-fold increase of mobile genetic element transfer (resistance genes). No data were found for benzalkonium chloride, octenidine dihydrochloride, hydrogen peroxide and povidone iodine.

18.2.6 Antibiotic Resistance Gene Expression

In a vanA *E. Faecium,* chlorhexidine digluconate was able to induce a $\geq$10-fold increase of vanHAX encoding VanA-type vancomycin resistance. No data were found for benzalkonium chloride, octenidine dihydrochloride, hydrogen peroxide and povidone iodine.

18.2.7 Other Risks Associated with Commonly Used Biocidal Agents

Other risks may also be relevant when the agents are used in alcohol-based skin antiseptics. They are not covered in detail. Sensitization to the agent may occur possibly resulting in local or systemic allergic reactions up to anaphylactic reactions. This has been described at least for chlorhexidine digluconate [15]. Some agents are cationic surfactants possibly resulting in a higher degree of skin irritation [15].

18.3 Effect on Biofilm

18.3.1 Biofilm Development

Biofilm is of clinical relevance, e.g. in catheter-associated bloodstream infection [9]. Typical biocidal agents in skin antiseptics show a different effect on biofilm development (Fig. 18.2). For povidone iodine biofilm formation can be inhibited in three species (*S. aureus, S. epidermidis, C. albicans.* A decrease of biofilm formation by *S. aureus* and *P. aeruginosa* was described for octenidine dihydrochloride but only at concentrations of $\geq$0.31% which has no relevance in skin antiseptics. Chlorhexidine digluconate exposure resulted in a decrease of biofilm formation in six species

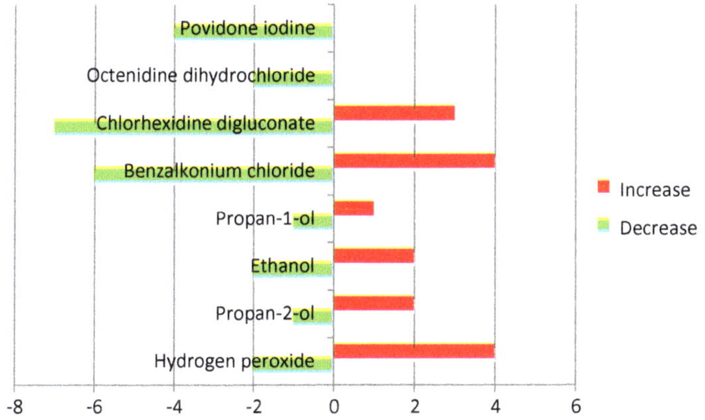

Fig. 18.2 Number of species with a decrease or increase of biofilm formation caused by biocidal agents that may be found in skin antiseptics

(*B. cepacia*, *C. albicans*, *E. faecalis*, *E. coli*, *S. aureus*, *S mutans*) and mixed biofilm. An increase, however, was observed in *K. pneumoniae*, *S. marcescens* and *S. epidermidis*. A similar result was seen for benzalkonium chloride with a decrease in 6 species (*L. monocytogenes*, *E.* Enteritidis, *E. coli*, *S. epidermidis*, *S. aureus*, *P. aeruginosa*) and an increase in *E. coli* and *S. epidermidis*. For the three alcohols, the effect seems to be equal regarding increase or decrease. For hydrogen peroxide, more species reacted with an increase (*A. oleivorans*, *P. aeruginosa*, *S. epidermidis*, *S. parasanguinis*) rather than a decrease of biofilm formation (*Candida* spp., *S. epidermidis*).

Despite some studies with evidence on enhanced biofilm formation by alcohols, the risk for a clinically significant effect remains low because the contact time in skin antisepsis is often ≤ 3 min. In addition, the evaporation time of the alcohols is probably too short for a relevant biofilm formation enhancement that may have been caused by a low concentration during alcohol evaporation at the end of the application of the antiseptic.

18.3.2 Biofilm Fixation

No data were found to assess the biofilm fixation potential of propan-2-ol, propan-1-ol, ethanol, povidone iodine, octenidine dihydrochloride, hydrogen peroxide, benzalkonium chloride or chlorhexidine digluconate.

18.3.3 Biofilm Removal

The propanols had a rather poor biofilm removal capacity. Chlorhexidine digluconate, ethanol and benzalkonium chloride showed mostly a poor or moderate biofilm removal. Octenidine dihydrochloride could equally show a poor, moderate and strong biofilm removal. Hydrogen peroxide removed mostly biofilm to a

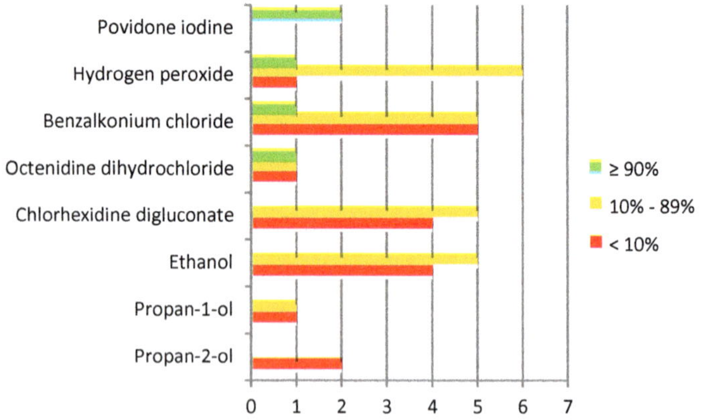

Fig. 18.3 Number of species with a strong ($\geq$90%), moderate (10–89%) or poor biofilm removal (<10%) by biocidal agents that may be found in skin antiseptics

moderate extent and with one species even to a strong extent. Povidone iodine has so far only shown to have a strong biofilm removal potential (Fig. 18.3).

18.4 Health Benefit of Commonly Used Biocidal Agents in Skin Antiseptics

A health benefit for alcohols alone in skin antiseptics (e.g. prevention of surgical site infection) has so far not been proven although they are considered to be the first choice biocidal agent for skin antisepsis [25]. The main benefit of the alcohols is the strong and immediate bactericidal and yeasticidal activity [12].

A health benefit can be expected for skin antiseptics based on alcohol and chlorhexidine digluconate for the prevention of central line-associated bloodstream infections. A meta-analysis of randomized controlled trials published in 2002 showed that treatment of the puncture site of central venous catheters with chlorhexidine digluconate instead of povidone iodine resulted in a risk ratio of 0.49 suggesting that approximately 50% of the infections could be prevented [4]. A recent study from France on 11 intensive care units with 1,181 patients and 2,457 vascular catheters described a 84% reduction of catheter-associated bloodstream infections when 70% propan-2-ol in combination with 2% chlorhexidine digluconate was used compared to 69% ethanol in combination with 5% povidone iodine [19]. Even though the selection of skin antiseptics was not ideal (ideally it would have been the same type of alcohol at the same concentration with two different types of additional biocidal agents), it nevertheless suggests a health benefit for patients treated with the propan-2-ol-chlorhexidine digluconate combination.

A similar but not significant health benefit has been shown for the combination of 45% propan-2-ol and 30% propan-1-ol with 0.1% octenidine dihydrochloride [8]. The rate of central line-associated bloodstream infections was 4.1% in the alcohol-octenidine group of 194 patients and 8.3% in the alcohol control group (74% ethanol, 10% propan-2-ol) with 194 patients. Some in vivo efficacy has been shown for an alcohol-based skin antiseptic (45% iso-propanol, 30% n-propanol) containing in addition 0.1% octenidine dihydrochloride. It significantly reduced bacterial re-growth over 24 or 48 h at the insertion site of central venous lines or epidural catheters [7, 16]. 0.1% octenidine dihydrochloride in 70% propan-2-ol also had a sustained efficacy on the resident skin flora of the upper arm when measured 10 min, 3 h or 6 h after application. It was equivalent to the effect of 2% chlorhexidine digluconate but superior to the effect of 0.5% chlorhexidine digluconate and 1% povidone iodine after 3 and 6 h [18].

A health benefit can probably also be expected for skin antiseptics based on alcohol and chlorhexidine digluconate for the prevention of surgical site infections as some studies suggest, but others do not [5, 20, 24]. No studies are currently available to evaluate a health benefit (prevention of surgical site infections) for the combination of alcohol and octenidine dihydrochloride.

Since 2016, the WHO recommends to use skin antiseptics based on alcohol and chlorhexidine digluconate for the prevention of surgical site infections. The reason is that their meta-analysis of available studies resulted in a significant health benefit for patients [25]. This recommendation has been described by some authors as "premature" because some other studies were not included in the meta-analysis which may have changed the overall result [17]. The WHO reviewed their recommendations once more and reconfirmed it [2]. The CDC recommended in 2017 alcohol-based skin antiseptics prior to surgery and did not recommend any additional biocidal agents [3]. The Commission for Hospital Hygiene and Infection Prevention at the Robert Koch-Institute in Germany recommends since 2018 to use alcohol-based skin antiseptics and state that additional biocidal agents may have a health benefit but the evidence is not strong enough to favour specific biocidal agents [14]. The variability of current recommendations seems to reflect the scientific uncertainty on the expectable health benefit associated with a combination of alcohol and chlorhexidine digluconate or other biocidal agents for the prevention of surgical site infections.

A health benefit has not been shown for any other biocidal agents used in alcohol-based skin antiseptics such as benzalkonium chloride, povidone iodine or hydrogen peroxide. A persistent antimicrobial activity within 48 h at the insertion site of central venous lines or epidural catheters has not been shown for a skin antiseptic based on 63% propan-2-ol and benzalkonium chloride, probably at 0.025% [6, 16]. A persistent efficacy of hydrogen peroxide is unlikely anyway because many bacterial species such as *S. epidermidis* have catalase activity decomposing hydrogen peroxide to water and oxygen.

18.5 Antiseptic Stewardship Implications

Overall, the probability for a clinically relevant selection pressure caused by low-level exposure to alcohols is small. The main reason is the volatility of the alcohols. An appropriate volume for skin antisepsis may be low with 1–2 ml (e.g. before an injection) or high with 20–25 ml (e.g. before major surgery) depending on the size and type of treated skin surface and the recommended exposure time. The alcohol(s) should be completely evaporated from the skin before the intervention begins. The contact time between the alcohols at the use concentration (63–75%) and the microorganisms may vary between 15 s and 10 min. The evaporation time at the end of the contact time is probably too short for any adaptive response possibly resulting in a lower susceptibility to the alcohols that may have been caused by a low concentration during alcohol evaporation at the end of the application of the antiseptic. In that respect alcohols are the preferred choice as the main biocidal agent in skin antiseptics.

Alcohol-based skin antiseptics with additional chlorhexidine digluconate seem to have a health benefit, at least for the prevention of catheter-associated blood stream infections and probably also for the prevention of surgical site infections although the last one is still under controversial debate in the scientific community. At the same time, low-level chlorhexidine digluconate exposure has the risk for a strong and stable adaptive response in various nosocomial pathogens such as *E. coli*, *S. marcescens*, *P. aeruginosa*, *K. pneumoniae* and *S. aureus* resulting in tolerance to chlorhexidine digluconate and some other biocidal agents such as triclosan, benzalkonium chloride or hydrogen peroxide. Cross-tolerance between chlorhexidine digluconate and selected antibiotics can occur in numerous species. Transporter and efflux pump genes can be up-regulated in *B. fragilis* and *B. cepacia complex*, horizontal gene transfer can be successfully induced in *E. coli*, and vancomycin resistance can be induced by a $\geq$ 10-fold increase of vanHAX in a vanA *E. faecium*. Biofilm formation is rather decreased than increased. Despite all risks, there are obvious health benefits. In addition, treatment of the skin with an alcohol-based antiseptic is often carried out only once in a patient, e.g. before an operation. It may well be carried out more often, e.g. once every two days in central venous catheters covered with a gauze dressing or once every seven days in central venous catheters covered with a transparent dressing [10]. But overall, it is a rather seldom type of treatment compared to hand disinfection with up to 179 applications per person and day [10]. The associated health benefit currently allows accepting the obvious chlorhexidine digluconate-associated risks.

Alcohol-based skin antiseptics with additional octenidine dihydrochloride may also have a health benefit, at least for the prevention of catheter-associated blood stream infections although the evidence is much weaker compared to chlorhexidine digluconate. Prevention of surgical site infections by additional octenidine dihydrochloride has not been described. Low-level exposure indicates that at least in *P. aeruginosa* a strong and stable adaptive response including a cross-tolerance to chlorhexidine digluconate can be induced. Cross tolerance between octenidine dihydrochloride and selected antibiotics can occur in *P. aeruginosa*. Biofilm formation is mainly decreased. The expectable health benefit of octenidine dihydrochloride in alcohol-based skin antiseptics is not as convincing compared to chlorhexidine digluconate, but the associated risks associated with selection pressure and biofilm formation seem to be smaller.

There is currently no evidence to support benzalkonium chloride as an additional biocidal agent in alcohol-based skin antiseptics. On the contrary, there are some relevant risks such as the possibility of a strong and stable adaptive response in various nosocomial pathogens such as *E. coli*, *S. aureus*, *P. aeruginosa* and *A. baumannii* resulting in tolerance to benzalkonium chloride and some other biocidal agents (chlorhexidine digluconate or triclosan) and selected antibiotics, up-regulation of transporter and efflux pump genes in *B. cepacia complex*, *E. coli* and *L. monocytogenes* and an increase in biofilm formation in *S. epidermidis*. That is why alcohol-based skin antiseptics containing benzalkonium chloride should be replaced by alcohol-based skin antiseptics without benzalkonium chloride. They should have at least equal efficacy and local tolerance. They may also be replaced by alcohol-based skin antiseptics of superior efficacy, e.g. with additional chlorhexidine digluconate or possibly octenidine dihydrochloride.

There is currently also no evidence for a health benefit to support hydrogen peroxide or povidone iodine as additional biocidal agents in alcohol-based skin antiseptics. The risks are, however, rather small. A strong adaptive response to low-level exposure has not been reported yet for any of the two biocidal agents. That is why the associated selection pressure can be regarded as substantially lower compared to benzalkonium chloride.

References

1. Allegranzi B, Bischoff P, de Jonge S, Kubilay NZ, Zayed B, Gomes SM, Abbas M, Atema JJ, Gans S, van Rijen M, Boermeester MA, Egger M, Kluytmans J, Pittet D, Solomkin JS (2016) New WHO recommendations on preoperative measures for surgical site infection prevention: an evidence-based global perspective. Lancet Infect Dis 16(12):e276–e287. https://doi.org/10.1016/s1473-3099(16)30398-x
2. Allegranzi B, Egger M, Pittet D, Bischoff P, Nthumba P, Solomkin J (2017) WHO's recommendation for surgical skin antisepsis is premature - Authors' reply. Lancet Infect Dis 17(10):1024–1025. https://doi.org/10.1016/s1473-3099(17)30526-1
3. Berrios-Torres SI, Umscheid CA, Bratzler DW, Leas B, Stone EC, Kelz RR, Reinke CE, Morgan S, Solomkin JS, Mazuski JE, Dellinger EP, Itani KMF, Berbari EF, Segreti J, Parvizi J, Blanchard J, Allen G, Kluytmans J, Donlan R, Schecter WP (2017) Centers for disease control and prevention guideline for the prevention of surgical site infection. JAMA Surg 152(8):784–791. https://doi.org/10.1001/jamasurg.2017.0904
4. Chaiyakunapruk N, Veenstra DL, Lipsky BA, Saint S (2002) Chlorhexidine compared with povidone-iodine solution for vascular catheter-site care: a meta-analysis. Ann Intern Med 136 (11):792–801
5. Darouiche RO, Wall MJ, Itani KM, Otterson MF, Webb AL, Carrick MM, Miller HJ, Awad SS, Crosby CT, Mosier MC, Alsharif A, Berger DH (2010) Chlorhexidine-alcohol versus povidone-iodine for surgical-site antisepsis. N Engl J Med 362(1):18–26
6. Debreceni G, Meggyesi R, Mestyan G (2007) Efficacy of spray disinfection with a 2-propanol and benzalkonium chloride containing solution before epidural catheter insertion–a prospective, randomized, clinical trial. Br J Anaesth 98(1):131–135. https://doi.org/10.1093/bja/ael288
7. Dettenkofer M, Jonas D, Wiechmann C, Rossner R, Frank U, Zentner J, Daschner FD (2002) Effect of skin disinfection with octenidine dihydrochloride on insertion site colonization of intravascular catheters. Infection 30(5):282–285
8. Dettenkofer M, Wilson C, Gratwohl A, Schmoor C, Bertz H, Frei R, Heim D, Luft D, Schulz S, Widmer AF (2010) Skin disinfection with octenidine dihydrochloride for central venous catheter site care: a double-blind, randomized, controlled trial. Clin Microbiol Infect 16(6):600–606. https://doi.org/10.1111/j.1469-0691.2009.02917.x

9. Hoiby N, Bjarnsholt T, Moser C, Bassi GL, Coenye T, Donelli G, Hall-Stoodley L, Hola V, Imbert C, Kirketerp-Moller K, Lebeaux D, Oliver A, Ullmann AJ, Williams C (2015) ESCMID guideline for the diagnosis and treatment of biofilm infections 2014. Clin Microbiol Infect 21(Suppl 1):S1–25. https://doi.org/10.1016/j.cmi.2014.10.024
10. Kampf G (2016) Acquired resistance to chlorhexidine – is it time to establish an "antiseptic stewardship" initiative? J Hosp Infect 94(3):213–227
11. Kampf G (2017) Hautantiseptik - neue Erkenntnisse und Empfehlungen. Krankenhaushygiene Up2date 12(2):143–155
12. Kampf G, Kramer A (2004) Epidemiologic background of hand hygiene and evaluation of the most important agents for scrubs and rubs. Clin Microbiol Rev 17(4):863–893
13. KRINKO am Robert Koch Institut (2017) Prävention von Infektionen, die von Gefäßkathetern ausgehen. Bundesgesundheitsblatt 60(2):171–215
14. KRINKO am Robert Koch Institut (2018) Prävention postoperativer Wundinfektionen. Bundesgesundheitsblatt 61(4):448–473
15. Lachapelle JM (2014) A comparison of the irritant and allergenic properties of antiseptics. Eur J Dermatol: Ejd 24(1):3–9. https://doi.org/10.1684/ejd.2013.2198
16. Lutz JT, Diener IV, Freiberg K, Zillmann R, Shah-Hosseini K, Seifert H, Berger-Schreck B, Wisplinghoff H (2016) Efficacy of two antiseptic regimens on skin colonization of insertion sites for two different catheter types: a randomized, clinical trial. Infection 44(6):707–712. https://doi.org/10.1007/s15010-016-0899-6
17. Maiwald M, Widmer AF (2017) WHO's recommendation for surgical skin antisepsis is premature. Lancet Infect Dis 17(10):1023–1024. https://doi.org/10.1016/s1473-3099(17)30448-6
18. Melichercikova V, Urban J, Goroncy-Bermes P (2010) Residual effect of antiseptic substances on human skin. J Hosp Infect 75(3):238–239. https://doi.org/10.1016/j.jhin.2009.12.010
19. Mimoz O, Lucet JC, Kerforne T, Pascal J, Souweine B, Goudet V, Mercat A, Bouadma L, Lasocki S, Alfandari S, Friggeri A, Wallet F, Allou N, Ruckly S, Balayn D, Lepape A, Timsit JF (2015) Skin antisepsis with chlorhexidine-alcohol versus povidone iodine-alcohol, with and without skin scrubbing, for prevention of intravascular-catheter-related infection (CLEAN): an open-label, multicentre, randomised, controlled, two-by-two factorial trial. Lancet 386(10008):2069–2077. https://doi.org/10.1016/s0140-6736(15)00244-5
20. Ngai IM, Van Arsdale A, Govindappagari S, Judge NE, Neto NK, Bernstein J, Bernstein PS, Garry DJ (2015) Skin preparation for prevention of surgical site infection after cesarean delivery: a randomized controlled trial. Obstet Gynecol 126(6):1251–1257. https://doi.org/10.1097/aog.0000000000001118
21. O'Grady NP, Alexander M, Burns LA, Dellinger EP, Garland J, Heard SO, Lipsett PA, Masur H, Mermel LA, Pearson ML, Raad, II, Randolph AG, Rupp ME, Saint S (2011) Guidelines for the prevention of intravascular catheter-related infections. Am J Infect Control 39(4 Suppl 1):S1–S34. doi:S0196-6553(11)00085-X [pii]. https://doi.org/10.1016/j.ajic.2011.01.003 [doi]
22. Rochon-Edouard S, Pons JL, Veber B, Larkin M, Vassal S, Lemeland JF (2004) Comparative in vitro and in vivo study of nine alcohol-based handrubs. Am J Infect Control 32(4):200–204. https://doi.org/10.1016/j.ajic.2003.08.003
23. Rosas-Ledesma P, Mariscal A, Carnero M, Munoz-Bravo C, Gomez-Aracena J, Aguilar L, Granizo JJ, Lafuente A, Fernandez-Crehuet J (2009) Antimicrobial efficacy in vivo of a new formulation of 2-butanone peroxide in n-propanol: comparison with commercial products in a cross-over trial. J Hosp Infect 71(3):223–227. https://doi.org/10.1016/j.jhin.2008.11.007
24. Tuuli MG, Liu J, Stout MJ, Martin S, Cahill AG, Odibo AO, Colditz GA, Macones GA (2016) A randomized trial comparing skin antiseptic agents at cesarean delivery. N Engl J Med 374(7):647–655. https://doi.org/10.1056/NEJMoa1511048
25. WHO (2016) Global guidelines for the prevention of surgical site infections. WHO, Geneva

Antiseptic Stewardship for Surface Disinfectants

19

19.1 Composition and Intended Use

Surface disinfectants can be based on different types of biocidal agents such as benzalkonium chloride, DDAC, glutaraldehyde, alcohols, hydrogen peroxide, silver (mostly in combination with hydrogen peroxide), peracetic acid and sodium hypochlorite [3, 4, 6, 9, 10, 12]. Many products contain two or more of them as a formulation. They often contain additional auxiliary agents, e.g. for adjustment of the pH value or as anticorrosives.

Surface disinfectants are used in health care, nursing homes, veterinary medicine, food processing and manufacturing and occasionally also in the domestic setting. In health care, areas with different risks for infection from contaminated surfaces have been described. A "possible risk" for infection exists in some areas such as general wards, whereas a "special risk" for infection exists in other areas such as operating theatres, intensive care units, transplant units or haemato-oncology units [13]. The summary below is an extract of previous book chapters on the biocidal agents.

19.2 Selection Pressure Associated with Commonly Used Biocidal Agents

19.2.1 Change of Susceptibility by Low-Level Exposure

Any adaptive effects were classified as "no MIC increase", "weak MIC increase" with a $\leq$4-fold MIC increase and "strong MIC increase" with a >4-fold MIC increase. The last category was divided into an unstable or stable MIC increase, and sometimes the stability was unknown. A species may be found in two or more categories indicating that the adaptive response depends on the type of isolate and not primarily on the species itself. Most data on different adaptive effects caused by low-level exposure were found for benzalkonium chloride (78 species), DDAC (48 species) and

G. Kampf, *Antiseptic Stewardship*, https://doi.org/10.1007/978-3-319-98785-9_19

silver (20 species). Only few data were found for hydrogen peroxide (8 species), sodium hypochlorite (7 species), peracetic acid (3 species) and glutaraldehyde and propan-1-ol (both 1 species). No data were found for ethanol and propan-2-ol.

Figure 19.1 shows the distribution of adaptive response categories for the different biocidal agents. The majority of species did not show any MIC increase or only a weak MIC increase ($\leq$ 4-fold). A strong adaptive response was most frequently seen in benzalkonium chloride (44% of the evaluated species), followed by silver (40%) and DDAC (15%). With peracetic acid, one species showed a strong adaptive response. The strong MIC increase was stable in 50% (silver), 41% (benzalkonium chloride) and 14% (DDAC) of species. The strong and stable adaptive response to silver was mostly dependent on the presence of sil genes (see also Chap. 15).

A strong and stable MIC increase after low-level exposure is probably the most critical adaptive response. Some species can be found in this group that have certainly a high relevance for infection control (Table 19.1). Most of them are among the group of Gram-negative species.

19.2.2 Cross-Tolerance to Other Biocidal Agents

Other risks may also be relevant when the agents are used in surface disinfectants. The most common effect is cross-tolerance to other biocidal agents. Isolates of 22 primarily benzalkonium chloride-tolerant species were cross-tolerant to chlorhexidine digluconate and triclosan. An isolate of a benzalkonium chloride-tolerant *E. coli* was cross-tolerant to DDAC, and a benzalkonium chloride-tolerant *L. monocytogenes* was cross-tolerant to another quaternary ammonium compound,

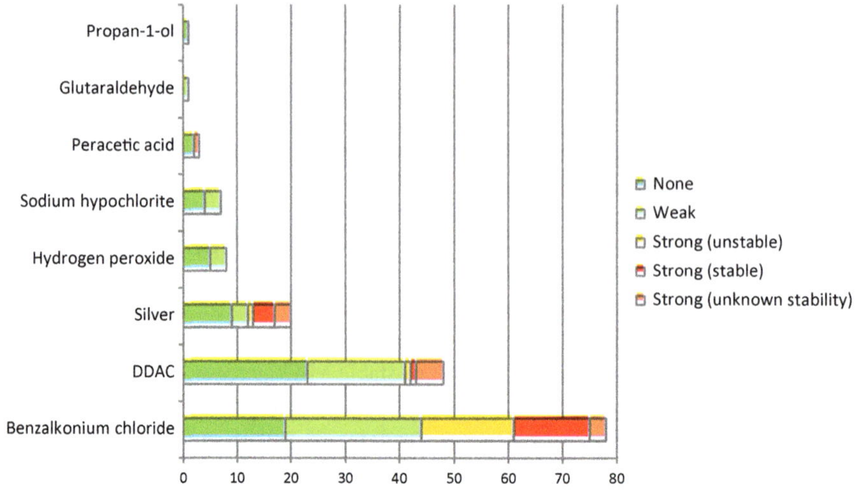

Fig. 19.1 Number of species with no, a weak or a strong adaptive MIC increase after low-level exposure to biocidal agents typically found in surface disinfectants

Table 19.1 Examples for healthcare-associated bacterial species with a strong (>4-fold MIC increase) and stable adaptive response after low-level exposure to selected biocidal agents

Biocidal agent	Bacterial species with a strong and stable adaptive MIC increase
Benzalkonium chloride	*Enterobacter* spp. ($\leq$300-fold)
	E. coli ($\leq$100-fold)
	S. aureus ($\leq$39-fold)
	P. aeruginosa ($\leq$33-fold)
	A. baumannii ($\leq$31-fold)
Silver	*E. coli* (128-fold)[a]
	E. cloacae ($\geq$32-fold)[a]
	K. pneumoniae ($\geq$32-fold)[a]
	K. oxytoca ($\geq$16-fold)[a]
DDAC	*P. aeruginosa* ($\geq$18-fold)

[a]Mainly sil-positive isolates or strains

alkylamine and sodium hypochlorite. Isolates of primarily DDAC-tolerant *E. coli and P. fluorescens* can be cross-tolerant to benzalkonium chloride, and isolates of *E. faecium*, *E. hirae*, *E. coli*, *P. putida*, *S. enteritidis* and *C. argentea* can be cross-tolerant to copper via specific efflux pumps. In addition, isolates of primarily peracetic acid-tolerant *B. subtilis* can be cross-tolerant to other oxidising agents, isolates of primarily hydrogen peroxide-tolerant *E. coli* can be cross-tolerant to aldehyde, and isolates of primarily hydrogen peroxide-tolerant *S. cerevisiae* can be cross-tolerant to ethanol. Isolates of primarily sodium hypochlorite-tolerant *E. coli* can be cross-tolerant to hydrogen peroxide, and cross-tolerance to benzalkonium chloride, another quaternary ammonium compound and alkylamine can occur in *L. monocytogenes*. Primarely glutaraldehyde-tolerant *E. coli*, *Halomonas* spp. and *B. cepacia* can be cross-tolerant to other aldehydes. And ethanol-adapted isolates of *L. monocytogenes* can be cross-tolerant to hydrogen peroxide. No cross-tolerance has so far been described between propan-1-ol and propan-2-ol and other biocidal agents. Especially, the rather frequently observed cross-tolerance between benzalkonium chloride, triclosan and chlorhexidine is a clear indication to carefully select biocidal agents in order to reduce this type of cross-tolerance to a minimum.

19.2.3 Cross-Tolerance to Antibiotics

Ethanol, propan-1-ol, propan-2-ol, peracetic acid, hydrogen peroxide and sodium hypochlorite have so far never been described with a cross-tolerance to antibiotics. A cross-tolerance between selected antibiotics and the biocidal agents benzalkonium chloride and silver can occur in numerous species. Occasional cross-resistance between DDAC and selected antibiotics was found in *C. coli, E. coli, L. monocytogenes* and *S. enterica*. Cross resistances to rifampicin and sometimes also to isoniazid have been reported in glutaraldehyde-resistant *M. chelonae*.

19.2.4 Efflux Pump Genes

Transporter and efflux pump genes were up-regulated after benzalkonium chloride exposure in *B. cepacia complex*, *E. coli* and *L. monocytogenes*, after silver exposure in *A. baumannii*, *E. coli*, *E. hirae* and *C. albicans*, and after exposure to glutaraldehyde in *Pseudomonas* spp.

19.2.5 Resistance Gene Plasmids

Plasmids with silver resistance genes can be found in *A. baumannii*, *C. metallidurans*, *E. cloacae*, *Klebsiella* spp., *P. stutzeri*, *Salmonella* spp. and *S. marcescens*. A plasmid with resistance to glutaraldehyde was detected in *S. aureus*.

19.2.6 Viable But Not Culturable

Sodium hypochlorite is able to induce the VBNC state with enhanced antibiotic tolerance in *E. coli*. Peracetic acid is able to induce the VBNC state in *S. Typhimurium*.

19.2.7 Horizontal Gene Transfer

Mobile genetic element transfer (resistance genes) can be successfully induced 5-fold by ethanol in *B. subtilis*.

19.2.8 Other Risks Associated with Biocidal Agents in Surface Disinfectants

Occupational exposure risks, material compatibility, stability, user acceptance and may be other risks can be found with different biocidal agents and products [18]. Some biocidal agents such as benzalkonium chloride may bind to some types of fibre such as white pulp or cotton towels so that the strength of the disinfectant solution is not sufficient anymore to ensure an adequate antimicrobial activity [2, 8]. Use solutions may become contaminated especially when disinfectants are based on quaternary ammonium compounds when the tissue dispensers are not reprocessed adequately [10, 11]. These aspects should also been taken into account.

19.3 Effect of Commonly Used Biocidal Agents on Biofilm

19.3.1 Biofilm Development

Surface-attached cells are likely to be common on dry hospital surfaces, and there is evidence that they also harbour established biofilms [14]. Reduced susceptibility to biocides combined with protection from physical removal through cleaning is likely

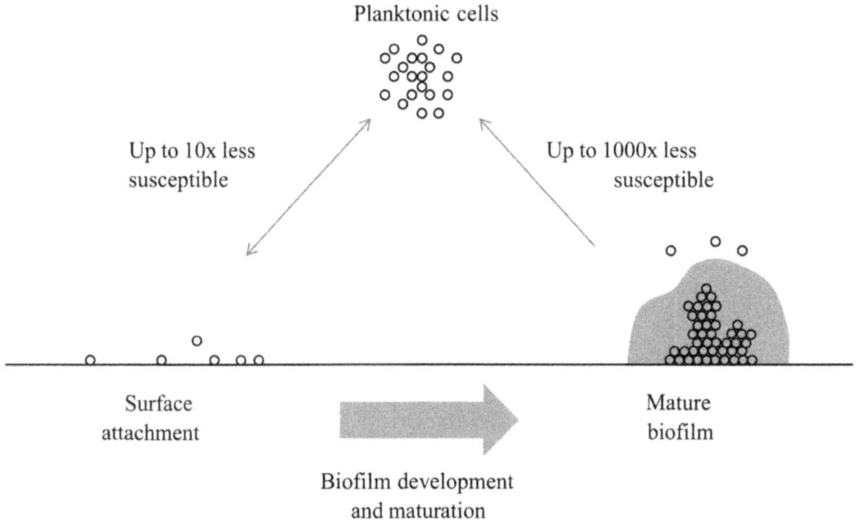

Fig. 19.2 Schematic of surface attachment, biofilm formation and biocide susceptibility [17]. Reprinted from the Journal of Hospital Infection, Volume number 89, Issue number 1, Authors Otter JA, Vickery K, Walker JT, deLancey Pulcini E, Stoodley P, Goldenberg SD et al., Surface-attached cells, biofilms and biocide susceptibility: implications for hospital cleaning and disinfection, Pages 16–27, Copyright 2015, with permission from Elsevier

to contribute to failures in hospital cleaning and disinfection (Fig. 19.2). Biofilms may explain why vegetative bacteria can survive for unusually long periods (weeks to months) on dry hospital surfaces. Also, the presence of surface-attached bacteria and biofilms is likely to interfere with attempts to recover bacteria from hospital surfaces, and may lead to underestimation of both the prevalence of contamination with pathogens and the number of bacteria that are on surfaces. This has important implications, particularly for hospital outbreak investigation [17].

Typical biocidal agents in surface disinfectants show a different effect on biofilm development (Fig. 19.3). For silver, biofilm formation can be inhibited in six species (*C. albicans*, *C. parapsilosis*, *C. tropicalis*, *E. coli*, *S. epidermidis*, *S. aureus*). Similar results are found for peracetic acid with an inhibition of biofilm formation in three species (*C. sakazakii*, *Candida* spp., *S. aureus*). For benzalkonium chloride, biofilm formation can be inhibited (*L. monocytogenes*, *S. Enteritidis*, *P. aeruginosa*, *E. coli*, *S. aureus* and *S. epidermidis*) or enhanced (*S. agalactiae*, *E. coli*, *S. aureus* and *S. epidermidis*). For the alcohols, the effect on biofilm formation seems to be both an increase and a decrease. Sodium hypochlorite and hydrogen peroxide can rather inhibit than enhance biofilm formation.

19.3.2 Biofilm Fixation

For most biocidal agents, no data were found to assess the biofilm fixation potential (silver, propan-2-ol, propan-1-ol, ethanol, sodium hypochlorite, hydrogen peroxide,

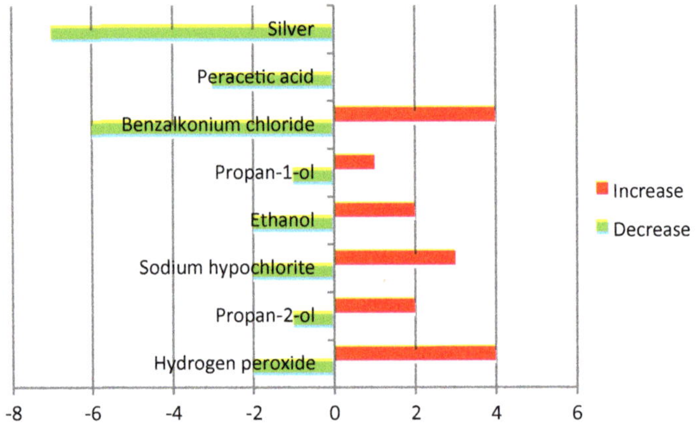

Fig. 19.3 Number of species with a decrease or increase of biofilm formation caused by biocidal agents that may be found in surface disinfectants

triclosan). Glutaraldehyde typically results in a moderate to strong biofilm fixation, whereas peracetic acid typically causes a poor or moderate biofilm fixation. Benzalkonium chloride was able to increase biofilm mechanical stability in *P. fluorescens* suggesting some biofilm fixation.

19.3.3 Biofilm Removal

Propan-2-ol and glutaraldehyde had a rather poor biofilm removal capacity. Ethanol, propan-1-ol and benzalkonium chloride could remove biofilm poorly or moderately. Silver, hydrogen peroxide, sodium hypochlorite and peracetic acid showed mostly a moderate and rarely a poor or strong biofilm removal (Fig. 19.4).

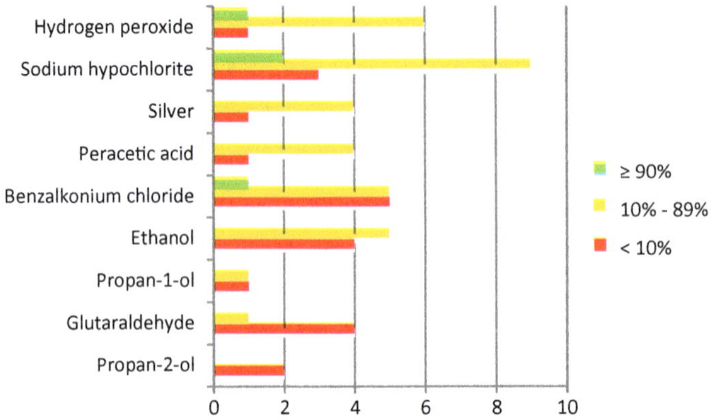

Fig. 19.4 Number of species with a strong ($\geq 90\%$), moderate (10–89%) or poor biofilm removal (<10%) by biocidal agents that may be found in surface disinfectants

19.4 Health Benefits of Biocidal Agents in Surface Disinfectants

The expected benefit of the surface disinfectants is a major reduction of the surface contamination with the aim to lower the risk of pathogen transmission from the contaminated surface to the patient or the healthcare worker. Some biocidal agents are typically used as single agents, e.g. sodium hypochlorite. Other agents are used as mixtures, e.g. different quaternary ammonium compounds (sometimes also in combination with aldehydes), peracetic acid in combination with hydrogen peroxide, or hydrogen peroxide in combination with silver. It is therefore often difficult to evaluate the effect of a single biocidal agent on surfaces.

A health benefit by surface disinfection has been questioned for routine use in hospitals suggesting that a patient health benefit often depends on the epidemiological situation and the target micro-organism [7]. In an outbreak situation or on special care units, however, a health benefit has been described for various biocidal agents such as hydrogen peroxide (multiresistant *A. baumannii*) [5] or sodium hypochlorite (*C. difficile*) [1, 16]. In addition, a surface disinfection with an inappropriate concentration of the biocidal agent (e.g. 0.08% instead of 0.5% sodium hypochlorite) resulted in an outbreak of imipenem-resistant *A. baumannii* on an intensive care unit [15] suggesting a health benefit for regular use.

19.5 Antiseptic Stewardship Implications

Biofilm containing multiresistant micro-organisms can persist on clinical surfaces from an intensive care unit despite terminal cleaning, suggesting that current cleaning practices may be inadequate to control biofilm development. The presence of multiresistant micro-organisms being protected within these biofilms may be the mechanism by which they persist within the hospital environment [19]. That is why the effect on biofilm development and removal is considered a major aspect for antiseptic stewardship in addition to selection pressure.

A low adaptive response in combination with mostly an inhibition of biofilm formation and removal of existing biofilm can be attributed to none of the evaluated biocidal agents. Nevertheless, peracetic acid showed only rarely a strong adaptive response, mainly decreased biofilm formation and moderately removed existing biofilm. No relevant adaptive response was found with sodium hypochlorite and hydrogen peroxide, and biofilm removal was overall favourable for both agents but they mostly increased biofilm formation. Silver could also cause a strong and stable adaptive response in a few mainly sil-positive species. It mainly decreased biofilm formation and was as nanoparticles able to moderately remove biofilm in the majority of species. Peracetic acid, hydrogen peroxide and sodium hypochlorite have so far not been associated with antibiotic cross-resistance. Benzalkonium chloride was the substance causing a strong and stable adaptive response in numerous species including cross-tolerance to other biocidal agents and selected antibiotics. Biofilm formation could be inhibited or enhanced, and biofilm removal was moderate or poor.

Data for the other biocidal agents are less comprehensive. Glutaraldehyde and propan-1-ol did not cause a strong adaptive response. The effect of propan-1-ol on biofilm formation and removal was inconsistent, glutaraldehyde showed only poor biofilm removal. For ethanol, propan-2-ol and DDAC only one or two of the parameter could be described so that a further evaluation was not done.

Overall, on surfaces where biofilm formation should be inhibited the use or peracetic acid seems to be the most appropriate option (low selection pressure). Hydrogen peroxide and sodium hypochlorite have also a low selection pressure and can moderately remove biofilm in many species. They seem to be appropriate on surfaces where enhancement of biofilm formation is of minor relevance because they can enhance biofilm formation in a few species. Benzalkonium chloride seems to be the least suitable biocidal agent taking into account the observed strong and stable adaptive response in many species and the inconclusive effect on biofilm formation and removal.

References

1. Apisarnthanarak A, Zack JE, Mayfield JL, Freeman J, Dunne WM, Little JR, Mundy LM, Fraser VJ (2004) Effectiveness of environmental and infection control programs to reduce transmission of *Clostridium difficile*. Clin Infect Dis: Off Publ Infect Dis Soc Am 39:601–602
2. Bloß R, Meyer S, Kampf G (2010) Adsorption of active ingredients from surface disinfectants to different types of fabrics. J Hosp Infect 75:56–61
3. Boyce JM (2016) Modern technologies for improving cleaning and disinfection of environmental surfaces in hospitals. Antimicrob Resist Infect Control 5:10. https://doi.org/10.1186/s13756-016-0111-x
4. Boyce JM (2018) Alcohols as surface disinfectants in healthcare settings. Infect Control Hosp Epidemiol 39(3):323–328. https://doi.org/10.1017/ice.2017.301
5. Chmielarczyk A, Higgins PG, Wojkowska-Mach J, Synowiec E, Zander E, Romaniszyn D, Gosiewski T, Seifert H, Heczko P, Bulanda M (2012) Control of an outbreak of Acinetobacter baumannii infections using vaporized hydrogen peroxide. J Hosp Infect 81(4):239–245. https://doi.org/10.1016/j.jhin.2012.05.010
6. Cobrado L, Pinto Silva A, Pina-Vaz C, Rodrigues A (2018) Effective Disinfection of a Burn Unit after Two Cases of Sepsis Caused by Multi-Drug-Resistant Acinetobacter baumannii. Surg Infect. https://doi.org/10.1089/sur.2017.311
7. Dettenkofer M, Wenzler S, Amthor S, Antes G, Motschall E, Daschner FD (2004) Does disinfection of environmental surfaces influence nosocomial infection rates? A systematic review. Am J Infect Control 32(2):84–89
8. Engelbrecht K, Ambrose D, Sifuentes L, Gerba C, Weart I, Koenig D (2013) Decreased activity of commercially available disinfectants containing quaternary ammonium compounds when exposed to cotton towels. Am J Infect Control 41(10):908–911. https://doi.org/10.1016/j.ajic.2013.01.017
9. Horejsh D, Kampf G (2011) Efficacy of three surface disinfectants against spores of Clostridium difficile ribotype 027. Int J Hyg Environ Health 214(2):172–174. S1438-4639(10)00128-8 [pii]. https://doi.org/10.1016/j.ijheh.2010.10.004 [doi]

10. Kampf G, Degenhardt S, Lackner S, Jesse K, von Baum H, Ostermeyer C (2014) Poorly processed reusable surface disinfection tissue dispensers may be a source of infection. BMC Infect Dis 14(1):37. 1471-2334-14-37 [pii]. https://doi.org/10.1186/1471-2334-14-37 [doi]
11. Kampf G, Degenhardt S, Lackner S, Ostermeyer C (2014) Effective processing or reusable dispensers for surface disinfection tissues - the devil is in the details. GMS Hyg Infect Control 9(1):DOC09
12. Kean R, Sherry L, Townsend E, McKloud E, Short B, Akinbobola A, Mackay WG, Williams C, Jones BL, Ramage G (2018) Surface disinfection challenges for Candida auris: an in-vitro study. J Hosp Infect 98(4):433–436. https://doi.org/10.1016/j.jhin.2017.11.015
13. KRINKO am Robert Koch Institut (2004) Anforderungen an die Hygiene bei der Reinigung und Desinfektion von Flächen. Bundesgesundheitsblatt 47(1):51–61
14. Ledwoch K, Dancer SJ, Otter JA, Kerr K, Roposte D, Maillard JY (2018) Beware Biofilm! Dry biofilms containing bacterial pathogens on multiple healthcare surfaces; a multicentre study. J Hosp Infect. https://doi.org/10.1016/j.jhin.2018.06.028
15. Liu WL, Liang HW, Lee MF, Lin HL, Lin YH, Chen CC, Chang PC, Lai CC, Chuang YC, Tang HJ (2014) The impact of inadequate terminal disinfection on an outbreak of imipenem-resistant Acinetobacter baumannii in an intensive care unit. PLoS ONE 9(9): e107975. https://doi.org/10.1371/journal.pone.0107975
16. Mayfield JM, Leet T, Miller J, Mundy LM (2000) Environmental control to reduce transmission of *Clostridium difficile*. Clin Infect Dis 31:995–1000
17. Otter JA, Vickery K, Walker JT, deLancey Pulcini E, Stoodley P, Goldenberg SD, Salkeld JA, Chewins J, Yezli S, Edgeworth JD (2015) Surface-attached cells, biofilms and biocide susceptibility: implications for hospital cleaning and disinfection. J Hosp Infect 89 (1):16–27. https://doi.org/10.1016/j.jhin.2014.09.008
18. Quinn MM, Henneberger PK, Braun B, Delclos GL, Fagan K, Huang V, Knaack JL, Kusek L, Lee SJ, Le Moual N, Maher KA, McCrone SH, Mitchell AH, Pechter E, Rosenman K, Sehulster L, Stephens AC, Wilburn S, Zock JP (2015) Cleaning and disinfecting environmental surfaces in health care: Toward an integrated framework for infection and occupational illness prevention. Am J Infect Control 43(5):424–434. https://doi.org/10.1016/j.ajic.2015.01.029
19. Vickery K, Deva A, Jacombs A, Allan J, Valente P, Gosbell IB (2012) Presence of biofilm containing viable multiresistant organisms despite terminal cleaning on clinical surfaces in an intensive care unit. J Hosp Infect 80(1):52–55. https://doi.org/10.1016/j.jhin.2011.07.007

Antiseptic Stewardship for Instrument Disinfectants 20

20.1 Composition and Intended Use

Instrument disinfectants can be based on different types of biocidal agents such as benzalkonium chloride, DDAC, glutaraldehyde, hydrogen peroxide, peracetic acid and sodium hypochlorite [5–7]. Many products contain two or more of them in a formulation. They often contain additional auxiliary agents, e.g. for adjustment of the pH value or as anticorrosives.

Instrument disinfectants are mainly used in health care and veterinary medicine, occasionally also in nursing homes. As part of the reprocessing of instruments, they are used for disinfection of non-critical medical devices such as stethoscopes, semi-critical medical devices such as flexible endoscopes and critical medical devices such as surgical instruments [4, 6]. Treatment of semi-critical items requires high-level disinfection typically with glutaraldehyde or peracetic acid. Low-level disinfection is sufficient for non-critical items where different biocidal agents may be used [5]. The summary below is an extract of previous book chapters on the biocidal agents.

20.2 Selection Pressure Associated with Commonly Used Biocidal Agents

20.2.1 Change of Susceptibility by Low-Level Exposure

The adaptive effects were classified as "no MIC increase", "weak MIC increase" with a $\leq$4-fold MIC increase and "strong MIC increase" with a >4-fold MIC increase. The last category was divided into an unstable or stable MIC increase; sometimes the stability was unknown. A species may be found in two or more categories indicating that the adaptive response depends on the type of isolate. Most data on different adaptive effects caused by low-level exposure were found for benzalkonium chloride (78 species) and DDAC (48 species). Only few data were

G. Kampf, *Antiseptic Stewardship*, https://doi.org/10.1007/978-3-319-98785-9_20

found for hydrogen peroxide (8 species), sodium hypochlorite (7 species), peracetic acid (3 species) and glutaraldehyde (1 species).

Figure 20.1 shows the distribution of adaptive response categories for the different biocidal agents. The majority of species did not show any MIC increase or only a weak MIC increase (≤ 4-fold). A strong adaptive response was most frequently seen in benzalkonium chloride (44% of the evaluated species), DDAC (15%) and peracetic acid (one species). The strong MIC increase was stable in 41% (benzalkonium chloride) and 14% (DDAC) of species. Hydrogen peroxide, sodium hypochlorite and glutaraldehyde have so far not shown a strong adaptive MIC increase.

A strong and stable MIC increase after low-level exposure is the most critical adaptive response. Some species can be found in this group that have certainly a high relevance for infection control (Table 20.1). Most of them are among the Gram-negative species.

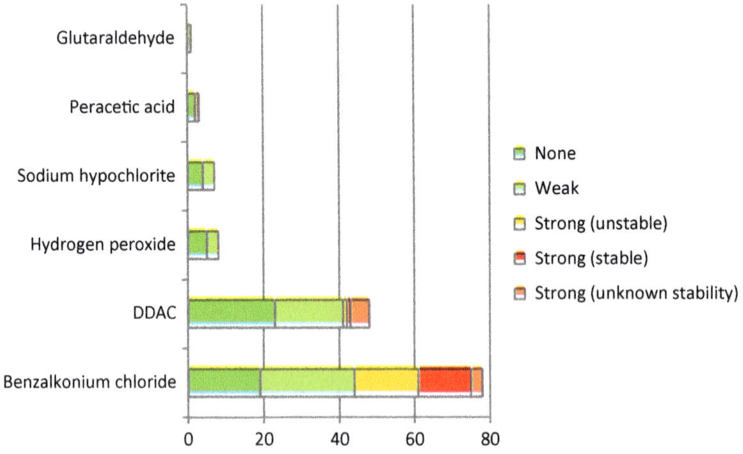

Fig. 20.1 Number of species with no, a weak or a strong adaptive MIC increase after low-level exposure to biocidal agents that may be found in instrument disinfectants

Table 20.1 Bacterial species with a strong (>4-fold MIC increase) and stable adaptive response after low-level exposure to selected biocidal agents sometimes found in instrument disinfectants

Biocidal agent	Bacterial species with a strong and stable adaptive MIC increase
Benzalkonium chloride	*Enterobacter* spp. (≤ 300-fold)
	E. coli (≤ 100-fold)
	S. aureus (≤ 39-fold)
	P. aeruginosa (≤ 33-fold)
	A. baumannii (≤ 31-fold)
DDAC	*P. aeruginosa* (≥ 18-fold)

20.2.2 Cross-Tolerance to Other Biocidal Agents

Isolates of 22 primarily benzalkonium chloride-tolerant species were cross-tolerant to chlorhexidine digluconate and triclosan. An isolate of a benzalkonium chloride-tolerant *E. coli* was cross-tolerant to DDAC, and a benzalkonium chloride-tolerant *L. monocytogenes* was cross-tolerant to another quaternary ammonium compound, alkylamine and sodium hypochlorite. Isolates of primarily DDAC-tolerant *E. coli and P. fluorescens* can be cross-tolerant to benzalkonium chloride, and isolates of *E. faecium, E. hirae, E. coli, P. putida, S. enteritidis* and *C. argentea* can be cross-tolerant to copper via specific efflux pumps. In addition, isolates of primarily peracetic acid-tolerant *B. subtilis* can be cross-tolerant to other oxidizing agents. Isolates of primarily hydrogen peroxide-tolerant *E. coli* can be cross-tolerant to aldehyde, and isolates of primarily hydrogen peroxide-tolerant *S. cerevisiae* can be cross-tolerant to ethanol. Isolates of primarily sodium hypochlorite-tolerant *E. coli* can be cross-tolerant to hydrogen peroxide, and in *L. monocytogenes* cross-tolerance to benzalkonium chloride, another quaternary ammonium compound and alkylamine can occur. Primarily glutaraldehyde-tolerant *E. coli, Halomonas* spp. and *B. cepacia* can be cross-tolerant to other aldehydes.

20.2.3 Cross-Tolerance to Antibiotics

Peracetic acid, hydrogen peroxide and sodium hypochlorite have so far never been described with a cross-tolerance to antibiotics. A cross-tolerance between selected antibiotics and benzalkonium chloride can occur in numerous species. Occasional cross-resistance between DDAC and selected antibiotics was found in *C. coli, E. coli, L. monocytogenes* and *S. enterica*. Cross resistances to rifampicin and sometimes also to isoniazid have been reported in glutaraldehyde-resistant *M. chelonae*.

20.2.4 Efflux Pump Genes

Transporter and efflux pump genes were up-regulated after BAC exposure in *B. cepacia complex, E. coli* and *L. monocytogenes* and after exposure to glutaraldehyde in *Pseudomonas* spp.

20.2.5 Resistance Gene Plasmids

A plasmid with resistance to glutaraldehyde was detected in *S. aureus*.

20.2.6 Viable but not Culturable

Sodium hypochlorite is able to induce the VBNC state with enhanced antibiotic tolerance in *E. coli*. In *S. Typhimurium* peracetic acid is able to induce the VBNC state.

20.2.7 Other Risks Associated with Biocidal Agents in Instrument Disinfectants

Occupational exposure risks, material compatibility, stability, user acceptance, corrosiveness and may be other risks vary between biocidal agents and products [2].

20.3 Effect of Commonly Used Biocidal Agents on Biofilm

20.3.1 Biofilm Development

Biocidal agents in instrument disinfectants have different effects on biofilm development (Fig. 20.2). With peracetic acid, biofilm formation can be inhibited in three species (*C. sakazakii*, *Candida* spp. and *S. aureus*). For benzalkonium chloride, biofilm formation can be inhibited (*C. albicans*, *C. parapsilosis*, *C. tropicalis*, *E. coli*, *S. aureus* and *S. epidermidis*) or enhanced (*S. agalactiae*, *E. coli*, *S. aureus* and *S. epidermidis*). Sodium hypochlorite and hydrogen peroxide can rather enhance than inhibit biofilm formation. No data were found for glutaraldehyde and DDAC.

20.3.2 Biofilm Fixation

Glutaraldehyde usually results in a moderate to strong biofilm fixation, whereas the effect of peracetic acid is typically a poor to moderate biofilm fixation. Benzalkonium chloride was able to increase the mechanical stability of a *P. fluorescens* biofilm suggesting some fixation potential. No data were found to assess the biofilm fixation potential of other biocidal agents typically used for instrument disinfection (DDAC, sodium hypochlorite, hydrogen peroxide).

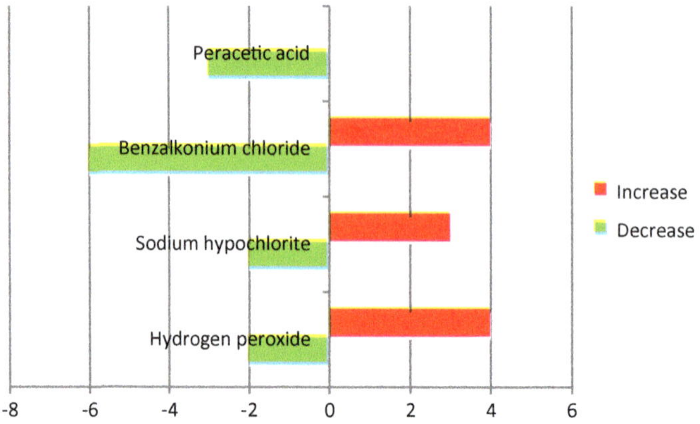

Fig. 20.2 Number of species with a decrease or increase of biofilm formation caused by biocidal agents that may be found in instrument disinfectants

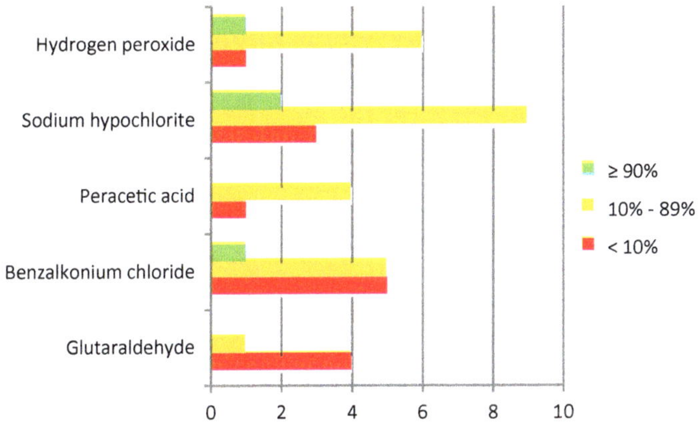

Fig. 20.3 Number of species with a strong (≥ 90%), moderate (10–89%) or poor biofilm removal (<10%) by biocidal agents that may be found in instrument disinfectants

20.3.3 Biofilm Removal

Glutaraldehyde has mostly a poor biofilm removal capacity. It is mostly poor or moderate with benzalkonium chloride. Peracetic acid, sodium hypochlorite and hydrogen peroxide revealed mostly a moderate biofilm removal (Fig. 20.3). No data were found for DDAC.

20.4 Expected Health Benefit of Biocidal Agents in Instrument Disinfectants

Disinfection of instruments has the aim to reduce the microbial load on medical devices after use. For uncritical and semi-critical instruments, it is not followed by sterilization so that a health benefit is expected for the patient (e.g. no infection after a bronchoscopy). The reprocessing, however, includes various steps including cleaning so that it is often not possible to attribute the health benefit to one part of the reprocessing, e.g. the disinfection. The entire process should be validated which includes the use of an instrument disinfectant. As long as the entire reprocessing is validated for the respective medical device, there will be basically no advantage or disadvantage of the biocidal agents used for the disinfection step. For critical instruments, it is expected to be a health benefit for the person who performs functionality tests of cleaned and disinfected medical devices prior to sterilization resulting in a reduction of microbial exposure during work.

20.5 Antiseptic Stewardship Implications

Critical instruments require cleaning and disinfection followed by sterilization. It is overall unlikely that biofilm can remain on instruments or that bacteria can survive in biofilm after validated reprocessing including sterilization has been performed. But the presence of biofilm on semi-critical medical devices such as flexible endoscopes is of relevance for patient safety. It may serve as a continuous source of microbial spread, and it may result in disinfection failure [1, 3]. That is why the effect of biocidal agents on biofilm development and removal is of major relevance for the reprocessing of some semi-critical medical devices.

A low adaptive response in combination with an inhibition of biofilm formation and removal of existing biofilm in the majority of species could not be attributed to any of the biocidal agents. Nevertheless, peracetic acid showed only rarely a strong adaptive response, mainly decreased biofilm formation and moderately removed existing biofilm. No relevant adaptive response was found with sodium hypochlorite and hydrogen peroxide; biofilm removal was overall favourable for both agents but they increased biofilm formation in more species. All three biocidal agents have so far not been associated with antibiotic cross-resistance. Benzalkonium chloride was the substance causing most frequently a strong and stable adaptive response including cross-tolerance to other biocidal agents and selected antibiotics. Biofilm formation could be inhibited or enhanced, and biofilm removal was often poor or moderate.

Data for the other biocidal agents are less comprehensive. Glutaraldehyde did not cause a strong adaptive response and showed only poor biofilm removal. For DDAC, only one or two of the parameter could be described so that a further evaluation is not done.

Overall, on surfaces of instruments where biofilm formation should be inhibited (e.g. flexible endoscopes), the use or peracetic acid seems to be the most appropriate option (low selection pressure). Hydrogen peroxide and sodium hypochlorite have also a low selection pressure and can moderately remove biofilm. They seem to be appropriate on surfaces where enhancement of biofilm formation is of minor relevance. All three biocidal agents have so far not been associated with antibiotic cross-resistance. Benzalkonium chloride seems to be the least suitable biocidal agent taking into account the frequently observed strong and stable adaptive response, the inconclusive effect on biofilm formation and removal and the occurrence of cross-resistance with other biocidal agents and selected antibiotics.

References

1. Akinbobola AB, Sherry L, McKay WG, Ramage G, Williams C (2017) Tolerance of Pseudomonas aeruginosa in in-vitro biofilms to high-level peracetic acid disinfection. J Hosp Infect 97(2):162–168. https://doi.org/10.1016/j.jhin.2017.06.024
2. Dettenkofer M, Block C (2005) Hospital disinfection: efficacy and safety issues. Curr Opin Infect Dis 18(4):320–325

3. Kovaleva J, Peters FT, van der Mei HC, Degener JE (2013) Transmission of infection by flexible gastrointestinal endoscopy and bronchoscopy. Clin Microbiol Rev 26(2):231–254. https://doi.org/10.1128/cmr.00085-12
4. KRINKO am Robert Koch Institut (2012) Anforderungen an die Hygiene bei der Aufbereitung von Medizinprodukten. Bundesgesundheitsblatt 55(10):1244–1310
5. Rutala WA, Weber DJ (2004) Disinfection and sterilization in health care facilities: what clinicians need to know. Clin Infect Dis: Off Publ Infect Dis Soc Am 39(5):702–709. https://doi.org/10.1086/423182
6. Rutala WA, Weber DJ (2016) Disinfection, sterilization, and antisepsis: an overview. Am J Infect Control 44(5 Suppl):e1–e6. https://doi.org/10.1016/j.ajic.2015.10.038
7. Silva e Souza AC, Pereira MS, Rodrigues MA (1998) [Disinfection of medical and surgical equipment: efficacy of chemical disinfectants and water and soap]. Revista latino-americana de enfermagem 6(3):95–105

21 Antiseptic Stewardship for Antimicrobial Soaps

21.1 Composition and Intended Use

Antimicrobial soaps can be based on different types of biocidal agents such as chlorhexidine digluconate, povidone iodine, triclosan, benzalkonium chloride, DDAC, octenidine dihydrochloride, polihexanide or sodium hypochlorite. Most products contain a single biocidal agent although combinations may be found. They often contain auxiliary agents, e.g. detergents, emollients or sometimes also fragrances.

They are used in health care, veterinary medicine, food processing and manufacturing and occasionally also in the domestic setting. A typical application in healthcare is for surgical hand scrubbing or for a hygienic hand wash [29]. They are also used in some departments as a patient preoperative antiseptic body wash with the aim to reduce the risk for surgical site infections [13, 20]. On intensive care units, they may be used for a patient antiseptic body wash with the aim to reduce catheter-associated bloodstream infections [23]. Another aim of using antimicrobial soaps is decolonization of the patients or healthcare workers skin in case of colonization with specific species such as MRSA or VRE, often in combination with other antiseptic treatments such as nasal decolonization [5, 9, 19, 21, 24]. In some parts of the world, antimicrobial soaps were also used at home for regular hand washing [1]. The summary below is an extract of previous book chapters on the biocidal agents.

21.2 Selection Pressure Associated with Commonly Used Biocidal Agents

21.2.1 Change of Susceptibility by Low-Level Exposure

The adaptive effects were classified as "no MIC increase", "weak MIC increase" with a $\leq$4-fold MIC increase and "strong MIC increase" with a >4-fold MIC

G. Kampf, *Antiseptic Stewardship*, https://doi.org/10.1007/978-3-319-98785-9_21

increase. The last category was divided into an unstable or stable MIC increase, and sometimes the stability was unknown. A species may be found in two or more categories indicating that the adaptive response depends on the type of isolate. Most data on different adaptive effects caused by low-level exposure were found for triclosan (90 species), chlorhexidine digluconate and benzalkonium chloride (both 78 species), polihexanide (55 species) and DDAC (48 species). Only few data were found for povidone iodine (5 species) and octenidine dihydrochloride (3 species).

Figure 21.1 shows the distribution of adaptive response categories for the different biocidal agents. The majority of species did not show any MIC increase or only a weak MIC increase ($\leq$4-fold). A strong adaptive response was most frequently seen in benzalkonium chloride (44% of the evaluated species), followed by triclosan (34%), chlorhexidine digluconate (26%), DDAC (15%) and polihexanide (11%). The strong MIC increase was stable in 42% (triclosan), 41% (benzalkonium chloride) and 40% (chlorhexidine digluconate) of species. With octenidine dihydrochloride, one species was found with a strong and stable adaptive response. Povidone iodine and sodium hypochlorite have so far not shown a strong MIC increase.

A strong and stable MIC increase after low-level exposure is probably the most critical adaptive response. Some species can be found in this group that have certainly a high relevance for infection control (Table 21.1). The strongest adaptive MIC increase was found with triclosan (up to 8,192-fold), whereas the changes observed with polihexanide were rather moderate (5-fold–8-fold) and only found in Gram-positive species.

The effect on biofilm is not covered in this chapter because it was assumed that is has only minor relevance for antimicrobial soaps.

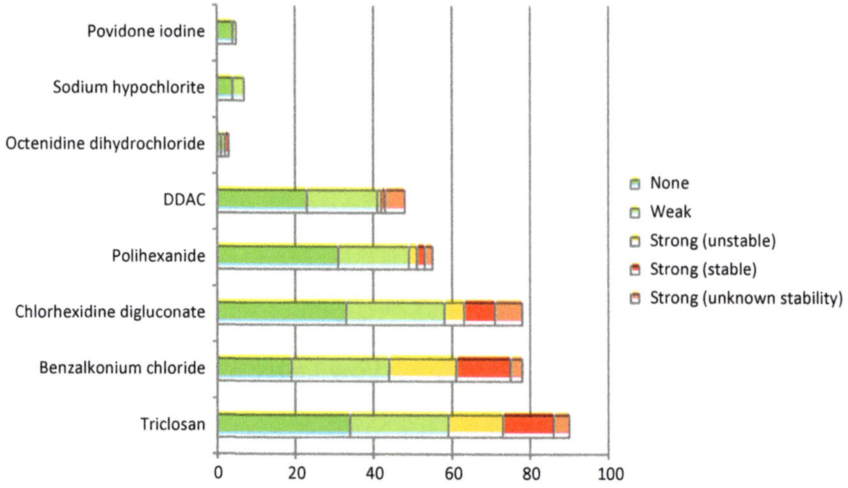

Fig. 21.1 Number of species with no, a weak or a strong adaptive MIC increase after low-level exposure to biocidal agents that may be found in antiseptic soaps

Table 21.1 Bacterial species with a strong (>4-fold MIC increase) and stable adaptive response after low-level exposure to selected biocidal agents used in antiseptic soaps

Biocidal agent	Bacterial species with a strong and stable adaptive MIC increase
Triclosan	*E. coli* (up to 8,192-fold)
	S. aureus (up to 313-fold)
	Staphylococcus spp. (up to 150-fold)
Chlorhexidine digluconate	*E. coli* ($\leq$ 500-fold)
	S. marcescens ($\leq$ 128-fold)
	P. aeruginosa ($\leq$ 32-fold)
	K. pneumoniae ($\leq$ 16-fold)
	S. aureus ($\leq$ 16-fold)
Benzalkonium chloride	*Enterobacter* spp. ($\leq$ 300-fold)
	E. coli ($\leq$ 100-fold)
	S. aureus ($\leq$ 39-fold)
	P. aeruginosa ($\leq$ 33-fold)
	A. baumannii ($\leq$ 31-fold)
Polihexanide	*E. faecalis* (8-fold)
	S. aureus (8-fold)
	S. epidermidis (4.8-fold)
Octenidine dihydrochloride	*P. aeruginosa* ($\leq$ 32-fold)
DDAC	*P. aeruginosa* ($\geq$ 18-fold)

21.2.2 Cross-Tolerance to Other Biocidal Agents

Cross-tolerance to other biocidal agents is quite common in some biocidal agents. Isolates of 22 primarily benzalkonium chloride-tolerant species were cross-tolerant to chlorhexidine digluconate and triclosan. An isolate of a benzalkonium chloride-tolerant *E. coli* was cross-tolerant to DDAC, and a benzalkonium chloride-tolerant *L. monocytogenes* was cross-tolerant to another QAC, alkylamine and sodium hypochlorite. Similar results were found with triclosan. Isolates of 17 primarily triclosan-tolerant species were cross-tolerant to chlorhexidine digluconate and 13 species to benzalkonium chloride. Primarily chlorhexidine digluconate tolerant isolates of *E. coli* and *S. Virchow* were cross-tolerant to triclosan, isolates of *S. Tyhimurium* were cross-tolerant to benzalkonium chloride, and isolates of *A. baylyi* were cross-tolerant to hydrogen peroxide. Isolates of primarily DDAC-tolerant *E. coli and P. fluorescens* can be cross-tolerant to benzalkonium chloride, and isolates of primarily octenidine dihydrochloride-tolerant *P. aeruginosa* can be cross-tolerant to chlorhexidine digluconate. Isolates of primarily sodium hypochlorite-tolerant *E. coli* can be cross-tolerant to hydrogen peroxide, and in *L. monocytogenes* cross-tolerance to benzalkonium chloride, another quaternary ammonium compound and alkylamine can occur. No cross-tolerance to other biocidal agents has been reported for povidone iodine and polihexanide. Especially, the rather frequently observed cross-tolerance between benzalkonium chloride, triclosan and chlorhexidine digluconate is a clear indication to carefully select biocidal agents in order to reduce this type of cross-tolerance to a minimum.

21.2.3 Cross-Tolerance to Antibiotics

Povidone iodine, sodium hypochlorite and polihexanide have so far never been described with a cross-tolerance to antibiotics. A cross-tolerance between triclosan, chlorhexidine digluconate and benzalkonium chloride and selected antibiotics can occur in numerous species. Occasional cross-resistance between DDAC and selected antibiotics was found in *C. coli*, *E. coli*, *L. monocytogenes* and *S. enterica*. Cross-tolerance between octenidine dihydrochloride and selected antibiotics can occur in *P. aeruginosa*.

21.2.4 Efflux Pump Genes

Transporter and efflux pump genes were up-regulated after benzalkonium chloride exposure in *B. cepacia complex*, *E. coli* and *L. monocytogenes*, and after chlorhexidine digluconate exposure in *B. fragilis* and *B. cepacia complex*.

21.2.5 Horizontal Gene Transfer

Horizontal gene transfer can be successfully induced by chlorhexidine digluconate and triclosan in *E. coli* (sulphonamide resistance by conjugation).

21.2.6 Antibiotic Resistance Gene Expression

In a vanA, *E. faecium* chlorhexidine digluconate was able to induce a $\geq$10-fold increase of vanHAX encoding VanA-type vancomycin resistance.

21.2.7 Other Risks Associated with Biocidal Agents in Antimicrobial Soaps

Other risks may also be relevant in antimicrobials soaps. They are not covered in detail. Sensitization to the agent may occur possibly resulting in local or systemic allergic reactions up to anaphylactic reactions [14, 22]. This has been described at least for chlorhexidine digluconate and polihexanide. Some agents are cationic surfactants possibly resulting in a higher degree of skin irritation [14]. Some antimicrobial soaps have been described with a bacterial contamination mainly with Gram-negative species (see Chaps. 10 and 13).

21.3 Expected Health Benefit of Biocidal Agents in Antimicrobial Soaps

Most antimicrobial soaps are based on a single biocidal ingredient so that an expected health benefit is rather dependent on the type of use, the target population and other factors such as additional antiseptic treatments or possible sources for dermal recontamination.

21.3.1 Antiseptic Body Wash Before Surgery

Use of antimicrobial soaps for an antiseptic body wash before surgery may have a health benefit in combination with nasal mupirocin [6]. In cardiac surgery, the bundle reduced the rate of superficial but not deep or organ space surgical site infections [13]. Among 3,924 patients undergoing ventral hernia repair, however, the prehospital chlorhexidine digluconate baths were associated with a significantly higher incidence of surgical site infections [20]. There is currently no general recommendation for a routine preference of antiseptic soaps over plain soaps before surgery [2].

21.3.2 Antiseptic Body Wash for Patients on Intensive Care Units

Universal decolonization with chlorhexidine digluconate bathing and potentially nasal mupirocin may be more effective than vertical strategies that include active surveillance and isolation [10]. Studies support the recently published recommendation that ICU patients over 2 months of age should be bathed with chlorhexidine digluconate on a daily basis to prevent central line-associated bloodstream infections as basic practice [23].

21.3.3 Antiseptic Body Wash for Decolonization of MRSA

Another indication for antiseptic body washing is to limit the spread of MRSA. Some studies suggest that routine daily bathing of MRSA-positive patients on intensive care units with soaps based on octenidine dihydrochloride in combination with other measures can significantly decrease acquisition of MRSA [5, 19, 24]. Other studies, however, did not demonstrate an effect [9, 21]. For eradication of MRSA on healthcare workers (skin and nasal cavity), the application of three products based on octenidine dihydrochloride over 5 d (antiseptic body wash once per day, antiseptic nasal gel thrice per day, antiseptic mouth rinse thrice per day) was effective only in 3 of 40 healthcare workers [21].

The evidence for chlorhexidine digluconate bathing includes studies with a health benefit but also some without a health benefit [4, 17, 26, 27]. Based on the different types of interventions, it is almost impossible to predict a health benefit when patients are colonized with MRSA or VRE and washed daily with an antiseptic soap. Although povidone iodine has broad-spectrum properties, it is considered not to be ideal for topical decolonization due to a lack of evidence for persistence and inferior outcomes compared with chlorhexidine digluconate [23].

21.3.4 Surgical Scrubbing

Surgical scrubbing is usually performed with soaps based on povidone iodine or chlorhexidine digluconate. Using either type of soap resulted in an equivalent surgical site infection rate compared to the use of alcohol-based hand rubs for surgical hand disinfection for 5 min [18]. A lower microbial density on the hands of the surgeons will result in a lower microbial count in the glove juice which is expected to be relevant for the prevention of surgical site infections in case of glove punctures [16]. Some residual effect of povidone iodine has been described for a soap used for surgical scrubbing but this effect seems doubtful considering that the overall efficacy even with a "residual effect" is mostly inferior to alcohol-based hand rubs [28].

21.3.5 Hygienic Hand Wash

In health care, there is basically no indication for a hygienic hand wash. If hands are clean, it is recommended to use an alcohol-based hand rub when an indication for hand hygiene occurs. Visibly soiled hands should be washed either with plain soap or an antimicrobial soap [29]. There is apparently no health benefit associated with the use of antimicrobial soap for the decontamination of soiled hands of healthcare workers.

For food processing and manufacturing, antimicrobial soaps may have some effect although it has been acknowledged that it is difficult to quantify [8].

At home, there is no health benefit to be expected when antiseptic soaps are used instead of plain soap for regular hand washing [15]. Such an effect is unlikely anyway because the time spent for lathering hands when soap is used is between 2.6 and 5.6 s, e.g. in public restrooms, indicating that the effect of an antimicrobial soap can only be minimal in such a short exposure time. Rinsing hands after lathering was always longer [25].

21.4 Antiseptic Stewardship Implications

Chlorhexidine digluconate, benzalkonium chloride and triclosan showed most frequently a strong and also stable adaptive response including a cross-tolerance to other agents. Other biocidal agents were less adaptive such as povidone iodine or sodium hypochlorite (no strong adaptive response), octenidine dihydrochloride (1 species with a strong and stable adaptive response) and polihexanide (2 species with a strong and stable adaptive response). Especially with polihexanide, it is noteworthy that the MIC change was rather moderate (5-fold–8-fold) and only found in Gram-positive species.

Under the assumption that all biocidal agents used in antimicrobial soaps have an equivalent bactericidal activity at appropriate concentrations, it seems that the lowest adaptive reaction is found with povidone iodine, sodium hypochlorite, octenidine dihydrochloride and polihexanide. In order to reduce selection pressure, they should be preferred biocidal agents in antimicrobial soaps when a health benefit is likely or proven, e.g. for antiseptic body wash on ICU patients or for decolonization of MRSA in combination with other antiseptic measures.

Some applications of antimicrobial soaps could be stopped completely. Especially, the domestic use of antimicrobial soaps, e.g. based on triclosan, is seen critically. Giuliano and Rybak recently reviewed the evidence evaluating the use of triclosan as an antimicrobial soap and its association with antimicrobial resistance. They concluded that there was no beneficial effect of triclosan over non-antimicrobial soap, and triclosan resistance has been demonstrated. They concluded that the risks outweigh the benefits of triclosan use [7]. The Canadian Paediatric Society promotes hand hygiene using plain soap and water in the vast majority of domestic settings [3]. A similar recommendation exists for Germany [12]. That is why antimicrobial soaps should not be routinely used in the domestic setting.

Another simple option is to ban soaps based on chlorhexidine digluconate, triclosan or benzalkonium chloride for hand hygiene in health care. One possible use of these soaps is in direct patient care. Based on the WHO recommendation for hand hygiene from 2009, it is recommended to wash hands when they are visibly soiled. The use of plain soap is adequate. Treatment of clean hands should preferably be done with alcohol-based hand rubs [29]. In the surgical theatre, the use of antimicrobial soaps, e.g. based on chlorhexidine digluconate, is one option recommended by the WHO. The scrubbing usually lasts for 6–10 min and consumes between 5 and 20 l water per scrub [11]. They may only be effective with additional postscrub water-based chlorhexidine digluconate treatments of the hands which pose an additional contamination and selection pressure risk [11]. Alcohol-based hand rubs have a stronger effect on the resident hand flora, require typically 1.5 min for application, cause less skin irritation and do not pose any relevant selection pressure to bacterial species due to their volatility [30, 31]. That is why surgical scrubbing has more disadvantages than advantages, especially regarding the possible selection pressure by chlorhexidine digluconate as the principal antimicrobial agent.

References

1. Aiello AE, Larson EL, Levy SB (2007) Consumer antibacterial soaps: effective or just risky? Clin Infect Dis: Off Publ Infect Dis Soc Am 45(Suppl 2):S137–S147. https://doi.org/10.1086/519255
2. Allegranzi B, Bischoff P, de Jonge S, Kubilay NZ, Zayed B, Gomes SM, Abbas M, Atema JJ, Gans S, van Rijen M, Boermeester MA, Egger M, Kluytmans J, Pittet D, Solomkin JS (2016) New WHO recommendations on preoperative measures for surgical site infection prevention:

an evidence-based global perspective. Lancet Infect Dis 16(12):e276–e287. https://doi.org/10.1016/s1473-3099(16)30398-x
3. Allen UD (2006) Antimicrobial products in the home: the evolving problem of antibiotic resistance. Paediatr Child Health 11(3):169–173
4. Frost SA, Alogso MC, Metcalfe L, Lynch JM, Hunt L, Sanghavi R, Alexandrou E, Hillman KM (2016) Chlorhexidine bathing and health care-associated infections among adult intensive care patients: a systematic review and meta-analysis. Crit Care (London, England) 20(1):379. https://doi.org/10.1186/s13054-016-1553-5
5. Gastmeier P, Kampf KP, Behnke M, Geffers C, Schwab F (2016) An observational study of the universal use of octenidine to decrease nosocomial bloodstream infections and MDR organisms. J Antimicrob Chemother 71(9):2569–2576. https://doi.org/10.1093/jac/dkw170
6. George S, Leasure AR, Horstmanshof D (2016) Effectiveness of decolonization with chlorhexidine and mupirocin in reducing surgical site infections: a systematic review. Dimension Crit Care Nurs: DCCN 35(4):204–222. https://doi.org/10.1097/dcc.0000000000000192
7. Giuliano CA, Rybak MJ (2015) Efficacy of triclosan as an antimicrobial hand soap and its potential impact on antimicrobial resistance: a focused review. Pharmacotherapy 35(3):328–336. https://doi.org/10.1002/phar.1553
8. Haas CN, Marie JR, Rose JB, Gerba CP (2005) Assessment of benefits from use of antimicrobial hand products: reduction in risk from handling ground beef. Int J Hyg Environ Health 208(6):461–466. https://doi.org/10.1016/j.ijheh.2005.04.009
9. Harris PN, Le BD, Tambyah P, Hsu LY, Pada S, Archuleta S, Salmon S, Mukhopadhyay A, Dillon J, Ware R, Fisher DA (2015) Antiseptic body washes for reducing the transmission of methicillin-resistant staphylococcus aureus: a cluster crossover study. Open Forum Infect Dis 2(2):ofv051. https://doi.org/10.1093/ofid/ofv051
10. Huang SS, Septimus E, Kleinman K, Moody J, Hickok J, Avery TR, Lankiewicz J, Gombosev A, Terpstra L, Hartford F, Hayden MK, Jernigan JA, Weinstein RA, Fraser VJ, Haffenreffer K, Cui E, Kaganov RE, Lolans K, Perlin JB, Platt R (2013) Targeted versus universal decolonization to prevent ICU infection. N Engl J Med 368(24):2255–2265. https://doi.org/10.1056/nejmoa1207290 [doi]
11. Kampf G (2018) Aqueous chlorhexidine for surgical hand disinfection? J Hosp Infect 98(4):378–379. https://doi.org/10.1016/j.jhin.2017.11.012
12. Kampf G, Dettenkofer M (2011) Desinfektionsmaßnahmen im häuslichen Umfeld – was macht wirklich Sinn? Hyg Med 36(1–2):8–11
13. Kohler P, Sommerstein R, Schonrath F, Ajdler-Schaffler E, Anagnostopoulos A, Tschirky S, Falk V, Kuster SP, Sax H (2015) Effect of perioperative mupirocin and antiseptic body wash on infection rate and causative pathogens in patients undergoing cardiac surgery. Am J Infect Control 43(7):e33–e38. https://doi.org/10.1016/j.ajic.2015.04.188
14. Lachapelle JM (2014) A comparison of the irritant and allergenic properties of antiseptics. Eur J Dermatol: EJD 24(1):3–9. https://doi.org/10.1684/ejd.2013.2198
15. Larson E, Aiello A, Lee LV, Della-Latta P, Gomez-Duarte C, Lin S (2003) Short- and long-term effects of handwashing with antimicrobial or plain soap in the community. J Commun Health 28(2):139–150
16. Misteli H, Weber WP, Reck S, Rosenthal R, Zwahlen M, Füglistaler P, Bolli MK, Örtli D, Widmer AF, Marti WR (2009) Surgical glove perforation and the risk of surgical site infection. Arch Surg 144(6):553–558
17. Musuuza JS, Sethi AK, Roberts TJ, Safdar N (2017) Implementation of daily chlorhexidine bathing to reduce colonization by multidrug-resistant organisms in a critical care unit. Am J Infect Control 45(9):1014–1017. https://doi.org/10.1016/j.ajic.2017.02.038
18. Parienti JJ, Thibon P, Heller R, Le Roux Y, von Theobald P, Bensadoun H, Bouvet A, Lemarchand F, Le Coutour X (2002) Hand-rubbing with an aqueous alcoholic solution versus traditional surgical hand-scrubbing and 30-day surgical site infection rates - a randomized equivalence study. JAMA 288(6):722–727

19. Pichler G, Pux C, Babeluk R, Hermann B, Stoiser E, De Campo A, Grisold A, Zollner-Schwetz I, Krause R, Schippinger W (2018) MRSA prevalence rates detected in a tertiary care hospital in Austria and successful treatment of MRSA positive patients applying a decontamination regime with octenidine. Eur J Clin Microbiol Infect Dis 37(1):21–27. https://doi.org/10.1007/s10096-017-3095-4
20. Prabhu AS, Krpata DM, Phillips S, Huang LC, Haskins IN, Rosenblatt S, Poulose BK, Rosen MJ (2017) Preoperative chlorhexidine gluconate use can increase risk for surgical site infections after ventral hernia repair. J Am Coll Surg 224(3):334–340. https://doi.org/10.1016/j.jamcollsurg.2016.12.013
21. Richter A, Eder I, Konig B, Lutze B, Rodloff AC, Thome UH, Weiss M, Chaberny IF (2018) [Decolonization of health care workers in a neonatal intensive care unit carrying a methicillin-susceptible Staphylococcus aureus Isolate]. Gesundheitswesen (Bundesverband der Arzte des Offentlichen Gesundheitsdienstes (Germany)) 80(1):54–58. https://doi.org/10.1055/s-0043-122277
22. Schunter JA, Stocker B, Brehler R (2017) A case of severe Anaphylaxis to Polyhexanide: cross-reactivity between Biguanide Antiseptics. Int Arch Allergy Immunol 173(4):233–6. https://doi.org/10.1159/000478700
23. Septimus EJ, Schweizer ML (2016) Decolonization in prevention of health care-associated infections. Clin Microbiol Rev 29(2):201–222. https://doi.org/10.1128/cmr.00049-15
24. Spencer C, Orr D, Hallam S, Tillmanns E (2013) Daily bathing with octenidine on an intensive care unit is associated with a lower carriage rate of meticillin-resistant Staphylococcus aureus. J Hosp Infect 83(2):156–159. https://doi.org/10.1016/j.jhin.2012.10.007
25. Toshima Y, Ojima M, Yamada H, Mori H, Tonomura M, Hioki Y, Koya E (2001) Observation of everyday hand-washing behavior of Japanese, and effects of antibacterial soap. Int J Food Microbiol 68(1–2):83–91
26. Urbancic KF, Martensson J, Glassford N, Eyeington C, Robbins R, Ward PB, Williams D, Johnson PD, Bellomo R (2018) Impact of unit-wide chlorhexidine bathing in intensive care on bloodstream infection and drug-resistant organism acquisition. Crit Care Resuscitation: J Australas Acad Crit Care Med 20(2):109–116
27. Velazquez-Meza ME, Mendoza-Olazaran S, Echaniz-Aviles G, Camacho-Ortiz A, Martinez-Resendez MF, Valero-Moreno V, Garza-Gonzalez E (2017) Chlorhexidine whole-body washing of patients reduces methicillin-resistant Staphylococcus aureus and has a direct effect on the distribution of the ST5-MRSA-II (New York/Japan) clone. J Med Microbiol 66(6):721–728. https://doi.org/10.1099/jmm.0.000487
28. Wade JJ, Casewell MW (1991) The evaluation of residual antimicrobial activity on hands and its clinical relevance. J Hosp Infect 18(Suppl. B):23–28
29. WHO (2009) WHO guidelines on hand hygiene in health care. First Global Patient Safety Challenge Clean Care is Safer Care, WHO, Geneva
30. Widmer AF (2013) Surgical hand hygiene: scrub versus rub. J Hosp Infect 83(suppl. 1):S35–S39
31. Widmer AF, Rotter M, Voss A, Nthumba P, Allegranzi B, Boyce J, Pittet D (2010) Surgical hand preparation: state-of-the-art. J Hosp Infect 74(2):112–122

22 Antiseptic Stewardship for Wound and Mucous Membrane Antiseptics

22.1 Composition and Intended Use

Wound and mucous membrane antiseptics can be based on different types of biocidal agents such as chlorhexidine digluconate, polihexanide, hydrogen peroxide, sodium hypochlorite, povidone iodine or octenidine dihydrochloride [1, 2]. In addition, silver may be used as an antimicrobial agent for wound treatment, e.g. in wound dressings. Most products contain a single biocidal agent.

They are used in health care, veterinary medicine and occasionally also in the domestic setting. Wound antiseptics are indicated for infected or critically colonized wounds [2]. Depending on a risk score, wound antiseptics may also be indicated for other types of wounds [2]. Mucous membrane antiseptics are typically applied prior to surgery, e.g. to the genitourinary or oral mucosa [3]. The summary below is an extract of previous book chapters on the biocidal agents.

22.2 Selection Pressure Associated with Commonly Used Biocidal Agents

22.2.1 Change of Susceptibility by Low-Level Exposure

The adaptive effects were classified as "no MIC increase", "weak MIC increase" with a $\leq$4-fold MIC increase and "strong MIC increase" with a >4-fold MIC increase. The last category was divided into an unstable or stable MIC increase; sometimes the stability was unknown. A species may be found in two or more categories indicating that the adaptive response depends on the type of isolate. Most data on different adaptive effects caused by low-level exposure were found for chlorhexidine digluconate (78 species), polihexanide (55 species) and silver (20 species). Only few data were found for hydrogen peroxide (8 species), sodium hypochlorite (7 species), povidone iodine (5 species) and octenidine dihydrochloride (3 species).

G. Kampf, *Antiseptic Stewardship*, https://doi.org/10.1007/978-3-319-98785-9_22

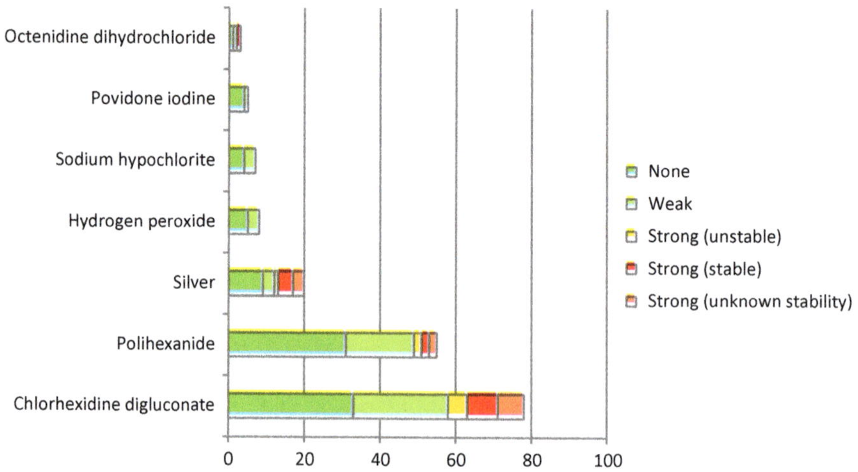

Fig. 22.1 Number of species with no, a weak or a strong adaptive MIC increase after low-level exposure to biocidal agents that may be found in wound or mucous membrane antiseptics

Figure 22.1 shows the distribution of adaptive response categories for the different biocidal agents. The majority of species did not show any MIC increase or only a weak MIC increase ($\leq$4-fold). A strong adaptive response was most frequently seen in silver (40%), chlorhexidine digluconate (26%) and polihexanide (11%). The strong MIC increase was stable in 50% (silver, mainly in sil-positive strains), 40% (chlorhexidine digluconate) and 33% (polihexanide) of species. With octenidine dihydrochloride, one species was found with a strong and stable adaptive response. Hydrogen peroxide, sodium hypochlorite and povidone iodine have so far not shown a strong MIC increase.

A strong and stable MIC increase after low-level exposure is probably the most critical adaptive response. Some species can be found in this group that have certainly a high relevance for infection control (Table 22.1). Most of them are among the Gram-negative species. It is noteworthy that the changes observed with polihexanide were rather moderate (5-fold–8-fold) and only found in Gram-positive species.

22.2.2 Cross-Tolerance to Other Biocidal Agents

Primarily chlorhexidine digluconate-tolerant isolates of *E. coli* and *S. Virchow* can be cross-tolerant to triclosan, isolates of *S. Tyhimurium* can be cross-tolerant to benzalkonium chloride, and isolates of *A. baylyi* can be cross-tolerant to hydrogen peroxide. Isolates of primarily octenidine dihydrochloride-tolerant *P. aeruginosa* can be cross-tolerant to chlorhexidine digluconate. Isolates of primarily sodium hypochlorite-tolerant *E. coli* can be cross-tolerant to hydrogen peroxide, and in *L. monocytogenes* cross-tolerance to benzalkonium chloride, another quaternary

Table 22.1 Bacterial species with a strong (>4-fold MIC increase) and stable adaptive response after low-level exposure to selected biocidal agents sometimes found in wound or mucous membrane antiseptics

Biocidal agent	Bacterial species with a strong and stable adaptive MIC increase
Chlorhexidine digluconate	*E. coli* ($\leq$ 500-fold)
	S. marcescens ($\leq$ 128-fold)
	P. aeruginosa ($\leq$ 32-fold)
	K. pneumoniae ($\leq$ 16-fold)
	S. aureus ($\leq$ 16-fold)
Silver	*E. coli* (128-fold)[a]
	E. cloacae ($\geq$ 32-fold)[a]
	K. pneumoniae ($\geq$ 32-fold)[a]
	K. oxytoca ($\geq$ 16-fold)[a]
Polihexanide	*E. faecalis* (8-fold)
	S. aureus (8-fold)
	S. epidermidis (4.8-fold)
Octenidine dihydrochloride	*P. aeruginosa* ($\leq$ 32 fold)

[a]Mainly sil-positive isolates or strains

ammonium compound and alkylamine can occur. Isolates of primarily hydrogen peroxide-tolerant *E. coli* can be cross-tolerant to aldehyde, and isolates of primarily hydrogen peroxide-tolerant *S. cerevisiae* can be cross-tolerant to ethanol. No cross-tolerance to other biocidal agents has been reported for povidone iodine and polihexanide.

22.2.3 Cross-Tolerance to Antibiotics

Povidone iodine, sodium hypochlorite, hydrogen peroxide and polihexanide have so far never been described with a cross-tolerance to antibiotics. A cross-tolerance between both silver and chlorhexidine digluconate and selected antibiotics can occur in numerous species. Cross-tolerance between octenidine dihydrochloride and selected antibiotics can occur in *P. aeruginosa*.

22.2.4 Efflux Pump Genes

Transporter and efflux pump genes were up-regulated after chlorhexidine digluconate exposure in *B. fragilis* and *B. cepacia complex*.

22.2.5 Horizontal Gene Transfer

Horizontal gene transfer can be successfully induced by chlorhexidine digluconate in *E. coli* (sulphonamide resistance by conjugation).

22.2.6 Antibiotic Resistance Gene Expression

In a vanA *E. faecium*, chlorhexidine digluconate was able to induce a ≥ 10-fold increase of vanHAX encoding VanA-type vancomycin resistance.

22.2.7 Other Risks Associated with Biocidal Agents in Wound and Mucous Membrane Antiseptics

Other risks may also be relevant in wound and mucous membrane antiseptics. They are not covered here in detail. Local tolerability including its possible toxic effect on cartilage, any favorable or negative effect on wound healing, its efficacy in the presence of organic load, the potential for sensitization and any systemic risk should also be evaluated [2].

22.3 Effect of Commonly Used Biocidal Agents on Biofilm

22.3.1 Biofilm Development

Typical biocidal agents in wound and mucous membrane antiseptics show a different effect on biofilm development (Fig. 22.2). For silver, often as nanoparticles, biofilm formation can be inhibited in *C. parapsilosis*, *C. tropicalis*, *C. albicans*, *E. coli*, *P. fluorescens*, *S. epidermidis* and *S. aureus*. Similar results are found for povidone iodine with an inhibition of biofilm formation in four species: *E. faecalis*, *S. aureus*, *S. epidermidis* and *C. albicans*. A decrease of biofilm formation was described for octenidine dihydrochloride but only at concentrations of $\geq 0.31\%$ which has no relevance in wound and mucous membrane antiseptics. Chlorhexidine digluconate exposure resulted in a decrease of biofilm formation in the majority of species. Sodium hypochlorite and hydrogen peroxide can rather enhance than inhibit biofilm formation. No data were found for polihexanide.

22.3.2 Biofilm Fixation

No data were found to assess the biofilm fixation potential of octenidine dihydrochloride, silver, chlorhexidine digluconate, povidone iodine, polihexanide, sodium hypochlorite or hydrogen peroxide.

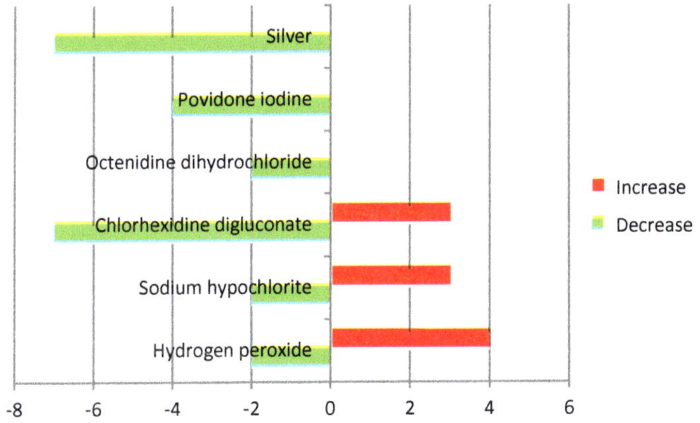

Fig. 22.2 Number of species with a decrease or increase of biofilm formation caused by biocidal agents that may be found in wound or mucous membrane antiseptics

22.3.3 Biofilm Removal

Povidone iodine has so far only been described with a strong biofilm removal. Silver, sodium hypochlorite and hydrogen peroxide have a mostly moderate biofilm removal capacity. Octenidine dihydrochloride could equally show a poor, moderate and strong biofilm removal. It is poor or moderate with polihexanide and chlorhexidine digluconate (Fig. 22.3).

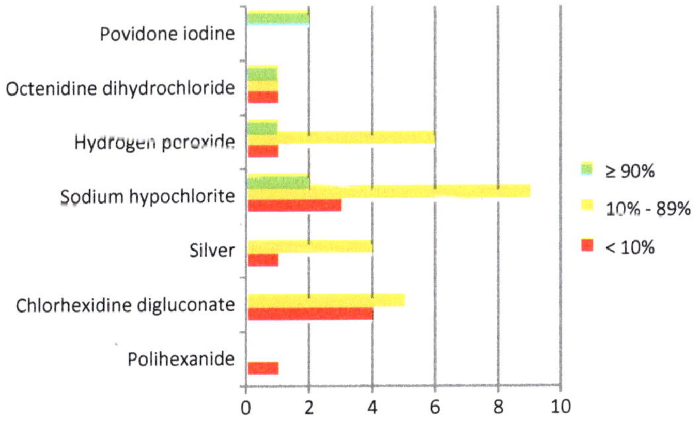

Fig. 22.3 Number of species with a strong ($\geq 90\%$), moderate (10–89%) or poor biofilm removal (<10%) by biocidal agents that may be found in wound or mucous membrane antiseptics

22.4 Health Benefits of Biocidal Agents in Wound and Mucous Membrane Antiseptics

For patients with wounds, a health benefit is already the prevention of a wound infection, e.g. after soft tissue traumatic injuries or in wounds after cardiothoracic surgery. There is some evidence for polihexanide and sodium hypochlorite to suggest that a targeted antiseptic wound treatment is able to reduce infection rates [2]. The use of mucous membrane antiseptics is recommended prior to surgery for prevention of surgical site infections, e.g. in urology, gynaecology or ophthalmology [3].

22.5 Antiseptic Stewardship Implications

A low adaptive response in combination with a frequently observed inhibition of biofilm formation and a rather strong removal of existing biofilm can be attributed only to povidone iodine. Sodium hypochlorite and hydrogen peroxide also revealed a low adaptive response but can enhance biofilm formation in a few more species and have only a moderate biofilm removal capacity. Limited data with octenidine dihydrochloride suggest an inconsistent adaptive effect and also an inconsistent effect on biofilm removal. Polihexanide can exhibit a strong adaptive response which is in comparison to other biocidal agents quite low (5-fold to 8-fold) and only described in Gram-positive species. Its biofilm removal capacity is poor. Chlorhexidine digluconate and silver may both show quite frequently a strong adaptive response mainly among Gram-negative species. The effect caused by silver depends largely on the presence of sil-genes in the strains. Silver can inhibit biofilm formation where the effect of chlorhexidine digluconate is inconsistent. For biofilm removal, silver nanoparticles have mostly a moderate effect, whereas the effect of chlorhexidine digluconate is mostly poor.

The indication for wound or mucous membrane antiseptics depends on multiple factors and can not only rely on the potential for selection pressure. Nevertheless, povidone iodine seems to exhibit the lowest selection pressure and chlorhexidine digluconate the highest one.

References

1. Assadian O (2016) Octenidine dihydrochloride: chemical characteristics and antimicrobial properties. J Wound Care 25(3 Suppl):S3–S6. https://doi.org/10.12968/jowc.2016.25.Sup3.S3
2. Kramer A, Dissemond J, Kim S, Willy C, Mayer D, Papke R, Tuchmann F, Assadian O (2017) Consensus on wound antisepsis: update 2018. Skin Pharmacol Physiol 31(1):28–58. https://doi.org/10.1159/000481545
3. KRINKO am Robert Koch Institut (2018) Prävention postoperativer Wundinfektionen. Bundesgesundheitsblatt 61(4):448–473

Printed by Printforce, the Netherlands